Birkhäuser

Monographs in Mathematics
Vol. 96

Wolfgang Arendt
Charles J.K. Batty
Matthias Hieber
Frank Neubrander

Vector-valued Laplace Transforms and Cauchy Problems

Second Edition

Birkhäuser

Wolfgang Arendt
Angewandte Analysis
Universität Ulm
89069 Ulm
Germany
wolfgang.arendt@uni-ulm.de

Charles J.K. Batty
St. John's College
Oxford OX1 3JP
UK
charles.batty@sjc.ox.ac.uk

Matthias Hieber
Fachbereich Mathematik
TU Darmstadt
Schlossgartenstr. 7
64289 Darmstadt
Germany
hieber@mathematik.tu-darmstadt.de

Frank Neubrander
Department of Mathematics
Louisiana State University
Baton Rouge, LA 70803
USA
neubrand@math.lsu.edu

2010 Mathematics Subject Classification: 35A22, 46F12, 35K25

ISBN 978-3-0348-0327-4 ISBN 978-3-0348-0087-7 (eBook)
DOI 10.1007/978-3-0348-0087-7

Cover design: deblik, Berlin

Printed on acid-free paper

Springer Basel AG is part of Springer Science+Business Media

www.birkhauser-science.com

Contents

Prefaces

Preface to the First Edition

Linear evolution equations in Banach spaces have seen important developments in the last two decades. This is due to the many different applications in the theory of partial differential equations, probability theory, mathematical physics, and other areas, and also to the development of new techniques. One important technique is given by the Laplace transform. It played an important role in the early development of semigroup theory, as can be seen in the pioneering monograph by Hille and Phillips [HP57]. But many new results and concepts have come from Laplace transform techniques in the last 15 years. In contrast to the classical theory, one particular feature of this method is that functions with values in a Banach space have to be considered.

The aim of this book is to present the theory of linear evolution equations in a systematic way by using the methods of vector-valued Laplace transforms.

It is simple to describe the basic idea relating these two subjects. Let A be a closed linear operator on a Banach space X. The *Cauchy problem* defined by A is the initial value problem

$$(CP) \quad \begin{cases} u'(t) = & Au(t) \quad (t \geq 0), \\ u(0) = & x, \end{cases}$$

where $x \in X$ is a given initial value. If u is an exponentially bounded, continuous function, then we may consider the Laplace transform

$$\hat{u}(\lambda) = \int_0^\infty e^{-\lambda t} u(t)\, dt$$

of u for large real λ. It turns out that u is a (mild) solution of (CP) if and only if

$$(\lambda - A)\hat{u}(\lambda) = x \qquad (\lambda \text{ large}). \tag{1}$$

Thus, if λ is in the resolvent set of A, then

$$\hat{u}(\lambda) = (\lambda - A)^{-1}x. \tag{2}$$

Now it is a typical feature of concrete evolution equations that no explicit information on the solution is known and only in exceptional cases can the solution be given by a formula. On the other hand, in many cases much can be said about the resolvent of the given operator. The fact that the Laplace transform allows us to reduce the Cauchy problem (CP) to the characteristic equation (1) explains its usefulness. The Laplace transform is the link between solutions and resolvents, between Cauchy problems and spectral properties of operators.

There are two important themes in the theory of Laplace transforms. The first concerns representation theorems; i.e., results which give criteria to decide whether a given function is a Laplace transform. Clearly, in view of (2), such results, applied to the resolvent of an operator, give information on the solvability of the Cauchy problem.

The other important subject is asymptotic behaviour, where the most challenging and delicate results are *Tauberian theorems* which allow one to deduce asymptotic properties of a function from properties of its transform. Since in the case of solutions of (CP) the transform is given by the resolvent, such results may allow one to deduce results of asymptotic behaviour from spectral properties of A.

These two themes describe the essence of this book, which is divided into three parts. In the first, representation theorems for Laplace transforms are given, and corresponding to this, well-posedness of the Cauchy problem is studied. The second is a systematic study of asymptotic behaviour of Laplace transforms first of arbitrary functions, and then of solutions of (CP). The last part contains applications and illustrative examples. Each part is preceded by a detailed introduction where we describe the interplay between the diverse subjects and explain how the sections are related.

We have assumed that the reader is already familiar with the basic topics of functional analysis and the theory of bounded linear operators, Lebesgue integration and functions of a complex variable. We require some standard facts from Fourier analysis and slightly more advanced areas of functional analysis for which we give references in the text. There are also four appendices (A, B, C and E) which collect together background material on other standard topics for use in various places in the book, while Appendix D gives a proof of a technical result in the geometry of Banach spaces which is needed in Section 4.6.

Finally, a few words should be said about the realization of the book. The collaboration of the authors is based on two research activities: the common work of W. Arendt, M. Hieber and F. Neubrander on integrated semigroups and the work of W. Arendt and C. Batty on asymptotic behaviour of semigroups over many years. Laplace transform methods are common to both.

The actual contributions are as follows.

Part I: All four authors wrote this part.

Part II was written by W. Arendt and C. Batty.

Part III was written by W. Arendt (Chapters 6 and 7) and M. Hieber (Chapter 8).

C. Batty undertook the coordination needed to make the material into a consistent book.

The authors are grateful to many colleagues and friends with whom they had a fruitful cooperation, frequently over many years, which allowed them to discuss the material presented in the book. We would especially like to acknowledge among them H. Amann, B. Bäumer, Ph. Bénilan, J. van Casteren, R. Chill, O. El-Mennaoui, J. Goldstein, H. Kellermann, V. Keyantuo, R. deLaubenfels, G. Lumer, R. Nagel, J. van Neerven, J. Prüss, F. Räbiger, A. Rhandi, W. Ruess, Q.P. Vũ, and L. Weis

Special thanks go to S. Bu, R. Chill, M. Haase, R. Nagel and R. Schnaubelt, who read parts of the manuscript and gave very useful comments.

The enormous technical work on the computer, in particular typing large parts of the manuscript and unifying 4 different TEX dialects, was done with high competence in a most reliable and efficient way by Mahamadi Warma. To him go our warmest thanks.

The authors are grateful to Professor H. Amann, editor of "Monographs in Mathematics", for his support. The cooperation with Birkhäuser Verlag, and with Dr. T. Hintermann in particular, was most enjoyable and efficient.

Ulm, Oxford, Darmstadt, Baton Rouge
August, 2000

Wolfgang Arendt
Charles Batty
Matthias Hieber
Frank Neubrander

Preface to the Second Edition

Ten years after the publication of the first edition of this monograph, it is clear that vector-valued Laplace transform methods continue to play an important role in the analysis of partial differential equations and other disciplines of analysis. Among the most notable new achievements of this period are the characterization of generators of cosine functions on Hilbert space due to Crouzeix, and quantitative Tauberian theorems for Laplace transforms with applications to energy estimates for wave equations.

In this second edition, the new developments have been taken into account by updating the Notes on each Chapter and the Bibliography. For example, the characterization of generators of cosine functions on Hilbert space by a purely geometric condition on the numerical range is precisely stated in Theorem 3.17.5. The main text has not been substantially changed, except in Section 4.4 where some results are now presented in quantitative forms. Their applications in the study of damped wave equations are explained in detail in the Notes of the section.

A few minor mathematical gaps and typographical errors have been corrected, and we are grateful to M. Haase, J. van Neerven, R. Schumann and D. Seifert for alerting us to some of them.

September 2010 The Authors

Part I

Laplace Transforms and Well-Posedness of Cauchy Problems

As a guide-line for Part I, as well as for the entire book, we have in mind the formula

$$\hat{u}(\lambda) = R(\lambda, A)x \tag{3}$$

saying that a mild solution of the Cauchy problem, $u'(t) = Au(t)$ with initial value x, is given by the resolvent of the underlying operator A evaluated at x. Thus, if we want to find solutions, we first have to characterize those functions which are Laplace transforms; i.e., we study representation theorems. Correspondingly, on the side of evolution equations, we investigate existence and uniqueness of solutions of the Cauchy problem. Other subjects treated here include regularity and positivity.

Part I contains three chapters as follows:

1. The Laplace Integral
2. The Laplace Transform
3. Cauchy Problems

We start with an introduction to the vector-valued Lebesgue integral; i.e., the Bochner integral. For our purposes it suffices to consider functions defined on the real line. Then we introduce the Laplace integral and investigate its analytic properties, giving special attention to its diverse abscissas. This will play an important role when solutions of the Cauchy problem are considered, as the abscissas give information about the asymptotic behaviour for large time. Operational properties of the Laplace integral are also discussed. Finally, we introduce functions of (semi) bounded variation defined on the half-line and the Laplace-Stieltjes transform. They will be needed when we study resolvent positive operators (Section 3.11) and Hille-Yosida operators (Section 3.5).

The vector-valued Fourier transform on the line is introduced in Section 1.6 and we prove the Paley-Wiener theorem for functions with values in a Hilbert space. This is the first of several representation theorems for Laplace transforms which we present in this book.

In Chapter 2, real representation theorems are the central subject. We prove a vector-valued version of Widder's classical theorem which describes those functions which are Laplace transforms of bounded measurable functions. The vector-valued version (Section 2.2) will lead directly to generation theorems in Chapter 3 for semigroups and integrated semigroups (Section 3.3) and for cosine functions (Section 3.15). A particularly simple representation theorem is valid for holomorphic functions (Section 2.6). The Laplace transform is an isomorphism between certain classes of holomorphic functions defined on sectors in the complex plane. This will lead directly to the generation theorem for holomorphic semigroups in Section 3.7. The third representation theorem is a vector-valued version of Bernstein's theorem describing Laplace-Stieljes transforms of monotonic functions (Section 2.7). It has its counterpart for Cauchy problems in Section 3.11 where resolvent positive operators are considered.

The uniqueness theorem for Laplace transforms is easy to prove (Section 1.7), but it has striking consequences. It gives directly an approximation result for sequences of Laplace transforms. In Chapter 3 we find its counterpart for Cauchy problems in the form of the Trotter-Kato theorem (Section 3.6).

For Cauchy problems, the most satisfying situation is when there corresponds exactly one (mild) solution to each initial value. This notion of well-posedness is equivalent to existence of a C_0-semigroup (Section 3.1). We also consider weaker forms of well-posedness which are characterized by the existence of integrated semigroups. In applications, they allow one to describe precise regularity for certain partial differential equations in $L^p(\mathbb{R}^n)$, and Chapter 8 is devoted to this. Here in Part I, there are three situations where integrated semigroups occur in a natural way. Operators satisfying the Hille-Yosida condition generate locally Lipschitz continuous integrated semigroups. Using convolution properties established in Section 1.3, we prove a beautiful existence and uniqueness theorem due to Da Prato and Sinestrari for the inhomogeneous Cauchy problem defined by such operators. The second interesting class of examples are resolvent positive operators which always generate twice integrated semigroups. This will be proved in Section 3.11. In Chapter 6 a resolvent positive operator will provide an elegant transition from elliptic to parabolic problems. Finally, in Section 3.14 we show that the second order Cauchy problem is well-posed on a space X if and only if the associated canonical system generates an integrated semigroup on the product space $X \times X$.

In Section 3.10 we show that integrated semigroups and semigroups are equivalent, up to the choice of the underlying Banach space. This choice is particularly interesting in the context of the second order Cauchy problem. In Section 3.14 we will show the remarkable result that the space of well-posedness is unique, and we find the phase space associated to the second order problem. In the applications to the wave equation given in Chapter 7 we will see how this space is well adapted to perturbation theory, allowing us to prove well-posedness of the wave equation defined by very general second order elliptic operators.

Special attention is given to C_0-groups; i.e., to Cauchy problems allowing unique mild solutions on the line. In Section 3.9 we study when a holomorphic semigroup of angle $\pi/2$ has a boundary group. This problem will occur again in Section 3.16 where we investigate which cosine functions allow a square root reduction. A striking theorem due to Fattorini shows that on UMD-spaces a square root reduction is always possible; i.e., each generator A of a cosine function is of the form $A = B^2 - \omega$ where B generates a C_0-group and $\omega \geq 0$. This beautiful result concludes the three sections on the second order Cauchy problem, applications of which will be given in Chapters 7 and 8.

Chapter 1

The Laplace Integral

The first three sections of this chapter are of a preliminary nature. There, we collect properties of the Bochner integral of functions of a real variable with values in a Banach space X. We then concentrate on the basic properties of the Laplace integral

$$\hat{f}(\lambda) := \int_0^\infty e^{-\lambda t} f(t)\, dt := \lim_{\tau\to\infty} \int_0^\tau e^{-\lambda t} f(t)\, dt$$

for locally Bochner integrable functions $f : \mathbb{R}_+ \to X$. In Section 1.4 we describe the set of complex numbers λ for which the Laplace integral converges. It will be shown that the domain of convergence is non-empty if and only if the antiderivative of f is of exponential growth. In Section 1.5 we discuss the holomorphy of $\lambda \mapsto \hat{f}(\lambda)$ and in Section 1.7 we show that f is uniquely determined by the Laplace integrals $\hat{f}(\lambda)$ (uniqueness and inversion). In Section 1.6 we prove the operational properties of the Laplace integral which are essential in applications to differential and integral equations. In particular, we show that the Laplace integral of the convolution $k * f : t \mapsto \int_0^t k(t-s) f(s)\, ds$ of a scalar-valued function k with a vector-valued function f is given by

$$\widehat{(k * f)}(\lambda) = \hat{k}(\lambda)\hat{f}(\lambda)$$

if $\hat{f}(\lambda)$ exists and $\hat{k}(\lambda)$ exists as an absolutely convergent integral. In Section 1.8 we consider vector-valued Fourier transforms and we show that Plancherel's theorem and the Paley-Wiener theorem extend to functions with values in a Hilbert space. Finally, after introducing the basic properties of the Riemann-Stieltjes integral in Section 1.9, we extend in Section 1.10 the basic properties of Laplace integrals to Laplace-Stieltjes integrals

$$\widehat{dF}(\lambda) := \int_0^\infty e^{-\lambda t}\, dF(t) := \lim_{\tau\to\infty} \int_0^\tau e^{-\lambda t}\, dF(t)$$

of functions F of bounded semivariation.

If f is Bochner integrable, then the normalized antiderivative $t \mapsto F(t) := \int_0^t f(s)\,ds$ is of bounded variation. We will see that $\hat{f}(\lambda)$ exists if and only if $\widehat{dF}(\lambda)$ exists, and in this case $\hat{f}(\lambda) = \widehat{dF}(\lambda)$. Thus, the Laplace-Stieltjes integral is a natural extension of the Laplace integral. This extension is crucial for our discussion of the Laplace transform in Chapter 2 since there are many functions $r : (\omega, \infty) \to X$ which can be represented as a Laplace-Stieltjes integral, but not as a Laplace integral of a Bochner integrable function. Examples are, among others, Dirichlet series $r(\lambda) = \sum_{n=1}^{\infty} a_n e^{-\lambda n} = \widehat{dF}(\lambda)$, where F is the step function $\sum_{n=1}^{\infty} a_n \chi_{(n,\infty)}$, or any function $r(\lambda) = \widehat{dF}(\lambda)$, where F is of bounded semivariation, but not the antiderivative of a Bochner integrable function.

1.1 The Bochner Integral

This section contains some properties of the Bochner integral of vector-valued functions. We shall consider only those properties which are used in later sections, and we shall assume that the reader is familiar with the basic facts about measure and integration of scalar-valued functions.

Let X be a complex Banach space, and let I be an interval (bounded or unbounded) in $\mathbb{R}$, or a rectangle in $\mathbb{R}^2$. A function $f : I \to X$ is *simple* if it is of the form $f(t) = \sum_{r=1}^{n} x_r \chi_{\Omega_r}(t)$ for some $n \in \mathbb{N} := \{1, 2, \ldots\}, x_r \in X$ and Lebesgue measurable sets $\Omega_r \subset I$ with finite Lebesgue measure $m(\Omega_r)$; f is a *step function* when each Ω_r can be chosen to be an interval, or a rectangle in $\mathbb{R}^2$. Here χ_Ω denotes the characteristic (indicator) function of Ω. In the representation of a simple function, the sets Ω_r may always be arranged to be disjoint, and then

$$f(t) = \begin{cases} x_r & (t \in \Omega_r; r = 1, 2, \ldots, n) \\ 0 & \text{otherwise.} \end{cases}$$

A function $f : I \to X$ is *measurable* if there is a sequence of simple functions g_n such that $f(t) = \lim_{n\to\infty} g_n(t)$ for almost all $t \in I$. Since any χ_Ω for Ω measurable is a pointwise almost everywhere (a.e.) limit of a sequence of step functions, it is not difficult to see that the functions g_n can be chosen to be step functions. When $X = \mathbb{C}$, this definition agrees with the usual definition of (Lebesgue) measurable functions. It is easy to see that if $f : I \to X$, $g : I \to X$ and $h : I \to \mathbb{C}$ are measurable, then $f + g$ and $h \cdot f$ are measurable. Moreover, if $k : X \to Y$ is continuous (where Y is any Banach space), then $k \circ f$ is measurable whenever f is measurable. In particular, $\|f\|$ is measurable. If X is a closed subspace of Y, and f is measurable as a Y-valued function, then f is also measurable as an X-valued function.

To verify measurability of a function we often use the characterization given by Pettis's theorem below. We say that $f : I \to X$ is *countably valued* if there is a countable partition $\{\Omega_n : n \in \mathbb{N}\}$ of I into subsets Ω_n such that f is constant on each Ω_n; it is easy to see that f is measurable if each Ω_n is measurable (and conversely, $\{t \in I : f(t) = x\}$ is measurable whenever f is measurable and $x \in X$). Also, $f : I \to X$ is called *almost separably valued* if there is a null set Ω_0 in I such that $f(I \setminus \Omega_0) := \{f(t) : t \in I \setminus \Omega_0\}$ is separable (equivalently, $f(I \setminus \Omega_0)$ is contained in a separable closed subspace of X); f is *weakly measurable* if $x^* \circ f : t \mapsto \langle f(t), x^* \rangle$ is Lebesgue measurable for each x^* in the dual space X^* of X.

Here and subsequently, $\langle \cdot, \cdot \rangle$ denotes the duality between X and X^*. For a subset D of X, we shall denote the closure of D in X by $\overline{D}$. For $x \in X$ and $\varepsilon > 0$, we shall let $B_X(x, \varepsilon) := B(x, \varepsilon) := \{y \in X : \|y - x\| < \varepsilon\}$ and $\overline{B}(x, \varepsilon) := \{y \in X : \|y - x\| \leq \varepsilon\}$. We shall also use this notation when $X = \mathbb{R}^n$ or $X = \mathbb{C}$, when it will be implicit that the norm is the Euclidean norm.

Theorem 1.1.1 (Pettis). *A function $f : I \to X$ is measurable if and only if it is weakly measurable and almost separably valued.*

Proof. If f is measurable, then there exist a null set Ω_0 and simple functions g_n such that $g_n \to f$ pointwise on $I \setminus \Omega_0$. The simple functions $x^* \circ g_n$ converge to $x^* \circ f$ on $I \setminus \Omega_0$ for all $x^* \in X^*$. Therefore, f is weakly measurable. The values taken by the functions g_n form a countable set D and $f(I \setminus \Omega_0) \subset \overline{D}$. Thus, f is almost separably valued.

To prove the converse statement one may replace X by the smallest closed subspace which contains $f(I \setminus \Omega_0)$ and then choose a countable dense set $\{x_n : n \in \mathbb{N}\}$. By the Hahn-Banach theorem, there are unit vectors $x_n^* \in X^*$ with $|\langle x_n, x_n^* \rangle| = \|x_n\|$. For any $\varepsilon > 0$ and $x \in X$ there exists x_k such that $\|x - x_k\| < \varepsilon$. Hence,

$$\begin{aligned} \sup_n |\langle x, x_n^* \rangle| &\leq \|x\| \leq \|x_k\| + \varepsilon = |\langle x_k, x_k^* \rangle| + \varepsilon \\ &\leq |\langle x - x_k, x_k^* \rangle| + |\langle x, x_k^* \rangle| + \varepsilon \\ &\leq \sup_n |\langle x, x_n^* \rangle| + 2\varepsilon. \end{aligned}$$

So

$$\|x\| = \sup_n |\langle x, x_n^* \rangle| \quad \text{for all } x \in X. \tag{1.1}$$

This implies that $t \mapsto \|f(t) - x\| = \sup_n |\langle f(t) - x, x_n^* \rangle|$ is measurable for all $x \in X$. Let

$$\Delta := \{t \in I \setminus \Omega_0 : \|f(t)\| > 0\} \text{ and } \Delta_{n,\varepsilon} := \{t \in \Delta : \|f(t) - x_n\| < \varepsilon\}$$

for $\varepsilon > 0$ and $n \in \mathbb{N}$. The sets $\Delta_{n,\varepsilon}$ are measurable and $\bigcup_n \Delta_{n,\varepsilon} = \Delta$. For fixed $\varepsilon > 0$, the sets $\Omega_{1,\varepsilon} := \Delta_{1,\varepsilon}$ and $\Omega_{n,\varepsilon} := \Delta_{n,\varepsilon} \setminus \bigcup_{k<n} \Delta_{k,\varepsilon}$ $(n \geq 2)$ form a measurable partition of Δ. Define a measurable, countably valued function g_ε on I by $g_\varepsilon := \sum_{i=1}^\infty x_i \chi_{\Omega_{i,\varepsilon}}$. Let $t \in I \setminus \Omega_0$. If $t \notin \Delta$, then $f(t) = 0 = g_\varepsilon(t)$. If $t \in \Delta$, then there exists $n \in \mathbb{N}$ such that $t \in \Omega_{n,\varepsilon}$. Hence,

$$\|f(t) - g_\varepsilon(t)\| < \varepsilon \quad \text{for all } t \in I \setminus \Omega_0.$$

This shows that f is the uniform limit almost everywhere of a sequence of measurable, countably valued functions.

Let (I_n) be an increasing sequence of bounded subintervals of I with $I = \bigcup_n I_n$. For each $n \in \mathbb{N}$, define a simple function $h_n := g_{2^{-n}}\chi_{H_n}$, where $H_n := I_n \cap \bigcup_{i=1}^{k_n} \Omega_{i,2^{-n}}$ and k_n is chosen such that the Lebesgue measure $m(I_n \setminus H_n) < 2^{-n}$. If $t \in \bigcap_{n=k}^{\infty} H_n$ for some $k \geq 1$, then

$$\|f(t) - h_n(t)\| = \|f(t) - g_{2^{-n}}(t)\| < 2^{-n}$$

for all $n \geq k$. Thus, $\lim_{n\to\infty} h_n(t) = f(t)$ for all $t \in \bigcup_{k=1}^{\infty} \bigcap_{n=k}^{\infty} H_n$. For $k \geq j$,

$$m\left(I_j \setminus \bigcap_{n=k}^{\infty} H_n\right) \leq \sum_{n=k}^{\infty} m(I_n \setminus H_n) < 2^{-k+1}.$$

Hence, $I_j \setminus \bigcup_{k=1}^{\infty} \bigcap_{n=k}^{\infty} H_n$ is null, for each j. This shows that $\lim_{n\to\infty} h_n(t) = f(t)$ for almost all $t \in I$. □

Corollary 1.1.2. *Let $f : I \to X$. Then the following statements hold:*

a) *The function f is measurable if and only if it is the uniform limit almost everywhere of a sequence of measurable, countably valued functions.*

b) *If X is separable, then f is measurable if and only if it is weakly measurable.*

c) *If f is continuous, then it is measurable.*

d) *If $f_n : I \to X$ are measurable functions and $f_n \to f$ pointwise a.e., then f is measurable.*

Proof. The statement b) is an immediate consequence of Pettis's Theorem 1.1.1. For d), observe first that f is weakly measurable. Define $\Omega_0 := \cup_n \Omega_n$ where Ω_n is a null set such that $f_n(I \setminus \Omega_n)$ is separable. Then $m(\Omega_0) = 0$ and $\Delta := \bigcup_n f_n(I \setminus \Omega_0)$ is separable. Since the least closed subspace containing Δ is separable and contains $f(I \setminus \Omega_0)$ it follows that f is almost separably valued. Thus, f is measurable. If f is continuous, then f is weakly measurable and the countable set $\{f(t) : t \in \mathbb{Q}\}$ is dense in the range of f. Again by Pettis's theorem, f is measurable. One implication of a) was established in the proof of Pettis's theorem and the converse follows from d). □

Pettis's theorem can be improved considerably in the following way. A subset W of X^* is called *separating* if for all $x \in X \setminus \{0\}$ there exists $x^* \in W$ such that $\langle x, x^* \rangle \neq 0$.

Corollary 1.1.3. *Let $f : I \to X$ be an almost separably valued function. Assume that $x^* \circ f$ is measurable for all x^* in a separating subset W of X^*. Then f is measurable.*

Proof. Changing f on a set of measure 0 and replacing X by a subspace, we can assume that X is separable. Let

$$Y := \{x^* \in X^* : \; x^* \circ f \; \text{ is measurable}\}.$$

Then Y is a subspace of X^* which contains W. By the Hahn-Banach theorem, Y is weak* dense in X^*. Let $Y_1 = Y \cap \overline{B}_{X^*}(0,1)$. We show that Y_1 is weak* closed. Let x^* be in the weak* closure of Y_1. Since X is separable, the weak* topology on $\overline{B}_{X^*}(0,1)$ is metrizable (see [Meg98, Theorem 2.6.23], for example). Thus, there exists a sequence $(x_n^*)_{n\in\mathbb{N}}$ in Y_1 converging to x^*. Hence, $x_n^* \circ f \to x^* \circ f$ as $n \to \infty$ pointwise on I. Thus, $x^* \circ f$ is measurable; i.e., $x^* \in Y_1$. This proves the claim. It follows from the Krein-Smulyan theorem (Theorem A.6) that Y is weak* closed. Since Y is weak* dense, we have $Y = X^*$; i.e., f is weakly measurable. Now the result follows from Theorem 1.1.1. □

For a simple function $g : I \to X$, $g = \sum_{i=1}^n x_i \chi_{\Omega_i}$, we define

$$\int_I g(t)\, dt := \sum_{i=1}^n x_i m(\Omega_i)$$

where $m(\Omega)$ is the Lebesgue measure of Ω. It is routine to verify that the definition is independent of the representation $g = \sum_{i=1}^n x_i \chi_{\Omega_i}$, and the integral so defined is linear.

A function $f : I \to X$ is called *Bochner integrable* if there exist simple functions g_n such that $g_n \to f$ pointwise a.e., and $\lim_{n\to\infty} \int_I \|f(t) - g_n(t)\|\, dt = 0$. If $f : I \to X$ is Bochner integrable, then the *Bochner integral of f on I* is

$$\int_I f(t)\, dt := \lim_{n\to\infty} \int_I g_n(t)\, dt.$$

It is easy to see that this limit exists and is independent of the choice of the sequence (g_n). If Ω is measurable with finite measure, then χ_Ω can be approximated in L^1-norm by step functions, and it follows that the functions g_n can always be chosen to be step functions. The integral $\int_I f(t)\, dt$ lies in the closed linear span of $\{f(t) : \; t \in I\}$. The set of all Bochner integrable functions from I to X is a linear space and the Bochner integral is a linear mapping. When $X = \mathbb{C}$, the definitions of Bochner integrability and integrals agree with those of Lebesgue integration theory.

When I is a rectangle, we may denote a Bochner integral by $\int_I f(s,t)\, d(s,t)$.

It is one of the great virtues of the Bochner integral that the class of Bochner integrable functions is easily characterized.

Theorem 1.1.4 (Bochner). *A function $f : I \to X$ is Bochner integrable if and only if f is measurable and $\|f\|$ is integrable. If f is Bochner integrable, then*

$$\left\| \int_I f(t)\, dt \right\| \le \int_I \|f(t)\|\, dt. \tag{1.2}$$

Proof. If f is Bochner integrable, then there exists an approximating sequence of simple functions g_n. Thus f and $\|f\|$ are measurable. The integrability of $\|f\|$ follows from

$$\int_I \|f(t)\|\,dt \le \int_I \|g_n(t)\|\,dt + \int_I \|f(t) - g_n(t)\|\,dt.$$

Moreover,

$$\begin{aligned} \left\|\int_I f(t)\,dt\right\| = \lim_{n\to\infty} \left\|\int_I g_n(t)\,dt\right\| &\le \lim_{n\to\infty} \int_I \|g_n(t)\|\,dt \\ &= \int_I \|f(t)\|\,dt. \end{aligned}$$

To prove the converse statement, let (h_n) be a sequence of simple functions approximating f pointwise on $I \setminus \Omega_0$, where $m(\Omega_0) = 0$. Define simple functions by

$$g_n(t) := \begin{cases} h_n(t) & \text{if } \|h_n(t)\| \le \|f(t)\|(1 + n^{-1}), \\ 0 & \text{otherwise.} \end{cases}$$

Then $\|g_n(t)\| \le \|f(t)\|(1 + n^{-1})$ and $\lim_{n\to\infty} \|g_n(t) - f(t)\| = 0$ for all $t \in I \setminus \Omega_0$. Because the functions $\|f\|$ and $\|g_n - f\|$ are integrable and $\|g_n(t) - f(t)\| \le 3\|f(t)\|$, we can apply the scalar dominated convergence theorem and obtain that $\lim_{n\to\infty} \int_I \|g_n(t) - f(t)\|\,dt = 0$. □

Example 1.1.5. a) Let X be the Lebesgue space $L^\infty(0,1)$ of all (equivalence classes of) bounded measurable functions from $(0,1)$ to $\mathbb{C}$. Let $f : (0,1) \to L^\infty(0,1)$ be given by $f(t) := \chi_{(0,t)}$. Then f is not almost separably valued since $\|f(t) - f(s)\| = 1$ for $t \ne s$. Thus, f is not measurable and therefore not Bochner integrable.

b) Let X be the Banach space c_0 of all complex sequences $x = (x_n)_{n\in\mathbb{N}}$ such that $\lim_{n\to\infty} x_n = 0$, with $\|x\| = \sup_n |x_n|$. Identify X^* with the space ℓ^1 of all complex sequences $a = (a_n)_{n\in\mathbb{N}}$ such that $\|a\| := \sum_{n=1}^\infty |a_n| < \infty$. Let $f : [0,1] \to c_0$ be given by $f(t) := (f_n(t))_{n\in\mathbb{N}}$ where $f_n(t) := n\chi_{[0,\frac{1}{n}]}(t)$. Let $x^* = (a_n)_{n\in\mathbb{N}} \in \ell^1$. Then $t \mapsto \langle f(t), x^*\rangle = \sum_{n=1}^\infty n a_n \chi_{[0,\frac{1}{n}]}(t)$ is measurable on $[0,1]$. Since c_0 is separable, it follows from Pettis's theorem that f is measurable. Moreover,

$$\int_0^1 |\langle f(t), x^*\rangle|\,dt \le \sum_{n=1}^\infty |a_n| = \|x^*\| < \infty.$$

However, $\|f(t)\| = n$ for $t \in (\frac{1}{n+1}, \frac{1}{n}]$, so $t \mapsto \|f(t)\|$ is not integrable on $[0,1]$. Thus, f is not Bochner integrable on $[0,1]$. □

Now we will consider the behaviour of the Bochner integral under linear operators. The following result is a straightforward consequence of the definition of the Bochner integral, and we shall use it frequently without comment, especially in the case of a linear functional ($Y = \mathbb{C}$).

Proposition 1.1.6. *Let $T : X \to Y$ be a bounded linear operator between Banach spaces X and Y, and let $f : I \to X$ be Bochner integrable. Then $T \circ f : t \mapsto T(f(t))$ is Bochner integrable and $T \int_I f(t)\,dt = \int_I T(f(t))\,dt$.*

We shall also need a version of Proposition 1.1.6 for a closed operator A on X (see Appendix B for notation and terminology).

Proposition 1.1.7. *Let A be a closed linear operator on X. Let $f : I \to X$ be Bochner integrable. Suppose that $f(t) \in D(A)$ for all $t \in I$ and $A \circ f : I \to X$ is Bochner integrable. Then $\int_I f(t)\,dt \in D(A)$ and*

$$A \int_I f(t)\,dt = \int_I A(f(t))\,dt.$$

Proof. Consider $X \times X$ as a Banach space in the norm $\|(x,y)\| = \|x\| + \|y\|$. The graph $G(A)$ of A is a closed subspace of $X \times X$. Define $g : I \to G(A) \subset X \times X$ by $g(t) = (f(t), A(f(t)))$. It is easy to see that g is measurable and

$$\int_I \|g(t)\|\,dt = \int_I \|f(t)\|\,dt + \int_I \|A(f(t))\|\,dt < \infty.$$

By Theorem 1.1.4, g is Bochner integrable. Moreover, $\int_I g(t)\,dt \in G(A)$. Applying Proposition 1.1.6 to the two projection maps of $X \times X$ onto X shows that

$$\int_I g(t)\,dt = \left(\int_I f(t)\,dt\,,\ \int_I A(f(t))\,dt \right).$$

This gives the result. □

Now we give vector-valued versions of two classical theorems of integration theory.

Theorem 1.1.8 (Dominated Convergence). *Let $f_n : I \to X$ $(n \in \mathbb{N})$ be Bochner integrable functions. Assume that $f(t) := \lim_{n\to\infty} f_n(t)$ exists a.e. and there exists an integrable function $g : I \to \mathbb{R}$ such that $\|f_n(t)\| \le g(t)$ a.e. for all $n \in \mathbb{N}$. Then f is Bochner integrable and $\int_I f(t)\,dt = \lim_{n\to\infty} \int_I f_n(t)\,dt$. Furthermore, $\int_I \|f(t) - f_n(t)\|\,dt \to 0$ as $n \to \infty$.*

Proof. The function f is Bochner integrable since it is measurable (by Corollary 1.1.2) and since $\|f\|$ is integrable (because $\|f(t)\| \le g(t)$ a.e.). Define $h_n(t) := \|f(t) - f_n(t)\|$ for $t \in I$. Since $|h_n(t)| \le 2g(t)$ and $h_n(t) \to 0$ a.e., the scalar dominated convergence theorem implies that $\int_I \|f(t) - f_n(t)\|\,dt \to 0$ as $n \to \infty$. By (1.2),

$$\left\| \int_I f(t)\,dt - \int_I f_n(t)\,dt \right\| \to 0.$$

□

Theorem 1.1.9 (Fubini's Theorem). *Let $I = I_1 \times I_2$ be a rectangle in $\mathbb{R}^2$, let $f : I \to X$ be measurable, and suppose that*

$$\int_{I_1} \int_{I_2} \|f(s,t)\| \, dt \, ds < \infty.$$

Then f is Bochner integrable and the repeated integrals

$$\int_{I_1} \int_{I_2} f(s,t) \, dt \, ds, \qquad \int_{I_2} \int_{I_1} f(s,t) \, ds \, dt$$

exist and are equal, and they coincide with the double integral $\int_I f(s,t) \, d(s,t)$.

Proof. Since any measurable function is almost separably valued, we may assume that X is separable.

The scalar-valued case of Fubini's theorem implies that $\|f\|$ is integrable on I, $\int_{I_2} \|f(s,t)\| \, dt$ exists for almost all $s \in I_1$, and for each $x^* \in X^*$ the repeated integrals

$$\int_{I_1} \int_{I_2} \langle f(s,t), x^* \rangle \, dt \, ds, \qquad \int_{I_2} \int_{I_1} \langle f(s,t), x^* \rangle \, ds \, dt$$

exist and are equal. It follows from Theorem 1.1.4 that $f : I \to X$ is Bochner integrable and $\int_{I_2} f(s,t) \, dt$ exists for almost all $s \in I_1$, and from Theorem 1.1.1 that $s \mapsto \int_{I_2} f(s,t) \, dt$ is measurable. Moreover,

$$\int_{I_1} \left\| \int_{I_2} f(s,t) \, dt \right\| \, ds \le \int_{I_1} \int_{I_2} \|f(s,t)\| \, dt \, ds < \infty,$$

so Theorem 1.1.4 shows that $\int_{I_1} \left(\int_{I_2} f(s,t) \, dt \right) \, ds$ exists. Since

$$\int_{I_2} \int_{I_1} \|f(s,t)\| \, ds \, dt = \int_{I_1} \int_{I_2} \|f(s,t)\| \, dt \, ds,$$

it follows similarly that $\int_{I_2} \left(\int_{I_1} f(s,t) \, ds \right) \, dt$ exists. For any $x^* \in X^*$,

$$\begin{aligned} \left\langle \int_{I_1} \left(\int_{I_2} f(s,t) \, dt \right) \, ds, x^* \right\rangle &= \int_{I_1} \int_{I_2} \langle f(s,t), x^* \rangle \, dt \, ds \\ &= \int_I \langle f(s,t), x^* \rangle \, d(s,t) \\ &= \left\langle \int_I f(s,t) \, d(s,t), x^* \right\rangle \\ &= \int_{I_2} \int_{I_1} \langle f(s,t), x^* \rangle \, ds \, dt \\ &= \left\langle \int_{I_2} \left(\int_{I_1} f(s,t) \, ds \right) \, dt, x^* \right\rangle. \end{aligned}$$

The Hahn-Banach theorem implies that

$$\int_{I_1} \left(\int_{I_2} f(s,t)\, dt \right) ds = \int_I f(s,t)\, d(s,t) = \int_{I_2} \left(\int_{I_1} f(s,t)\, ds \right) dt. \qquad \square$$

Let $L^1(I,X)$ denote the space of all Bochner integrable functions $f : I \to X$, and let

$$\|f\|_1 := \int_I \|f(t)\|\, dt.$$

In the usual way, we shall identify functions which differ only on sets of measure zero. Then $\|\cdot\|_1$ is a norm on $L^1(I,X)$.

Theorem 1.1.10. *The space $L^1(I,X)$ is a Banach space.*

Proof. Let (f_n) be a sequence in $L^1(I,X)$ with $\sum \|f_n\|_1 < \infty$. By the monotone convergence theorem for series of positive scalar-valued functions, $\sum \|f_n(t)\| < \infty$ a.e., $\sum_{n=1}^{\infty} \|f_n(\cdot)\|$ is integrable, and

$$\int_I \sum_{n=1}^{\infty} \|f_n(t)\|\, dt = \sum_{n=1}^{\infty} \int_I \|f_n(t)\|\, dt.$$

Hence, $\sum_{n=1}^{\infty} f_n(t)$ converges a.e. to a sum $g(t)$ in the Banach space X. By Corollary 1.1.2, g is measurable. Moreover, $\|g(t)\| \le \sum_{n=1}^{\infty} \|f_n(t)\|$, so $\|g\|$ is integrable. By Theorem 1.1.4, g is integrable. Finally,

$$\left\| g - \sum_{n=1}^{k} f_n \right\|_1 \le \int_I \|g(t) - \sum_{n=1}^{k} f_n(t)\|\, dt \le \int_I \sum_{n=k+1}^{\infty} \|f_n(t)\|\, dt \to 0$$

as $k \to \infty$. Thus, $L^1(I,X)$ is a Banach space. $\square$

By the definition of Bochner integrability, the simple functions are dense in $L^1(I,X)$ and, by the remarks following the definition, the step functions are dense. It follows easily that the infinitely differentiable functions of compact support are also dense in $L^1(I,X)$.

We shall be particularly interested in the case when $I = \mathbb{R}_+ := [0,\infty)$. If $f \in L^1(\mathbb{R}_+,X)$, an application of the Dominated Convergence Theorem shows that

$$\int_0^{\infty} f(t)\, dt = \lim_{\tau\to\infty} \int_0^{\tau} f(t)\, dt. \tag{1.3}$$

When $f \in L^1_{loc}(\mathbb{R}_+,X)$ (i.e., f is Bochner integrable on $[0,\tau]$ for every $\tau \in \mathbb{R}_+$), the limit in (1.3) may exist without f being Bochner integrable on $\mathbb{R}_+$. If the limit exists, we say that $\int_0^{\infty} f(t)\, dt$ *converges as an improper* (or *principal value*) *integral*, and we define

$$\int_0^{\infty} f(t)\, dt := \lim_{\tau\to\infty} \int_0^{\tau} f(t)\, dt.$$

When $f \in L^1(\mathbb{R}_+, X)$, i.e. $\int_0^\infty \|f(t)\|\, dt < \infty$, we say that the integral is *absolutely convergent*.

For $1 < p < \infty$, let $L^p(I, X)$ denote the space of all measurable functions $f : I \to X$ such that

$$\|f\|_p := \left(\int_I \|f(t)\|^p\, dt \right)^{1/p} < \infty.$$

Let $L^\infty(I, X)$ be the space of all measurable functions $f : I \to X$ such that

$$\|f\|_\infty := \operatorname*{ess\,sup}_{t \in I} \|f(t)\| < \infty.$$

Note that the spaces $L^p(I, \mathbb{C})$ $(1 \le p \le \infty)$ are the usual Lebesgue spaces which we shall denote simply by $L^p(I)$. With the usual identifications, each of the spaces $L^p(I, X)$ becomes a Banach space. The proofs of completeness are similar to the scalar-valued cases.

The proof of Theorem 1.1.4 shows that the simple functions are dense in $L^p(I, X)$ for $1 < p < \infty$ (so $L^p(I, X)$ can also be defined in a similar way to the Bochner integrable functions). It follows that the step functions, and the infinitely differentiable functions of compact support, are also dense. By considering such functions first and then approximating, one may show as in the scalar-valued case, that if $f \in L^p(I, X)$ and

$$f_t(s) := \begin{cases} f(s-t) & \text{if } s - t \in I, \\ 0 & \text{otherwise,} \end{cases}$$

then $t \mapsto f_t$ is continuous from $\mathbb{R}$ into $L^p(I, X)$ for $1 \le p < \infty$.

We have presented the theory above in the case when I is an interval in $\mathbb{R}$ (or, for Fubini's theorem, I is a rectangle in $\mathbb{R}^2$). Almost all the integrals which appear in this book will indeed be over intervals in $\mathbb{R}$ (or repeated integrals in $\mathbb{R}^2$). However, the entire theory works, with no changes in the proofs, when I is a measurable set in $\mathbb{R}^n$ (or in $\mathbb{R}^m \times \mathbb{R}^n$, in Fubini's theorem). Since the step functions are dense in each of the spaces $L^p(I \times J, X)$ for $1 \le p < \infty$, it is easy to see from Fubini's theorem that there is an isometric isomorphism between $L^p(I \times J, X)$ and $L^p(I, L^p(J, X))$ given by $f \mapsto g$, where

$$(g(s))(t) := f(s, t).$$

This enables many properties of the spaces $L^p(I, X)$ when I is a rectangle in $\mathbb{R}^n$ to be deduced from the case $n = 1$.

Finally in this section, we introduce notation for spaces of continuous and differentiable functions. Let I be an interval in $\mathbb{R}$. We denote by $C(I, X)$ the vector space of all continuous functions $f : I \to X$. For $k \in \mathbb{N}$, we denote by $C^k(I, X)$ the space of all k-times differentiable functions with continuous kth derivative, and we put $C^\infty(I,X) := \bigcap_{k=1}^\infty C^k(I,X)$. When I is compact, $C(I,K)$ is a closed subspace of $L^\infty(I,X)$ and therefore a Banach space with respect to the supremum norm $\|\cdot\|_\infty$.

When I is not compact, we let $C_c(I,X)$ be the space of all functions in $C(I,X)$ with compact support, and $C_c^\infty(I,X) := C_c(I,X) \cap C^\infty(I,X)$. Thus $C_c^\infty(I,X)$ is a dense subspace of $L^p(I,X)$ for $1 \le p < \infty$. When $I = \mathbb{R}_+$ or $I = \mathbb{R}$, we shall also consider the space $C_0(I,X)$ of all continuous functions $f : I \to X$ such that $\lim_{|t|\to\infty, t\in I} \|f(t)\| = 0$ and the space $\mathrm{BUC}(I,X)$ of all bounded, uniformly continuous functions $f : I \to X$. These are both Banach spaces with respect to $\|\cdot\|_\infty$, and $C_0(I,X) \subset \mathrm{BUC}(I,X) \subset L^\infty(I,X)$.

When $X = \mathbb{C}$, we shall write $C(I)$ in place of $C(I,\mathbb{C})$, etc., and we shall extend this notation to cases when I is replaced by an open subset Ω of $\mathbb{R}^n$. Note that $C_c^\infty(\Omega)$ coincides with the space $\mathcal{D}(\Omega)$ of *test functions* on Ω (see Appendix E), and we shall use both notations according to context. Furthermore, when Ω is any locally compact space, we shall let $C_0(\Omega)$ be the Banach space of all continuous complex-valued functions on Ω which vanish at infinity, with the supremum norm. When K is any compact space, we shall let $C(K)$ be the Banach space of all continuous complex-valued functions on K, with the supremum norm.

1.2 The Radon-Nikodym Property

In this section we consider properties of functions F obtained as indefinite integrals. If $f : [a,b] \to X$ is Bochner integrable, we say that $F : [a,b] \to X$ is an *antiderivative* or *primitive* of f if

$$F(t) = F(a) + \int_a^t f(s)\,ds \qquad (t \in [a,b]).$$

Given a function $F : [a,b] \to X$ and a partition π, $a = t_0 < t_1 < \ldots < t_n = b$, let

$$V(\pi,F) := \sum_{i=1}^n \|F(t_i) - F(t_{i-1})\|.$$

Then F is said to be *of bounded variation* if

$$V(F) := V_{[a,b]}(F) := \sup_\pi V(\pi,F) < \infty,$$

where the supremum is taken over all partitions π of $[a,b]$.

We say that F is *absolutely continuous* on $[a,b]$ if for every $\varepsilon > 0$ there exists $\delta > 0$ such that $\sum_i \|F(b_i) - F(a_i)\| < \varepsilon$ for every finite collection $\{(a_i,b_i)\}$ of disjoint intervals in $[a,b]$ with $\sum_i (b_i - a_i) < \delta$. We say that F is *Lipschitz continuous* if there exists M such that $\|F(t) - F(s)\| \le M|t-s|$ for all $s,t \in [a,b]$. Clearly, any Lipschitz continuous function is absolutely continuous.

Proposition 1.2.1. *Let $F : [a,b] \to X$ be absolutely continuous. Then F is of bounded variation. Moreover, if $G(t) := V_{[a,t]}(F)$, then G is absolutely continuous on $[a,b]$.*

Proof. Take $\varepsilon > 0$, and let δ be as in the definition of absolute continuity of F. Then $\sum_i V_{[a_i,b_i]}(F) \le \varepsilon$ whenever $\{(a_i,b_i)\}$ is a finite collection of disjoint subintervals of $[a,b]$ with $\sum_i (b_i - a_i) < \delta$. In particular, F is of bounded variation on any subinterval of length less than δ. Since $[a,b]$ is a finite union of such intervals, F is of bounded variation on $[a,b]$. Moreover,

$$\sum_i |G(b_i) - G(a_i)| = \sum_i V_{[a_i,b_i]}(F) < \varepsilon.$$

Thus, G is absolutely continuous. □

A point $t \in [a,b]$ is said to be a *Lebesgue point* of $f \in L^1([a,b],X)$ if $\lim_{h\to 0} \frac{1}{h}\int_t^{t+h} \|f(s) - f(t)\|\, ds = 0$. It is easy to see that any point of continuity is a Lebesgue point, and the following proposition shows that almost all points are Lebesgue points.

Proposition 1.2.2. *Let $f : [a,b] \to X$ be Bochner integrable and $F(t) := \int_a^t f(s)\, ds$ $(t \in [a,b])$. Then*

a) *F is differentiable a.e. and $F' = f$ a.e.*

b) *$\lim_{h\to 0} \frac{1}{h}\int_t^{t+h} \|f(s) - f(t)\|\, ds = 0$ t-a.e.*

c) *F is absolutely continuous.*

d) *$V_{[a,b]}(F) = \int_a^b \|f(s)\|\, ds$.*

Proof. To show a) and b) let g_n be step functions such that

$$f(t) = \lim_{n\to\infty} g_n(t) \text{ a.e. and } \lim_{n\to\infty} \int_a^b \|f(t) - g_n(t)\|\, dt = 0.$$

For $h > 0$,

$$\begin{aligned}\left\| \frac{1}{h}\int_t^{t+h} f(s)\, ds - f(t) \right\| &\le \frac{1}{h}\int_t^{t+h} \|f(s) - f(t)\|\, ds \\ &\le \frac{1}{h}\int_t^{t+h} \|f(s) - g_n(s)\|\, ds \\ &\quad + \frac{1}{h}\int_t^{t+h} \|g_n(s) - g_n(t)\|\, ds + \|g_n(t) - f(t)\|.\end{aligned}$$

Since g_n is a step function and $s \mapsto \|f_n(s) - g_n(s)\|$ is Lebesgue integrable, it follows from Lebesgue's theorem [Rud87, Theorem 8.17] that

$$\begin{aligned}\limsup_{h\downarrow 0}\left\|\frac{1}{h}\int_t^{t+h} f(s)\,ds - f(t)\right\| &\le \limsup_{h\downarrow 0}\frac{1}{h}\int_t^{t+h}\|f(s)-f(t)\|\,ds \\ &\le 2\|g_n(t) - f(t)\|\end{aligned}$$

for all $t \in [a,b] \setminus \Omega_n$ and some null set Ω_n. Taking the limit as $n \to \infty$ yields the right-differentiability of F and

$$\lim_{h\downarrow 0}\frac{1}{h}\int_t^{t+h}\|f(s)-f(t)\|\,ds = 0$$

for all $t \in [a,b] \setminus \bigcup_{n\in\mathbb{N}} \Omega_n$. The left-hand limits are similar.

For c), let $\varepsilon > 0$. There exists $\delta > 0$ such that $\int_\Omega \|f(s)\|\,ds < \varepsilon$ whenever $\mu(\Omega) < \delta$. If $\{(a_i,b_i)\}$ is a finite collection of disjoint subintervals of $[a,b]$ with $\sum_i (b_i - a_i) < \delta$, then taking $\Omega = \bigcup_i (a_i,b_i)$, we deduce that

$$\sum_i \|F(b_i) - F(a_i)\| = \sum_i \left\|\int_{a_i}^{b_i} f(s)\,ds\right\| \le \int_\Omega \|f(s)\|\,ds < \varepsilon.$$

To prove the statement d), observe first that, for any partition π of $[a,b]$,

$$V(\pi,F) = \sum_i \left\|\int_{t_{i-1}}^{t_i} f(s)\,ds\right\| \le \int_a^b \|f(s)\|\,ds.$$

Thus, $V(F) \le \int_a^b \|f(s)\|\,ds$. Conversely, given $\varepsilon > 0$, we may choose a step function g such that $\int_a^b \|f(s) - g(s)\|\,ds < \varepsilon$. There is a partition π of $[a,b]$ such that g is constant on each interval (t_{i-1},t_i). Then

$$\begin{aligned}\int_a^b \|f(s)\|\,ds - V(F) &\le \int_a^b \|f(s)\|\,ds - V(\pi,F) \\ &\le \int_a^b \|g(s)\|\,ds + \varepsilon - \sum_i \left\|\int_{t_{i-1}}^{t_i} f(s)\,ds\right\| \\ &= \sum_i \left(\left\|\int_{t_{i-1}}^{t_i} g(s)\,ds\right\| - \left\|\int_{t_{i-1}}^{t_i} f(s)\,ds\right\|\right) + \varepsilon \\ &\le \int_a^b \|f(s) - g(s)\|\,ds + \varepsilon \\ &\le 2\varepsilon.\end{aligned}$$

Since $\varepsilon > 0$ is arbitrary, this completes the proof of d). □

In the scalar case, the fundamental theorem of calculus [Rud87, Theorem 8.18] states that any absolutely continuous function $F : [a, b] \to \mathbb{C}$ is differentiable a.e., $f := F'$ is Lebesgue integrable, and $F(t) - F(a) = \int_a^t f(s)\,ds$ for all $t \in [a, b]$. We will see below (Example 1.2.8) that the fundamental theorem does not hold for Lipschitz continuous functions with values in arbitrary Banach spaces. The following weaker statement holds for all Banach spaces.

Proposition 1.2.3. *Let $F : [a, b] \to X$ be absolutely continuous, and suppose that $f(t) := F'(t)$ exists a.e. Then f is Bochner integrable and $F(t) = F(a) + \int_a^t f(s)\,ds$ for all $t \in [a, b]$.*

Proof. Since $f(t) = \lim_{n\to\infty} n(F(t + 1/n) - F(t))$, it follows from Corollary 1.1.2 that f is measurable. Let $G(t) := V_{[a,t]}(F)$, so $G : [a, b] \to \mathbb{R}$ is absolutely continuous by Proposition 1.2.1. Hence G is differentiable a.e. and $G' \in L^1[a, b]$. Since

$$\|F(t + h) - F(t)\| \le V_{[t,t+h]}(F) = G(t + h) - G(t),$$

$\|f(t)\| \le G'(t)$ a.e. Hence $\|f\| \in L^1[a, b]$, so f is Bochner integrable by Theorem 1.1.4. For $x^* \in X^*$,

$$\begin{aligned}\langle F(t), x^*\rangle &= \langle F(a), x^*\rangle + \int_a^t \langle f(s), x^*\rangle\,ds \\ &= \left\langle F(a) + \int_a^t f(s)\,ds, x^*\right\rangle\end{aligned}$$

by the scalar fundamental theorem of calculus. By the Hahn-Banach theorem, $F(t) = F(a) + \int_a^t f(s)\,ds$. □

Let I be any interval in $\mathbb{R}$. A function $F : I \to X$ is said to be *absolutely continuous* if it is absolutely continuous on each compact interval of I. We now consider the property that every absolutely continuous function $F : I \to X$ is differentiable a.e. It is easy to see that this property is independent of the interval I, so it is a property of X alone.

Proposition 1.2.4. *For any Banach space X the following are equivalent:*

(i) *Every absolutely continuous function $F : \mathbb{R}_+ \to X$ is differentiable a.e.*

(ii) *Every Lipschitz continuous function $F : \mathbb{R}_+ \to X$ is differentiable a.e.*

Proof. Clearly, (i) implies (ii). Assume that statement (ii) holds and let $F : \mathbb{R}_+ \to X$ be absolutely continuous. By Proposition 1.2.1, F is locally of bounded variation and G is absolutely continuous where $G(t) := V_{[0,t]}(F)$. Let $h(t) := G(t) + t$. Then h is strictly increasing, $h(0) = 0$, and $h(\mathbb{R}_+) = \mathbb{R}_+$. Moreover,

$$\|F(t) - F(s)\| \le G(t) - G(s) \le h(t) - h(s)$$

for all $t \geq s \geq 0$. Hence, $F \circ h^{-1} : \mathbb{R}_+ \to X$ is Lipschitz continuous. By assumption, $F \circ h^{-1}$ is differentiable a.e. Since $|h(t) - h(s)| \geq |t - s|$, h^{-1} maps null sets to null sets. Moreover, h is differentiable a.e., since G is absolutely continuous. It follows from the chain rule that F is differentiable a.e. □

Definition 1.2.5. *A Banach space X is said to have the* Radon-Nikodym property *if the equivalent conditions of Proposition* 1.2.4 *are satisfied.*

By the remarks above, the space X has the Radon-Nikodym property if and only if every Lipschitz continuous function $F : [0, 1] \to X$ is differentiable almost everywhere. It is clear that a closed subspace of a space with the Radon-Nikodym property has the Radon-Nikodym property.

Next we exhibit a class of spaces having the Radon-Nikodym property.

Theorem 1.2.6 (Dunford-Pettis). *Let $X = Y^*$ where Y is a Banach space, and suppose that X is separable. Then X has the Radon-Nikodym property.*

Proof. By Proposition 1.2.4, it suffices to show that every Lipschitz function $F : \mathbb{R}_+ \to X$ is differentiable a.e. We may assume that $F(0) = 0$ and the Lipschitz constant is 1, so that $\|F(t) - F(s)\| \leq |t - s|$. For $y \in Y$, the function $\langle y, F(\cdot)\rangle : \mathbb{R}_+ \to \mathbb{C}$ is Lipschitz with Lipschitz constant $\|y\|$, so there exists $g_y \in L^\infty(\mathbb{R}_+)$ such that $\|g_y\|_\infty \leq \|y\|$ and

$$\langle y, F(t)\rangle = \int_0^t g_y(s)\, ds \quad (t \in \mathbb{R}_+).$$

Moreover, g_y is unique up to null sets.

Since Y^* is separable, Y is also separable (see [Meg98, Theorem 1.12.11]). Let D be a countable dense subset of Y. Suppose that $y = \sum_{i=1}^n \alpha_i y_i$ for some $n \in \mathbb{N}$, $\alpha_i \in \mathbb{Q} + i\mathbb{Q}$ and $y_i \in D$. Then

$$\langle y, F(t)\rangle = \sum_{i=1}^n \alpha_i \langle y_i, F(t)\rangle = \int_0^t \sum_{i=1}^n \alpha_i g_{y_i}(s)\, ds,$$

so $g_y(s) = \sum_{i=1}^n \alpha_i g_{y_i}(s)$ a.e. Hence,

$$\left| \sum_{i=1}^n \alpha_i g_{y_i}(s) \right| \leq \left\| \sum_{i=1}^n \alpha_i y_i \right\| \tag{1.4}$$

for almost all $s \in \mathbb{R}_+$. This holds for all possible $n \in \mathbb{N}$, $\alpha_i \in \mathbb{Q} + i\mathbb{Q}$ and $y_i \in D$, but there are only countably many such possibilities. Hence there is a null subset Ω of $\mathbb{R}_+$ such that (1.4) holds for all $s \in \mathbb{R}_+ \setminus \Omega$ and all $n \in \mathbb{N}, \alpha_i \in \mathbb{Q} + i\mathbb{Q}$ and $y_i \in D$. It follows immediately that (1.4) holds for all $\alpha_i \in \mathbb{C}$. This shows that for $s \in \mathbb{R}_+ \setminus \Omega$, the map $y \mapsto g_y(s)$ from D to $\mathbb{C}$ extends to a unique $f(s) \in Y^* = X$ with $\|f(s)\| \leq 1$. For $y \in D$, $\langle y, f(\cdot)\rangle$ is measurable and bounded, and

$$\langle y, F(t)\rangle = \int_0^t \langle y, f(s)\rangle\, ds. \tag{1.5}$$

By density of D and the dominated convergence theorem, (1.5) is valid for $y \in Y$. Since Y is weak* dense in $X^* = Y^{**}$ and f is separably valued, it follows from Corollary 1.1.3 that $f : \mathbb{R}_+ \to X$ is measurable. Since f is bounded, f is locally Bochner integrable and it follows from (1.5) that

$$F(t) = \int_0^t f(s)\,ds.$$

By Proposition 1.2.2 a), F is differentiable a.e. □

Corollary 1.2.7. *Every reflexive space has the Radon-Nikodym property.*

Proof. Since a continuous function has separable range, it suffices to show that every separable reflexive space has the Radon-Nikodym property. This follows from Theorem 1.2.6. □

Next we give some examples of spaces which do not have the Radon-Nikodym property.

Example 1.2.8. a) Let $X = C[0,1]$ and define $F : [0,1] \to C[0,1]$ by $F(t)(s) := (t-s)\chi_{[0,t]}(s)$. For $0 \le t_1 \le t_2 \le 1$,

$$F(t_2)(s) - F(t_1)(s) = \begin{cases} t_2 - t_1 & (0 \le s \le t_1), \\ t_2 - s & (t_1 < s \le t_2), \\ 0 & (t_2 < s \le 1). \end{cases}$$

Thus $\|F(t_2) - F(t_1)\|_\infty = |t_2 - t_1|$, so F is Lipschitz continuous. However, $\lim_{h\to 0} \frac{1}{h}(F(t+h) - F(t))$ does not exist in the norm topology, so $C[0,1]$ does not have the Radon-Nikodym property.

b) It follows from a) that $L^\infty(0,1)$ does not have the Radon-Nikodym property either. However, when F is as in a) and $L^\infty(0,1)$ is identified with $L^1(0,1)^*$, the weak* derivative $F'_{w*}(t) := \text{w*-}\lim_{h\to 0} \frac{1}{h}(F(t+h) - F(t)) = \chi_{[0,t]}$ exists in $L^\infty(0,1)$. Note that F'_{w*} is not measurable and therefore not Bochner integrable (Example 1.1.5 a)). We shall see in Section 1.9 that F'_{w*} is Riemann integrable and $F(t) = \int_0^t F'_{w*}(s)\,ds$ in the sense of Riemann integrals. □

More generally, it follows from the next result that every Banach space containing a closed subspace isomorphic to c_0 (see Example 1.1.5 b) and Appendix D) does not have the Radon-Nikodym property.

Proposition 1.2.9. *The space c_0 does not have the Radon-Nikodym property.*

Proof. Let $F(t) := (F_n(t))_{n\in\mathbb{N}}$, where $F_n(t) := \frac{1}{n}\sin(nt)$ $(n \in \mathbb{N})$. Then $F : \mathbb{R}_+ \to c_0$ is Lipschitz continuous since

$$|F_n(t) - F_n(s)| = \left| \int_s^t \cos(nr)\,dr \right| \le |t-s| \quad (t, s \ge 0,\ n \in \mathbb{N}).$$

However, F is nowhere differentiable, since $F_n'(t) = \cos(nt)$ and $(\cos(nt))_{n\in\mathbb{N}} \notin c_0$. Thus, c_0 does not have the Radon-Nikodym property. □

It follows from Theorem 1.2.6 that the space $l^1 = c_0^*$ has the Radon-Nikodym property. However, $L^1(0,1)$ does not, even though c_0 is not a closed subspace of $L^1(0,1)$ (see Appendix D).

Proposition 1.2.10. *The space $L^1(0,1)$ does not have the Radon-Nikodym property.*

Proof. In fact, define $F : [0,1] \to L^1(0,1)$ by $F(t) = \chi_{[0,t]}$. Then F is clearly Lipschitz continuous. Let $0 < t < 1$. Then F is not differentiable at t. In fact,

$$\left\| \frac{1}{h}\left(F(t+h) - F(t)\right) - \frac{1}{2h}\left(F(t+2h) - F(t)\right) \right\|_1 = 1$$

for $0 < h < (1-t)/2$. □

1.3 Convolutions

For $k, h \in L^1(\mathbb{R})$, standard arguments with Fubini's theorem (see [Rud87, Theorem 7.14]) show that the *convolution*

$$(k * h)(t) := \int_{\mathbb{R}} k(t-s)h(s)\,ds$$

exists a.e. and $k*h \in L^1(\mathbb{R})$. Moreover, convolution is commutative and associative.

In this section, we consider convolutions involving vector and operator-valued functions. If $k : \mathbb{R} \to \mathbb{C}$ and $f : \mathbb{R} \to X$ are measurable, we define the *convolution* by

$$(k * f)(t) = \int_{\mathbb{R}} k(t-s)f(s)\,ds \tag{1.6}$$

whenever this exists (as a Bochner integral). Since $(k * f)(t) = \int_{\mathbb{R}} k(s)f(t-s)\,ds$, we may write $f * k$ in place of $k * f$.

Proposition 1.3.1. *Let $k, h \in L^1(\mathbb{R})$ and $f \in L^1(\mathbb{R}, X)$. Then*

a) *$(k * f)(t)$ exists for almost all $t \in \mathbb{R}$ and $k * f \in L^1(\mathbb{R}, X)$.*

b) *$h * (k * f) = (h * k) * f$ a.e.*

Proof. These results may be deduced from the vector-valued version of Fubini's theorem (Theorem 1.1.9) in the same way as in the scalar case. (Alternatively, they may be deduced from their scalar cases, using Theorem 1.1.4 and the Hahn-Banach theorem). □

Many standard facts about convolutions of scalar-valued functions extend to the vector-valued case. We summarize some of them here.

Proposition 1.3.2. a) (**Young's inequality**) *Let* $1 \le p,q,r \le \infty$ *satisfy* $1/p + 1/q = 1 + 1/r$. *If* $k \in L^p(\mathbb{R})$ *and* $f \in L^q(\mathbb{R}, X)$, *then* $k * f \in L^r(\mathbb{R}, X)$ *and*

$$\|k * f\|_r \le \|k\|_p \|f\|_q.$$

b) *Let* $1 < p, p' < \infty$ *satisfy* $1/p + 1/p' = 1$. *If* $k \in L^p(\mathbb{R})$ *and* $f \in L^{p'}(\mathbb{R}, X)$, *then* $k * f \in C_0(\mathbb{R}, X)$.

c) *If* $k \in L^1(\mathbb{R})$ *and* $f \in L^\infty(\mathbb{R}, X)$, *or if* $k \in L^\infty(\mathbb{R})$ *and* $f \in L^1(\mathbb{R}, X)$, *then* $k * f \in \mathrm{BUC}(\mathbb{R}, X)$.

d) *If* $k \in L^1(\mathbb{R})$ *and* $f \in C_0(\mathbb{R}, X)$, *or if* $k \in C_0(\mathbb{R}, X)$ *and* $f \in L^1(\mathbb{R})$, *then* $k * f \in C_0(\mathbb{R}, X)$.

Proof. Under the assumptions of a), the scalar-valued version of Young's inequality [RS72, Section IX.4] shows that $(|k| * \|f\|)(t)$ exists a.e., and $|k| * \|f\| \in L^r(\mathbb{R})$. If $(|k| * \|f\|)(t)$ exists, then $(k * f)(t)$ exists and

$$\|(k * f)(t)\| \le (|k| * \|f\|)(t).$$

Moreover, $k * f$ is weakly measurable by the scalar-valued theory, and hence measurable by Pettis's theorem (Theorem 1.1.1). Thus, $k * f \in L^r(\mathbb{R}, X)$.

The proofs of the remaining parts are similar to the scalar-valued case [Rud62, Theorem 1.1.6]. Suppose that $1 \le p < \infty$ and $1/p + 1/p' = 1$. For $t, h \in \mathbb{R}$,

$$\begin{aligned} \|(k * f)(t + h) - (k * f)(t)\| &\le \left\| \int_{\mathbb{R}} (k(t + h - s) - k(t - s)) f(s)\, ds \right\| \\ &\le \left(\int_{\mathbb{R}} |k(s + h) - k(s)|^p\, ds \right)^{1/p} \|f\|_{p'} \\ &\to 0 \end{aligned}$$

as $h \to 0$. This shows that $k * f$ is uniformly continuous. When $k \in L^\infty(\mathbb{R})$ and $f \in L^1(\mathbb{R}, X)$, uniform continuity is established in a similar way with the roles of k and f reversed. Boundedness follows from a). This proves c). For b) and d), observe that $k * f$ has compact support if both k and f have compact support, and then the results follow by density arguments. □

When k and f are defined on $\mathbb{R}_+$, or on $[0, \tau]$ where $\tau > 0$, then $k * f$ may be defined by (1.6) by taking $k(t)$ and $f(t)$ to be zero for other values of t. Then

$$(k * f)(t) = \int_0^t k(t - s) f(s)\, ds \qquad (t \ge 0).$$

It is immediate that Propositions 1.3.1 and 1.3.2 remain valid in these contexts. Note that if $k \in L^1_{loc}(\mathbb{R}_+)$ and $f \in L^1_{loc}(\mathbb{R}_+, X)$, then $k * f \in L^1_{loc}(\mathbb{R}_+, X)$.

As a tool, we shall need the notion of regularization from harmonic analysis. A *mollifier* is a sequence $(\rho_n)_{n\in\mathbb{N}}$ in $L^1(\mathbb{R})$ of the following form. The function $\rho_1 \in L^1(\mathbb{R})$ satisfies $\int_{\mathbb{R}} \rho_1(t)\,dt = 1$, and $\rho_n \in L^1(\mathbb{R})$ is given by $\rho_n(t) = n\rho_1(nt)$ $(t \in \mathbb{R},\ n \in \mathbb{N})$. The next lemma shows that any mollifier acts as an approximate unit on various function spaces.

Lemma 1.3.3. *Let $(\rho_n)_{n\in\mathbb{N}}$ be a mollifier.*

a) *Let $f \in \mathrm{BUC}(\mathbb{R}, X)$. Then $\lim_{n\to\infty} \| f * \rho_n - f\|_\infty = 0$.*

b) *Let $f \in L^1(\mathbb{R}, X)$. Then $\lim_{n\to\infty} \| f * \rho_n - f\|_1 = 0$.*

Proof. a) Let $\varepsilon > 0$. There exists $c > 0$ such that $2\| f \|_\infty \int_{|s|\geq c} |\rho_1(s)|\,ds \leq \varepsilon$. Then for $t \in \mathbb{R}$, $n \in \mathbb{N}$,

$$\begin{aligned}
\|(f * \rho_n)(t) - f(t)\| &= \left\| \int_{\mathbb{R}} (f(t-s) - f(t))\rho_n(s)\,ds \right\| \\
&= \left\| \int_{\mathbb{R}} (f(t-\frac{s}{n}) - f(t))\rho_1(s)\,ds \right\| \\
&\leq \int_{|s|\leq c} \left\| f\left(t - \frac{s}{n}\right) - f(t) \right\| |\rho_1(s)|\,ds \\
&\quad + \int_{|s|>c} \left\| f\left(t - \frac{s}{n}\right) - f(t) \right\| |\rho_1(s)|\,ds \\
&\leq \sup_{|h|\leq c/n} \| f(t-h) - f(t) \| \; \|\rho_1\|_1 + \varepsilon \\
&\leq 2\varepsilon
\end{aligned}$$

for all sufficiently large n, since f is uniformly continuous.

b) Let $f \in L^1(\mathbb{R}, X)$. Then

$$\begin{aligned}
\| f * \rho_n - f \|_1 &= \int_{\mathbb{R}} \left\| \int (f(t-s) - f(t))\rho_n(s)\,ds \right\| dt \\
&= \int_{\mathbb{R}} \left\| \int_{\mathbb{R}} \left(f(t - \frac{s}{n}\right) - f(t))\rho_1(s)\,ds \right\| dt \\
&\leq \int_{\mathbb{R}} \int_{\mathbb{R}} \left\| f\left(t - \frac{s}{n}\right) - f(t) \right\| dt\, |\rho_1(s)|\,ds.
\end{aligned}$$

As observed in Section 1.1, $\lim_{n\to\infty} \int_{\mathbb{R}} \| f(t - \frac{s}{n}) - f(t) \|\,dt = 0$ for all $s \in \mathbb{R}$. Now the claim follows from the dominated convergence theorem. □

Lemma 1.3.3 is also valid for the spaces $L^p(\mathbb{R}, X)$ $(1 < p < \infty)$ (see Remark 1.3.8 b)). The notion of mollifier can be extended to a family $\{\rho_\varepsilon :\ 0 < \varepsilon \leq 1\}$ where $\rho_\varepsilon(t) = \varepsilon^{-1}\rho(t/\varepsilon)$, and Lemma 1.3.3 remains valid in that case.

The theory of vector-valued convolutions and mollifiers extends, with almost no changes, to the case of functions on $\mathbb{R}^n$ for $n \geq 1$.

Now we move on to consider convolutions of vector-valued functions with operator-valued functions.

The space of all bounded linear operators from a Banach space X into a Banach space Y is denoted by $\mathcal{L}(X,Y)$, or simply by $\mathcal{L}(X)$ when $Y = X$. A function $T : \mathbb{R}_+ \to \mathcal{L}(X,Y)$ is *strongly continuous* if $t \mapsto T(t)x$ is continuous for all $x \in X$. By the uniform boundedness principle, a strongly continuous function T is locally bounded. Note also that $\|T\|$ is lower semi-continuous and hence measurable.

We state the convolution results for strongly continuous functions $T : \mathbb{R}_+ \to \mathcal{L}(X,Y)$, but they are also valid for $T : (0,\infty) \to \mathcal{L}(X,Y)$ if T is strongly continuous on $(0,\infty)$ and bounded on $(0,1)$. There are similar results for compact intervals $[0,\tau]$.

Proposition 1.3.4. *Let $f \in L^1_{loc}(\mathbb{R}_+,X)$ and let $T : \mathbb{R}_+ \to \mathcal{L}(X,Y)$ be strongly continuous. Then the convolution*

$$(T * f)(t) := \int_0^t T(t-s)f(s)\,ds$$

exists (as a Bochner integral) and defines a continuous function $T * f : \mathbb{R}_+ \to Y$.

Proof. Fix $t \geq 0$. First, we show that $s \mapsto T(t-s)f(s)$ is measurable on $[0,t]$. When $f(s) = \chi_\Omega(s)x$ for some measurable $\Omega \subset \mathbb{R}_+$ and $x \in X$, then

$$T(t-s)f(s) = \chi_\Omega(s) \cdot T(t-s)x.$$

This is measurable, being the product of a measurable scalar-valued function and a continuous vector-valued function. By linearity, $T(t-\cdot)f(\cdot)$ is measurable when f is a simple function. For measurable f, there is a sequence of simple functions $f_n \to f$ a.e. and then $T(t-s)f_n(s) \to T(t-s)f(s)$ s-a.e., so $T(t-\cdot)f(\cdot)$ is measurable.

Since

$$\|T(t-s)f(s)\| \leq \|T(t-s)\|\,\|f(s)\|$$

it follows from Theorem 1.1.4 that $(T * f)(t)$ exists. Continuity of $T * f$ follows from the dominated convergence theorem (Theorem 1.1.8). $\square$

Now we state the analogue of Proposition 1.3.2 for operator-valued functions on $\mathbb{R}_+$.

Proposition 1.3.5. *Let $f \in L^1_{loc}(\mathbb{R}_+,X)$ and $T : \mathbb{R}_+ \to \mathcal{L}(X,Y)$ be strongly continuous.*

a) (**Young's inequality**) *Let $1 \leq p,q,r \leq \infty$ satisfy $1/p + 1/q = 1 + 1/r$. If $\int_0^\infty \|T(t)\|^p\,dt < \infty$ and $f \in L^q(\mathbb{R}_+,X)$, then $T * f \in L^r(\mathbb{R}_+,Y)$ and*

$$\|T * f\|_r \leq \|f\|_q \left(\int_0^\infty \|T(t)\|^p\,dt\right)^{1/p}.$$

b) *Let* $1 < p, p' < \infty$ *satisfy* $1/p + 1/p' = 1$. *If* $\int_0^\infty \|T(t)\|^p\, dt < \infty$ *and* $f \in L^{p'}(\mathbb{R}_+, X)$, *then* $T * f \in C_0(\mathbb{R}_+, Y)$.

c) *If* $\int_0^\infty \|T(t)\|\, dt < \infty$ *and* $f \in \mathrm{BUC}(\mathbb{R}_+, X)$, *or if* T *is bounded and* $f \in L^1(\mathbb{R}_+, X)$, *then* $T * f \in \mathrm{BUC}(\mathbb{R}_+, Y)$.

d) *If* $\int_0^\infty \|T(t)\|\, dt < \infty$ *and* $f \in C_0(\mathbb{R}_+, X)$, *or if* $\lim_{t\to\infty} \|T(t)\| = 0$ *and* $f \in L^1(\mathbb{R}_+, X)$, *then* $T * f \in C_0(\mathbb{R}_+, Y)$.

Proof. The proofs are similar to Proposition 1.3.2, with the exception of the uniform continuity of $T * f$ when $\|T\|$ is integrable and f is bounded and uniformly continuous. Then, for $0 \le t \le t + h$,

$$\begin{aligned}
&\|(T * f)(t+h) - (T * f)(t)\| \\
&\le \left\| \int_t^{t+h} T(s)f(t+h-s)\, ds \right\| + \left\| \int_0^t T(s)(f(t+h-s) - f(t-s))\, ds \right\| \\
&\le \left(\int_t^{t+h} \|T(s)\|\, ds \right) \|f\|_\infty + \left(\int_0^\infty \|T(s)\|\, ds \right) \sup_{s\ge 0} \|f(s+h) - f(s)\| \\
&\to 0
\end{aligned}$$

uniformly in t as $h \to 0$. □

If f or T is more regular, then $T * f$ is continuously differentiable. We give two such results.

Proposition 1.3.6. *Let* $T : \mathbb{R}_+ \to \mathcal{L}(X, Y)$ *be strongly continuous and bounded,* $x \in X$, $f' \in L^1_{loc}(\mathbb{R}_+, X)$, $f(t) = x + \int_0^t f'(s)\, ds$ $(t \ge 0)$. *Then* $T * f \in C^1(\mathbb{R}_+, Y)$ *and*

$$(T * f)'(t) = (T * f')(t) + T(t)x.$$

Proof. Let $u(t) := (T * f')(t) + T(t)x$. Then by Proposition 1.3.4, $u \in C(\mathbb{R}_+, Y)$. By Fubini's theorem we have

$$\begin{aligned}
\int_0^t u(s)\, ds &= \int_0^t \int_0^s T(r)f'(s-r)\, dr\, ds + \int_0^t T(s)x\, ds \\
&= \int_0^t \int_r^t T(r)f'(s-r)\, ds\, dr + \int_0^t T(s)x\, ds \\
&= \int_0^t \int_0^{t-r} T(r)f'(s)\, ds\, dr + \int_0^t T(s)x\, ds \\
&= \int_0^t T(r)(f(t-r) - x)\, dr + \int_0^t T(s)x\, ds \\
&= (T * f)(t) \qquad (t \ge 0).
\end{aligned}$$

By Proposition 1.2.2 a), this proves the claim. □

Proposition 1.3.7. *Let* $T : [0,\tau] \to \mathcal{L}(X,Y)$ *be Lipschitz continuous with* $T(0) = 0$, *and let* $f \in L^1([0,\tau],X)$. *Then* $T * f \in C^1([0,\tau],Y)$.

Proof. First, suppose that $f \in C^1([0,\tau],X)$. By Proposition 1.3.6, $T * f$ has a derivative $g \in C([0,\tau],Y)$. For $0 \le r \le t \le \tau$,

$$\begin{aligned}\left\| \int_r^t g(s)\,ds \right\| &= \|(T*f)(t) - (T*f)(r)\| \\ &\le \int_r^t \|T(t-s)f(s)\|\,ds + \int_0^r \|(T(t-s) - T(r-s))f(s)\|\,ds \\ &\le \int_r^t L\,(t-s)\|f(s)\|\,ds + \int_0^r L\,(t-r)\|f(s)\|\,ds \\ &\le L\,(t-r)\|f\|_1,\end{aligned}$$

where L is a Lipschitz constant for T, so that $\|T(t_2) - T(t_1)\| \le L|t_2 - t_1|$ and, in particular, $\|T(s)\| \le Ls$ since $T(0) = 0$. It follows that $\|g(s)\| \le L\|f\|_1$ for all $s \in [0,\tau]$.

Now, consider $f \in L^1([0,\tau],X)$. There is a sequence (f_n) in $C^1([0,\tau],X)$ such that $\|f_n - f\|_1 \to 0$. By Proposition 1.3.5, $\|(T*(f_n - f))(t)\| \to 0$. By the first paragraph,

$$\begin{aligned}\|(T*f_n)' - (T*f_m)'\|_\infty &\le L\,\|f_n - f_m\|_1 \\ &\to 0,\end{aligned}$$

so $((T*f_n)')$ converges uniformly to a function $g \in C([0,\tau],Y)$. Since

$$(T*f)(t) = \lim_{n\to\infty}(T*f_n)(t) = \lim_{n\to\infty}\int_0^t (T*f_n)'(s)\,ds = \int_0^t g(s)\,ds,$$

it follows that $T * f \in C^1([0,\tau],Y)$. □

Remark 1.3.8. a) There is an analogous theory of operator-valued convolutions on $\mathbb{R}$. One may define

$$(T*f)(t) := \int_{\mathbb{R}} T(t-s)f(s)\,ds$$

for almost all t, if $T : \mathbb{R} \to \mathcal{L}(X,Y)$ is strongly continuous, $\int_{\mathbb{R}} \|T(t)\|^p\,dt < \infty$ and $f \in L^q(\mathbb{R},X)$ where $1/p + 1/q \ge 1$, and Proposition 1.3.5 is valid in this case.

b) Convolutions of scalar, vector or operator-valued functions can sometimes be considered as Bochner integrals with values in a function space. Suppose that $0 < \tau \le \infty$, $k \in L^1(0,\tau)$ and $f \in L^p((0,\tau),X)$ where $1 \le p < \infty$. Define $G : (0,\tau) \to L^p((0,\tau),X)$ by

$$G(s)(t) := \begin{cases} f(t-s) & (0 < s < t < \tau) \\ 0 & \text{otherwise.}\end{cases}$$

Then G is continuous and $\|G(s)\|_p \le \|f\|_p$. We can therefore form the Bochner integral $\int_0^\tau k(s)G(s)\,ds$ in $L^1((0,\tau),X)$. Then

$$\int_0^\tau k(s)G(s)\,ds = k * f \quad \text{a.e. in } (0,\tau).$$

This can be proved by first considering step functions, or by considering integrals of the form $\int_0^\tau \langle (k*f)(t), x^*\rangle g(t)\,dt$ for arbitrary $x^* \in X^*$ and $g \in L^{p'}(0,\tau)$. Thus, $\|k * f\|_p \le \|k\|_1 \|f\|_p$, a special case of Young's inequality.

This approach can also be used for convolutions on $\mathbb{R}$, when the spaces $L^p(\mathbb{R},X)$ can be replaced by $\mathrm{BUC}(\mathbb{R},X)$ or $C_0(\mathbb{R},X)$. For example, this leads to a very short proof of Lemma 1.3.3 for all these spaces.

The same idea can be used in the operator-valued case. Suppose that $0 < \tau \le \infty$, $T : (0,\tau) \to \mathcal{L}(X,Y)$ is strongly continuous and $f \in L^p((0,\tau),X)$, where $1 \le p < \infty$. Let

$$H(s)(t) := \begin{cases} T(s)f(t-s) & (0 < s < t < \tau), \\ 0 & \text{otherwise.} \end{cases}$$

Then $H(s) \in L^p((0,\tau),Y)$, $\|H(s)\|_p \le \|T(s)\|\,\|f\|_p$, and $H : (0,\tau) \to L^p((0,\tau),Y)$ is continuous. If $\int_0^\tau \|T(s)\|\,ds < \infty$, one may form the Bochner integral $\int_0^\tau H(s)\,ds$ in $L^p((0,\tau),Y)$ and it coincides a.e. with $T * f$. □

1.4 Existence of the Laplace Integral

Let X be a complex Banach space and $L^1_{loc}(\mathbb{R}_+,X) := \{f : \mathbb{R}_+ \to X\colon f$ is Bochner integrable on $[0,\tau]$ for all $\tau > 0\}$. This section is concerned with the existence of *the Laplace integral*

$$\hat{f}(\lambda) := \int_0^\infty e^{-\lambda t} f(t)\,dt := \lim_{\tau\to\infty} \int_0^\tau e^{-\lambda t} f(t)\,dt$$

for $f \in L^1_{loc}(\mathbb{R}_+,X)$ and $\lambda \in \mathbb{C}$. Note that $\int_0^\tau e^{-\lambda t} f(t)\,dt$ exists as a Bochner integral, and if $\int_0^\infty e^{-\lambda t} f(t)\,dt$ exists as a Bochner integral then it agrees with the definition above, by the dominated convergence theorem.

Of special interest will be *the abscissa of convergence of* $\hat{f}$, given by

$$\mathrm{abs}(f) := \inf\{\mathrm{Re}\,\lambda : \hat{f}(\lambda) \text{ exists}\}.$$

It will be shown that the set of those $\lambda \in \mathbb{C}$ for which the Laplace integral converges is either empty or a right half-plane whose left boundary point $\mathrm{abs}(f)$ coincides with the exponential growth bound of the antiderivative

$$t \mapsto F(t) - F_\infty \qquad (t \ge 0),$$

where $F(t) := \int_0^t f(s)\,ds$, $F_\infty := \lim_{t\to\infty} F(t)$ if the limit exists, and $F_\infty := 0$ otherwise. Therefore, one particular result of this section will be that a locally Bochner integrable function f is Laplace transformable if and only if its *normalized antiderivative* $F : t \to \int_0^t f(s)\,ds$ is exponentially bounded.

Proposition 1.4.1. *Let $f \in L^1_{loc}(\mathbb{R}_+, X)$. Then the Laplace integral $\hat{f}(\lambda)$ converges if $\operatorname{Re}\lambda > \operatorname{abs}(f)$ and diverges if $\operatorname{Re}\lambda < \operatorname{abs}(f)$.*

Proof. Clearly, $\hat{f}(\lambda)$ does not exist if $\operatorname{Re}\lambda < \operatorname{abs}(f)$. For $\lambda_0 \in \mathbb{C}$ define $G_0(t) := \int_0^t e^{-\lambda_0 s} f(s)\,ds \;\; (t \geq 0)$. Then, for all $\lambda \in \mathbb{C}$ and $t \geq 0$, integration by parts gives

$$\begin{aligned}\int_0^t e^{-\lambda s} f(s)\,ds &= \int_0^t e^{-(\lambda-\lambda_0)s} e^{-\lambda_0 s} f(s)\,ds \\ &= e^{-(\lambda-\lambda_0)t} G_0(t) + (\lambda-\lambda_0)\int_0^t e^{-(\lambda-\lambda_0)s} G_0(s)\,ds. \qquad (1.7)\end{aligned}$$

If $\hat{f}(\lambda_0)$ exists, then G_0 is bounded. Moreover, it follows from (1.7) that $\hat{f}(\lambda)$ exists if $\operatorname{Re}\lambda > \operatorname{Re}\lambda_0$ and

$$\hat{f}(\lambda) = (\lambda-\lambda_0)\int_0^\infty e^{-(\lambda-\lambda_0)s} G_0(s)\,ds \qquad (\operatorname{Re}\lambda > \operatorname{Re}\lambda_0). \qquad (1.8)$$

This shows that $\hat{f}(\lambda)$ exists if $\operatorname{Re}\lambda > \operatorname{abs}(f)$. □

If $\hat{f}(\lambda)$ converges for all $\lambda \in \mathbb{C}$, then $\operatorname{abs}(f) := -\infty$. If the domain of convergence is empty, then $\operatorname{abs}(f) := \infty$. A function f is called *Laplace transformable* if $\operatorname{abs}(f) < \infty$.

It follows from Proposition 1.4.1 that the interior of the domain of convergence of $\hat{f}(\lambda)$ is the open right half-plane $\{\operatorname{Re}\lambda > \operatorname{abs}(f)\}$. As the following example shows, the domain of convergence may or may not include points on the boundary $\operatorname{abs}(f) + i\mathbb{R}$.

Example 1.4.2. Let $f(t) := (1+t)^{-1}$. Then $\operatorname{abs}(f) = 0$ since the Laplace integral $\hat{f}(\lambda)$ converges for $\lambda > 0$ but not for $\lambda = 0$. If $\lambda = ir$ $(r \neq 0)$, then integration by parts implies that $\hat{f}(ir)$ converges and

$$\hat{f}(ir) = \frac{1}{ir} - \frac{1}{ir}\int_0^\infty \frac{e^{-irt}}{(1+t)^2}\,dt.$$

Thus, the domain of convergence of $\hat{f}(\lambda)$ is $\{\operatorname{Re}\lambda \geq 0, \lambda \neq 0\}$. For $f(t) := 1$, the domain of convergence of $\hat{f}(\lambda)$ is the open half-plane $\{\operatorname{Re}\lambda > 0\}$; for $f(t) := (1+t^2)^{-1}$ it is the closed half-plane $\{\operatorname{Re}\lambda \geq 0\}$. □

From the proof of Proposition 1.4.1 and the uniform boundedness principle one also obtains the remarkable result that the abscissa of convergence of $\hat{f}(\lambda)$ is given by

$$\operatorname{abs}(f) = \inf \left\{ \lambda \in \mathbb{R} : \sup_{t>0} \left| \int_0^t e^{-\lambda s} \langle f(s), x^* \rangle \, ds \right| < \infty \text{ for all } x^* \in X^* \right\}. \tag{1.9}$$

To see this we denote the right-hand side of (1.9) by $\operatorname{abs}_w(f)$. Clearly, $\operatorname{abs}_w(f) \leq \operatorname{abs}(f)$. Assume that $\operatorname{abs}_w(f) < \operatorname{abs}(f)$. Then there exists λ_0 such that $\operatorname{abs}_w(f) < \lambda_0 < \operatorname{abs}(f)$ and $\sup_{t>0} |\langle G_0(t), x^* \rangle| < \infty$ $(x^* \in X^*)$, where G_0 is as in the proof of Proposition 1.4.1. It follows from the uniform boundedness principle that G_0 is bounded. Thus, by (1.7), $\hat{f}(\lambda)$ exists for all $\lambda > \lambda_0$. Since this contradicts $\lambda_0 < \operatorname{abs}(f)$ one obtains that $\operatorname{abs}_w(f) = \operatorname{abs}(f)$.

Next $\operatorname{abs}(f)$ will be described by the exponential growth of f and its antiderivatives. For $f : \mathbb{R}_+ \to X$ the *exponential growth bound* is given by

$$\omega(f) := \inf \left\{ \omega \in \mathbb{R} : \sup_{t \geq 0} \|e^{-\omega t} f(t)\| < \infty \right\}.$$

It is obvious that

$$\operatorname{abs}(f) \leq \operatorname{abs}(\|f\|) \leq \omega(f). \tag{1.10}$$

It will be shown in Example 1.4.4 below that there are cases in which one has strict inequalities in (1.10). In fact, it is possible that $\operatorname{abs}(f) = -\infty$ and $\omega(f) = \infty$. It will be shown next that $\operatorname{abs}(f)$ is determined by the exponential growth of the antiderivative $t \mapsto F(t) - F_\infty$ $(t \geq 0)$, where $F(t) := \int_0^t f(s)\, ds$, $F_\infty := \lim_{t\to\infty} F(t)$ if the limit exists, and $F_\infty := 0$ otherwise.

Theorem 1.4.3. *Let* $f \in L^1_{loc}(\mathbb{R}_+, X)$. *Then* $\operatorname{abs}(f) = \omega(F - F_\infty)$.

Proof. Suppose that $\operatorname{abs}(f) < \infty$. For $\lambda_0 > \operatorname{abs}(f)$ define $G_0(t) := \int_0^t e^{-\lambda_0 s} f(s)\, ds$ $(t \geq 0)$. Then G_0 is continuous and convergent as $t \to \infty$, so G_0 is bounded. To prove that $\operatorname{abs}(f) \geq \omega(F - F_\infty)$ one considers the three cases $\operatorname{abs}(f) > 0, \operatorname{abs}(f) = 0$, and $\operatorname{abs}(f) < 0$. First, let $\operatorname{abs}(f) > 0$. Then $F_\infty = 0$ and, for $\lambda_0 > \operatorname{abs}(f)$, integration by parts gives

$$F(t) = \int_0^t f(s)\, ds = \int_0^t e^{\lambda_0 s} e^{-\lambda_0 s} f(s)\, ds = e^{\lambda_0 t} G_0(t) \; - \; \lambda_0 \int_0^t e^{\lambda_0 s} G_0(s)\, ds.$$

It follows that $\|F(t)\| \leq Ce^{\lambda_0 t} + C(e^{\lambda_0 t} - 1) \leq 2Ce^{\lambda_0 t}$ for $t \geq 0$, where $C := \sup_{s \geq 0} \|G_0(s)\|$. This shows that $\operatorname{abs}(f) \geq \omega(F - F_\infty)$ if $\operatorname{abs}(f) > 0$. Second, let $\operatorname{abs}(f) = 0$. If $F_\infty = 0$, then the same procedure as above yields $\operatorname{abs}(f) \geq \omega(F - F_\infty)$. If $\operatorname{abs}(f) = 0$ and $\lim_{t\to\infty} F(t) = F_\infty$ exists, then it follows from the continuity of F that $\sup_{t \geq 0} \|F(t) - F_\infty\| < \infty$. Thus, $\operatorname{abs}(f) = 0 \geq \omega(F - F_\infty)$. This shows that $\operatorname{abs}(f) \geq \omega(F - F_\infty)$ if $\operatorname{abs}(f) = 0$. Third, let $\operatorname{abs}(f) < 0$. Choose $\operatorname{abs}(f) < \lambda_0 < 0$. For $r \geq t \geq 0$ one has

$$\begin{aligned} F(r) - F(t) &= \int_t^r e^{\lambda_0 s} e^{-\lambda_0 s} f(s)\, ds \\ &= e^{\lambda_0 r} G_0(r) - e^{\lambda_0 t} G_0(t) \; - \; \lambda_0 \int_t^r e^{\lambda_0 s} G_0(s)\, ds. \end{aligned}$$

Since G_0 is bounded and $\lim_{r\to\infty} F(r) = F_\infty$, it follows that

$$\|F_\infty - F(t)\| = \left\| e^{\lambda_0 t} G_0(t) + \lambda_0 \int_t^\infty e^{-|\lambda_0|s} G_0(s)\,ds \right\| \le 2Ce^{\lambda_0 t}$$

for $t \ge 0$. This proves that $\operatorname{abs}(f) \ge \omega(F - F_\infty)$.

To show the reverse inequality, suppose that $\omega(F - F_\infty) < \infty$ and let $\omega > \omega(F - F_\infty)$. Since F is continuous, there exists $M \ge 0$ such that $\|F(t) - F_\infty\| \le Me^{\omega t}$ for all $t \ge 0$. Let $\lambda > \omega > \omega(F - F_\infty)$. Using the fact that $F - F_\infty$ is an antiderivative of f, integration by parts yields

$$\int_0^t e^{-\lambda s} f(s)\,ds = e^{-\lambda t}(F(t) - F_\infty) + F_\infty + \lambda \int_0^t e^{-\lambda s}(F(s) - F_\infty)\,ds.$$

Hence, $\hat{f}(\lambda)$ exists for $\lambda > \omega(F - F_\infty)$ and is given by $\hat{f}(\lambda) = F_\infty + \lambda(\widehat{F - F_\infty})(\lambda)$. This shows that $\operatorname{abs}(f) \le \omega(F - F_\infty)$. Since $\lambda\widehat{F_\infty}(\lambda) = \lambda\int_0^\infty e^{-\lambda t} F_\infty\,dt = F_\infty$ if $\operatorname{Re}\lambda > 0$,

$$\hat{f}(\lambda) = \lambda\hat{F}(\lambda) \quad \text{if} \quad \operatorname{Re}\lambda > \max\{\operatorname{abs}(f), 0\}. \tag{1.11}$$

□

If $\omega \ge 0$, then the triangle inequality implies that $\omega(F) \le \omega$ if and only if $\omega(F - F_\infty) \le \omega$. Thus, a locally Bochner integrable function f is Laplace transformable if and only if its antiderivative $F(t) = \int_0^t f(s)\,ds$ is exponentially bounded; and,

$$\operatorname{abs}(f) \le \omega \iff \omega(F) \le \omega \quad (\text{if } \omega \ge 0). \tag{1.12}$$

The following is a first example of a function which is Laplace transformable but not exponentially bounded; i.e., $\operatorname{abs}(f) < \omega(f) = \infty$.

Example 1.4.4. For $t \ge 0$ let $f(t) := e^t e^{e^t} \cos(e^{e^t})$. Then $\omega(f) = \infty$. Since

$$F(t) = \int_0^t f(s)\,ds = \int_0^t e^s e^{e^s} \cos\left(e^{e^s}\right) ds = \int_e^{e^{e^t}} \cos(u)\,du = \sin e^{e^t} - \sin(e),$$

it follows that $F_\infty = 0$. Thus, by Theorem 1.4.3, $\operatorname{abs}(f) = \omega(F) = 0$. □

Finally, the results of this section will be formulated for strongly continuous operator-valued functions $T : \mathbb{R}_+ \to \mathcal{L}(X, Y)$. We define the *exponential growth bound* of T by

$$\omega(T) := \omega(\|T\|) = \inf\left\{\omega \in \mathbb{R} : \sup_{t\ge0} \|e^{-\omega t}T(t)\| < \infty\right\}.$$

By the uniform boundedness principle,

$$\omega(T) = \sup\{\omega(u_x) : x \in X\},$$

where $u_x(t) := T(t)x$.

If $T : \mathbb{R}_+ \to \mathcal{L}(X,Y)$ is strongly continuous, and $\lambda \in \mathbb{C}$, then $\int_0^t e^{-\lambda s}T(s)\,ds$ denotes the bounded operator $x \mapsto \int_0^t e^{-\lambda s}T(s)x\,ds$, and we define

$$\begin{aligned}\operatorname{abs}(T) &:= \inf\left\{\operatorname{Re}\lambda : \int_0^t e^{-\lambda s}T(s)\,ds \text{ converges strongly as } t\to\infty\right\}\\ &= \sup\{\operatorname{abs}(u_x) : x\in X\}.\end{aligned}$$

Here and in what follows, to say $S : \mathbb{R}_+ \to \mathcal{L}(X,Y)$ *converges strongly* as $t \to \infty$ refers to the strong operator topology; i.e., it means that $\lim_{t\to\infty} S(t)x$ exists in Y for all $x \in X$.

Proposition 1.4.5. *Let $T : \mathbb{R}_+ \to \mathcal{L}(X,Y)$ be strongly continuous, let $S(t) = \int_0^t T(s)\,ds$, and S_∞ be the strong limit of $S(t)$ as $t\to\infty$ if it exists, and $S_\infty := 0$ otherwise. Then*

a) $\lim_{t\to\infty}\int_0^t e^{-\lambda s}T(s)\,ds$ *exists in operator norm whenever* $\operatorname{Re}\lambda > \operatorname{abs}(T)$.

b) $$\operatorname{abs}(T) = \inf\left\{\lambda\in\mathbb{R} : \sup_{t>0}\left|\int_0^t e^{-\lambda s}\langle T(s)x,y^*\rangle\,ds\right| < \infty \quad \textit{for all } x\in X \textit{ and } y^*\in Y^*\right\}.$$

c) $\operatorname{abs}(T) = \omega(S - S_\infty)$.

Proof. If $\int_0^t e^{-\lambda_0 s}T(s)\,ds$ converges strongly, then it is uniformly bounded in operator norm. Thus, a) follows from (1.7), while b) is immediate from (1.9).

To prove c), it is possible to repeat the proof of Theorem 1.4.3. Alternatively, one may deduce c) from Theorem 1.4.3 as follows. Let $u_x(t) := T(t)x$, $\widetilde{v_x}(t) := S(t)x - S_\infty x$, and

$$v_x(t) := \begin{cases} S(t)x - \lim\limits_{s\to\infty} S(s)x & \text{if the limit exists,}\\ S(t)x & \text{otherwise.}\end{cases}$$

By Theorem 1.4.3, $\operatorname{abs}(u_x) = \omega(v_x)$. Moreover, $\omega(v_x)$ and $\omega(\widetilde{v_x})$ coincide if either of them is strictly positive, since $v_x - \widetilde{v_x}$ is a constant function.

First suppose that $\operatorname{abs}(T) < 0$. Then $\omega(v_x) = \operatorname{abs}(u_x) \le \operatorname{abs}(T) < 0$, so $\lim_{t\to\infty} v_x(t)$ exists for all $x \in X$. Thus, $S(t)$ converges strongly, so $v_x = \widetilde{v_x}$ and

$$\begin{aligned}\operatorname{abs}(T) &= \sup\{\operatorname{abs}(u_x) : x\in X\}\\ &= \sup\{\omega(v_x) : x\in X\}\\ &= \sup\{\omega(\widetilde{v_x}) : x\in X\}\\ &= \omega(S - S_\infty).\end{aligned}$$

Next, suppose that $\omega(S - S_\infty) < 0$. Then again, $S(t)$ converges strongly and, as above, $\operatorname{abs}(T) = \omega(S - S_\infty)$.

Next, suppose that $\text{abs}(T) > 0$. Then

$$\begin{aligned}\text{abs}(T) &= \sup\{\text{abs}(u_x) : x \in X,\ \text{abs}(u_x) > 0\}\\ &= \sup\{\omega(v_x) : x \in X,\ \omega(v_x) > 0\}\\ &= \sup\{\omega(\widetilde{v_x}) : x \in X,\ \omega(\widetilde{v_x}) > 0\}\\ &= \omega(S - S_\infty).\end{aligned}$$

Finally, suppose that $\omega(S - S_\infty) > 0$. Then the same argument as in the previous paragraph can be applied to show that $\text{abs}(T) = \omega(S - S_\infty)$. □

Remark 1.4.6. If $\widetilde{S}_\infty$ is the norm-limit of $S(t)$ as $t \to \infty$ if this exists and $\widetilde{S}_\infty \in \mathcal{L}(X,Y)$ is arbitrary otherwise, and S_∞ is as in Proposition 1.4.5, then it is trivial that $\omega(S - \widetilde{S}_\infty) = \omega(S - S_\infty)$. □

If $T : \mathbb{R}_+ \to \mathcal{L}(X,Y)$ is strongly continuous and $\text{abs}(T) < \infty$, we define the *Laplace integral* of T by

$$\hat{T}(\lambda) := \int_0^\infty e^{-\lambda s} T(s)\, ds := \lim_{t\to\infty} \int_0^t e^{-\lambda s} T(s)\, ds \qquad (\text{Re}\,\lambda > \text{abs}(T)),$$

where the right-hand integral is interpreted as above and the limit exists in operator norm (Proposition 1.4.5).

1.5 Analytic Behaviour

Let $f \in L^1_{loc}(\mathbb{R}_+, X)$. In this section it will be shown that $\lambda \mapsto \hat{f}(\lambda)$ is holomorphic for $\text{Re}\,\lambda > \text{abs}(f)$. In general, $\hat{f}$ need not have a singularity on the boundary $\text{abs}(f) + i\mathbb{R}$ of its domain of convergence and may be extended holomorphically to a strictly larger half-plane. However, it will be shown that $\text{abs}(f)$ is a singularity of $\hat{f}$ if X is an ordered Banach space with a normal positive cone and $f(t) \geq 0$ a.e. (see Appendix C). We put $\mathbb{N}_0 := \mathbb{N} \cup \{0\}$.

Theorem 1.5.1. *Let* $f \in L^1_{loc}(\mathbb{R}_+, X)$ *with* $\text{abs}(f) < \infty$. *Then* $\lambda \mapsto \hat{f}(\lambda)$ *is holomorphic for* $\text{Re}\,\lambda > \text{abs}(f)$ *and, for all* $n \in \mathbb{N}_0$ *and* $\text{Re}\,\lambda > \text{abs}(f)$,

$$\hat{f}^{(n)}(\lambda) = \int_0^\infty e^{-\lambda t}(-t)^n f(t)\, dt \tag{1.13}$$

(as an improper Bochner integral).

Proof. Define $q_k : \mathbb{C} \to X \ \ (k \in \mathbb{N}_0)$ by

$$q_k(\lambda) := \int_0^k e^{-\lambda t} f(t)\, dt = \lim_{N\to\infty} \sum_{n=0}^N \frac{\lambda^n}{n!} \int_0^k (-t)^n f(t)\, dt.$$

The limits exist uniformly for λ in bounded subsets of $\mathbb{C}$. By the Weierstrass convergence theorem (a simple special case of Vitali's Theorem A.5), the functions q_k are entire and $q_k^{(j)}(\lambda) = \int_0^k e^{-\lambda t}(-t)^j f(t)\,dt$ for all $j \in \mathbb{N}_0$.

Let $\operatorname{Re}\lambda > \lambda_0 > \operatorname{abs}(f)$ and define $G_0(t) := \int_0^t e^{-\lambda_0 s} f(s)\,ds$. By Proposition 1.4.1, G_0 is bounded and integration by parts gives

$$\begin{aligned}\hat{f}(\lambda) - q_k(\lambda) &= \int_k^\infty e^{-(\lambda-\lambda_0)s} e^{-\lambda_0 s} f(s)\,ds \\ &= -e^{-(\lambda-\lambda_0)k} G_0(k) + (\lambda - \lambda_0)\int_k^\infty e^{-(\lambda-\lambda_0)s} G_0(s)\,ds.\end{aligned}$$

It follows that q_k converges to $\hat{f}$ uniformly on compact subsets of $\{\lambda : \operatorname{Re}\lambda > \operatorname{abs}(f)\}$. Again by the Weierstrass convergence theorem, $\hat{f}$ is holomorphic and $q_k^{(j)}(\lambda) \to \hat{f}^{(j)}(\lambda)$ as $k \to \infty$, for $\operatorname{Re}\lambda > \operatorname{abs}(f)$. □

If $\operatorname{abs}(f) < \infty$, then the *abscissa of holomorphy* of $\hat{f}$ is denoted by

$$\operatorname{hol}(\hat{f}) := \inf\{\omega \in \mathbb{R} : \hat{f} \text{ extends holomorphically for } \operatorname{Re}\lambda > \omega\}.$$

By Theorem 1.5.1,

$$\operatorname{hol}(\hat{f}) \le \operatorname{abs}(f). \tag{1.14}$$

In general, equality does not hold in (1.14) (see Example 1.5.2 below). However we shall see that equality does hold for positive functions on ordered spaces (Theorem 1.5.3) and for exponentially bounded functions which extend holomorphically into a sector $\{|\arg(\lambda)| < \alpha\}$ for some $0 < \alpha < \frac{\pi}{2}$ (see Section 2.6). Furthermore, in Chapter 4 (Theorem 4.4.19) we shall see that

$$\operatorname{abs}(f) \le \operatorname{hol}_0(\hat{f})$$

whenever f is exponentially bounded, where

$$\operatorname{hol}_0(\hat{f}) := \inf\{\omega \in \mathbb{R} : \hat{f} \text{ has a bounded holomorphic extension for } \operatorname{Re}\lambda > \omega\}$$

is the *abscissa of boundedness* of $\hat{f}$. The following example shows that it may happen that $\operatorname{hol}(\hat{f}) < \operatorname{abs}(f) < \omega(f)$.

Example 1.5.2. Let $f(t) = e^t \sin e^t$ $(t \ge 0)$. Obviously, $\omega(f) = 1$. Since $F(t) = \int_0^t f(s)\,ds = \cos 1 - \cos e^t$, one obtains from Theorem 1.4.3 that $\operatorname{abs}(f) = 0$. It follows from

$$\begin{aligned}\int_0^t e^{-\lambda s} f(s)\,ds &= \int_0^t e^{-\lambda s}\frac{d}{ds}(-\cos e^s)\,ds\\ &= -e^{-\lambda t}\cos e^t + \cos 1 - \int_0^t \lambda e^{-\lambda s}\cos e^s\,ds\\ &= -e^{-\lambda t}\cos e^t + \cos 1 - \lambda\int_0^t e^{-(\lambda+1)s}\frac{d}{ds}(\sin e^s)\,ds\\ &= -e^{-\lambda t}\cos e^t + \cos 1 - \lambda e^{-(\lambda+1)t}\sin e^t\\ &\quad +\lambda\sin 1 - \lambda(\lambda+1)\int_0^t e^{-(\lambda+2)s}e^s\sin e^s\,ds\end{aligned}$$

that

$$\hat{f}(\lambda) = \cos 1 + \lambda\sin 1 - \lambda(\lambda+1)\hat{f}(\lambda+2)$$

if $\operatorname{Re}\lambda > 0$. Thus, $\hat{f}$ has a holomorphic extension to $\operatorname{Re}\lambda > -2$, then to $\operatorname{Re}\lambda > -4$, etc. This shows that $\operatorname{hol}(\hat{f}) = -\infty$.

Another remarkable property of this function is that $\operatorname{abs}(f) < \operatorname{abs}(|f|)$. Clearly, $\int_0^\infty e^{-\lambda t}|f(t)|dt = \int_1^\infty u^{-\lambda}|\sin u|\,du$ converges for $\lambda > 1$. Since

$$\int_1^{n\pi}\frac{1}{u}|\sin u|\,du \ge \frac{\sqrt{2}}{2}\sum_{j=1}^{n-1}\int_{(4j+1)\pi/4}^{(4j+3)\pi/4}\frac{1}{u}\,du \ge \frac{\sqrt{2}}{2}\sum_{j=1}^{n-1}\frac{2\pi}{4j+3},$$

it follows that $\operatorname{abs}(|f|) = 1$. □

Theorem 1.5.3. *Let X be an ordered Banach space with a normal positive cone. Assume that* $\operatorname{abs}(f) < \infty$ *and* $f(t) \ge 0$ *a.e. Then* $\operatorname{hol}(\hat{f}) = \operatorname{abs}(f)$. *If, in addition,* $\operatorname{abs}(f) > -\infty$, *then $\hat{f}$ has a singularity at* $\operatorname{abs}(f)$.

Proof. If $\operatorname{abs}(f) = -\infty$, then the statement follows from (1.14). If $\operatorname{abs}(f) > -\infty$, then one may assume that $\operatorname{abs}(f) = 0$. Otherwise replace f by $t \mapsto e^{-\omega t}f(t)$, where $\omega := \operatorname{abs}(f)$. By (1.14), $\lambda \mapsto \hat{f}(\lambda)$ is holomorphic for $\operatorname{Re}\lambda > 0$. Assume that $\hat{f}$ is holomorphic at $\lambda = 0$. Then there exists a full circle with centre at 1 and radius $1 + 2\lambda_0$ for some $\lambda_0 > 0$ such that $\hat{f}$ is holomorphic on it (the extension of $\hat{f}$ to points in the complex plane with $\operatorname{Re}\lambda \le 0$ is denoted by the same symbol). Hence,

$$\hat{f}(-\lambda_0) = \sum_{k=0}^\infty (-1)^k(1+\lambda_0)^k\frac{1}{k!}\hat{f}^{(k)}(1).$$

Next it will be shown that $\int_0^\infty e^{\lambda_0 t}\langle f(t), x^*\rangle\,dt$ converges for all $x^* \in X^*$, which is a contradiction to (1.9). Let $g(t) := \langle f(t), x^*\rangle$. Since X_+^* is generating (Proposition C.2), one can assume that $x^* \ge 0$ and thus $g(t) \ge 0$. Then, by Theorem 1.5.1 and the monotone convergence theorem,

$$\begin{aligned}\hat{g}(-\lambda_0) &= \sum_{k=0}^\infty (1+\lambda_0)^k\frac{1}{k!}\int_0^\infty e^{-t}t^k g(t)\,dt\\ &= \int_0^\infty e^{-t}e^{(1+\lambda_0)t}g(t)\,dt = \int_0^\infty e^{\lambda_0 t}g(t)\,dt.\end{aligned}$$

□

The following example shows that there are positive functions with $\mathrm{hol}(\hat{f}) < \omega(f)$.

Example 1.5.4. Let $\Omega := \bigcup_{n\in\mathbb{N}}[n, n+e^{-n^3})$ and $f(t) := e^{t^2}\chi_\Omega(t)$. Then $f(t) \geq 0$ and $\omega(f) = \infty$. However, $\int_0^\infty e^{-\lambda s} f(s)\,ds = \sum_{n=0}^\infty \int_n^{n+e^{-n^3}} e^{-\lambda s} e^{s^2}\,ds$ converges for all $\lambda \in \mathbb{R}$. Hence, $\mathrm{hol}(\hat{f}) = \mathrm{abs}(f) = -\infty$. □

Finally in this section, we consider operator-valued functions. Let $T : \mathbb{R}_+ \to \mathcal{L}(X,Y)$ be strongly continuous, where X and Y are arbitrary Banach spaces, and assume that $\mathrm{abs}(T) < \infty$.

As in Section 1.4, let $u_x(t) := T(t)x \quad (x \in X,\ t \geq 0)$, and

$$\hat{T}(\lambda) := \lim_{t\to\infty} \int_0^t e^{-\lambda s} T(s)\,ds \qquad (\mathrm{Re}\,\lambda > \mathrm{abs}(T)),$$

where the integral is interpreted as in Section 1.4 and the limit exists in operator norm (Proposition 1.4.5).

Since $\hat{T}(\lambda)x = \widehat{u_x}(\lambda)$, it follows from Theorem 1.5.1, (1.8) and Proposition A.3 that $\hat{T} : \{\mathrm{Re}\,\lambda > \mathrm{abs}(T)\} \to \mathcal{L}(X,Y)$ is holomorphic. We define

$$\mathrm{hol}(\hat{T}) := \inf\Big\{\omega \in \mathbb{R} : \hat{T} \text{ extends to a holomorphic function from } \{\mathrm{Re}\,\lambda > \omega\} \text{ into } \mathcal{L}(X,Y)\Big\}.$$

Proposition 1.5.5. *Let $T : \mathbb{R}_+ \to \mathcal{L}(X,Y)$ be strongly continuous with $\mathrm{abs}(T) < \infty$. Then*

$$\mathrm{hol}(\hat{T}) = \sup\{\mathrm{hol}(\widehat{u_x}) : x \in X\}.$$

Proof. Since $\hat{T}(\lambda)x = \widehat{u_x}(\lambda)$ for $\mathrm{Re}\,\lambda > \mathrm{abs}(T)$, it is clear that $\mathrm{hol}(\hat{T}) \geq \mathrm{hol}(\widehat{u_x})$ for all $x \in X$. We have to show that if $\omega < \mathrm{abs}(T)$ and each $\widehat{u_x}$ extends holomorphically to $\{\mathrm{Re}\,\lambda > \omega\}$ and $\hat{T}(\lambda)x = \widehat{u_x}(\lambda)$ $(\mathrm{Re}\,\lambda > \omega)$, then $\hat{T}$ is a holomorphic function of $\{\mathrm{Re}\,\lambda > \omega\}$ into $\mathcal{L}(X,Y)$.

Take λ_0 with $\mathrm{Re}\,\lambda_0 > \mathrm{abs}(T)$. For $|\lambda - \lambda_0| < \mathrm{abs}(T) - \omega$,

$$\hat{T}(\lambda)x = \sum_{n=0}^\infty \frac{\widehat{u_x}^{(n)}(\lambda_0)}{n!}(\lambda - \lambda_0)^n.$$

Moreover, for $0 < r < \mathrm{abs}(T) - \omega$,

$$\sup\left\{\left\|\sum_{n=0}^m \frac{\widehat{u_x}^{(n)}(\lambda_0)}{n!}(\lambda - \lambda_0)^n\right\| : m \in \mathbb{N},\ |\lambda - \lambda_0| < r\right\} < \infty$$

for each $x \in X$. By the uniform boundedness principle, $\hat{T}(\lambda) \in \mathcal{L}(X,Y)$ and $\hat{T}$ is bounded on $\{|\lambda - \lambda_0| < r\}$. It follows from Proposition A.3 that $\hat{T} : \{\mathrm{Re}\,\lambda > \omega\} \to \mathcal{L}(X,Y)$ is holomorphic. □

1.6 Operational Properties

The importance of Laplace integrals in applications to differential equations lies in the fact that they transform the analytic operations of differentiation, integration and convolution into algebraic operations of multiplication. In this section, we establish these and other basic properties of Laplace transforms.

Proposition 1.6.1. *Let $f \in L^1_{loc}(\mathbb{R}_+, X)$, $\mu \in \mathbb{C}$ and $s \in \mathbb{R}_+$. Let*

$$\begin{aligned} g(t) &:= e^{-\mu t} f(t) \quad (t \geq 0), \\ f_s(t) &:= f(s+t) \quad (t \geq 0), \\ h_s(t) &:= \begin{cases} f(t-s) & (t \geq s), \\ 0 & (0 \leq t < s). \end{cases} \end{aligned}$$

Let $\lambda \in \mathbb{C}$. Then

a) *$\hat{g}(\lambda)$ exists if and only if $\hat{f}(\lambda + \mu)$ exists, and then $\hat{g}(\lambda) = \hat{f}(\lambda + \mu)$.*

b) *$\widehat{f_s}(\lambda)$ exists if and only if $\hat{f}(\lambda)$ exists, and then*

$$\widehat{f_s}(\lambda) = e^{\lambda s}\left(\hat{f}(\lambda) - \int_0^s e^{-\lambda t} f(t)\, dt\right).$$

c) *$\widehat{h_s}(\lambda)$ exists if and only if $\hat{f}(\lambda)$ exists, and then $\widehat{h_s}(\lambda) = e^{-\lambda s}\hat{f}(\lambda)$.*

Proof. These results follow immediately from the formulae:

$$\begin{aligned} \int_0^\tau e^{-\lambda t} g(t)\, dt &= \int_0^\tau e^{-(\lambda+\mu)t} f(t)\, dt, \\ \int_0^\tau e^{-\lambda t} f_s(t)\, dt &= e^{\lambda s}\left(\int_0^{s+\tau} e^{-\lambda t} f(t)\, dt - \int_0^s e^{-\lambda t} f(t)\, dt\right), \\ \int_0^\tau e^{-\lambda t} h_s(t)\, dt &= e^{-\lambda s} \int_0^{\tau - s} e^{-\lambda t} f(t)\, dt \quad (\tau > s). \end{aligned}$$

□

Proposition 1.6.2. *Let $f \in L^1_{loc}(\mathbb{R}_+, X)$ and $T \in \mathcal{L}(X,Y)$, and let $(T \circ f)(t) = T(f(t))$. Then $T \circ f \in L^1_{loc}(\mathbb{R}_+, Y)$. If $\hat{f}(\lambda)$ exists, then $\widehat{T \circ f}(\lambda)$ exists and equals $T(\hat{f}(\lambda))$.*

Proof. By Proposition 1.1.6, $T \circ f \in L^1_{loc}(\mathbb{R}_+, Y)$ and

$$\int_0^\tau e^{-\lambda t}(T \circ f)(t)\, dt = T \int_0^\tau e^{-\lambda t} f(t)\, dt.$$

The second statement follows on letting $\tau \to \infty$. □

Proposition 1.6.3. *Let $f \in L^1_{loc}(\mathbb{R}_+, X)$ and A be a closed operator on X. Suppose that $f(t) \in D(A)$ a.e. and $A \circ f \in L^1_{loc}(\mathbb{R}_+, X)$. Let $\lambda \in \mathbb{C}$. If $\hat{f}(\lambda)$ and $\widehat{A \circ f}(\lambda)$ both exist, then $\hat{f}(\lambda) \in D(A)$ and $\widehat{A \circ f}(\lambda) = A(\hat{f}(\lambda))$.*

Proof. By Proposition 1.1.7,

$$\int_0^\tau e^{-\lambda t}(A \circ f)(t)\, dt = A \int_0^\tau e^{-\lambda t} f(t)\, dt.$$

Since A is closed, the second statement follows on letting $\tau \to \infty$. $\square$

Now we consider convolutions.

Proposition 1.6.4. *Let $k \in L^1_{loc}(\mathbb{R}_+)$, $f \in L^1_{loc}(\mathbb{R}_+, X)$, $\lambda \in \mathbb{C}$, and suppose that $\operatorname{Re}\lambda > \max(\operatorname{abs}(|k|), \operatorname{abs}(f))$. Then $(\widehat{k * f})(\lambda)$ exists and $(\widehat{k * f})(\lambda) = \hat{k}(\lambda)\hat{f}(\lambda)$.*

Proof. Replacing $k(t)$ by $e^{-\lambda t}k(t)$ and $f(t)$ by $e^{-\lambda t}f(t)$, and using Proposition 1.6.1, we may assume that $\lambda = 0$.

First, we give the simple proof in the case when $f \in L^1(\mathbb{R}_+, X)$. Then Fubini's theorem gives that $k * f \in L^1(\mathbb{R}_+, X)$ and

$$\begin{aligned} (\widehat{k * f})(0) &= \int_0^\infty (k * f)(t)\, dt \\ &= \int_0^\infty \int_0^t k(t-s) f(s)\, ds\, dt \\ &= \int_0^\infty \left(\int_s^\infty k(t-s)\, dt \right) f(s)\, ds \\ &= \hat{k}(0)\hat{f}(0). \end{aligned}$$

Now, assume only that $\hat{f}(0)$ exists. Replacing $f(t)$ by $f(t) - e^{-t}\hat{f}(0)$ (and using the previous case), we may assume that $\hat{f}(0) = 0$. Let $\varepsilon > 0$. There exists K such that $\left\| \int_0^\tau f(s)\, ds \right\| < \varepsilon$ whenever $\tau > K$. Then

$$\int_0^\tau (k * f)(t)\, dt = (1 * (k * f))(\tau) = (k * (1 * f))(\tau).$$

Hence,

$$\begin{aligned} \left\| \int_0^\tau (k * f)(t)\, dt \right\| &\le \left\| \int_0^K k(\tau - t) \left(\int_0^t f(s)\, ds \right) dt \right\| \\ &\quad + \left\| \int_K^\tau k(\tau - t) \left(\int_0^t f(s)\, ds \right) dt \right\| \\ &\le M \int_{\tau-K}^\tau |k(s)|\, ds + \varepsilon \int_0^{\tau-K} |k(s)|\, ds, \end{aligned}$$

where $M := \sup_{t\geq 0} \left\| \int_0^t f(s)\, ds \right\| < \infty$. Letting $\tau \to \infty$,

$$\limsup_{\tau\to\infty} \left\| \int_0^\tau (k * f)(t)\, dt \right\| \leq \varepsilon \|k\|_1.$$

Since $\varepsilon > 0$ is arbitrary, it follows that $(\widehat{k * f})(0) = 0$, as required. □

As a corollary, we recover a simple result which was already observed in (1.11).

Corollary 1.6.5. *Let $f \in L^1_{loc}(\mathbb{R}_+, X)$ and let $F(t) = \int_0^t f(s)\, ds$. If $\operatorname{Re}\lambda > 0$ and $\hat{f}(\lambda)$ exists, then $\hat{F}(\lambda)$ exists and $\hat{F}(\lambda) = \hat{f}(\lambda)/\lambda$.*

Proof. This is immediate from Proposition 1.3.1 with $k(t) = 1$. □

Corollary 1.6.6. *Let $f : \mathbb{R}_+ \to X$ be absolutely continuous and differentiable a.e. If $\operatorname{Re}\lambda > 0$ and $\widehat{f'}(\lambda)$ exists, then $\hat{f}(\lambda)$ exists and $\widehat{f'}(\lambda) = \lambda\hat{f}(\lambda) - f(0)$.*

Proof. By Proposition 1.2.3, $f' \in L^1_{loc}(\mathbb{R}_+, X)$ and $f(t) - f(0) = \int_0^t f'(s)\, ds$. The result follows from Corollary 1.6.5. □

Now we want to consider the substitution of $\sqrt{\lambda}$ for λ; we will find a function h such that $\hat{h}(\lambda) = \hat{f}(\sqrt{\lambda})$. For this we first calculate the Laplace integral of a special function.

Lemma 1.6.7. *Let $s > 0$ and*

$$\begin{aligned} \phi_s(t) &= \frac{e^{-s^2/4t}}{\sqrt{\pi t}}, \\ \psi_s(t) &= \frac{s e^{-s^2/4t}}{2\sqrt{\pi} t^{3/2}} \qquad (t > 0). \end{aligned}$$

Then

$$\begin{aligned} \widehat{\phi_s}(\lambda) &= \frac{1}{\sqrt{\lambda}} e^{-s\sqrt{\lambda}}, \\ \widehat{\psi_s}(\lambda) &= e^{-s\sqrt{\lambda}} \qquad (\operatorname{Re}\lambda > 0). \end{aligned}$$

Proof. First, we show that, for $\alpha > 0$,

$$\int_0^\infty e^{-((\alpha/u)-u)^2}\, du = \int_0^\infty \frac{\alpha}{v^2} e^{-(v-\alpha/v)^2}\, dv = \frac{\sqrt{\pi}}{2}.$$

The first equality follows from the change of variable $v := \alpha/u$. Taking the average and making the change of variable $w := u - \alpha/u$, this gives

$$\begin{aligned}\int_0^\infty e^{-((\alpha/u)-u)^2}\,du &= \frac{1}{2}\int_0^\infty \left(1+\frac{\alpha}{u^2}\right)e^{-(u-\alpha/u)^2}\,du\\ &= \frac{1}{2}\int_{-\infty}^\infty e^{-w^2}\,dw\\ &= \frac{\sqrt{\pi}}{2}.\end{aligned}$$

Now, for $\lambda > 0$,

$$\begin{aligned}\widehat{\phi_s}(\lambda) &= e^{-s\sqrt{\lambda}}\int_0^\infty \frac{1}{\sqrt{\pi t}}e^{-(\sqrt{\lambda t}-s/(2\sqrt{t}))^2}\,dt\\ &= e^{-s\sqrt{\lambda}}\frac{2}{\sqrt{\pi\lambda}}\int_0^\infty \frac{\sqrt{\lambda}}{2u^2}e^{-(s\sqrt{\lambda}/(2u)-u)^2}\,du\\ &= e^{-s\sqrt{\lambda}}\frac{2}{\sqrt{\pi\lambda}}\frac{\sqrt{\pi}}{2}\\ &= \frac{e^{-s\sqrt{\lambda}}}{\sqrt{\lambda}},\\ \widehat{\psi_s}(\lambda) &= e^{-s\sqrt{\lambda}}\int_0^\infty \frac{s}{2\sqrt{\pi}t^{3/2}}e^{-(\sqrt{\lambda t}-s/(2\sqrt{t}))^2}\,dt\\ &= e^{-s\sqrt{\lambda}}\frac{2}{\sqrt{\pi}}\int_0^\infty e^{-(s\sqrt{\lambda}/(2u)-u)^2}\,du\\ &= e^{-s\sqrt{\lambda}}.\end{aligned}$$

For $\operatorname{Re}\lambda > 0$, the results follow by uniqueness of holomorphic extensions. □

Proposition 1.6.8. *Let $f \in L^1_{loc}(\mathbb{R}_+, X)$ with $\omega(f) < \infty$, and let*

$$\begin{aligned}g(t) &= \int_0^\infty \frac{e^{-s^2/4t}}{\sqrt{\pi t}}f(s)\,ds,\\ h(t) &= \int_0^\infty \frac{se^{-s^2/4t}}{2\sqrt{\pi}t^{3/2}}f(s)\,ds.\end{aligned}$$

Then $\hat{g}(\lambda) = \hat{f}(\sqrt{\lambda})/\sqrt{\lambda}$ and $\hat{h}(\lambda) = \hat{f}(\sqrt{\lambda})$, whenever $\operatorname{Re}\lambda > (\max\{\omega(f), 0\})^2$.

Proof. If $\operatorname{Re}\lambda > (\max\{\omega(f), 0\})^2$, then $\operatorname{Re}\sqrt{\lambda} > \max\{\omega(f), 0\}$, and Fubini's theorem and Lemma 1.6.7 give

$$\begin{aligned}\hat{g}(\lambda) &= \int_0^\infty\int_0^\infty \frac{e^{-s^2/4t}}{\sqrt{\pi t}}e^{-\lambda t}f(s)\,dt\,ds\\ &= \int_0^\infty \frac{e^{-s\sqrt{\lambda}}}{\sqrt{\lambda}}f(s)\,ds\\ &= \hat{f}(\sqrt{\lambda})/\sqrt{\lambda}.\end{aligned}$$

Similarly, $\hat{h}(\lambda) = \hat{f}(\sqrt{\lambda})$. □

1.7 Uniqueness, Approximation and Inversion

In this section we shall show that any $f \in L^1_{loc}(\mathbb{R}_+, X)$ with $\operatorname{abs}(f) < \infty$ is uniquely determined by its Laplace transform $\hat{f}$ and we shall give the Post-Widder inversion formula (Theorem 1.7.7). Other inversion theorems appear in Section 2.3. The following elementary statement will be used in the proofs of many results in this and the following sections.

Lemma 1.7.1. *Let $a, b > 0$ and define $\lambda_n := a + nb$, $e_{-\lambda_n}(t) := e^{-\lambda_n t}$ ($n \in \mathbb{N}_0$, $t \geq 0$). Then $\{e_{-\lambda_n} : n \in \mathbb{N}_0\}$ is total in $L^1(\mathbb{R}_+)$.*

Proof. By the Stone-Weierstrass theorem, the linear span of the set $P := \{t \mapsto a^{-1}t^{bn/a} : n \in \mathbb{N}_0\}$ is dense in $C[0,1]$ and thus in $L^1(0,1)$. Now the statement follows from the fact that $\Phi : L^1(0,1) \to L^1(\mathbb{R}_+)$ defined by $(\Phi g)(t) := ae^{-at}g(e^{-at})$ is an isometric isomorphism which maps P onto the exponential functions $\{e_{-\lambda_n} : n \in \mathbb{N}_0\}$. □

Proposition 1.7.2. *Let $f \in L^1_{loc}(\mathbb{R}_+, X)$ with $\operatorname{abs}(f) < \infty$, let $a > \operatorname{abs}(f)$, $b > 0$ and $\lambda_n := a + nb$. If $\hat{f}(\lambda_n) = 0$ for all $n \in \mathbb{N}$, then $f(t) = 0$ a.e.*

Proof. One can assume that $a > \max\{\operatorname{abs}(f), 0\}$. Define $F(t) := \int_0^t f(s)\, ds$ ($t \geq 0$). Then $0 = \hat{f}(\lambda_n) = \lambda_n \hat{F}(\lambda_n)$ for all $n \in \mathbb{N}_0$ (see Corollary 1.6.5). It follows from Theorem 1.4.3 that $a > \omega(F)$. Thus $G(t) := e^{-at}F(t)$ is continuous and bounded on $\mathbb{R}_+$, and $\hat{G}(nb) = \hat{F}(a+nb) = 0$ ($n \in \mathbb{N}$). For $x^* \in X^*$ define $g_{x^*}(t) := \langle G(t), x^* \rangle$ ($t \geq 0$). Then $g_{x^*} \in L^\infty(\mathbb{R}_+) = L^1(\mathbb{R}_+)^*$ and

$$\langle e_{-nb}, g_{x^*} \rangle = \int_0^\infty e^{-nbt} g_{x^*}(t)\, dt = \langle \hat{G}(nb), x^* \rangle = 0.$$

Since $\{e_{-nb} : n \in \mathbb{N}\}$ is a total subset of $L^1(\mathbb{R}_+)$ by Lemma 1.7.1, it follows that $g_{x^*}(t) = 0$ for all $t \geq 0$ and $x^* \in X^*$. This implies that $F(t) = e^{at}G(t) = 0$ for all $t \geq 0$ and thus $f(t) = 0$ a.e. □

Because of its importance we reformulate this result in the following form which is used frequently in the book.

Theorem 1.7.3 (Uniqueness Theorem). *Let $f, g \in L^1_{loc}(\mathbb{R}_+, X)$ with $\operatorname{abs}(f) < \infty$ and $\operatorname{abs}(g) < \infty$, and let $\lambda_0 > \max(\operatorname{abs}(f), \operatorname{abs}(g))$. Suppose that $\hat{f}(\lambda) = \hat{g}(\lambda)$ whenever $\lambda > \lambda_0$. Then $f(t) = g(t)$ a.e.*

Remark 1.7.4. A sequence $(\lambda_n)_{n\in\mathbb{N}}$ of complex numbers is called *a uniqueness sequence* for the Laplace transform if $f = 0$ a.e. whenever $f \in L^1_{loc}(\mathbb{R}_+, X)$, $\operatorname{abs}(f) < \operatorname{Re}\lambda_n$ for all n, and $\hat{f}(\lambda_n) = 0$ for all n. It was shown in Proposition 1.7.2 that equidistant sequences $\lambda_n = a + nb$ $(b > 0)$ are examples of uniqueness sequences. In particular, this shows that a function of the form $\lambda \mapsto q(\lambda)\sin(\lambda)$ cannot have a representation as a Laplace transform. A finite sequence $(\lambda_1, \cdots, \lambda_n)$ is not a uniqueness sequence since the function $\lambda \mapsto (\lambda+\mu)^{-2n}(\lambda-\lambda_1)\cdots(\lambda-\lambda_n)$ $(\operatorname{Re}\lambda > -\mu)$ is the Laplace transform of the convolution product of the functions $t \mapsto (1-(\lambda_i+\mu)t)e^{-\mu t}$ $(1 \le i \le n)$. A characterization of uniqueness sequences will be given in the Notes to this section. □

One can deduce the following fundamental result on approximation from the uniqueness theorem by a quotient argument (see also the proof of Vitali's theorem in Appendix A).

Theorem 1.7.5 (Approximation). *Let $f_n \in C(\mathbb{R}_+, X)$ with $\|f_n(t)\| \le Me^{\omega t}$ for some $M > 0$, $\omega \in \mathbb{R}$ and all $n \in \mathbb{N}$. Let $\lambda_0 \ge \omega$. The following are equivalent:*

(i) *The Laplace transforms $\hat{f}_n$ converge pointwise on (λ_0, ∞) and the sequence $(f_n)_{n\in\mathbb{N}}$ is equicontinuous on $\mathbb{R}_+$.*

(ii) *The functions f_n converge uniformly on compact subsets of $\mathbb{R}_+$.*

Moreover, if (ii) *holds, then $\hat{f}(\lambda) = \lim_{n\to\infty}\hat{f}_n(\lambda)$ for all $\lambda > \lambda_0$, where $f(t) := \lim_{n\to\infty} f_n(t)$.*

Proof. The space $c(X) := \{(x_n)_{n\in\mathbb{N}} : x_n \in X \text{ and } \lim_{n\to\infty} x_n \text{ exists}\}$ is a closed subspace of $l^\infty(X) := \{(x_n)_{n\in\mathbb{N}} : x_n \in X \text{ and } \sup_{n\in\mathbb{N}}\|x_n\| < \infty\}$. Define $w : \mathbb{R}_+ \to l^\infty(X)$ by $w(t) = (f_n(t))_{n\in\mathbb{N}}$. Assume that (i) holds. The equicontinuity of f_n implies the continuity of w, and the convergence of $\hat{f}_n(\lambda)$ implies that $\hat{w}(\lambda) = (\hat{f}_n(\lambda))_{n\in\mathbb{N}} \in c(X)$ for all $\lambda > \lambda_0$. Consider the quotient mapping $q : l^\infty(X) \to l^\infty(X)/c(X)$. Then $\widehat{(q \circ w)}(\lambda) = q(\hat{w}(\lambda)) = 0$ for all $\lambda > \lambda_0$. Since $q \circ w : \mathbb{R}_+ \to l^\infty(X)/c(X)$ is continuous, it follows from the uniqueness theorem that $q \circ w(t) = 0$ for all $t \ge 0$; i.e., $w(t) \in c(X)$ for all $t \ge 0$. Hence $(f_n)_{n\in\mathbb{N}}$ converges pointwise. Since $(f_n)_{n\in\mathbb{N}}$ is equicontinuous, this implies uniform convergence on each compact subset of $\mathbb{R}_+$.

Conversely, assume that (ii) holds. Clearly, uniform convergence implies equicontinuity. Let $f(t) = \lim_{n\to\infty} f_n(t)$. Then $\lim_{n\to\infty}\widehat{f_n}(\lambda) = \hat{f}(\lambda)$ for all $\lambda > \lambda_0$ by the dominated convergence theorem. □

We point out that one cannot omit the condition that the sequence $(f_n)_{n\in\mathbb{N}}$ is equicontinuous. To give an example, let $X = \mathbb{C}$ and $f_n(t) := e^{int}$ $(t \ge 0, n \in \mathbb{N})$. Then $\widehat{f_n}(\lambda) = \frac{1}{\lambda - in}$ converges to 0 as $n \to \infty$ for all $\lambda > 0$. But $f_n(t)$ does not converge as $n \to \infty$ if $t \in \mathbb{R}_+ \setminus 2\pi\mathbb{Z}$.

Another quotient argument enables us to deduce the converse of Proposition 1.6.3 from the uniqueness theorem.

Proposition 1.7.6. *Let A be a closed linear operator on X, let $f, g \in L^1_{loc}(\mathbb{R}_+, X)$ such that $\operatorname{abs}(f) < \infty$ and $\operatorname{abs}(g) < \infty$, and let $\omega > \max\{\operatorname{abs}(f), \operatorname{abs}(g)\}$. Then the following assertions are equivalent:*

(i) $f(t) \in D(A)$ *and* $Af(t) = g(t)$ *a.e. on* $\mathbb{R}_+$.

(ii) $\hat{f}(\lambda) \in D(A)$ *and* $A\hat{f}(\lambda) = \hat{g}(\lambda)$ *for all* $\lambda > \omega$.

Proof. The implication (i) $\Rightarrow$ (ii) has already been proved in Proposition 1.6.3.

(ii) $\Rightarrow$ (i): Let $G(A)$ be the graph of A, which is a closed subspace of $X \times X$, and let $q : X \times X \to (X \times X)/G(A)$ be the quotient map. Define $h : \mathbb{R}_+ \to (X \times X)/G(A)$ by $h(t) = q(f(t), g(t))$. Then $\hat{h}(\lambda) = q(\hat{f}(\lambda), \hat{g}(\lambda)) = 0$ for all $\lambda > \omega$, by (ii). By the uniqueness theorem, $h(t) = 0$ a.e. This proves that (i) is true. □

Recall from Proposition 1.2.2 that t is a Lebesgue point of $f \in L^1_{loc}(\mathbb{R}_+, X)$ if $\lim_{h\to 0} \frac{1}{h} \int_t^{t+h} \|f(s) - f(t)\|\, ds = 0$ and that almost all points t are Lebesgue points of f. We now prove the Post-Widder inversion formula.

Theorem 1.7.7 (Post-Widder). *Let* $f \in L^1_{loc}(\mathbb{R}_+, X)$. *Assume that* $\operatorname{abs}(f) < \infty$ *and that* $t > 0$ *is a Lebesgue point of* f. *Then*

$$f(t) = \lim_{k\to\infty} (-1)^k \frac{1}{k!} \left(\frac{k}{t}\right)^{k+1} \hat{f}^{(k)}\left(\frac{k}{t}\right).$$

Proof. To explain the structure of the proof, we first consider the special case when f is a bounded continuous function. By Theorem 1.5.1,

$$(-1)^k \frac{1}{k!} \left(\frac{k}{t}\right)^{k+1} \hat{f}^{(k)}\left(\frac{k}{t}\right) = \int_0^\infty \rho_k(s) f(s)\, ds,$$

where $\rho_k(s) := \frac{1}{k!} \left(\frac{k}{t}\right)^{k+1} e^{-ks/t} s^k$ $(s > 0)$. The functions ρ_k are "approximate Dirac δ-functions"; i.e., $\rho_k \geq 0$, $\int_0^\infty \rho_k(s)\, ds = 1$, and for all $\varepsilon > 0$ and all open intervals $I \subset \mathbb{R}_+$ containing t we have $\int_{s\notin I} \rho_k(s)\, ds < \varepsilon$ for all sufficiently large k (see below). Since f is assumed to be bounded and continuous, it follows from

$$\begin{aligned} \left\| \int_0^\infty \rho_k(s) f(s)\, ds - f(t) \right\| &= \left\| \int_0^\infty \rho_k(s)(f(s) - f(t))\, ds \right\| \\ &\leq 2\|f\|_\infty \int_{s\notin I} \rho_k(s)\, ds + \sup_{s\in I} \|f(s) - f(t)\| \end{aligned}$$

that $\int_0^\infty \rho_k(s) f(s)\, ds \to f(t)$ as $k \to \infty$. This proves the statement for bounded and continuous functions f.

Now let $f \in L^1_{loc}(\mathbb{R}_+, X)$ and $\max(\operatorname{abs}(f), 0) < \omega < \infty$. By (1.12), $\omega > \omega(F)$, where $F(s) := \int_0^s f(r)\, dr$. In the following let $t > 0$ be a fixed Lebesgue point of f, and let $k \in \mathbb{N}$ such that $k > \omega t$. Let

$$G(s) := \int_t^s (f(r) - f(t))\, dr = F(s) - F(t) - f(t)(s - t) \quad (s \geq 0).$$

Since $\omega > \omega(F)$, there exists $M > 0$ such that $\|G(s)\| \le Me^{\omega s}$ for all $s \ge 0$. Since $\frac{1}{k!}\lambda^{k+1}\int_0^\infty e^{-\lambda s}s^k\,ds = 1$ for $\lambda > 0$ (by induction and integration by parts), it follows from Theorem 1.5.1 and integration by parts that

$$\begin{aligned} J_k &:= (-1)^k \frac{1}{k!}\left(\frac{k}{t}\right)^{k+1} \hat{f}^{(k)}\left(\frac{k}{t}\right) - f(t) \\ &= \frac{1}{k!}\left(\frac{k}{t}\right)^{k+1} \int_0^\infty e^{-ks/t}s^k f(s)\,ds - f(t) \\ &= \frac{1}{k!}\left(\frac{k}{t}\right)^{k+1} \int_0^\infty e^{-ks/t}s^k (f(s) - f(t))\,ds \\ &= \frac{1}{k!}\left(\frac{k}{t}\right)^{k+1} k\int_0^\infty e^{-ks/t}s^{k-1}\left(\frac{s}{t} - 1\right) G(s)\,ds \\ &= \frac{k^{k+2}}{k!t}\int_0^\infty e^{-ku}u^{k-1}(u-1)G(ut)\,du. \end{aligned}$$

Let $\varepsilon > 0$. Since t is a Lebesgue point of f, there exists $0 < \delta < 1$ such that

$$\frac{1}{t}\|G(ut)\| = \left\|\frac{1}{t}\int_t^{tu} (f(r) - f(t))\,dr\right\| \le \frac{\varepsilon}{3}|u-1|$$

if $|u-1| \le \delta$. Define

$$J_{1,k} := \frac{k^{k+2}}{k!t}\int_{1-\delta}^{1+\delta} e^{-ku}u^{k-1}(u-1)G(ut)\,du.$$

Then

$$\begin{aligned} \|J_{1,k}\| &\le \frac{\varepsilon}{3}\frac{k^{k+2}}{k!}\int_{1-\delta}^{1+\delta} e^{-ku}u^{k-1}(u-1)^2\,du \\ &\le \frac{\varepsilon}{3}\frac{k^{k+2}}{k!}\int_0^\infty e^{-ku}\left(u^{k+1} - 2u^k + u^{k-1}\right)du \\ &= \frac{\varepsilon}{3}\frac{k^{k+2}}{k!}\left(\frac{(k+1)!}{k^{k+2}} - \frac{2k!}{k^{k+1}} + \frac{(k-1)!}{k^k}\right) = \frac{\varepsilon}{3} \end{aligned}$$

for all $k \in \mathbb{N}$ with $k > \omega t$. Let

$$J_{2,k} := \frac{k^{k+2}}{k!t}\int_0^{1-\delta} e^{-ku}u^{k-1}(u-1)G(ut)\,du.$$

If $k > 1/\delta$, the function $u \mapsto e^{-ku}u^{k-1}$ is increasing on $(0, 1-\delta)$. Thus,

$$\|J_{2,k}\| \le \frac{k^{k+2}}{k!t}e^{-k(1-\delta)}(1-\delta)^{k-1}\int_0^{1-\delta}(1-u)\|G(ut)\|\,du =: b_k.$$

Since $b_{k+1}/b_k = \left(1+\frac{1}{k}\right)^{k+2} e^{\delta-1}(1-\delta) \to e^{\delta}(1-\delta) < 1$ as $k \to \infty$, one obtains that $\|J_{2,k}\| < \varepsilon/3$ for all sufficiently large k. Finally let

$$J_{3,k} := \frac{1}{t}\frac{1}{k!}k^{k+2}\int_{1+\delta}^{\infty} e^{-ku}u^{k-1}(u-1)G(ut)\,du.$$

The function $u \mapsto e^{-mu}u^m$ is decreasing on $(1+\delta,\infty)$ for all $m \in \mathbb{N}$. Choose $k_0 > t\omega$ and let $k > k_0$. Then

$$\begin{aligned}\|J_{3,k}\| &= \left\|\frac{k^{k+2}}{k!t}\int_{1+\delta}^{\infty} e^{-(k-k_0)u}u^{k-k_0}e^{-k_0u}u^{k_0-1}(u-1)G(ut)\,du\right\| \\ &\le \frac{k^{k+2}}{k!t}e^{-(k-k_0)(1+\delta)}(1+\delta)^{k-k_0}\int_{1+\delta}^{\infty} e^{-k_0u}u^{k_0-1}(u-1)Me^{\omega ut}\,du \\ &=: c_k.\end{aligned}$$

Since $c_{k+1}/c_k = \left(1+\frac{1}{k}\right)^{k+2} e^{-1-\delta}(1+\delta) \to e^{-\delta}(1+\delta) < 1$, one obtains that $\|J_{3,k}\| < \varepsilon/3$ for all sufficiently large k. It follows from $J_k = J_{1,k} + J_{2,k} + J_{3,k}$ that $J_k \to 0$ as $k \to \infty$. □

1.8 The Fourier Transform and Plancherel's Theorem

In this section, we give a brief summary of the properties of vector-valued Fourier transforms, and we extend Plancherel's theorem and the Paley-Wiener theorem (characterizing the Laplace transforms of L^2-functions) to functions with values in a Hilbert space.

For $f \in L^1(\mathbb{R}, X)$, the *Fourier transform* of f is the function $\mathcal{F}f : \mathbb{R} \to X$ defined by

$$(\mathcal{F}f)(s) := \int_{-\infty}^{\infty} e^{-ist}f(t)\,dt.$$

We also define

$$(\overline{\mathcal{F}}f)(s) := \int_{-\infty}^{\infty} e^{ist}f(t)\,dt = (\mathcal{F}f)(-s) = (\mathcal{F}\check{f})(s),$$

where $\check{f}(t) := f(-t)$.

Many properties of the Fourier transform on $L^1(\mathbb{R}, X)$ can be proved in exactly the same way as for the scalar-valued case (some can also be proved by applying linear functionals and using the scalar-valued results and the Hahn-Banach theorem), and we quote some of them here. Proofs of the scalar-valued cases may be found in standard textbooks such as [Rud87], [RS72], [Yos80], [Rud91].

Theorem 1.8.1. *Let $f \in L^1(\mathbb{R}, X)$ and $g \in L^1(\mathbb{R})$.*

a) $\mathcal{F}(g * f)(s) = (\mathcal{F}g)(s)(\mathcal{F}f)(s)$.

b) $\int_{-\infty}^{\infty} g(t)(\mathcal{F}f)(t)\,dt = \int_{-\infty}^{\infty} (\mathcal{F}g)(t)f(t)\,dt$.

c) **(Riemann-Lebesgue Lemma)** $\mathcal{F}f \in C_0(\mathbb{R}, X)$.

d) **(Inversion Theorem)** *If* $\mathcal{F}f \in L^1(\mathbb{R}, X)$, *then* $f = \frac{1}{2\pi}\overline{\mathcal{F}}(\mathcal{F}f)$ *a.e.*

For scalar-valued functions, Plancherel's theorem [Rud87, Theorem 19.2] shows that $\mathcal{F}f \in L^2(\mathbb{R})$ and $\|\mathcal{F}f\|_2 = \sqrt{2\pi}\|f\|_2$ whenever $f \in L^1(\mathbb{R}) \cap L^2(\mathbb{R})$, and hence the restriction of $\mathcal{F}$ to $L^1(\mathbb{R}) \cap L^2(\mathbb{R})$ has a unique extension to a bounded linear operator on $L^2(\mathbb{R})$ (also denoted by $\mathcal{F}$), $\frac{1}{\sqrt{2\pi}}\mathcal{F}$ is a unitary operator on $L^2(\mathbb{R})$, and Theorem 1.8.1 b) holds for $f, g \in L^2(\mathbb{R})$. Hence, $\mathcal{F}^{-1} = (2\pi)^{-1}\overline{\mathcal{F}}$, where $(\overline{\mathcal{F}}f)(t) = (\mathcal{F}f)(-t)$.

Plancherel's theorem is not true for vector-valued functions, except when the space X is a Hilbert space (see the Notes). If X is a Hilbert space with inner product $(\cdot|\cdot)_X$, then $L^2(\mathbb{R}, X)$ is also a Hilbert space with inner product

$$(f|g)_{L^2(\mathbb{R},X)} := \int_{-\infty}^{\infty} (f(t)|g(t))_X \, dt.$$

As observed in Section 1.1, the simple functions are dense in $L^2(\mathbb{R}_+, X)$. In this Hilbert space context, this can also be shown by computing the orthogonal complement of the simple functions.

Theorem 1.8.2 (Plancherel's Theorem). *Let X be a Hilbert space. Then $\mathcal{F}f \in L^2(\mathbb{R}, X)$ and $\|\mathcal{F}f\|_{L^2(\mathbb{R},X)} = \sqrt{2\pi}\|f\|_{L^2(\mathbb{R},X)}$ for all $f \in L^1(\mathbb{R}, X) \cap L^2(\mathbb{R}, X)$. The restriction of $\mathcal{F}$ to $L^1(\mathbb{R}, X) \cap L^2(\mathbb{R}, X)$ extends to a bounded linear operator $\mathcal{F}$ on $L^2(\mathbb{R}, X)$ and $\frac{1}{\sqrt{2\pi}}\mathcal{F}$ is a unitary operator on the Hilbert space $L^2(\mathbb{R}, X)$. Moreover,*

$$\int_{-\infty}^{\infty} ((\mathcal{F}f)(t)|g(t))_X \, dt = \int_{-\infty}^{\infty} (f(t)|(\mathcal{F}g)(-t))_X \, dt \tag{1.15}$$

for all $f, g \in L^2(\mathbb{R}, X)$.

Proof. Let $f \in L^1(\mathbb{R}, X) \cap L^2(\mathbb{R}, X)$. To prove that

$$\|\mathcal{F}f\|_{L^2(\mathbb{R},X)} = \sqrt{2\pi}\|f\|_{L^2(\mathbb{R},X)},$$

it suffices to assume that X is separable, since f is almost separably valued. Let $\{e_n : n \in \mathbb{N}\}$ be an orthonormal basis of X, and let $f_n(t) := (f(t)|e_n)_X$. Then $f_n \in L^1(\mathbb{R}) \cap L^2(\mathbb{R})$ and $((\mathcal{F}f)(s)|e_n)_X = (\mathcal{F}f_n)(s)$. Now, using the scalar-valued Plancherel theorem,

$$\begin{aligned}
\int_{-\infty}^{\infty} \|(\mathcal{F}f)(s)\|^2 \, ds &= \int_{-\infty}^{\infty} \sum_{n=1}^{\infty} |(\mathcal{F}f_n)(s)|^2 \, ds \\
&= 2\pi \int_{-\infty}^{\infty} \sum_{n=1}^{\infty} |f_n(t)|^2 \, dt \\
&= 2\pi \int_{-\infty}^{\infty} \|f(t)\|^2 \, dt.
\end{aligned}$$

This proves the first part of the result. Since $L^1(\mathbb{R}_+,X) \cap L^2(\mathbb{R}_+,X)$ is dense in $L^2(\mathbb{R}_+,X)$, $\mathcal{F}$ extends uniquely to a bounded linear operator on $L^2(\mathbb{R},X)$ such that $\frac{1}{\sqrt{2\pi}}\mathcal{F}$ is an isometry.

One may prove (1.15) in a similar way, using Parseval's formula and the corresponding scalar-valued result (Theorem 1.8.1 b)). This implies that the adjoint operator of $\mathcal{F}$ in the sense of Hilbert spaces is $\overline{\mathcal{F}}$, where $(\overline{\mathcal{F}}f)(t) = (\mathcal{F}f)(-t)$. Hence, $\overline{\mathcal{F}}\mathcal{F} = 2\pi I$ so $\overline{\mathcal{F}}$ is surjective, and it follows that $\mathcal{F}$ is surjective. □

We remark that Plancherel's theorem extends to functions of several variables with values in a Hilbert space X: the normalized Fourier transform $(2\pi)^{-n/2}\mathcal{F}$ is a unitary operator on $L^2(\mathbb{R}^n,X)$. This can be deduced from the scalar-valued case as in the proof of Theorem 1.8.2. Alternatively, it may be deduced from Theorem 1.8.2 by induction, using the identification $L^2(\mathbb{R}^{n+1},X) = L^2(\mathbb{R},L^2(\mathbb{R}^n,X))$.

When $f \in L^1(\mathbb{R}_+,X)$, we consider f as being a member of $L^1(\mathbb{R},X)$ with $f(t) = 0$ for $t < 0$, so

$$(\mathcal{F}f)(s) = \int_0^\infty e^{-ist} f(t)\,dt = \hat{f}(is),$$

where, as usual, $\hat{f}$ is the Laplace transform of f. Similarly, if $t \mapsto e^{-at}f(t)$ belongs to $L^1(\mathbb{R}_+,X)$ (for example, if $f \in L^2(\mathbb{R}_+,X)$ and $a > 0$), then its Fourier transform is $s \mapsto \hat{f}(a+is)$. Thus, Plancherel's theorem can be used to study Laplace transforms of functions in $L^2(\mathbb{R}_+,X)$ when X is a Hilbert space.

Let $\mathbb{C}_+ := \{\lambda \in \mathbb{C} : \operatorname{Re}\lambda > 0\}$ and $H^2(\mathbb{C}_+,X)$ be the space of all holomorphic functions $g : \mathbb{C}_+ \to X$ such that

$$\|g\|^2_{H^2(\mathbb{C}_+,X)} := \sup_{\alpha>0} \int_{-\infty}^\infty \|g(\alpha+is)\|^2\,ds < \infty.$$

For scalar-valued functions, the Paley-Wiener theorem [Rud87, Theorem 9.13] shows that $g \in H^2(\mathbb{C}_+) := H^2(\mathbb{C}_+,\mathbb{C})$ if and only if $g = \hat{f}|_{\mathbb{C}_+}$ for some $f \in L^2(\mathbb{R}_+)$. Then

$$\sup_{\alpha>0} \int_{-\infty}^\infty |g(\alpha+is)|^2\,ds = \lim_{\alpha\downarrow 0} \int_{-\infty}^\infty |g(\alpha+is)|^2\,ds.$$

Moreover, g has $\mathcal{F}f$ as a boundary function in the sense that $g(\alpha+is) \to (\mathcal{F}f)(s)$ s-a.e. and in L^2-norm, as $\alpha \downarrow 0$. In addition, g is the Poisson integral of $\mathcal{F}f$:

$$g(\alpha+is) = \frac{\alpha}{\pi} \int_{-\infty}^\infty \frac{(\mathcal{F}f)(r)}{\alpha^2+(s-r)^2}\,dr$$

[Dur70, Chapter 11], [Koo80, Chapter VI].

Again, these results are not true for vector-valued functions in general, but they are true in the case of Hilbert spaces.

Theorem 1.8.3 (Paley-Wiener Theorem). *Let X be a Hilbert space. Then the map $f \mapsto \hat{f}|_{\mathbb{C}_+}$ is an isometric isomorphism of $L^2(\mathbb{R}_+, X)$ onto $H^2(\mathbb{C}_+, X)$. Moreover, for $f \in L^2(\mathbb{R}_+, X)$,*

$$\hat{f}(\alpha + is) = \frac{\alpha}{\pi} \int_{-\infty}^{\infty} \frac{(\mathcal{F}f)(r)}{\alpha^2 + (s-r)^2}\, dr. \tag{1.16}$$

As $\alpha \downarrow 0$, $\|\hat{f}(\alpha+is)-(\mathcal{F}f)(s)\| \to 0$ (s)-a.e. and $\int_{-\infty}^{\infty} \|\hat{f}(\alpha+is)-(\mathcal{F}f)(s)\|^2\, ds \to 0$.

Proof. Let $f \in L^2(\mathbb{R}_+, X)$. For $\alpha > 0$,

$$\hat{f}(\alpha + is) = \int_0^{\infty} e^{-ist} \left(e^{-\alpha t} f(t)\right) dt.$$

By Plancherel's Theorem 1.8.2,

$$\int_{-\infty}^{\infty} \left\|\hat{f}(\alpha + is)\right\|^2 ds = 2\pi \int_0^{\infty} e^{-2\alpha t} \|f(t)\|^2\, dt \le 2\pi \|f\|_2^2.$$

Thus $\hat{f} \in H^2(\mathbb{C}_+, X)$. Moreover, Plancherel's theorem and the dominated convergence theorem give

$$\int_{-\infty}^{\infty} \|\hat{f}(\alpha + is) - (\mathcal{F}f)(s)\|^2\, ds = 2\pi \int_0^{\infty} |e^{-\alpha t} - 1|^2 \|f(t)\|^2\, dt \to 0$$

as $\alpha \downarrow 0$. For $\alpha > 0$ and $x \in X$,

$$(\hat{f}(\alpha + is)|x)_X = \frac{\alpha}{\pi} \int_{-\infty}^{\infty} \frac{((\mathcal{F}f)(r)|x)_X}{\alpha^2 + (s-r)^2}\, dr = \left(\int_{-\infty}^{\infty} \frac{(\mathcal{F}f)(r)}{\alpha^2 + (s-r)^2}\, dr \,\middle|\, x \right)_X,$$

which establishes (1.16). The proof that $\|\hat{f}(\alpha + is) - (\mathcal{F}f)(s)\| \to 0$ a.e. is similar to the scalar-valued case.

Conversely, let $g \in H^2(\mathbb{C}_+, X)$. Then $\hat{g}$ is separably valued, so we may assume that X is separable. Let $\{e_n : n \in \mathbb{N}\}$ be an orthonormal basis of X, and let $g_n(\lambda) = (g(\lambda)|e_n)_X$. Then $g_n \in H^2(\mathbb{C}_+)$, so the scalar-valued case implies that there exists $f_n \in L^2(\mathbb{R}_+)$ such that $g_n = \widehat{f_n}|_{\mathbb{C}_+}$. Moreover,

$$\begin{aligned}
\int_0^{\infty} \sum_{n=1}^{\infty} |f_n(t)|^2\, dt &= \lim_{\alpha\downarrow 0} \sum_{n=1}^{\infty} \int_0^{\infty} e^{-2\alpha t} |f_n(t)|^2\, dt \\
&= \frac{1}{2\pi} \lim_{\alpha\downarrow 0} \sum_{n=1}^{\infty} \int_{-\infty}^{\infty} |g_n(\alpha + is)|^2\, ds \\
&= \frac{1}{2\pi} \lim_{\alpha\downarrow 0} \int_{-\infty}^{\infty} \|g(\alpha + is)\|^2\, ds < \infty.
\end{aligned}$$

Hence $\sum_{n=1}^{\infty} |f_n(t)|^2$ converges, and therefore $\sum_{n=1}^{\infty} f_n(t) e_n$ converges to a sum $f(t)$ in X, for almost all t. Now, $f \in L^2(\mathbb{R}_+, X)$ and, for $\lambda \in \mathbb{C}_+$,

$$\hat{f}(\lambda) = \sum_{n=1}^{\infty} \widehat{f_n}(\lambda) e_n = g(\lambda).$$

□

1.9 The Riemann-Stieltjes Integral

This section is an introduction to the Riemann-Stieltjes integral $\int_a^b g(t)\,dF(t)$ of a vector-valued function F and a scalar-valued function g on $[a,b]$. Such integrals play an important role in the approach to Laplace transform theory taken in Chapter 2. More precisely, let $f : \mathbb{R}_+ \to X$ be a bounded measurable function. Then $t \mapsto F(t) := \int_0^t f(s)\,ds$ is Lipschitz continuous and we will show in Sections 1.10 and 2.1 that

$$\hat{f}(\lambda) = \int_0^\infty e^{-\lambda t} f(t)\,dt = \int_0^\infty e^{-\lambda t}\,dF(t) = T_F(e_{-\lambda})$$

for all $\lambda > 0$, where $T_F : g \mapsto \int_0^\infty g(t)\,dF(t)$ is a bounded linear operator from $L^1(\mathbb{R}_+)$ to X and $e_{-\lambda}(t) := e^{-\lambda t}$. Thus, the Laplace integrals $\hat{f}(\lambda)$ are evaluations of a bounded linear operator T_F at the exponential function $e_{-\lambda}$. Since the map $\Phi_S : F \mapsto T_F$ turns out to be an isometric isomorphism between the Lipschitz continuous functions $F : \mathbb{R}_+ \to X$ and $\mathcal{L}(L^1(\mathbb{R}_+), X)$ (see Section 2.1), many functional analytic arguments can be applied to Laplace transform theory.

A function $F : [a,b] \to X$ is of *bounded semivariation* if there exists $M \geq 0$ such that $\|\sum_i (F(t_i) - F(s_i))\| \leq M$ for every choice of a finite number of non-overlapping intervals (s_i, t_i) in $[a,b]$. Recall from Section 1.2 that F is of *bounded variation* if there exists $M \geq 0$ such that $\sum_i \|F(t_i) - F(t_{i-1})\| \leq M$ for every finite partition $a = t_0 < t_1 < \ldots < t_n = b$ of $[a,b]$. Further, F is of *weak bounded variation* if $x^* \circ F : t \mapsto \langle F(t), x^* \rangle$ is of bounded variation for every $x^* \in X^*$. The set of functions $F : [a,b] \to X$ of bounded semivariation is denoted by $\mathrm{BSV}([a,b], X)$. A function $F : \mathbb{R}_+ \to X$ is in $\mathrm{BSV}_{loc}(\mathbb{R}_+, X)$ if it is of bounded semivariation on every compact subinterval of $\mathbb{R}_+$.

As remarked in Section 1.2, any Lipschitz function is of bounded variation, and it is easy to see that any function of bounded variation is of bounded semivariation. We show in the following proposition that the function $F : [0,1] \to L^\infty[0,1]$ defined by $F(t) = \chi_{[0,t]}$ is of bounded semivariation. Since $\|F(t) - F(s)\| = 1$ for all $t \neq s$, F is not of bounded variation, not separably valued, and not measurable (see Example 1.1.5). Hence, functions of bounded semivariation may not be measurable.

Proposition 1.9.1. *Let X be an ordered Banach space with normal cone. Let $F : [a,b] \to X$ be increasing. Then F is of bounded semivariation.*

Proof. Let (s_i, t_i) $(i = 1, 2, \ldots, n)$ be disjoint intervals in $[a,b]$. Then

$$0 \leq \sum_{i=1}^n (F(t_i) - F(s_i)) \leq F(t_n) - F(s_1) \leq F(b) - F(a).$$

Hence

$$\left\| \sum_{i=1}^n (F(t_i) - F(s_i)) \right\| \leq c\|F(b) - F(a)\|,$$

where c is a constant associated with the normal cone (see Appendix C). □

In the context of Proposition 1.9.1, it is easy to see directly that F is of weak bounded variation. In fact, there is the following general result.

Proposition 1.9.2. *A function $F : [a,b] \to X$ is of bounded semivariation if and only if it is of weak bounded variation.*

Proof. Assume that F is of weak bounded variation. Let $S_\Omega := \sum_i (F(t_i) - F(s_i))$, where Ω is the union of finitely many disjoint intervals (s_i, t_i) in $[a,b]$. For each $x^* \in X^*$, there exists $M_{x^*} := V_{[a,b]}(x^* \circ F)$ such that $|\langle S_\Omega, x^* \rangle| \le M_{x^*}$ for all such Ω. It follows from the uniform boundedness principle that F is of bounded semivariation.

Now, let F be of bounded semivariation. Then there exists a constant $M \ge 0$ such that $\|\sum_i (F(t_i) - F(s_i))\| \le M$ for any choice of a finite number of disjoint intervals (s_i, t_i) in $[a,b]$. To obtain the weak bounded variation, one writes $x^* = x_1^* + i x_2^*$, where x_j^* is a real-linear functional, and distinguishes between the subintervals on which the numbers $\langle F(t_i) - F(t_{i-1}), x_j^* \rangle$ are either positive or negative. □

Let F, g be two functions defined on an interval $[a,b]$, one with values in X and the other with values in $\mathbb{C}$. If π denotes a finite partition $a = t_0 < t_1 < \ldots < t_n = b$ of $[a,b]$ with *partitioning points* t_i and with some *intermediate points* $s_i \in [t_{i-1}, t_i]$ $(i = 1, \ldots, n)$, we denote by $|\pi| = \max_i (t_i - t_{i-1})$ the *norm* of π, and by

$$S(g, F, \pi) := \sum_{i=1}^{n} g(s_i) \, (F(t_i) - F(t_{i-1}))$$

the *Riemann-Stieltjes sum* associated with g, F and π. We say that g is *Riemann-Stieltjes integrable with respect to F* if

$$\int_a^b g(t) \, dF(t) := \lim_{|\pi| \to 0} S(g, F, \pi)$$

exists in the norm topology of X. Here π runs through all partitions of $[a,b]$ with intermediate points, and the limit must be independent of the choice of intermediate points.

It is immediate from the definition that the set of all functions g which are Riemann-Stieltjes integrable with respect to a fixed function F is a linear space, and the Riemann-Stieltjes integral is linear in g (and also in F). If F is of bounded variation, g is bounded and $\int_a^b g(t) \, dF(t)$ exists, then

$$\left\| \int_a^b g(t) \, dF(t) \right\| \le \sup_{t \in [a,b]} \|g(t)\| \, V_{[a,b]}(F). \tag{1.17}$$

If F is of bounded semivariation, then it follows from the proof of Proposition 1.9.2 that

$$\left\| \int_a^b g(t)\, dF(t) \right\| \leq 4M \sup_{t\in[a,b]} \|g(t)\|, \tag{1.18}$$

where

$$M := \sup \left\{ \left\| \sum_i (F(t_i) - F(s_i)) \right\| : (s_i, t_i) \text{ disjoint subintervals of } [a,b] \right\}. \tag{1.19}$$

The Riemann-Stieltjes integral respects closed operators; there are easy analogues of Propositions 1.1.6 and 1.1.7, both when g is scalar-valued and F is vector-valued and in the alternative case.

When $F(t) = t$, we write $S(g, \pi)$ for $S(g, F, \pi)$ and call it the *Riemann sum* associated with g and π. We say that g is *Riemann integrable* on $[a,b]$ if g is Riemann-Stieltjes integrable with respect to $F(t) = t$, and we write

$$\int_a^b g(t)\, dt := \lim_{|\pi|\to 0} S(g, \pi).$$

In the scalar-valued case, $g : [a,b] \to \mathbb{C}$ is Riemann integrable if and only if g is bounded and continuous a.e., and the Riemann and Lebesgue integrals are then equal [Rud76, Theorem 11.33]. By applying linear functionals, it follows that if $g : [a,b] \to X$ is Riemann integrable, then it is bounded and weakly measurable. If X is separable, then g is measurable by Pettis's theorem 1.1.1 and hence Bochner integrable by Theorem 1.1.4. Whenever g is both Riemann and Bochner integrable, the two integrals coincide (so our notation should not cause confusion). However, Riemann integrable functions with values in an inseparable space may be nowhere continuous and not even measurable (see Example 1.9.7 below).

Now we return to Riemann-Stieltjes sums and integrals of two functions F and g. Let π be a partition of $[a,b]$ with partitioning points $a = t_0 < t_1 < \ldots < t_n = b$ and intermediate points $s_i \in [t_{i-1}, t_i]$. If one chooses $s_0 = a$ and $s_{n+1} = b$, then we obtain a partition π' with partitioning points $a = s_0 \leq s_1 \leq \ldots \leq s_{n+1} = b$, with intermediate points $t_i \in [s_i, s_{i+1}]$, and $|\pi'| \leq 2|\pi|$. Moreover,

$$S(F, g, \pi) = g(b)F(b) - g(a)F(a) - S(g, F, \pi').$$

It follows that if g is Riemann-Stieltjes integrable with respect to F, then so is F with respect to g (and vice versa, by symmetry) and the following formula of integration by parts holds:

$$\int_a^b g(t)\, dF(t) = g(b)F(b) - g(a)F(a) - \int_a^b F(t)\, dg(t). \tag{1.20}$$

Example 1.9.3. Let $a \le c \le d \le b$ and let I be an interval with endpoints c and d. Let $F : [a, b] \to X$. If $a \notin I$ and $b \notin I$, then F and χ_I are Riemann-Stieltjes integrable with respect to each other if and only if F is continuous at c and d, and then

$$\int_a^b \chi_I(t)\, dF(t) = -\int_a^b F(t)\, d\chi_I(t) = F(d) - F(c).$$

If $a = c \in I$ and $b \notin I$, then F and χ_I are Riemann-Stieltjes integrable with respect to each other if and only if F is continuous at d, and then

$$\int_a^b \chi_I(t)\, dF(t) = F(d) - F(a), \qquad \int_a^b F(t)\, d\chi_I(t) = -F(d).$$

If $a \notin I$ and $b = d \in I$, then F and χ_I are Riemann-Stieltjes integrable with respect to each other if and only if F is continuous at c, and then

$$\int_a^b \chi_I(t)\, dF(t) = F(b) - F(c), \qquad \int_a^b F(t)\, d\chi_I(t) = F(c). \qquad \square$$

Proposition 1.9.4. *Let $F : [a, b] \to X$ and $g : [a, b] \to \mathbb{C}$. If one function is continuous and the other is of bounded semivariation, then F and g are Riemann-Stieltjes integrable with respect to each other.*

Proof. a) Assume that F is of bounded semivariation and that g is continuous. Let $\varepsilon > 0$. Then there exists $\delta > 0$ such that $|g(s_1) - g(s_2)| < \varepsilon$ whenever $|s_1 - s_2| < \delta$.

Let π_j, $(j = 1, 2)$, be two partitions of $[a, b]$ with $|\pi_j| < \delta/2$. Let $a = t_0 < t_1 < \ldots < t_n = b$ be the partitioning points of π_1 and π_2 together. Then

$$S(g, F, \pi_j) = \sum_{i=1}^{n} g(s_{j,i})(F(t_i) - F(t_{i-1}))$$

where $s_{j,i}$, t_i and t_{i-1} belong to the same subinterval of π_j. In particular, $|s_{1,i} - s_{2,i}| < \delta$. For $x^* \in X^*$,

$$\begin{aligned} |\langle S(g, F, \pi_1) - S(g, F, \pi_2), x^* \rangle| &= \left| \sum_{i=1}^{n} (g(s_{1,i}) - g(s_{2,i})) \langle F(t_i) - F(t_{i-1}), x^* \rangle \right| \\ &< \varepsilon \sum_{i=1}^{n} |\langle F(t_i) - F(t_{i-1}), x^* \rangle| \\ &\le 4\varepsilon M \|x^*\|, \end{aligned}$$

where M is defined by (1.19). Now, Cauchy's convergence criterion implies that $\int_a^b g(t)\, dF(t)$ exists.

b) Assume that F is continuous and g is of bounded semivariation. By Proposition 1.9.2, g is of bounded variation. Given $\varepsilon > 0$ there exists $\delta > 0$ such that $\|F(s_1) - F(s_2)\| < \varepsilon$ whenever $|s_1 - s_2| < \delta$. Similarly to a), one shows that

$$\|S(F, g, \pi_1) - S(F, g, \pi_2)\| < \varepsilon V_{[a,b]}(g)$$

whenever $|\pi_1| < \delta/2$ and $|\pi_2| < \delta/2$, where $V_{[a,b]}(g)$ is the total variation of g. Hence $\int_a^b F(t)\, dg(t)$ exists. □

Corollary 1.9.5. *Let $g : [a,b] \to \mathbb{C}$ be piecewise continuous, and $F : [a,b] \to X$ be continuous and of bounded semivariation. Then F and g are Riemann-Stieltjes integrable with respect to each other.*

Proof. Since g is piecewise continuous, $g = g_1 + g_2$ where g_1 is continuous and g_2 is a step function. Now the result follows from Proposition 1.9.4 and Example 1.9.3. □

Corollary 1.9.6. *Let X be an ordered Banach space with normal cone. Let $f : [a,b] \to X$ be increasing. Then f is Riemann integrable.*

Proof. This is immediate from Propositions 1.9.1 and 1.9.4. □

Example 1.9.7. Let $f : [0,1] \to L^\infty[0,1]$ be defined by $f(t) := \chi_{[0,t]}$, so f is increasing, nowhere continuous and not measurable (Example 1.1.5 a)). By Corollary 1.9.6, f is Riemann integrable and

$$\int_0^1 f(t)\, dt = \lim_{n\to\infty} \frac{1}{n} \sum_{i=1}^n \chi_{[0,i/n]}.$$

This shows that $\int_0^1 f(t)\, dt$ is the function $s \mapsto 1-s$. Similarly, if $F(t) := \int_0^t f(r)\, dr$, then

$$F(t)(s) = (t-s)\chi_{[0,t]}(s)$$

(see Example 1.2.8 b)). □

Proposition 1.9.8. *Let $F : [a,b] \to X$ be of bounded semivariation and $g : [a,b] \to \mathbb{C}$ be of bounded variation. Then gF is of bounded semivariation.*

Proof. There exists $M > 0$ such that $|g(t)| \le M$ and $\|F(t)\| \le M$ for all $t \in [a,b]$. The function gF is of bounded semivariation since it is of weak bounded variation. This follows from the assumptions and the estimates:

$$\begin{aligned}
&\sum_i |\langle g(t_i)F(t_i) - g(t_{i-1})F(t_{i-1}), x^*\rangle| \\
&\quad\le \sum_i |\langle (g(t_i) - g(t_{i-1}))F(t_i), x^*\rangle| \\
&\qquad + \sum_i |\langle g(t_{i-1})(F(t_i) - F(t_{i-1})), x^*\rangle| \\
&\quad\le M\|x^*\| \sum_i |g(t_i) - g(t_{i-1})| + M \sum_i |\langle F(t_i) - F(t_{i-1}), x^*\rangle|,
\end{aligned}$$

for all $x^* \in X^*$. □

In the remainder of this section, we give some results which reduce Riemann-Stieltjes integrals to Riemann or Bochner integrals when g or F has a derivative in an appropriate sense.

Proposition 1.9.9. *Let $F : [a,b] \to X$ be of bounded semivariation and $g \in C^1[a,b]$. Then Fg' is Riemann integrable and*

$$\int_a^b F(t)\,dg(t) = \int_a^b F(t)g'(t)\,dt.$$

Proof. a) We show that hF is Riemann integrable for each $h \in C[a,b]$. If h is a step function, then hF is of bounded semivariation and hence Riemann integrable, by Proposition 1.9.4. Since each continuous function h on $[a,b]$ is a uniform limit of step functions and F is bounded, the claim follows since a uniform limit of Riemann integrable functions is Riemann integrable.

b) There exists $M > 0$ such that $|g'(t)| \le M$ and $\|F(t)\| \le M$ for all $t \in [a,b]$. For $\varepsilon > 0$ there exists $\delta > 0$ such that $|g'(s) - g'(s')| < \varepsilon$ whenever $|s - s'| < \delta$. Let π be a partition of $[a,b]$ with $|\pi| < \delta$ and with partitioning points t_i and intermediate points s_i. By the mean value theorem, there exist $s_i' \in (t_{i-1}, t_i)$ such that $g(t_i) - g(t_{i-1}) = g'(s_i')(t_i - t_{i-1})$. Let π' be the partition with partitioning points t_i and intermediate points s_i', so $|\pi'| = |\pi|$. Then

$$\begin{aligned} S(F, g, \pi') &= \sum_i F(s_i')g'(s_i')(t_i - t_{i-1}) \\ &= S(Fg', \pi'). \end{aligned}$$

Letting $|\pi| \to 0$, it follows that $\int_a^b F(t)\,dg(t) = \int_a^b F(t)g'(t)\,dt$. □

Proposition 1.9.10. *Let $F : [a,b] \to X$ be of bounded semivariation and $g, h \in C[a,b]$. Then $G(t) := \int_a^t h(s)\,dF(s)$ is of bounded semivariation on $[a,b]$ and*

$$\int_a^b g(t)\,dG(t) = \int_a^b g(t)h(t)\,dF(t).$$

Proof. Let M be such that $|h(t)| \le M$ for all $t \in [a,b]$. Let π be a partition of $[a,b]$ with partitioning points t_i and intermediate points s_i. By (1.17), for $x^* \in X^*$,

$$\begin{aligned} \sum_i |\langle G(t_i) - G(t_{i-1}), x^* \rangle| &= \sum_i \left| \int_{t_{i-1}}^{t_i} h(s)\,d\langle F(s), x^* \rangle \right| \\ &\le M V_{[a,b]}(x^* \circ F). \end{aligned}$$

It follows from Proposition 1.9.2 that G is of bounded semivariation. By Proposition 1.9.4, $\int_a^b g(t)\,dG(t)$ and $\int_a^b g(t)h(t)\,dF(t)$ both exist.

For $\varepsilon > 0$ there exists $\delta > 0$ such that $|g(s') - g(s)| < \varepsilon$ whenever $|s' - s| < \delta$.

If $|\pi| < \delta$, then

$$\begin{aligned}
&\left|\left\langle S(g,G,\pi) - \int_a^b g(t)h(t)\,dF(t), x^*\right\rangle\right| \\
&= \left|\sum_i \int_{t_{i-1}}^{t_i} (g(s_i) - g(t))h(t)\,d\langle F(t), x^*\rangle\right| \\
&\leq \varepsilon M V_{[a,b]}(x^* \circ F).
\end{aligned}$$

It follows that

$$\left\langle \int_a^b g(t)\,dG(t) - \int_a^b g(t)h(t)\,dF(t), x^* \right\rangle = 0.$$

The result follows from the Hahn-Banach theorem. □

The following result gives analogues of a special case of Proposition 1.9.10 and of Proposition 1.9.9, with Riemann integrals replaced by Bochner integrals.

Proposition 1.9.11. *Let $g : [a,b] \to \mathbb{C}$ and $F : [a,b] \to X$. If F is an antiderivative of a Bochner integrable function f and if g is continuous, then $\int_a^b g(s)\,dF(s)$ exists and equals the Bochner integral $\int_a^b g(s)f(s)\,ds$. If F is continuous and g is absolutely continuous, then $\int_a^b F(s)\,dg(s)$ equals the Bochner integral $\int_a^b F(s)g'(s)\,ds$.*

Proof. Assume that g is continuous and that there exists a Bochner integrable function f such that $F(t) = F(a) + \int_a^t f(s)\,ds$ for all $t \in [a,b]$. Then F is of bounded variation (Proposition 1.2.2) and the Riemann-Stieltjes integral $\int_a^b g(s)\,dF(s)$ exists by Proposition 1.9.4. The Bochner integral $\int_a^b g(s)f(s)\,ds$ exists by Theorem 1.1.4, since g is bounded and measurable. For $\varepsilon > 0$ there exists $\delta > 0$ such that $|g(s') - g(s)| < \varepsilon$ whenever $|s' - s| < \delta$. For any partition π with $|\pi| < \delta$,

$$\begin{aligned}
\left\| S(g,F,\pi) - \int_a^b g(s)f(s)\,ds \right\| &= \left\| \sum_i \int_{t_{i-1}}^{t_i} (g(s_i) - g(s))f(s)\,ds \right\| \\
&\leq \varepsilon \int_a^b \|f(s)\|\,ds.
\end{aligned}$$

Letting $|\pi| \to 0$ and $\varepsilon \to 0$, the result follows.

The proof of the second statement is analogous and is omitted. □

Combining Propositions 1.9.4, 1.9.8 and 1.9.9 with the integration by parts formula (1.20) one obtains the following statement which will be used frequently in later sections. Let $F : [0,t] \to X$ be of bounded semivariation. Then

$$\int_0^t e^{-\lambda s}\,dF(s) = e^{-\lambda t}F(t) - F(0) + \lambda \int_0^t e^{-\lambda s}F(s)\,ds \tag{1.21}$$

for all $\lambda \in \mathbb{C}$. One should notice that the integral $\int_0^t e^{-\lambda s} F(s)\, ds$ is a Riemann integral if F is of bounded semivariation, by Propositions 1.9.4 and 1.9.9. If F is also continuous, then the integral can be taken in the Bochner sense.

1.10 Laplace-Stieltjes Integrals

This section contains the essential properties of the *Laplace-Stieltjes integral*

$$\widehat{dF}(\lambda) := \int_0^\infty e^{-\lambda t}\, dF(t) := \lim_{\tau\to\infty} \int_0^\tau e^{-\lambda t} dF(t),$$

where $F \in \mathrm{BSV}_{loc}(\mathbb{R}_+, X)$; i.e., $F : \mathbb{R}_+ \to X$ is of bounded semivariation on each compact subinterval of $\mathbb{R}_+$. First we observe that the Laplace-Stieltjes integral is a generalization of the Laplace integral.

Proposition 1.10.1. *Let $f \in L^1_{loc}(\mathbb{R}_+, X)$ and $F(t) := \int_0^t f(s)\, ds$. For $\lambda \in \mathbb{C}$, $\widehat{dF}(\lambda)$ exists if and only if $\hat{f}(\lambda)$ exists, and then $\widehat{dF}(\lambda) = \hat{f}(\lambda)$.*

Proof. By Proposition 1.9.11,

$$\int_0^\tau e^{-\lambda t}\, dF(t) = \int_0^\tau e^{-\lambda t} f(t)\, dt \qquad (\tau \ge 0),$$

and the result follows by letting $\tau \to \infty$. □

The results obtained in Sections 1.4, 1.5 and 1.6 for Laplace integrals carry over to Laplace-Stieltjes integrals with only minor modifications of the proofs, and we give these below, starting with elementary properties.

Throughout this section, integrals over $\mathbb{R}_+$ are to be understood as improper Riemann-Stieltjes (or Riemann) integrals. Thus

$$\begin{aligned} \int_0^\infty g(t)\, dF(t) &:= \lim_{\tau\to\infty} \int_0^\tau g(t)\, dF(t), \\ \int_0^\infty F(t)\, dt &:= \lim_{\tau\to\infty} \int_0^\tau F(t)\, dt. \end{aligned}$$

It is easy to see that $\int_0^\infty F(t)\, dt$ exists if $F \in \mathrm{BSV}_{loc}(\mathbb{R}_+, X)$ and $\|F(t)\| \le h(t)$ $(t \ge 0)$ for some $h \in L^1(\mathbb{R}_+)$.

Recall from Section 1.4 that the *exponential growth bound of* a function $F \in \mathrm{BSV}_{loc}(\mathbb{R}_+, X)$ is defined by

$$\omega(F) := \inf\left\{ \omega \in \mathbb{R} : \sup_{t\ge 0} \|e^{-\omega t} F(t)\| < \infty \right\}.$$

It follows from (1.21) that $\mathrm{abs}(dF) \le \omega(F)$ and

$$\widehat{dF}(\lambda) = -F(0) + \lambda \int_0^\infty e^{-\lambda t} F(t)\, dt \quad (\mathrm{Re}\,\lambda > \omega(F)). \tag{1.22}$$

Note that (1.22) is a generalization of both (1.11) and Corollary 1.6.5.

Proposition 1.10.2. *Let* $f \in \mathrm{BSV}_{loc}(\mathbb{R}_+, X)$ *and* $F(t) := \int_0^t f(s)\,ds$. *Then* F *is locally Lipschitz continuous, and*

$$\widehat{df}(\lambda) = -f(0) + \lambda\widehat{dF}(\lambda) = -f(0) + \lambda^2\hat{F}(\lambda)$$

whenever $\mathrm{Re}\,\lambda > \omega(f)$.

Proof. Since f is locally bounded, (1.17) implies that F is locally Lipschitz continuous and $\omega(F) \le \omega(f)$. In particular, $F \in L^1_{loc}(\mathbb{R}_+, X) \cap \mathrm{BSV}_{loc}(\mathbb{R}_+, X)$. By (1.21) and Proposition 1.9.10,

$$\int_0^\tau e^{-\lambda s}\,df(s) = e^{-\lambda\tau}f(\tau) - f(0) + \lambda\int_0^\tau e^{-\lambda s}\,dF(s).$$

Letting $\tau \to \infty$ gives

$$\widehat{df}(\lambda) = -f(0) + \lambda\widehat{dF}(\lambda)$$

whenever $\mathrm{Re}\,\lambda > \omega(f)$. By (1.22), $\widehat{dF}(\lambda) = \lambda\hat{F}(\lambda)$. □

Now we give a generalization of Proposition 1.6.1 a).

Proposition 1.10.3. *Let* $F \in \mathrm{BSV}_{loc}(\mathbb{R}_+, X)$, $\mu \in \mathbb{C}$ *and let* $G(t) := \int_0^t e^{-\mu s}\,dF(s)$ $(t \ge 0)$. *For* $\lambda \in \mathbb{C}$, $\widehat{dG}(\lambda)$ *exists if and only if* $\widehat{dF}(\lambda + \mu)$ *exists, and then* $\widehat{dG}(\lambda) = \widehat{dF}(\lambda + \mu)$.

Proof. By Proposition 1.9.10,

$$\int_0^\tau e^{-\lambda t}\,dG(t) = \int_0^\tau e^{-(\lambda+\mu)t}\,dF(t),$$

and the result follows on letting $\tau \to \infty$. □

For $F \in \mathrm{BSV}_{loc}(\mathbb{R}_+, X)$, let

$$\mathrm{abs}(dF) := \inf\left\{\mathrm{Re}\,\lambda : \widehat{dF}(\lambda) \text{ exists}\right\}.$$

Proposition 1.10.4. *Let* $F \in \mathrm{BSV}_{loc}(\mathbb{R}_+, X)$. *Then* $\widehat{dF}(\lambda)$ *converges if* $\mathrm{Re}\,\lambda > \mathrm{abs}(dF)$ *and diverges if* $\mathrm{Re}\,\lambda < \mathrm{abs}(dF)$.

Proof. Clearly, $\widehat{dF}(\lambda)$ does not exist if $\mathrm{Re}\,\lambda < \mathrm{abs}(dF)$. For $\lambda_0 \in \mathbb{C}$ define $G_0(t) := \int_0^t e^{-\lambda_0 s}\,dF(s)$ $(\lambda \in \mathbb{C},\ t \ge 0)$. Then, by Proposition 1.9.10,

$$\int_0^t e^{-\lambda s}\,dF(s) = \int_0^t e^{-(\lambda-\lambda_0)s}\,dG_0(s) \quad (\lambda \in \mathbb{C}, t \ge 0).$$

Integration by parts (1.20), and Proposition 1.9.9, yield

$$\int_0^t e^{-\lambda s}\,dF(s) = e^{-(\lambda-\lambda_0)t}G_0(t) + (\lambda - \lambda_0)\int_0^t e^{-(\lambda-\lambda_0)s}G_0(s)\,ds. \tag{1.23}$$

If $\widehat{dF}(\lambda_0)$ exists, then G_0 is bounded. Therefore, $\widehat{dF}(\lambda)$ exists if $\operatorname{Re}\lambda > \operatorname{Re}\lambda_0$ and

$$\widehat{dF}(\lambda) = (\lambda - \lambda_0)\int_0^\infty e^{-(\lambda-\lambda_0)s} G_0(s)\,ds \quad (\operatorname{Re}\lambda > \operatorname{Re}\lambda_0). \tag{1.24}$$

This shows that $\widehat{dF}(\lambda)$ exists if $\operatorname{Re}\lambda > \operatorname{abs}(dF)$ and, as for the Laplace integral (see (1.9)),

$$\operatorname{abs}(dF) = \inf\left\{\lambda \in \mathbb{R} : \sup_{t>0}\left|\int_0^t e^{-\lambda s}\, d\langle F(s), x^*\rangle\right| < \infty \text{ for all } x^* \in X^*\right\}. \tag{1.25}$$

□

Theorem 1.10.5. *Let* $F \in \mathrm{BSV}_{loc}(\mathbb{R}_+, X)$ *and let* $F_\infty := \lim_{t\to\infty} F(t)$ *if the limit exists,* $F_\infty := 0$ *otherwise. Then* $\operatorname{abs}(dF) = \omega(F - F_\infty)$.

Proof. For $\lambda_0 > \operatorname{abs}(dF)$ define $G_0(t) := \int_0^t e^{-\lambda_0 s} dF(s)$ $(t \geq 0)$. Then G_0 is bounded. To prove that $\operatorname{abs}(dF) \geq \omega(F - F_\infty)$ one considers the two cases $\operatorname{abs}(dF) \geq 0$ and $\operatorname{abs}(dF) < 0$. First, let $\operatorname{abs}(dF) \geq 0$ and $\lambda_0 > \operatorname{abs}(dF)$. It follows from Proposition 1.9.10, integration by parts (1.20), and Proposition 1.9.9 that

$$F(t) = F(0) + \int_0^t e^{\lambda_0 s}\, dG_0(s) = F(0) + e^{\lambda_0 t} G_0(t) - \lambda_0 \int_0^t e^{\lambda_0 s} G_0(s)\,ds$$

for all $t \geq 0$, so $\sup_{t\geq 0} \|e^{-\lambda_0 t}(F(t) - F_\infty)\| < \infty$. Thus $\omega(F - F_\infty) \leq \operatorname{abs}(dF)$ if $\operatorname{abs}(dF) \geq 0$. Second, let $\operatorname{abs}(dF) < 0$. Choose $\operatorname{abs}(dF) < \lambda_0 < 0$. For $r \geq t \geq 0$ one has

$$F(r) - F(t) = \int_t^r e^{\lambda_0 s}\, dG_0(s) = e^{\lambda_0 r} G_0(r) - e^{\lambda_0 t} G_0(t) - \lambda_0 \int_t^r e^{\lambda_0 s} G_0(s)\,ds.$$

Thus,

$$\lim_{r\to\infty} F(r) = F_\infty = F(t) - e^{\lambda_0 t} G_0(t) - \lambda_0 \int_t^\infty e^{\lambda_0 s} G_0(s)\,ds$$

exists and $\sup_{t\geq 0} \|e^{-\lambda_0 t}(F(t) - F_\infty)\| < \infty$. Therefore, $\omega(F - F_\infty) \leq \operatorname{abs}(dF)$ if $\operatorname{abs}(dF) < 0$.

To show the reverse inequality let $\omega > \omega(F - F_\infty)$. Then there exists $M \geq 0$ such that $\|F(t) - F_\infty\| \leq M e^{\omega t}$ for all $t \geq 0$. Let $\lambda > \omega > \omega(F - F_\infty)$. Integration by parts (see (1.21)) yields

$$\int_0^t e^{-\lambda s}\, dF(s) = e^{-\lambda t}(F(t) - F_\infty) + F_\infty - F(0) + \lambda \int_0^t e^{-\lambda s}(F(s) - F_\infty)\,ds.$$

Hence, $\widehat{dF}(\lambda)$ exists for $\lambda > \omega(F - F_\infty)$ and is given by

$$\widehat{dF}(\lambda) = F_\infty - F(0) + \lambda \int_0^\infty e^{-\lambda s}(F(s) - F_\infty)\,ds. \tag{1.26}$$

This shows that $\operatorname{abs}(dF) \leq \omega(F - F_\infty)$. □

Note that (1.26) is a generalization of (1.22).

Theorem 1.10.6. *Let* $F \in \mathrm{BSV}_{loc}(\mathbb{R}_+, X)$ *and assume that* $\mathrm{abs}(dF) < \infty$. *Then* $\lambda \mapsto \widehat{dF}(\lambda)$ *is holomorphic for* $\mathrm{Re}\,\lambda > \mathrm{abs}(dF)$, *and*

$$\widehat{dF}^{(n)}(\lambda) = \int_0^\infty e^{-\lambda t}(-t)^n \, dF(t) \quad (\mathrm{Re}\,\lambda > \mathrm{abs}(dF),\ n \in \mathbb{N}_0)$$

(as an improper Riemann-Stieltjes integral).

Proof. Let $q_k(\lambda) := \int_0^k e^{-\lambda t}\, dF(t)$. It follows from (1.18) that

$$q_k(\lambda) = \lim_{N\to\infty} \sum_{n=0}^{N} \frac{\lambda^n}{n!} \int_0^k (-t)^n \, dF(t).$$

By the Weierstrass convergence theorem (a special case of Vitali's Theorem A.5), q_k is entire and

$$q_k^{(j)}(\lambda) = \int_0^k e^{-\lambda t}(-t)^j \, dF(t)$$

for all $j \in \mathbb{N}_0$. Let $\mathrm{Re}\,\lambda > \lambda_0 > \mathrm{abs}(dF)$, and define

$$G_0(t) := \int_0^t e^{-\lambda_0 s} \, dF(s).$$

Then G_0 is bounded and it follows from Proposition 1.9.10, (1.20) and Proposition 1.9.9 that

$$\begin{aligned} \widehat{dF}(\lambda) - q_k(\lambda) &= \int_k^\infty e^{-(\lambda-\lambda_0)s} \, dG_0(s) \\ &= -e^{-(\lambda-\lambda_0)k} G_0(k) + (\lambda - \lambda_0) \int_k^\infty e^{-(\lambda-\lambda_0)s} G_0(s) \, ds. \end{aligned}$$

Hence q_k converges to $\widehat{dF}$ uniformly on compact subsets of $\{\lambda : \mathrm{Re}\,\lambda > \mathrm{abs}(dF)\}$. Again by the Weiestrass convergence theorem, $\widehat{dF}$ is holomorphic and $q_k^{(j)}(\lambda) \to \widehat{dF}^{(j)}(\lambda)$ as $k \to \infty$, if $\mathrm{Re}\,\lambda > \mathrm{abs}(dF)$. □

Finally in this section, we consider operator-valued Laplace-Stieltjes integrals. Let $S : \mathbb{R}_+ \to \mathcal{L}(X, Y)$ be a function. By the uniform boundedness principle, $S \in \mathrm{BSV}_{loc}(\mathbb{R}_+, \mathcal{L}(X, Y))$ if and only if $v_x := S(\cdot)x \in \mathrm{BSV}(\mathbb{R}_+, Y)$ for all $x \in X$. When $S \in \mathrm{BSV}_{loc}(\mathbb{R}_+, \mathcal{L}(X, Y))$, we let

$$\begin{aligned} \omega(S) &:= \inf\left\{\omega \in \mathbb{R} : \sup_{t\ge 0} \|e^{-\omega t} S(t)\| < \infty\right\}, \\ \mathrm{abs}(dS) &:= \inf\left\{\mathrm{Re}\,\lambda : \int_0^t e^{-\lambda s} \, dS(s) \text{ converges strongly as } t \to \infty\right\} \\ &= \sup\{\mathrm{abs}(dv_x) : x \in X\}. \end{aligned}$$

The following analogue of Proposition 1.4.5 follows from Theorems 1.10.5 and 1.10.6.

Proposition 1.10.7. *Let $S \in \mathrm{BSV}_{loc}(\mathbb{R}_+, \mathcal{L}(X,Y))$ and let S_∞ be the strong limit of $S(t)$ as $t \to \infty$ if it exists, and $S_\infty := 0$ otherwise. Then*

a) $\lim_{t\to\infty} \int_0^t e^{-\lambda s}\, dS(s)$ *exists in operator norm whenever* $\operatorname{Re}\lambda > \operatorname{abs}(dS)$,

b) $\operatorname{abs}(dS) = \omega(S - S_\infty)$.

Proof. If $\int_0^t e^{-\lambda_0 s}\, dS(s)$ converges strongly as $t \to \infty$, then it is uniformly bounded in operator norm. Thus, a) follows from (1.23). Hence,

$$\begin{aligned} \operatorname{abs}(dS) &= \inf\left\{ \operatorname{Re}\lambda : \int_0^t e^{-\lambda s}\, dS(s) \text{ converges in norm as } t \to \infty \right\} \\ &= \omega(S - \widetilde{S}_\infty), \end{aligned}$$

by Theorem 1.10.6, where $\widetilde{S}_\infty$ is the norm limit of $S(t)$ as $t \to \infty$, if this exists, $\widetilde{S}_\infty := 0$ otherwise. It is trivial that $\omega(S - \widetilde{S}_\infty) = \omega(S - S_\infty)$, so b) is proved. $\square$

1.11 Notes

Section 1.1

The Bochner integral is an extension of the Lebesgue integral to functions with values in Banach spaces. Introduced around 1930 by Bochner, it has become a widely used integral in infinite dimensional applications. Much of Section 1.1 follows the treatment of the Bochner integral in Chapter III of [HP57], where many references to the original literature can be found. Comprehensive treatments of the Bochner integral and vector-valued measures, as well as references to the literature are contained in the monograph [DU77] by Diestel and Uhl. Corollary 1.1.3 is taken from [Are01].

One reason for introducing the Riemann integral here is that increasing functions with values in an ordered Banach space with normal cone are always Riemann integrable (Corollary 1.9.6). However, if the space is not separable, Riemann integrable functions are not necessarily Bochner integrable and their antiderivative may be nowhere differentiable. One way to circumvent these difficulties is to consider generalizations of the Riemann integral. This allows a version of the fundamental theorem of calculus where all continuous functions $f : [0,1] \to X$ (where X is a Banach space) with $f(0) = 0$ are differentiable in the mean and coincide with the generalized Riemann integral of their derivatives (see [BLN99]).

Section 1.2

The Radon-Nikodym property was identified in the 1970s as an important property in the theory of vector measures and also in the geometry of Banach spaces. Our treatment is based on [DU77].

Section 1.3

Most of the results are vector-valued versions of standard material. Proposition 1.3.7 is contained in [KH89].

Section 1.4

The Laplace transform has a long history, dating back to Euler's paper *'De constructione aequationum'* from 1737, Lagrange's *'Mémoire sur l'utilité de la méthode de prendre le milieux entre les résultats de plusieurs observations'* from 1773, and Laplace's *'Mémoire sur les approximations des formules qui sont finctions de très grands nombres'* from 1785. Since then it has been widely used in mathematics and engineering (in particular in ordinary differential, difference and functional equations, electrical engineering and applications to signal processing problems). Modern Laplace transform theory began to emerge at the end of the 19th century when Heaviside popularized a user-friendly and powerful *operational calculus* within the engineering community in connection with his research in electromagnetism [Hea93]. Since his methods were to a large degree based on purely formal operations with few mathematical justifications, many mathematicians at the beginning of the 20th century began to strive for a solid mathematical foundation of Heaviside's operational calculus by virtue of the Laplace transform. These efforts culminated in Widder's books *'The Laplace Transform'* [Wid41] and *'An Introduction to Transform Theory'* [Wid71] as well as Doetsch's *'Theorie und Anwendung der Laplace-Transformation'* [Doe37] and his monumental, three volumed *'Handbuch der Laplace-Transformation'* [Doe50]. These monographs have been among the best introductions to the subject and have become classic texts. A first comprehensive look at Laplace transform theory for functions with values in a Banach space X is contained in Hille's monograph *'Functional Analysis and Semi-Groups'* from 1948 [Hil48]. Many historical notes on Laplace transform theory can be found in the books of Doetsch and in survey articles by Deakin [Dea81], [Dea82], and Martis in Biddau [Bid33].

One weakness of Laplace transform theory—compared to Heaviside's operational calculus—are the restrictions concerning the growth of the functions at infinity. To remove these restrictions, Vignaux introduced in 1939 an asymptotic version of the Laplace transform [Vig39], [VC44]. For an extension of the asymptotic Laplace transform to vector-valued functions, and references to the literature, see [LN99] and [LN01].

The characterization of the abscissa of convergence by the exponential growth of the antiderivative are vector-valued versions of classical results due to Landau (1906) and Pincherle (1913) that can be found in [Doe50, Volume I,Theorems 2.2.7 and 2.2.8], or [HP57, Section 1.6.2].

Section 1.5
Theorem 1.5.1 is due to Pincherle and Landau (1905); Theorem 1.5.3 is due to Landau (1906). The proofs given here, as well as Example 1.5.2, follow [Doe50, Volume I, Sections 3.2–3.4] where further references to the classical literature can be found.

Section 1.6
The results of this section are straightforward vector-valued versions of standard results in the classical theory of Laplace transforms (see [Doe50, Volume I, Sections 2.14, 2.15], for example).

Section 1.7
Theorem 1.7.5 has been proved by Ti-Jun Xiao and Jin Liang [XL00], but a special case was given by Kurtz [Kur69] and the general result was mentioned by Chernoff [Che74, p.106]. In fact, the proof of [Che68, Proposition] carries over to the case considered in Theorem 1.7.5. The short proof given here appeared in [Bob97b] and [Are01].

The Inversion Theorem 1.7.7 is due to Post (1930) and Widder (1934) (see [Wid41, Section 7.6], and [Doe50, Volume I, Section 8.2]). Numerically more efficient Post-Widder type inversion formulas for the Laplace transform can be found in [Jar08, Theorem 4.1] and [JNO08]. They are derived from rational approximation methods for operator semigroups developed by Hersh and Kato [HK79] and Brenner and Thomée [BT79] (see also [Kov07]).

The Uniqueness Theorem 1.7.3 was mentioned first by Pastor [Pas19] in 1919 and is a special case of the following result of Shen [She47] (see also [Doe50, Volume I, Section 2.9], [BN94], [Mih09]).

Theorem 1.11.1. *Let $f \in L^1_{loc}(\mathbb{R}_+, X)$ with $\operatorname{abs}(f) < \infty$. Let (λ_n) be an infinite sequence with no accumulation point and $\operatorname{Re}\lambda_n \geq a > 0$ for all $n \in \mathbb{N}$ and some $a > \operatorname{abs}(f)$. If*

$$\sum_{n=1}^{\infty} 1 - \left|\frac{\lambda_n - 1}{\lambda_n + 1}\right| = \infty,$$

then $(\lambda_n)_{n\in\mathbb{N}}$ is a uniqueness sequence; i.e., $\hat{f}(\lambda_n) = 0$ $(n \in \mathbb{N})$ implies that $f = 0$.

Conversely, let (λ_n) be a sequence with $\lambda_n \neq 0$ and $|\arg(\lambda_n)| \leq \theta < \pi/2$ $(n \in \mathbb{N})$. If (λ_n) has no accumulation point and the sum above is finite, then there exists $0 \neq f \in L^1_{loc}(\mathbb{R}_+, X)$ with $\hat{f}(\lambda_n) = 0$ for all $n \in \mathbb{N}$.

Consider the horizontal sequences $\lambda_n := a + n^\gamma b$ for $a, b > 0$. If $0 < \gamma \leq 1$ or $\gamma < 0$, then (λ_n) is a uniqueness sequence. If $\gamma > 1$ or $a = 0$ and $\gamma < -1$, then $\{\lambda_n\}$ is the set of zeros of a non-trivial Laplace transform. For example, if $f(t) = \frac{1}{\sqrt{t}} \sin(\frac{1}{t})$, then $\hat{f}(\lambda) = \sqrt{\frac{\pi}{\lambda}} e^{-\sqrt{2\lambda}} \sin(\sqrt{2\lambda})$ which has zeros for $\lambda_n = 2n^2\pi^2$ $(n \in \mathbb{N})$. The vertical sequences $\lambda_n = 1 + in^\gamma$ are uniqueness sequences if $0 < \gamma \leq \frac{1}{2}$. If $\gamma > \frac{1}{2}$, then $\{\lambda_n\}$ is the set of zeros of a non-trivial Laplace transform.

Uniqueness sequences are important in the discussion of Cauchy problems which are well-posed in the regularized sense (see [Bäu01] or [LN99] for definitions and references).

Section 1.8

For $1 \leq p \leq 2$, a Banach space X is said to have *Fourier type* p if the Fourier transform on $L^1(\mathbb{R}, X) \cap L^p(\mathbb{R}, X)$ extends to a bounded linear operator of $L^p(\mathbb{R}, X)$ into $L^{p'}(\mathbb{R}, X)$, where $1/p + 1/p' = 1$. The Hausdorff-Young inequalities show that $\mathbb{C}$ has Fourier type for every $p \in [1, 2]$. Every Banach space has Fourier type 1, and a space with Fourier type p also has Fourier type q whenever $1 \leq q \leq p$. Theorem 1.8.2 shows that Hilbert spaces have Fourier type 2, and conversely Kwapień [Kwa72] showed that a space with Fourier type 2 is isomorphic to a Hilbert space. A space of the form $L^p(\Omega, \mu)$, where $1 \leq p < \infty$ and (Ω, μ) is any measure space, has Fourier type $\min(p, p')$. The spaces with non-trivial Fourier type (i.e., Fourier type p for some $p > 1$) have been characterized by Bourgain [Bou82], [Bou88] (see also [Pis86]). Every superreflexive space (a Banach space with an equivalent uniformly convex norm) has non-trivial Fourier type, but there exist reflexive spaces which do not have non-trivial Fourier type and there exist non-reflexive spaces which do have non-trivial Fourier type.

A Banach space X is said to have the *analytic Radon-Nikodym property* (ARNP) if each function $g \in H^2(\mathbb{R}, X)$ has a boundary function, i.e. $\lim_{\alpha\downarrow 0} g(\alpha + is)$ exists s-a.e. This property was first considered by Bukhvalov [Buk81], [BD82] using functions on the unit disc rather than $\mathbb{C}_+$, and H^p-spaces for $p \neq 2$, but this formulation is equivalent. Thus, Theorem 1.8.3 shows in particular that Hilbert spaces have the (ARNP). Every reflexive space has the (ARNP), and more generally, any space with the Radon-Nikodym property, and also any space of the form $L^1(\Omega, \mu)$, has the (ARNP). On the other hand,

c_0 does not have the (ARNP), and there exist spaces with non-trivial Fourier type which do not have the (ARNP) (see [HN99]).

Section 1.9
This section contains some of the basic properties of the Riemann-Stieltjes integral; see [HP57] and [Wid41] for further details and references to the original literature.

Section 1.10
In the classical Laplace transform literature, many authors preferred Laplace-Stieltjes integrals $\int_0^\infty e^{-\lambda t}\,dF(t)$ since they include Laplace integrals $\int_0^\infty e^{-\lambda t} f(t)\,dt$ (when F is differentiable a.e.) and Dirichlet series $\sum_{i=1}^\infty a_i e^{-\lambda t_i}$ (when F is a step function). The importance of the Laplace-Stieltjes integral for our purposes is that many classical results for Laplace-Stieltjes integrals of complex-valued functions F can be extended to functions with values in arbitrary Banach spaces X, whereas vector-valued extensions of the corresponding Laplace transform results often require additional assumptions on X (see, for example, [Zai60]). All results of this section are vector-valued versions of classical results for Laplace-Stieltjes transforms in [Wid41] or [Wid71].

Chapter 2

The Laplace Transform

In this chapter the emphasis of the discussion shifts from Laplace integrals $\hat{f}(\lambda)$ and $\widehat{dF}(\lambda)$ to the Laplace transform $\mathcal{L} : f \mapsto \hat{f}$ and to the Laplace-Stieltjes transform $\mathcal{L}_S : F \mapsto \widehat{dF}$. The Laplace transform is considered first as an operator acting on $L^\infty(\mathbb{R}_+, X)$ and the Laplace-Stieltjes transform as an operator on

$$\mathrm{Lip}_0(\mathbb{R}_+, X) := \Big\{ F : \mathbb{R}_+ \to X : F(0) = 0,\ \|F\|_{\mathrm{Lip}_0(\mathbb{R}_+,X)} := \sup_{t,s\geq 0} \frac{\|F(t) - F(s)\|}{|t-s|} < \infty \Big\}.$$

These domains of $\mathcal{L}$ and $\mathcal{L}_S$ are relatively easy to deal with and have immediate and important applications to abstract differential and integral equations.

The following observation is the key to one of the basic structures of Laplace transform theory. If $f \in L^\infty(\mathbb{R}_+, X)$, then $t \mapsto F(t) := \int_0^t f(s)\, ds$ belongs to $\mathrm{Lip}_0(\mathbb{R}_+, X)$ and

$$\mathcal{L}(f)(\lambda) = \int_0^\infty e^{-\lambda t} f(t)\, dt = \int_0^\infty e^{-\lambda t}\, dF(t) = T_F(e_{-\lambda}),$$

where $T_F : g \mapsto \int_0^\infty g(s) dF(s)$ is a bounded linear operator from $L^1(\mathbb{R}_+)$ into X, and where $e_{-\lambda}$ denotes the exponential function $t \mapsto e^{-\lambda t}$. The operator T_F is fundamental to Laplace transform theory. In Section 2.1 it is shown that $\Phi_S : F \mapsto T_F$ is an isometric isomorphism between $\mathrm{Lip}_0(\mathbb{R}_+, X)$ and $\mathcal{L}(L^1(\mathbb{R}_+), X)$ (Riesz-Stieltjes representation theorem). This representation is crucial for the following reason. The main purpose of Laplace transform theory is to translate properties of the generating function F into properties of the resulting function $\lambda \mapsto r(\lambda) = \int_0^\infty e^{-\lambda t}\, dF(t)$ and vice versa. Since $F(t) = T_F \chi_{[0,t]} = \int_0^\infty \chi_{[0,t]}(s)\, dF(s)$ and $r(\lambda) = T_F e_{-\lambda} = \int_0^\infty e^{-\lambda s}\, dF(s)$, the generating function F as well as the resulting function r are evaluations of the *same* bounded linear operator acting on different total subsets of $L^1(\mathbb{R}_+)$.

In Section 2.2, the range of the Laplace-Stieltjes transform acting on $\mathrm{Lip}_0(\mathbb{R}_+, X)$ is characterized. It is shown that a function $r : \mathbb{R}_+ \to X$ has a Laplace-Stieltjes representation $r = \mathcal{L}_S(F)$ for some $F \in \mathrm{Lip}_0(\mathbb{R}_+, X)$ if and only if r is a C^∞-function whose Taylor coefficients satisfy the estimate

$$\|r\|_W := \sup_{n\in\mathbb{N}_0} \sup_{\lambda>0} \frac{\lambda^{n+1}}{n!} \|r^{(n)}(\lambda)\| < \infty. \tag{2.1}$$

This can be rephrased by saying that the Laplace-Stieltjes transform is an isometric isomorphism between the Banach spaces $\mathrm{Lip}_0(\mathbb{R}_+, X)$ and

$$C_W^\infty((0,\infty), X) := \{r \in C^\infty((0,\infty), X) : \|r\|_W < \infty\}.$$

If the Banach space X has the Radon-Nikodym property (see Section 1.2), then (and only then) "Widder's growth conditions" (2.1) are necessary and sufficient for r to have a Laplace representation $r = \mathcal{L}(f)$ for some $f \in L^\infty(\mathbb{R}_+, X)$; i.e., Banach spaces with the Radon-Nikodym property are precisely those Banach spaces in which the Laplace transform is an isometric isomorphism between $L^\infty(\mathbb{R}_+, X)$ and $C_W^\infty((0,\infty), X)$. For $X = \mathbb{C}$, this is a classical result usually known as "Widder's Theorem".

If $r = \mathcal{L}_S(F)$ for some $F \in \mathrm{Lip}_0(\mathbb{R}_+, X)$, then the inverse Laplace-Stieltjes transform has many different representations. A few of them, such as

$$\begin{aligned} F(t) &= \frac{1}{2\pi i}\int_\Gamma e^{\lambda t}\frac{r(\lambda)}{\lambda}\,d\lambda = \lim_{n\to\infty}\sum_{j=1}^\infty (-1)^{j+1} e^{tnj} r(nj) \\ &= \lim_{k\to\infty} (-1)^k \frac{1}{k!}\left(\frac{k}{t}\right)^{k+1} \frac{d^k}{dt^k}\left(\frac{r(\lambda)}{\lambda}\right)\Big|_{\lambda=k/t}, \end{aligned}$$

will be proved in Section 2.3.

In Section 2.4, the results of the previous sections are extended to functions with exponential growth at infinity; i.e., we investigate the Laplace transform acting on functions f with $\operatorname{ess\,sup}_{t\geq 0} \|e^{-\omega t} f(t)\| < \infty$.

In applications it is usually impossible to verify whether or not a given function r satisfies Widder's growth conditions (2.1). Thus, in Sections 2.5 and 2.6 some complex growth conditions are discussed which are necessary (and in a certain sense sufficient) for a holomorphic function $r : \{\mathrm{Re}\,\lambda > \omega\} \to X$ to have a Laplace representation. In Section 2.5, the growth condition considered is $\sup_{\mathrm{Re}\,\lambda>\omega} \|\lambda^{1+b} r(\lambda)\| < \infty$ for some $b > 0$.

In Section 2.6, we discuss functions r which are holomorphic in a sector $\Sigma := \{|\arg(\lambda)| < \frac{\pi}{2} + \varepsilon\}$ and satisfy $\sup_{\lambda\in\Sigma} \|\lambda r(\lambda)\| < \infty$. We will see that any such r is the Laplace transform of a function which is holomorphic in the sector $\{|\arg(\lambda)| < \varepsilon\}$. The final class of functions which we will consider are the completely monotonic ones; i.e., C^∞-functions r with values in an ordered Banach space such that $(-1)^n r^{(n)}(\lambda) \geq 0$ for all $n \in \mathbb{N}_0$ and $\lambda > \omega$. In the scalar case,

Bernstein's theorem states that a function r is completely monotonic if and only if it is the Laplace-Stieltjes transform of an increasing function. In Section 2.7 we investigate for which ordered Banach spaces Bernstein's theorem holds.

2.1 Riesz-Stieltjes Representation

In the following sections the emphasis will be on the properties of the Laplace transform $\mathcal{L} : f \mapsto \hat{f}$ and the Laplace-Stieltjes transform $\mathcal{L}_S : F \mapsto \widehat{dF}$. As is the case with all linear operators, the choice of the domain is crucial. For the Laplace-Stieltjes transform $\mathcal{L}_S$ the most convenient choice of the domain space is

$$\mathrm{Lip}_0(\mathbb{R}_+, X) := \Big\{ F : \mathbb{R}_+ \to X : \ F(0) = 0, \ \|F\|_{\mathrm{Lip}_0(\mathbb{R}_+,X)} := \sup_{t,s \geq 0} \frac{\|F(t) - F(s)\|}{|t-s|} < \infty \Big\}.$$

If $F(t) = \int_0^t f(s)\, ds$ for $f \in L^\infty(\mathbb{R}_+, X)$, then $F \in \mathrm{Lip}_0(\mathbb{R}_+, X)$ and

$$\int_0^\infty e^{-\lambda t}\, dF(t) = \int_0^\infty e^{-\lambda t} f(t)\, dt \quad (\lambda > 0),$$

by Proposition 1.10.1. Thus, any result for $\mathcal{L}_S$ acting on $\mathrm{Lip}_0(\mathbb{R}_+, X)$ translates into one for $\mathcal{L}$ acting on $L^\infty(\mathbb{R}_+, X)$. However, since there are Banach spaces in which not every Lipschitz continuous function is the antiderivative of an L^∞-function (see Section 1.2), the Laplace-Stieltjes transform is a true generalization of the Laplace transform. It is the generalization needed to deal effectively with Laplace transforms of vector-valued functions.

In this section we investigate the Riesz-Stieltjes operator Φ_S which assigns to $F \in \mathrm{Lip}_0(\mathbb{R}_+, X)$ a bounded linear operator $T_F : L^1(\mathbb{R}_+) \to X$ such that

$$T_F f := \int_0^\infty f(s)\, dF(s) := \lim_{\tau \to \infty} \int_0^\tau f(s)\, dF(s),$$

when $f \in L^1(\mathbb{R}_+)$ is continuous. It will be shown that Φ_S is an isometric isomorphism between $\mathrm{Lip}_0(\mathbb{R}_+, X)$ and $\mathcal{L}(L^1(\mathbb{R}_+), X)$, the space of all bounded linear operators from the Banach space $L^1(\mathbb{R}_+)$ into X (Riesz-Stieltjes representation). This observation is fundamental for the whole chapter. To see why the Riesz-Stieltjes representation is such an important tool, observe that

$$F(t) = T_F \chi_{[0,t]} \ (t \geq 0), \quad \text{and} \quad \widehat{dF}(\lambda) = T_F e_{-\lambda} \ (\lambda > 0).$$

Thus, if one knows F, then the operator T_F is specified on the set of characteristic functions $\chi_{[0,t]}$ $(t > 0)$, which is total in $L^1(\mathbb{R}_+)$. Therefore, T_F and, in particular, the Laplace integrals $T_F e_{-\lambda} = \widehat{dF}(\lambda)$ $(\lambda > 0)$ are completely determined.

Conversely, the Laplace integrals $\widehat{dF}(\lambda)$ determine T_F on the set of exponential functions $e_{-\lambda}$ ($\lambda > 0$), which is also total in $L^1(\mathbb{R}_+)$ (Lemma 1.7.1). Hence, the Laplace integrals $\widehat{dF}(\lambda)$ determine the properties of T_F and, in particular, the properties of $F(t) = T_F\chi_{[0,t]}$ $(t \geq 0)$.

Theorem 2.1.1 (Riesz-Stieltjes Representation). *There exists a unique isometric isomorphism* $\Phi_S : F \mapsto T_F$ *from* $\mathrm{Lip}_0(\mathbb{R}_+, X)$ *onto* $\mathcal{L}(L^1(\mathbb{R}_+), X)$ *such that*

$$T_F\chi_{[0,t]} = F(t) \tag{2.2}$$

for all $t \geq 0$ *and* $F \in \mathrm{Lip}_0(\mathbb{R}_+, X)$. *Moreover,*

$$T_F g = \lim_{t\to\infty} \int_0^t g(s)\, dF(s) := \int_0^\infty g(s)\, dF(s) \tag{2.3}$$

for all continuous functions $g \in L^1(\mathbb{R}_+)$.

Note that it is part of the claim that the improper integral in (2.3) converges. We shall call the isomorphism Φ_S the *Riesz-Stieltjes operator*.

Proof. Let $D := \mathrm{span}\{\chi_{[0,t)} : t > 0\}$, the space of step functions, which is dense in $L^1(\mathbb{R}_+)$. For each $f \in D$ there exists a unique representation

$$f = \sum_{i=1}^n \alpha_i \chi_{[t_{i-1},t_i)},$$

where $0 = t_0 < t_1 < \ldots < t_n$, $\alpha_i \in \mathbb{C}$ $(i = 1, \ldots, n)$. Let $F \in \mathrm{Lip}_0(\mathbb{R}_+, X)$. Define $T_F : D \to X$ by

$$T_F(f) = T_F\left(\sum_{i=1}^n \alpha_i \chi_{[t_{i-1},t_i)}\right) := \sum_{i=1}^n \alpha_i (F(t_i) - F(t_{i-1})).$$

Then,

$$\|T_F(f)\| \leq \|F\|_{\mathrm{Lip}_0(\mathbb{R}_+,X)} \sum_{i=1}^n |\alpha_i|(t_i - t_{i-1}) = \|F\|_{\mathrm{Lip}_0(\mathbb{R}_+,X)} \|f\|_1.$$

Hence, T_F has a unique extension $T_F \in \mathcal{L}(L^1(\mathbb{R}_+), X)$. Moreover,

$$\|T_F\| \leq \|F\|_{\mathrm{Lip}_0(\mathbb{R}_+,X)}.$$

Conversely, if $T \in \mathcal{L}(L^1(\mathbb{R}_+), X)$, let $F(t) := T\chi_{[0,t)}$ for $t \geq 0$. Then for $t > s \geq 0$,

$$\|F(t) - F(s)\| = \|T\chi_{[s,t)}\| \leq \|T\|\ \|\chi_{[s,t)}\|_1 = \|T\|(t - s).$$

Thus, $F \in \mathrm{Lip}_0(\mathbb{R}_+, X)$ and $\|F\|_{\mathrm{Lip}_0(\mathbb{R}_+,X)} \leq \|T\|$. It follows from the definitions that $T = T_F$ and if $T = T_G$ then $F = G$. This shows that $F \mapsto T_F$ is an isometric isomorphism.

Finally, let $g \in L^1(\mathbb{R}_+)$ be a continuous function and let $F \in \mathrm{Lip}_0(\mathbb{R}_+, X)$. Take $t > 0$, and let π be a partition of $[0,t]$ with partitioning points $0 = t_0 < t_1 < \ldots < t_n = t$ and intermediate points $s_i \in [t_{i-1}, t_i]$. Let

$$f_\pi := \sum_{i=1}^{n} g(s_i)\chi_{[t_{i-1},t_i)}.$$

Thus, $S(g, F, \pi) = T_F(f_\pi)$. As $|\pi| \to 0$, $\|f_\pi - g\chi_{[0,t)}\|_1 \to 0$, so

$$\int_0^t g(s)\,dF(s) = T_F(g\chi_{[0,t)}).$$

As $t \to \infty$, $\|g\chi_{[0,t)} - g\|_1 \to 0$, so

$$\int_0^\infty g(s)\,dF(s) = T_F(g). \qquad \square$$

We conclude this section by discussing convergence of functions and their Laplace-Stieltjes transforms. In fact, the Laplace-Stieltjes transform allows us to give a purely operator-theoretic proof of the following approximation theorem. Note, however, that the essential implication (i) $\Rightarrow$ (iv) can also be obtained with the help of Theorem 1.7.5 (which may easily be strengthened by merely considering convergence on a sequence of equidistant points).

Theorem 2.1.2. *Let $M > 0$, $F_n \in \mathrm{Lip}_0(\mathbb{R}_+, X)$ with $\|F_n\|_{\mathrm{Lip}_0(\mathbb{R}_+,X)} \leq M$ for all $n \in \mathbb{N}$, and $r_n = \mathcal{L}_S(F_n)$. The following are equivalent:*

(i) *There exist $a, b > 0$ such that $\lim_{n\to\infty} r_n(a + kb)$ exists for all $k \in \mathbb{N}_0$.*

(ii) *There exists $r \in C^\infty((0,\infty), X)$ such that $r_n \to r$ uniformly on compact subsets of $(0,\infty)$.*

(iii) *$\lim_{n\to\infty} F_n(t)$ exists for all $t \geq 0$.*

(iv) *There exists $F \in \mathrm{Lip}_0(\mathbb{R}_+, X)$ such that $F_n \to F$ uniformly on compact subsets of $\mathbb{R}_+$.*

Moreover, if r and F are as in (ii) *and* (iv), *then $r = \mathcal{L}_S(F)$.*

Proof. By the Riesz-Stieltjes Representation Theorem 2.1.1, there exist $T_n \in \mathcal{L}(L^1(\mathbb{R}_+), X)$ such that $\|T_n\| = \|F_n\|_{\mathrm{Lip}_0(\mathbb{R}_+,X)} \leq M$, $T_n e_{-\lambda} = r_n(\lambda)$, and $T_n\chi_{[0,t]} = F_n(t)$ ($n \in \mathbb{N}$, $t \geq 0$, $\lambda > 0$). Each of the statements imply that the uniformly bounded family of operators T_n converges on a total subset of $L^1(\mathbb{R}_+)$

(see also Lemma 1.7.1). By equicontinuity (see Proposition B.15), for any uniformly bounded sequence of operators, the topology of simple convergence on a total subset equals the topology of simple convergence and the topology of uniform convergence on compact subsets. Thus there exists $T \in \mathcal{L}(L^1(\mathbb{R}_+), X)$ such that $T_n g \to Tg$ as $n \to \infty$ for all $g \in L^1(\mathbb{R}_+)$ (simple convergence). For all $b > 0$ the sets $K_b := \{\chi_{[0,t]} : 0 \le t \le b\}$ and $E_b := \{e_{-\lambda} : \frac{1}{b} \le \lambda \le b\}$ are compact in $L^1(\mathbb{R}_+)$ (continuous images of compact sets are compact). Hence, $T_n \to T$ uniformly on K_b and E_b (uniform convergence on compact subsets). Now the statements follow from the Riesz-Stieltjes representation. □

2.2 A Real Representation Theorem

In this section the range of the Laplace-Stieltjes transform $\mathcal{L}_S : F \mapsto \widehat{dF}$ acting on $\mathrm{Lip}_0(\mathbb{R}_+, X)$ will be characterized. Since $\lambda \mapsto \widehat{dF}(\lambda) = \lambda\widehat{F}(\lambda)$ is holomorphic and, by Proposition 1.7.2, functions like $\lambda \mapsto (\sin\lambda)x$ $(x \in X)$ cannot be in the range of $\mathcal{L}_S$, the range must be a proper subset of $C^\infty((0,\infty), X)$. The following observations will lead to a complete description of the range.

Let $F \in \mathrm{Lip}_0(\mathbb{R}_+, X)$ and $T_F := \Phi_S(F)$, where Φ_S is the Riesz-Stieltjes operator of Section 2.1. Define

$$r(\lambda) := \widehat{dF}(\lambda) = \int_0^\infty e^{-\lambda t}\, dF(t) \quad (\lambda > 0).$$

Then, by Theorem 1.10.6, $r \in C^\infty((0,\infty), X)$ and

$$r^{(n)}(\lambda) = \int_0^\infty e^{-\lambda t}(-t)^n\, dF(t) = T_F k_{n,\lambda},$$

where $k_{n,\lambda}(t) := e^{-\lambda t}(-t)^n$ $(t \ge 0,\ \lambda > 0,\ n \in \mathbb{N}_0)$. Since $\|k_{n,\lambda}\|_1 = \int_0^\infty e^{-\lambda t} t^n\, dt = n!/\lambda^{n+1}$ and $\|T_F\| = \|F\|_{\mathrm{Lip}_0(\mathbb{R}_+,X)}$, it follows that

$$\|r^{(n)}(\lambda)\| \le \|F\|_{\mathrm{Lip}_0(\mathbb{R}_+,X)} n!/\lambda^{n+1}$$

for all $n \in \mathbb{N}_0$ and $\lambda > 0$. Thus, r is a C^∞-function whose Taylor coefficients satisfy

$$\|r\|_W := \sup_{\lambda>0, k\in\mathbb{N}_0} \frac{\lambda^{k+1}}{k!}\|r^{(k)}(\lambda)\| \le \|F\|_{\mathrm{Lip}_0(\mathbb{R}_+,X)}.$$

This shows that the Laplace-Stieltjes transform $\mathcal{L}_S : F \to \widehat{dF}$ maps $\mathrm{Lip}_0(\mathbb{R}_+, X)$ into the space

$$C_W^\infty((0,\infty), X) := \{r \in C^\infty((0,\infty), X) : \|r\|_W < \infty\}.$$

In 1936, Widder showed that the Laplace transform maps $L^\infty(\mathbb{R}_+, \mathbb{R})$ onto $C_W^\infty((0,\infty), \mathbb{R})$. The following result is the vector-valued version of Widder's classical theorem.

Theorem 2.2.1 (Real Representation Theorem). *The Laplace-Stieltjes transform* $\mathcal{L}_S$ *is an isometric isomorphism between* $\mathrm{Lip}_0(\mathbb{R}_+, X)$ *and* $C_W^\infty((0,\infty), X)$.

Proof. We have already shown that $\mathcal{L}_S$ maps $\mathrm{Lip}_0(\mathbb{R}_+, X)$ into $C_W^\infty((0,\infty), X)$ and that $\|\mathcal{L}_S(F)\|_W \le \|F\|_{\mathrm{Lip}_0(\mathbb{R}_+,X)}$. If $\mathcal{L}_S(F) = \widehat{dF} = 0$ for some $F \in \mathrm{Lip}_0(\mathbb{R}_+, X)$, then $T_F e_{-\lambda} = \int_0^\infty e^{-\lambda t}\, dF(t) = \widehat{dF}(\lambda) = 0$ for all $\lambda > 0$. Since the exponential functions $e_{-\lambda}$ ($\lambda > 0$) are total in $L^1(\mathbb{R}_+)$ (Lemma 1.7.1), it follows that $T_F = 0$. In particular, $T_F \chi_{[0,t]} = F(t) = 0$ for all $t \ge 0$. Thus, $\mathcal{L}_S$ is one-to-one.

The hard part of the proof is to show that $\mathcal{L}_S$ is onto. Let $r \in C_W^\infty((0,\infty), X)$. Define $T_k \in \mathcal{L}(L^1(\mathbb{R}_+), X)$ by

$$T_k f \;:=\; \int_0^\infty f(t)(-1)^k \frac{1}{k!}\left(\frac{k}{t}\right)^{k+1} r^{(k)}\left(\frac{k}{t}\right) dt \qquad (k \in \mathbb{N}_0).$$

The operators T_k are uniformly bounded by $\|r\|_W$ since $\|T_k f\| \le \|r\|_W \|f\|_1$ for all $f \in L^1(\mathbb{R}_+)$. We will show below that $T_k e_{-\lambda} \to r(\lambda)$ as $k \to \infty$ for all $\lambda > 0$. Since the exponential functions $e_{-\lambda}$ ($\lambda > 0$) are total in $L^1(\mathbb{R}_+)$ it then follows from Proposition B.15 that there exists $T \in \mathcal{L}(L^1(\mathbb{R}_+), X)$ with $\|T\| \le \|r\|_W$ such that $T_k f \to Tf$ for all $f \in L^1(\mathbb{R}_+)$. In particular,

$$r(\lambda) \;=\; \lim_{k\to\infty} T_k e_{-\lambda} \;=\; T e_{-\lambda}.$$

The Riesz-Stieltjes Representation Theorem 2.1.1 then yields the existence of some $F \in \mathrm{Lip}_0(\mathbb{R}_+, X)$ with $\|F\|_{\mathrm{Lip}_0(\mathbb{R}_+,X)} = \|T\| \le \|r\|_W$ such that $Tg = \int_0^\infty g(t)\, dF(t)$ for all continuous functions $g \in L^1(\mathbb{R}_+)$. Hence, for all $\lambda > 0$,

$$r(\lambda) \;=\; Te_{-\lambda} \;=\; \int_0^\infty e^{-\lambda t}\, dF(t) \;=\; \widehat{dF}(\lambda).$$

Thus, $\mathcal{L}_S$ is onto and $\|\mathcal{L}_S(F)\|_W = \|\widehat{dF}\|_W = \|F\|_{\mathrm{Lip}_0(\mathbb{R}_+,X)}$ for $F \in \mathrm{Lip}_0(\mathbb{R}_+, X)$.

It remains to be shown that $T_k e_{-\lambda} \to r(\lambda)$ as $k \to \infty$ for all $\lambda > 0$. Observe that

$$\begin{aligned}
T_k e_{-\lambda} \;&=\; \int_0^\infty e^{-\lambda t}(-1)^k \frac{1}{k!}\left(\frac{k}{t}\right)^{k+1} r^{(k)}\left(\frac{k}{t}\right) dt \\
&=\; (-1)^k \frac{1}{(k-1)!}\int_0^\infty \left(e^{-\lambda k/u} u^{k-1}\right) r^{(k)}(u)\, du \\
&=\; (-1)^k \frac{1}{(k-1)!}\left[\sum_{j=0}^{k-1} (-1)^j \frac{d^j}{du^j}\left(e^{-\lambda k/u} u^{k-1}\right) r^{(k-j-1)}(u)\Big|_{u=0}^{\infty} \right.\\
&\qquad \left. +(-1)^k \int_0^\infty \frac{d^k}{du^k}\left(e^{-\lambda k/u} u^{k-1}\right) r(u)\, du\right].
\end{aligned}$$

To discuss the derivatives of $u \mapsto e^{-\lambda k/u} u^{k-1}$, define $G(x,u) \;:=\; e^{-x/u}\left(\frac{u}{x}\right)^{k-1}$. Then $G(sx, su) = G(x,u)$ for all $s > 0$. Differentiating both sides of the last equality with respect to s and then setting $s = 1$ yields $x\frac{\partial G}{\partial x}(x,u)+u\frac{\partial G}{\partial u}(x,u) = 0$ or $\frac{1}{x}\frac{\partial G}{\partial u}(x,u) = -\frac{1}{u}\frac{\partial G}{\partial x}(x,u)$. This implies that

$$\frac{\partial}{\partial u}\left(e^{-x/u}\frac{u^{k-1}}{x^k}\right) = -\frac{\partial}{\partial x}\left(e^{-x/u}\frac{u^{k-2}}{x^{k-1}}\right).$$

By induction on j, it follows that

$$\frac{\partial^j}{\partial u^j}\left(e^{-x/u}\frac{u^{k-1}}{x^k}\right) = (-1)^j\frac{\partial^j}{\partial x^j}\left(e^{-x/u}\frac{u^{k-j-1}}{x^{k-j}}\right) \qquad (0 \le j \le k),$$

or

$$\frac{\partial^j}{\partial u^j}\left(e^{-x/u}u^{k-1}\right) = (-1)^j x^k u^{k-j-1}\frac{\partial^j}{\partial x^j}\left(\frac{e^{-x/u}}{x^{k-j}}\right). \tag{2.4}$$

Hence,

$$\begin{aligned} h(u) &:= \sum_{j=0}^{k-1}(-1)^j\frac{\partial^j}{\partial u^j}\left(e^{-x/u}u^{k-1}\right)r^{(k-j-1)}(u) \\ &= \sum_{j=0}^{k-1}x^k\frac{\partial^j}{\partial x^j}\left(\frac{e^{-x/u}}{x^{k-j}}\right)u^{k-j-1}r^{(k-j-1)}(u). \end{aligned}$$

Since

$$\|u^{k-j-1}r^{(k-j-1)}(u)\| \le \frac{\|r\|_W(k-j-1)!}{u},$$

one obtains that

$$\|h(u)\| \le \sum_{j=0}^{k-1}\frac{\|r\|_W(k-j-1)!}{u}x^k\left|\frac{\partial^j}{\partial x^j}\left(\frac{e^{-x/u}}{x^{k-j}}\right)\right|.$$

It follows that $\lim_{u\to\infty} h(u) = 0 = \lim_{u\to 0} h(u)$. Therefore, letting $x = \lambda k$,

$$T_k e_{-\lambda} = \frac{1}{(k-1)!}\int_0^\infty \frac{d^k}{du^k}\left(e^{-\lambda k/u}u^{k-1}\right)r(u)\,du.$$

Since by (2.4),

$$\frac{\partial^k}{\partial u^k}\left(e^{-x/u}u^{k-1}\right) = (-1)^k\frac{x^k}{u}\frac{\partial^k}{\partial x^k}\left(e^{-x/u}\right) = \frac{x^k}{u^{k+1}}e^{-x/u},$$

it follows that

$$\begin{aligned} T_k e_{-\lambda} &= \frac{\lambda^k k^k}{(k-1)!}\int_0^\infty e^{-\lambda k/u}\frac{1}{u^{k+1}}r(u)\,du \\ &= \frac{\lambda^k k^{k+1}}{k!}\int_0^\infty e^{-\lambda k t}t^{k-1}r\left(\frac{1}{t}\right)\,dt. \end{aligned}$$

Define $f(t) := \frac{1}{t}r(\frac{1}{t})$ and $s := \frac{1}{\lambda}$. Then

$$\begin{aligned} T_k e_{-\lambda} &= \frac{s}{k!}\left(\frac{k}{s}\right)^{k+1}\int_0^\infty e^{-kt/s}t^k f(t)\,dt \\ &= s(-1)^k\frac{1}{k!}\left(\frac{k}{s}\right)^{k+1}\hat{f}^{(k)}\left(\frac{k}{s}\right). \end{aligned}$$

Finally, one concludes from the Post-Widder Inversion Theorem 1.7.7 that

$$\lim_{k\to\infty} T_k e_{-\lambda} = sf(s) = r\left(\frac{1}{s}\right) = r(\lambda)$$

for all $\lambda > 0$. □

For later use in Section 2.5, we observe that in the Widder conditions it is not necessary to consider all values of k.

Proposition 2.2.2. *Let $r \in C^\infty((0,\infty),X)$, and suppose that $\lim_{\lambda\to\infty} r(\lambda) = 0$ and there exist $M > 0$ and infinitely many integers m such that $\sup_{\lambda>0}\|\frac{\lambda^{m+1}}{m!}r^{(m)}(\lambda)\| \le M$. Then $r \in C_W^\infty((0,\infty),X)$ and $\|r\|_W \le M$.*

Proof. It suffices to show that if $\|r^{(m)}(\lambda)\| \le Mm!/\lambda^{m+1}$, for all $\lambda > 0$, then $\|r^{(k)}(\lambda)\| \le Mk!/\lambda^{k+1}$ for all $\lambda > 0$ and $0 \le k < m$. Let

$$\tilde{r}(\lambda) := \frac{(-1)^m}{(m-1)!}\int_\lambda^\infty (\lambda-\mu)^{m-1}r^{(m)}(\mu)\,d\mu.$$

Note that the integral is absolutely convergent, $\tilde{r}^{(m)}(\lambda) = r^{(m)}(\lambda)$, and the substitution $t = \lambda/\mu$ gives

$$\|\tilde{r}(\lambda)\| \le Mm\int_\lambda^\infty \frac{(\mu-\lambda)^{m-1}}{\mu^{m+1}}\,d\mu = \frac{Mm}{\lambda}\int_0^1 (1-t)^{m-1}\,dt = \frac{M}{\lambda}.$$

Hence $r - \tilde{r}$ is a polynomial and $\lim_{\lambda\to\infty}(r-\tilde{r})(\lambda) = 0$, so $r = \tilde{r}$. It follows that

$$\begin{aligned} \|r^{(k)}(\lambda)\| &= \left\|\frac{(-1)^m}{(m-k-1)!}\int_\lambda^\infty (\lambda-\mu)^{m-k-1}r^{(m)}(\mu)\,d\mu\right\| \\ &\le \frac{Mm!}{(m-k-1)!}\int_\lambda^\infty \frac{(\mu-\lambda)^{m-k-1}}{\mu^{m+1}}\,d\mu \\ &= \frac{Mk!}{\lambda^{k+1}} \end{aligned}$$

for $\lambda > 0$ and $0 \le k < m$. □

Now it will be shown that the Laplace transform is an isometric isomorphism between $L^\infty(\mathbb{R}_+,X)$ and $C_W^\infty((0,\infty),X)$ if and only if the Banach space X has the

Radon-Nikodym property. Recall from Section 1.2 that X has the *Radon-Nikodym property* if every $F \in \mathrm{Lip}_0(\mathbb{R}_+, X)$ is differentiable a.e., or equivalently if every absolutely continuous function $F : \mathbb{R}_+ \to X$ is differentiable a.e. As shown in Theorem 1.2.6 and Corollary 1.2.7, every separable dual space (for example, l^1) and every reflexive Banach space have the Radon-Nikodym property. However, $L^1(\mathbb{R}_+)$ and c_0 do not have the property (Propositions 1.2.9 and 1.2.10).

Theorem 2.2.3. *Let X be a Banach space. The following are equivalent:*

(i) *X has the Radon-Nikodym property.*

(ii) *The Laplace transform $\mathcal{L} : f \mapsto \hat{f}$ is an isometric isomorphism between $L^\infty(\mathbb{R}_+, X)$ and $C_W^\infty((0,\infty), X)$.*

(iii) *The Riesz operator $\Phi : f \mapsto R_f$, $R_f g := \int_0^\infty g(t)f(t)\,dt$ is an isometric isomorphism between $L^\infty(\mathbb{R}_+, X)$ and $\mathcal{L}(L^1(\mathbb{R}_+), X)$.*

Proof. Define the normalized antiderivative $I : L^\infty(\mathbb{R}_+, X) \to \mathrm{Lip}_0(\mathbb{R}_+, X)$ by $I(f) := F$, $F(t) := \int_0^t f(s)\,ds$ $(t \geq 0)$. Then I is one-to-one and $\|I(f)\|_{\mathrm{Lip}_0(\mathbb{R}_+,X)} \leq \|f\|_\infty$ for all $f \in L^\infty(\mathbb{R}_+, X)$. If I is onto, then X has the Radon-Nikodym property (see Proposition 1.2.2). Conversely, if X has the Radon-Nikodym property and $F \in \mathrm{Lip}_0(\mathbb{R}_+, X)$ then $f(t) := F'(t)$ exists for almost all $t \geq 0$. Since $f(t) = \lim_{h\to 0} \frac{F(t+h)-F(t)}{h}$ a.e., one concludes that $\|f\|_\infty \leq \|F\|_{\mathrm{Lip}_0(\mathbb{R}_+,X)}$. In particular, $f \in L^\infty(\mathbb{R}_+, X)$ and by Proposition 1.2.3, $F = I(f)$. Thus X has the Radon-Nikodym property if and only if I is an isometric isomorphism.

The Riesz-Stieltjes operator $\Phi_S : F \mapsto T_F$, where $T_F g = \int_0^\infty g(t)\,dF(t)$ for all continuous $g \in L^1(\mathbb{R}_+)$, is an isometric isomorphism between $\mathrm{Lip}_0(\mathbb{R}_+, X)$ and $\mathcal{L}(L^1(\mathbb{R}_+), X)$, and the Laplace-Stieltjes transform

$$\mathcal{L}_S : F \mapsto \widehat{dF}, \quad \widehat{dF}(\lambda) = \int_0^\infty e^{-\lambda t}\,dF(t),$$

is an isometric isomorphism between $\mathrm{Lip}_0(\mathbb{R}_+, X)$ and $C_W^\infty((0,\infty), X)$. When $F = I(f)$, $T_F g = \int_0^\infty g(t)f(t)\,dt$ for all $g \in L^1(\mathbb{R}_+)$, by Proposition 1.9.11 and continuity in L^1-norm. Now the statements follow from the fact that $\Phi = \Phi_S \circ I$ and $\mathcal{L} = \mathcal{L}_S \circ I$ on $L^\infty(\mathbb{R}_+, X)$. □

Example 2.2.4. a) Consider $X = L^1(\mathbb{R}_+)$. Let $F(t) := \chi_{[0,t]}$ $(t \geq 0)$ and $r(\lambda) := e_{-\lambda}$ $(\mathrm{Re}\,\lambda > 0)$, where $e_{-\lambda}(t) = e^{-\lambda t}$. Then $F \in \mathrm{Lip}_0(\mathbb{R}_+, L^1(\mathbb{R}_+))$ and

$$r(\lambda) = \int_0^\infty e^{-\lambda t}\,dF(t) = \widehat{dF}(\lambda).$$

Since F is nowhere differentiable (see Proposition 1.2.10), there does not exist $f \in L^\infty(\mathbb{R}_+, L^1(\mathbb{R}_+))$ such that

$$r(\lambda) = \int_0^\infty e^{-\lambda t} f(t)\,dt.$$

b) Consider $C_0(\mathbb{R}_+)$ as a subspace of $L^\infty(\mathbb{R}_+)$. Define $F : \mathbb{R}_+ \to C_0(\mathbb{R}_+)$ by $F(t)(s) := (t-s)\chi_{[0,t]}(s)$, and $f : \mathbb{R}_+ \to L^\infty(\mathbb{R}_+)$ by $f(t) := \chi_{[0,t]}$. Then $F \in \mathrm{Lip}_0(\mathbb{R}_+, C_0(\mathbb{R}_+))$ and $F(t) = \int_0^t f(s)\,ds$ as a Riemann integral in $L^\infty(\mathbb{R}_+)$, but F is nowhere differentiable and f is not measurable (see Examples 1.2.8 and 1.9.7). Moreover,

$$\frac{1}{\lambda}e_{-\lambda} = \int_0^\infty e^{-\lambda t}\,dF(t) = \int_0^\infty e^{-\lambda t} f(t)\,dt$$

as (improper) Riemann-Stieltjes and Riemann integrals, but $\lambda \mapsto \frac{1}{\lambda}e_{-\lambda}$ is not the Laplace transform of any function in $L^1(\mathbb{R}_+, L^\infty(\mathbb{R}_+))$. □

2.3 Real and Complex Inversion

We have shown in Section 2.2 that the Laplace-Stieltjes transform $\mathcal{L}_S$ is an isometric isomorphism between $\mathrm{Lip}_0(\mathbb{R}_+, X)$ and $C_W^\infty((0,\infty), X)$. In this section we will derive several representations of the inverse Laplace-Stieltjes transform $\mathcal{L}_S^{-1}$.

Theorem 2.3.1 (Post-Widder Inversion). *Let $F \in \mathrm{Lip}_0(\mathbb{R}_+, X)$, $r = \mathcal{L}_S(F)$, and $t > 0$. Then*

$$F(t) = \lim_{k\to\infty} (-1)^k \frac{1}{k!}\left(\frac{k}{t}\right)^{k+1} \frac{d^k}{d\lambda^k}\left(\frac{r(\lambda)}{\lambda}\right)\Big|_{\lambda=k/t}.$$

Proof. Since $\omega(F) \le 0$ and $F(0) = 0$, it follows from (1.22) that

$$\frac{r(\lambda)}{\lambda} = \int_0^\infty e^{-\lambda t} F(t)\,dt$$

for all $\lambda > 0$, where the integral is an absolutely convergent Bochner integral. Now the statement follows from Theorem 1.7.7. □

Applying Leibniz's rule $(f \cdot r)^{(k)} = \sum_{j=0}^k \binom{k}{j} f^{(k-j)} r^{(j)}$ to $f(\lambda) := \frac{1}{\lambda}$ and r one can rewrite the Post-Widder inversion of the Laplace-Stieltjes transform as

$$F(t) = \lim_{k\to\infty} \sum_{j=0}^k (-1)^j \frac{1}{j!}\left(\frac{k}{t}\right)^j r^{(j)}\left(\frac{k}{t}\right) \qquad (t > 0). \tag{2.5}$$

Compared to the Post-Widder inversion, it is remarkable that in the following Phragmén-Doetsch inversion formula only the values $r(k)$ for large $k \in \mathbb{N}$ are needed and that the convergence is uniform for all $t \ge 0$.

Theorem 2.3.2 (Phragmén-Doetsch Inversion). *Let $F \in \mathrm{Lip}_0(\mathbb{R}_+, X)$ and $r = \mathcal{L}_S(F)$. Then*

$$\left\| F(t) - \sum_{j=1}^\infty \frac{(-1)^{j+1}}{j!} e^{tkj} r(kj) \right\| \le \frac{c}{k}\|r\|_W$$

for all $t \ge 0$ and $k \in \mathbb{N}$, where $c \approx 1.0159...$, and $\|r\|_W = \|F\|_{\mathrm{Lip}_0(\mathbb{R}_+,X)}$.

Proof. By the Riesz-Stieltjes Representation Theorem 2.1.1 and the Real Representation Theorem 2.2.1, there exists $T \in \mathcal{L}(L^1(\mathbb{R}_+), X)$ such that $r(\lambda) = \int_0^\infty e^{-\lambda t}\, dF(t) = Te_{-\lambda}$ $(\lambda > 0)$, $T\chi_{[0,t]} = F(t)$ $(t \geq 0)$ and $\|T\| = \|r\|_W = \|F\|_{\mathrm{Lip}_0(\mathbb{R}_+,X)}$. Thus,

$$\left\| F(t) - \sum_{j=1}^{\infty} \frac{(-1)^{j+1}}{j!} e^{tkj} r(kj) \right\| \leq \|T\| \left\| \chi_{[0,t]} - \sum_{j=1}^{\infty} (-1)^{j+1} \frac{1}{j!} e^{tkj} e_{-kj} \right\|_1 .$$

Define $p_{k,t}(s) := 1 - e^{-e^{k(t-s)}} = \sum_{j=1}^\infty (-1)^{j+1} \frac{1}{j!} e^{tkj} e_{-kj}(s)$. Then,

$$\begin{aligned}
\|\chi_{[0,t]} - p_{k,t}\|_1 &= \int_0^t |p_{k,t}(s) - 1|\, ds + \int_t^\infty |p_{k,t}(s)|\, ds \\
&= \int_0^t e^{-e^{k(t-s)}}\, ds + \int_t^\infty \left(1 - e^{-e^{k(t-s)}}\right) ds \\
&= \frac{1}{k} \int_1^{e^{kt}} \frac{e^{-u}}{u}\, du + \frac{1}{k} \int_0^1 \frac{1 - e^{-u}}{u}\, du \\
&\leq \frac{1}{k} \left(\int_1^\infty \frac{e^{-u}}{u}\, du + \int_0^1 \frac{1 - e^{-u}}{u}\, du \right)
\end{aligned}$$

for all $t \geq 0$ and $k \in \mathbb{N}$. Now the claim follows from the fact that $\int_1^\infty \frac{1}{u} e^{-u}\, du + \int_0^1 \frac{1-e^{-u}}{u}\, du = -2\,\mathrm{Ei}(-1) + \gamma \approx 1.0159...$, where $\mathrm{Ei}(z)$ is the exponential integral and γ is Euler's constant (see [Leb72, Section 3.1]). □

The following corollary shows that the Phragmén-Doetsch inversion is invariant under exponentially decaying perturbations for small values of t.

Corollary 2.3.3. *Let $F \in \mathrm{Lip}_0(\mathbb{R}_+, X)$, $r = \mathcal{L}_S(F)$, and $q(\lambda) = r(\lambda) + a(\lambda)$ $(\lambda > 0)$, where $a : (0, \infty) \to X$ is a function such that $\limsup_{n\to\infty} \frac{1}{n} \log \|a(n)\| \leq -T$ for some $T > 0$. Then*

$$F(t) = \lim_{k\to\infty} \sum_{j=1}^{\infty} \frac{(-1)^{j+1}}{j!} e^{tkj} q(kj)$$

for all $0 \leq t < T$.

Proof. Let $0 < T_0 < T$ and choose k_0 such that $\|a(k)\| \leq e^{-T_0 k}$ for all $k \geq k_0$. Then,

$$\begin{aligned}
&\left\| F(t) - \sum_{j=1}^{\infty} \frac{(-1)^{j+1}}{j!} e^{tkj} q(kj) \right\| \\
&\leq \left\| F(t) - \sum_{j=1}^{\infty} \frac{(-1)^{j+1}}{j!} e^{tkj} r(kj) \right\| + \left\| \sum_{j=1}^{\infty} \frac{(-1)^{j+1}}{j!} e^{tkj} a(kj) \right\| \\
&\leq \frac{2}{k} \|r\|_W + \sum_{j=1}^{\infty} \frac{1}{j!} e^{tkj} e^{-T_0 kj} \leq \frac{2}{k} \|r\|_W + e^{e^{-(T_0 - t)k}} - 1.
\end{aligned}$$

□

The Post-Widder inversion and the Phragmén-Doetsch inversion are called real inversions of the Laplace-Stieltjes transform since they use only properties of $r(\lambda)$ for large real λ. For the following complex inversion formula we use the fact that if $r(\lambda) = \int_0^\infty e^{-\lambda t}\, dF(t)$ $(\lambda > 0)$ for some $F \in \mathrm{Lip}_0(\mathbb{R}_+, X)$, then r admits a holomorphic extension for $\operatorname{Re}\lambda > 0$ which we denote by the same symbol (see Theorem 1.10.6). We shall give here a proof based on the Riesz-Stieltjes representation, but we shall give another, rather simple, proof in Section 4.2.

Theorem 2.3.4 (Complex Inversion). *Let $F \in \mathrm{Lip}_0(\mathbb{R}_+, X)$ and $r = \mathcal{L}_S(F)$. Then*

$$F(t) = \lim_{k\to\infty} \frac{1}{2\pi i} \int_{c-ik}^{c+ik} e^{\lambda t} \frac{r(\lambda)}{\lambda}\, d\lambda,$$

where the limit is uniform for $t \in [0, a]$ for any $a > 0$, and $c > 0$ is arbitrary.

Proof. By the Riesz-Stieltjes Representation Theorem 2.1.1, there exists $T \in \mathcal{L}(L^1(\mathbb{R}_+), X)$ such that $r(\lambda) = Te_{-\lambda}$ $(\operatorname{Re}\lambda > 0)$ and $F(t) = T\chi_{[0,t]}$ $(t \geq 0)$. Thus,

$$\left\| F(t) - \frac{1}{2\pi i} \int_{c-ik}^{c+ik} e^{\lambda t} \frac{r(\lambda)}{\lambda}\, d\lambda \right\| \leq \|T\| \left\| \chi_{[0,t]} - \frac{1}{2\pi i} \int_{c-ik}^{c+ik} e^{\lambda t} \frac{e_{-\lambda}}{\lambda}\, d\lambda \right\|_1 .$$

Now the statement follows from the next lemma. □

Lemma 2.3.5. *Let $t \geq 0$ and $a, c > 0$. Then the functions*

$$h_{k,t} := \frac{1}{2\pi i} \int_{c-ik}^{c+ik} e^{\lambda t} \frac{e_{-\lambda}}{\lambda}\, d\lambda$$

converge towards $\chi_{[0,t]}$ in $L^1(\mathbb{R}_+)$ as $n \to \infty$, uniformly for $t \in [0, a]$.

Proof. Let $\|h_{k,t} - \chi_{[0,t]}\|_1 = A_k + B_k$, where $A_k := \int_0^t |h_{k,t}(s) - 1|\, ds$ and $B_k := \int_t^\infty |h_{k,t}(s)|\, ds$. We show first that $\lim_{k\to\infty} A_k = 0$. The residue of the function $\lambda \mapsto e^{\lambda(t-s)}/\lambda$ at the point 0 is 1. By Cauchy's theorem,

$$h_{k,t}(s) - 1 = \frac{1}{2\pi i} \left(\int_{\Gamma_+} - \int_{\Gamma_-} - \int_{\Gamma_0} \right) \frac{e^{\lambda(t-s)}}{\lambda}\, d\lambda,$$

where $\Gamma_\pm := \{\lambda : \lambda = u \pm ik;\ 0 \leq u \leq c\}$, $\Gamma_0 := \{\lambda : \lambda = ke^{iu};\ \pi/2 \leq u \leq 3\pi/2\}$. Along Γ_+, and similarly along Γ_-, it follows from $0 \leq s \leq t$ that

$$\left| \int_{\Gamma_+} \frac{e^{\lambda(t-s)}}{\lambda}\, d\lambda \right| = \left| \int_0^c \frac{e^{(u+ik)(t-s)}}{u+ik}\, du \right| \leq c\frac{e^{c(t-s)}}{k}.$$

Along Γ_0, for $0 \leq s < t$,

$$\left| \int_{\Gamma_0} \frac{e^{\lambda(t-s)}}{\lambda}\, d\lambda \right| = \left| \int_{\pi/2}^{3\pi/2} e^{k(t-s)e^{iu}}\, du \right| \leq \int_{\pi/2}^{3\pi/2} e^{k(t-s)\cos u}\, du.$$

Hence,

$$\begin{aligned} A_k &= \int_0^t |h_{k,t}(t-s) - 1|\, ds \\ &\le \int_0^t \left(\frac{ce^{cs}}{\pi k} + \frac{1}{2\pi} \int_{\pi/2}^{3\pi/2} e^{ks\cos u}\, du \right) ds \\ &\to 0 \end{aligned}$$

as $k \to \infty$, uniformly for $t \in [0, a]$ for all $a > 0$, by the monotone convergence theorem, or by explicit estimation.

In order to estimate B_k, we define $\widetilde{\Gamma}_\pm := \{\lambda : \lambda = u \pm ik \; ; \; c \le u \le k\}$, $\widetilde{\Gamma}_0 := \{\lambda : \lambda = k\sqrt{2}e^{iu} \; ; \; -\pi/4 \le u \le \pi/4\}$. By Cauchy's theorem,

$$h_{k,t}(s) = \frac{1}{2\pi i} \left(-\int_{\widetilde{\Gamma}_+} + \int_{\widetilde{\Gamma}_-} + \int_{\widetilde{\Gamma}_0} \right) \frac{e^{\lambda(t-s)}}{\lambda}\, d\lambda.$$

Along $\widetilde{\Gamma}_+$, and similarly along $\widetilde{\Gamma}_-$, it follows from $s - t \ge 0$ that

$$\begin{aligned} \left| \int_{\widetilde{\Gamma}_+} \frac{e^{\lambda(t-s)}}{\lambda}\, d\lambda \right| &= \left| \int_c^k \frac{e^{(u+ik)(t-s)}}{u+ik}\, du \right| \le \frac{1}{k} \int_c^k e^{-u(s-t)}\, du \\ &= \frac{e^{-c(s-t)} - e^{-k(s-t)}}{k(s-t)}. \end{aligned}$$

Along $\widetilde{\Gamma}_0$,

$$\begin{aligned} \left| \int_{\widetilde{\Gamma}_0} \frac{e^{\lambda(t-s)}}{\lambda}\, d\lambda \right| &= \left| \int_{-\pi/4}^{\pi/4} e^{k\sqrt{2}(t-s)e^{iu}}\, du \right| \le \int_{-\pi/4}^{\pi/4} e^{k\sqrt{2}(t-s)\cos u}\, du \\ &= 2\int_0^{\pi/4} e^{k\sqrt{2}(t-s)\cos(u)}\, du \le \frac{\pi}{2} e^{k\sqrt{2}(t-s)\cos(\pi/4)} = \frac{\pi}{2} e^{-k(s-t)}. \end{aligned}$$

Hence, for all $t \ge 0$,

$$\begin{aligned} \int_t^\infty |h_{k,t}(s)|\, ds &\le \frac{1}{\pi} \int_t^\infty \frac{e^{-c(s-t)} - e^{-k(s-t)}}{k(s-t)}\, ds + \frac{1}{4} \int_t^\infty e^{-k(s-t)}\, ds \\ &= \frac{1}{\pi} \int_0^\infty z_k(s)\, ds + \frac{1}{4k}, \end{aligned}$$

where $z_k(s) := \frac{1}{ks}(e^{-cs} - e^{-ks}) \le e^{-cs}$ for $k \ge c$ by the mean value theorem applied to e^{-x} over $[cs, ks]$. By the dominated convergence theorem, or by explicit estimation, $B_k \to 0$ as $k \to \infty$, uniformly for $t \in [0, a]$ for all $a > 0$. $\square$

2.4 Transforms of Exponentially Bounded Functions

So far in this chapter, Laplace transforms have been considered for bounded or globally Lipschitz continuous functions. We shall now adapt the results of the previous sections to functions with exponential growth at infinity, by an elementary "shifting" procedure (see Proposition 1.6.1 a) and Proposition 1.10.3). More precisely, for $\omega \in \mathbb{R}$ we consider the Laplace-Stieltjes transform acting on

$$\mathrm{Lip}_\omega(\mathbb{R}_+, X) := \Big\{ G : \mathbb{R}_+ \to X : \ G(0) = 0, \\ \|G\|_{\mathrm{Lip}_\omega(\mathbb{R}_+,X)} := \sup_{t>s\geq 0} \frac{\|G(t) - G(s)\|}{\int_s^t e^{\omega r}\, dr} < \infty \Big\}$$

and the Laplace transform acting on

$$L^\infty_\omega(\mathbb{R}_+, X) := \Big\{ g \in L^1_{loc}(\mathbb{R}_+, X) : \ \|g\|_{\omega,\infty} := \operatorname*{ess\,sup}_{t\geq 0} \|e^{-\omega t} g(t)\| < \infty \Big\}.$$

It is easy to see that

$$\|G\|_{\mathrm{Lip}_\omega(\mathbb{R}_+,X)} = \begin{cases} \sup_{0\leq s<t} \dfrac{\|G(t) - G(s)\|}{(t-s)e^{\omega t}} & \text{if } \omega \geq 0, \\ \sup_{0\leq s<t} \dfrac{\|G(t) - G(s)\|}{(t-s)e^{\omega s}} & \text{if } \omega \leq 0. \end{cases}$$

It is clear that the multiplication operator $M_\omega : g \mapsto e^{-\omega\cdot} g(\cdot)$ is an isometric isomorphism between $L^\infty_\omega(\mathbb{R}_+, X)$ and $L^\infty(\mathbb{R}_+, X)$, and we now set up the corresponding isomorphism between $\mathrm{Lip}_\omega(\mathbb{R}_+, X)$ and $\mathrm{Lip}_0(\mathbb{R}_+, X)$.

For $G \in \mathrm{Lip}_\omega(\mathbb{R}_+, X)$ and $f \in \mathrm{BSV}_{loc}(\mathbb{R}_+)$, it follows from the definition of the Riemann-Stieltjes integral that

$$\left\| \int_a^b f(t)\, dG(t) \right\| \leq \|G\|_{\mathrm{Lip}_\omega(\mathbb{R}_+,X)} \int_a^b |f(t)| e^{\omega t}\, dt \quad (0 \leq a \leq b). \tag{2.6}$$

Let

$$(I_\omega G)(t) := \int_0^t e^{-\omega s}\, dG(s).$$

Then (2.6) implies that

$$I_\omega G \in \mathrm{Lip}_0(\mathbb{R}_+, X) \ \text{ and } \ \|I_\omega G\|_{\mathrm{Lip}_0(\mathbb{R}_+,X)} \leq \|G\|_{\mathrm{Lip}_\omega(\mathbb{R}_+,X)}.$$

Similarly if $F \in \mathrm{Lip}_0(\mathbb{R}_+, X)$ and

$$(J_\omega F)(t) := \int_0^t e^{\omega s}\, dF(s),$$

then $J_\omega F \in \mathrm{Lip}_\omega(\mathbb{R}_+, X)$ and $\|J_\omega F\|_{\mathrm{Lip}_\omega(\mathbb{R}_+,X)} \le \|F\|_{\mathrm{Lip}_0(\mathbb{R}_+,X)}$. Moreover, $J_\omega I_\omega G = G$ and $I_\omega J_\omega F = F$, by Proposition 1.9.10. Hence, I_ω is an isometric isomorphism of $\mathrm{Lip}_\omega(\mathbb{R}_+, X)$ onto $\mathrm{Lip}_0(\mathbb{R}_+, X)$.

Note that if $G \in L^\infty_\omega(\mathbb{R}_+, X)$ then $\omega(G) \le \omega$ and $\mathrm{abs}(dG) \le \omega$ by Theorem 1.10.5. Thus, the Laplace-Stieltjes transform

$$(\mathcal{L}_{S,\omega} G)(\lambda) := \widehat{dG}(\lambda) = \int_0^\infty e^{-\lambda t}\, dG(t)$$

exists for $\lambda > \omega$. By Proposition 1.10.3,

$$(\mathcal{L}_{S,\omega} G)(\lambda) = (\mathcal{L}_S I_\omega G)(\lambda - \omega). \tag{2.7}$$

Let

$$C^\infty_W((\omega,\infty), X) := \Big\{ r \in C^\infty((\omega,\infty), X) : \\ \|r\|_W := \sup_{\lambda>\omega, k\in\mathbb{N}_0} \frac{(\lambda-\omega)^{k+1}}{k!} \|r^{(k)}(\lambda)\| < \infty \Big\}.$$

This is a Banach space, and it is clear that the shift $S_\omega : r \mapsto r(\cdot - \omega)$ is an isometric isomorphism of $C^\infty_W((0,\infty), X)$ onto $C^\infty_W((\omega,\infty), X)$. The equation (2.7) may be written as $\mathcal{L}_{S,\omega} = S_\omega \circ \mathcal{L}_S \circ I_\omega$.

Now we can give the following reformulation of the Real Representation Theorem 2.2.1.

Theorem 2.4.1. *Let $\omega \in \mathbb{R}$. The Laplace-Stieltjes transform is an isometric isomorphism of $\mathrm{Lip}_\omega(\mathbb{R}_+, X)$ onto $C^\infty_W((\omega,\infty), X)$. In particular, for $M > 0$ and $r \in C^\infty_W((\omega,\infty), X)$, the following are equivalent:*

(i) $\|(\lambda-\omega)^{k+1}\frac{1}{k!}r^{(k)}(\lambda)\| \le M \quad (\lambda > \omega,\ k \in \mathbb{N}_0)$.

(ii) *There exists $G : \mathbb{R}_+ \to X$ satisfying $G(0) = 0$ and $\|G(t+h) - G(t)\| \le M \int_t^{t+h} e^{\omega r}\, dr \quad (t, h \ge 0)$, such that $r(\lambda) = \int_0^\infty e^{-\lambda t}\, dG(t)$ for all $\lambda > \omega$.*

Proposition 1.6.1 a) gives

$$\mathcal{L}_\omega = S_\omega \circ \mathcal{L} \circ M_\omega$$

where $\mathcal{L}$ and $\mathcal{L}_\omega$ are the Laplace transforms on $L^\infty(\mathbb{R}_+, X)$ and $L^\infty_\omega(\mathbb{R}_+, X)$. Hence Theorem 2.2.3 can be reformulated as follows.

Theorem 2.4.2. *Let $M > 0$, $\omega \in \mathbb{R}$. If X has the Radon-Nikodym property then for any $r \in C^\infty_W((\omega,\infty), X)$ the following are equivalent:*

(i) $\|(\lambda-\omega)^{k+1}\frac{1}{k!}r^{(k)}(\lambda)\| \le M \quad (\lambda > \omega,\ k \in \mathbb{N}_0)$.

(ii) *There exists $g \in L^1_{loc}(\mathbb{R}_+, X)$ with $\|g(t)\| \le Me^{\omega t}$ for almost all $t \ge 0$ such that $r(\lambda) = \int_0^\infty e^{-\lambda t} g(t)\, dt$ for all $\lambda > \omega$.*

As in Theorem 2.1.1 one shows that there exists an isometric isomorphism $\Phi_{S,\omega}$ between the spaces $\mathrm{Lip}_\omega(\mathbb{R}_+, X)$ and $\mathcal{L}(L^1_\omega(\mathbb{R}_+), X)$, where

$$L^1_\omega(\mathbb{R}_+) := \left\{ h \in L^1_{loc}(\mathbb{R}_+) : \ \|h\|_{\omega,1} := \int_0^\infty e^{\omega t} |h(t)| \, dt < \infty \right\}.$$

The isomorphism $\Phi_{S,\omega}$ assigns to every function $G \in \mathrm{Lip}_\omega(\mathbb{R}_+, X)$ an operator $T \in \mathcal{L}(L^1_\omega(\mathbb{R}_+), X)$ with $\|T\| = \|G\|_{\mathrm{Lip}_\omega(\mathbb{R}_+,X)}$ such that

$$Th = \int_0^\infty h(t) \, dG(t)$$

for all continuous functions $h \in L^1_\omega(\mathbb{R}_+)$, $T\chi_{[0,t]} = G(t)$ for all $t \geq 0$, and $Te_{-\lambda} = \widehat{dG}(\lambda)$ if $\mathrm{Re}\,\lambda > \omega$.

The inversion theorems in Section 2.3 all remain valid, with almost no changes in the proofs (the version of Theorem 2.3.4 for $\mathrm{Lip}_\omega(\mathbb{R}_+, X)$ can be deduced directly from the case $\omega = 0$ by using the isomorphism I_ω). Thus, if $r = \widehat{dF}$ for some $F \in \mathrm{Lip}_\omega(\mathbb{R}_+, X)$, then

$$F(t) = \lim_{k\to\infty} (-1)^k \frac{1}{k!} \left(\frac{k}{t}\right)^{k+1} \frac{d^k}{d\lambda^k} \left(\frac{r(\lambda)}{\lambda}\right)\Bigg|_{\lambda=k/t}. \tag{2.8}$$

If $c > \max(\omega, 0)$, then

$$F(t) = \lim_{k\to\infty} \frac{1}{2\pi i} \int_{c-ik}^{c+ik} e^{\lambda t} \frac{r(\lambda)}{\lambda} \, d\lambda, \tag{2.9}$$

where the limit exists uniformly on compact subsets of $\mathbb{R}_+$. Finally,

$$F(t) = \lim_{k\to\infty} \sum_{j=1}^\infty (-1)^{j+1} \frac{1}{j!} e^{tkj} r(kj), \tag{2.10}$$

where the limit exists uniformly on $\mathbb{R}_+$.

The following is a consequence of the Phragmén-Doetsch inversion (2.10).

Proposition 2.4.3. *Let* $\varepsilon > 0$ *and* $f \in L^1_{loc}(\mathbb{R}_+, X)$ *with* $\mathrm{abs}(f) < \infty$. *The following are equivalent.*

(i) $\limsup_{\lambda\to\infty} \frac{1}{\lambda} \log \|\hat{f}(\lambda)\| \leq -\varepsilon$.

(ii) $f = 0$ *a.e. on* $[0, \varepsilon]$.

Proof. Let $F(t) := \int_0^t f(s) \, ds$ and $G(t) := \int_0^t F(s) \, ds$. Since $\mathrm{abs}(f) < \infty$, $\omega(F) < \infty$ by Theorem 1.4.3 and hence $G \in \mathrm{Lip}_\omega(\mathbb{R}_+, X)$ for some $\omega \in \mathbb{R}$. By Corollary 1.6.5 and Proposition 1.10.1,

$$\hat{f}(\lambda) = \lambda \widehat{F}(\lambda) = \lambda \widehat{dG}(\lambda) = \lambda^2 \widehat{G}(\lambda)$$

for $\operatorname{Re}\lambda > \omega$. Define

$$r(\lambda) := \frac{1}{\lambda}\hat{f}(\lambda) = \widehat{F}(\lambda) = \widehat{dG}(\lambda)$$

for $\lambda > \omega$. If (i) holds, then $\limsup_{\lambda\to\infty} \frac{1}{\lambda}\log\|r(\lambda)\| \le -\varepsilon$. Let $0 < \xi < \varepsilon$. Then there exist $M, \lambda_0 > 0$ such that $\|r(\lambda)\| \le Me^{-\lambda\xi}$ for all $\lambda > \lambda_0$. Let $t \in [0, \xi)$. Then, for $\lambda_0 < k \in \mathbb{N}$,

$$\left\| \sum_{j=1}^{\infty} \frac{(-1)^{j+1}}{j!} e^{tkj} r(kj) \right\| \le M \sum_{j=1}^{\infty} \frac{1}{j!} e^{(t-\xi)kj} = M\left(e^{e^{(t-\xi)k}} - 1\right) \to 0$$

as $k \to \infty$. Since $r = \widehat{dG}$, it follows from (2.10) that $G = 0$ on $[0, \xi)$ for all $0 < \xi < \varepsilon$. Thus, $G = 0$ on $[0, \varepsilon]$ and hence $f = 0$ a.e. on $[0, \varepsilon]$, by Proposition 1.2.2. This proves that (i) $\Rightarrow$ (ii).

Suppose that (ii) holds. Then $F = 0$ on $[0, \varepsilon]$. Thus

$$r(\lambda) = \int_0^\infty e^{-\lambda t} F(t)\, dt = \int_\varepsilon^\infty e^{-\lambda t} F(t)\, dt = e^{-\lambda\varepsilon} \int_0^\infty e^{-\lambda t} F(t+\varepsilon)\, dt.$$

Since $t \mapsto F(t+\varepsilon)$ is exponentially bounded, it follows that $\| \int_0^\infty e^{-\lambda t} F(t+\varepsilon)\, dt\| \le C$ for some $C > 0$ and therefore $\|r(\lambda)\| \le Ce^{-\varepsilon\lambda}$ for all sufficiently large λ. This proves that (ii) $\Rightarrow$ (i). ☐

If $f \in L^1_{loc}(\mathbb{R}_+, X)$ with $\operatorname{abs}(f) < \infty$, then it follows from Corollary 1.6.5 and the exponential boundedness of F that there exist $M, \lambda_0 > 0$ such that $\|\hat{f}(\lambda)\| \le M$ for all $\lambda > \lambda_0$. Thus, $\limsup_{\lambda\to\infty} \frac{1}{\lambda}\log\|\hat{f}(\lambda)\| \le 0$. This and the previous proposition yield the following corollary.

Corollary 2.4.4. *Let* $f \in L^1_{loc}(\mathbb{R}_+, X)$ *with* $\operatorname{abs}(f) < \infty$. *Then the following are equivalent:*

(i) $\limsup_{\lambda\to\infty} \frac{1}{\lambda}\log\|\hat{f}(\lambda)\| = 0$.

(ii) *For every* $\varepsilon > 0$, *the restriction of* f *to* $[0, \varepsilon]$ *does not vanish a.e.*

2.5 Complex Conditions

It was shown in the previous section that a holomorphic function $q : \{\operatorname{Re}\lambda > \omega\} \to X$ has a Laplace-Stieltjes or multiplied Laplace representation

$$q(\lambda) = \int_0^\infty e^{-\lambda t}\, dF(t) = \lambda \int_0^\infty e^{-\lambda t} F(t)\, dt$$

if there exists a constant $M > 0$ such that the Taylor coefficients $\frac{1}{k!}q^{(k)}(\lambda)$ are bounded by $M/(\lambda-\omega)^{k+1}$ for all $\lambda > \omega$ and $k \in \mathbb{N}_0$. Since only properties of the function q along the real half-line (ω, ∞) are involved, Widder's growth conditions are also referred to as "real conditions". In many instances, these real conditions are too difficult to be checked because all derivatives of q have to be considered, whereas the growth of q in a complex half-plane $\operatorname{Re}\lambda > \omega$ can be estimated. In these cases one can apply the following representation theorem.

Theorem 2.5.1 (Complex Representation). *Let $\omega \geq 0$, let $q : \{\operatorname{Re}\lambda > \omega\} \to X$ be a holomorphic function with $\sup_{\operatorname{Re}\lambda>\omega} \|\lambda q(\lambda)\| < \infty$ and let $b > 0$. Then there exists $f \in C(\mathbb{R}_+, X)$ with $\sup_{t>0} \|e^{-\omega t} t^{-b} f(t)\| < \infty$ such that $q(\lambda) = \lambda^b \hat{f}(\lambda)$ for $\operatorname{Re}\lambda > \omega$.*

Proof. Let $\alpha > \omega$ and define

$$f(t) := \lim_{R\to\infty} \frac{1}{2\pi i} \int_{\alpha-iR}^{\alpha+iR} e^{\lambda t} \frac{q(\lambda)}{\lambda^b}\, d\lambda = \frac{1}{2\pi} \int_{-\infty}^{\infty} e^{(\alpha+ir)t} \frac{q(\alpha+ir)}{(\alpha+ir)^b}\, dr.$$

Observe that the latter integral is absolutely convergent, by the assumption on q, so the limit exists uniformly for t in compact subsets of $\mathbb{R}_+$. Hence, f is continuous on $\mathbb{R}_+$. By applying Cauchy's theorem over rectangles with vertices $\alpha \pm iR$, $\beta \pm iR$, and using the assumption on q, it is easy to see that the definition of f is independent of $\alpha > \omega$.

For $\alpha > \omega$ and $R > 0$, let $\Gamma_{\alpha,R}$ be the path consisting of the vertical half-line $\{\alpha + ir : r < -R\}$, the semicircle $\{\alpha + Re^{i\theta} : \frac{-\pi}{2} \leq \theta \leq \frac{\pi}{2}\}$, and the half-line $\{\alpha + ir : r > R\}$. By Cauchy's theorem,

$$\begin{aligned} f(t) &= \frac{1}{2\pi i} \int_{\Gamma_{\alpha,R}} e^{\lambda t} \frac{q(\lambda)}{\lambda^b}\, d\lambda \\ &= \frac{1}{2\pi} \int_{-\infty}^{-R} e^{(\alpha+ir)t} \frac{q(\alpha+ir)}{(\alpha+ir)^b}\, dr \\ &\quad + \frac{1}{2\pi} \int_{-\pi/2}^{\pi/2} e^{(\alpha+Re^{i\theta})t} \frac{q(\alpha+Re^{i\theta})}{(\alpha+Re^{i\theta})^b} Re^{i\theta}\, d\theta \\ &\quad + \frac{1}{2\pi} \int_{R}^{\infty} e^{(\alpha+ir)t} \frac{q(\alpha+ir)}{(\alpha+ir)^b}\, dr. \end{aligned}$$

Hence,

$$\begin{aligned} \|f(t)\| &\leq \frac{Me^{\alpha t}}{\pi} \int_R^\infty \frac{dr}{r^{b+1}} + \frac{M}{2\pi} \int_{-\pi/2}^{\pi/2} \frac{e^{(\alpha+R\cos\theta)t}}{R^b}\, d\theta \\ &= \frac{Me^{\alpha t}}{\pi b R^b} + \frac{Me^{\alpha t}}{\pi R^b} \int_0^{\pi/2} e^{Rt\cos\theta}\, d\theta, \end{aligned}$$

where $M := \sup_{\operatorname{Re}\lambda>\omega} \|\lambda q(\lambda)\|$. Choosing $R = 1/t$, we obtain that $\|f(t)\| \leq Ct^b e^{\alpha t}$ for some C independent of $\alpha > \omega$. Hence, $\|f(t)\| \leq Ct^b e^{\omega t}$.

Given λ with $\operatorname{Re}\lambda > \omega$, choose $\omega < \alpha < \operatorname{Re}\lambda$. By the dominated convergence theorem and Fubini's theorem,

$$\begin{aligned}\int_0^\infty e^{-\lambda t} f(t)\,dt &= \lim_{R\to\infty} \int_0^\infty e^{-\lambda t} \frac{1}{2\pi i} \int_{\alpha-iR}^{\alpha+iR} e^{zt} \frac{q(z)}{z^b}\,dz\,dt \\ &= \lim_{R\to\infty} \frac{1}{2\pi i} \int_{\alpha-iR}^{\alpha+iR} \frac{q(z)}{(\lambda-z)z^b}\,dz.\end{aligned}$$

By Cauchy's residue theorem around the path consisting of the semicircle $\{\alpha + Re^{i\theta} : -\pi/2 \le \theta \le \pi/2\}$ and the line-segment $\{\alpha + ir : -R \le r \le R\}$,

$$\begin{aligned}\frac{1}{2\pi i} \int_{\alpha-iR}^{\alpha+iR} \frac{q(z)}{(\lambda-z)z^b}\,dz &= \frac{1}{2\pi} \int_{-\pi/2}^{\pi/2} \frac{q(\alpha+Re^{i\theta})Re^{i\theta}}{(\lambda-\alpha-Re^{i\theta})(\alpha+Re^{i\theta})^b}\,d\theta + \frac{q(\lambda)}{\lambda^b} \\ &\to \frac{q(\lambda)}{\lambda^b}\end{aligned}$$

as $R \to \infty$, using the assumption on q. □

We mention that Theorem 2.5.1 does not hold for $b = 0$. In fact, Desch and Prüss [DP93] construct a scalar-valued holomorphic function q on $\mathbb{C}_+$ satisfying

$$\sup_{\operatorname{Re}\lambda>0} \|q(\lambda)\|(1+|\lambda|) < \infty$$

such that q is not the Laplace transform of a function $f \in L^\infty_{loc}(0,\infty)$.

On the other hand, if $\lambda q(\lambda)$ and $\lambda^2 q'(\lambda)$ are bounded on the right half-plane, then q is the Laplace transform of a bounded continuous function, as we show in the following corollary.

Corollary 2.5.2 (Prüss). *Let $q : \{\operatorname{Re}\lambda > 0\} \to X$ be holomorphic. If there exists $M > 0$ such that $\|\lambda q(\lambda)\| \le M$ and $\|\lambda^2 q'(\lambda)\| \le M$ for $\operatorname{Re}\lambda > 0$, then there exists a bounded function $f \in C((0,\infty), X)$ such that $q(\lambda) = \int_0^\infty e^{-\lambda t} f(t)\,dt$ for $\operatorname{Re}\lambda > 0$. In particular, $q \in C_W^\infty((0,\infty), X)$.*

Proof. It follows from Theorem 2.5.1 that there are functions $f_i \in C(\mathbb{R}_+, X)$ ($i = 0, 1$) and $C > 0$ such that $\|f_i(t)\| \le Ct$ for $t > 0$,

$$q(\lambda) = \lambda \int_0^\infty e^{-\lambda t} f_0(t)\,dt, \quad \text{and} \quad \lambda q'(\lambda) = \lambda \int_0^\infty e^{-\lambda t} f_1(t)\,dt$$

for $\operatorname{Re}\lambda > 0$. By Theorem 1.5.1,

$$q'(\lambda) = \int_0^\infty e^{-\lambda t} f_0(t)\,dt - \lambda \int_0^\infty e^{-\lambda t} t f_0(t)\,dt = \int_0^\infty e^{-\lambda t} f_1(t)\,dt.$$

Integration by parts (or Corollary 1.6.5) yields

$$\lambda \int_0^\infty e^{-\lambda t} \left(\int_0^t f_0(s)\,ds - t f_0(t) \right) dt = \lambda \int_0^\infty e^{-\lambda t} \int_0^t f_1(s)\,ds\,dt.$$

Since the Laplace transform is one-to-one, it follows that $tf_0(t) = \int_0^t f_0(s)\,ds - \int_0^t f_1(s)\,ds$. Thus, $f_0 \in C^1((0,\infty),X)$ and $tf_0'(t) = -f_1(t)$. Therefore, $\|f_0'(t)\| \le C$ for all $t > 0$ and

$$q(\lambda) = \lambda \int_0^\infty e^{-\lambda t} f_0(t)\,dt = \int_0^\infty e^{-\lambda t} f_0'(t)\,dt \quad (\mathrm{Re}\,\lambda > 0). \qquad \square$$

Remark 2.5.3. If $f \in L^\infty((0,\infty),X)$, then $r = \hat{f}$ is holomorphic on the right half-plane and

$$\begin{aligned} \|\lambda r(\lambda)\| &\le \frac{|\lambda|}{\mathrm{Re}\,\lambda}\|f\|_\infty, \\ \|\lambda^2 r'(\lambda)\| &\le \left(\frac{|\lambda|}{\mathrm{Re}\,\lambda}\right)^2 \|f\|_\infty \qquad (\mathrm{Re}\,\lambda > 0). \end{aligned}$$

In particular, $\lambda r(\lambda)$ and $\lambda^2 r'(\lambda)$ are bounded on each sector $\Sigma_\alpha = \{re^{i\gamma} : r > 0,\ |\gamma| < \alpha\}$ where $\alpha \in (0,\pi/2)$. In Corollary 2.5.2 the estimate is required uniformly on the right half-plane, which is more. On the other hand, continuity is obtained as an additional result. $\square$

We close this section with a characterization of Laplace transforms of functions in $L^1_{loc}(\mathbb{R}_+,X)$ with $\|f(t)\| \le Mt^n$ for some $M, n \ge 0$ and almost all $t \ge 0$ (if X has the Radon-Nikodym property) or the Laplace-Stieltjes transforms of functions $H : \mathbb{R}_+ \to X$ with $H(0) = 0$ and $\|H(t) - H(s)\| \le M\int_s^t r^n\,dr$ for some $M > 0$ and all $0 \le s \le t$ (for general X).

Corollary 2.5.4. *Let $M > 0$, $n \in \mathbb{N}_0$, and $r \in C^\infty((0,\infty),X)$. The following are equivalent:*

(i) $\|\frac{\lambda^{k+n+1}}{(k+n)!} r^{(k)}(\lambda)\| \le M \quad (\lambda > 0,\ k \in \mathbb{N}_0)$.

(ii) *There exists $H : \mathbb{R}_+ \to X$ satisfying $H(0) = 0$ and $\|H(t) - H(s)\| \le M\int_s^t r^n\,dr$ $(0 \le s \le t)$, such that $r(\lambda) = \int_0^\infty e^{-\lambda t}\,dH(t)$ for all $\lambda > 0$.*

Proof. By the Real Representation Theorem 2.2.1, the statement holds for $n = 0$. Therefore, let $n \ge 1$. To show that (i) implies (ii), define

$$m(\lambda) := (-1)^n \int_\lambda^\infty \frac{1}{(n-1)!}(u-\lambda)^{n-1} r(u)\,du$$

for $\lambda > 0$. Then, $m^{(k)}(\lambda) = r^{(k-n)}(\lambda)$ for all $k \ge n$ and $\lambda > 0$. Since $\|\frac{\lambda^{k+1}}{k!} m^{(k)}(\lambda)\| = \|\frac{\lambda^{k+1}}{k!} r^{(k-n)}(\lambda)\| \le M$ for all $\lambda > 0$ and $k \ge n$, it follows from Proposition 2.2.2 that $m \in C_W^\infty((0,\infty),X)$ and $\|m\|_W \le M$. By Theorem 2.2.1, there exists $G : \mathbb{R}_+ \to X$ with $G(0) = 0$ and $\|G(t) - G(s)\| \le M|t-s|$ for all $t, s \ge 0$ such that $m(\lambda) = \int_0^\infty e^{-\lambda t}\,dG(t)$ for all $\lambda > 0$. By Theorem 1.5.1 and Proposition 1.9.10,

$$r(\lambda) = m^{(n)}(\lambda) = \int_0^\infty e^{-\lambda t}(-t)^n\,dG(t) = \int_0^\infty e^{-\lambda t}\,dH(t),$$

where $H(t) := \int_0^t (-s)^n \, dG(s)$. Now the statement (ii) follows from $\|H(t) - H(s)\| = \| \int_s^t (-r)^n \, dG(r)\| \le M \int_s^t r^n \, dr$ for all $0 \le s \le t$.

To show that (ii) implies (i), let $x^* \in X^*$. The function $x^* \circ H$ is locally Lipschitz continuous, hence absolutely continuous and differentiable a.e. If $h(t) := \frac{d}{dt}\langle H(t), x^*\rangle$, then $|h(t)| \le Mt^n \|x^*\|$ and $\langle r(\lambda), x^*\rangle = \int_0^\infty e^{-\lambda t} h(t)\, dt$, by Proposition 1.9.11. Hence,

$$\begin{aligned} \left| \left\langle \frac{\lambda^{k+n+1}}{(k+n)!} r^{(k)}(\lambda), x^* \right\rangle \right| &= \left| \frac{\lambda^{k+n+1}}{(k+n)!} \int_0^\infty e^{-\lambda t} (-t)^k h(t)\, dt \right| \\ &\le M\|x^*\|. \end{aligned}$$

Now (i) follows from the Hahn-Banach theorem. □

2.6 Laplace Transforms of Holomorphic Functions

In this section those functions are characterized which are Laplace transforms of holomorphic, exponentially bounded functions defined on some open sector $\Sigma_\alpha := \{re^{i\gamma} : r > 0, -\alpha < \gamma < \alpha\}$ for some $0 < \alpha \le \pi/2$. The closure of Σ_α is denoted by $\overline{\Sigma}_\alpha$. We shall use the same notation for $0 < \alpha < \pi$. Note that $\Sigma_{\frac{\pi}{2}} = \mathbb{C}_+ := \{\operatorname{Re} \lambda > 0\}$.

Theorem 2.6.1 (Analytic Representation). *Let $0 < \alpha \le \frac{\pi}{2}$, $\omega \in \mathbb{R}$ and $q : (\omega, \infty) \to X$. The following are equivalent:*

(i) *There is a holomorphic function $f : \Sigma_\alpha \to X$ such that $\sup_{z \in \Sigma_\beta} \|e^{-\omega z} f(z)\| < \infty$ for all $0 < \beta < \alpha$ and $q(\lambda) = \hat{f}(\lambda)$ for all $\lambda > \omega$.*

(ii) *The function q has a holomorphic extension $\tilde{q} : \omega + \Sigma_{\alpha + \frac{\pi}{2}} \to X$ such that $\sup_{\lambda \in \omega + \Sigma_{\gamma + \frac{\pi}{2}}} \|(\lambda - \omega)\tilde{q}(\lambda)\| < \infty$ for all $0 < \gamma < \alpha$.*

Proof. Assume that (i) holds. Let $0 < \beta < \alpha$. Then there exists $M > 0$ such that $\|f(z)\| \le M|e^{\omega z}|$ for all $z \in \overline{\Sigma}_\beta \setminus \{0\}$. Define paths $\Gamma_\pm$ by $\Gamma_\pm := \{se^{\pm i\beta} : 0 \le s < \infty\}$. By Cauchy's theorem,

$$\begin{aligned} q(\lambda) &= \int_0^\infty e^{-\lambda t} f(t)\, dt = \int_{\Gamma_\pm} e^{-\lambda z} f(z) dz \\ &= e^{\pm i\beta} \int_0^\infty e^{-\lambda s e^{\pm i\beta}} f(se^{\pm i\beta})\, ds \qquad (2.11) \end{aligned}$$

for all $\lambda > \omega$. Let $0 < \varepsilon < \frac{\pi}{2} - \beta$, and let $\lambda \in \mathbb{C}$ with $-\frac{\pi}{2} - \beta + \varepsilon < \arg(\lambda - \omega) < \frac{\pi}{2} - \beta - \varepsilon$. Then $-\frac{\pi}{2} + \varepsilon < \arg((\lambda - \omega)e^{i\beta}) < \frac{\pi}{2} - \varepsilon$, so $\operatorname{Re}((\lambda - \omega)e^{i\beta}) \ge |\lambda - \omega| \sin \varepsilon$. It follows that

$$\|e^{-\lambda s e^{i\beta}} f(se^{i\beta})\| \le Me^{-s|\lambda - \omega| \sin \varepsilon}.$$

Consequently, the integral

$$q_+(\lambda) := e^{i\beta} \int_0^\infty e^{-\lambda s e^{i\beta}} f(se^{i\beta})\, ds$$

is absolutely convergent and defines a holomorphic function in the region $-\frac{\pi}{2} - \beta + \varepsilon < \arg(\lambda - \omega) < \frac{\pi}{2} - \beta - \varepsilon$, with $\|(\lambda - \omega)q_+(\lambda)\| \leq M/\sin\varepsilon$. Similarly,

$$q_-(\lambda) := e^{-i\beta} \int_0^\infty e^{-\lambda s e^{-i\beta}} f(se^{-i\beta})\, ds$$

defines a holomorphic function in the region $-\frac{\pi}{2} + \beta + \varepsilon < \arg(\lambda - \omega) < \frac{\pi}{2} + \beta - \varepsilon$, with $\|(\lambda - \omega)q_-(\lambda)\| \leq M/\sin\varepsilon$. By (2.11), both q_+ and q_- are extensions of q, and together they define a holomorphic extension $\tilde{q}$ to $\omega + \Sigma_{\frac{\pi}{2}+\beta-\varepsilon}$, satisfying $\|(\lambda - \omega)\tilde{q}(\lambda)\| \leq M/\sin\varepsilon$ in the sector. Since $\beta < \alpha$ and $0 < \varepsilon < \frac{\pi}{2} - \beta$ are arbitrary, this proves (ii).

Assume that (ii) holds. Let $0 < \gamma < \alpha$ and $\delta > 0$. There exists $M > 0$ such that $\|(\lambda - \omega)\tilde{q}(\lambda)\| \leq M$ for all $\lambda \in (\omega + \overline{\Sigma}_{\gamma+\frac{\pi}{2}}) \setminus \{\omega\}$. Consider an oriented path Γ (depending on γ and δ) consisting of

$$\Gamma_\pm := \left\{\omega + re^{\pm i(\gamma+\pi/2)} : \delta \leq r\right\} \quad \text{and} \quad \Gamma_0 := \left\{\omega + \delta e^{i\theta} : -\gamma - \tfrac{\pi}{2} \leq \theta \leq \gamma + \tfrac{\pi}{2}\right\}.$$

Let $0 < \varepsilon < \gamma$ and consider $z \in \Sigma_{\gamma-\varepsilon}$. For $\lambda = \omega + re^{\pm i(\gamma+\pi/2)} \in \Gamma_\pm$,

$$\begin{aligned} \operatorname{Re}(\lambda z) &= \omega \operatorname{Re} z + r|z| \cos(\arg z \pm (\gamma + \pi/2)) \\ &\leq \omega \operatorname{Re} z - r|z| \sin\varepsilon. \end{aligned}$$

Hence,

$$\|e^{\lambda z}\tilde{q}(\lambda)\| \leq e^{\omega \operatorname{Re} z} e^{-r|z|\sin\varepsilon} \frac{M}{r} \quad (\lambda \in \Gamma_\pm). \tag{2.12}$$

This shows that

$$f(z) := \frac{1}{2\pi i} \int_\Gamma e^{\lambda z} \tilde{q}(\lambda)\, d\lambda \tag{2.13}$$

is absolutely convergent, uniformly for z in compact subsets of Σ_γ, and therefore defines a holomorphic function in Σ_γ. By Cauchy's theorem, this function is independent of $\delta > 0$, and also independent of $\gamma < \alpha$ so long as $\arg z < \gamma$ (here we use the assumption on $\tilde{q}$ to estimate the integral of $e^{\lambda z}\tilde{q}(\lambda)$ over arcs $\{\omega + Re^{i\theta} : \gamma_1 + \frac{\pi}{2} \leq |\theta| \leq \gamma_2 + \frac{\pi}{2}\}$). Hence (2.13) defines a holomorphic function $f \in \Sigma_\alpha$.

To estimate $f(z)$, we choose $\delta = |z|^{-1}$, and choose γ and ε such that $\gamma < \alpha$ and $|\arg z| < \gamma - \varepsilon$. On Γ_0, $\lambda = \omega + |z|^{-1}e^{i\theta}$ $(-\gamma - \pi/2 \leq \theta \leq \gamma + \pi/2)$, so

$$\begin{aligned} \left\| \frac{1}{2\pi i} \int_{\Gamma_0} e^{\lambda z} \tilde{q}(\lambda)\, d\lambda \right\| &\leq \frac{1}{2\pi} \int_{-\gamma-\pi/2}^{\gamma+\pi/2} e^{\omega \operatorname{Re} z} e^{\cos(\arg z + \theta)} M\, d\theta \\ &\leq M e^{1+\omega \operatorname{Re} z}. \end{aligned} \tag{2.14}$$

On $\Gamma_\pm$, $\lambda = \omega + re^{\pm i(\gamma+\pi/2)}$, and the estimate (2.12) gives

$$\begin{aligned} \left\| \frac{1}{2\pi i} \int_{\Gamma_\pm} e^{\lambda z} \tilde{q}(\lambda)\, d\lambda \right\| &\leq \frac{1}{2\pi} \int_{|z|^{-1}}^{\infty} e^{\omega \operatorname{Re} z} e^{-r|z| \sin \varepsilon} \frac{M}{r}\, dr \\ &= \frac{M e^{\omega \operatorname{Re} z}}{2\pi} \int_1^\infty \frac{e^{-r \sin \varepsilon}}{r}\, dr \\ &\leq \frac{M e^{\omega \operatorname{Re} z}}{2\pi \sin \varepsilon}. \end{aligned} \tag{2.15}$$

Now (2.14) and (2.15) establish that

$$\sup_{z \in \Sigma_{\gamma-\varepsilon}} \|e^{-\omega z} f(z)\| < \infty$$

for any $0 < \varepsilon < \gamma < \alpha$.

Next we will show that $\hat{f}(\lambda) = q(\lambda)$ whenever $\lambda > \omega$. Given such λ, choose $0 < \delta < \lambda - \omega$, and $0 < \gamma < \alpha$. Then λ is to the right of the path Γ, and Fubini's theorem and Cauchy's residue theorem imply that

$$\begin{aligned} \hat{f}(\lambda) &= \int_0^\infty e^{-\lambda t} \frac{1}{2\pi i} \int_\Gamma e^{\mu t} \tilde{q}(\mu)\, d\mu\, dt \\ &= \frac{1}{2\pi i} \int_\Gamma \frac{\tilde{q}(\mu)}{\lambda - \mu}\, d\mu \\ &= \tilde{q}(\lambda) + \lim_{R\to\infty} \frac{1}{2\pi i} \int_{\widetilde{\Gamma}_R} \frac{\tilde{q}(\mu)}{\lambda - \mu}\, d\mu, \end{aligned}$$

where $\widetilde{\Gamma}_R := \{\omega + Re^{i\theta} : \; -\gamma - \pi/2 \leq \theta \leq \gamma + \pi/2\}$. Then

$$\begin{aligned} \left\| \int_{\widetilde{\Gamma}_R} \frac{\tilde{q}(\mu)}{\lambda - \mu}\, d\mu \right\| &\leq \int_{-\gamma-\pi/2}^{\gamma+\pi/2} \frac{M}{|\omega + Re^{i\theta} - \lambda|}\, d\theta \\ &\to 0 \end{aligned}$$

as $R \to \infty$. This proves that $\hat{f}(\lambda) = q(\lambda)$. □

When f is as in Theorem 2.6.1 (i), it is an easy consequence of Cauchy's integral formula for derivatives that

$$\sup_{z \in \Sigma_\beta} \left\| z^k e^{-\omega z} f^{(k)}(z) \right\| < \infty$$

for all $0 < \beta < \alpha$.

Recall from Sections 1.4 and 1.5 that, for $f \in L^1_{loc}(\mathbb{R}_+, X)$,

$$\begin{aligned} \omega(f) &:= \inf \left\{ \omega \in \mathbb{R} : \sup_{t \geq 0} \|e^{-\omega t} f(t)\| < \infty \right\}, \\ \operatorname{abs}(f) &:= \inf \left\{ \operatorname{Re} \lambda : \; \hat{f}(\lambda) \; \text{ exists} \right\}, \\ \operatorname{hol}(\hat{f}) &:= \inf \left\{ \omega \in \mathbb{R} : \; \hat{f} \; \text{ has a holomorphic extension for } \; \operatorname{Re} \lambda > \omega \right\}. \end{aligned}$$

Moreover, $\mathrm{hol}(\hat{f}) \le \mathrm{abs}(f) \le \omega(f)$. We will now show that equalities hold when f is holomorphic and exponentially bounded on a sector.

Theorem 2.6.2. *Let $0 < \alpha < \pi/2$, let $f : \Sigma_\alpha \to X$ be holomorphic, and suppose that there exists $\omega \in \mathbb{R}$ such that $\sup_{z\in\Sigma_\alpha} \|e^{-\omega z} f(z)\| < \infty$. Then $\mathrm{hol}(\hat{f}) = \mathrm{abs}(f) = \omega(f)$.*

Proof. By Theorem 2.6.1, there exists $\gamma > 0$ such that $\hat{f}$ has a holomorphic extension (also denoted by $\hat{f}$) to $\omega + \Sigma_{\gamma+\pi/2}$ and $C := \sup_{\lambda\in\Sigma_{\gamma+\pi/2}} \|(\lambda-\omega)\hat{f}(\lambda)\| < \infty$. By definition of $\mathrm{hol}(\hat{f})$, $\hat{f}$ also has a holomorphic extension to $\mathrm{hol}(\hat{f}) + \Sigma_{\pi/2} = \{\mathrm{Re}\,\lambda > \mathrm{hol}(\hat{f})\}$.

Let $\omega' > \mathrm{hol}(\hat{f})$. There exists $\gamma' > 0$ such that

$$\omega' + \overline{\Sigma}_{\gamma'+\pi/2} \subseteq (\omega + \Sigma_{\gamma+\pi/2}) \cup (\mathrm{hol}(\hat{f}) + \Sigma_{\pi/2}).$$

Hence, $\hat{f}$ is holomorphic on $\omega' + \Sigma_{\gamma'+\pi/2}$ and continuous on the closure. Let

$$U := \left\{\lambda \in (\omega' + \overline{\Sigma}_{\gamma'+\pi/2}) \cap (\omega + \Sigma_{\gamma+\pi/2}) : \ |\lambda - \omega'| < 2|\lambda - \omega|\right\}.$$

If $\lambda \in U$, then $\|(\lambda - \omega')\hat{f}(\lambda)\| \le 2C$. Moreover, $(\omega' + \overline{\Sigma}_{\gamma'+\pi/2}) \setminus U$ is compact. Hence, $\sup_{\lambda\in\omega'+\Sigma_{\gamma'+\pi/2}} \|(\lambda-\omega')\hat{f}(\lambda)\| < \infty$. It follows from Theorem 2.6.1, and the fact that the Laplace transform is one-to-one, that $\sup_{z\in\Sigma_\beta} \|e^{-\omega' z} f(z)\| < \infty$ for some $\beta > 0$, and in particular, $\omega(f) \le \omega'$. Since this holds whenever $\omega' > \mathrm{hol}(\hat{f})$, it follows that $\omega(f) \le \mathrm{hol}(\hat{f})$, completing the proof. □

In the remainder of this section we will consider asymptotic behaviour of $f(t)$ as $t \to \infty$ and as $t \to 0$. In the case of holomorphic functions defined on a sector it can be described completely in terms of the Laplace transform. This is not the case in general, and in Chapter 4 a systematic treatment of this question will be given. However, here we can use contour arguments directly on the basis of the representation formula (2.13).

First we show that asymptotic behaviour along one ray and on the whole sector are equivalent. This is a consequence of Vitali's theorem (Theorem A.5).

Proposition 2.6.3. *Let $0 < \alpha \le \pi$ and let $f : \Sigma_\alpha \to X$ be holomorphic such that*

$$\sup_{z\in\Sigma_\beta} \|f(z)\| < \infty$$

for all $0 < \beta < \alpha$. Let $x \in X$.

a) *If $\lim_{t\to\infty} f(t) = x$, then $\lim_{\substack{|z|\to\infty \\ z\in\Sigma_\beta}} f(z) = x$ for all $0 < \beta < \alpha$.*

b) *If $\lim_{t\downarrow 0} f(t) = x$, then $\lim_{\substack{|z|\to 0 \\ z\in\Sigma_\beta}} f(z) = x$ for all $0 < \beta < \alpha$.*

Proof. a) Let $f_k(z) = f(kz)$. It follows from Vitali's theorem that $\lim_{k\to\infty} f_k(z) = x$ uniformly on compact subsets of Σ_α. Let $0 < \beta < \alpha$. Let $\varepsilon > 0$. There exists $k_0 \in \mathbb{N}$ such that $\|f_k(z) - x\| \le \varepsilon$ whenever $z \in \Sigma_\beta$, $1 \le |z| \le 2$, $k \ge k_0$. Let $z \in \Sigma_\beta$, $|z| \ge k_0$. Choose $k \in \mathbb{N}$ such that $k \le |z| < k+1$. Then

$$\|f(z) - x\| = \|f_k(z/k) - x\| \le \varepsilon.$$

This proves a).

b) This follows by applying a) to the function $z \mapsto f(z^{-1})$. □

Now we consider the asymptotic behaviour of $f(t)$ as $t \to \infty$ and $t \downarrow 0$.

Theorem 2.6.4 (Tauberian Theorem). *Consider the situation of Theorem* 2.6.1, *and let* $x \in X$.

a) *One has* $\lim_{t\downarrow 0} f(t) = x$ *if and only if* $\lim_{\lambda\to\infty} \lambda q(\lambda) = x$.

b) *Assume that* $\omega = 0$. *Then* $\lim_{t\to\infty} f(t) = x$ *if and only if* $\lim_{\lambda\downarrow 0} \lambda q(\lambda) = x$.

Proof. We can assume that $\omega = 0$ for both cases a) and b) by replacing $f(z)$ by $e^{-\omega z} f(z)$ otherwise. Replacing $f(t)$ by $f(t) - x$, we can also assume that $x = 0$. For simplicity, we shall denote the function $\tilde{q}$ of Theorem 2.6.1 by q.

Assume that $\lim_{\lambda\to\infty} \lambda q(\lambda) = x$. Let $0 < \gamma < \alpha$. By Proposition 2.6.3, $\lim_{\substack{|\lambda|\to\infty \\ \lambda\in\Sigma_{\gamma+\pi/2}}} \lambda q(\lambda) = x$. Let $\varepsilon > 0$. There exists $\delta_0 > 0$ such that $\|\lambda q(\lambda)\| \le \varepsilon$ whenever $|\lambda| \ge \delta_0$, $\lambda \in \Sigma_{\gamma+\frac{\pi}{2}}$. Let $0 < t \le 1/\delta_0$. Now we choose the contour Γ as in the proof of Theorem 2.6.1, (ii) $\Rightarrow$ (i), with $\delta = 1/t$. Then

$$\begin{aligned}
\left\|\frac{1}{2\pi i}\int_{\Gamma_0} e^{\lambda t} q(\lambda)\, d\lambda\right\| &= \left\|\frac{1}{2\pi i}\int_{-\gamma-\pi/2}^{\gamma+\pi/2} e^{e^{i\theta}} q\left(\frac{e^{i\theta}}{t}\right)\frac{ie^{i\theta}}{t}\, d\theta\right\| \\
&\le \frac{\varepsilon}{2\pi}\int_{-\gamma-\pi/2}^{\gamma+\pi/2} e^{\cos\theta}\, d\theta \\
&\le \varepsilon\, e,
\end{aligned}$$

and

$$\begin{aligned}
&\left\|\frac{1}{2\pi i}\int_{\Gamma_\pm} e^{\lambda t} q(\lambda)\, d\lambda\right\| \\
&= \left\|\frac{1}{2\pi i}\int_{1/t}^{\infty} e^{t\cdot re^{\pm i(\gamma+\pi/2)}} q(re^{\pm i(\gamma+\pi/2)}) re^{\pm i(\gamma+\pi/2)} \frac{dr}{r}\right\| \\
&= \left\|\frac{1}{2\pi i}\int_{1}^{\infty} e^{se^{\pm i(\gamma+\pi/2)}} q(\frac{s}{t}e^{\pm(\gamma+\pi/2)})\frac{s}{t}e^{\pm i(\gamma+\pi/2)} \frac{ds}{s}\right\| \\
&\to 0
\end{aligned}$$

as $t \downarrow 0$ by the dominated convergence theorem. It follows from the representation (2.13) that $\limsup_{t\downarrow 0} \|f(t)\| \le \varepsilon\, e$.

The converse implication is easy and does not depend on holomorphy. Assume that $\lim_{t\downarrow 0}\|f(t)\| = 0$. Let $\varepsilon > 0$. There exists $\tau > 0$ such that $\|f(t)\| \le \varepsilon$ for all $t \in [0,\tau]$. Then

$$\begin{aligned}\limsup_{\lambda\to\infty}\|\lambda q(\lambda)\| &\le \limsup_{\lambda\to\infty}\left\{\|\lambda\int_0^\tau e^{-\lambda t}f(t)\,dt\| + \|\lambda\int_\tau^\infty e^{-\lambda t}f(t)\,dt\|\right\}\\ &\le \varepsilon + \limsup_{\lambda\to\infty}\lambda\int_\tau^\infty e^{-\lambda t}Me^{\omega t}\,dt\\ &= \varepsilon + \limsup_{\lambda\to\infty} M\frac{\lambda}{\lambda-\omega}e^{-(\lambda-\omega)\tau} = \varepsilon,\end{aligned}$$

where $\omega > \omega(f)$ and M is suitable. This completes the proof of a).

The assertion b) is proved in the same way as a). □

2.7 Completely Monotonic Functions

Throughout this section, X will be an ordered Banach space with normal cone (see Appendix C). Let $f : \mathbb{R}_+ \to X$ be increasing. Then f is of bounded semivariation on each interval $[0,\tau]$, by Proposition 1.9.1. Assume that $\omega(f) < \infty$. Then the Laplace-Stieltjes transform

$$\widehat{df}(\lambda) = \lim_{\tau\to\infty}\int_0^\tau e^{-\lambda t}\,df(t) =: \int_0^\infty e^{-\lambda t}\,df(t) \tag{2.16}$$

converges on the half-plane $\{\operatorname{Re}\lambda > \operatorname{abs}(df)\}$, and defines a holomorphic function $\widehat{df}$ on $\{\operatorname{Re}\lambda > \operatorname{abs}(df)\}$. Recall from Theorem 1.10.5 that $\operatorname{abs}(df) < \infty$ if and only if $\omega(f) < \infty$.

Theorem 2.7.1. *Let $f : \mathbb{R}_+ \to X$ be an increasing function. Assume that $-\infty < \operatorname{abs}(df) < \infty$. Then $\operatorname{abs}(df)$ is a singularity of $\widehat{df}$.*

Proof. Replacing $f(t)$ by $\int_0^t e^{-\operatorname{abs}(df)s}\,df(s)$, we can assume that $\operatorname{abs}(df) = 0$. Assume that $\widehat{df}$ has a holomorphic extension to a neighbourhood of 0. Then there exists $\delta > 0$ such that

$$\widehat{df}(-\delta) = \sum_{n=0}^\infty (-1)^n(1+\delta)^n\frac{(\widehat{df})^{(n)}(1)}{n!}.$$

Let $x^* \in X_+^*$. Then

$$\langle\widehat{df}(-\delta), x^*\rangle = \sum_{n=0}^\infty \frac{(1+\delta)^n}{n!}\int_0^\infty e^{-t}t^n\,d\langle f(t), x^*\rangle.$$

Since all expressions are positive we may interchange the sum and the integral and obtain

$$\begin{aligned}\int_0^\infty e^{\delta t}\, d\langle f(t), x^*\rangle &= \int_0^\infty e^{-t} e^{(1+\delta)t}\, d\langle f(t), x^*\rangle \\ &= \sum_{n=0}^\infty \frac{(1+\delta)^n}{n!} \int_0^\infty e^{-t} t^n\, d\langle f(t), x^*\rangle \\ &= \langle \widehat{df}(-\delta), x^*\rangle < \infty.\end{aligned}$$

Since X_+^* spans X^* (see Proposition C.2), it follows that $\operatorname{abs}(x^* \circ f) \le -\delta$ for all $x^* \in X^*$. It follows from (1.25) that $\operatorname{abs}(df) \le -\delta$, which contradicts the assumption. □

Corollary 2.7.2. *Let $f \in L^1_{loc}(\mathbb{R}_+, X)$ such that $f(t) \ge 0$ a.e. Assume that $-\infty <$ $\operatorname{abs}(f) < \infty$. Then* $\operatorname{abs}(f)$ *is a singularity of $\hat{f}$. Hence,* $\operatorname{hol}(\hat{f}) = \operatorname{abs}(f)$.

Proof. This is immediate from Proposition 1.10.1 and Theorem 2.7.1. □

Our aim is to characterize functions of the form $\widehat{df}$ where $f : \mathbb{R}_+ \to X$ is increasing. Then

$$(-1)^n \widehat{df}^{(n)}(\lambda) = \int_0^\infty e^{-\lambda t} t^n\, df(t) \ge 0$$

for all $n \in \mathbb{N}_0$, $\lambda > \omega$. Thus $\widehat{df}$ is completely monotonic in the sense of the following definition.

Definition 2.7.3. *A function $r : (\omega, \infty) \to X$ is* completely monotonic *if r is infinitely differentiable and*

$$(-1)^n r^{(n)}(\lambda) \ge 0 \quad \textit{for all } \lambda > \omega,\ n \in \mathbb{N}_0. \tag{2.17}$$

In the following, we shall assume that $\omega = 0$ for simplicity (otherwise, we can replace $r(\lambda)$ by $r(\lambda + \omega)$ and $f(t)$ by $\int_0^t e^{-\omega s}\, df(s)$). Recall that by Theorem 1.10.5 $\operatorname{abs}(df) \le 0$ if and only if $\omega(f) \le 0$.

Definition 2.7.4. *We say that* Bernstein's theorem holds in X *if for every completely monotonic function $r : (0, \infty) \to X$ there exists an increasing function $f : \mathbb{R}_+ \to X$ such that $\omega(f) \le 0$ and $r(\lambda) = \widehat{df}(\lambda)$ for all $\lambda > 0$.*

Bernstein's theorem does hold in $X = \mathbb{R}$; this is just Bernstein's classical theorem from 1928 [Ber28]. Here we will prove it, as a special case of Theorem 2.7.7, with the help of the Real Representation Theorem 2.2.1.

Definition 2.7.5. *The space X has the* interpolation property *if, given two sequences $(x_n)_{n\in\mathbb{N}}$, $(y_n)_{n\in\mathbb{N}}$ in X such that*

$$x_n \le x_{n+1} \le y_{n+1} \le y_n \quad (n \in \mathbb{N}) \tag{2.18}$$

there exists $z \in X$ *such that*

$$x_n \leq z \leq y_n \quad \textit{for all } n \in \mathbb{N}. \tag{2.19}$$

Examples 2.7.6. a) *Assume that* $X = Y^*$ *where* Y *is an ordered Banach space with normal cone. Then* X *has the interpolation property.*

Proof. Let $x_n^* \leq x_{n+1}^* \leq y_{n+1}^* \leq y_n^*$ $(n \in \mathbb{N})$. Replacing x_n^* by $x_n^* - x_1^*$ and y_n^* by $y_n^* - x_1^*$ we can assume that $x_n^* \geq 0$. Define $z^* \in X^*$ by $\langle x, z^* \rangle = \sup_{n \in \mathbb{N}} \langle x, x_n^* \rangle$. Then z^* is linear and positive, and hence continuous (see Appendix C).

b) If X is reflexive, then X has the interpolation property. This follows from a).

c) Each von Neumann algebra (i.e., a $*$-subalgebra of $\mathcal{L}(H)$ which is closed in the strong operator topology, where H is a Hilbert space) has the interpolation property. This follows from a) and [Ped89, Theorem 4.6.17].

d) Every σ-order complete Banach lattice (i.e., a Banach lattice in which each countable order-bounded set has a supremum) has the interpolation property.

e) If X has order continuous norm (i.e., each decreasing positive sequence converges) then X has the interpolation property.

f) The space $C[0, 1]$ does not have the interpolation property.

See the Notes for further comments on the interpolation property. □

Now we can formulate the following characterization, which is the main result of this section.

Theorem 2.7.7. *Bernstein's theorem holds in* X *if and only if* X *has the interpolation property.*

The proof of Theorem 2.7.7 will be carried out in several steps. On the way we will prove a characterization of completely monotonic functions which is valid without restrictions on the space. First, we study convex functions.

Let $J \subset \mathbb{R}$ be an interval. A function $F : J \to X$ is called *convex* if

$$F(\lambda s + (1 - \lambda)t) \leq \lambda F(s) + (1 - \lambda)F(t)$$

for all $s, t \in J$, $0 < \lambda < 1$. Many order properties of convex functions carry over from the scalar case since for $x \in X$ we have

$$x \geq 0 \quad \text{if and only if} \quad \langle x, x^* \rangle \geq 0 \quad \text{for all } x^* \in X_+^*.$$

For example, a twice differentiable function F is convex if and only if $F'' \geq 0$.

Lemma 2.7.8. *Let* $[a, b]$ *be a closed interval in the interior of* J *and let* $F : J \to X$ *be convex. Then* F *is Lipschitz continuous on* $[a, b]$. *Moreover, if* $F(J) \subset X_+$ *and* $F(a) = 0$, *then* F *is increasing on* $[a, b]$.

Proof. Let $c < a$, $d > b$ such that $[c,d] \subset J$. Then for $a \le t < s \le b$,

$$\frac{F(a)-F(c)}{a-c} \le \frac{F(s)-F(t)}{s-t} \le \frac{F(d)-F(b)}{d-b}.$$

Since the cone is normal this implies that F is Lipschitz continuous on $[a,b]$. The second assertion is easy to see. □

We notice in particular that every convex function defined on an open interval is continuous.

Let $-\infty < a < b \le \infty$ and let $f : [a,b) \to X_+$ be increasing. Then f is Riemann integrable on $[a,t]$ whenever $a \le t < b$ (see Corollary 1.9.6). Let

$$F(t) := \int_a^t f(s)\,ds \quad (a \le t < b). \tag{2.20}$$

Then $F : [a,b) \to X_+$ is convex.

If X has the interpolation property, then the following converse result holds.

Proposition 2.7.9. *Assume that X has the interpolation property. Let $F : [a,b) \to X_+$ be convex such that $F(a) = 0$, where $-\infty < a < b \le \infty$. Then there exists an increasing function $f : [a,b) \to X_+$ such that (2.20) holds.*

Proof. The following two properties follow from convexity:

a) Let $a \le s < b$. Then the difference quotient

$$\frac{1}{h}\left(F(s+h) - F(s)\right)$$

is positive and increasing for $h \in (0, b-s)$.

b) Let $a \le s < s+h \le t < t+k < b$. Then

$$\frac{1}{h}\left(F(s+h) - F(s)\right) \le \frac{1}{k}\left(F(t+k) - F(t)\right). \tag{2.21}$$

Put $f(a) = 0$. It follows from the interpolation property, a) and b) that for each $t \in (a,b)$ there exists $f(t) \in X$ such that

$$\frac{1}{h}\left(F(s+h) - F(s)\right) \le f(t) \le \frac{1}{k}\left(F(t+k) - F(t)\right) \tag{2.22}$$

whenever $a \le s < s+h \le t < t+k < b$. It follows from (2.21) and (2.22) that $f : [a,b) \to X_+$ is increasing.

Let $G(t) := \int_a^t f(s)\,ds$. We show that $F = G$. Let $a < t < b$. Let $a \le t_0 < t_1 < \ldots < t_n = t$ be a partition of $[a,t]$. Setting $h_i := t_i - t_{i-1}$, we obtain from

(2.22) that

$$\begin{aligned}\sum_{i=1}^{n} f(t_{i-1})(t_i - t_{i-1}) &\leq \sum_{i=1}^{n} \frac{1}{h_i} \left(F(t_{i-1} + h_i) - F(t_{i-1})\right) h_i \\ &= \sum_{i=1}^{n} \left(F(t_i) - F(t_{i-1})\right) \\ &= F(t) - F(a) = F(t).\end{aligned}$$

It follows from the definition of the Riemann integral that $G(t) \leq F(t)$. Also by (2.22),

$$\begin{aligned}\sum_{i=1}^{n} f(t_i)(t_i - t_{i-1}) &\geq \sum_{i=1}^{n} \frac{F(t_i) - F(t_{i-1})}{t_i - t_{i-1}} (t_i - t_{i-1}) \\ &= F(t).\end{aligned}$$

Hence $G(t) \geq F(t)$. □

Next, we prove a converse version of Proposition 2.7.9.

Proposition 2.7.10. *Assume that for every convex function $F : \mathbb{R}_+ \to X_+$ with $F(0) = 0$ and $\omega(F) = 0$ there exists an increasing function $f : \mathbb{R}_+ \to X_+$ such that $F(t) = \int_0^t f(s)\, ds$ $(t \geq 0)$. Then X has the interpolation property.*

Proof. Let $x_n \leq x_{n+1} \leq y_{n+1} \leq y_n$ $(n \in \mathbb{N})$. We can assume that $x_1 \geq 0$ (replacing x_n by $x_n - x_1$ and y_n by $y_n - x_1$ otherwise). Define $f : \mathbb{R}_+ \to X$ by

$$f(t) := \begin{cases} x_n & \text{if } t \in [\frac{n-1}{n}, \frac{n}{n+1});\ n \geq 1, \\ y_n & \text{if } t \in [\frac{n+1}{n}, \frac{n}{n-1});\ n \geq 2, \\ y_1 & \text{if } t \in [2, \infty), \\ 0 & \text{if } t = 1. \end{cases}$$

Then $f \in L^1_{loc}(\mathbb{R}_+, X)$. Let $F(t) := \int_0^t f(s)\, ds$. Then $F : \mathbb{R}_+ \to X_+$ is convex and $F(0) = 0$. By assumption, there exists an increasing function $g : \mathbb{R}_+ \to X$ such that $F(t) = \int_0^t g(s)\, ds$ $(t \geq 0)$. Then

$$\frac{F(t-h) - F(t)}{-h} \leq g(t) \leq \frac{F(t+h) - F(t)}{h}$$

for all $t > 0$ and $h > 0$ small enough. It follows that $g(t) = F'(t)$ whenever F is differentiable at t. Consequently, $g(t) = x_n$ if $t \in (\frac{n-1}{n}, \frac{n}{n+1})$ and $g(t) = y_n$ if $t \in (\frac{n+1}{n}, \frac{n}{n-1})$. Hence, $x_n \leq g(1) \leq y_n$. Thus, $z := g(1)$ interpolates between the two sequences. □

For completeness, we also give the usual representation of convex functions as a corollary of Proposition 2.7.9.

Corollary 2.7.11. *Assume that X has the interpolation property. Let $F : (a,b) \to X$ be convex, and let $c \in (a,b)$. Then there exist $x \in X$ and an increasing function $f : (a,b) \to X$ such that*

$$F(t) = F(c) + (t-c)x + \int_c^t f(s)\,ds$$

for all $t \in (a,b)$.

Proof. We may assume that $c = 0$. It follows from convexity that

$$\frac{1}{t}\left(F(0) - F(-t)\right) \leq \frac{1}{s}\left(F(s) - F(0)\right)$$

whenever $0 < s < b$, $0 < t < -a$. Moreover, the left-hand difference quotient is decreasing in t, and the right-hand one is increasing in s. By the interpolation property, there exists $x \in X$ such that

$$\frac{1}{t}\left(F(0) - F(-t)\right) \leq x \leq \frac{1}{s}\left(F(s) - F(0)\right)$$

for all $0 < t < -a$, $0 < s < b$. In particular, the function

$$G(t) := F(t) - F(0) - tx \quad (t \in (a,b))$$

is positive, convex and satisfies $G(0) = 0$.

By Proposition 2.7.9, there exist increasing functions $f_1 : [0,b) \to X_+$ and $f_2 : [0,-a) \to X_+$ such that

$$\begin{aligned} G(t) &= \int_0^t f_1(s)\,ds \quad \text{for } t \in [0,b) \quad \text{and} \\ G(-t) &= \int_0^t f_2(s)\,ds \quad \text{for } t \in [0,-a). \end{aligned}$$

We can assume that $f_1(0) = f_2(0) = 0$. Let $f(t) := f_1(t)$ for $t \in [0,b)$ and $f(t) := -f_2(-t)$ for $t \in (a,0)$. Then f is increasing and $G(t) = \int_0^t f(s)\,ds$ for all $t \in (a,b)$. ☐

Now we will study completely monotonic functions. We need the following formulas (2.23) and (2.24) (the latter is merely needed for $n = 1$ and $n = 2$). In the remainder of this section we shall sometimes use loose notation such as $\frac{r(\lambda)}{\lambda}$ to denote the function $\lambda \mapsto \frac{r(\lambda)}{\lambda}$, and $\left(\frac{r(\lambda)}{\lambda}\right)'$ and $\left(\frac{r(\lambda)}{\lambda}\right)^{(n)}$ to denote its derivatives of orders 1 and n.

Lemma 2.7.12. *Let $r \in C^\infty((0,\infty), X)$. Then*

$$\frac{(-1)^n}{n!}\lambda^{n+1}\left(\frac{r(\lambda)}{\lambda}\right)^{(n)} = \sum_{m=0}^{n} \frac{(-1)^m}{m!}\lambda^m r^{(m)}(\lambda) \tag{2.23}$$

and

$$\left(\lambda^{k+n}\left(\frac{r(\lambda)}{\lambda^n}\right)^{(k)}\right)^{(n)} = \lambda^k r^{(k+n)}(\lambda) \tag{2.24}$$

for all $\lambda > 0$, $k, n \in \mathbb{N}_0$. *In particular, if* r *is completely monotonic, then* $\lambda \mapsto r(\lambda)/\lambda$ *is also completely monotonic.*

Proof. The first formula (2.23) is immediate from Leibniz's rule. It follows that if r is completely monotonic, then $\lambda \mapsto r(\lambda)/\lambda$ is also completely monotonic.

We show by induction over n that (2.24) holds for all $k \in \mathbb{N}_0$. It is obvious for $n = 0$. Moreover,

$$\begin{aligned} \lambda^k r^{(k+1)}(\lambda) = \lambda^k \left(\lambda \frac{r(\lambda)}{\lambda}\right)^{(k+1)} &= \lambda^k \left\{\lambda \left(\frac{r(\lambda)}{\lambda}\right)^{(k+1)} + (k+1)\left(\frac{r(\lambda)}{\lambda}\right)^{(k)}\right\} \\ &= \left(\lambda^{k+1}\left(\frac{r(\lambda)}{\lambda}\right)^{(k)}\right)' \end{aligned}$$

for $\lambda > 0$. This shows that (2.24) holds for $n = 1$.

Now assume that (2.24) holds for a fixed $n \in \mathbb{N}$ and $k \in \mathbb{N}_0$. Then, applying (2.24) to r' yields

$$\lambda^k r^{(k+n+1)}(\lambda) = \left(\lambda^{k+n}\left(\frac{r'(\lambda)}{\lambda^n}\right)^{(k)}\right)^{(n)} \tag{2.25}$$

for $\lambda > 0$. Observe that

$$\begin{aligned} &\left(\lambda^{k+n+1}(r(\lambda)/\lambda^{n+1})^{(k)}\right)' \\ &= \left(\lambda^n \cdot \lambda^{k+1}(r(\lambda)/\lambda^{n+1})^{(k)}\right)' \\ &= n\lambda^{n-1}\left(\lambda^{k+1}(r(\lambda)/\lambda^{n+1})^{(k)}\right) + \lambda^n\left(\lambda^{k+1}(r(\lambda)/\lambda^{n+1})^{(k)}\right)' \\ &= n\lambda^{n-1}\left(\lambda^{k+1}(r(\lambda)/\lambda^{n+1})^{(k)}\right) + \lambda^n\lambda^k(r(\lambda)/\lambda^n)^{(k+1)}, \end{aligned}$$

by applying (2.24) for $n = 1$ to the function $r(\lambda)/\lambda^n$ instead of r. Hence,

$$\begin{aligned} &\left(\lambda^{k+n+1}(r(\lambda)/\lambda^{n+1})^{(k)}\right)' \\ &= n\lambda^{n+k}\left(r(\lambda)/\lambda^{n+1}\right)^{(k)} + \lambda^{n+k}\left(r'(\lambda)/\lambda^n - nr(\lambda)/\lambda^{n+1}\right)^{(k)} \\ &= \lambda^{n+k}\left(r'(\lambda)/\lambda^n\right)^{(k)} \end{aligned}$$

for $\lambda > 0$. It follows from (2.25) that

$$\left(\lambda^{k+n+1}(r(\lambda)/\lambda^{n+1})^{(k)}\right)^{(n+1)} = \left(\lambda^{n+k}(r'(\lambda)/\lambda^n)^{(k)}\right)^{(n)} = \lambda^k r(\lambda)^{(k+n+1)}.$$

Thus, (2.24) holds when n is replaced by $n + 1$. □

Proposition 2.7.13. *Let $F \in \mathrm{Lip}_0(\mathbb{R}_+, X)$ and let*

$$r(\lambda) = \lambda \widehat{dF}(\lambda) = \lambda \int_0^\infty e^{-\lambda t}\, dF(t) \quad (\lambda > 0).$$

Then r is completely monotonic if and only if F is convex and $F(t) \geq 0$ $(t \geq 0)$.

Proof. Assume that r is completely monotonic. Note that $\frac{r(\lambda)}{\lambda} = \int_0^\infty e^{-\lambda t}\, dF(t)$. Thus, by the Post-Widder formula (Theorem 2.3.1), for $t > 0$ we have $F(t) = \lim_{k\to\infty} F_k(t)$, where

$$F_k(t) := G_k(k/t), \quad G_k(\lambda) := \frac{(-1)^k}{k!}\lambda^{k+1}\left(r(\lambda)/\lambda^2\right)^{(k)}.$$

By Lemma 2.7.12, $\lambda \mapsto r(\lambda)/\lambda^2$ is completely monotonic, and it follows that $F_k(t) \geq 0$. We show that F_k is convex; i.e., that $F_k''(t) = -\left(kt^{-2}G_k'(k/t)\right)' \geq 0$.

Let $H(\lambda) := -\lambda^2 k G_k'(k\lambda)$. Then $F_k''(t) = \frac{d}{dt}H(1/t) = -t^{-2}H'(1/t)$. Thus it suffices to show that $H'(\lambda) \leq 0$ or equivalently $2\lambda k G_k'(k\lambda) + \lambda^2 k^2 G_k''(k\lambda) \geq 0$ for $\lambda > 0$. Letting $\mu := k\lambda$ we have to show that

$$(\mu G_k(\mu))'' = 2G_k'(\mu) + \mu G_k''(\mu) \geq 0 \quad (\mu > 0).$$

This is true since (2.24) for $n = 2$ gives

$$(\mu G_k(\mu))'' = \frac{(-1)^k}{k!}\left(\mu^{k+2}(r(\mu)/\mu^2)^{(k)}\right)'' = \frac{(-1)^k}{k!}\mu^k r^{(k+2)}(\mu) \geq 0 \quad (\mu > 0).$$

This proves one implication.

Conversely, suppose that F is convex and $F(t) \geq 0$ for all $t \geq 0$. Let $x^* \in X_+^*$. Then $x^* \circ F$ is convex, positive and Lipschitz continuous. There is an increasing, bounded function $g : \mathbb{R}_+ \to \mathbb{R}_+$ such that $g(t) = \frac{d}{dt}\langle F(t), x^*\rangle$ a.e., and $\langle F(t), x^*\rangle = \int_0^t g(s)\, ds$ for all $t \geq 0$ (see Proposition 2.7.9). We may assume that $g(0) = 0$. By Proposition 1.10.1 and (1.22),

$$\langle r(\lambda), x^*\rangle = \lambda\langle \widehat{dF}(\lambda), x^*\rangle = \lambda\hat{g}(\lambda) = \widehat{dg}(\lambda) \quad (\lambda > 0).$$

Hence, $x^* \circ r$ is completely monotonic for all $x^* \in X_+^*$ and therefore r is completely monotonic. □

Next we prove a representation theorem for completely monotonic functions defined on $\mathbb{R}_+$ (and not merely $(0, \infty)$).

Proposition 2.7.14. *Let $r \in C^\infty(\mathbb{R}_+, X)$ such that $(-1)^n r^{(n)}(\lambda) \geq 0$ $(\lambda \geq 0)$. Then there exists a convex function $F \in \mathrm{Lip}_0(\mathbb{R}_+, X)$ such that $F(t) \geq 0$ $(t \geq 0)$ and*

$$r(\lambda) = \lambda \widehat{dF}(\lambda) \quad (\lambda > 0). \tag{2.26}$$

Proof. It follows from (2.23) that for $k \in \mathbb{N}$ and $\lambda > 0$,

$$p_k(\lambda) := \frac{(-1)^k}{k!}\lambda^{k+1}\left(\frac{r(\lambda)}{\lambda}\right)^{(k)} = \sum_{m=0}^{k} \frac{(-1)^m}{m!}\lambda^m r^{(m)}(\lambda) \geq 0.$$

Moreover, $\lim_{\lambda\downarrow 0} p_k(\lambda) = r(0)$. It follows from (2.24) for $n = 1$ that

$$p_k'(\lambda) = \frac{(-1)^k}{k!}\lambda^k r^{(k+1)}(\lambda) \leq 0 \quad (\lambda > 0).$$

Thus $0 \leq p_k(\lambda) \leq r(0)$ for all $\lambda > 0$. Since the cone is normal, this implies that the function $\frac{r(\lambda)}{\lambda}$ is in $C_W^\infty((0,\infty), X)$. By Theorem 2.2.1, there exists $F \in \mathrm{Lip}_0(\mathbb{R}_+, X)$ such that $\frac{r(\lambda)}{\lambda} = \widehat{dF}(\lambda)$ $(\lambda > 0)$. It follows from Proposition 2.7.13 that F is positive and convex. □

Theorem 2.7.15. *A function $r : (0,\infty) \to X$ is completely monotonic if and only if there exists a convex function $F : \mathbb{R}_+ \to X_+$ satisfying $F(0) = 0$ and $\omega(F) \leq 0$ such that*

$$r(\lambda) = \lambda \int_0^\infty e^{-\lambda t}\, dF(t) \quad (\lambda > 0). \tag{2.27}$$

In that case, F is uniquely determined by r.

Proof. a) Assume that r is of the form (2.27). Let $x^* \in X_+^*$. Then there exists an increasing function $f : \mathbb{R}_+ \to \mathbb{R}_+$ such that $f(0) = 0$ and

$$\langle F(t), x^*\rangle = \int_0^t f(s)\, ds \quad (t \geq 0).$$

Thus

$$\langle r(\lambda), x^*\rangle = \int_0^\infty e^{-\lambda t}\, df(t) \quad (\lambda > 0).$$

Hence, $\langle r(\cdot), x^*\rangle$ is completely monotonic and

$$\langle (-1)^n r^{(n)}(\lambda), x^*\rangle = (-1)^n \left(\frac{d}{d\lambda}\right)^n \langle r(\lambda), x^*\rangle \geq 0.$$

Since $x^* \in X_+^*$ is arbitrary, it follows that r is completely monotonic.

b) Conversely, let r be completely monotonic. By Proposition 2.7.14, there exists a convex function $G \in \mathrm{Lip}_0(\mathbb{R}_+, X)$ such that $G(t) \geq 0$ $(t \geq 0)$ and $r(\lambda+1) = \lambda \int_0^\infty e^{-\lambda t}\, dG(t)$ $(\lambda > 0)$. Let

$$F(t) := \int_0^t (1 - (t-s))e^s\, dG(s).$$

Then F is positive and convex. In fact, let $x^* \in X_+^*$. Then there exists an increasing function $g : \mathbb{R}_+ \to \mathbb{R}_+$ such that $\langle G(t), x^* \rangle = \int_0^t g(s)\,ds$ and $g(0) = 0$. By Proposition 1.9.10, Fubini's Theorem and (1.20),

$$\begin{aligned}\langle F(t), x^* \rangle &= \int_0^t e^s g(s)\,ds - \int_0^t (t-s)e^s g(s)\,ds \\ &= \int_0^t \left(e^s g(s) - \int_0^s e^r g(r)\,dr \right) ds \\ &= \int_0^t \int_0^s e^r\,dg(r)\,ds \quad (t \geq 0).\end{aligned}$$

Thus $x^* \circ F$ is positive and convex for all $x^* \in X_+^*$, so F is positive and convex. By Proposition 1.10.1 and (1.22),

$$\langle r(\lambda+1), x^* \rangle = \lambda \hat{g}(\lambda) = \int_0^\infty e^{-\lambda t}\,dg(t)$$

for $\lambda > 0$. By Proposition 1.10.3, for $\lambda > 1$,

$$\langle r(\lambda), x^* \rangle = \int_0^\infty e^{-\lambda t} e^t\,dg(t) = \int_0^\infty e^{-\lambda t}\,df(t),$$

where

$$f(t) := \int_0^t e^s\,dg(s) = e^t g(t) - \int_0^t e^s g(s)\,ds,$$

by (1.20). Since $\langle F(t), x^* \rangle = \int_0^t f(s)\,ds$, it follows that $r(\lambda) = \lambda \int_0^\infty e^{-\lambda t}\,dF(t)$ for $\lambda > 1$. By Theorem 2.7.1, $\operatorname{abs}(dF)$ is a singularity of $\widehat{dF}$. Moreover, applying Proposition 2.7.14 to $r(\cdot + \delta)$ shows that r has a holomorphic extension to $\{\lambda \in \mathbb{C} : \operatorname{Re}\lambda > \delta\}$ for all $\delta > 0$, and hence to $\{\lambda \in \mathbb{C} : \operatorname{Re}\lambda > 0\}$. It follows from the uniqueness of holomorphic extensions that $\operatorname{abs}(dF) \leq 0$ and $\widehat{dF}(\lambda) = r(\lambda)$ for $\lambda > 0$. By Theorem 1.10.5, $\omega(F) \leq 0$ (actually, $\omega(F) = 0$ unless $r \equiv 0$). Finally, uniqueness of F follows from the Post-Widder formula (Theorem 2.3.1). □

Theorem 2.7.16. *Assume that X has the interpolation property. Let $r : (0,\infty) \to X$ be completely monotonic. Then there exists an increasing function $f : \mathbb{R}_+ \to X_+$ such that $f(0) = 0$, $\omega(f) \leq 0$ and*

$$r(\lambda) = \int_0^\infty e^{-\lambda t}\,df(t) \quad (\lambda > 0).$$

Proof. By Theorem 2.7.15, there exists a convex function $F : \mathbb{R}_+ \to X_+$ satisfying $F(0) = 0$ and $\omega(F) \leq 0$ such that $r(\lambda) = \lambda \int_0^\infty e^{-\lambda t}\,dF(t)$ for all $\lambda > 0$. By Proposition 2.7.9, there exists an increasing function $f : \mathbb{R}_+ \to X_+$ such that

$F(t) = \int_0^t f(s)\,ds$ $(t \geq 0)$. We can assume that $f(0) = 0$. Let $\omega > 0$. There exists $M \geq 0$ such that $\|F(t)\| \leq Me^{\omega t}$. Since f is increasing we have

$$\frac{t}{2} f(t/2) \leq \int_{t/2}^{t} f(s)\,ds \leq F(t).$$

It follows that $\omega(f) \leq 0$ (actually, $\omega(f) = 0$ unless $r \equiv 0$). By Proposition 1.10.2,

$$\int_0^\infty e^{-\lambda t}\,df(t) = \lambda \int_0^\infty e^{-\lambda t}\,dF(t) = r(\lambda) \quad (\lambda > 0). \qquad \square$$

Now we can prove Theorem 2.7.7.

Proof of Theorem 2.7.7. One direction is given by Theorem 2.7.16. In order to prove the other, assume that Bernstein's theorem holds in X. We show that X has the interpolation property. Let $F : \mathbb{R}_+ \to X_+$ be convex such that $F(0) = 0$ and $\omega(F) = 0$. By Proposition 2.7.10, it suffices to show that $F(t) = \int_0^t f(s)\,ds$ $(t \geq 0)$ for some increasing function $f : \mathbb{R}_+ \to X$. By Proposition 2.7.13, $r(\lambda) := \lambda \int_0^\infty e^{-\lambda t}\,dF(t)$ defines a completely monotonic function on $(0, \infty)$. By assumption, there exists an increasing function $f : \mathbb{R}_+ \to X$ such that

$$r(\lambda) = \int_0^\infty e^{-\lambda t}\,df(t).$$

We may assume that $f(0) = 0$. Let $H(t) := \int_0^t f(s)\,ds$. Using Proposition 1.10.2 and (1.22), $\lambda^2 \widehat{H}(\lambda) = \widehat{df}(\lambda) = r(\lambda) = \lambda^2 \widehat{F}(\lambda)$ for all $\lambda > 0$. It follows from the uniqueness theorem that $H(t) = F(t)$ for all $t \geq 0$. $\square$

If $r : (0, \infty) \to X$ is completely monotonic, there may be many increasing functions $f : \mathbb{R}_+ \to X_+$ such that $r = \widehat{df}$. However, if X has order continuous norm, then we may pick out a normalized version of f.

Let $f : \mathbb{R}_+ \to X$ be increasing and assume that X has order continuous norm. For $t \geq 0$ we define $f(t+) = \lim_{s \downarrow t} f(s)$, and for $t > 0$ we let $f(t-) = \lim_{s \uparrow t} f(s)$. We say that f *has a jump at* $t > 0$ if $f(t+) \neq f(t-)$.

Lemma 2.7.17. *Assume that X has order continuous norm and that $f : \mathbb{R}_+ \to X$ is increasing. Then the number of jumps of f is countable.*

Proof. Let $\tau > 0$ and $J := \{t \in (0, \tau) :\ f(t+) \neq f(t-)\}$. Let $\varepsilon > 0$ and $J_\varepsilon := \{t \in J :\ \|f(t+) - f(t-)\| \geq \varepsilon\}$. We claim that J_ε is finite. Otherwise there exist $t_n \in J_\varepsilon$ $(n \in \mathbb{N})$, $t_n \neq t_m$ for $n \neq m$. Let $x_n = f(t_n+) - f(t_n-)$. Then $\sum_{n=1}^m x_n \leq f(\tau) - f(0)$ for all $m \in \mathbb{N}$. Since X has order continuous norm, the sum $\sum_{n=1}^\infty x_n$ converges. Hence, $\|x_n\| \to 0$ as $n \to \infty$. This is a contradiction. Since $J = \bigcup_{n \in \mathbb{N}} J_{1/n}$, it follows that J is countable. $\square$

We continue to assume that X has order continuous norm. Let $f : \mathbb{R}_+ \to X$ be increasing. We define the *normalization* $f^* : \mathbb{R}_+ \to X$ of f by

$$f^*(t) = \begin{cases} f(0+) & \text{if } t = 0, \\ \frac{1}{2}\left(f(t+) + f(t-)\right) & \text{if } t > 0. \end{cases}$$

The function f is called *normalized* if $f = f^*$.

It follows from the definition of the Riemann-Stieltjes integral that

$$\int_0^t g(s)\, df(s) = \int_0^t g(s)\, df^*(s)$$

for every $t > 0$ and every continuous function $g : [0,t] \to \mathbb{C}$. In fact, one may take a sequence of partitions $(\pi_n)_{n\in\mathbb{N}}$ with intermediate points which avoid the jumps of f). Then $S(g, f, \pi_n) = S(g, f^*, \pi_n)$ for all $n \in \mathbb{N}$, and so

$$\int_0^t g(s)\, df(s) = \lim_{n\to\infty} S(g, f, \pi_n) = \lim_{n\to\infty} S(g, f^*, \pi_n) = \int_0^t g(s)\, df^*(s).$$

In conclusion, we obtain the following result.

Theorem 2.7.18 (Bernstein's theorem). *Assume that X has order continuous norm. Let $r : (0,\infty) \to X$ be completely monotonic. Then there exists a unique normalized increasing function $f : \mathbb{R}_+ \to X$ such that $f(0) = 0$, $\omega(f) \le 0$ and*

$$r(\lambda) = \int_0^\infty e^{-\lambda t}\, df(t) \quad (\lambda > 0).$$

Proof. Since X has the interpolation property, existence follows from Theorem 2.7.16. For uniqueness, suppose that $r(\lambda) = \int_0^\infty e^{-\lambda t}\, df(t)$ $(\lambda > 0)$. By Proposition 1.10.2, $r(\lambda) = \lambda \int_0^\infty e^{-\lambda t}\, dF(t)$ $(\lambda > 0)$ where $F(t) := \int_0^t f(s)\, ds$. It follows from Theorem 2.7.15 that F is uniquely determined by r. Since

$$F'(t+) := \lim_{h\downarrow 0} \frac{1}{h}\left(F(t+h) - F(t)\right) = f(t+)$$

if $t \ge 0$, and

$$F'(t-) := \lim_{h\downarrow 0} \frac{1}{h}\left(F(t) - F(t-h)\right) = f(t-)$$

if $t > 0$, the normalized function f is also unique. □

2.8 Notes

Section 2.1

Representation of operators from a space of the form $L^1(\Omega, \mu)$ into a Banach space

X by vector measures is a classical subject (see [DU77, Section III.1]). In view of the applications to Cauchy problems, Stieltjes integrals seem more appropriate than vector measures in our context. In the context of Laplace transform theory, the Riesz-Stieltjes Representation Theorem 2.1.1 appeared in a paper of Hennig and Neubrander [HN93] (see also [Neu94] and [BN94]). For a discussion of the representation of bounded linear operators in $\mathcal{L}(L^p(\mathbb{R}_+), X)$ as functions of bounded p'-variation $(1/p + 1/p' = 1, p' > 1)$, see the work of Weis [Wei93] and Vieten [Vie95].

Section 2.2

For real-valued functions, Theorem 2.2.1 was proved by Widder in 1936 [Wid36] (see also [Wid41]). In trying to extend scalar-valued Laplace transform theory to vector-valued functions, Hille [Hil48] remarks on several occasions that Widder's theorem can be lifted to infinite dimensions if the space is reflexive, but not in general (see [Hil48, p.213] or [Miy56]). In fact, it was shown by Zaidman [Zai60] (see also [Are87b] or Theorem 2.2.3) that Widder's theorem extends to a Banach space X if and only if X has the Radon-Nikodym property (for example, if X is reflexive). In 1965, Berens and Butzer [BB65] gave necessary and sufficient complex conditions for the Laplace-Stieltjes representability of functions in reflexive and uniformly convex Banach spaces. However, these results were of limited applicability. In general, important classes of Banach spaces that appear in studying evolution equations do not possess the Radon-Nikodym property. As a consequence, in the 1960s and 1970s Laplace transform methods were applied mainly to special vector-valued functions, like resolvents and semigroups, which have nice additional algebraic properties. In the theory of C_0-semigroups the link between the generator A and the semigroup T is given via the Laplace transform

$$(\lambda - A)^{-1}x = \int_0^\infty e^{-\lambda t} T(t)x \, dt \quad (x \in X).$$

The crucial algebraic property which made it possible to extend classical Laplace transform results to this abstract setting is the algebraic semigroup law $T(t+s) = T(t)T(s)$, $(t, s \geq 0)$. Hille and Phillips comment in the foreword to [HP57] that "*.... in keeping with the spirit of the times the algebraic tools now play a major role....*" and that "*.... the Laplace-Stieltjes transform methods..... have not been replaced but rather supplemented by the new tools.*" The major disadvantage of the "algebraic approach" to linear evolution equations becomes obvious if one compares the mathematical theories associated with them (for example, semigroup theories, cosine families, the theory of integro-differential equations, etc.). It is striking how similar the results and techniques are. Still, without a Laplace transform theory for functions with values in arbitrary Banach spaces, every type of linear evolution equation required its own theory because the algebraic properties of the operator families changed from one case to another. In the late 1970s, in search of a general analytic principle behind all these theories, the study of Laplace transforms of functions with values in arbitrary Banach spaces was revitalized by Sova (see [Sov77] up to [Sov82]). An important result for Laplace transforms in Banach spaces is Theorem 2.6.1, proved by Sova in 1979 [Sov79b], [Sov79c]. This analytic representation theorem is behind every generation result for analytic solution families of linear evolution equations.

The Real Representation Theorem 2.2.1 shows that the statement of Widder's Theorem extends to arbitrary Banach spaces if the Laplace transform is replaced by the Laplace-Stieltjes transform. It is due to [Are87b] where it was deduced from the scalar result by Widder [Wid41] by duality arguments. The proof of Theorem 2.2.1 given here is a modification of Widder's original proof given in [Wid41]; see [HN93]. Further extensions of these results are given in [Bob97a], [Bob97b], [Kis00], [Bob01] and [Cho02].

The characterization of the range of the Laplace-Stieltjes transform acting on $\mathrm{Lip}_0(\mathbb{R}_+, X)$ given in Theorem 2.2.1 is based on the Post-Widder inversion formula in Theorem 1.7.7. Corresponding to other inversion formulas, equivalent descriptions can be formulated. Employing the complex inversion formula (see [Sov80b], [BN94]), or the Phragmén-Doetsch inversion (see [PC98]), one can prove that the following growth and regularity conditions are equivalent.

Theorem 2.8.1. *Let $r : (0, \infty) \to X$ be continuous. The following are equivalent:*

(i) $r \in C^\infty((0,\infty), X)$ *and*

$$\sup_{\substack{\lambda>0\\ k\in\mathbb{N}_0}} \left\| \frac{\lambda^{k+1}}{k!} r^{(k)}(\lambda) \right\| < \infty.$$

(ii) $\lim_{\lambda\to\infty} r(\lambda) = 0$ *and r has an extension to a holomorphic function $r : \{\mathrm{Re}\,\lambda > 0\} \to X$ such that, for all $\gamma > 0$, $\sup_{\mathrm{Re}\,\lambda>\gamma} \|r(\lambda)\| < \infty$ and*

$$\sup_{\substack{s>0\\ k\in\mathbb{N}_0}} \left\| \frac{1}{2\pi} \int_{-\infty}^{\infty} \frac{r(\gamma+it)}{(1-ist)^{k+2}}\, dt \right\| < \infty.$$

(iii) $\sup_{\lambda>0} \|\lambda r(\lambda)\| < \infty$ *and*

$$\sup_{\substack{\lambda>0\\ k\in\mathbb{N}}} \left\| \sum_{j=1}^{\infty} \frac{(-1)^{j-1}}{(j-1)!} e^{jk} \lambda r(j\lambda) \right\| < \infty.$$

For a discussion of the L^p-conditions

$$\int_0^\infty \left\| \left(\frac{k}{t}\right)^{k+1} \frac{1}{k!} r^{(k)} \left(\frac{k}{t}\right) \right\|^p dt \le M \qquad \text{for all } k \ge 0,$$

and their connection to the representability of r as the Laplace transform of a function of bounded p-variation ($p > 1$), see [Wid41, Chapter VII], [Lev69], [Sov81a], [Wei93], and [Vie95]. It is shown in [KMV03] that a function $r \in C^\infty((0,\infty), X)$ is the finite Laplace-Stieltjes transform $r(\lambda) = \int_0^\tau e^{-\lambda t}\, dF(t)$ of a Lipschitz continuous function $F : [0,\tau] \to X$ with $\|F(t) - F(s)\| \le M|t-s|$ for all $0 \le t, s \le \tau$ if and only if

$$\sup_{k\in\mathbb{N}_0} \sup_{\lambda>k/\tau} \left\| \frac{\lambda^{k+1}}{k!} r^{(k)}(\lambda) \right\| \le M$$

and

$$\sup_{k\in\mathbb{N}} \sup_{\lambda\in(0,k/\tau)} \left\| \tau^{-k} e^{\lambda\tau} r^{(k)}(\lambda) \right\| < \infty.$$

Section 2.3

Theorem 2.3.2 goes back to Phragmén's proof of the Uniqueness Theorem 1.7.3 (see [Phr04]), and to Doetsch [Doe37] who recognized the usefulness of the formula as an inversion procedure (see also [Doe50, Volume I, Section 8.1]). The Phragmén-Doetsch inversion formula shows that a Laplace transformable function f is determined by the values of $\hat{f}(\lambda_n)$, where $\lambda_n = n \geq n_0$. An extension of the Phragmén-Doetsch inversion to arbitrary Müntz sequences $(\lambda_n) \subset \mathbb{R}_+$ (i.e., $\lambda_{n+1} - \lambda_n \geq 1$ and $\sum_{n=1}^{\infty} \lambda_n^{-1} = \infty$), has been obtained by Bäumer [Bäu03] (see also [BLN99]). There does not seem to be any inversion formula that holds for arbitrary uniqueness sequences (see Theorem 1.11.1). Corollary 2.3.3 is taken from [BN96] and is one of the key ingredients in the theory of asymptotic Laplace transforms (see [LN99], [LN01]). Whereas the complex inversion formula in Theorem 2.3.4 (the proof given here is from [HN93]) is in general affected by exponentially decaying perturbations of the Laplace transform, the following modification, due to Lyubich [Lyu66], gives a complex inversion formula which holds locally even if the transform undergoes such perturbations.

Theorem 2.8.2. *Let* $\tau > 0$, $\omega > 0$, $F \in \mathrm{Lip}_0(\mathbb{R}_+, X)$, *and* $q(\lambda) = \int_0^\infty e^{-\lambda t}\, dF(t) + a(\lambda)$ $(\lambda > 0)$, *where* $a \in L^1_{loc}(\mathbb{R}_+, X)$ *is a function with* $\limsup_{\lambda\to\infty} \frac{1}{\lambda} \log \|a(\lambda)\| \leq -\tau$. *Then*

$$H(\mu) := \frac{1}{2\pi i} \int_\omega^\infty e^{\mu t} \frac{q(t)}{t}\, dt$$

is well defined for $\mathrm{Re}\,\mu < 0$, *has a holomorphic continuation to the sliced half-plane* $\{\mu : \mathrm{Re}\,\mu < \tau\} \setminus [0, \tau)$, *and*

$$F(t) = \lim_{\varepsilon\to 0}(H(t + i\varepsilon) - H(t - i\varepsilon)) \qquad \textit{for all } t \in [0, \tau).$$

Haase [Haa08] has given a different approach to Theorem 2.3.4 and Lemma 2.3.5.

Section 2.4

With the exception of Proposition 2.4.3 which is due to Doetsch (see [Doe50, Volume I, Section 14.3]) and Corollary 2.4.4, the results are straightforward reformulations of the main theorems of the sections 2.1–2.3. Using a Phragmén-Doetsch type inversion formula along sequences $(\lambda_n) \subset \mathbb{R}_+$ with $\lambda_{n+1} - \lambda_n \geq 1$ and $\sum_{n=1}^{\infty} \lambda_n^{-1} = \infty$ (Müntz sequences), one can strengthen the statement of Proposition 2.4.3 as follows (see [Bäu03]).

Theorem 2.8.3. *Let* $0 \leq \tau$ *and let* $f \in L^1_{loc}(\mathbb{R}_+, X)$ *with* $\mathrm{abs}(f) < \infty$. *Then the following are equivalent:*

(i) $f(t) = 0$ *almost everywhere on* $[0, \tau]$ *and* $\tau \in \mathrm{supp}(f)$.

(ii) *Every Müntz sequence* (β_n) *satisfies* $\limsup_{n\to\infty} \frac{1}{\beta_n} \log \|\hat{f}(\beta_n)\| = -\tau$.

(iii) *For every Müntz sequence* (β_n) *there exists a Müntz subsequence* (β_{n_k}) *such that*

$$\lim_{k\to\infty} \frac{1}{\beta_{n_k}} \log \|\hat{f}(\beta_{n_k})\| = -\tau.$$

(iv) *There exists a Müntz sequence* (β_n) *with* $\limsup_{n\to\infty} \frac{1}{\beta_n} \log \|\hat{f}(\beta_n)\| = -\tau$.

(v) $\limsup_{\lambda\to\infty} \frac{1}{\lambda} \log \|\hat{f}(\lambda)\| = -\tau$.

As a consequence of these equivalences one obtains the following short proof of Titchmarsh's theorem (see [Bäu03], [BLN99] or [MB87, Section VI.7]).

Corollary 2.8.4 (Titchmarsh's Theorem). *Let* $k \in L^1[0, \tau]$ *with* $0 \in \mathrm{supp}(k)$ *and* $f \in L^1([0, \tau], X)$. *If* $k \star f = 0$ *on* $[0, \tau]$, *then* $f = 0$.

Proof. We extend k and f by zero to $\mathbb{R}_+$. Then, by Proposition 2.4.3 and Corollary 2.4.4, $\limsup_{\lambda\to\infty} \frac{1}{\lambda}\log|\hat{k}(\lambda)| = 0$ and $\limsup_{\lambda\to\infty}\frac{1}{\lambda}\log\|\widehat{k\star f}(\lambda)\| \le -T$. By taking subsequences, it follows from the theorem above that there exists a Müntz sequence (β_n) such that $\lim_{n\to\infty}\frac{1}{\beta_n}\log|\hat{k}(\beta_n)| = 0$ and

$$\begin{aligned} -\tau \ge \lim_{n\to\infty}\frac{1}{\beta_n}\log\|\widehat{k\star f}(\beta_n)\| &= \lim_{n\to\infty}\frac{1}{\beta_n}\log|\hat{k}(\beta_n)| + \lim_{n\to\infty}\frac{1}{\beta_n}\log\|\hat{f}(\beta_n)\| \\ &= \lim_{n\to\infty}\frac{1}{\beta_n}\log\|\hat{f}(\beta_n)\|. \end{aligned}$$

Thus, $f = 0$ on $[0,\tau]$. □

A function $k \in L^1_{loc}(\mathbb{R}_+)$ with $\mathrm{abs}(k) < \infty$ is a *regularizing function* if

$$\limsup_{\lambda\to\infty}\frac{1}{\lambda}\log|\hat{k}(\lambda)| = 0,$$

or, equivalently, if $0 \in \mathrm{supp}(k)$ (by Corollary 2.4.4). By the Titchmarsh-Foiaş theorem (see [BLN99]), the condition $0 \in \mathrm{supp}(k)$ is necessary and sufficient for the convolution operator $\mathcal{K} : f \to k * f$, $(k*f)(t) := \int_0^t k(t-s)f(s)\,ds$ to be an injective operator on $C(\mathbb{R}_+, X)$ with dense range in the Fréchet space $C_*(\mathbb{R}_+, X)$ of all continuous functions $g : \mathbb{R}_+ \to X$ such that $g(0) = 0$, equipped with the seminorms $\|g\|_n := \sup_{t\in[0,n]}\|g(t)\|$. Moreover, $\|f\|_{\mathcal{K},n} := \sup_{t\in[0,n]}\|\mathcal{K}f(t)\|$ defines a family of seminorms on $C(\mathbb{R}_+, X)$ and $\mathcal{K}$ extends to an isomorphism between the Fr echet completion $C^{[k]}(\mathbb{R}_+, X)$ of $C(\mathbb{R}_+, X)$ with respect to that family of seminorms and the Fréchet space $C_*(\mathbb{R}_+, X)$. Typical examples of regularizing functions are

$$k(t) = \frac{t^{b-1}}{\Gamma(b)} \quad \text{with} \quad \hat{k}(\lambda) = \frac{1}{\lambda^b}\ (b > 0), \quad \text{or}$$

$$k_\delta(t) = \frac{1}{2\pi i}\int_{\omega+i\mathbb{R}} e^{t\lambda-\lambda^\delta}\,d\lambda \quad \text{with} \quad \hat{k}_\delta(\lambda) = e^{-\lambda^\delta}\ (0 < \delta < 1).$$

Note that $k_{1/2}(t) = \frac{1}{2\sqrt{\pi}}t^{-3/2}e^{-1/4t}$ (see Lemma 1.6.7).

If k is a regularizing function, then the elements of the Fréchet space $C^{[k]}(\mathbb{R}_+, X)$ are called *k-generalized functions.* A k-generalized function u is said to be Laplace transformable if the continuous function $f := k * u \in C_*(\mathbb{R}_+, X)$ is Laplace transformable and the Laplace transform of u is defined as

$$\hat{u}(\lambda) := \frac{\hat{f}(\lambda)}{\hat{k}(\lambda)}.$$

Let $\mathcal{H} = \{\lambda : \mathrm{Re}\,\lambda > \omega\}$ and $m : \mathcal{H} \to \mathbb{C}$ be holomorphic. A meromorphic function $q : \mathcal{H} \to X$ is said to have an m-multiplied Laplace representation if there exists $f \in C_*(\mathbb{R}_+, X)$ with $\mathrm{abs}(f) \le \omega$ such that $mq = \hat{f}$ on $\mathcal{H}$. If $m = \hat{k}$ for some regularizing function k, then the meromorphic function q has a Laplace representation $q = \hat{u}$ for $u = \mathcal{K}^{-1}f \in C^{[k]}(\mathbb{R}_+, X)$ (see [Bäu97], [BLN99], and [LN99]).

Section 2.5

Theorem 2.5.1 is a standard result of Laplace transform theory. Corollary 2.5.2 is due to

Prüss [Prü93], the proof given here is from [BN94]. Corollary 2.5.4 is a special case of results in [DVW02] (see also [DHW97]).

Theorem 2.5.1 can be interpreted in terms of k-generalized functions and Laplace transforms (see the Notes of Section 2.4; we use the same notation here). Let $q : \mathcal{H} \to X$ be holomorphic with $\sup_{\lambda \in \mathcal{H}} \|\lambda q(\lambda)\| < \infty$. As shown in Theorem 2.5.1, for all $b > 0$ there exists $f \in C_*(\mathbb{R}_+, X)$ such that $q(\lambda) = \lambda^b \hat{f}(\lambda)$ on $\mathcal{H}$. Thus, $q(\lambda) = \hat{u}(\lambda) = \frac{\hat{f}(\lambda)}{\hat{k}(\lambda)}$, where $k(t) = \frac{1}{\Gamma(b)} t^{b-1}$ and $u = \mathcal{K}^{-1} f \in C^{[k]}(\mathbb{R}_+, X)$ coincides with the b-th (distributional) derivative of f. More generally, if q is a meromorphic function on some half-plane $\mathcal{H}$ with values in X for which $\lambda \mapsto \lambda \hat{k}_0(\lambda) q(\lambda)$ is holomorphic on $\mathcal{H}$ and

$$\sup_{\lambda \in \mathcal{H}} \|\lambda \hat{k}_0(\lambda) q(\lambda)\| < \infty$$

for some regularizing function k_0, then it follows from Theorem 2.5.1 that there exists $f \in C_*(\mathbb{R}_+, X)$ such that $\frac{1}{\lambda}\hat{k}_0(\lambda) q(\lambda) = \hat{k}(\lambda) q(\lambda) = \hat{f}(\lambda)$ or $q(\lambda) = \hat{u}(\lambda)$, where $k := 1 * k_0$ and $u \in C^{[k]}(\mathbb{R}_+, X)$ is a generalized function such that $f = k * u$. Notice that if k_i are regularizing functions and $k_1 * k_2 = k_3$, then $C^{[k_1]}(\mathbb{R}_+, X)$ is continuously embedded in $C^{[k_3]}(\mathbb{R}_+, X)$. Thus, a faster growing q will have a less regular u such that $q = \hat{u}$.

Section 2.6

Theorem 2.6.1 is due to Sova [Sov79b] and Theorem 2.6.2 is taken from [Neu89b].

Section 2.7.

In 1893, Stieltjes proved in a letter to Hermite that a bounded continuous function $f : \mathbb{R}_+ \to \mathbb{R}$ is positive if and only if $\hat{f}^{(n)}(\lambda) \geq 0$ for all $n \in \mathbb{N}_0$ and all λ sufficiently large (see [BB05]). Bernstein proved his theorem in 1928 [Ber28].

The characterization of those ordered Banach spaces in which Bernstein's theorem (Theorem 2.7.7) holds is due to Arendt [Are94a].

The interpolation property is of particular interest for spaces of the form $C(K)$, where K is a compact space. Then it can be described in terms of K: the space $C(K)$ has the interpolation property if and only if K is an F-space (i.e., if $A, B \subset K$ are open and disjoint F_σ-sets, then $\overline{A} \cap \overline{B} = \emptyset$). Note that $C(K)$ is σ-order complete if and only if K is quasi-stonean (i.e., if $A \subset K$ is an open F_σ-set, then $\bar{A}$ is open). For example, $K := \beta\mathbb{N} \setminus \mathbb{N}$ is a F-space which is not quasi-stonean (where $\beta\mathbb{N}$ denotes the Stone-Čech compactification of $\mathbb{N}$). Whereas every quasi-stonean space K is totally disconnected (i.e. the connected component of each point x is $\{x\}$), there exist connected compact F-spaces. One reason why these spaces have been studied is that $C(K)$ has the Grothendieck property (see Section 4.3) if K is an F-space. We refer to the article by Seever [See68] for this and further information.

The interpolation property is also equivalent to two other vector-valued versions of classical theorems; namely, Riesz's representation theorem for positive functionals on $C[0,1]$ and Hausdorff's theorem on the moment problem. More precisely, the following is proved in [Are94a].

Theorem 2.8.5. *Let X be an ordered Banach space with normal cone. The following are equivalent:*

(i) *X has the interpolation property.*

(ii) *For every positive $T \in \mathcal{L}(C[0,1], X)$ there exists an increasing function $f : [0,1] \to X$ such that $Tg = \int_0^1 g(t)\, df(t)$ for all $g \in C[0,1]$.*

(iii) *For each completely monotonic sequence* $(x_n)_{n\in\mathbb{N}}$ *in* X *there exists an increasing function* $f : [0,1] \to X$ *such that* $x_n = \int_0^1 t^n \, df(t)$ $(n \in \mathbb{N})$.

Here, a sequence $x = (x_n)_{n\in\mathbb{N}}$ is called *completely monotonic* if $(-\Delta)^k x \geq 0$ for all $k \in \mathbb{N}$ where $\Delta : X^{\mathbb{N}} \to X^{\mathbb{N}}$ is given by $\Delta x = (x_{n+1} - x_n)_{n\in\mathbb{N}}$.

Bernstein's theorem in ordered Banach spaces with order continuous norm (Theorem 2.7.18) is proved in [Are87a] with the help of the classical scalar theorem. A first vector-valued version of Bernstein's theorem is due to Bochner [Boc42]. But Bochner considered convergence in order, whereas for our purposes norm convergence of Riemann-Stieltjes sums and improper integrals is essential in order to make the results applicable to operator theory. Here we deduce Bernstein's theorem from the Real Representation Theorem 2.2.1.

One can obtain Widder's theorem (the scalar case of Theorem 2.2.1) as an easy corollary of Bernstein's classical result (see [Wid71, Section 6.8]). However this argument is restricted to the scalar case. On the other hand, it is possible to deduce the vector-valued version of Theorem 2.2.1 from the scalar case by a duality argument (see [Are87b] and the Notes of Section 2.2).

Chapter 3

Cauchy Problems

In this chapter we study systematically well-posedness of the Cauchy problem. Given a closed operator A on a Banach space X we will see in Section 3.1 that the abstract Cauchy problem

$$\begin{cases} u'(t) = & Au(t) \quad (t \geq 0), \\ u(0) = & x, \end{cases}$$

is mildly well-posed (i.e., for each $x \in X$ there exists a unique mild solution) if and only if the resolvent of A is a Laplace transform; and this in turn is the same as saying that A generates a C_0-semigroup. Well-posedness in a weaker sense will lead to generators of integrated semigroups (Section 3.2). The real representation theorem from Section 2.2 will give us directly the characterization of generators of C_0-semigroups in terms of a resolvent estimate; namely, the Hille-Yosida theorem. When the operators are not densely defined, we obtain Hille-Yosida operators which are studied in detail in Section 3.5. Also for results on approximation of semigroups in Section 3.6 we can use corresponding results on Laplace transforms from Section 1.7. Much attention is given to holomorphic semigroups which are particularly simple to characterize by means of the results of Section 2.6. We consider not only holomorphic semigroups which are strongly continuous at 0, but more general holomorphic semigroups which will be useful in applications to the heat equation with Dirichlet boundary conditions in Chapter 6. When the holomorphic semigroup exists on the right half-plane, the boundary behaviour is of special interest. If the semigroup is locally bounded, then a boundary C_0-group is obtained on the imaginary axis. This case is particularly important for fractional powers (see also the Notes of Section 3.7) and for the second order problem (Section 3.16). When the holomorphic semigroup is polynomially bounded we obtain k-times integrated semigroups where the k depends on the degree of the polynomial. A typical example is the Gaussian semigroup. Its boundary is governed by the Schrödinger operator $i\Delta$, which we study in Section 3.9 and in Chapter 8. The last three sections are devoted to the second order Cauchy problem; i.e., to the theory of cosine functions. A central result will be to establish a unique phase space on which the associated system is well-posed. This is a particularly interesting special case of the intermediate spaces which are constructed in Section 3.10 for integrated semigroups.

In two places we will give results for UMD-spaces which are not valid in general Banach spaces: in Section 3.12 where we establish a particularly simple complex inversion formula for semigroups, and in Section 3.16 where we prove Fattorini's remarkable theorem on the square root reduction.

There is no special section on perturbation theory, but we prove perturbation results for Hille-Yosida operators, integrated semigroups and generators of cosine functions in the corresponding sections. For holomorphic semigroups we consider not only "relatively small perturbations" but also "compact perturbations" with respect to A. This chapter contains some interesting examples of holomorphic semigroups in Sections 3.7 and 3.9, but for real applications we refer to Part III.

Throughout this chapter we will make extensive use (sometimes without comment) of notation, terminology and basic properties of closed operators which may be found in Appendix B. In some examples we shall use some basic notions of distributions and Sobolev spaces which may be found in Appendix E.

3.1 C_0-semigroups and Cauchy Problems

Let A be a closed operator on a Banach space X. We consider the abstract Cauchy problem

$$(ACP_0) \quad \begin{cases} u'(t) = & Au(t) \quad (t \geq 0), \\ u(0) = & x, \end{cases}$$

where $x \in X$. By a *classical solution* of (ACP_0) we understand a function $u \in C^1(\mathbb{R}_+, X)$ such that $u(t) \in D(A)$ for all $t \geq 0$ and (ACP_0) holds.

If a classical solution exists, then it follows that $x = u(0) \in D(A)$. It will be useful to find a weaker notion of solution where x may be arbitrary. This can be done by integrating the equation. Assume that u is a classical solution. Since A is closed, it follows from Proposition 1.1.7 that

$$\int_0^t u(s)\,ds \in D(A) \quad \text{and} \quad A\int_0^t u(s)\,ds = u(t) - x \quad (t \geq 0). \tag{3.1}$$

Definition 3.1.1. *A function $u \in C(\mathbb{R}_+, X)$ is called a* mild solution *of (ACP_0) if* (3.1) *holds.*

The following assertion shows that mild and classical solutions differ merely by regularity.

Proposition 3.1.2. *A mild solution u of (ACP_0) is a classical solution if and only if $u \in C^1(\mathbb{R}_+, X)$.*

Proof. Assume that $u \in C^1(\mathbb{R}_+, X)$. Let $t \geq 0$. Then

$$\frac{1}{h}(u(t+h) - u(t)) = \frac{1}{h} A \int_t^{t+h} u(s)\, ds$$

for all $h \neq 0$ small enough ($h > 0$ if $t = 0$). Since A is closed, it follows that

$$\begin{aligned} u(t) &= \lim_{h \to 0} \frac{1}{h} \int_t^{t+h} u(s)\, ds \in D(A) \text{ and} \\ u'(t) &= Au(t). \end{aligned}$$

□

Next we want to characterize mild solutions with the help of Laplace transforms. Let $u \in C(\mathbb{R}_+, X)$. Recall from (1.12) that $\operatorname{abs}(u) < \infty$ if and only if

$$\left\| \int_0^t u(s)\, ds \right\| \leq M e^{\omega t} \quad (t \geq 0) \tag{3.2}$$

for some $M, \omega \geq 0$. As before, we denote by

$$\hat{u}(\lambda) := \int_0^\infty e^{-\lambda t} u(t)\, dt \quad (\lambda > \omega)$$

the Laplace transform of u.

Theorem 3.1.3. *Let $u \in C(\mathbb{R}_+, X)$ such that (3.2) holds. Then the following assertions are equivalent:*

(i) *u is a mild solution of (ACP_0).*

(ii) *$\hat{u}(\lambda) \in D(A)$ and $\lambda \hat{u}(\lambda) - A\hat{u}(\lambda) = x$ for all $\lambda > \omega$.*

Proof. (i) ⇒ (ii): Let u be a mild solution. Let $\lambda > \omega$. We know from (1.11) that

$$\hat{u}(\lambda) = \lambda \int_0^\infty e^{-\lambda t} \int_0^t u(s)\, ds\, dt.$$

Since A is closed, it follows from Proposition 1.6.3 that $\hat{u}(\lambda) \in D(A)$ and

$$\begin{aligned} A\hat{u}(\lambda) &= \lambda \int_0^\infty e^{-\lambda t} A \int_0^t u(s)\, ds\, dt \\ &= \lambda \int_0^\infty e^{-\lambda t} (u(t) - x)\, dt \\ &= \lambda \hat{u}(\lambda) - x. \end{aligned}$$

(ii) ⇒ (i): Let $v(t) := \int_0^t u(s)\, ds$. Then by (1.11), $\hat{v}(\lambda) = \hat{u}(\lambda)/\lambda \in D(A)$ and

$$A\hat{v}(\lambda) = A\hat{u}(\lambda)/\lambda = \hat{u}(\lambda) - x/\lambda = \hat{f}(\lambda) \quad (\lambda > \omega),$$

where $f(t) := u(t) - x$ $(t \geq 0)$. It follows from Proposition 1.7.6 that $v(t) \in D(A)$ and $Av(t) = f(t) = u(t) - x$ for all $t \geq 0$; i.e., u is a mild solution of (ACP_0). $\square$

Let u be a mild solution of (ACP_0) satisfying (3.2). Assume that $\omega < \lambda \in \rho(A)$. Then it follows from Theorem 3.1.3 that $\hat{u}(\lambda) = R(\lambda, A)x$. Thus the Laplace transform of a mild solution is the resolvent applied to the initial value. This leads us to consider operators whose resolvent exists on a half-line and is a Laplace transform.

For this, let $T : \mathbb{R}_+ \to \mathcal{L}(X)$ be strongly continuous. Recall from Proposition 1.4.5 that $\operatorname{abs}(T) < \infty$ if and only if

$$\left\| \int_0^t T(s)x\,ds \right\| \leq Me^{\omega t}\|x\| \qquad (t \geq 0, x \in X) \tag{3.3}$$

for some $\omega \geq 0$, $M \geq 0$. In that case, $\operatorname{abs}(T) \leq \omega$ and

$$\hat{T}(\lambda)x := \lim_{t\to\infty} \int_0^t e^{-\lambda s}T(s)x\,ds$$

defines a bounded operator $\hat{T}(\lambda) \in \mathcal{L}(X)$ whenever $\operatorname{Re}\lambda > \operatorname{abs}(T)$. Moreover, $\hat{T} : \{\operatorname{Re}\lambda > \operatorname{abs}(T)\} \to \mathcal{L}(X)$ is holomorphic (see Section 1.5).

As in Sections 1.4 and 1.5, we denote by $\omega(T)$ the exponential growth bound of T, and by $\operatorname{hol}(\hat{T})$ the abscissa of holomorphy of $\hat{T}$. Recall that $\operatorname{hol}(\hat{T}) \leq \operatorname{abs}(T) \leq \omega(T)$.

We will consider Laplace transforms of operator-valued functions on many occasions in this and subsequent chapters. The following definition will be helpful.

Definition 3.1.4. *Let $\lambda_0 \in \mathbb{R}$ and let $R : (\lambda_0, \infty) \to \mathcal{L}(X)$ be a function. We say that R is a* Laplace transform *if there exists a strongly continuous function $T : \mathbb{R}_+ \to \mathcal{L}(X)$ such that* $\operatorname{abs}(T) \leq \lambda_0$ *and*

$$R(\lambda) = \hat{T}(\lambda) \qquad (\lambda > \lambda_0).$$

The following proposition is a simple consequence of the uniqueness theorem.

Proposition 3.1.5. *Let $T : \mathbb{R}_+ \to \mathcal{L}(X)$ be strongly continuous such that* $\operatorname{abs}(T) < \infty$. *Let $\omega >$* $\operatorname{abs}(T)$. *Then the following hold:*

a) *If $B \in \mathcal{L}(X)$ such that $B\hat{T}(\lambda) = \hat{T}(\lambda)B$ for all $\lambda > \omega$, then $BT(t) = T(t)B$ for all $t \geq 0$.*

b) *In particular, if $\hat{T}(\mu)\hat{T}(\lambda) = \hat{T}(\lambda)\hat{T}(\mu)$ for all $\lambda, \mu > 0$, then $T(t)T(s) = T(s)T(t)$ for all $t, s \geq 0$.*

Proof. a) For $x \in X$ and $\lambda > \omega$, one has

$$\int_0^\infty e^{-\lambda t}T(t)Bx\,dt = \hat{T}(\lambda)Bx = B\hat{T}(\lambda)x = \int_0^\infty e^{-\lambda t}BT(t)x\,dt.$$

It follows from the uniqueness theorem that $T(t)Bx = BT(t)x$ for all $t \geq 0$.

b) Let $\mu > \omega$. It follows from a) that $\hat{T}(\mu)T(t) = T(t)\hat{T}(\mu)$ for all $t \geq 0$. Fixing $t \geq 0$ and applying a) to $B := T(t)$ shows that $T(s)T(t) = T(t)T(s)$ for all $s \geq 0$. □

Now we introduce C_0-semigroups.

Definition 3.1.6. *A* C_0-semigroup *is a strongly continuous function* $T : \mathbb{R}_+ \to \mathcal{L}(X)$ *such that*

$$\begin{aligned} T(t+s) &= T(t)T(s) \quad (t, s \geq 0), \\ T(0) &= I. \end{aligned}$$

In the next theorem we show that C_0-semigroups are exactly those strongly continuous operator-valued functions whose Laplace transforms are resolvents. It is remarkable that C_0-semigroups are automatically exponentially bounded.

Theorem 3.1.7. *Let* $T : \mathbb{R}_+ \to \mathcal{L}(X)$ *be a strongly continuous function. The following assertions are equivalent:*

(i) $\operatorname{abs}(T) < \infty$ *and there exists an operator* A *such that* $(\lambda_0, \infty) \subset \rho(A)$ *and*

$$\hat{T}(\lambda) = R(\lambda, A) \quad (\lambda > \lambda_0)$$

for some $\lambda_0 > \operatorname{abs}(T)$.

(ii) T *is a* C_0*-semigroup.*

In that case, $\omega(T) < \infty$, $\{\operatorname{Re}\lambda > \operatorname{hol}(\hat{T})\} \subset \rho(A)$ *and* $\hat{T}(\lambda) = R(\lambda, A)$ *whenever* $\operatorname{Re}\lambda > \operatorname{hol}(\hat{T})$.

Proof. a) Let T be a C_0-semigroup. We show first that $\omega(T) < \infty$. Let $M := \sup_{0 \leq t \leq 1} \|T(t)\|$. Then $M < \infty$ by the uniform boundedness principle. Let $\omega = \log M$. Let $t \in \mathbb{R}_+$. Take $n \in \mathbb{N}_0$ and $s \in [0, 1)$ such that $t = n + s$. Then $\|T(t)\| = \|T(s)T(1)^n\| \leq MM^n = Me^{\omega n} \leq Me^{\omega t}$.

b) Assume that $\operatorname{abs}(T) < \infty$. Let $\mu > \lambda > \operatorname{abs}(T)$. Then integration by parts yields for $x \in X$,

$$\begin{aligned} \frac{\hat{T}(\lambda)x - \hat{T}(\mu)x}{\mu - \lambda} &= \int_0^\infty e^{(\lambda-\mu)t}\hat{T}(\lambda)x\,dt - \int_0^\infty \frac{1}{\mu - \lambda} e^{(\lambda-\mu)t} e^{-\lambda t} T(t)x\,dt \\ &= \int_0^\infty e^{(\lambda-\mu)t} \int_0^\infty e^{-\lambda s} T(s)x\,ds\,dt \\ &\quad - \int_0^\infty e^{(\lambda-\mu)t} \int_0^t e^{-\lambda s} T(s)x\,ds\,dt \\ &= \int_0^\infty e^{(\lambda-\mu)t} \int_t^\infty e^{-\lambda s} T(s)x\,ds\,dt \end{aligned}$$

$$= \int_0^\infty e^{-\mu t} \int_t^\infty e^{-\lambda(s-t)} T(s)x \, ds \, dt$$
$$= \int_0^\infty e^{-\mu t} \int_0^\infty e^{-\lambda s} T(s+t)x \, ds \, dt.$$

On the other hand,

$$\hat{T}(\mu)\hat{T}(\lambda)x = \int_0^\infty e^{-\mu t} \int_0^\infty e^{-\lambda s} T(s)T(t)x \, ds \, dt.$$

So it follows from the uniqueness theorem (Theorem 1.7.3) that $(\hat{T}(\lambda))_{\lambda > \mathrm{abs}(T)}$ is a pseudo-resolvent (see Appendix B) if and only if T satisfies $T(s+t) = T(s)T(t)$ $(s, t \geq 0)$.

Now assume that $\hat{T}$ is a pseudo-resolvent. Then $T(0)$ is a projection. Moreover, $T(0)x = 0$ if and only if $T(t)x = T(t)T(0)x = 0$ for all $t \geq 0$. Thus by the uniqueness theorem, $T(0)x = 0$ if and only if $\hat{T}(\lambda)x = 0 \quad (\lambda > \omega(T))$. By Proposition B.6, $(\hat{T}(\lambda))_{\lambda > \mathrm{abs}(T)}$ is a resolvent if and only if $T(0) = I$. This proves that (i) $\Leftrightarrow$ (ii).

c) It follows from (i) and Proposition B.5 that $\{\mathrm{Re}\,\lambda > \mathrm{hol}(\hat{T})\} \subset \rho(A)$ and $\hat{T}(\lambda) = R(\lambda, A)$ whenever $\mathrm{Re}\,\lambda > \mathrm{hol}(\hat{T})$. □

Definition 3.1.8. *Let T be a C_0-semigroup. The* generator *of T is defined as the operator A on X such that $(\omega(T), \infty) \subset \rho(A)$ and $\hat{T}(\lambda) = R(\lambda, A)$ for all $\lambda > \omega(T)$.*

Thus, an operator A is the generator of a C_0-semigroup if and only if its resolvent is a Laplace transform in the sense of Definiton 3.1.4.

In the following proposition we collect the diverse relations of a C_0-semigroup and its generator. These properties will be used frequently without further reference.

Proposition 3.1.9. *Let T be a C_0-semigroup on X and let A be its generator. Then the following properties hold:*

a) $\lim_{\lambda\to\infty} \lambda R(\lambda, A)x = x$ *for all $x \in X$; in particular, A is densely defined.*

b) *For all $x \in X$, the function $u_x(t) := T(t)x$ is a mild solution of (ACP_0).*

c) $R(\lambda, A)T(t) = T(t)R(\lambda, A)$ *for all $\lambda \in \rho(A)$ and $t \geq 0$.*

d) *$x \in D(A)$ implies $T(t)x \in D(A)$ and $AT(t)x = T(t)Ax$.*

e) $\int_0^t T(s)x \, ds \in D(A)$ *and* $A \int_0^t T(s)x \, ds = T(t)x - x$ *for all $x \in X$ and $t \geq 0$.*

f) *Let $x, y \in X$. Then $x \in D(A)$ and $Ax = y$ if and only if $\int_0^t T(s)y \, ds = T(t)x - x$ for all $t \geq 0$.*

g) *Let $x \in X$. Then $x \in D(A)$ if and only if $y = \lim_{t\downarrow 0} \frac{1}{t}(T(t)x - x)$ exists. In that case, $Ax = y$.*

h) *$T(\cdot)x$ is a classical solution of (ACP_0) if and only if $x \in D(A)$.*

i) *If $\lambda \in \mathbb{C}$ then $(e^{\lambda t}T(t))_{t\geq 0}$ is a C_0-semigroup and $A + \lambda$ is its generator.*

j) *Let $x \in X$ and $\lambda \in \mathbb{C}$. Then $x \in D(A)$ and $Ax = \lambda x$ if and only if $T(t)x = e^{\lambda t}x$ for all $t \geq 0$.*

Proof. a) follows from the following Abelian argument. There exist $M \geq 0$ and $\omega \in \mathbb{R}$ such that

$$\|T(t)\| \leq Me^{\omega t} \qquad (t \geq 0).$$

Let $\varepsilon > 0$ and $x \in X$. There exists $\tau > 0$ such that $\|T(t)x - x\| \leq \varepsilon$ for all $t \in [0, \tau]$. Therefore

$$\begin{aligned}
&\limsup_{\lambda\to\infty} \|\lambda R(\lambda, A)x - x\| \\
&\leq \limsup_{\lambda\to\infty} \left\|\lambda \int_0^\infty e^{-\lambda t}(T(t)x - x)\,dt\right\| \\
&\leq \limsup_{\lambda\to\infty} \left\{\lambda \int_0^\tau e^{-\lambda t}\varepsilon\,dt + \lambda \int_\tau^\infty e^{-\lambda t}(Me^{\omega t} + 1)\,dt\,\|x\|\right\} \\
&= \varepsilon.
\end{aligned}$$

b) By Theorem 3.1.3, a function $u \in C(\mathbb{R}_+, X)$ is a mild solution if and only if $\hat{u}(\lambda) = R(\lambda, A)x = \hat{T}(\lambda)x$ for all $\lambda > \omega(T)$. So the claim follows from the uniqueness theorem.

c) follows from Proposition 3.1.5.

d) follows from c) (by Proposition B.7).

e) follows from b).

f) Let $x \in D(A)$. Then by d), $T(s)Ax = AT(s)x$. Hence by e), $\int_0^t T(s)Ax\,ds = A\int_0^t T(s)x\,ds = T(t)x - x$. Conversely, let $x, y \in X$ such that $\int_0^t T(s)y\,ds = T(t)x - x$ for all $t \geq 0$. Then

$$\begin{aligned}
R(\lambda, A)y &= \lambda \int_0^\infty e^{-\lambda t} \int_0^t T(s)y\,ds\,dt \\
&= \lambda \int_0^\infty e^{-\lambda t}(T(t)x - x)\,dt \\
&= \lambda R(\lambda, A)x - x.
\end{aligned}$$

Thus $x \in D(A)$ and $y = \lambda x - (\lambda - A)x = Ax$.

g) Let $x \in D(A)$. Then by f), $\frac{1}{t}(T(t)x - x) = \frac{1}{t}\int_0^t T(s)Ax\,ds \to Ax$ as $t \to 0$. Conversely, let $x, y \in X$ such that $y = \lim_{t\downarrow 0} \frac{1}{t}(T(t)x - x) = \lim_{t\downarrow 0} \frac{1}{t}A\int_0^t T(s)x\,ds$. Since A is closed, it follows that $x \in D(A)$ and $Ax = y$.

h) Let $x \in D(A)$. Then $T(t)x = x + \int_0^t T(s)Ax\,ds$ by e). Thus, $T(\cdot)x \in C^1(\mathbb{R}_+, X)$ and the claim follows from Proposition 3.1.2. Conversely, if $T(\cdot)x$ is a classical solution, then $x = T(0)x \in D(A)$ by definition.

i) follows from Theorem 3.1.7.

j) Replacing A by $A - \lambda$ we may assume that $\lambda = 0$. Now the claim follows from f). □

Property g) is sometimes expressed by saying that A is the *infinitesimal generator* of T.

Since almost all the C_0-semigroups which arise naturally from differential operators cannot be written down explicitly, we do not give examples in this section. However the reader who wishes to see explicit examples may look already to Examples 3.3.10, 3.4.8, 3.7.5, 3.7.6 and 3.7.9, and to various examples in Chapter 5.

We note here that if $T : \mathbb{R}_+ \to \mathcal{L}(X)$ satisfies $T(t+s) = T(t)T(s)$ $(t, s \geq 0)$ and $\lim_{t\downarrow 0} \|T(t)x - x\| = 0$ $(x \in X)$, then T is a C_0-semigroup. To see this, we have to show that T is strongly continuous at $t > 0$. Right-continuity follows immediately from the estimate $\|T(t+h)x - T(t)x\| \leq \|T(t)\| \, \|T(h)x - x\|$. For left-continuity, note that the assumptions imply that there exist $M > 0$ and $\delta > 0$ such that $\|T(h)\| \leq M$ whenever $0 < h < \delta$ (otherwise, there exist $t_n \downarrow 0$ such that $\|T(t_n)\| \to \infty$ and, by the uniform boundedness theorem, there exists $x \in X$ such that $(T(t_n)x)$ is unbounded, which is a contradiction). Hence for $0 < h < \delta$, we have $\|(T(t-h)x - T(t)x\| \leq \|T(\delta - h)\| \, \|T(t-\delta)\| \, \|x - T(h)x\| \to 0$ as $h \downarrow 0$. Since $T(0)T(t)x = T(t)x$, letting $t \downarrow 0$ shows that $T(0) = I$.

The following result characterizes C_0-semigroups which are norm-continuous on $\mathbb{R}_+$. It also describes the situation when the generator A of a C_0-semigroup T is bounded. Since A is closed, this is equivalent to saying that $D(A) = X$.

Theorem 3.1.10. *Let A be the generator of a C_0-semigroup T. The following assertions are equivalent:*

(i) *The operator A is bounded; i.e., $D(A) = X$.*

(ii) $\lim_{t\downarrow 0} \|T(t) - I\| = 0$.

In that case, $T(t) = e^{tA} := \sum_{k=0}^{\infty} \frac{t^k A^k}{k!}$ $(t \geq 0)$.

Proof. (i) ⇒ (ii): Assume that A is bounded. Then clearly, $T(t) := \sum_{k=0}^{\infty} \frac{t^k A^k}{k!}$ defines a continuous mapping $T : \mathbb{R}_+ \to \mathcal{L}(X)$ such that $T(0) = I$ and $\|T(t)\| \leq e^{t\|A\|}$. Let $\lambda > \|A\|$. Then

$$\int_0^\infty e^{-\lambda t} T(t)\, dt = \sum_{k=0}^{\infty} \frac{A^k}{k!} \int_0^\infty e^{-\lambda t} t^k \, dt = \sum_{k=0}^{\infty} A^k \lambda^{-(k+1)} = R(\lambda, A).$$

Thus, T is a C_0-semigroup and A is its generator by Definition 3.1.8.

(ii) ⇒ (i): It follows from Proposition 4.1.3 or direct computation as in the proof of Proposition 3.1.9 a) that $\lim_{\lambda\to\infty} \|\lambda R(\lambda, A) - I\| = 0$. Thus, there exists $\lambda > \omega(T)$ such that $\|\lambda R(\lambda, A) - I\| < 1/2$. This implies that $\lambda R(\lambda, A)$ is invertible in $\mathcal{L}(X)$. In particular, $D(A) = \lambda R(\lambda, A)X = X$. □

Now we consider uniqueness of mild solutions of $(ACP)_0$.

Proposition 3.1.11. *Let T be a C_0-semigroup and A be its generator. Let $\tau > 0$, $x \in X$. Let $u \in C([0,\tau], X)$ such that $\int_0^t u(s)\,ds \in D(A)$ and*

$$A \int_0^t u(s)\,ds = u(t) - x$$

for all $t \in [0,\tau]$. Then $u(t) = T(t)x$.

Proof. Let $v(t) = \int_0^t (u(s) - T(s)x)\,ds$. Then by hypothesis and by Proposition 3.1.9 e), $v(t) \in D(A)$ $(0 \le t \le \tau)$. Moreover, $v'(t) = Av(t)$ $(0 \le t \le \tau)$ and $v(0) = 0$. We show that $v \equiv 0$. Let $S(t)y := \int_0^t T(s)y\,ds$. Then $S(t)y \in D(A)$ and $AS(t)y = T(t)y - y$ for all $y \in X$, by Proposition 3.1.9 e). Let $0 < t \le \tau$, $w(s) := S(t-s)v(s)$, $0 \le s \le t$. Then

$$\begin{aligned} w'(s) &= -T(t-s)v(s) + S(t-s)v'(s) \\ &= -T(t-s)v(s) + S(t-s)Av(s) \\ &= -T(t-s)v(s) + AS(t-s)v(s) \\ &= -v(s). \end{aligned}$$

Since $w(t) = w(0) = 0$, we conclude that

$$0 = w(t) = \int_0^t w'(s)\,ds = -\int_0^t v(s)\,ds.$$

Since $t \in (0,\tau]$ is arbitrary, it follows that $v(s) = 0$ for $s \in [0,\tau]$. □

Proposition 3.1.9 and Proposition 3.1.11 show in particular that the abstract Cauchy problem (ACP_0) is well-posed (in the sense of mild solutions) whenever the operator A generates a C_0-semigroup T. Moreover, the orbits are given by $T(\cdot)x$ where x is the initial value.

Now we show the converse assertion. If (ACP_0) is mildly well-posed (i.e., for each x there exists a unique mild solution), then the operator generates a C_0-semigroup. More precisely, we have the following result.

Theorem 3.1.12. *Let A be a closed operator. The following assertions are equivalent:*

(i) *For all $x \in X$ there exists a unique mild solution of (ACP_0).*

(ii) *The operator A generates a C_0-semigroup.*

(iii) *$\rho(A) \neq \emptyset$ and for all $x \in D(A)$ there exists a unique classical solution of (ACP_0).*

When these assertions hold, the mild solution of (ACP_0) is given by $u(t) = T(t)x$.

Proof. (i) $\Rightarrow$ (ii): Let u_x be the mild solution for the initial value $x \in X$. It follows from uniqueness that $u_x(t)$ is linear in x. So for each $t \geq 0$ there exists a linear mapping $T(t) : X \to X$ such that $T(t)x = u_x(t)$ for all $x \in X$. We show that $T(t)$ is continuous. Denote by $\Phi : X \to C(\mathbb{R}_+, X)$ the mapping $\Phi(x) = u_x$. Note that $C(\mathbb{R}_+, X)$ is a Fréchet space for the topology of uniform convergence on intervals of the form $[0, \tau]$ where $\tau > 0$. The mapping Φ is linear. We show that Φ has a closed graph. In fact, let $x_n \to x$ in X and $u_{x_n} \to u$ in $C(\mathbb{R}_+, X)$. Let $t > 0$. Then $\int_0^t u_{x_n}(s)\,ds$ converges to $\int_0^t u(s)\,ds$ as $n \to \infty$. Since $A \int_0^t u_{x_n}(s)\,ds = u_{x_n}(t) - x_n$ and since A is closed, it follows that $\int_0^t u(s)\,ds \in D(A)$ and $A \int_0^t u(s)\,ds = \lim_{n\to\infty} u_{x_n}(t) - x_n = u(t) - x$. Thus $u(t) = T(t)x$; i.e., $u = \Phi(x)$. It follows from the closed graph theorem that Φ is continuous. This implies that $T(t) \in \mathcal{L}(X)$ for all $t \geq 0$.

Let u be a mild solution of (ACP_0) with initial value x. Then it is easy to see that $u(\cdot + s)$ is a mild solution for the initial value $u(s)$. So uniqueness implies that $T(t+s)x = T(t)T(s)x$.

We have shown that T is a C_0-semigroup. Let B be the generator of T. Then $R(\lambda, B) = \hat{T}(\lambda)$ $(\lambda > \omega(T))$. On the other hand, by Theorem 3.1.3, $\hat{T}(\lambda)x \in D(A)$ and $(\lambda - A)\hat{T}(\lambda)x = x$ for all $x \in X$ and $\lambda > \omega(T)$. Thus, $D(B) \subset D(A)$ and $(\lambda - A)R(\lambda, B)x = x$ $(x \in X)$ if $\lambda > \omega(T)$. If we show that $(\lambda - A)$ is injective, then it follows that $\lambda \in \rho(A)$ and $R(\lambda, A) = R(\lambda, B)$. Thus, $A = B$.

Assume that $\lambda > \omega(T)$ and let $x \in D(A)$ such that $(\lambda - A)x = 0$. Then $u(t) := e^{\lambda t}x$ is a mild solution. Thus, $T(t)x = e^{\lambda t}x$. Since $\omega(T) < \lambda$, it follows that $x = 0$.

(ii) $\Rightarrow$ (iii) follows from Proposition 3.1.9 b) and Proposition 3.1.11.

(iii) $\Rightarrow$ (i): Let $\lambda \in \rho(A)$. Let $x \in X$. There exists a classical solution v of (ACP_0) with initial value $R(\lambda, A)x$. It is easy to check that $u(t) := (\lambda - A)v(t)$ defines a mild solution of (ACP_0) with initial value x. This shows existence. In order to show uniqueness, let u be a mild solution for the initial value $x = 0$. Then $v(t) := \int_0^t u(s)\,ds$ defines a classical solution for the initial value 0. Hence $v(t) = 0$ for all $t \geq 0$ by assumption. It follows that $u(t) = 0$ $(t \geq 0)$. ☐

As a corollary of Theorem 3.1.12 we show that one can also characterize C_0-semigroups and their generator by property e) of Proposition 3.1.9. This will be useful later.

Corollary 3.1.13. *Let A be a closed operator on X and $T : \mathbb{R}_+ \to \mathcal{L}(X)$ be strongly continuous such that $\int_0^t T(s)x\,ds \in D(A)$ and*

$$A \int_0^t T(s)x\,ds = T(t)x - x$$

for all $x \in X$, $t \geq 0$. Assume that $T(t)x \in D(A)$ and $AT(t)x = T(t)Ax$ for all $x \in D(A)$, $t \geq 0$. Then T is a C_0-semigroup and A is its generator.

Proof. Let $x \in X$. Then $u(t) := T(t)x$ defines a mild solution of (ACP_0). As in the proof of Proposition 3.1.11, u is the unique mild solution. Now the claim follows from Theorem 3.1.12. □

We show by an example that the condition that A has non-empty resolvent set cannot be omitted in assertion (iii) of Theorem 3.1.12; i.e., it can happen that the abstract Cauchy problem (ACP_0) is well-posed in the sense of classical solutions without A being the generator of a C_0-semigroup.

Example 3.1.14. Let B be a densely defined closed operator on a Banach space Y such that $D(B) \neq Y$. Consider the operator A on $X := Y \times Y$ given by

$$A = \begin{pmatrix} 0 & B \\ 0 & 0 \end{pmatrix}$$

with domain $Y \times D(B)$. Then A is closed and densely defined. Moreover, for all $(x, y) \in D(A)$,

$$u(t) = (x + tBy, y) \quad (t \geq 0)$$

is the unique classical solution of (ACP_0). However, there does not exist a mild solution for an initial value (x, y) if $y \in Y \setminus D(B)$. This is easy to see.

Now we consider the inhomogeneous Cauchy problem. In contrast to the homogeneous case, we consider this on a bounded interval $[0, \tau]$ where $\tau \in (0, \infty)$, but results on $\mathbb{R}_+$ can be deduced by letting τ vary. We shall apply Laplace transform techniques to inhomogeneous Cauchy problems on $\mathbb{R}_+$ in Section 5.6.

Let A be a closed operator and let $f \in L^1([0, \tau], X)$ where $\tau > 0$. We consider the inhomogeneous Cauchy problem

$$(ACP_f) \quad \begin{cases} u'(t) = & Au(t) + f(t) \quad (t \in [0, \tau]), \\ u(0) = & x, \end{cases}$$

where $x \in X$. A function $u \in C([0, \tau], X)$ is called a *mild solution* of (ACP_f) if $\int_0^t u(s)\,ds \in D(A)$ and

$$u(t) = x + A\int_0^t u(s)\,ds + \int_0^t f(s)\,ds \quad (t \in [0, \tau]).$$

Assume that $f \in C([0, \tau], X)$. Then we define a *classical solution* as a function $u \in C^1([0, \tau], X)$ such that $u(t) \in D(A)$ for all $t \in [0, \tau]$ and such that (ACP_f) is valid. Note that in that case, since $Au(t) = u'(t) - f(t)$ $(t \in [0, \tau])$, one has $u \in C([0, \tau], D(A))$, where $D(A)$ is seen as a Banach space with the graph norm. Since A is closed, the proof of Proposition 3.1.2 is also valid in the inhomogeneous case, so the following holds.

Proposition 3.1.15. *Let $f \in C([0, \tau], X)$ and $u \in C([0, \tau], X)$ be a mild solution of (ACP_f). Then u is a classical solution if and only if $u \in C^1([0, \tau], X)$.*

In the case when A generates a C_0-semigroup there always exists a mild solution.

Proposition 3.1.16. *Let A be the generator of a C_0-semigroup T on X. Then for every $f \in L^1([0,\tau],X)$ the problem (ACP_f) has a unique mild solution u given by*

$$u(t) = T(t)x + \int_0^t T(t-s)f(s)\,ds \quad (t \in [0,\tau]). \tag{3.4}$$

Sometimes, (3.4) *is called* the variation of constants formula *for the solution.*

Proof. Uniqueness: Let $u_1, u_2 \in C([0,\tau],X)$ be two mild solutions of (ACP_f). Then $u := u_1 - u_2 \in C([0,\tau],X)$, $u(0) = 0$ and $A\int_0^t u(s)\,ds = u(t)$ for all $t \in [0,\tau]$. It follows from Proposition 3.1.11 that $u \equiv 0$.

Existence: We have seen that $T(\cdot)x$ is a mild solution of the homogeneous Cauchy problem. It remains to show that $v(t) := \int_0^t T(t-s)f(s)\,ds$ is a mild solution of (ACP_f) with initial value $x = 0$. Extending f by 0 to $\mathbb{R}_+$, Proposition 1.3.4 shows that $v \in C([0,\tau],X)$. Using Proposition 3.1.9 e) and Fubini's theorem we obtain

$$\begin{aligned}
A\int_0^t v(s)\,ds &= A\int_0^t\int_0^s T(s-r)f(r)\,dr\,ds \\
&= A\int_0^t\int_r^t T(s-r)f(r)\,ds\,dr \\
&= \int_0^t A\int_0^{t-r} T(s)f(r)\,ds\,dr \\
&= \int_0^t (T(t-r)f(r) - f(r))\,dr \\
&= v(t) - \int_0^t f(r)\,dr.
\end{aligned}$$

This proves the claim. ☐

Corollary 3.1.17. *Let A be the generator of a C_0-semigroup. Let $x \in D(A)$, $f_0 \in X$, $f(t) = f_0 + \int_0^t f'(s)\,ds$ $(t \in [0,\tau])$ for some function $f' \in L^1([0,\tau],X)$. Then the function u defined by* (3.4) *is a classical solution of (ACP_f).*

This follows from Proposition 3.1.16, Proposition 1.3.6 and Proposition 3.1.15. This result will later be extended to a class of operators which are not densely defined (Theorem 3.5.2).

Finally, given a closed operator A on X, we consider the Cauchy problem on the line

$$ACP_0(\mathbb{R}) \begin{cases} u'(t) = & Au(t) \quad (t \in \mathbb{R}), \\ u(0) = & x, \end{cases}$$

where $x \in X$. A function $u \in C(\mathbb{R}, X)$ is called a *mild solution* of $ACP_0(\mathbb{R})$ if $\int_0^t u(s)\,ds \in D(A)$ and

$$A\int_0^t u(s)\,ds = u(t) - x \qquad \text{for all } t \in \mathbb{R}.$$

Proposition 3.1.18. *Assume that A is an operator such that A generates a C_0-semigroup T_+ and $-A$ generates a C_0-semigroup T_-. Define*

$$U(t) = \begin{cases} T_+(t) & \text{if } t \geq 0, \\ T_-(-t) & \text{if } t < 0. \end{cases} \tag{3.5}$$

Then $U : \mathbb{R} \to \mathcal{L}(X)$ is strongly continuous, $U(0) = I$ and $U(t+s) = U(t)U(s)$ $(t, s \in \mathbb{R})$.

Proof. Note first that $T_+(t)T_-(s) = T_-(s)T_+(t)$ for $s, t \geq 0$, by Proposition 3.1.5. The only assertion which is not obvious is to show that $U(t-s) = U(t)U(-s)$ if $t \geq 0,\ s \geq 0$. We can assume that $0 \leq s \leq t$ (replacing A by $-A$ for the other case). Let $x \in X$, $t \geq 0$, $v(s) := T_+(t-s)x$ for $s \in [0,t]$. Then for $0 \leq r \leq t$,

$$\begin{aligned} -A\int_0^r v(s)\,ds &= -A\int_0^r T_+(t-s)x\,ds = -A\int_{t-r}^t T_+(s)x\,ds \\ &= T_+(t-r)x - T_+(t)x = v(r) - T_+(t)x. \end{aligned}$$

Thus, v is a mild solution of the problem

$$\begin{cases} v'(s) = & -Av(s) \quad (0 < s \leq t), \\ v(0) = & T_+(t)x. \end{cases}$$

It follows from Proposition 3.1.11 that $v(s) = T_-(s)T_+(t)x$. Hence,

$$U(t-s)x = T_+(t-s)x = v(s) = T_-(s)T_+(t)x = U(-s)U(t)x. \qquad \square$$

Definition 3.1.19. *An operator A on X is said* to generate a C_0-group *if A and $-A$ generate C_0-semigroups. In that case, the function $U : \mathbb{R} \to \mathcal{L}(X)$ defined by* (3.5) *is called* the C_0-group generated by A.

Proposition 3.1.20. *A closed operator A generates a C_0-group if and only if for every $x \in X$ there exists a unique mild solution u of $ACP_0(\mathbb{R})$. In that case, $u(t) = U(t)x$ $(t \in \mathbb{R})$, where U is the C_0-group generated by A. If $x \in D(A)$, then $U(\cdot)x \in C^1(\mathbb{R}, X)$, $U(t)x \in D(A)$ for all $t \in \mathbb{R}$ and $\frac{d}{dt}U(t)x = AU(t)x$ $(t \in \mathbb{R})$.*

Proof. Assume that $ACP_0(\mathbb{R})$ is mildly well-posed; i.e., for all $x \in X$ there exists a unique mild solution of $ACP_0(\mathbb{R})$. Then it is clear that for each $x \in X$ there exists a mild solution of

$$(CP)_\pm \quad \begin{cases} u'(t) = & \pm Au(t) \quad (t \geq 0), \\ u(0) = & x. \end{cases}$$

The solutions of $(CP)_+$ and $(CP)_-$ are both unique. In fact, let $u \in C(\mathbb{R}_+, X)$ be a mild solution of $(CP)_+$ with initial value $u(0) = 0$. Then extending u by 0 on $(-\infty, 0)$ one obtains a mild solution of $ACP_0(\mathbb{R})$. Hence $u \equiv 0$ by assumption. The same argument is valid for $(CP)_-$. Now it follows from Theorem 3.1.12 that A and $-A$ are both generators of C_0-semigroups.

Conversely, if A generates a C_0-group, then it is easy to see that $U(\cdot)x$ is a mild solution of $ACP_0(\mathbb{R})$. The remaining properties follow directly from the corresponding results for semigroups. □

If A generates a C_0-semigroup T, then mild solutions of $ACP_0(\mathbb{R})$ can be described differently.

Definition 3.1.21. *A function $u \in C(\mathbb{R}, X)$ is called a* complete orbit of T *if*

$$u(t+s) = T(t)u(s) \quad \textit{for all } t \geq 0,\ s \in \mathbb{R}.$$

Proposition 3.1.22. *Let A be the generator of a C_0-semigroup T and let $u \in C(\mathbb{R}, X)$, $x = u(0)$. Then u is a mild solution of $ACP_0(\mathbb{R})$ if and only if u is a complete orbit.*

Proof. Assume that u is a complete orbit. Let $x = u(0)$. Since $T(t)x = u(t)$ for $t \geq 0$, we have

$$A \int_0^t u(s)\, ds = u(t) - x \quad (t \geq 0).$$

For $t < 0$ we have $T(-t)u(t) = u(0) = x$ and

$$\begin{aligned} A \int_0^t u(r)\, dr &= A \int_{-t}^0 u(r+t)\, dr = -A \int_0^{-t} u(r+t)\, dr \\ &= -A \int_0^{-t} T(r)u(t)\, dr = u(t) - T(-t)u(t) = u(t) - x. \end{aligned}$$

Thus u is a mild solution of $ACP_0(\mathbb{R})$.

Conversely, assume that u is a mild solution of $ACP_0(\mathbb{R})$ with $x = u(0)$. Let $s \in \mathbb{R}$. Then for $t \geq 0$,

$$A \int_0^t u(r+s)\, dr = A \int_s^{s+t} u(r)\, dr = u(t+s) - u(s).$$

Thus $u(\cdot + s)$ is a mild solution of (ACP_0) for $x = u(s)$. Hence $u(t+s) = T(t)u(s)$ for $t \geq 0$. □

If a C_0-semigroup T extends to a C_0-group, then $T(t)$ is invertible for all $t > 0$. The following converse statement is sometimes useful (and will be needed in Proposition 4.7.2, for example).

Proposition 3.1.23. *Let A be the generator of a C_0-semigroup T. If there exists $t_0 > 0$ such that $T(t_0)$ is invertible, then A generates a C_0-group.*

Proof. a) Let $t > 0$. We show that $T(t)$ is invertible. Let $T(t)x = 0$. Choose $n \in \mathbb{N}$ such that $nt_0 > t$. Then $T(nt_0)x = T(nt_0 - t)T(t)x = 0$. Since $T(nt_0) = T(t_0)^n$ is invertible, it follows that $x = 0$. Thus, $T(t)$ is injective. Let $y \in X$. Let $x := T(nt_0 - t)T(nt_0)^{-1}y$. Then $T(t)x = y$. Thus, $T(t)$ is surjective.

b) Define $U(t) := T(t)$ for $t \geq 0$ and $U(t) := T(-t)^{-1}$ for $t < 0$. Then $U : \mathbb{R} \to \mathcal{L}(X)$ satisfies $U(t+s) = U(t)U(s)$ for all $t, s \in \mathbb{R}$. Let $x \in X$, $t_0 \in \mathbb{R}$. Let $t_1 > \max\{-t_0, 0\}$. Then

$$\lim_{t \to t_0} U(t)x = T(t_1)^{-1} \lim_{t \to t_0} T(t + t_1)x = U(t_0)x.$$

Thus U is strongly continuous.

c) We show that the generator of $(U(-t))_{t \geq 0}$ is $-A$. Let $x \in D(A)$. Then

$$\begin{aligned} \lim_{t \downarrow 0} \frac{1}{t}\left(U(-t)x - x\right) &= \lim_{t \downarrow 0} U(-1)\left(\frac{1}{t}T(1-t)x - T(1)x\right) \\ &= T(1)^{-1}(-AT(1)x) = -Ax. \end{aligned}$$

Conversely, if $x \in X$ such that $y := \lim_{t \downarrow 0} \frac{1}{t}(U(-t)x - x)$ exists, it follows as above that $-y = \lim_{t \downarrow 0} \frac{1}{t}(T(t)x - x)$. Thus, $x \in D(A)$ and $-Ax = y$. Now the claim follows from Proposition 3.1.9. □

3.2 Integrated Semigroups and Cauchy Problems

Let T be a C_0-semigroup on a Banach space X with generator A. For $k \in \mathbb{N}$ we define $S : \mathbb{R}_+ \to \mathcal{L}(X)$ by

$$S(t)x := \int_0^t \frac{(t-s)^{k-1}}{(k-1)!} T(s)x \, ds \qquad (t \geq 0, x \in X).$$

By Theorem 3.1.7, there exist $M, \omega \geq 0$ such that $\| T(t) \| \leq Me^{\omega t}$ for all $t \geq 0$. Taking Laplace transforms and integrating by parts yields

$$R(\lambda, A) = \lambda^k \int_0^\infty e^{-\lambda t} S(t) \, dt \qquad (\lambda > \omega). \tag{3.6}$$

Here, the Laplace integral is understood in the sense of Section 1.4, but, for each $x \in X$ and $\lambda > \omega$, one has $R(\lambda, A)x = \lambda^k \int_0^\infty e^{-\lambda t} S(t)x \, dt$ as an absolutely convergent Bochner integral.

The above formula (3.6) is the basic idea behind the following definition. Consider an arbitrary strongly continuous function $S : \mathbb{R}_+ \to \mathcal{L}(X)$. We recall from Proposition 1.4.5 that $\operatorname{abs}(S) < \infty$ if and only if there exist constants $M, \omega \geq 0$ such that

$$\left\| \int_0^t S(s) \, ds \right\| \leq Me^{\omega t} \qquad (t \geq 0). \tag{3.7}$$

In that case, the Laplace integral

$$\hat{S}(\lambda)x := \int_0^\infty e^{-\lambda t} S(t)x\,dt := \lim_{\tau\to\infty} \int_0^\tau e^{-\lambda t} S(t)x\,dt$$

exists for all $\lambda \in \mathbb{C}$ with $\operatorname{Re}\lambda > \omega$ and all $x \in X$ and defines a bounded operator $\hat{S}(\lambda)$ on X. Hence, the following definition is meaningful.

Definition 3.2.1. *Let A be an operator on a Banach space X and $k \in \mathbb{N}_0$. We call A the* generator of a k-times integrated semigroup *if there exist $\omega \geq 0$ and a strongly continuous function $S : \mathbb{R}_+ \to \mathcal{L}(X)$ such that* $\operatorname{abs}(S) \leq \omega$, $(\omega, \infty) \subset \rho(A)$ *and*

$$R(\lambda, A) = \lambda^k \int_0^\infty e^{-\lambda t} S(t)\,dt \qquad (\lambda > \omega). \tag{3.8}$$

In this case, S is called the k-times integrated semigroup generated by A. *If $k = 1$ we also use the notion* once integrated semigroup.

By Theorem 3.1.7, a 0-times integrated semigroup is the same as a C_0-semigroup. The discussion above shows that if A generates a 0-times integrated semigroup, then A generates a k-times integrated semigroup for every $k \in \mathbb{N}$. The same argument shows that if A generates a k-times integrated semigroup, then A generates an n-times integrated semigroup for every $n > k$.

As in the situation of C_0-semigroups we collect diverse relations of an integrated semigroup and its generator.

Lemma 3.2.2. *Let $k \in \mathbb{N}$ and let S be a k-times integrated semigroup on X with generator A. Then the following hold:*

a) $R(\mu, A)S(t) = S(t)R(\mu, A) \quad (t \geq 0, \mu \in \rho(A))$.

b) *If $x \in D(A)$, then $S(t)x \in D(A)$ and $AS(t)x = S(t)Ax$ for all $t \geq 0$.*

c) *Let $x \in D(A)$ and $t \geq 0$. Then*

$$\int_0^t S(s)Ax\,ds = S(t)x - \frac{t^k}{k!}x.$$

In particular, $\frac{d}{dt}(S(t)x) = S(t)Ax + \frac{t^{k-1}}{(k-1)!}x$.

d) *Let $x \in X$ and $t \geq 0$. Then $\int_0^t S(s)x\,ds \in D(A)$ and*

$$A\int_0^t S(s)x\,ds = S(t)x - \frac{t^k}{k!}x.$$

In particular, $S(0) = 0$.

e) *Let $x, y \in X$ such that $\int_0^t S(s)y\,ds = S(t)x - \frac{t^k}{k!}x$ for all $t \geq 0$. Then $x \in D(A)$ and $Ax = y$.*

Proof. By definition and assertion (3.7), there exist constants $M, \omega \geq 0$ such that $(\omega, \infty) \subset \rho(A)$ and $\| \int_0^t S(s)\, ds \| \leq M e^{\omega t}$ for $t \geq 0$. In the following let $\lambda > \omega$.

a) follows from Proposition 3.1.5.

b) is implied by a) (by Proposition B.7).

c) Let $x \in D(A)$. Integrating by parts yields

$$\begin{aligned}
\lambda^{k+1} \int_0^\infty e^{-\lambda t} \frac{t^k}{k!} x\, dt &= R(\lambda, A)(\lambda - A)x \\
&= \lambda^{k+1} \int_0^\infty e^{-\lambda t} S(t)x\, dt - \lambda^k \int_0^\infty e^{-\lambda t} S(t)Ax\, dt \\
&= \lambda^{k+1} \int_0^\infty e^{-\lambda t} S(t)x\, dt \\
&\quad -\lambda^{k+1} \int_0^\infty e^{-\lambda t} \int_0^t S(s)Ax\, ds\, dt.
\end{aligned}$$

The uniqueness theorem implies the assertion.

d) Let $\mu \in \rho(A)$ and $x \in X$. By a) and c) we have

$$\int_0^t S(s)x\, ds = \mu R(\mu, A) \int_0^t S(s)x\, ds - R(\mu, A)S(t)x + \frac{t^k}{k!} R(\mu, A)x.$$

Hence $\int_0^t S(s)x\, ds \in D(A)$ and

$$(\mu - A) \int_0^t S(s)x\, ds = \mu \int_0^t S(s)x\, ds - S(t)x + \frac{t^k}{k!} x.$$

e) Let $x, y \in X$ such that $\int_0^t S(s)y\, ds = S(t)x - \frac{t^k}{k!} x$ for all $t \geq 0$. Then

$$\begin{aligned}
R(\lambda, A)(\lambda x - y) &= \lambda^{k+1} \int_0^\infty e^{-\lambda t} S(t)x\, dt - \lambda^k \int_0^\infty e^{-\lambda t} S(t)y\, dt \\
&= \lambda^{k+1} \int_0^\infty e^{-\lambda t} S(t)x\, dt - \lambda^{k+1} \int_0^\infty e^{-\lambda t} \int_0^t S(s)y\, ds\, dt \\
&= x.
\end{aligned}$$

Hence $x \in D(A)$ and $\lambda x - y = \lambda x - Ax$, which implies that $Ax = y$. □

Remark 3.2.3. Observe that in contrast to the situation of C_0-semigroups, generators of k-times integrated semigroups for $k \geq 1$ need not be densely defined. However, assertion d) of Lemma 3.2.2 implies that $S(t)x \in \overline{D(A)}$ for $t \geq 0$ and $x \in X$. □

We saw in Theorem 3.1.7 that C_0-semigroups are precisely those operator-valued functions whose Laplace transforms are resolvents $R(\lambda, A)$ of operators A. By definition, k-times integrated semigroups are exactly those operator-valued functions whose Laplace transforms are $\lambda^{-k} R(\lambda, A)$ for operators A. In the following proposition we show that this property corresponds to the functional equation (3.9) for S.

Proposition 3.2.4. *Let $S : \mathbb{R}_+ \to \mathcal{L}(X)$ be a strongly continuous function satisfying $\| \int_0^t S(s)\, ds \| \le Me^{\omega t}$ $(t \ge 0)$ for some $M, \omega \ge 0$. Let $k \in \mathbb{N}$. For $\lambda > \omega$ set*

$$R(\lambda) := \lambda^k \int_0^\infty e^{-\lambda t} S(t)\, dt.$$

Then the following assertions are equivalent:

(i) *There exists an operator A such that $(\omega, \infty) \subset \rho(A)$ and $R(\lambda) = (\lambda - A)^{-1}$ for $\lambda > \omega$.*

(ii) *For $s, t \ge 0$,*

$$S(t)S(s) = \frac{1}{(k-1)!} \left[\int_t^{t+s} (s+t-r)^{k-1} S(r)\, dr - \int_0^s (s+t-r)^{k-1} S(r)\, dr \right], \tag{3.9}$$

and $S(t)x = 0$ for all $t \ge 0$ implies $x = 0$.

Proof. We first claim that $\{R(\lambda) : \lambda > \omega\}$ is a pseudo-resolvent if and only if (3.9) holds. Since

$$\frac{R(\lambda)}{\lambda^k} \frac{R(\mu)}{\mu^k} = \int_0^\infty e^{-\lambda t} \int_0^\infty e^{-\mu s} S(t)S(s)\, ds\, dt \qquad (\lambda, \mu > \omega),$$

the claim follows from the uniqueness theorem (Theorem 1.7.3) provided we are able to prove that

$$\frac{1}{\mu - \lambda} \frac{1}{\lambda^k} \frac{1}{\mu^k} (R(\lambda) - R(\mu)) \tag{3.10}$$

equals the term

$$\int_0^\infty e^{-\lambda t} \int_0^\infty e^{-\mu s} \frac{1}{(k-1)!} \int_t^{s+t} (s+t-r)^{k-1} S(r)\, dr\, ds\, dt - \int_0^\infty e^{-\lambda t} \int_0^\infty e^{-\mu s} \frac{1}{(k-1)!} \int_0^s (s+t-r)^{k-1} S(r)\, dr\, ds\, dt. \tag{3.11}$$

Notice that (3.10) equals

$$\frac{1}{\mu^k} \frac{1}{\mu - \lambda} \left(\frac{R(\lambda)}{\lambda^k} - \frac{R(\mu)}{\mu^k} \right) + \frac{1}{\mu - \lambda} \left(\frac{1}{\mu^k} - \frac{1}{\lambda^k} \right) \frac{R(\mu)}{\mu^k} =: I + II.$$

As in the proof of Theorem 3.1.7 we see that term I equals

$$\begin{aligned} &\frac{1}{\mu^k} \int_0^\infty e^{-\mu t} \int_0^\infty e^{-\lambda s} S(t+s)\, ds\, dt \\ &= \int_0^\infty e^{-\lambda t} \int_0^\infty e^{-\mu s} \int_t^{t+s} \frac{(t+s-r)^{k-1}}{(k-1)!} S(r)\, dr\, ds\, dt. \end{aligned}$$

Hence, it remains to show that term II is equal to the second term in (3.11). This follows from the following computation:

$$\begin{aligned}
-II &= \sum_{j=0}^{k-1} \lambda^{-(j+1)} \mu^{(j-k)} \hat{S}(\mu) \\
&= \sum_{j=0}^{k-1} \lambda^{-(j+1)} \int_0^\infty e^{-\mu s} \int_0^s \frac{(s-r)^{k-j-1}}{(k-j-1)!} S(r)\, dr\, ds \\
&= \sum_{j=0}^{k-1} \int_0^\infty e^{-\lambda t} \frac{t^j}{j!}\, dt \int_0^\infty e^{-\mu s} \int_0^s \frac{(s-r)^{k-j-1}}{(k-j-1)!} S(r)\, dr\, ds \\
&= \int_0^\infty e^{-\lambda t} \int_0^\infty e^{-\mu s} \int_0^s \frac{(s+t-r)^{k-1}}{(k-1)!} S(r)\, dr\, ds\, dt.
\end{aligned}$$

Finally, recall that $\{R(\lambda) : \lambda > \omega\}$ is the resolvent of an operator A in X if and only if $\operatorname{Ker} R(\lambda) = \{0\}$ (see Proposition B.6). This is equivalent to the fact that $S(t)x = 0$ for all $t \geq 0$ implies $x = 0$. □

In contrast to the situation for semigroups the functional equation (3.9) in Proposition 3.2.4 does not imply that $\operatorname{abs}(S) < \infty$. This will be shown at the end of this section in Remark 3.2.15.

It follows from Proposition 3.1.5 or from the above Proposition 3.2.4 that for a k-times integrated semigroup S we have

$$S(t)S(s) = S(s)S(t) \qquad (s, t \geq 0). \tag{3.12}$$

The above elementary properties of integrated semigroups will be used in the following without further notice.

A particular example of an integrated semigroup is the antiderivative of a semigroup which is not necessarily strongly continuous at 0. In order to make this more precise, consider a strongly continuous function $T : (0, \infty) \to \mathcal{L}(X)$ satisfying

a) $T(t+s) = T(t)T(s) \quad (s, t > 0)$,

b) there exists $c > 0$ such that $||T(t)|| \leq c$ for all $t \in (0, 1]$,

c) $T(t)x = 0$ for all $t > 0$ implies $x = 0$.

Then by the proof of Theorem 3.1.7, there exist constants $M, \omega \geq 0$ such that $||T(t)|| \leq Me^{\omega t}$ for all $t > 0$. For $t \geq 0$ set

$$S(t) := \int_0^t T(s)\, ds.$$

Then $(S(t))_{t \geq 0}$ satisfies condition (ii) of Proposition 3.2.4 with $k = 1$. Hence, there exists an operator A such that $(\omega, \infty) \subset \rho(A)$ and

$$R(\lambda, A) = \lambda \int_0^\infty e^{-\lambda t} S(t)\, dt = \int_0^\infty e^{-\lambda t} T(t)\, dt \qquad (\lambda > \omega). \tag{3.13}$$

Definition 3.2.5. *Let $T : (0,\infty) \to \mathcal{L}(X)$ be a strongly continuous function satisfying assumptions* a), b) *and* c) *above. Let A be defined as in* (3.13). *Then T is called a* semigroup *and A is called its* generator.

We will see in the following Section 3.3 that a semigroup T on X is a C_0-semigroup on X if and only if $\overline{D(A)} = X$.

Proposition 3.2.6. *Let A be the generator of a k-times integrated semigroup S on X for some $k \in \mathbb{N}$ and let $a \in \mathbb{C}$. Then $A - a$ generates a k-times integrated semigroup S_a on X which is given by*

$$S_a(t) = e^{-at}S(t) + \sum_{j=1}^{k} \binom{k}{j} a^j \int_0^t \frac{(t-s)^{j-1}}{(j-1)!} e^{-as} S(s)\, ds.$$

Proof. Taking Laplace transforms of S_a we obtain, for μ sufficiently large

$$\begin{aligned} \int_0^\infty e^{-\mu t} S_a(t)\, dt &= \int_0^\infty e^{-(\mu+a)t} S(t)\, dt \\ &\quad + \int_0^\infty e^{-(\mu+a)t} S(t)\, dt \sum_{j=1}^{k} \binom{k}{j} a^j \mu^{-j} \\ &= \frac{R(\mu+a, A)}{(\mu+a)^k} \frac{1}{\mu^k} \sum_{j=0}^{k} \binom{k}{j} a^j \mu^{k-j} \\ &= \frac{R(\mu+a, A)}{(\mu+a)^k} \frac{(\mu+a)^k}{\mu^k} = \frac{R(\mu, A-a)}{\mu^k}. \end{aligned}$$

Hence, the assertion follows directly from Definition 3.2.1. □

Proposition 3.2.7. *Let A be an operator on X and let $\mu \in \rho(A), k \in \mathbb{N}$. Then A generates a k-times integrated semigroup S on X if and only if there exists $\omega \in \mathbb{R}$ such that $(\omega,\infty) \subset \rho(A)$ and $R(\cdot, A)R(\mu, A)^k$ is a Laplace transform $\hat{T}$ in the sense of Definition* 3.1.4. *In that case, $\omega(T) < \infty$ if and only if $\omega(S) < \infty$.*

Proof. By Proposition 3.2.6, the operator A generates a k-times integrated semigroup if and only if $A - \mu$ does so. By Definition 3.2.1 and Proposition 1.6.1 a), this is equivalent to $\lambda \mapsto (\lambda - \mu)^{-k} R(\lambda, A)$ being a Laplace transform. The resolvent equation implies that

$$R(\lambda, A)R(\mu, A)^k = \frac{R(\lambda, A)}{(\mu-\lambda)^k} - \frac{R(\mu, A)}{(\mu-\lambda)^k} - \frac{R(\mu, A)^2}{(\mu-\lambda)^{k-1}} - \cdots - \frac{R(\mu, A)^k}{(\mu-\lambda)} \tag{3.14}$$

for $\lambda, \mu \in \rho(A), \lambda \neq \mu$. The first assertion follows easily. Moreover, each step in the passage between S and T preserves exponential boundedness. □

In the following we characterize those operators which generate a k-times integrated semigroup for some $k \in \mathbb{N}$ simply by the fact that the resolvent is polynomially bounded on a half-plane. The real characterization given in the next section determines precisely the order of integration.

Theorem 3.2.8. *Let A be an operator and let $k \in \mathbb{N}$.*

a) *Assume that there exists $\omega \geq 0$, $M \geq 0$, $b > 0$ such that $\lambda \in \rho(A)$ and $\|R(\lambda, A)\| \leq M|\lambda|^{k-1-b}$ whenever $\operatorname{Re}\lambda > \omega$. Then A generates a k-times integrated semigroup S satisfying $\omega(S) \leq \omega$.*

b) *Conversely, if A generates a k-times integrated semigroup S such that $\omega(S) < \infty$, then for $\omega > \max\{\omega(S), 0\}$ there exists M such that $\lambda \in \rho(A)$ and $\|R(\lambda, A)\| \leq M|\lambda|^k$ whenever $\operatorname{Re}\lambda > \omega$.*

Proof. a) Apply Theorem 2.5.1 to $q(\lambda) := \lambda^b R(\lambda, A)/\lambda^k$.

b) Let $\max\{\omega(S), 0\} < \omega_1 < \omega$. There exists $M_1 \geq 0$ such that $\|S(t)\| \leq M_1 e^{\omega_1 t}$ $(t \geq 0)$. Hence, by Proposition B.5, $\lambda \in \rho(A)$ and

$$\begin{aligned}\|R(\lambda, A)\| &= \left\|\lambda^k \int_0^\infty e^{-\lambda t} S(t)\,dt\right\| \\ &\leq |\lambda|^k M_1 (\operatorname{Re}\lambda - \omega_1)^{-1} \leq |\lambda|^k M_1(\omega - \omega_1)^{-1}\end{aligned}$$

whenever $\operatorname{Re}\lambda > \omega$. □

We now turn our attention to the inhomogenous Cauchy problem

$$(ACP_f) \quad \begin{cases} u'(t) = & Au(t) + f(t) \quad (t \in [0, \tau]), \\ u(0) = & x, \end{cases} \tag{3.15}$$

where $\tau > 0, f \in L^1([0, \tau], X), x \in X$ and A is assumed to be the generator of a k-times integrated semigroup S on X for some $k \in \mathbb{N}$. Recall from Section 3.1 that by a *mild solution* of (ACP_f) we understand a function $u \in C([0, \tau], X)$ such that $\int_0^t u(s)\,ds \in D(A)$ and $u(t) = A\int_0^t u(s)\,ds + x + \int_0^t f(s)\,ds$ for all $t \in [0, \tau]$ and that by a *classical solution* of (ACP_f) we mean a function $u \in C^1([0, \tau], X) \cap C([0, \tau], D(A))$ satisfying (ACP_f) for all $t \in [0, \tau]$. For $x \in X$ consider the function v given by

$$v(t) := S(t)x + \int_0^t S(s) f(t - s)\,ds \qquad (t \in [0, \tau]). \tag{3.16}$$

It follows from Proposition 1.3.4 that $v \in C([0, \tau], X)$.

Lemma 3.2.9. a) *If there is a mild solution u of (ACP_f), then $v \in C^k([0, \tau], X)$ and $u = v^{(k)}$.*

b) *If there exists a classical solution u of (ACP_f), then $v \in C^{k+1}([0, \tau], X)$ and $u = v^{(k)}$.*

Proof. a) For $0 \leq s \leq t \leq \tau$ set $w(s) := S(t-s)\int_0^s u(r)dr$. Since $\int_0^s u(r)dr \in D(A)$ for $s \in [0, \tau]$, it follows from Lemma 3.2.2 c) that

$$\begin{aligned} w'(s) &= -S(t-s)A\int_0^s u(r)\,dr - \frac{(t-s)^{k-1}}{(k-1)!}\int_0^s u(r)\,dr + S(t-s)u(s) \\ &= -\frac{(t-s)^{k-1}}{(k-1)!}\int_0^s u(r)\,dr + S(t-s)\left(x + \int_0^s f(r)\,dr\right) \qquad (s \in [0, t]).\end{aligned}$$

Since $0 = w(0) - w(t) = -\int_0^t w'(s)\,ds$ we have

$$\int_0^t S(t-s)\left(x + \int_0^s f(r)\,dr\right)ds = \int_0^t \frac{(t-s)^{k-1}}{(k-1)!}\int_0^s u(r)\,dr\,ds.$$

Using this and Proposition 1.3.6, it follows that

$$\begin{aligned} u(t) &= \frac{d^{k+1}}{dt^{k+1}}\left(\int_0^t \frac{(t-s)^{k-1}}{(k-1)!}\int_0^s u(r)\,dr\,ds\right) \\ &= \frac{d^k}{dt^k}\left(S(t)x + \int_0^t S(s)f(t-s)\,ds\right) = v^{(k)}(t). \end{aligned}$$

b) This follows immediately from Proposition 3.1.15. □

Lemma 3.2.10. *Let v be defined by (3.16). Assume that $v \in C^k([0,\tau],X)$. Then $u := v^{(k)}$ is a mild solution of (ACP_f). Moreover, if $v \in C^{k+1}([0,\tau],X)$, then $u := v^{(k)}$ is a classical solution of (ACP_f).*

Proof. By Fubini's theorem,

$$\int_0^t v(s)\,ds = \int_0^t S(s)x\,ds + \int_0^t \int_0^{t-r} S(s)f(r)\,ds\,dr$$

for $t \in [0,\tau]$. By Lemma 3.2.2 d), $\int_0^t S(s)x\,ds \in D(A)$, $\int_0^{t-r} S(s)f(r)\,ds \in D(A)$ and $A\int_0^{t-r} S(s)f(r)\,ds = S(t-r)f(r) - \frac{(t-r)^k}{k!}f(r)$. By Proposition 1.1.7, $\int_0^t v(s)\,ds \in D(A)$ and

$$A\int_0^t v(s)\,ds = v(t) - \frac{t^k}{k!}x - \int_0^t \frac{(t-r)^k}{k!}f(r)\,dr. \tag{3.17}$$

Since A is closed and $v \in C^k([0,\tau],X)$, it follows from (3.17) that $v^{(j-1)}(t) \in D(A)$ for $t \in [0,\tau]$ and that

$$Av^{(j-1)}(t) = v^{(j)}(t) - \frac{t^{k-j}}{(k-j)!}x - \int_0^t \frac{(t-r)^{k-j}}{(k-j)!}f(r)\,dr \tag{3.18}$$

for $j = 1,\dots,k$. Since $v(0) = 0$, this implies that $v^{(j)}(0) = 0$ for $j = 1,2,\dots,k-1$. It now follows from (3.18) for $j = k$ that $u := v^{(k)}$ is a mild solution of (ACP_f). If $v \in C^{k+1}([0,\tau],X)$, we may differentiate (3.18) once more and see that $v^{(k)}(t) \in D(A)$ and that

$$Av^{(k)}(t) = v^{(k+1)}(t) - f(t) \quad (t \in [0,\tau]). \tag{3.19}$$

Hence $u := v^{(k)}$ satisfies $u'(t) = Au(t) + f(t)$ for $t \in [0,\tau]$. Also, by (3.18) for $j = k$, $u(0) = v^{(k)}(0) = x$. □

Combining the above Lemmas 3.2.9 and 3.2.10 with Lemma 3.2.2 c) and Proposition 1.3.6, we obtain the following corollary.

Corollary 3.2.11. *Let A be the generator of a k-times integrated semigroup on X for some $k \in \mathbb{N}$.*

a) *Assume that $x \in D(A^{k+1})$. Then there exists a unique classical solution of (ACP_0).*

b) *Assume that $f \in C^{k+1}([0,\tau],X)$ and that there exist $x_j \in D(A)$ for $j = 0,1,\dots,k$ satisfying $x_0 = x, x_{j+1} = Ax_j + f^{(j)}(0)$ $(j = 0,1,\dots,k)$. Then there exists a unique classical solution of (ACP_f).*

c) *Assume that $f \in C^{k}([0,\tau],X)$ and that there exist $x_j \in D(A)$ for $j = 0,1,\dots,k-1$ satisfying $x_0 = x$, $x_{j+1} = Ax_j + f^{(j)}(0)$ $(j = 0,1,\dots,k-1)$. Then there exists a unique mild solution of (ACP_f).*

Remark 3.2.12. We remark that in contrast to the case of a C_0-semigroup (see Corollary 3.1.17) a mere regularity condition on the function f does not suffice to ensure the existence of a classical solution of (ACP_f). Indeed, let A be the generator of a once integrated semigroup S on X such that A does not generate a C_0-semigroup on X. Then there exists $y \in X$ such that $S(\cdot)y \notin C^1([0,\tau],X)$. Consider the function f defined by $f(t) := y$ for $t \in [0,\tau]$. If $x = 0$, then $v(t) = \int_0^t S(s)y\,ds$, but $v \notin C^2([0,\tau],X)$. □

In the following we turn our attention to the converse of Corollary 3.2.11; i.e., we are aiming to show that A is the generator of an integrated semigroup whenever the associated Cauchy problem (ACP_0) admits a unique classical solution for all initial data x belonging to the domain of some power of A. To this end, we restrict ourselves to the case of generators of *exponentially bounded k-times integrated semigroups*; i.e., we assume that the function S in Definition 3.2.1 satisfies in addition the property that $\|S(t)\| \le Me^{\omega t}$ for all $t \ge 0$ and some suitable constants $M,\omega \ge 0$.

For $x \in X$ consider then the "$(k+1)$-times integrated version" of (ACP_0), which is to find $v \in C^1(\mathbb{R}_+,X) \cap C(\mathbb{R}_+,D(A))$ satisfying

$$(ACP_{k+1}) \begin{cases} v'(t) = & Av(t) + \dfrac{t^k}{k!}x \qquad (t \ge 0), \\ v(0) = & 0. \end{cases} \tag{3.20}$$

Assume that A generates an exponentially bounded k-times integrated semigroup S on X and define v by $v(t) := \int_0^t S(s)x\,ds$ for $t \ge 0$. Then by Lemma 3.2.2 d),

$$v'(t) = S(t)x = A\int_0^t S(s)x\,ds + \frac{t^k}{k!}x \qquad (t \ge 0) \tag{3.21}$$

and $v(0) = 0$. Hence v is a classical solution of (ACP_{k+1}). It is unique by Lemma 3.2.9 and exponentially bounded since S is so. We have therefore proved the implication (i) ⇒ (ii) of the following result. Recall that a 0-times integrated semigroup is the same as a C_0-semigroup, so the following may be compared with Theorem 3.1.10.

Theorem 3.2.13. *Let A be a closed operator on X and let $k \in \mathbb{N}_0$. The following assertions are equivalent:*

(i) *A generates an exponentially bounded k-times integrated semigroup on X.*

(ii) *For all $x \in X$ there exists a unique classical solution of (ACP_{k+1}) which is exponentially bounded.*

(iii) *$\rho(A) \neq \emptyset$ and for every $x \in D(A^{k+1})$ there exists a unique classical solution of (ACP_0) which is exponentially bounded.*

Proof. The remarks before Theorem 3.2.13 imply the assertion (i) $\Rightarrow$ (ii). Moreover, the implication (i) $\Rightarrow$ (iii) follows from Corollary 3.2.11. For the proof of the implication (ii) $\Rightarrow$ (i) we need the following "uniform exponential boundedness principle".

Lemma 3.2.14. *Let X, Y be Banach spaces and let $V : \mathbb{R}_+ \to \mathcal{L}(X, Y)$ be a function. Assume that $V(\cdot)x$ is exponentially bounded for all $x \in X$. Then there exist constants $M \geq 0, \omega \in \mathbb{R}$ such that*

$$\| V(t) \| \leq M e^{\omega t} \qquad (t \geq 0).$$

Proof. Observe that, for $n \in \mathbb{N}$, the set X_n defined by

$$X_n := \left\{ x \in X : \| V(t)x \| \leq n e^{nt} \text{ for all } t \geq 0 \right\}$$

is a closed subset of X. The hypothesis implies that $X = \bigcup_{n \in \mathbb{N}} X_n$. Hence, by Baire's theorem, there exists $n_0 \in \mathbb{N}$ such that X_{n_0} has non-empty interior. It follows that there exist $z \in X, \varepsilon > 0, M \geq 0$ and $\omega \in \mathbb{R}$ such that

$$\| V(t)x \| \leq M e^{\omega t} \qquad (t \geq 0)$$

provided $\| x - z \| \leq \varepsilon$. For $\| y \| \leq 1$ we have

$$\varepsilon e^{-\omega t} \| V(t)y \| \leq e^{-\omega t} \| V(t)(\varepsilon y + z) \| + e^{-\omega t} \| V(t)z \| \leq 2M$$

for $t \geq 0$. Thus $\| V(t) \| \leq \frac{2M}{\varepsilon} e^{\omega t}$ for $t \geq 0$. □

The above Lemma 3.2.14 enables us now to prove the implication (ii) $\Rightarrow$ (i) in Theorem 3.2.13.

(ii) $\Rightarrow$ (i): Denote by $V(\cdot)x$ the solution of (ACP_{k+1}). The mapping $V(t) : X \to D(A)$ is linear by uniqueness. We even have that $V(t) \in \mathcal{L}(X, D(A))$. Indeed, the space $C(\mathbb{R}_+, D(A))$ is a Fréchet space for the seminorms

$$p_m(f) := \sup_{0 \leq t \leq m} \|f(t)\|_{D(A)}.$$

Define a mapping $\Phi : X \to C(\mathbb{R}_+, D(A))$ by $\Phi(x) = V(\cdot)x$. Then Φ is closed and the closed graph theorem implies that Φ is continuous (see the proof of Theorem

3.1.12). In particular, the mapping $X \to D(A), x \mapsto V(t)x$ is continuous for $t \geq 0$. The hypothesis together with Lemma 3.2.14 implies that $\| V(t) \| \leq Me^{\omega t}$ $(t \geq 0)$ for suitable constants $M, \omega \geq 0$. Therefore $Q(\lambda)x := \lambda^{k+1} \int_0^\infty e^{-\lambda t} V(t)x\, dt$ is well defined for $\lambda > \omega$. Since

$$\left\| \int_0^t AV(s)x\, ds \right\| = \left\| V(t)x - \frac{t^{k+1}}{(k+1)!}x \right\| \leq \left(Me^{\omega t} + \frac{t^{k+1}}{(k+1)!} \right) \|x\|,$$

it follows from Theorem 1.4.3 that $\operatorname{abs}(AV(\cdot)x) \leq \omega$. By Proposition 1.6.3, $Q(\lambda)x \in D(A)$ for all $x \in X$, all $\lambda > \omega$ and

$$\begin{aligned}(\lambda - A)Q(\lambda)x &= \lambda^{k+2} \int_0^\infty e^{-\lambda t} V(t)x\, dt - \lambda^{k+1} \int_0^\infty e^{-\lambda t} AV(t)x\, dt \\ &= \lambda^{k+2} \int_0^\infty e^{-\lambda t} V(t)x\, dt - \lambda^{k+1} \int_0^\infty e^{-\lambda t} \frac{d}{dt} V(t)x\, dt \\ &\quad + \lambda^{k+1} \int_0^\infty \frac{t^k}{k!} e^{-\lambda t} x\, dt \\ &= x\end{aligned}$$

for $\lambda > \omega$. In order to show that $\lambda - A$ is injective for $\lambda > \omega$ assume that $(\lambda - A)x = 0$ for some $x \in D(A)$ and $\lambda > \omega$. Then the solution $V(t)x$ of (ACP_{k+1}) is given by $V(t)x = \left(\int_0^t \frac{(t-s)^k}{k!} e^{\lambda s}\, ds \right) x$. Since $\| V(t)x \| \leq Me^{\omega t}\|x\|$ for all $t \geq 0$, it follows that $x = 0$. Hence, $R(\lambda, A) = Q(\lambda)$ for $\lambda > \omega$ and V is a $(k+1)$-times integrated semigroup generated by A. By hypothesis $S(t)x := \frac{d}{dt} V(t)x = AV(t)x + \frac{t^k}{k!}x$ exists for all $t \geq 0$ and all $x \in X$ and $V(0) = 0$. Integrating by parts shows that $R(\lambda, A) = \lambda^k \hat{S}(\lambda)$, so S is a k-times integrated semigroup generated by A.

(iii) $\Rightarrow$ (ii): For the time being, assume that $0 \in \rho(A)$. Let $x \in X$ and let u be the solution of (ACP_0) with initial value $u(0) = A^{-k-1}x$. Then v given by

$$v(t) := u(t) - A^{-k-1}x - tA^{-k}x - \ldots - \frac{t^{k-1}}{(k-1)!} A^{-2}x - \frac{t^k}{k!} A^{-1}x$$

is an exponentially bounded solution of (ACP_{k+1}). Let $\bar{v}$ be another solution of (ACP_{k+1}). Then $u = v - \bar{v}$ solves (ACP_0) with initial value $u(0) = 0$. Hence, $u \equiv 0$ and we have proved (ii) provided $0 \in \rho(A)$. In the case where $0 \notin \rho(A)$, we have that $0 \in \rho(A - \mu)$ for some $\mu \in \mathbb{C}$. Hence, the preceding argument shows that (ii) holds for $(A - \mu)$. The implication (ii) $\Rightarrow$ (i) and Proposition 3.2.6 imply that A generates a k-times integrated semigroup which, as we have seen, implies (ii). $\square$

Remark 3.2.15. a) The assumption that $\rho(A) \neq \emptyset$ in Theorem 3.2.13 (iii) cannot be omitted even if $k = 0$ (see Example 3.1.11).

b) The assumption of exponential boundedness in Theorem 3.2.13 (ii) and (iii) cannot be omitted as the following example shows: Let $1 \leq p < \infty$ and X be the space ℓ^p of all complex sequences $x = (x_n)_{n\in\mathbb{N}}$ such that $\|x\| := (\sum_{n=1}^{\infty} |x_n|^p)^{1/p} < \infty$. Define the closed operator A on X by

$$D(A) := \{x \in X : (a_n x_n) \in X\}, \quad Ax := (a_n x_n)_{n\in\mathbb{N}}$$

where $a_n := n + ie^{n^2}$ for $n \in \mathbb{N}$. For $t \geq 0$ set $S(t)x := \left((\int_0^t e^{sa_n}\, ds)x_n\right)_n$. Then $S(t) \in \mathcal{L}(X)$ for $t \geq 0$ and $S(\cdot)x$ is strongly continuous. For $x \in X$ let $v(t) := \int_0^t S(s)x\, ds$. We verify that $v'(t) = Av(t) + tx$ for $t \geq 0$; i.e., v is the unique solution of (ACP_2). However, v is not exponentially bounded if $x_n \neq 0$ for all $n \in \mathbb{N}$. Observe also that S satisfies $S(s)S(t) = \int_s^{s+t} S(r)dr - \int_0^t S(r)dr$. However, S is not Laplace transformable since v is not exponentially bounded. □

3.3 Real Characterization

In Section 3.1 (respectively, 3.2) we proved that the Cauchy problem

$$u'(t) = Au(t) \quad (t \geq 0), \qquad u(0) = x,$$

possesses a unique classical solution for all $x \in D(A)$ (respectively, $x \in D(A^{k+1})$) provided the operator A generates a C_0-semigroup (respectively, k-times integrated semigroup) on X. It is therefore interesting to characterize generators of C_0-semigroups (respectively, integrated semigroups) by properties of the operators A or their resolvents.

In the following we characterize generators of C_0-semigroups (respectively, exponentially bounded integrated semigroups) in terms of estimates for the resolvents and all their powers for real λ. Recall that an operator A was defined to be the generator of a k-times integrated semigroup S on X for some $k \in \mathbb{N}_0$ if $(\omega, \infty) \subset \rho(A)$ for some $\omega \geq 0$ and there exists a strongly continuous function $S : \mathbb{R}_+ \to \mathcal{L}(X)$ satisfying $\mathrm{abs}(S) \leq \omega$ and

$$R(\lambda, A) = \lambda^k \int_0^\infty e^{-\lambda t} S(t)\, dt \qquad (\lambda > \omega).$$

By applying the Real Representation Theorem 2.4.1 to the special case of resolvents, we obtain the following characterization. Here and in what follows, we use the notation $(R(\lambda, A)/\lambda^k)^{(n)}$ to denote the nth derivative of the function $\lambda \mapsto R(\lambda, A)/\lambda^k$. Note that the first derivative of $R(\lambda, A)$ is $-R(\lambda, A)^2$ (see Corollary B.3).

Theorem 3.3.1. *Let A be a linear operator on X. Let $M \geq 0, \omega \in \mathbb{R}$ and $k \in \mathbb{N}_0$. Then the following assertions are equivalent:*

(i) $(\omega, \infty) \subset \rho(A)$ *and*

$$\sup_{n\in\mathbb{N}_0} \sup_{\lambda>\omega} \|(\lambda - \omega)^{n+1} (R(\lambda, A)/\lambda^k)^{(n)}/n!\| \leq M.$$

(ii) *A generates a $(k+1)$-times integrated semigroup S_{k+1} on X satisfying*

$$\|S_{k+1}(t) - S_{k+1}(s)\| \leq M \int_s^t e^{\omega r}\, dr \qquad (0 \leq s \leq t).$$

Proof. The implication (i) $\Rightarrow$ (ii) follows from Theorem 2.4.1 and assertion (1.22). Conversely, assume that (ii) holds. By Definition 3.2.1, there exists $\omega' \geq \omega$ such that $\frac{R(\lambda,A)}{\lambda^k} = \lambda \int_0^\infty e^{-\lambda t} S_{k+1}(t)\, dt$ for all $\lambda > \omega'$. By Proposition B.5, $(\omega, \infty) \subset \rho(A)$ and (i) follows from Theorem 2.4.1 and (1.22). □

When $k > 0$ and $X \neq \{0\}$, conditions (i) and (ii) of Theorem 3.3.1 cannot be satisfied for $\omega < 0$.

Note that, given a linear operator A satisfying condition (i) above, one cannot improve the order of integration of S_{k+1}, in general (see Example 3.3.10 below). However, if A is densely defined, we obtain the following characterization.

Theorem 3.3.2. *Let A be a densely defined operator on X. Let $M \geq 0, \omega \in \mathbb{R}$ and $k \in \mathbb{N}_0$. Then the following assertions are equivalent:*

(i) *$(\omega, \infty) \subset \rho(A)$ and*

$$\sup_{n\in\mathbb{N}_0} \sup_{\lambda>\omega} \|(\lambda-\omega)^{n+1}(R(\lambda,A)/\lambda^k)^{(n)}/n!\| \leq M.$$

(ii) *A generates a k-times integrated semigroup S_k on X satisfying*

$$\|S_k(t)\| \leq Me^{\omega t} \quad (t \geq 0).$$

The following lemma will be useful in the proof of Theorem 3.3.2. In addition to the space $\mathrm{Lip}_\omega(\mathbb{R}_+, X)$ defined in Section 2.4, we also set

$$\begin{aligned} C^1_\omega(\mathbb{R}_+, X) &:= \left\{ f \in C^1(\mathbb{R}_+, X) : f(0) = 0, \sup_{t\geq 0} \|e^{-\omega t} f'(t)\| < \infty \right\} \\ &= C^1(\mathbb{R}_+, X) \cap \mathrm{Lip}_\omega(\mathbb{R}_+, X). \end{aligned}$$

Lemma 3.3.3. *For $\omega \in \mathbb{R}$, the space $C^1_\omega(\mathbb{R}_+, X)$ is a closed subspace of $\mathrm{Lip}_\omega(\mathbb{R}_+, X)$. In particular, if $S \in \mathrm{Lip}_\omega(\mathbb{R}_+, \mathcal{L}(X))$, then $\{x \in X : S(\cdot)x \in C^1(\mathbb{R}_+, X)\}$ is a closed subspace of X.*

Proof. Let $(f_n) \subset C^1_\omega(\mathbb{R}_+, X)$ such that (f_n) converges to f in $\mathrm{Lip}_\omega(\mathbb{R}_+, X)$. Then $f_n(t) = \int_0^t f_n'(s)\, ds$ for $t \geq 0$. Since $\sup_{s\in[0,t]} \|f_n'(s) - f_m'(s)\| \leq e^{\omega t}\|f_n - f_m\|_{\mathrm{Lip}_\omega(\mathbb{R}_+,X)}$ for $n, m \in \mathbb{N}_0$, it follows that (f_n') converges uniformly on compact sets to a function $g \in C(\mathbb{R}_+, X)$ and that $f(t) = \int_0^t g(s)\, ds$ for $t \geq 0$. Hence $f \in C^1_\omega(\mathbb{R}_+, X)$. The final statement follows, since $x \mapsto S(\cdot)x$ is continuous from X into $\mathrm{Lip}_\omega(\mathbb{R}_+, X)$. □

Proof of Theorem 3.3.2. Assume that (ii) holds. Then A also generates a $(k+1)$-times integrated semigroup S_{k+1} on X which in addition satisfies assertion (ii) of Theorem 3.3.1. Hence assertion (i) follows from that theorem.

Conversely, assume that (i) holds. By Theorem 3.3.1, A generates a $(k+1)$-times integrated semigroup S_{k+1} on X such that $S_{k+1} \in \mathrm{Lip}_\omega(\mathbb{R}_+, \mathcal{L}(X))$ with $\|S_{k+1}\|_{\mathrm{Lip}_\omega(\mathbb{R}_+,\mathcal{L}(X))} \leq M$. By Lemma 3.2.2 c),

$$S_k(t)x := \frac{d}{dt} S_{k+1}(t)x \tag{3.22}$$

exists for all $x \in D(A)$ and $t \mapsto S_k(t)x$ is continuous. By Lemma 3.3.3, the definition of $S_k(t)x$ given in (3.22) is also meaningful for $x \in \overline{D(A)}$ and $t \mapsto S_k(t)x$ is also continuous for $x \in \overline{D(A)}$. By assumption $\overline{D(A)} = X$ and therefore A is the generator of the k-times integrated semigroup S_k on X which clearly satisfies

$$\|S_k(t)\| \leq Me^{\omega t} \quad (t \geq 0).$$

□

Notice that the special case $k = 0$ in Theorem 3.3.2 is precisely the classical Hille-Yosida theorem (in the general form presented here due to Hille, Yosida, Feller, Miyadera and Phillips), which we state explicitly due to its special importance.

Theorem 3.3.4 (Hille-Yosida). *Let A be a densely defined operator on X. Then A generates a C_0-semigroup on X if and only if there exist constants $M \geq 0, \omega \in \mathbb{R}$ such that $(\omega, \infty) \subset \rho(A)$ and*

$$\|\, (\lambda - \omega)^{n+1} R(\lambda, A)^{(n)}/n! \,\| \leq M \qquad (\lambda > \omega, n \in \mathbb{N}_0). \tag{3.23}$$

It is immediate from Theorem 3.3.2 and the relation

$$(-1)^n R(\lambda, A)^{(n)}/n! = R(\lambda, A)^{n+1} \qquad (\lambda \in \rho(A), n \in \mathbb{N}_0) \tag{3.24}$$

(see Corollary B.3) that the generator of a C_0-semigroup T of contractions may be characterized as follows.

Corollary 3.3.5. *Let A be a densely defined operator on X. Then A generates a C_0-semigroup on X satisfying $\|T(t)\| \leq 1$ for all $t \geq 0$ if and only if $(0, \infty) \subset \rho(A)$ and*

$$\|\, \lambda R(\lambda, A) \,\| \leq 1 \quad (\lambda > 0). \tag{3.25}$$

It is possible to express the semigroup in terms of the resolvent via "Euler's formula" for exponentials, which is well known in the scalar case.

Corollary 3.3.6. *Let A be the generator of a C_0-semigroup T. Then*

$$T(t)x = \lim_{n\to\infty} \left(I - \tfrac{t}{n}A\right)^{-n} x \tag{3.26}$$

for $t > 0$ and $x \in X$.

Proof. By (3.24) we have

$$\left(I - \tfrac{t}{n}A\right)^{-n} = \lambda^n R(\lambda, A)^n = \frac{(-1)^{n-1}}{(n-1)!}\lambda^n R(\lambda, A)^{(n-1)}$$

where $\lambda = \frac{n}{t}$. Thus assertion (3.26) is precisely the Post-Widder inversion formula proved in Theorem 1.7.7. □

For a densely defined operator A on X denote its adjoint by A^*. Then $R(\lambda, A)^* = R(\lambda, A^*)$ for all $\lambda \in \rho(A) = \rho(A^*)$ (see Proposition B.11). As a direct consequence of Theorem 3.3.1 and Theorem 3.3.4 we obtain the following result.

Corollary 3.3.7. *Let A be the generator of a C_0-semigroup on X. Then the adjoint A^* of A generates a once integrated semigroup on X^*.*

We remark that if the underlying Banach space X is reflexive, then the adjoint A^* of A even generates a C_0-semigroup on X^*. This follows from Theorem 3.3.4, since A^* is densely defined (see Proposition B.10). In fact, the following proposition shows that operators satisfying the Hille-Yosida condition (3.23) acting on reflexive spaces are necessarily densely defined.

Proposition 3.3.8. *Let A be a linear operator on a reflexive Banach space X. Assume that there exist constants $M, \omega \geq 0$ such that $(\omega, \infty) \subset \rho(A)$ and $\|\lambda R(\lambda, A)\| \leq M$ $(\lambda > \omega)$. Then A is densely defined.*

Proof. Let $x \in X$ and for $n \in \mathbb{N}$ with $n > \omega$ set $a_n := R(n, A)x$. By assumption, $(na_n)_{n\in\mathbb{N}, n>\omega}$ is a bounded sequence. Let z be a weak limit point of the relatively weakly compact set $\{na_n : n \in \mathbb{N}\}$. Since $na_n - Aa_n = x$ and $\lim_{n\to\infty} a_n = 0$, the closedness of A implies that $x = z$. But z is in the weak closure of $D(A)$ and hence in the norm closure of $D(A)$. □

Corollary 3.3.9. *Let X be a reflexive Banach space and assume that A generates a C_0-semigroup on X. Then the adjoint A^* of A generates a C_0-semigroup on X^*.*

Example 3.3.10. Let $1 \leq p < \infty$ and $X := L^p(\mathbb{R})$. Consider the operator $A_p f := f'$ on $L^p(\mathbb{R})$ with domain $D(A_p) := W^{1,p}(\mathbb{R})$. For the definition of the Sobolev space $W^{1,p}(\mathbb{R})$, see Appendix E. Then A_p generates the C_0-semigroup T_p on $L^p(\mathbb{R})$ given by

$$(T_p(t)f)(x) = f(x+t) \qquad (x \in \mathbb{R}, t \geq 0).$$

Identifying $L^p(\mathbb{R})^*$ with $L^{p'}(\mathbb{R})$ where $1/p + 1/p' = 1$, Corollary 3.3.9 implies that $-A_{p'}$ generates a C_0-semigroup on $L^{p'}(\mathbb{R})$, provided $p > 1$. In fact, this is evident since $T_{p'}$ extends to a C_0-group. If $p = 1$, then by Corollary 3.3.7, $-A_\infty$ generates a once integrated semigroup S on $L^\infty(\mathbb{R})$, where $A_\infty f := -f'$ with domain $D(A_\infty) := W^{1,\infty}(\mathbb{R})$. This is given by $(S(t)g)(x) = \int_0^t g(x-s)\,ds$. □

As a further consequence of Theorem 3.3.1 and Theorem 3.3.2 we note the following corollary.

Corollary 3.3.11. *Let T be a semigroup on X in the sense of Definition* 3.2.5 *and let A be the generator of T. Then T is a C_0-semigroup on X if and only if $D(A)$ is dense in X.*

Finally, given an operator A in X and a closed subspace Y of X, we define the *part A_Y of A in Y* by

$$\begin{aligned} D(A_Y) &:= \{y \in Y \cap D(A) : Ay \in Y\} \\ A_Y y &:= Ay. \end{aligned} \tag{3.27}$$

If $D(A) \subset Y$, then $\rho(A) \subset \rho(A_Y)$ and $R(\lambda, A_Y) = R(\lambda, A)|_Y$ for all $\lambda \in \rho(A)$. An important case is $Y = \overline{D(A)}$ when A is not densely defined. Then it may well happen that A_Y is not densely defined either (a concrete example is the Poisson operator considered in Section 6.1). Nevertheless the following holds true.

Lemma 3.3.12. *Let A be an operator on X such that $(\omega, \infty) \subset \rho(A)$ and*

$$M := \sup_{\lambda > \omega} \|\lambda R(\lambda, A)\| < \infty$$

for some $\omega \in \mathbb{R}$. Let $Y = \overline{D(A)}$. Then

a) $\lim_{\lambda\to\infty} \lambda R(\lambda, A)x = x$ *for all $x \in Y$.*

b) *$D(A_Y)$ is dense in Y.*

c) *If A satisfies the Hille-Yosida condition* (3.23), *then A_Y generates a C_0-semigroup on Y.*

Proof. a) The assumption implies that $\lim_{\lambda\to\infty} R(\lambda, A)x = 0$ for all $x \in X$. Hence, $\lim_{\lambda\to\infty} \lambda R(\lambda, A)x = \lim_{\lambda\to\infty}(x + R(\lambda, A)Ax) = x$ if $x \in D(A)$. Consequently, $\lim_{\lambda\to\infty} \lambda R(\lambda, A)x = x$ if $x \in \overline{D(A)} = Y$.

b) Since $R(\lambda, A)x \in D(A_Y)$ if $x \in Y$, this follows from a).

c) This follows from the Hille-Yosida theorem. □

Although the part of an operator A in $\overline{D(A)}$ may not be densely defined, we obtain the following result from the proof of Theorem 3.3.2.

Corollary 3.3.13. *Let A be an operator satisfying the equivalent conditions of Theorem* 3.3.1 *for some $k \in \mathbb{N}_0$ and let $Y = \overline{D(A)}$. Then the part A_Y of A in Y generates a k-times integrated semigroup on Y.*

Proof. By Lemma 3.2.2, $S_{k+1}(\cdot)x \in C^1(\mathbb{R}_+, X)$ provided $x \in D(A)$. It follows from Lemma 3.3.3 that $S_{k+1}(\cdot)x \in C^1_\omega(\mathbb{R}_+, X)$ for all $x \in \overline{D(A)} = Y$. For $t > 0$ and $x \in Y$ let $S_k(t)x = \frac{d}{dt}S_{k+1}(t)x$. Since $S_{k+1}(t)D(A) \subset D(A)$ for $t > 0$, it follows that $S_k(t)x \in Y$ for all $x \in Y$. Thus, $S_k : \mathbb{R}_+ \to \mathcal{L}(Y)$ is strongly continuous and

$$R(\lambda, A)x = \lambda^{k+1} \int_0^\infty e^{-\lambda t} S_{k+1}(t)x \, dt = \lambda^k \int_0^\infty e^{-\lambda t} S_k(t)x \, dt$$

for all $x \in Y$ and λ sufficiently large. □

Proposition 3.3.14. *Let T be a C_0-semigroup on X with generator A. Let $X^{\odot} := \overline{D(A^*)}$. Then $T(t)^{\odot} := T(t)^*|_{X^{\odot}}$ defines a C_0-semigroup whose generator $A^{\odot}$ is the part of A^* in $X^{\odot}$.*

The C_0-semigroup $T^{\odot}$ is known as the *sun-dual* of T.

Proof. It follows from Lemma 3.3.12 that the part of A^* in $X^{\odot}$ generates a C_0-semigroup $T^{\odot}$ on $X^{\odot}$. For $x \in X$ and $x^* \in X^{\odot}$, Corollary 3.3.6 shows that

$$\begin{aligned}
\langle x, T(t)^{\odot} x^* \rangle &= \lim_{n\to\infty} \langle x, (I - \tfrac{t}{n}A^{\odot})^{-n} x^* \rangle \\
&= \lim_{n\to\infty} \langle x, ((I - \tfrac{t}{n}A)^{-n})^* x^* \rangle \\
&= \lim_{n\to\infty} \langle (I - \tfrac{t}{n}A)^{-n} x, x^* \rangle \\
&= \langle T(t)x, x^* \rangle \\
&= \langle x, T(t)^* x^* \rangle.
\end{aligned}$$

It follows that $T(t)^{\odot} = T(t)^*|_{X^{\odot}}$ for all $t > 0$. □

3.4 Dissipative Operators

In the previous section we saw that, by the Hille-Yosida characterization, the generator of a contraction semigroup may be characterized in terms of a resolvent estimate for real λ. It is the aim of this section to give a second characterization of such semigroups, which turns out to be quite useful in particular when dealing with differential operators. In order to do so, we define for $x \in X$ the *subdifferential* $\mathrm{dN}(x)$ of the norm $N : X \to \mathbb{R}_+, N(x) = \|x\|$ at x by

$$\mathrm{dN}(x) := \{x^* \in X^* : \|x^*\| \le 1, \langle x, x^* \rangle = \|x\|\}. \tag{3.28}$$

The Hahn-Banach theorem implies that $\mathrm{dN}(x) \neq \emptyset$ for all $x \in X$.

Definition 3.4.1. *An operator A on X is called* dissipative *if for every $x \in D(A)$ there exists $x^* \in \mathrm{dN}(x)$ such that*

$$\mathrm{Re}\langle Ax, x^* \rangle \le 0. \tag{3.29}$$

A useful characterization of dissipative operators is the following.

Lemma 3.4.2. *An operator A on X is dissipative if and only if*

$$\|(\lambda - A)x\| \ge \lambda \|x\| \qquad (x \in D(A),\ \lambda > 0) \tag{3.30}$$

or equivalently,

$$\|x - tAx\| \ge \|x\| \qquad (x \in D(A),\ t > 0).$$

Proof. Assume that A is dissipative. Let $x \in D(A), t > 0$. Let $x^* \in \mathrm{dN}(x)$ satisfy (3.29). Then for $t > 0$,

$$\begin{aligned} \|x - tAx\| &\geq \operatorname{Re}\langle x - tAx, x^*\rangle \\ &= \|x\| - t\operatorname{Re}\langle Ax, x^*\rangle \\ &\geq \|x\|. \end{aligned}$$

Conversely, let $x \in D(A)$ and assume that $\|x - tAx\| \geq \|x\|$ $(t > 0)$. Choose $x_t^* \in \mathrm{dN}(x - tAx)$ and let x^* be a weak* accumulation point of x_t^* as $t \downarrow 0$. Then $\|x^*\| \leq 1$. Since $\|x - tAx\| = \langle x - tAx, x_t^*\rangle$, letting $t \downarrow 0$ shows that $\|x\| = \langle x, x^*\rangle$. Thus, $x^* \in \mathrm{dN}(x)$. Moreover,

$$\begin{aligned} \|x\| \leq \|x - tAx\| &= \operatorname{Re}\langle x, x_t^*\rangle - t\operatorname{Re}\langle Ax, x_t^*\rangle \\ &\leq \|x\| - t\operatorname{Re}\langle Ax, x_t^*\rangle. \end{aligned}$$

Thus, $\operatorname{Re}\langle Ax, x_t^*\rangle \leq 0$. Letting $t \downarrow 0$ shows that $\operatorname{Re}\langle Ax, x^*\rangle \leq 0$. □

Example 3.4.3. a) Dissipative operators acting on Hilbert spaces or L^p-spaces may be characterized as follows:

(i) Let A be an operator on a Hilbert space H. Denote by $(\cdot|\cdot)$ the inner product in H. Then A is dissipative if and only if $\operatorname{Re}(Ax|x) \leq 0$ for all $x \in D(A)$.

(ii) Let $\Omega \subset \mathbb{R}^n$ be open, $1 < p < \infty$, set $X := L^p(\Omega)$ and identify X^* with $L^{p'}(\Omega)$ where $1/p + 1/p' = 1$. For $f \in X \setminus \{0\}$ we define $\operatorname{sign} f \in L^\infty(\Omega)$ by

$$(\operatorname{sign} f)(x) := \begin{cases} 0 & \text{if } f(x) = 0, \\ \dfrac{f(x)}{|f(x)|} & \text{if } f(x) \neq 0. \end{cases}$$

Then $\mathrm{dN}(f) = \|f\|_p^{-(p-1)} \{\operatorname{sign} \bar{f} \cdot |f|^{p-1}\}$, where $\bar{f}$ denotes the complex conjugate function of f. Therefore an operator A on X is dissipative if and only if

$$\operatorname{Re} \int_\Omega Af \cdot \operatorname{sign} \bar{f} \cdot |f|^{p-1}\, dx \leq 0 \tag{3.31}$$

for all $f \in D(A)$.

b) If A is dissipative, then cA is dissipative for all $c > 0$.

c) If $B \in \mathcal{L}(X)$, then $B - \|B\|$ is dissipative.

Lemma 3.4.4. *Let A be a densely defined dissipative operator on X. Then A is closable and $\overline{A}$ is dissipative.*

Proof. Let $(x_n) \subset D(A)$ such that $x_n \to 0$ and $Ax_n \to y$ for some $y \in X$ as $n \to \infty$. We show that $y = 0$. To this end, let $z \in D(A)$. It follows from Lemma

3.4.2 that $\|(I - tA)x\| \geq \|x\|$ for all $t > 0$ and all $x \in D(A)$. Hence $\| x_n + tz \| \leq \| x_n + tz - tA(x_n + tz) \|$ for all $n \in \mathbb{N}$ and all $t > 0$. This implies that $\| tz \| \leq \| tz - ty - t^2 Az \|$ and hence $\| z \| \leq \| z - y - tAz \|$ for all $t > 0$. Letting $t \to 0$ we obtain $\| z \| \leq \| z - y \|$ for all $z \in D(A)$. Since $D(A)$ is dense, it follows that $y = 0$ which means that A is closable. Taking limits in (3.30) shows that $\overline{A}$ is dissipative. □

The following theorem due to Lumer and Phillips characterizes generators A of C_0-semigroups of contractions in terms of dissipativity of A.

Theorem 3.4.5 (Lumer-Phillips). *Let A be a densely defined operator on X. Then A generates a C_0-semigroup of contractions on X if and only if*

a) *A is dissipative, and*

b) *$(\lambda - A)D(A) = X$ for some (or all) $\lambda > 0$.*

Proof. Let A be the generator of a C_0-semigroup of contractions. Then assertion b) holds by the Hille-Yosida theorem (Corollary 3.3.5). Moreover, the Hille-Yosida theorem combined with Lemma 3.4.2 implies assertion a).

In order to prove the converse implication note that by Lemma 3.4.2 we have

$$\| (\lambda - A)x \| \geq \lambda \| x \| \qquad (x \in D(A), \lambda > 0). \tag{3.32}$$

Since $(\lambda_0 - A)D(A) = X$ for some $\lambda_0 > 0$, it follows from (3.32) that $\lambda_0 - A$ is invertible and that $\| R(\lambda_0, A) \| \leq \lambda_0^{-1}$. We show that this property holds for all $\lambda > 0$. In fact, let $\Lambda := \rho(A) \cap (0, \infty)$. Then $\Lambda \neq \emptyset$ and therefore A is closed. Furthermore let $(\lambda_n) \subset \Lambda$ such that $\lim_{n\to\infty} \lambda_n = \lambda > 0$. By Corollary B.3, $\operatorname{dist}(\lambda_n, \sigma(A)) \geq \| R(\lambda_n, A) \|^{-1} \geq \lambda_n$ for all $n \in \mathbb{N}$ and it follows that $\lambda \in \Lambda$. This shows that Λ is closed in $(0, \infty)$. Since Λ is obviously open, it follows that $\Lambda = (0, \infty)$ and therefore $(0, \infty) \subset \rho(A)$. Inequality (3.32) implies that $\|R(\lambda, A)\| \leq \lambda^{-1}$ for all $\lambda > 0$ and the Hille-Yosida theorem finally implies the assertion. □

By the same proof with Theorem 3.3.1 replacing the Hille-Yosida theorem we obtain the following characterization in the case when $D(A)$ is not necessarily dense.

Corollary 3.4.6. *Let A be an operator on X. The following assertions are equivalent:*

(i) *A is dissipative and $(\lambda - A)D(A) = X$ for some (or all) $\lambda > 0$.*

(ii) *A generates a once integrated semigroup S satisfying*

$$\|S(t) - S(s)\| \leq |t - s| \qquad (t, s \geq 0).$$

In concrete examples, dissipativity is often relatively easy to verify whereas the range condition b) in Theorem 3.4.5 is hard to show. However, in the following example of the Dirichlet-Laplacian, the range condition is just a consequence of the Riesz-Fréchet lemma.

Example 3.4.7 (The Laplacian with Dirichlet boundary conditions). Let $\Omega \subset \mathbb{R}^n$ be an open set. Consider the operator $A : D(A) \to L^2(\Omega)$ defined by

$$\begin{aligned} D(A) &:= \{u \in H_0^1(\Omega) : \Delta u \in L^2(\Omega)\}, \\ Au &:= \Delta u. \end{aligned}$$

Here, $H_0^1(\Omega)$ is the Sobolev space defined in Appendix E, and Δu is defined to be $\sum_{j=1}^n D_j^2 u$ in the sense of distributions (see also Appendix E). Denoting by $(\cdot|\cdot)$ the inner product in $L^2(\Omega)$, we see that $(Au|u) = \int_\Omega (\Delta u)\bar{u} = -\int_\Omega |\nabla u|^2 \le 0$ for $u \in D(A)$. Hence, by Example 3.4.3 a), the operator A is dissipative. In order to prove the range condition b) of Theorem 3.4.5, let $f \in L^2(\Omega)$. Then the mapping $\Phi : v \mapsto \int_\Omega fv$ defines a continuous linear functional on the Hilbert space $H_0^1(\Omega)$. By the Riesz-Fréchet lemma, there exists a unique $u \in H_0^1(\Omega)$ such that $\Phi(v) = (v|\bar{u})_{H_0^1(\Omega)}$ for all $v \in H_0^1(\Omega)$. Here,

$$(v|\bar{u})_{H_0^1(\Omega)} := \int_\Omega uv + \sum_{j=1}^n \int_\Omega D_j u D_j v$$

denotes the inner product in $H_0^1(\Omega)$. Considering in particular $v \in \mathcal{D}(\Omega)$, it follows that $u - \Delta u = f$ in $\mathcal{D}(\Omega)'$. This implies that $u \in D(A)$ and $u - Au = f$. Obviously, $D(A)$ is dense in $L^2(\Omega)$ and by the Lumer-Phillips theorem, A generates a contraction semigroup on $L^2(\Omega)$. We call A the *Laplacian with Dirichlet boundary conditions on* $L^2(\Omega)$, and we denote it by $\Delta_{L^2(\Omega)}$. We remark that in the case where Ω is a bounded domain with boundary of class C^2, it can be shown (see [Bre83, Théorème IX.25]) that

$$D(\Delta_{L^2(\Omega)}) = H^2(\Omega) \cap H_0^1(\Omega). \qquad \square$$

Example 3.4.8. Consider the Hilbert space $X := L^2(0,1)$ and the operator

$$\begin{aligned} D(A) &:= \{u \in H^1(0,1) : u(0) = 0\}, \\ Au &:= u'. \end{aligned}$$

Then $\operatorname{Re}\,(u|Au) = \frac{1}{2}\int_0^1 (u(x)\overline{u'(x)} + u'(x)\overline{u(x)})\,dx = \frac{1}{2}|u(1)|^2 \ge 0$. For $f \in L^2(0,1)$ and $\lambda \in \mathbb{C}$, define u by

$$u(x) := \int_0^x e^{-\lambda(x-y)} f(y)\,dy \qquad (x \in (0,1)).$$

Then $u \in D(A)$ and $\lambda u + u' = f$. Hence, the range condition $(\lambda + A)D(A) = X$ is fulfilled for all $\lambda \in \mathbb{C}$ and by the Lumer-Phillips theorem, $-A$ generates a contraction semigroup T on X. It is not difficult to see that T is given by

$$(T(t)f)(x) = \begin{cases} f(x-t) & (t \le x), \\ 0 & (t > x). \end{cases} \tag{3.33}$$

In fact, the mapping $x \mapsto \int_0^\infty e^{-\lambda t}(T(t)f)(x)\,dt = \int_0^x e^{-\lambda(x-y)}f(y)\,dy$ belongs to $D(A)$ and $(\lambda + \frac{\partial}{\partial x})\int_0^\infty e^{-\lambda t}(T(t)f)(x)\,dt = f(x)$ for $x \in (0,1)$. Since $(\lambda + A)$ is invertible for all $\lambda \in \mathbb{C}$, we see that $(\lambda + A)^{-1} = \int_0^\infty e^{-\lambda t}T(t)\,dt$.

The representation (3.33) implies that T also defines a C_0-semigroup of positive contractions on $L^p(0,1)$ for $1 \le p < \infty$. Its generator is given by $-A_p$ where

$$D(A_p) = \{u \in W^{1,p}(0,1) : u(0) = 0\} \qquad \text{and} \quad A_p u = u'. \tag{3.34}$$

□

3.5 Hille-Yosida Operators

In this section we consider operators which satisfy the Hille-Yosida condition (3.23) but are not necessarily densely defined.

Definition 3.5.1. *A linear operator A on X is called a* Hille-Yosida operator *if there exist $\omega \in \mathbb{R}, M \ge 0$ such that $(\omega,\infty) \subset \rho(A)$ and*

$$\| (\lambda - \omega)^n R(\lambda, A)^n \| \le M \qquad (n \in \mathbb{N}_0, \lambda > \omega). \tag{3.35}$$

We note that by the Hille-Yosida theorem and the identity

$$(-1)^n R(\lambda, A)^{(n)}/n! = R(\lambda, A)^{n+1},$$

the class of densely defined Hille-Yosida operators coincides with the class of generators of C_0-semigroups on X. We also observe that by Theorem 3.3.1 an operator A on X is a Hille-Yosida operator if and only if A generates a once integrated semigroup S on X satisfying

$$\| S(t) - S(s) \| \le M \int_s^t e^{\omega r}\,dr \qquad (0 \le s \le t) \tag{3.36}$$

for some $\omega \in \mathbb{R}, M \ge 0$. The above estimate (3.36) implies in particular that S is a locally Lipschitz continuous function on $\mathbb{R}_+$. This fact will be of crucial importance in the proof of the following result on the inhomogeneous Cauchy problem for operators which are not necessarily densely defined. More precisely, consider the problem

$$(ACP_f)\begin{cases} u'(t) = & Au(t) + f(t) \quad (t \in [0,\tau]), \\ u(0) = & x, \end{cases} \tag{3.37}$$

where $f : [0,\tau] \to X$ is a given function and $x \in X$. When $\overline{D(A)} = X$, then the inhomogeneous problem can be solved in the classical sense by means of the variation of constants formula provided $x \in D(A)$ and $f \in C^1([0,\tau],X)$ (see Corollary 3.2.11 b)). Note that this method cannot be used when $\overline{D(A)} \ne X$ and when $f(t) \notin \overline{D(A)}$. The method which we use in the following to treat (ACP_f) is based on the fact that a Hille-Yosida operator generates a once integrated semigroup which is locally Lipschitz continuous. Employing the results of Section 3.2 in the present situation, we see that existence and uniqueness results for (ACP_f) are equivalent to the fact that v given by

$$v(t) = S(t)x + \int_0^t S(t)f(t-s)ds$$

is sufficiently regular. More precisely, the following holds true.

Theorem 3.5.2 (Da Prato-Sinestrari). *Let A be a Hille-Yosida operator on X and let $\tau > 0$.*

a) *Let $f \in L^1([0,\tau], X)$ and $x \in \overline{D(A)}$. Then there exists a unique mild solution of (ACP_f).*

b) *Let $f(t) = f_0 + \int_0^t g(s)\,ds$ where $f_0 \in X$ and $g \in L^1([0,\tau], X)$. Let $x \in D(A)$ and assume that $Ax + f_0 \in \overline{D(A)}$. Then there exists a unique classical solution of (ACP_f).*

Remark 3.5.3. a) Note that $x \in \overline{D(A)}$ is a necessary condition for a mild solution u to exist, because $\lim_{t\to 0} \frac{1}{t}\int_0^t u(s)\,ds = x$ and $\int_0^t u(s)\,ds \in D(A)$ by definition of a mild solution.

b) If a classical solution u of (ACP_f) exists, then $x \in D(A)$ and

$$Ax + f(0) \in \overline{D(A)}. \tag{3.38}$$

In fact,

$$Au(0) + f(0) = u'(0) = \lim_{t\to 0} \tfrac{1}{t}(u(t) - u(0)) \in \overline{D(A)}.$$ □

Proof of Theorem 3.5.2. Let A be a Hille-Yosida operator on X. By Theorem 3.3.1, A generates a once integrated semigroup S on X which satisfies estimate (3.36). In order to prove assertion a), it suffices by Lemma 3.2.9, Proposition 1.3.7 and Lemma 3.2.10 to show that $t \mapsto S(t)x$ belongs to $C^1([0,\tau], X)$. This follows from the assumption that $x \in \overline{D(A)}$, Lemma 3.2.2 c) and Lemma 3.3.3. Thus, we have proved assertion a).

In order to prove assertion b), it suffices by Lemma 3.2.9 and Lemma 3.2.10 to verify that $v \in C^2([0,\tau], X)$. Since $x \in D(A)$ it follows from Lemma 3.2.2 c) and Proposition 1.3.6 that

$$v'(t) = x + S(t)Ax + S(t)f_0 + \int_0^t S(s)g(t-s)\,ds.$$

Now, by Proposition 1.3.7 the convolution term on the right hand side above belongs to $C^1([0,\tau], X)$ and $t \mapsto S(t)(Ax + f_0)$ belongs to $C^1([0,\tau], X)$ by the argument given in the proof of assertion a), since $Ax + f_0 \in \overline{D(A)}$ by assumption. □

Given a Hille-Yosida operator A, we consider now the problem whether or not the sum $A + B$ of A and some operator B is again a Hille-Yosida operator. We start with the following renorming lemma.

Lemma 3.5.4. *Let A be a Hille-Yosida operator satisfying estimate* (3.35) *for some $M > 0$ and $\omega = 0$. Then there exists an equivalent norm $|\cdot|$ on X such that $\|x\| \leq |x| \leq M\|x\|$ for $x \in X$ and*

$$|\lambda R(\lambda, A)x| \leq |x| \qquad (x \in X, \lambda > 0).$$

Proof. For $\mu > 0$ and $x \in X$ set

$$\|x\|_\mu := \sup_{n \geq 0} \|\mu^n R(\mu, A)^n x\|.$$

Then

$$\|x\| \leq \|x\|_\mu \leq M\|x\| \qquad \text{and} \qquad \|\mu R(\mu, A)\|_\mu \leq 1. \tag{3.39}$$

Let $\lambda \in (0, \mu]$ and set $y := R(\lambda, A)x$. It follows that $y = R(\mu, A)(x + (\mu - \lambda)y)$ and hence that $\|y\|_\mu \leq \frac{1}{\mu}\|x\|_\mu + (1 - \frac{\lambda}{\mu})\|y\|_\mu$. Therefore, $\|\lambda R(\lambda, A)\|_\mu \leq 1$ and it follows from (3.39) that

$$\|\lambda^n R(\lambda, A)^n x\| \leq \|\lambda^n R(\lambda, A)^n x\|_\mu \leq \|x\|_\mu \quad (0 < \lambda \leq \mu). \tag{3.40}$$

Hence $\|x\|_\lambda \leq \|x\|_\mu$ for $0 < \lambda \leq \mu$. Defining

$$|x| := \lim_{\mu \to \infty} \|x\|_\mu,$$

the assertion follows by taking $n = 1$ in (3.40) and letting $\mu \to \infty$. □

Theorem 3.5.5. *Let A be a Hille-Yosida operator on X and let $B \in \mathcal{L}(\overline{D(A)}, X)$. Then $A + B$ is a Hille-Yosida operator.*

Proof. Replacing A by $A - \omega$ we may assume that the estimate (3.35) is satisfied for A with $\omega = 0$. Denote by $|\cdot|$ the norm introduced in Lemma 3.5.4. It follows from that lemma that

$$|\lambda R(\lambda, A)| \leq 1 \quad (\lambda > 0).$$

Note that $\lambda - (A+B) = (I - BR(\lambda, A))(\lambda - A)$ for $\lambda > 0$. Since $|BR(\lambda, A)| \leq |B|/\lambda$, the operator $I - BR(\lambda, A)$ is invertible for $\lambda > |B|$ and

$$|(\lambda - (A + B))^{-1}| \leq |(\lambda - A)^{-1}|\, |(I - BR(\lambda))^{-1}| \leq \frac{1}{\lambda} \frac{1}{1 - |B|\lambda^{-1}} = \frac{1}{\lambda - |B|}$$

for those λ. Hence $|(\lambda - |B|)R(\lambda, A+B)| \leq 1$ for $\lambda > |B|$. Returning to the original norm we have for $x \in X$,

$$\|(\lambda - |B|)^n R(\lambda, A + B)^n x\| \leq |(\lambda - |B|)^n R(\lambda, A + B)^n x| \leq |x| \leq M\|x\|$$

for $\lambda > |B|$. Thus $A + B$ is a Hille-Yosida operator. □

Taking into account the Hille-Yosida theorem and the fact that generators of C_0-semigroups are densely defined we obtain the following corollary.

Corollary 3.5.6. *Let A be the generator of a C_0-semigroup on X and let $B \in \mathcal{L}(X)$. Then $A + B$ generates a C_0-semigroup.*

Next we consider perturbation by operators defined on the domain of the given operator. If A is a closed operator, we consider $D(A)$ with the graph norm $\|x\|_{D(A)} := \|x\| + \|Ax\|$ for which it is a Banach space. Two operators A, defined on X, and $\tilde{A}$, defined on a second Banach space $\tilde{X}$, are called *similar* if there exists an isomorphism $U : X \to \tilde{X}$ such that

$$D(\tilde{A}) = UD(A) \quad \text{and} \quad U^{-1}\tilde{A}Ux = Ax \quad \text{for all } x \in D(A).$$

In that case, A and $\tilde{A}$ have similar properties. For example, if A generates a C_0-semigroup T, then $\tilde{A}$ generates the C_0-semigroup $\tilde{T}$ on $\tilde{X}$ given by

$$\tilde{T}(t)y = UT(t)U^{-1}y \qquad (y \in \tilde{X},\ t \geq 0).$$

Similarly, if A generates an integrated semigroup S, then $\tilde{A}$ generates the integrated semigroup $\tilde{S}$ on $\tilde{X}$ given by

$$\tilde{S}(t)y = US(t)U^{-1}y \qquad (y \in \tilde{X},\ t \geq 0).$$

Theorem 3.5.7. *Let A be an operator such that $(\omega, \infty) \subset \rho(A)$ and $M := \sup_{\lambda > \omega} \|\lambda R(\lambda, A)\| < \infty$ for some $\omega \in \mathbb{R}$, and let $B \in \mathcal{L}(D(A))$. Then there exists a bounded operator $C \in \mathcal{L}(X)$ such that $A + B$ and $A + C$ are similar. In particular, if A is a Hille-Yosida operator, then $A + B$ is also a Hille-Yosida operator.*

For the proof we need the following well-known result.

Lemma 3.5.8. *Let $U, V \in \mathcal{L}(X)$. If $I - UV$ is invertible, then so is $I - VU$.*

Proof. One has $(I - VU)^{-1} = I + V(I - UV)^{-1}U$. □

Proof of Theorem 3.5.7. Choose $\lambda_0 > \omega$ and let $S := (\lambda_0 - A)BR(\lambda_0, A) \in \mathcal{L}(X)$. Choose $\lambda > \lambda_0$ such that $\|SR(\lambda, A)\| < 1$. Then $I - (\lambda_0 - A)BR(\lambda, A)R(\lambda_0, A) = I - SR(\lambda, A)$ is invertible. It follows from Lemma 3.5.8 that $I - BR(\lambda, A)$ is also invertible. Let $C := (\lambda - A)BR(\lambda, A) \in \mathcal{L}(X)$. We show that $A + B$ and $A + C$ are similar. Let $U := I - BR(\lambda, A)$. Then U is an isomorphism on X such that $UD(A) = D(A)$. Moreover,

$$\begin{aligned} U(A + C)U^{-1} &= U(A - \lambda + C)U^{-1} + \lambda \\ &= U[A - \lambda - (A - \lambda)BR(\lambda, A)]U^{-1} + \lambda \\ &= U(A - \lambda)[I - BR(\lambda, A)]U^{-1} + \lambda \\ &= U(A - \lambda) + \lambda = A - \lambda + B + \lambda = A + B. \end{aligned}$$

This proves the claim. Now the second assertion follows from Theorem 3.5.5. □

Finally, we collect several examples of Hille-Yosida operators.

Example 3.5.9. a) Let A be the generator of a C_0-semigroup on X. Then the adjoint A^* of A is a Hille-Yosida operator on X^*.

b) As a concrete example, consider $X = L^\infty(\mathbb{R})$ and define A by $Au := -u'$ with $D(A) := W^{1,\infty}(\mathbb{R})$. By Example 3.3.10, $(0,\infty) \subset \rho(A)$ and $\|R(\lambda, A)\| \le 1/\lambda$ for $\lambda > 0$.

c) Let A be the generator of a semigroup T in the sense of Definition 3.2.5. Since $S(t) := \int_0^t T(s)\,ds$ fulfills assumption (ii) of Theorem 3.3.1 for $k = 0$ it follows from that theorem that A is a Hille-Yosida operator.

d) Let $X := C[0,1]$ and define an operator A on X by

$$\begin{aligned} Au &:= -u' \\ D(A) &:= \{u \in C^1[0,1] : u(0) = 0\}. \end{aligned}$$

Then $\overline{D(A)} = \{u \in C[0,1] : u(0) = 0\} \neq X$, $(0,\infty) \subset \rho(A)$ and $\|R(\lambda, A)\| \le 1/\lambda$ for $\lambda > 0$. In fact, for $f \in X$ and $\lambda > 0$ set

$$u(x) := \int_0^x e^{-\lambda y} f(x-y)\,dy \quad (x \in [0,1]).$$

Then $u \in D(A)$, $(\lambda - A)u = f$ and

$$\sup_{x\in[0,1]} |u(x)| \le \|f\| \int_0^\infty e^{-\lambda y}\,dy = \frac{1}{\lambda}\|f\|,$$

which implies the assertions above. □

3.6 Approximation of Semigroups

In this section we study convergence of semigroups. It is interesting that we obtain the main result (Theorem 3.6.1) directly as a consequence of the approximation theorem for Laplace transforms given in Section 1.7, which in turn was proved by a simple functional analytic argument. At the end of the section we give a second proof of the Hille-Yosida theorem, as a simple corollary of the approximation theorem.

Let T_n be a C_0-semigroup on X with generator A_n $(n \in \mathbb{N})$. We suppose that

$$\|T_n(t)\| \le M \quad (t \ge 0,\ n \in \mathbb{N}). \tag{3.41}$$

If for each $x \in X$

$$\lim_{n\to\infty} T_n(t)x =: T(t)x$$

converges uniformly on $[0, \tau]$ for each $\tau > 0$, then it is easy to see that T is a C_0-semigroup. Denote its generator by A. Then it follows from the dominated convergence theorem that $\lim_{n\to\infty} R(\lambda, A_n)x = R(\lambda, A)x$ for all $x \in X$, $\lambda > 0$. We now show the converse assertion.

Theorem 3.6.1 (Trotter-Kato). *Let T_n be a C_0-semigroup on X with generator A_n $(n \in \mathbb{N})$ and suppose that (3.41) holds. Let A be a densely defined operator on X. Suppose that there exists $\omega \geq 0$ such that $(\omega, \infty) \subset \rho(A)$ and*

$$\lim_{n\to\infty} R(\lambda, A_n)x = R(\lambda, A)x \tag{3.42}$$

for all $x \in X$, $\lambda > \omega$. Then A is the generator of a C_0-semigroup T and

$$T(t)x = \lim_{n\to\infty} T_n(t)x \tag{3.43}$$

uniformly on $[0, \tau]$ for all $\tau > 0$ and all $x \in X$.

Proof. a) Let $x \in X$. We show that the sequence $(T_n(\cdot)x)_{n\in\mathbb{N}}$ is equicontinuous on $\mathbb{R}_+$. Since $D(A)$ is dense in X, we can assume that $x \in D(A)$. Let $\varepsilon > 0$. Let $\mu > \omega$, $y := (\mu - A)x$. Choose $n_0 \in \mathbb{N}$ such that $M\|R(\mu, A)y - R(\mu, A_n)y\| \leq \varepsilon/2$ for all $n \geq n_0$. Then (by Proposition 3.1.9 f)) for $n \geq n_0$,

$$\begin{aligned}
\|T_n(t)x - T_n(s)x\| &= \|T_n(t)R(\mu, A)y - T_n(s)R(\mu, A)y\| \\
&\leq \|T_n(t)R(\mu, A_n)y - T_n(s)R(\mu, A_n)y\| + \varepsilon/2 \\
&= \Big\| \int_0^t T_n(r)A_nR(\mu, A_n)y\,dr \\
&\qquad - \int_0^s T_n(r)A_nR(\mu, A_n)y\,dr \Big\| + \frac{\varepsilon}{2} \\
&\leq M|t-s|\,\|A_nR(\mu, A_n)y\| + \varepsilon/2 \\
&= M|t-s|\,\|\mu R(\mu, A_n)y - y\| + \varepsilon/2.
\end{aligned}$$

Since $\sup_{n\in\mathbb{N}} \|R(\mu, A_n)y\| < \infty$, there exists $\delta > 0$ such that $\|T_n(t)x - T_n(s)x\| \leq \varepsilon$ whenever $|t - s| \leq \delta$ and $n \geq n_0$. Since $T_n(\cdot)x : \mathbb{R}_+ \to X$ is continuous for $n < n_0$, this shows that the sequence is equicontinuous.

b) Now it follows from Theorem 1.7.5 that $T_n(t)x$ converges uniformly on $[0, \tau]$ as $n \to \infty$ for all $x \in X$ and all $\tau > 0$. It is clear that $T(t)x := \lim_{n\to\infty} T_n(t)x$ $(x \in X)$ defines a C_0-semigroup T on X. By the dominated convergence theorem one has

$$\begin{aligned}
\int_0^\infty e^{-\lambda t}T(t)x\,dt &= \lim_{n\to\infty} \int_0^\infty e^{-\lambda t}T_n(t)x\,dt \\
&= \lim_{n\to\infty} R(\lambda, A_n)x = R(\lambda, A)x \qquad (x \in X, \lambda > \omega).
\end{aligned}$$

By Definition 3.1.8 this means that A is the generator of T. □

In order to apply Theorem 3.6.1 it is useful to give other criteria equivalent to (3.42).

Proposition 3.6.2. *Let $\omega \in \mathbb{R}$. Let A and A_n be operators such that $(\omega, \infty) \subset \rho(A)$ and $(\omega, \infty) \subset \rho(A_n)$ for all $n \in \mathbb{N}$. Assume that $\sup_{n\in\mathbb{N}} \|R(\lambda, A_n)\| < \infty$ for all $\lambda > \omega$. Then the following assertions are equivalent:*

(i) $\lim_{n\to\infty} R(\mu, A_n)x = R(\mu, A)x$ *for all $x \in X$ and all $\mu > \omega$.*

(ii) $\lim_{n\to\infty} R(\mu, A_n)x = R(\mu, A)x$ *for all $x \in X$ and some $\mu > \omega$.*

(iii) *For all $x \in D(A)$ there exist $x_n \in D(A_n)$ such that $\lim_{n\to\infty} x_n = x$ and $\lim_{n\to\infty} A_n x_n = Ax$.*

(iv) *There exists a core D of A such that for all $x \in D$ there exist $x_n \in D(A_n)$ such that $\lim_{n\to\infty} x_n = x$ and $\lim_{n\to\infty} A_n x_n = Ax$.*

Proof. (i) $\Rightarrow$ (ii) is trivial.

(ii) $\Rightarrow$ (iii): Let $x \in D(A)$. Then $x_n := R(\mu, A_n)(\mu - A)x \in D(A_n)$, $x_n \to x$ by hypothesis, and $A_n x_n = \mu x_n - (\mu - A)x \to Ax$ as $n \to \infty$.

(iii) $\Rightarrow$ (iv) is trivial.

(iv) $\Rightarrow$ (i): Let $\mu > \omega$. Let $x \in D$. By hypothesis, there exist $x_n \in D(A_n)$ such that $x_n \to x$ and $A_n x_n \to Ax$. Let $y_n := (\mu - A_n)x_n$. Then $y_n \to y := (\mu - A)x$. Hence,

$$\begin{aligned}
&\limsup_{n\to\infty} \|R(\mu, A_n)y - R(\mu, A)y\| \\
&\quad\le \limsup_{n\to\infty} \Big(\|R(\mu, A_n)(y - y_n)\| + \|R(\mu, A_n)y_n - R(\mu, A)y\| \Big) \\
&\quad= \limsup_{n\to\infty} \|R(\mu, A_n)y_n - R(\mu, A)y\| \\
&\quad= \limsup_{n\to\infty} \|x_n - x\| = 0.
\end{aligned}$$

Since D is a core and $\mu - A$ is surjective, $\{(\mu - A)x : x \in D\}$ is dense in X (see Appendix B), and (i) follows by approximation. □

Corollary 3.6.3. *Let A be a densely defined operator on X. Let $A_n \in \mathcal{L}(X)$ such that*

$$\|e^{tA_n}\| \le M \quad (t \ge 0,\, n \in \mathbb{N}),$$

where $M \ge 0$. Assume that $(\omega, \infty) \subset \rho(A)$ and $\lim_{n\to\infty} A_n x = Ax$ for all $x \in D(A)$. Then A generates a C_0-semigroup T and for all $x \in X$, $\lim_{n\to\infty} e^{tA_n}x = T(t)x$ uniformly for $t \in [0, \tau]$ for all $\tau > 0$.

It is easy to deduce the Hille-Yosida theorem (Corollary 3.3.5) from Corollary 3.6.3. In fact, let A be a densely defined operator on X such that $(0, \infty) \subset \rho(A)$ and $\|\lambda R(\lambda, A)\| \le 1$ for all $\lambda > 0$. Denote by

$$A_n := n^2 R(n, A) - nI \quad (n \in \mathbb{N})$$

the Yosida approximation of A. Then

$$\|e^{tA_n}\| = e^{-nt}\|e^{tn^2R(n,A)}\| \le e^{-nt}e^{tn^2\|R(n,A)\|} \le 1 \qquad (t > 0,\ n \in \mathbb{N}).$$

Then $\lim_{n\to\infty} A_n x = Ax$ for all $x \in D(A)$. In fact, $A_n x - Ax = nR(n,A)Ax - Ax \to 0$ by Lemma 3.3.12. Now it follows from Corollary 3.6.3 that A generates a contractive C_0-semigroup.

3.7 Holomorphic Semigroups

This section is devoted to the study of holomorphic semigroups. This class of semigroups plays an important role in the theory of evolution equations. Indeed, the modern treatment of linear and nonlinear parabolic problems is based on the theory of holomorphic semigroups. When compared with arbitrary C_0-semigroups, holomorphic C_0-semigroups show many special properties. We only mention here

a) characterization results involving only a resolvent estimate (for example, Theorem 3.7.11 and Corollary 3.7.17),

b) regularity properties of solutions of the Cauchy problem (Corollary 3.7.21 and applications in Chapters 6 and 7),

c) determination of the asymptotic behaviour of the C_0-semigroup by spectral conditions on the generator (Theorem 5.1.12 and Theorem 5.6.5).

Throughout this section, let $\Sigma_\theta := \{z \in \mathbb{C} \setminus \{0\} : |\arg z| < \theta\}$ be the sector in the complex plane of angle $\theta \in (0, \pi]$. Recall from Definition 3.2.5 that a semigroup T was defined to be a strongly continuous mapping $(0,\infty) \to \mathcal{L}(X)$ satisfying a) $T(t+s) = T(t)T(s)$ for all $s, t > 0$; b) $\|T(t)\| \le c$ for all $t \in (0,1]$ and some $c > 0$; and c) $T(t)x = 0$ for all $t > 0$ implies $x = 0$.

Recall also from Corollary 3.3.11 that a semigroup is a C_0-semigroup if and only if its generator A is densely defined.

Definition 3.7.1. *Let $\theta \in (0, \frac{\pi}{2}]$. A semigroup T on X is called* holomorphic of angle θ *if it has a holomorphic extension to Σ_θ which is bounded on $\Sigma_{\theta'} \cap \{z \in \mathbb{C} : |z| \le 1\}$ for all $\theta' \in (0,\theta)$.*

If no confusion seems likely, we denote the extension of T to Σ_θ also by T. Also, if we do not want to specify the angle θ in Definition 3.7.1, we call a semigroup T *holomorphic* if it is holomorphic of angle θ for some $\theta \in (0, \frac{\pi}{2}]$.

Proposition 3.7.2. *Let $\theta \in (0, \frac{\pi}{2}]$ and let T be a semigroup on X with generator A. Assume that T is holomorphic of angle θ. Then the following hold:*

a) $T(z + z') = T(z)T(z') \quad (z, z' \in \Sigma_\theta)$.

b) *For all $\theta' \in (0,\theta)$ there exist $M \ge 0, \omega \ge 0$ such that $\|T(z)\| \le Me^{\omega \operatorname{Re} z}$ for all $z \in \Sigma_{\theta'}$.*

c) *Let* $\alpha \in (-\theta, \theta)$. *Denote by* T_α *the semigroup given by* $T_\alpha(t) := T(e^{i\alpha}t)$ $(t \geq 0)$. *Then* $e^{i\alpha}A$ *is the generator of* T_α.

d) *If* T *is a* C_0*-semigroup, then*

$$\lim_{z\to 0, z\in\Sigma_{\theta'}} T(z)x = x$$

for all $x \in X$ *and all* $\theta' \in (0, \theta)$.

Proof. a) For fixed $z' \in (0, \infty)$ consider the holomorphic functions $z \mapsto T(z + z')$ and $z \mapsto T(z)T(z')$ for $z \in \Sigma_\theta$. Since the two functions coincide on $(0, \infty)$, the identity theorem for holomorphic functions implies that $T(z + z') = T(z)T(z')$ for $z \in \Sigma_\theta$ and $z' \in (0, \infty)$. For fixed $z \in \Sigma_\theta$ the two holomorphic functions $z' \mapsto T(z + z')$ and $z' \mapsto T(z)T(z')$ coincide for $z' \in (0, \infty)$ and the assertion follows from the identity theorem for holomorphic functions.

b) Let $\theta' \in (0, \theta)$,

$$M := \sup_{z\in\Sigma_{\theta'}, |z|\leq 1} \|T(z)\|, \qquad \omega' := \max\{\log M, 0\}.$$

Then $\|T(z)\| \leq Me^{\omega'|z|}$ for all $z \in \Sigma_{\theta'}$. In fact, let $z = te^{i\beta}$ where $|\beta| \leq \theta'$. Applying the proof of Theorem 3.1.7 a) to T_β one obtains

$$\|T(z)\| = \|T_\beta(t)\| \leq Me^{\omega|z|}.$$

Since $|z| \leq \operatorname{Re} z/\cos\theta'$ for $z \in \Sigma_{\theta'}$, the claim follows with $\omega := \omega'/\cos\theta$.

c) Let A_α be the generator of T_α. For $R > 0$, let Γ_R be the contour consisting of the line segments $\{t : 0 \leq t \leq R\}$ and $\{te^{i\alpha} : 0 \leq t \leq R\}$ and the arc $\{Re^{i\varphi} : 0 \leq \varphi \leq \alpha\}$. Cauchy's theorem implies that $\int_{\Gamma_R} \exp(-\lambda e^{-i\alpha}z)T(z)x\,dz = 0$ for $\lambda > 0$ and $x \in X$. Letting $R \to \infty$, we obtain

$$\begin{aligned} e^{i\alpha}R(\lambda, A_\alpha)x = e^{i\alpha}\int_0^\infty e^{-\lambda t}T_\alpha(t)x\,dt &= \int_0^\infty \exp(-\lambda e^{-i\alpha}t)T(t)x\,dt \\ &= R(\lambda e^{-i\alpha}, A)x, \end{aligned}$$

by Theorem 3.1.7. This implies that $A_\alpha = e^{i\alpha}A$.

d) Let $\theta' \in (0, \theta)$. By b) there exist $\omega \geq 0$ and $M \geq 0$ such that $\|e^{-\omega z}T(z)\| \leq M$ for all $z \in \Sigma_{\theta'}$. It follows from Proposition 2.6.3 b) that

$$\lim_{z\to 0, z\in\Sigma_{\theta'}} e^{-\omega z}T(z)x = x$$

for all $x \in X$. This implies the claim. □

We note that in the situation of Proposition 3.7.2, T_α is a C_0-semigroup for each $\alpha \in (-\theta, \theta)$ whenever T is a C_0-semigroup; i.e., if $D(A)$ is dense.

Next, we define bounded holomorphic semigroups.

Definition 3.7.3. *Let $\theta \in (0, \frac{\pi}{2}]$. A semigroup T is called a* bounded holomorphic semigroup of angle θ *if T has a bounded holomorphic extension to $\Sigma_{\theta'}$ for each $\theta' \in (0, \theta)$.*

We denote the extension of T to Σ_θ by T again. If we do not want to specify the angle, we call T a *bounded holomorphic semigroup* if T is a bounded holomorphic semigroup of angle θ for some $\theta \in (0, \frac{\pi}{2}]$.

Some caution is required concerning this terminology. If T is a bounded semigroup which is holomorphic, then it is not necessarily a bounded holomorphic semigroup since it is just bounded on $\mathbb{R}_+$ and may not be bounded on a sector. For example, let $X = \mathbb{C}$ and $T(t) = e^{it}$ $(t \geq 0)$.

The following result is an immediate consequence of Proposition 3.7.2 b).

Proposition 3.7.4. *An operator A generates a holomorphic semigroup if and only if there exists $\omega \geq 0$ such that $A - \omega$ generates a bounded holomorphic semigroup.*

Next we give some examples of holomorphic semigroups.

Example 3.7.5 (Selfadjoint operators). *Let A be a selfadjoint operator on a Hilbert space H. Assume that A is bounded above by ω. Then A generates a bounded holomorphic C_0-semigroup of angle $\pi/2$ satisfying*

$$\|T(z)\| \leq e^{\omega \operatorname{Re} z} \qquad (\operatorname{Re} z > 0).$$

Proof. By the Spectral Theorem B.13, we can assume that $H = L^2(\Omega, \mu)$ and that A is given by

$$D(A) = \{f \in H : \ mf \in H\}, \qquad Af = m \cdot f,$$

where $m : \Omega \to (-\infty, \omega]$ is measurable. It is easy to see that

$$(T(z)f)(x) := e^{zm(x)} f(x) \qquad (x \in \Omega, \ \operatorname{Re} z > 0),$$

defines a holomorphic C_0-semigroup on H, whose generator is A. □

Example 3.7.6 (Gaussian semigroup). *Let X be one of the spaces $L^p(\mathbb{R}^n)$ $(1 \leq p < \infty)$, $C_0(\mathbb{R}^n)$ or $\mathrm{BUC}(\mathbb{R}^n)$. Then*

$$(G(t)f)(x) := (4\pi t)^{-n/2} \int_{\mathbb{R}^n} f(x-y) e^{-|y|^2/4t} \, dy \quad (t > 0, \ f \in X, \ x \in \mathbb{R}^n)$$

defines a bounded holomorphic C_0-semigroup of angle $\pi/2$ on X. Its generator is the Laplacian Δ_X on X with maximal domain; i.e.,

$$\begin{aligned} D(\Delta_X) &= \{f \in X : \ \Delta f \in X\}, \\ \Delta_X f &= \Delta f, \end{aligned}$$

where we identify X with a subspace of $\mathcal{D}(\mathbb{R}^n)'$, and $\Delta f = \sum_{j=1}^n D_j^2 f$ (see Appendix E).

Proof. a) Let $k_t \in \mathcal{S}(\mathbb{R}^n)$ be given by

$$k_t(x) = \frac{1}{(4\pi t)^{n/2}} e^{-|x|^2/4t} \qquad (t > 0, x \in \mathbb{R}^n).$$

Then $G(t)f = k_t * f \in X$. Note that $\mathcal{F}k_t = h_t$, where $h_t(x) := e^{-t|x|^2}$. Hence, $h_{t+s} = h_t h_s$. Recall that $\mathcal{F}$ is an isomorphism from $\mathcal{S}(\mathbb{R}^n)'$ onto $\mathcal{S}(\mathbb{R}^n)'$ such that $\mathcal{F}(\psi * f) = \mathcal{F}\psi \cdot \mathcal{F}f$ for all $f \in \mathcal{S}(\mathbb{R}^n)'$, $\psi \in \mathcal{S}(\mathbb{R}^n)$ (see Appendix E). Thus, $\mathcal{F}(G(t)f) = h_t \cdot \mathcal{F}f$. It follows that $G(t+s) = G(t)G(s) \quad (t, s > 0)$. Since $\{k_t : t > 0\}$ is an approximate identity (see Lemma 1.3.3), it follows that $\|k_t * f - f\|_X \to 0$ as $t \downarrow 0$ for all $f \in X$. We have shown that G is a C_0-semigroup.

b) The function k_z is also defined for $\operatorname{Re} z > 0$ and $z \mapsto k_z : \mathbb{C}_+ \to L^1(\mathbb{R}^n)$ is a holomorphic function satisfying $\sup_{z\in\Sigma_\theta} \|k_z\|_{L^1(\mathbb{R}^n)} < \infty$ for each $0 < \theta < \pi/2$. This shows that $G(z)f := k_z * f$ defines a holomorphic extension of G to $\mathbb{C}_+$ with values in $\mathcal{L}(X)$ such that $\sup_{z\in\Sigma_\theta} \|G(z)\| < \infty$ for each $0 < \theta < \pi/2$.

c) We identify the generator of G.

First step: Let $f \in X$ such that $\Delta f \in X$. We show that $\Delta(G(t)f) = G(t)(\Delta f)$. Let $m(x) = -|x|^2$. Then

$$\mathcal{F}(\Delta(G(t)f)) = m\mathcal{F}(G(t)f) = mh_t\mathcal{F}f = h_t m\mathcal{F}f = h_t\mathcal{F}(\Delta f) = \mathcal{F}(G(t)\Delta f).$$

This proves the claim.

Second step: Let $\psi \in \mathcal{S}(\mathbb{R}^n)$. Then

$$\int_0^t G_1(s)\Delta\psi \, ds = G_1(t)\psi - \psi,$$

where G_1 is the Gaussian semigroup on $L^1(\mathbb{R}^n)$. In fact,

$$\begin{aligned} \mathcal{F}\left(\int_0^t G_1(s)\Delta\psi \, ds\right)(x) &= \int_0^t \mathcal{F}(G_1(s)\Delta\psi)(x) \, ds \\ &= \int_0^t e^{-s|x|^2}(-|x|)^2(\mathcal{F}\psi)(x) \, ds \\ &= (e^{-t|x|^2} - 1)(\mathcal{F}\psi)(x) \\ &= (\mathcal{F}(G_1(t)\psi - \psi))(x). \end{aligned}$$

The claim follows from the uniqueness of Fourier transforms.

Third step: Let $f \in X$, $t > 0$. We show that

$$\Delta \int_0^t G(s)f \, ds = G(t)f - f.$$

Then it follows from Corollary 3.1.13 (using also the first step) that the generator of G is the Laplacian with maximal domain. Let $\psi \in \mathcal{S}(\mathbb{R}^n)$. Then Fubini's theorem gives $\langle \Delta\psi, G(s)f\rangle = \langle G_1(s)\Delta\psi, f\rangle$ and

$$\begin{aligned} \left\langle \psi, \Delta \int_0^t G(s)f\,ds \right\rangle &= \left\langle \Delta\psi, \int_0^t G(s)f\,ds \right\rangle \\ &= \int_0^t \langle \Delta\psi, G(s)f\rangle\,ds \\ &= \int_0^t \langle G_1(s)\Delta\psi, f\rangle\,ds \\ &= \left\langle \int_0^t G_1(s)\Delta\psi\,ds, f \right\rangle \\ &= \langle G_1(t)\psi - \psi, f\rangle \\ &= \langle \psi, G(t)f - f\rangle, \end{aligned}$$

by the second step. This proves the claim. □

Remark 3.7.7. a) Let $X = L^p(\mathbb{R}^n)$ $(1 < p < \infty)$. Then $D(\Delta_X) = W^{2,p}(\mathbb{R}^n)$. In fact, $R(1, \Delta_X)$ is given by the Fourier multiplier $x \mapsto (1 + |x|^2)^{-1}$ (see Appendix E). It is easy to see that the function $m_{jk}(x) := -x_j x_k(1 + |x|^2)^{-1}$ satisfies the condition of Mikhlin's theorem (Theorem E.3), so it is a Fourier multiplier for $L^p(\mathbb{R}^n)$ for $j, k \in \{1, \ldots, n\}$. One verifies easily that $\mathcal{F}(D_j D_k R(1, \Delta_X)f) = m_{jk}\mathcal{F}f$ for all $f \in \mathcal{S}(\mathbb{R}^n)$. Thus $D_j D_k R(1, \Delta_X) : \mathcal{S}(\mathbb{R}^n) \to \mathcal{S}(\mathbb{R}^n)$ has a bounded extension to $L^p(\mathbb{R}^n)$. This means that

$$D(\Delta_X) = R(1, \Delta_X)L^p(\mathbb{R}^n) \subset W^{2,p}(\mathbb{R}^n).$$

b) If $X = L^1(\mathbb{R}^n)$, $C_0(\mathbb{R}^n)$ or $\mathrm{BUC}(\mathbb{R}^n)$, then $D(\Delta_X)$ is not a classical function space. For example, if $X = L^1(\mathbb{R}^n)$, then $D(\Delta_X) \supsetneq W^{2,1}(\mathbb{R}^n)$. Similarly, if $X = C_0(\mathbb{R}^n)$, then $D(\Delta_X)$ contains functions which are not in $C^2(\mathbb{R}^n)$. See [DL90, Chapter II, Section 3, Remark 5]. □

Modifying the Banach space in Example 3.7.6 we obtain an example of a holomorphic semigroup which is not a C_0-semigroup. Another example will be given in Chapter 6. Let $C_b(\mathbb{R}^n)$ be the Banach space of all bounded continuous complex-valued functions on $\mathbb{R}^n$ with the supremum norm.

Example 3.7.8. Let $X = C_b(\mathbb{R}^n)$ or $L^\infty(\mathbb{R}^n)$. Define the Gaussian semigroup G on X as in Example 3.7.6. Then G is a bounded holomorphic semigroup which is not a C_0-semigroup. Its generator is the operator Δ_X defined as in Example 3.7.6. □

Example 3.7.9 (Poisson semigroup). Let X be one of the spaces considered in Example 3.7.6. Let

$$p_t(x) = c_n \frac{t}{(t^2 + |x|^2)^{(n+1)/2}} \qquad (x \in \mathbb{R}^n, t > 0)$$

where $c_n := \Gamma((n+1)/2)/\pi^{(n+1)/2}$. Then $p_t \in L^1(\mathbb{R}^n)$ and $(\mathcal{F}p_t)(x) = e^{-t|x|}$ $(t > 0)$. Similarly to Example 3.7.6, one shows that

$$T(t)f := p_t * f \qquad (t > 0)$$

defines a C_0-semigroup on X, which is called the Poisson semigroup. It is again a bounded holomorphic C_0-semigroup of angle $\pi/2$ on X. Its holomorphic extension to the sector $\Sigma_{\pi/2}$ is bounded on Σ_θ for $\theta < \pi/2$ and is given by

$$T(z)f := p_z * f,$$

where

$$p_z(x) := c_n \frac{z}{(z^2 + |x|^2)^{(n+1)/2}}$$

for $x \in \mathbb{R}^n$ and $\operatorname{Re} z > 0$. Its generator is the operator A_X defined by

$$\begin{aligned} A_X f &:= \mathcal{F}^{-1}(-|\cdot|\mathcal{F}f), \\ D(A_X) &:= \{f \in X : \mathcal{F}^{-1}(|\cdot|\mathcal{F}f) \in X\}, \end{aligned}$$

where $\mathcal{F}$ now denotes the Fourier transform in $\mathcal{S}(\mathbb{R}^n)'$.

A more explicit description of the operator $A_{L^p(\mathbb{R})}$ is of particular interest. Observing that

$$-(-i \operatorname{sign}(\xi))(i\xi) = -|\xi| \qquad (\xi \in \mathbb{R}),$$

it follows from (E.19) that

$$A_{L^p(\mathbb{R})} = -H\frac{\partial}{\partial x},$$

where H denotes the Hilbert transform defined by

$$(Hf)(x) := \lim_{\varepsilon \to 0, R \to \infty} \frac{1}{\pi} \int_{\varepsilon \le |y| \le R} \frac{f(x-y)}{y}\, dy.$$

Since the Hilbert transform acts as a bounded operator on $L^p(\mathbb{R})$ for $1 < p < \infty$ by Proposition E.5, it follows that the domain $D(A_{L^p(\mathbb{R})})$ of $A_{L^p(\mathbb{R})}$ coincides with $W^{1,p}(\mathbb{R})$. However, $A_{L^p(\mathbb{R})}$ is not a first-order differential operator. We shall return to the relationship between the Poisson and Gaussian semigroups in Example 3.8.5. □

Remark 3.7.10. Let G be the Gaussian semigroup on $L^1(\mathbb{R}^n)$. It follows from the explicit formula for k_z given in Example 3.7.6a) that for $\operatorname{Re} z > 0$

$$\|G(z)\|_{\mathcal{L}(L^1(\mathbb{R}^n))} = \|k_z\|_{L^1(\mathbb{R}^n)} = \left(\frac{|z|}{\operatorname{Re} z}\right)^{n/2}. \tag{3.44}$$

Thus, by Proposition 3.7.2c), G_α defined by $G_\alpha(t) := G(e^{i\alpha}t)$ for $|\alpha| < \frac{\pi}{2}$ is an example of a bounded C_0-semigroup, which due to (3.44), is not a contraction semigroup. □

The following characterization of a bounded holomorphic semigroup in terms of a single resolvent estimate for its generator is of fundamental importance.

Theorem 3.7.11. *Let A be an operator on X and $\theta \in (0, \frac{\pi}{2}]$. The following assertions are equivalent:*

(i) *A generates a bounded holomorphic semigroup of angle θ.*

(ii) *$\Sigma_{\theta+\frac{\pi}{2}} \subset \rho(A)$ and*

$$\sup_{\lambda \in \Sigma_{\theta+\frac{\pi}{2}-\varepsilon}} \|\lambda R(\lambda, A)\| < \infty \text{ for all } \varepsilon > 0.$$

Proof. In order to prove the assertion (i) $\Rightarrow$ (ii), note that if $\lambda_0 \in \rho(A)$ and $\lambda \mapsto R(\lambda, A)$ has a holomorphic extension to some open connected set Ω containing λ_0, then by Proposition B.5, $\Omega \subset \rho(A)$ and the extension is the resolvent. Thus (i) $\Rightarrow$ (ii) follows immediately from Theorem 2.6.1. In order to prove the converse implication (ii) $\Rightarrow$ (i), note that by Theorem 2.6.1 there exists a holomorphic function $T : \Sigma_\theta \to \mathcal{L}(X)$ which is bounded on Σ_α for $0 < \alpha < \theta$ such that

$$R(\lambda, A) = \int_0^\infty e^{-\lambda t} T(t)\, dt \qquad (\operatorname{Re} \lambda > 0). \tag{3.45}$$

The proof of Theorem 3.1.7 shows that T is a semigroup. This proves (i). □

In particular, a densely defined operator A satisfying (ii) of Theorem 3.7.11 is a Hille-Yosida operator, by Example 3.5.9 c). Moreover, the bounded holomorphic semigroup T generated by A is a C_0-semigroup if and only if $D(A)$ is dense in X, by Corollary 3.3.11.

We note from (2.13) that the semigroup T generated by A is given by

$$T(z) = \frac{1}{2\pi i} \int_\Gamma e^{\lambda z} R(\lambda, A)\, d\lambda \qquad (z \in \Sigma_\alpha), \tag{3.46}$$

if $0 < \alpha < \theta$, where the contour Γ consists of

$$\Gamma_\pm := \{re^{\pm\gamma} : \delta \le r\} \quad \text{and} \quad \Gamma_0 := \{\delta e^{i\theta'} : |\theta'| \le \gamma\}$$

with $\alpha + \frac{\pi}{2} < \gamma < \theta + \frac{\pi}{2}$ and $\delta > 0$.

If we do not want to specify the angle of holomorphy in the above theorem, then it suffices to verify condition (ii) above on a right half-plane. In fact, the following holds true.

Corollary 3.7.12. *For an operator A on X the following are equivalent:*

(i) *A generates a bounded holomorphic semigroup on X.*

(ii) $\{z \in \mathbb{C} : \operatorname{Re} z > 0\} \subset \rho(A)$ *and*

$$M := \sup_{\operatorname{Re}\lambda > 0} \|\lambda R(\lambda, A)\| < \infty.$$

Proof. By Theorem 3.7.11, we only have to prove that (ii) implies the second assertion of Theorem 3.7.11. Set $c := \frac{1}{2M}$. For $s \in \mathbb{R} \setminus \{0\}$ and $-c|s| < r \leq 0$, let $\lambda := c|s| + r + is$. Then $|\lambda - (r + is)| = c|s| \leq \frac{1}{2}\|R(\lambda, A)\|^{-1}$. By Corollary B.3, $r + is \in \rho(A)$ and

$$\| (r + is)R(r + is, A) \| \leq 2M \frac{|r + is|}{|s|} \leq 2M(c + 1).$$

Thus, (ii) is satisfied with $\theta = \arctan c$. □

Remark 3.7.13. Suppose that an operator A satisfies the equivalent conditions (i) and (ii) of the above corollary. For $Y := \overline{D(A)}$ let A_Y be the part of A defined as in (3.27). It follows from Lemma 3.3.12 and Corollary 3.7.12 that in this case A_Y generates a bounded holomorphic C_0-semigroup T_Y on Y. Moreover, $T_Y(t) = T(t)|_Y$ for $t \geq 0$, where T is the semigroup generated by A.

Corollary 3.7.14. *Let A be an operator on X such that $\sigma(A) \subset i\mathbb{R}$. Assume that there exists a constant $M > 0$ such that*

$$\|R(\lambda, A)\| \leq \frac{M}{|\operatorname{Re}\lambda|} \qquad (\operatorname{Re}\lambda \neq 0). \tag{3.47}$$

Then A^2 generates a bounded holomorphic semigroup of angle $\pi/2$ on X.

Proof. Let $\theta \in (0, \pi/2]$ and $\lambda \in \Sigma_{\theta+\pi/2}$. Then there exist $r > 0$ and $\varphi \in (0, \pi/2)$ such that $\lambda = r^2 e^{2i\varphi}$. Observe that

$$\lambda - A^2 = (re^{i\varphi} + A)(re^{i\varphi} - A).$$

The assumption implies that $\lambda \in \rho(A^2)$ and that

$$R(\lambda, A^2) = -R(re^{i\varphi}, A)R(-re^{i\varphi}, A).$$

The resolvent estimate (3.47) implies that

$$\|R(\lambda, A^2)\| \leq \frac{M^2}{(r\cos\varphi)^2} = \frac{M^2}{(\cos\varphi)^2}\frac{1}{|\lambda|} \qquad (\lambda \in \Sigma_{\theta+\pi/2}).$$

Hence, the assertion follows from Theorem 3.7.11. □

Applying Corollary 3.7.14 to the situation of generators of bounded C_0-groups we immediately obtain the following result.

Corollary 3.7.15. *Let A be the generator of a bounded C_0-group U on X. Then A^2 generates a bounded holomorphic C_0-semigroup T of angle $\pi/2$ on X. Moreover, for $t > 0$,*

$$T(t) = \int_{\mathbb{R}} k_t(s)U(s)\, ds,$$

where $k_t(s) = (4\pi t)^{-1/2}e^{-|s|^2/4t}$.

Proof. The fact that A^2 generates a bounded holomorphic C_0-semigroup of angle $\pi/2$ is immediate from Corollary 3.7.14. Define $T(0)x = x$ and

$$\begin{aligned} T(t)x = \int_{\mathbb{R}} k_t(s)U(s)x\, ds &= \int_{\mathbb{R}} k_1(s)U(s\sqrt{t})\, ds \\ &= \int_0^\infty k_t(s)(U(s)x + U(-s)x)\, ds. \end{aligned}$$

Then T is strongly continuous by the dominated convergence theorem. By Proposition 1.6.8,

$$\begin{aligned} \hat{T}(\lambda) &= \frac{1}{2\sqrt{\lambda}}R(\sqrt{\lambda}, A) + \frac{1}{2\sqrt{\lambda}}R(\sqrt{\lambda}, -A) \\ &= R(\lambda, A^2) \end{aligned}$$

for $\lambda > 0$. Thus T is a C_0-semigroup generated by A^2. □

Let X be any of the spaces considered in Example 3.7.6 in the case $n = 1$, and let U be the shift group: $(U(t)f)(x) = f(x - t)$ (see Example 3.3.10). Then the C_0-semigroup constructed in Corollary 3.7.15 is the Gaussian semigroup.

The following proposition will be useful in Chapter 6 when we are dealing with the holomorphic semigroup generated by the Laplacian subject to Dirichlet boundary conditions on spaces of continuous functions. Note that A is not necessarily densely defined.

Proposition 3.7.16. *Let A be a dissipative operator on X and assume that A generates a holomorphic semigroup T on X. Then $\|T(t)\| \le 1$ for all $t > 0$.*

Proof. By the remark after Theorem 3.7.11, A is a Hille-Yosida operator. Hence, there exists $\lambda_0 > 0$ with $\lambda_0 \in \rho(A)$. By the proof of the Lumer-Phillips theorem, we have $(0, \infty) \subset \rho(A)$ and $\|\lambda R(\lambda, A)\| \le 1$ for all $\lambda > 0$. By the proof of Corollary 3.3.6, we have $T(t)x = \lim_{n\to\infty} \|(I - \frac{t}{n}A)^{-n}x\|$ for $t > 0$ and $x \in X$. Hence,

$$\|T(t)\| \le \limsup_{n\to\infty} \left\| \left(I - \tfrac{t}{n}A\right)^{-n} \right\| \le 1$$

for $t > 0$. □

Applying Corollary 3.7.12 to the operator $A - \omega$ for suitable ω, in view of Proposition 3.7.4, we obtain the following characterization results for holomorphic semigroups.

Corollary 3.7.17. *Let A be an operator on X, and $a \in \mathbb{R}$. Then A generates a holomorphic semigroup if and only if there exists $r > 0$ such that $\{\lambda \in \mathbb{C} : \operatorname{Re}\lambda > a, |\lambda| > r\} \subset \rho(A)$ and*

$$\sup_{\substack{\operatorname{Re}\lambda > a\\ |\lambda|>r}} \|\lambda R(\lambda, A)\| < \infty.$$

Modifying the proof of Corollary 3.7.12 to the situation of holomorphic semigroups which are not necessarily bounded, we obtain the following result.

Corollary 3.7.18. *Let A be the generator of a semigroup T on X. Then T is holomorphic if and only if there exists $r > 0$ such that $\{is : s \in \mathbb{R}, |s| > r\} \subset \rho(A)$ and*

$$\sup_{|s|>r} \|sR(is, A)\| < \infty.$$

We now characterize bounded holomorphic semigroups in terms of the behaviour of $\|tAT(t)\|$ for positive t.

Theorem 3.7.19. *Let T be a bounded semigroup on X with generator A. Then T is a bounded holomorphic semigroup if and only if $T(t)x \in D(A)$ for all $t > 0$, $x \in X$, and*

$$\sup_{t>0} \| \, tAT(t) \, \| < \infty. \tag{3.48}$$

Proof. Suppose that T is a bounded holomorphic semigroup of angle θ. Then T is norm-differentiable on $(0, \infty)$, so $T(t)x \in D(A)$ for $t > 0$ and $x \in X$. By Cauchy's integral formula for the derivative,

$$AT(t) = T'(t) = \frac{1}{2\pi i} \int_{|z-t|=t\sin\theta/2} \frac{T(z)}{(z-t)^2}\, dz.$$

Hence,

$$\|tAT(t)\| \le (\sin\theta/2)^{-1} \sup_{z\in\Sigma_{\theta/2}} \|T(z)\| < \infty.$$

Conversely, let $M := \sup_{t>0} \{\| \, T(t) \, \|, \| \, tAT(t) \, \|\}$. By assumption, $T(t)x \in D(A)$ for all $t > 0$ and $x \in X$. Since $A^nT(t) = (AT(t/n))^n \in \mathcal{L}(X)$ for $n \in \mathbb{N}$, we have

$$\left\| \frac{A^nT(t)}{n!} \right\| = \left\| \frac{(AT(t/n))^n}{n!} \right\| \le \frac{(\frac{n}{t}M)^n}{n!} \le \left(\frac{eM}{t}\right)^n \quad \text{for all } n \in \mathbb{N}.$$

If $|z| < \frac{t}{2eM}$, then $\tilde{T}(t+z) := \sum_{n=0}^{\infty} z^n \frac{A^nT(t)}{n!}$ converges in norm and $\|\tilde{T}(t+z)\| \le 1 + M$. Then for $s \in [0, \frac{t}{2eM})$, $\tilde{T}(t+s) = T(t+s)$. In fact, let $x \in D(A)$ and $u(s) := \tilde{T}(t+s)x$. Then $u(s) \in D(A)$ and $u'(s) = Au(s)$ $(s \in [0, \frac{t}{2eM}))$. Since $u(0) = T(t)x$, it follows from Proposition 3.1.11 that $u(s) = T(s)T(t)x = T(t+s)x$. By uniqueness of analytic extensions, we obtain a bounded, holomorphic extension $\tilde{T}$ of T to the sector Σ_θ where $\theta = \arctan \frac{1}{2eM}$. □

Remark 3.7.20. A slight modification of the above proof implies the following assertion:
Let A be the generator of a bounded holomorphic semigroup T on X. Then $T(t)x \in D(A^n)$ for all $x \in X$, $t > 0$ and $n \in \mathbb{N}$, and we have

$$\sup_{t>0} ||t^n A^n T(t)|| < \infty \qquad (n \in \mathbb{N}).$$ □

Theorem 3.7.19 and Remark 3.7.20 have several important consequences for the regularity of the solution of the associated Cauchy problem. In contrast to the situation of C_0-semigroups where we obtain a classical solution of the Cauchy problem only if the initial condition x belongs to the domain $D(A)$ of the generator A, we see that if A generates a holomorphic C_0-semigroup, then for all $x \in X$ we obtain a solution which is differentiable for $t > 0$. More precisely, the following holds.

Corollary 3.7.21. *Let $x \in X$ and assume that A generates a holomorphic C_0-semigroup on X. Then there exists a unique function $u \in C^\infty((0,\infty), X) \cap C(\mathbb{R}_+, X) \cap C((0,\infty), D(A))$ satisfying*

$$\begin{cases} u'(t) = & Au(t) \quad (t > 0), \\ u(0) = & x. \end{cases}$$

The phenomenon described in Corollary 3.7.21 is frequently called the *smoothing effect* of holomorphic C_0-semigroups.

We now consider the inhomogeneous Cauchy problem

$$(ACP_f) \begin{cases} u'(t) = & Au(t) + f(t) \quad (t \in [0,\tau]), \\ u(0) = & x, \end{cases}$$

associated with the generator A of a holomorphic semigroup, and we establish that the variation of constants formula holds when $x \in \overline{D(A)}$.

Proposition 3.7.22. *Let A be the generator of a holomorphic semigroup T on X. Let $f \in L^1((0,\tau), X)$ and $x \in \overline{D(A)}$. Then (ACP_f) has a unique mild solution u which is given by*

$$u(t) = T(t)x + \int_0^t T(t-s)f(s)\,ds. \qquad (t > 0)$$

Proof. By the remark following Theorem 3.7.11, A is a Hille-Yosida operator. Hence, the first assertion follows from the Da Prato-Sinestrari Theorem 3.5.2. The formula for u follows from Lemma 3.2.9, since the once integrated semigroup generated by A is given by $S(t)x = \int_0^t T(s)x\,ds$ and the derivative of $S * f$ is easily seen to be $T * f$. □

The following perturbation result for generators of holomorphic semigroups is particularly useful when dealing with lower order perturbations of differential operators. Consider a closed operator A on X. Then a mapping $B : D(A) \to X$ is continuous (with respect to the graph norm on $D(A)$) if and only if

$$\|Bx\| \leq c\|Ax\| + b\|x\| \qquad (x \in D(A)),$$

for suitable $c, b \geq 0$. One frequently says that B is a *relatively bounded perturbation of* A in that case.

Theorem 3.7.23. *Let A be the generator of a holomorphic semigroup on X. Let $B : D(A) \to X$ be an operator such that for every $\varepsilon > 0$ there exists a constant $b \geq 0$ such that*

$$\| Bx \| \leq \varepsilon \| Ax \| + b \| x \| \qquad (x \in D(A)). \tag{3.49}$$

Then $A + B$ generates a holomorphic semigroup.

Proof. Assume first that A generates a bounded holomorphic semigroup on X. Corollary 3.7.12 implies that there exists $\theta \in (0, \frac{\pi}{2}]$ such that

$$\Sigma_{\theta+\pi/2} \subset \rho(A) \text{ and } \sup_{\lambda \in \Sigma_{\theta+\pi/2}} \|\lambda R(\lambda, A)\| =: M < \infty.$$

It follows from the assumption on B that given $\varepsilon > 0$, there exists $b \geq 0$ such that for $x \in X$

$$\begin{aligned} \| BR(\lambda, A)x \| &\leq \varepsilon \| AR(\lambda, A)x \| + b \| R(\lambda, A)x \| \\ &= \varepsilon \|\lambda R(\lambda, A)x - x\| + b\|R(\lambda, A)x\| \\ &\leq \varepsilon(M+1)\| x \| + \frac{bM}{|\lambda|}\| x \| \qquad (\lambda \in \Sigma_{\theta+\pi/2}). \end{aligned}$$

Choosing $\varepsilon := (2(M+1))^{-1}$, it follows that $\| BR(\lambda, A) \| < 3/4$ whenever $|\lambda| > 4bM$ and hence that $I - BR(\lambda, A)$ is invertible, with $\|(I - BR(\lambda, A))^{-1}\| < 4$. Since

$$\lambda - (A + B) = (I - BR(\lambda, A))(\lambda - A) \qquad (\lambda \in \Sigma_{\theta+\pi/2}),$$

it follows that $\lambda - (A + B)$ is invertible for $\lambda \in \Sigma_{\theta+\pi/2}$ with $|\lambda| > 4bM$ and that

$$\| R(\lambda, A + B) \| \leq \frac{4M}{|\lambda|} \qquad (\lambda \in \Sigma_{\theta+\pi/2}, \ |\lambda| > 4bM).$$

By Corollary 3.7.17, $A + B$ generates a holomorphic semigroup.

If A generates a holomorphic semigroup which is not bounded, choose $\omega \in \mathbb{R}$ such that $A - \omega$ generates a bounded holomorphic semigroup on X (see Proposition 3.7.4). The first part of the proof implies that $A + B - \omega$, and therefore also $A + B$, generates a holomorphic semigroup on X. □

Note that in the situation of Theorem 3.7.23, A and $A + B$ have the same domain. Thus, the semigroup generated by A is a C_0-semigroup if and only if the one generated by $A + B$ is a C_0-semigroup.

Example 3.7.24 (First order perturbations of the Laplacian). Consider the Gaussian semigroup G with generator Δ_X on any of the spaces X of Example 3.7.6. For $j = 1, 2, \ldots, n$, let U_j be the C_0-group on X defined by

$$(U_j(t)f)(x) = f(x_1, \ldots, x_j - t, \ldots, x_n)$$

with generator $-D_j$, and let T_j be the holomorphic C_0-semigroup with generator D_j^2 (see Corollary 3.7.15). Then $T_1, \ldots, T_n$ commute and $G(t) = T_1(t) \cdots T_n(t)$. Since

$$\begin{aligned}(T_j(t)f)(x) &= \int_{\mathbb{R}} (4\pi t)^{-1/2} e^{-(x_j-s)^2/4t} f(x_1, \ldots, s, \ldots, x_n)\, ds,\\ (D_jT_j(t)f)(x) &= -\int_{\mathbb{R}} \frac{(x_j - s)}{4\sqrt{\pi} t^{3/2}} e^{-(x_j-s)^2/4t} f(x_1, \ldots, s, \ldots, x_n)\, ds.\end{aligned}$$

Hence by Young's inequality (see Proposition 1.3.2),

$$\begin{aligned}\|D_jT_j(t)\|_{\mathcal{L}(X)} &\le \int_{\mathbb{R}} \frac{|s|}{4\sqrt{\pi} t^{3/2}} e^{-s^2/4t}\, ds\\ &= \frac{2}{\sqrt{\pi t}} \int_0^\infty u e^{-u^2}\, du\\ &= \frac{1}{\sqrt{\pi t}}.\end{aligned}$$

Since $\|T_j(t)\|_{\mathcal{L}(X)} = 1$, it follows that $\|D_jG(t)\|_{\mathcal{L}(X)} \le \frac{1}{\sqrt{\pi t}}$, and hence

$$\begin{aligned}\|D_jR(\lambda, \Delta_X)\|_{\mathcal{L}(X)} &= \left\| \int_0^\infty e^{-\lambda t} D_jG(t)\, dt \right\|_{\mathcal{L}(X)}\\ &\le \frac{1}{\sqrt{\pi}} \int_0^\infty \frac{e^{-\lambda t}}{\sqrt{t}}\, dt\\ &= \frac{c}{\sqrt{\lambda}}\end{aligned}$$

for all $\lambda > 0$, for some constant c.

Now let B be a first-order differential operator of the form

$$(Bf)(x) = \sum_{j=1}^n b_j(x)(D_jf)(x) + b_0(x)f(x),$$

for some $b_j \in L^\infty(\mathbb{R}^n)$ $(j = 0, 1, \ldots, n)$ (and b_j continuous if $X = C_0(\mathbb{R}^n)$; b_j

uniformly continuous if $X = \mathrm{BUC}(\mathbb{R}^n)$). For $f \in D(\Delta_X)$,

$$\begin{aligned} \|Bf\| &= \|BR(\lambda, \Delta_X)(\lambda - \Delta_X)f\| \\ &\leq \sum_{j=1}^{n} \|b_j\|_\infty \|D_j R(\lambda, \Delta_X)\|_{\mathcal{L}(X)} \, \|(\lambda - \Delta_X)f\| + \|b_0\|_\infty \|f\| \\ &\leq \sum_{j=1}^{n} \frac{\|b_j\|_\infty c}{\sqrt{\lambda}} \|\Delta_X f\| + \left(\sum_{j=1}^{n} \|b_j\|_\infty c\sqrt{\lambda} + \|b_0\|_\infty \right) \|f\|. \end{aligned}$$

Since λ may be chosen arbitrary large, this establishes (3.49), and Theorem 3.7.23 shows that $\Delta_X + B$ generates a holomorphic C_0-semigroup on X. □

We shall extend Example 3.7.24 to more general differential operators in Section 7.2.

We now prove a second perturbation theorem for holomorphic semigroups where the norm estimate (3.49) is replaced by compactness.

Theorem 3.7.25 (Desch-Schappacher). *Let A be the generator of a holomorphic C_0-semigroup T. Let $B : D(A) \to X$ be a compact linear operator where $D(A)$ carries the graph norm. Then $A + B$ generates a holomorphic C_0-semigroup S. Moreover, $T(t) - S(t)$ is compact for each $t > 0$.*

Proof. By Corollary 3.7.17, there exist $r > 0$, $M > 0$ such that $\lambda \in \rho(A)$ and $\|\lambda R(\lambda, A)\|_{\mathcal{L}(X)} \leq M$ whenever $|\lambda| \geq r$, $\operatorname{Re}\lambda > 0$. Since $D(A)$ is dense in X, it follows that $\lim_{\substack{|\lambda|\to\infty \\ \operatorname{Re}\lambda>0}} \lambda R(\lambda, A)x = x$ for all $x \in X$ (see Lemma 3.3.12). Since $\lambda R(\lambda, A)x - x = AR(\lambda, A)x$, it follows that $\lim_{\substack{|\lambda|\to\infty \\ \operatorname{Re}\lambda>0}} R(\lambda, A)x = 0$ in $D(A)$ for all $x \in X$. By Proposition B.15, the convergence is uniform on compact subsets of X. Since $B : D(A) \to X$ is compact, it follows that

$$\lim_{\substack{|\lambda|\to\infty \\ \operatorname{Re}\lambda>0}} \|R(\lambda, A)B\|_{\mathcal{L}(D(A))} = 0.$$

Consequently, there exists $r_1 \geq r$ such that $\|R(\lambda, A)B\|_{\mathcal{L}(D(A))} \leq \frac{1}{2}$ whenever $|\lambda| \geq r_1$, $\operatorname{Re}\lambda > 0$. Denote by $I_{D(A)}$ the identity map on $D(A)$. It follows that $(I_{D(A)} - R(\lambda, A)B)^{-1}$ exists in $\mathcal{L}(D(A))$ and

$$\|(I_{D(A)} - R(\lambda, A)B)^{-1}\|_{\mathcal{L}(D(A))} \leq 2 \qquad (|\lambda| \geq r_1, \ \operatorname{Re}\lambda > 0).$$

Thus $(\lambda - (A + B)) = (\lambda - A)(I - R(\lambda, A)B)$ is invertible and

$$R(\lambda, A + B) = (I_{D(A)} - R(\lambda, A)B)^{-1} R(\lambda, A) \qquad (\operatorname{Re}\lambda > 0, |\lambda| \geq r_1).$$

Moreover, for $|\lambda| \geq r_1$, $\operatorname{Re}\lambda > 0$,

$$\begin{aligned} \|R(\lambda, A + B)\|_{\mathcal{L}(X,D(A))} &\leq 2\|R(\lambda, A)\|_{\mathcal{L}(X,D(A))} \\ &\leq M_1 := 2\left(\frac{M}{r_1} + 1 + M\right), \end{aligned}$$

since

$$\begin{aligned}\|R(\lambda,A)x\|_{D(A)} &= \|R(\lambda,A)x\|_X + \|AR(\lambda,A)x\|_X\\ &\le \frac{M}{|\lambda|}\|x\|_X + \|\lambda R(\lambda,A)x - x\|_X\\ &\le \frac{M}{r_1}\|x\|_X + (M+1)\|x\|_X.\end{aligned}$$

Hence for $x \in X,\ |\lambda| \ge r_1,\ \operatorname{Re}\lambda > 0$,

$$\begin{aligned}\|\lambda R(\lambda,A+B)x\|_X &= \|(A+B)R(\lambda,A+B)x - x\|_X\\ &\le \|R(\lambda,A+B)x\|_{D(A)}\\ &\quad + \|B\|_{\mathcal{L}(D(A),X)}\|R(\lambda,A+B)x\|_{D(A)} + \|x\|_X\\ &\le \left(M_1(1+\|B\|_{\mathcal{L}(D(A),X)})+1\right)\|x\|_X.\end{aligned}$$

Now it follows from Corollary 3.7.17 that $A+B$ generates a holomorphic C_0-semigroup.

It remains to show the last assertion. Denote by $\mathcal{K}(X)$ the closed subspace of $\mathcal{L}(X)$ consisting of all compact operators and by $q : \mathcal{L}(X) \to \mathcal{L}(X)/\mathcal{K}(X)$ the quotient mapping. For $\lambda > \lambda_0 := \max\{\omega(T),\omega(S)\}$, we have

$$\begin{aligned}\int_0^\infty e^{-\lambda t}(S(t)-T(t))\,dt &= R(\lambda,A+B) - R(\lambda,A)\\ &= R(\lambda,A+B)[(\lambda-A)-(\lambda-A-B)]R(\lambda,A)\\ &= R(\lambda,A+B)BR(\lambda,A) \in \mathcal{K}(X).\end{aligned}$$

Since S and T are holomorphic, the function $U := S-T$ is norm-continuous on $(0,\infty)$. Since $\widehat{(q\circ U)}(\lambda) = q(\widehat{U}(\lambda)) = 0$ for all $\lambda > \lambda_0$, it follows from the uniqueness theorem that $q\circ U \equiv 0$; i.e., $U(t) \in \mathcal{K}(X)$ for all $t > 0$. □

We should point out that in the situation of Theorem 3.7.25 the norm estimate (3.49) is not true in general; see the Notes for further information.

3.8 Fractional Powers

A particularly interesting example of a holomorphic semigroup is the family of fractional powers of a sectorial operator. Consider an operator B on X for which $(-\infty,0] \subset \rho(B)$ and $\sup_{\lambda\le 0}(1-\lambda)\|R(\lambda,B)\| < \infty$. It follows from Corollary B.3 that B is *sectorial* in the sense that there exist constants $M > 0,\ \varphi \in (0,\pi)$ such that

$$\sigma(B) \subset \Sigma_\varphi \quad\text{and}\quad \|R(\lambda,B)\| \le \frac{M}{1+|\lambda|}, \qquad \lambda \in \mathbb{C}\backslash\Sigma_\varphi. \tag{3.50}$$

Let Γ be the downward path consisting of $\{se^{\pm i\varphi} : \ s \geq r\}$ and $\{re^{i\theta} : \ -\varphi \leq \theta \leq \varphi\}$, where $r > 0$ is chosen so small that $\sigma(B)$ is to the right of Γ. Then the *fractional powers* $(B^{-z})_{\operatorname{Re} z>0}$ of B are defined by

$$B^{-z} := \frac{1}{2\pi i} \int_\Gamma \lambda^{-z} R(\lambda, B)\, d\lambda, \qquad (\operatorname{Re} z > 0). \tag{3.51}$$

Here, $\lambda^{-z} = \exp(-z(\log|\lambda| + i\theta))$ if $\lambda = |\lambda|e^{i\theta}$, $-\pi < \theta < \pi$. Note that the integral is absolutely convergent, uniformly for z in compact subsets of $\mathbb{C}_+$, and therefore $z \mapsto B^{-z}$ is holomorphic from $\mathbb{C}_+$ to $\mathcal{L}(X)$. By Cauchy's theorem, the definition of B^{-z} is independent of the choices of φ and r. Moreover, when $z = n \in \mathbb{N}$, Γ may be replaced by a closed contour around 0 and then the residue theorem shows that

$$B^{-z} = -\frac{1}{(n-1)!} \left(\frac{d}{d\lambda}\right)^{n-1} R(\lambda, B)\bigg|_{\lambda=0} = (-1)^n R(0, B)^n = B^{-n}$$

in the usual sense.

Now, assume for the time being that $0 < \operatorname{Re} z < 1$. Then

$$\begin{aligned} B^{-z} &= \lim_{\substack{\varphi\uparrow\pi \\ r\downarrow 0}} \frac{1}{2\pi i} \int_\Gamma \lambda^{-z} R(\lambda, B)\, d\lambda \\ &= -\frac{e^{-i\pi z}}{2\pi i} \int_0^\infty s^{-z}(s+B)^{-1}\, ds + \frac{e^{i\pi z}}{2\pi i} \int_0^\infty s^{-z}(s+B)^{-1}\, ds \\ &= \frac{\sin \pi z}{\pi} \int_0^\infty s^{-z}(s+B)^{-1}\, ds. \end{aligned} \tag{3.52}$$

In the particular case when $X = \mathbb{C}$ and $B = 1$, (3.52) gives

$$\int_0^\infty s^{-z}(s+1)^{-1}\, ds = \frac{\pi}{\sin \pi z} \qquad (0 < \operatorname{Re} z < 1). \tag{3.53}$$

Hence,

$$\|B^{-z}\| \leq \frac{|\sin \pi z|}{\pi} \int_0^\infty s^{-\operatorname{Re} z} \frac{M}{1+s}\, ds = M \frac{|\sin \pi z|}{\sin(\pi \operatorname{Re} z)} \qquad (0 < \operatorname{Re} z < 1). \tag{3.54}$$

Theorem 3.8.1. *Let B be an operator on X such that $(-\infty, 0] \subset \rho(B)$ and $\sup_{\lambda\leq 0}(1-\lambda)\|R(\lambda, B)\| < \infty$. Then the family $(B^{-z})_{\operatorname{Re} z>0}$ defines a holomorphic semigroup on X of angle $\pi/2$. If B is densely defined, then $(B^{-z})_{\operatorname{Re} z>0}$ is a holomorphic C_0-semigroup.*

Proof. We have observed above that $z \mapsto B^{-z}$ is holomorphic for $\operatorname{Re} z > 0$, and it follows from (3.54) that B^{-z} is uniformly bounded in $\Sigma_\theta \cap \{z \in \mathbb{C} : \ |z| < 1\}$ for $\theta \in (0, \pi/2)$. To verify the semigroup property, let Γ and Γ' be two contours as in the definition of B^{-z}, with Γ to the left of Γ'. For $\operatorname{Re} z_1 > 0$ and $\operatorname{Re} z_2 > 0$, the resolvent identity and Fubini's theorem give

$$\begin{aligned}
B^{-z_1}B^{-z_2} &= \frac{1}{(2\pi i)^2}\int_\Gamma\int_{\Gamma'} \lambda^{-z_1}\mu^{-z_2}R(\lambda,B)R(\mu,B)\,d\mu\,d\lambda \\
&= \frac{1}{(2\pi i)^2}\int_\Gamma\int_{\Gamma'} \lambda^{-z_1}\mu^{-z_2}\left(\frac{R(\lambda,B)-R(\mu,B)}{\mu-\lambda}\right)d\mu\,d\lambda \\
&= \frac{1}{2\pi i}\int_\Gamma \lambda^{-z_1}\left(\frac{1}{2\pi i}\int_{\Gamma'}\frac{\mu^{-z_2}}{\mu-\lambda}\,d\mu\right)R(\lambda,B)\,d\lambda \\
&\quad+\frac{1}{2\pi i}\int_{\Gamma'} \mu^{-z_2}\left(\frac{1}{2\pi i}\int_{\Gamma}\frac{\lambda^{-z_1}}{\lambda-\mu}\,d\lambda\right)R(\mu,B)\,d\mu \\
&= \frac{1}{2\pi i}\int_\Gamma \lambda^{-z_1}\lambda^{-z_2}R(\lambda,B)\,d\lambda \\
&= B^{-(z_1+z_2)}.
\end{aligned}$$

Here, we have used Cauchy's integral formula (after changing to a closed contour around 0) to see that

$$\frac{1}{2\pi i}\int_{\Gamma'}\frac{\mu^{-z_2}}{\mu-\lambda}\,d\mu = \lambda^{-z_2} \quad\text{and}\quad \frac{1}{2\pi i}\int_{\Gamma}\frac{\lambda^{-z_1}}{\lambda-\mu}\,d\lambda = 0.$$

This proves the first assertion.

Now suppose that B is densely defined. We have to show that $\lim_{\substack{z\to 0\\ z\in\Sigma_\theta}} B^{-z}x = x$, for all $\theta \in (0,\pi/2)$ and all $x \in X$. Since B^{-z} is uniformly bounded on $\Sigma_\theta \cap \{z \in \mathbb{C} : |z| < 1\}$, we may assume that $x \in D(B)$ (see Proposition B.15). For $0 < \operatorname{Re} z < 1$, it follows from (3.52) and (3.53) that

$$\begin{aligned}
B^{-z}x - x &= \frac{\sin\pi z}{\pi}\int_0^\infty s^{-z}\left((s+B)^{-1}x-(s+1)^{-1}x\right)ds \\
&= \frac{\sin\pi z}{\pi}\int_0^\infty \frac{s^{-z}}{s+1}(s+B)^{-1}(I-B)x\,ds.
\end{aligned}$$

Hence,

$$\|B^{-z}x-x\| \le M\frac{|\sin\pi z|}{\pi}\int_0^\infty \frac{s^{-\operatorname{Re} z}}{(s+1)^2}\,ds\,\|(I-B)x\|.$$

It follows that $\lim_{\substack{z\to 0\\ z\in\Sigma_\theta}} \|B^{-z}x-x\| = 0$ for all $\theta\in(0,\pi/2)$. □

Now suppose that B is an operator on X such that $(-\infty,0) \subset \rho(B)$ and $M := \sup_{\lambda<0}\|\lambda R(\lambda,B)\| < \infty$. We shall show that B has a special type of square root, which has interesting properties for semigroup generators.

For $\varepsilon > 0$, $(-\infty,0] \subset \rho(\varepsilon+B)$ and

$$\sup_{\lambda\le 0}(1-\lambda)\|R(\lambda,\varepsilon+B)\| = \sup_{\lambda\le 0}(1-\lambda)\|R(\lambda-\varepsilon,B)\| \le M/\varepsilon.$$

Hence we can define $(\varepsilon + B)^{-1/2} \in \mathcal{L}(X)$ as above, and then $((\varepsilon + B)^{-1/2})^2 = (\varepsilon + B)^{-1}$. In particular, $(\varepsilon + B)^{-1/2}$ is injective. Let $(\varepsilon + B)^{1/2}$ be the algebraic inverse of $(\varepsilon + B)^{-1/2}$, so

$$\begin{aligned} D((\varepsilon + B)^{1/2}) &= \operatorname{Ran}((\varepsilon + B)^{-1/2}), \\ (\varepsilon + B)^{1/2}((\varepsilon + B)^{-1/2}y) &= y \qquad (y \in X). \end{aligned}$$

Then $(\varepsilon+B)^{1/2}$ is a closed operator on X, $((\varepsilon+B)^{1/2})^2 = \varepsilon+B$, and for $x \in D(B)$,

$$x = (\varepsilon + B)^{-1/2}((\varepsilon + B)^{-1/2}(\varepsilon + B)x),$$

so

$$\begin{aligned} (\varepsilon + B)^{1/2}x &= (\varepsilon + B)^{-1/2}(\varepsilon + B)x \\ &= \frac{1}{\pi}\int_0^\infty s^{-1/2}(s + \varepsilon + B)^{-1}(\varepsilon + B)x\, ds. \end{aligned} \tag{3.55}$$

Proposition 3.8.2. *Let B be a densely defined operator on X such that $(-\infty, 0) \subset \rho(B)$ and $\sup_{\lambda<0} \|\lambda R(\lambda, B)\| < \infty$. Then there is a unique closed operator $B^{1/2}$ such that*

a) *$(B^{1/2})^2 = B$, and*

b) *For $x \in D(B)$,*

$$B^{1/2}x = \lim_{\varepsilon\downarrow 0}(\varepsilon + B)^{1/2}x = \frac{1}{\pi}\int_0^\infty s^{-1/2}(s + B)^{-1}Bx\, ds.$$

Moreover, $D(B)$ is a core for $B^{1/2}$, and $D(B^{1/2}) = D((\varepsilon + B)^{1/2})$ for all $\varepsilon > 0$.

Proof. Since

$$\|s^{-1/2}(s + \varepsilon + B)^{-1}(\varepsilon + B)x\| \leq Ms^{-3/2}\left(\|x\| + \|Bx\|\right) \qquad (0 < \varepsilon < 1)$$

and

$$\begin{aligned} \|s^{-1/2}(s + \varepsilon + B)^{-1}(\varepsilon + B)x\| &= \|s^{-1/2}(x - s(s + \varepsilon + B)^{-1}x)\| \\ &\leq (M + 1)\|x\|s^{-1/2}, \end{aligned}$$

one can apply the dominated convergence theorem and take limits in (3.55) as $\varepsilon \downarrow 0$. Thus, we let

$$B^{1/2}x := \lim_{\lambda\downarrow 0}(\varepsilon + B)^{1/2}x = \frac{1}{\pi}\int_0^\infty s^{-1/2}(s + B)^{-1}Bx\, ds$$

for $x \in D(B)$. Moreover, for $\varepsilon > 0$,

$$\begin{aligned} & B^{1/2}x - (\varepsilon + B)^{1/2}x \\ = & \frac{1}{\pi} \int_0^\infty s^{-1/2} \left((s+B)^{-1}Bx - (s+\varepsilon+B)^{-1}(\varepsilon+B)x \right) ds \\ = & -\frac{\varepsilon}{\pi} \int_0^\infty s^{1/2}(s+\varepsilon+B)^{-1}(s+B)^{-1}x \, ds. \end{aligned}$$

Let

$$S_\varepsilon := -\frac{\varepsilon}{\pi} \int_0^\infty s^{1/2}(s+\varepsilon+B)^{-1}(s+B)^{-1} \, ds.$$

Then $S_\varepsilon \in \mathcal{L}(X)$,

$$\|S_\varepsilon\| \le \frac{\varepsilon}{\pi} \int_0^\infty \frac{M^2}{s^{1/2}(s+\varepsilon)} \, ds = M^2 \varepsilon^{1/2},$$

by (3.53), and

$$B^{1/2}x - (\varepsilon + B)^{1/2}x = S_\varepsilon x \qquad (x \in D(B), \ \varepsilon > 0).$$

We define

$$B^{1/2} := (\varepsilon + B)^{1/2} + S_\varepsilon \quad \text{with} \quad D(B^{1/2}) = D((\varepsilon + B)^{1/2}).$$

Since $(\varepsilon + B)^{1/2}$ is closed and S_ε is bounded, $B^{1/2}$ is closed. Moreover, b) holds. Since $(\varepsilon + B)^{1/2}$ is densely defined and invertible, $D(B) = D(((\varepsilon + B)^{1/2})^2)$ is a core for $(\varepsilon + B)^{1/2}$ (see Appendix B) and hence for $B^{1/2}$. It follows that the definition of $B^{1/2}$ is independent of ε.

Let $x \in D(B)$. Then $(\varepsilon + B)^{1/2}x \to B^{1/2}x$ and

$$B^{1/2}(\varepsilon + B)^{1/2}x = (\varepsilon + B)x + S_\varepsilon(\varepsilon + B)^{1/2}x \to Bx \quad \text{as} \quad \varepsilon \downarrow 0.$$

Since $B^{1/2}$ is closed, it follows that $B^{1/2}x \in D(B^{1/2})$ and $(B^{1/2})^2 x = Bx$.

Let $y \in D((B^{1/2})^2)$. By Lemma 3.3.12, $\varepsilon(\varepsilon + B)^{-1}y \to y$ and $B(\varepsilon(\varepsilon + B)^{-1}y) = \varepsilon(\varepsilon+B)^{-1}(B^{1/2})^2 y \to (B^{1/2})^2 y$ as $\varepsilon \to \infty$. Since B is closed, $y \in D(B)$. Thus $B = (B^{1/2})^2$.

Finally, let $\widetilde{B}$ be any closed operator such that $\widetilde{B}x = B^{1/2}x$ for all $x \in D(B)$ and $\widetilde{B}^2 = B$. Then $(\widetilde{B} + i)(\widetilde{B} - i) = \widetilde{B}^2 + I = B + I$, which is invertible, so $\widetilde{B} + i$ is invertible. Since $\widetilde{B}$ is densely defined and $\rho(\widetilde{B})$ is non-empty, $D(\widetilde{B}^2) = D(B)$ is a core for $\widetilde{B}$ (see Appendix B). Hence, $\widetilde{B}$ is the closure of $B^{1/2}|_{D(B)}$, and this proves uniqueness. □

Now suppose that A is the generator of a bounded C_0-semigroup T on X. If $0 \in \rho(A)$, then the theory above can be applied to $B := -A$, so $(-A)^{-z}$ is defined for $\operatorname{Re} z > 0$. Substituting $(s-A)^{-1} = \int_0^\infty e^{-st}T(t)\,dt$ into (3.52), it is not difficult to see that

$$(-A)^{-z} = \frac{1}{\Gamma(z)} \int_0^\infty t^{z-1} T(t)\,dt \tag{3.56}$$

for $0 < \operatorname{Re} z < 1$, and hence for $\operatorname{Re} z > 0$ by uniqueness of holomorphic extensions. We shall not use this.

Proposition 3.8.2 shows that $(-A)^{1/2}$ is defined whenever A generates a bounded C_0-semigroup.

Theorem 3.8.3. *Let A be the generator of a bounded C_0-semigroup T on X, and define*

$$S(t)x = \begin{cases} \displaystyle\int_0^\infty \frac{te^{-t^2/4s}}{2\sqrt{\pi}s^{3/2}} T(s)x\,ds & (t > 0), \\ x & (t = 0). \end{cases}$$

Then S is a bounded holomorphic C_0-semigroup of angle $\pi/4$, and the generator of S is $-(-A)^{1/2}$. Furthermore, for $x \in D(A)$, $u(t) := S(t)x$ is the unique bounded classical solution of the second order Cauchy problem

$$\begin{cases} u''(t) = & -Au(t) \qquad (t \geq 0), \\ u(0) = & x. \end{cases} \tag{3.57}$$

Moreover, if T is a bounded holomorphic C_0-semigroup of angle $\theta \in (0, \pi/2]$, then S is a bounded holomorphic C_0-semigroup of angle $(\frac{\theta}{2} + \frac{\pi}{4})$.

Proof. For $z \in \Sigma_{\pi/4}$, let $\psi_z(s) = \dfrac{ze^{-z^2/4s}}{2\sqrt{\pi}s^{3/2}}$ $(s > 0)$. Then $\psi_z \in L^1(\mathbb{R}_+)$, $\|\psi_z\|_1 = \frac{|z|}{\operatorname{Re}(z^2)}$, $\int_0^\infty \psi_z(s)\,ds = 1$, and $\lim_{\substack{z \in \Sigma_\theta \\ z \to 0}} \int_\delta^\infty |\psi_z(s)|\,ds = 0$, for $\delta > 0$ and $0 \leq \theta < \pi/4$.

For $t \in \mathbb{R}_+$, $S(t)x = \int_0^\infty \psi_t(s)T(s)x\,ds$ $(x \in X)$ and $\widehat{\psi_t}(\lambda) = e^{-t\sqrt{\lambda}}$ $(\lambda > 0)$ (Lemma 1.6.7). Hence $(\widehat{\psi_{t_1} * \psi_{t_2}})(\lambda) = \widehat{\psi_{t_1+t_2}}(\lambda)$. It follows from the uniqueness theorem that $\psi_{t_1} * \psi_{t_2} = \psi_{t_1+t_2}$ $(t_1, t_2 \geq 0)$. Now,

$$\begin{aligned} S(t_1)S(t_2)x &= \int_0^\infty \int_0^\infty \psi_{t_1}(s)\psi_{t_2}(r)T(s+r)x\,dr\,ds \\ &= \int_0^\infty \int_0^t \psi_{t_1}(s)\psi_{t_2}(t-s)\,ds\,T(t)x\,dt \\ &= S(t_1+t_2)x. \end{aligned}$$

For $z \in \Sigma_{\pi/4}$, let $S(z)x = \int_0^\infty \psi_z(s)T(s)x\,ds$. Then $S(\cdot)$ is holomorphic, $\|S(z)\| \leq \frac{|z|}{(\operatorname{Re} z)^2} \sup_{s \geq 0} \|T(s)\|$ and

$$\begin{aligned} \|S(z)x - x\| &= \left\| \int_0^\infty \frac{e^{-z^2/4s}}{2\sqrt{\pi}s^{3/2}} (T(s)z - z)\,ds \right\| \\ &\to 0 \qquad (z \in \Sigma_\theta,\ z \to 0). \end{aligned}$$

Thus S is a bounded holomorphic C_0-semigroup of angle $\pi/4$.

Let B be the generator of S and let $x \in D(A)$. For $t > 0$, we can differentiate through the integral sign and obtain

$$\begin{aligned} BS(t)x = \frac{d}{dt}(S(t)x) &= \int_0^\infty \frac{e^{-t^2/4s}}{2\sqrt{\pi}s^{3/2}}\left(1-\frac{t^2}{2s}\right)T(s)x\,ds, \\ 0 &= \int_0^\infty \frac{e^{-t^2/4s}}{2\sqrt{\pi}s^{3/2}}\left(1-\frac{t^2}{2s}\right)x\,ds. \end{aligned}$$

Hence

$$BS(t)x = \int_0^\infty \frac{e^{-t^2/4s}}{2\sqrt{\pi}s^{3/2}}\left(1-\frac{t^2}{2s}\right)(T(s)x - x)\,ds.$$

Since $\|T(s)x - x\| \le s\|Ax\|\sup_{t\ge 0}\|T(t)\|$, the dominated convergence theorem gives, on letting $t \downarrow 0$, that $x \in D(B)$ and

$$\begin{aligned} Bx &= \int_0^\infty \frac{T(s)x - x}{2\sqrt{\pi}s^{3/2}}\,ds \\ &= \int_0^\infty \frac{1}{\pi}\int_0^\infty \lambda^{1/2}e^{-s\lambda}\,d\lambda\,(T(s)x - x)\,ds \\ &= \frac{1}{\pi}\int_0^\infty \lambda^{1/2}(R(\lambda,A)x - \lambda^{-1}x)\,d\lambda \\ &= \frac{1}{\pi}\int_0^\infty \lambda^{-1/2}R(\lambda,A)Ax\,d\lambda \\ &= -(-A)^{1/2}x. \end{aligned}$$

Since $D(A)$ is a core for $-(-A)^{1/2}$, it follows that B extends $-(-A)^{1/2}$. However, $B + i$ is invertible (since B generates a bounded holomorphic semigroup) and $i - (-A)^{1/2}$ is invertible (since $I - A = ((-A)^{1/2} - i)((-A)^{1/2} + i)$), so $B = -(-A)^{1/2}$.

Let $x \in D(A) = D(((-A)^{1/2})^2)$. Then it is immediate that $u(t) := S(t)x$ is a bounded classical solution of (3.57).

Let u_1 be any bounded solution of (3.57). Take $\mu > 0$, and let

$$v(t) := R(\mu,A)(u(t) - u_1(t)) \qquad (t \ge 0).$$

Then $v(0) = 0$, v is bounded, and

$$v''(t) = -Av(t) = (I - \mu R(\mu,A))(u(t) - u_1(t)),$$

which is bounded. Since

$$v'(t) = v(t+1) - v(t) - \frac{1}{2}\int_t^{t+1}(t+1-s)v''(s)\,ds,$$

it follows that v' is bounded.

Let

$$w(t) := \begin{cases} v'(-t) + (-A)^{1/2}v(-t) & (t \le 0), \\ S(t)v'(0) & (t > 0). \end{cases}$$

Since $(-A)^{1/2}R(\mu, A)$ is bounded (by the closed graph theorem, or by direct calculation), w is bounded. For $t \le 0$,

$$\begin{aligned} w'(t) &= -v''(-t) - (-A)^{1/2}v'(-t) \\ &= Av(t) - (-A)^{1/2}v'(-t) \\ &= -(-A)^{1/2}w(t). \end{aligned}$$

Also, $w'(t) = -(-A)^{1/2}w(t)$ for all $t \ge 0$, since $-(-A)^{1/2}$ generates S. It follows from Proposition 3.1.11 that

$$w(t+s) = S(t)w(s) \qquad (t \ge 0, s \in \mathbb{R}).$$

Now we extend w to $\mathbb{C}$ by

$$w(\lambda + s) := S(\lambda)w(s) \qquad (\lambda \in \Sigma_{\pi/4}, s \in \mathbb{R}).$$

This is well defined since $S(\lambda_1 + \lambda_2) = S(\lambda_1)S(\lambda_2)$. Moreover, w is holomorphic, and bounded since λ may be chosen in $\Sigma_{\pi/8}$, where S is bounded. By Liouville's theorem, w is constant, so $S(t)v'(0) = v'(0)$ for all $t \ge 0$. Now,

$$\begin{aligned} v'(t) &= -(-A)^{1/2}v(t) + w(-t) \\ &= -(-A)^{1/2}v(t) + v'(0) \qquad (t \ge 0), \\ v(0) &= 0. \end{aligned}$$

By Proposition 3.1.16,

$$v(t) = \int_0^t S(t-s)v'(0)\,ds = tv'(0).$$

But v is bounded, so $v'(0) = 0$ and hence $v(t) = 0$. Since $R(\mu, A)$ is injective, it follows that $u_1(t) = u(t)$.

Now suppose that T is a bounded holomorphic C_0-semigroup of angle $\theta \in (0, \pi/2]$. Let $\alpha \in (-\theta, \theta)$. An application of Cauchy's theorem shows that

$$S(t) = \int_0^\infty \frac{te^{-t^2/4re^{i\alpha}}}{2\sqrt{\pi}r^{3/2}e^{i\alpha/2}} T(re^{i\alpha})\,dr \qquad (t > 0).$$

Now let

$$S(z) = \int_0^\infty \frac{ze^{-z^2/4re^{i\alpha}}}{2\sqrt{\pi}r^{3/2}e^{i\alpha/2}} T(re^{i\alpha})\,dr$$

for $\frac{\alpha}{2} - \frac{\pi}{4} < \arg z < \frac{\alpha}{2} + \frac{\pi}{4}$. This defines a holomorphic extension of S to this sector, and it is bounded in each proper subsector. Varying α provides a holomorphic extension of S to $\Sigma_{(\frac{\theta}{2}+\frac{\pi}{4})}$ which is bounded on $\Sigma_{\theta'}$ for $\theta' < (\frac{\theta}{2} + \frac{\pi}{4})$. □

Example 3.8.4. Let (Ω, μ) be a measure space, $X := L^2(\Omega, \mu)$, $m : \Omega \to \mathbb{R}_+$ be measurable, and A be defined by

$$\begin{aligned} D(A) &:= \{f \in X : mf \in X\} \\ Af &:= mf. \end{aligned}$$

Then $-A$ generates the C_0-semigroup $T(t)f = e^{-tm} f$. The operator $A^{1/2}$ of Proposition 3.8.2 is given by

$$D(A^{1/2}) = \{f \in X : m^{1/2} f \in X\} \quad \text{and} \quad A^{1/2} f = m^{1/2} f.$$

The semigroup S generated by $-A^{1/2}$, as in Theorem 3.8.3, is given by $S(t)f = e^{-tm^{1/2}} f$. □

Example 3.8.5. Let X be any of the spaces of Example 3.7.6, and let T be the Gaussian semigroup on X. Then $T(s)f = k_s * f$, where $k_s(x) = (4\pi s)^{-n/2} e^{-|x|^2/4s}$. Hence, the holomorphic semigroup S of Theorem 3.8.3 is given by $S(t)f = h_t * f$, where

$$h_t(x) = \int_0^\infty \frac{te^{-t^2/4s}}{2\sqrt{\pi} s^{3/2}} \frac{e^{-|x|^2/4s}}{(4\pi s)^{n/2}} \, ds.$$

Putting $r = (t^2 + |x|^2)/4s$ gives

$$h_t(x) = \frac{\Gamma(\frac{n+1}{2}) t}{\pi^{(n+1)/2} (t^2 + |x|^2)^{(n+1)/2}}.$$

Thus, S is the Poisson semigroup considered in Example 3.7.9. Note that although the generator Δ_X of T is a second order differential operator, the generator $A_X = -(-\Delta_X)^{1/2}$ of S is not a first order differential operator. □

It should be mentioned that (3.57) is an abstract *elliptic equation*. For example, if T is the Gaussian semigroup on $C_0(\mathbb{R}^n)$ (Example 3.8.5), then letting $u(t,x) := (S(t)f)(x)$ for $f \in C_0(\mathbb{R}^n)$, $x \in \mathbb{R}^n$ and $t > 0$, u is a solution of

$$u_{tt} + \Delta u := \frac{\partial^2 u}{\partial t^2} + \sum_{j=1}^n \frac{\partial^2 u}{\partial x_j^2} = 0 \quad \text{on } (0,\infty) \times \mathbb{R}^n;$$

i.e., u is a solution of the Laplace equation.

The *wave equation*

$$u_{tt} = \Delta u$$

will be treated in Sections 3.14–3.16 and Chapter 7.

3.9 Boundary Values of Holomorphic Semigroups

Let T be a holomorphic C_0-semigroup on X of angle $\frac{\pi}{2}$. In this section we are interested in the behaviour of $T(is+t)$ as t tends to 0 and we ask under what circumstances the "boundary value" $T(is)$ of T (which will be defined precisely below) exists and defines a C_0-group. We also address the converse problem: Which C_0-groups are obtained as boundary values of holomorphic C_0-semigroups?

As in Section 3.7, we let $\Sigma_\varphi := \{z \in \mathbb{C} \setminus \{0\} : |\arg z| < \varphi\}$ be the sector in the complex plane of angle $\varphi \in (0, \pi)$. Furthermore, we set $\Sigma_\varphi^+ := \Sigma_\varphi \cap \{z \in \mathbb{C} : \operatorname{Im} z \geq 0\}$, $\Sigma_\varphi^- := \Sigma_\varphi \cap \{z \in \mathbb{C} : \operatorname{Im} z \leq 0\}$ and define D by $D := \{z \in \Sigma_{\pi/2} : |z| \leq 1\}$. The following result gives an answer to our first question above.

Proposition 3.9.1. *Let A be the generator of a holomorphic C_0-semigroup T on X of angle $\varphi \in (0, \pi/2]$. Then the following are equivalent:*

(i) *$e^{i\varphi}A$ generates a C_0-semigroup $T(e^{i\varphi}\cdot)$ on X.*

(ii) $\sup_{z \in \Sigma_\varphi^+ \cap D} ||T(z)|| < \infty$.

In this case, the C_0-semigroup $T(e^{i\varphi}\cdot)$ is given by

$$T(e^{i\varphi}s)x = \lim_{t \downarrow 0} T(t + e^{i\varphi}s)x \qquad (x \in X, s \geq 0). \tag{3.58}$$

The C_0-semigroup $S(s) := T(e^{i\varphi}s)$ defined by (3.58) is called the *boundary semigroup* of T. The following lemma will be useful in the proof of Proposition 3.9.1.

Lemma 3.9.2. *Let A be the generator of a holomorphic C_0-semigroup T on X of angle $\varphi \in (0, \pi/2]$. Assume that $e^{i\varphi}A$ generates a C_0-semigroup S on X. Then*

$$T(t + e^{i\varphi}s) = T(t)S(s) \quad \textit{for all} \quad s, t \geq 0.$$

Proof. Obviously, the resolvents of A and $e^{i\varphi}A$ commute. By Proposition 3.1.5, $S(s)T(t) = T(t)S(s)$ for all $s, t \geq 0$. Fix $a, b \geq 0$ and denote by B the generator of the C_0-semigroup V on X defined by $V(t) := S(bt)T(at)$. For $x \in D(A)$ we have $\frac{d}{dt}V(t)x = (a + be^{i\varphi})AV(t)x$. Hence $\frac{d}{dt}V(t)x|_{t=0} = (a + be^{i\varphi})Ax$ and therefore B extends $(a + be^{i\varphi})A$. It follows that $V(t) = T((a + be^{i\varphi})t)$ for all $t \geq 0$ (see Proposition 3.7.2 c)). In particular, we have $V(1) = T(a + be^{i\varphi}) = S(b)T(a) = T(a)S(b)$ for all $a, b \geq 0$. □

Proof of Proposition 3.9.1. Assume that (i) holds. Let $z \in \Sigma_\varphi^+ \cap D$. There exist $a, b \in [0, 1]$ such that $z = a + be^{i\varphi}$. It follows from Lemma 3.9.2 that $T(z) = T(a + be^{i\varphi}) = T(a)T(be^{i\varphi})$. Hence, $\|T(z)\| \leq \|T(a)\| \, \|T(be^{i\varphi})\| \leq M$ for a suitable M and all $z \in \Sigma_\varphi^+ \cap D$. This implies (ii).

In order to prove the converse implication, fix $R > 0$. Then there exists $M_R > 0$ such that $\|T(z)\| \leq M_R$ whenever $z \in \Sigma_\varphi^+$ and $|z| \leq R$. For $x \in X$,

$0 < t < t' \leq 1$ and $s \geq 0$ satisfying $|t + e^{i\varphi}s| \leq R$, we have

$$\begin{aligned} \|T(t+e^{i\varphi}s)x - T(t'+e^{i\varphi}s)x\| &\leq \left\|T(t+e^{i\varphi}s)\big(x - T(t'-t)x\big)\right\| \\ &\leq M_R\|x - T(t'-t)x\|. \end{aligned}$$

It thus follows that $T(e^{i\varphi}s)x := \lim_{t\downarrow 0} T(t+e^{i\varphi}s)x$ exists uniformly in $s \in [0, R]$. Consequently, the mapping $\mathbb{R}_+ \to \mathcal{L}(X), s \mapsto T(e^{i\varphi}s)$ is strongly continuous. It is easy to see that $T(e^{i\varphi}\cdot)$ is a C_0-semigroup. Denote by B its generator. Let $x \in D(A)$ and $\tau > 0$. Then

$$\begin{aligned} \int_0^\tau T(e^{i\varphi}s)e^{i\varphi}Ax\,ds &= \lim_{t\downarrow 0}\int_0^\tau e^{i\varphi}AT(t+e^{i\varphi}s)x\,ds \\ &= \lim_{t\downarrow 0}\int_0^\tau \frac{d}{ds}\left(T(t+e^{i\varphi}s)x\right)\,ds \\ &= \lim_{t\downarrow 0}(T(t+e^{i\varphi}\tau)x - T(t)x) \\ &= T(e^{i\varphi}\tau)x - x. \end{aligned}$$

It follows from Proposition 3.1.9 f) that $x \in D(B)$ and $Bx = e^{i\varphi}Ax$. We have shown that B is an extension of $e^{i\varphi}A$. Since $\rho(B) \cap \rho(e^{i\varphi}A) \neq \emptyset$, it follows that both operators are equal. □

Remark 3.9.3. Note that the above result may be easily modified to the case where $e^{i\varphi}A$ and $e^{-i\varphi}A$ generate C_0-semigroups. Indeed, assuming that

$$\sup_{z\in\Sigma_\varphi\cap D}||T(z)|| < \infty, \tag{3.59}$$

it follows that $e^{\pm i\varphi}A$ generate C_0-semigroups on X which are given by $T(e^{\pm i\varphi}s)x := \lim_{t\downarrow 0} T(t+e^{\pm i\varphi}s)x$. In particular, if (3.59) is satisfied for $\varphi = \frac{\pi}{2}$, then iA and $-iA$ generate C_0-semigroups $T(\pm is)$ on X and we call $S(s) := T(is)$ $(s \in \mathbb{R})$, the *boundary group* of T. □

For $1 \leq p \leq \infty$, let Δ_p be the Laplacian on $L^p(\mathbb{R}^n)$ with maximal domain:

$$\begin{aligned} \Delta_p f &:= \Delta f, \\ D(\Delta_p) &:= \{f \in L^p(\mathbb{R}^n) : \Delta f \in L^p(\mathbb{R}^n)\}, \end{aligned} \tag{3.60}$$

where Δf is defined in the distributional sense. We proved in Example 3.7.6 that for $1 \leq p < \infty$ the operator Δ_p is the generator of the Gaussian semigroup T_p on $L^p(\mathbb{R}^n)$. Moreover, $\Delta_p^* = \Delta_{p'}$, where $1/p + 1/p' = 1$. Although Δ_∞ does not generate a C_0-semigroup on $L^\infty(\mathbb{R}^n)$, it does generate a holomorphic semigroup T_∞ given by

$$T_\infty(z)f := k_z * f \quad (f \in L^\infty(\mathbb{R}^n)),$$

(see Example 3.7.8).

We wish to determine whether $i\Delta_p$ generates a C_0-semigroup. By duality, we may restrict ourselves in the following to the case when $1 \le p \le 2$. Since $\mathcal{F}T_2(z)f = e^{-z|\cdot|^2}\mathcal{F}f$ and $(2\pi)^{-n/2}\mathcal{F}$ is unitary on $L^2(\mathbb{R}^n)$, we have $||T_2(z)||_{\mathcal{L}(L^2(\mathbb{R}^n))} = 1$ for all $z \in \mathbb{C}$ with $\operatorname{Re} z > 0$. Hence, by Proposition 3.9.1 the operator $i\Delta_2$ generates a C_0-semigroup on $L^2(\mathbb{R}^n)$. By Remark 3.7.10,

$$||T_1(z)||_{\mathcal{L}(L^1(\mathbb{R}^n))} = \left(\frac{|z|}{\operatorname{Re} z}\right)^{n/2} \qquad (\operatorname{Re} z > 0).$$

Hence, by the Riesz-Thorin interpolation theorem (see [Hör83, Theorem 7.1.12])

$$||T_p(z)||_{\mathcal{L}(L^p(\mathbb{R}^n))} \le \left(\frac{|z|}{\operatorname{Re} z}\right)^{n(1/p-1/2)} \qquad (1 \le p \le 2, \operatorname{Re} z > 0). \tag{3.61}$$

In fact, we will now show that a multiple of the above upper bound will also serve as a lower bound for $||T_p(z)||_{\mathcal{L}(L^p(\mathbb{R}^n))}$. Then we can apply Proposition 3.9.1 and obtain the following result.

Theorem 3.9.4 (Hörmander). *Let $1 \le p < \infty$. Then the operator $i\Delta_p$ generates a C_0-semigroup on $L^p(\mathbb{R}^n)$ if and only if $p = 2$.*

Proof. By Proposition 3.9.1, it suffices to show that a multiple of the above upper bound (3.61) will also serve as a lower bound for $||T_p(z)||_{\mathcal{L}(L^p(\mathbb{R}^n))}$. More precisely, we prove in the following that if $1 \le p < \infty$, then

$$||T_p(z)||_{\mathcal{L}(L^p(\mathbb{R}^n))} \ge 2^{-n/2p}\left(\frac{|z|}{\operatorname{Re} z}\right)^{n|1/p-1/2|} \qquad (\operatorname{Re} z > 0). \tag{3.62}$$

We already observed that it suffices to consider the case where $1 < p \le 2$.

Fix $z \in \mathbb{C}_+$ and consider the function $f : \mathbb{R}^n \to \mathbb{C}$ defined by

$$f(x) := \exp\left(\frac{-|x|^2}{\overline{z}}\right).$$

Taking p' such that $\frac{1}{p} + \frac{1}{p'} = 1$ we verify that

$$||f||_{p'} = \left(\frac{\pi}{p'}\right)^{n/2p'}\left(\frac{|z|^2}{\operatorname{Re} z}\right)^{n/2p'}. \tag{3.63}$$

Let $x, y \in \mathbb{R}^n$ and recall that $z \in \mathbb{C}_+$. Then

$$-\frac{|x-y|^2}{z} - \frac{|x|^2}{\overline{z}} = -2\operatorname{Re}\left(\frac{1}{z}\right)\left|x - \frac{y}{2}\right|^2 + 2i\left(x - \frac{y}{2}\right)y\operatorname{Im}\left(\frac{1}{z}\right) - \frac{|y|^2}{2}\operatorname{Re}\left(\frac{1}{z}\right).$$

Moreover,

$$
\begin{aligned}
&T_{p'}(z/4)f(y)\\
&= \frac{1}{(\pi z)^{n/2}} \int_{\mathbb{R}^n} \exp\left(-\frac{|x-y|^2}{z} - \frac{|x|^2}{\overline{z}}\right) dx\\
&= \frac{1}{(\pi z)^{n/2}} \int_{\mathbb{R}^n} \exp\left(-2|x|^2 \operatorname{Re}\left(\frac{1}{z}\right) + 2ixy \operatorname{Im}\left(\frac{1}{z}\right)\right) dx\\
&\qquad \times \exp\left(-\frac{|y|^2}{2} \operatorname{Re}\left(\frac{1}{z}\right)\right)\\
&= \frac{1}{(\pi z)^{n/2}} \left(\frac{1}{4\operatorname{Re}(z^{-1})}\right)^{n/2} \int_{\mathbb{R}^n} \exp\left(-\frac{|x|^2}{2}\right) \exp\left(ixy \frac{\operatorname{Im}(z^{-1})}{(\operatorname{Re}(z^{-1}))^{1/2}}\right) dx\\
&\qquad \times \exp\left(-\frac{|y|^2}{2} \operatorname{Re}\left(\frac{1}{z}\right)\right)\\
&= \frac{1}{(\pi z)^{n/2}} \left(\frac{1}{4\operatorname{Re}(z^{-1})}\right)^{n/2} (2\pi)^{n/2}\\
&\qquad \times \exp\left(-\frac{|y|^2}{2} \frac{(\operatorname{Im}(z^{-1}))^2}{\operatorname{Re}(z^{-1})}\right) \exp\left(-\frac{|y|^2}{2} \operatorname{Re}\left(\frac{1}{z}\right)\right),
\end{aligned}
$$

where in the last step we used the fact that the function $x \mapsto \exp(-|x|^2/2)$ is an eigenvector of the Fourier transform. Hence,

$$
|T_{p'}(z/4)f(y)| = \left(\frac{|z|}{2\operatorname{Re} z}\right)^{n/2} \exp\left(-\frac{|y|^2}{2\operatorname{Re} z}\right) \qquad (y \in \mathbb{R}^n),
$$

and since $\int_{\mathbb{R}^n} \exp(-|x|^2/2)\, dx = (2\pi)^{n/2}$, we obtain

$$
\|T_{p'}(z/4)f\|_{p'} = \left(\frac{|z|}{2\operatorname{Re} z}\right)^{n/2} \left(\frac{2\pi}{p'}\right)^{n/2p'} (\operatorname{Re} z)^{n/2p'}.
$$

Combining this equality with (3.63) we see that

$$
\|T_{p'}(z/4)\|_{\mathcal{L}(L^q(\mathbb{R}^n))} \geq \frac{\|T_{p'}(z/4)f\|_{p'}}{\|f\|_{p'}} = 2^{-n/2p} \left(\frac{|z|}{\operatorname{Re} z}\right)^{n(1/p-1/2)}.
$$

Finally, since $\|T_{p'}(z)\|_{\mathcal{L}(L^{p'}(\mathbb{R}^n))} = \|T_p(z)\|_{\mathcal{L}(L^p(\mathbb{R}^n))}$, it follows that

$$
\|T_p(z)\|_{\mathcal{L}(L^p(\mathbb{R}^n))} \geq 2^{-n/2p} \left(\frac{|z|}{\operatorname{Re} z}\right)^{n|1/p-1/2|} \qquad (\operatorname{Re} z > 0). \qquad \square
$$

Interesting examples of boundary values of holomorphic C_0-semigroups occur also in connection with fractional powers of operators and the so-called *Riemann-Liouville semigroup*. Indeed, consider in $L^p(0,1)$ the operator

$$
Au := u' \quad \text{with domain} \quad D(A) := \{u \in W^{1,p}(0,1) : u(0) = 0\}. \tag{3.64}
$$

We showed in Example 3.4.8 that $-A$ generates a C_0-semigroup T on $L^p(0,1)$ ($1 \le p < \infty$), which may be represented by

$$T(t)f(x) = \begin{cases} f(x-t) & (t \le x), \\ 0 & (t > x). \end{cases} \tag{3.65}$$

Since $T(t) = 0$ for $t \ge 1$, we see that $\operatorname{abs}(T) = -\infty$. Now inserting (3.65) in (3.56) we see that

$$A^{-z}f(x) = \frac{1}{\Gamma(z)} \int_0^x (x-y)^{z-1} f(y)\,dy \qquad (x \in (0,1), \operatorname{Re} z > 0, f \in L^p(0,1)).$$

By Theorem 3.8.1, $(A^{-z})_{\operatorname{Re} z > 0}$ is a holomorphic C_0-semigroup of angle $\pi/2$. This C_0-semigroup is called the *Riemann-Liouville semigroup*. By Proposition 3.9.1 and Remark 3.9.3, the question whether or not the Riemann-Liouville semigroup possesses a boundary group is equivalent to

$$\sup_{z \in \Sigma_{\pi/2} \cap D} \|A^{-z}\| < \infty. \tag{3.66}$$

When (3.66) holds true, we denote the boundary group by $(S(s)) := (A^{is})_{s \in \mathbb{R}}$. In order to show the estimate (3.66) we make use of the *transference principle* due to Coifman and Weiss [CW77]. We will use it in the following form [Ama95, Chapter III, Example 4.7.3 c)]. For a measure space (Ω, μ) and $1 \le p \le \infty$, we denote by $L^p(\Omega, \mu)$ the usual Banach space of equivalence classes of p-integrable functions (bounded functions when $p = \infty$).

Theorem 3.9.5 (Coifman-Weiss). *Let (Ω, μ) be a σ-finite measure space and let $1 < p < \infty$. Assume that $0 \in \rho(A)$ and that $-A$ generates a C_0-semigroup of positive contractions on $L^p(\Omega, \mu)$. For $t > 0$ and $s \in \mathbb{R}$ let A^{-t+is} be defined as in (3.51). Then $A^{is}f := \lim_{t \downarrow 0} A^{-t+is} f \in L^p(\Omega, \mu)$ for $f \in L^p(\Omega, \mu)$ and there exists a constant M, depending only on p, such that*

$$\|A^{is}\|_{\mathcal{L}(L^p(\Omega,\mu))} \le M(1+s^2)e^{\pi|s|/2} \quad (s \in \mathbb{R}).$$

Thus, for $1 < p < \infty$, the Riemann-Liouville semigroup admits a boundary group. For $p = 1$ the situation is different. Indeed,

$$\begin{aligned} \|A^{-z}\|_{\mathcal{L}(L^1(0,1))} &= \sup_{y \in [0,1]} \frac{1}{|\Gamma(z)|} \int_y^1 |(x-y)^{z-1}|\,dx \\ &= \frac{1}{|\Gamma(z)|} \sup_{y \in [0,1]} \int_y^1 (x-y)^{\operatorname{Re} z - 1}\,dx = \frac{1}{|\Gamma(z)|} \frac{1}{\operatorname{Re} z}, \end{aligned}$$

which by Proposition 3.9.1 implies that we do not have a boundary value for $p = 1$. In summary, we have proved the following result for the Riemann-Liouville semigroup on $L^p(0,1)$.

Theorem 3.9.6. *Let $1 \leq p < \infty$ and denote by G the generator of the Riemann-Liouville semigroup on $L^p(0,1)$. Then iG is the generator of a C_0-group on $L^p(0,1)$ provided $1 < p < \infty$. If $p = 1$, then iG does not generate a C_0-semigroup on $L^1(0,1)$.*

We now consider the converse to the situation described in Proposition 3.9.1; namely, we ask for conditions on the boundary group itself which imply that A generates a holomorphic C_0-semigroup. We begin with the following result.

Theorem 3.9.7. *Let A be an operator on X and let $\varphi \in (0, \pi/2)$. Assume that $e^{\pm i\varphi}A$ generate bounded C_0-semigroups on X. Then A generates a bounded holomorphic C_0-semigroup of angle φ.*

Our proof of Theorem 3.9.7 is based on the following version of the Phragmén-Lindelöf theorem.

Theorem 3.9.8 (Phragmén-Lindelöf). *Let $\varphi \in (0, \pi/2]$ and let $h : \overline{\Sigma}_\varphi \to X$ be continuous on $\overline{\Sigma}_\varphi$ and holomorphic in Σ_φ. Set $\alpha := \frac{\pi}{2\varphi}$. Assume that for all $\varepsilon > 0$ there exists a constant $C_\varepsilon > 0$ such that*

$$\|h(z)\| \leq C_\varepsilon e^{\varepsilon |z|^\alpha} \qquad (z \in \Sigma_\varphi).$$

If $\|h(re^{\pm i\varphi})\| \leq M$ for all $r > 0$, then $\|h(z)\| \leq M$ for all $z \in \Sigma_\varphi$.

For a proof of Theorem 3.9.8 we refer to [Con73, Cor.6.4.4].

Proof of Theorem 3.9.7. Denote by T_+, T_- the C_0-semigroups generated by $A_+ := e^{i\varphi}A$ and $A_- := e^{-i\varphi}A$, respectively. Let $M \geq 0$ such that $\|T_\pm(t)\| \leq M$ for all $t \geq 0$. Then

$$\|R(\lambda, A_\pm)\| = \left\| \int_0^\infty e^{-\lambda t} T_\pm(t)\, dt \right\| \leq \frac{M}{\operatorname{Re}\lambda} \qquad (\operatorname{Re}\lambda > 0).$$

For $\lambda \in \Sigma_\varphi^+$ this implies that

$$\|R(\lambda, A)\| = \|R(\lambda, e^{i\varphi}A_-)\| = \|R(\lambda e^{-i\varphi}, A_-)\| \leq \frac{M}{\operatorname{Re}(\lambda e^{-i\varphi})} \leq \frac{M}{|\lambda| \cos\varphi}.$$

Similarly, $\|R(\lambda, A)\| \leq \frac{M}{|\lambda|\cos\varphi}$ if $\lambda \in \Sigma_\varphi^-$. Thus we have

$$\|(I - zA)^{-1}\| = \|z^{-1}R(z^{-1}, A)\| \leq \frac{M}{\cos\varphi} \qquad (z \in \Sigma_\varphi).$$

For $n \in \mathbb{N}$ and $z \in \overline{\Sigma}_\varphi$ set $T_n(z) := (I - \frac{z}{n}A)^{-n}$. Then $\|T_n(z)\| \leq M$ for $z = re^{\pm i\varphi}$ and

$$\|T_n(z)\| \leq \left(\frac{M}{\cos\varphi}\right)^n \qquad \text{for} \quad z = re^{\pm i\alpha}, r \geq 0, |\alpha| < \varphi.$$

It follows from the Phragmén-Lindelöf Principle 3.9.8 that $\|T_n(z)\| \leq M$ for all $z \in \overline{\Sigma}_\varphi$ and all $n \in \mathbb{N}$. The Hille-Yosida Theorem 3.3.4 implies now that A generates a C_0-semigroup T. By Corollary 3.3.6, we have $\lim_{n\to\infty} T_n(t)x = T(t)x$ for $t \geq 0$ and $x \in X$. It thus follows from Vitali's theorem (see Theorem A.5 and Proposition A.3) that T has a holomorphic extension $\tilde{T}$ to Σ_φ satisfying $\|\tilde{T}(z)\| \leq M$ for all $z \in \Sigma_\varphi$. □

Consider now the case where $e^{\pm i\varphi}A$ generate C_0-semigroups $T_{\pm\varphi}$ which are not necessarily bounded. In this case, there exist constants $M, \omega \geq 0$ such that $\|T_{\pm\varphi}(t)\| \leq Me^{\omega t}$ for $t \geq 0$. It follows that $e^{\pm i\varphi}(A - \mu)$ generate bounded C_0-semigroups for $\mu := \frac{\omega}{\cos\varphi}$. Theorem 3.9.7 implies now that $A - \mu$ generates a bounded holomorphic C_0-semigroup of angle φ. We have thus proved the following result.

Corollary 3.9.9. *Let $\varphi \in (0, \pi/2)$ and assume that $e^{\pm i\varphi}A$ generate C_0-semigroups $T_{\pm\varphi}$ on X. Then A generates a holomorphic C_0-semigroup of angle φ with boundary semigroups $T_{\pm\varphi}$.*

The following result is a consequence of the above Corollary 3.9.9 and Proposition 3.9.1.

Corollary 3.9.10. *Assume that A generates a C_0-semigroup T and that iA generates a C_0-group U. Then T has a holomorphic extension to $\Sigma_{\pi/2}$ and U is the boundary group of T.*

Proof. By Corollary 3.9.9, the operator $e^{\pm i\pi/4}A$ generates a holomorphic C_0-semigroup of angle $\pi/4$. Thus $e^{i\theta}A$ generates a C_0-semigroup for all $\theta \in (-\pi/2, \pi/2)$. It follows from Corollary 3.9.9 again that A generates a holomorphic C_0-semigroup of angle $\pi/2$. By Proposition 3.9.1, U is its boundary group. □

Next we consider spectral conditions on A which imply that A generates a holomorphic C_0-semigroup on X under the assumption that iA generates a C_0-group U on X. An obvious necessary condition for this is that the spectrum of A is located in a left half-plane. However, this condition is not sufficient, in general. Indeed, consider the generator G of the Riemann-Liouville semigroup on $L^p(0,1)$ for $1 < p < \infty$ as introduced in Theorem 3.9.6 and let $A := -G$. Since G generates a C_0-semigroup T with $T(t) = 0$ for $t > 1$, $\sigma(A)$ is empty by Theorem 3.1.7. Moreover, iA generates a group by Theorem 3.9.6. However, A does not generate a C_0-semigroup. Nevertheless, if $\sigma(A)$ is contained in some left half-plane, then A generates a holomorphic C_0-semigroup on X provided U satisfies a certain growth condition.

Theorem 3.9.11. *Let A be an operator on X such that iA generates a C_0-group U on X. Assume that there exists a dense subspace Y of X such that for all $x \in Y$ there exist constants $C \geq 0$ and $k \in \mathbb{N}$ (depending on x) such that*

$$\|U(t)x\| \leq C(1 + |t|)^k \qquad (t \in \mathbb{R}).$$

If $\sigma(A) \subset \{\lambda \in \mathbb{C} : \operatorname{Re}\lambda \leq b\}$ for some $b \in \mathbb{R}$, then A generates a holomorphic C_0-semigroup of angle $\pi/2$ on X (whose boundary group is U).

The key of the proof of Theorem 3.9.11 is the following result of Phragmén-Lindelöf type.

Proposition 3.9.12. *Let $r : \overline{\Sigma_{\pi/2}} \to X$ be continuous. Assume that r is holomorphic in $\Sigma_{\pi/2}$ and that there exist constants $C, M \geq 0, R_0 > 0, k \in \mathbb{N}$ such that*

$$\begin{aligned} \|r(\lambda)\| &\leq \frac{C}{|\sin\varphi|^k} \qquad (\operatorname{Re}\lambda \geq 0, \operatorname{Im}\lambda \neq 0, |\lambda| \geq R_0, \arg\lambda = \varphi) \quad \textit{and} \\ \|r(is)\| &\leq M \qquad (s \in \mathbb{R}). \end{aligned}$$

Then $\|r(\lambda)\| \leq M$ for all $\lambda \in \mathbb{C}$ with $\operatorname{Re}\lambda \geq 0$.

Proof. For $R \geq R_0$ and $k \in \mathbb{N}$ consider the holomorphic function

$$\Phi : D_R^0 := \{z \in \Sigma_{\pi/2} : |z| < R\} \to X, \quad \lambda \mapsto \left(1 - \frac{\lambda^2}{R^2}\right)^k r(\lambda).$$

Let $\lambda := Re^{i\varphi}$ for $\varphi \in (-\pi/2, \pi/2)$. If $\varphi \neq 0$, we obtain

$$\|\Phi(\lambda)\| = |1 - e^{i2\varphi}|^k \|r(\lambda)\| = 2^k |\sin\varphi|^k \|r(\lambda)\| \leq 2^k C.$$

Moreover, $\Phi(R) = 0$ and

$$\|\Phi(is)\| = \left|\left(1 + \tfrac{s^2}{R^2}\right)^k\right| \|r(is)\| \leq 2^k M \quad (|s| \leq R).$$

The maximum principle implies that $\|\Phi(\lambda)\| \leq 2^k \max\{C, M\}$ for all $\lambda \in \overline{D_R^0}$. Letting $R \to \infty$, we deduce that $\|r(\lambda)\| \leq 2^k \max\{C, M\}$ provided $\operatorname{Re}\lambda \geq 0$. Now the Phragmén-Lindelöf Principle 3.9.8 implies that

$$\|r(\lambda)\| \leq M$$

for all $\lambda \in \mathbb{C}$ with $\operatorname{Re}\lambda \geq 0$. □

Proof of Theorem 3.9.11. Replacing A by $A - \omega$, we may assume that $\sigma(A) \subset \{\lambda \in \mathbb{C} : \operatorname{Re}\lambda < -\delta\}$ for some $\delta > 0$. By assumption, iA generates a group and we therefore have $\sup_{s\in\mathbb{R},|s|\geq w} \|sR(is, A)\| < \infty$ for suitable $w \geq 0$. It follows that

$$M := \sup_{s\in\mathbb{R}} \|sR(is, A)\| < \infty.$$

Thus the second assumption of Proposition 3.9.12 is satisfied for the function $\lambda \mapsto \lambda R(\lambda, A)x$ $(x \in X)$. In order to verify the first assumption in this proposition let $x \in Y$. Then, by hypothesis, there exist constants $k \in \mathbb{N}, C \geq 0$ such that

$\|U(t)x\| \leq C(1+|t|)^k$ $(t \in \mathbb{R})$. For λ of the form $\lambda = re^{i\varphi}$, where $r \geq 1$ and $\varphi \in (0, \pi/2]$ we therefore obtain

$$\begin{aligned}\|\lambda R(\lambda, A)x\| &= \left\|\lambda \int_0^\infty e^{i\lambda t} U(-t)x\,dt\right\| \leq |\lambda| C \int_0^\infty e^{-|\operatorname{Im}\lambda| t}(1+t^k)\,dt \\ &\leq C|\lambda| \left(\frac{1}{|\operatorname{Im}\lambda|} + \frac{k!}{|\operatorname{Im}\lambda|^{k+1}}\right) \leq \frac{C(1+k!)}{|\sin\varphi|^{k+1}}.\end{aligned}$$

Similarly, for $\lambda = re^{i\varphi}$, with $r \geq 1$ and $\varphi \in [-\pi/2, 0)$ we have

$$\|\lambda R(\lambda, A)x\| \leq \frac{C(1+k!)}{|\sin\varphi|^{k+1}}.$$

Hence both assumptions of Proposition 3.9.12 are satisfied for the function $\lambda \mapsto \lambda R(\lambda, A)x$ and it follows from that proposition that

$$\|\lambda R(\lambda, A)x\| \leq M\|x\| \qquad (\operatorname{Re}\lambda \geq 0, x \in Y).$$

Since Y is dense in X, we conclude that

$$\|\lambda R(\lambda, A)\| \leq M \qquad (\operatorname{Re}\lambda \geq 0).$$

It follows from Corollary 3.7.12 that A generates a bounded holomorphic C_0-semigroup and from Corollary 3.8.9 that the angle is $\pi/2$ and U is the boundary group. □

We finally turn our attention back to the question raised at the beginning of this section: under which conditions on the holomorphic C_0-semigroup T the "boundary value" of T exists and is again a C_0-semigroup. In the following, we weaken the sense of "boundary value" of T and also allow integrated semigroups as "boundary values" for holomorphic C_0-semigroups T. We say that an operator A on X generates a *k-times integrated group* on X for some $k \in \mathbb{N}_0$ if A and $-A$ generate k-times integrated semigroups on X. A k-times integrated group is called *exponentially bounded* if the integrated semigroups generated by A and $-A$ are exponentially bounded.

Theorem 3.9.13. *Let $\gamma \geq 0$ and $k \in \mathbb{N}$. Assume that A generates a holomorphic C_0-semigroup T of angle $\pi/2$. Then the following assertions hold:*

a) *Assume that there exist constants $M, \omega \geq 0$ such that*

$$\|T(z)\| \leq \frac{Me^{\omega|z|}}{(\operatorname{Re} z)^\gamma} \quad (\operatorname{Re} z > 0).$$

Then iA generates an exponentially bounded k-times integrated group provided $k > \gamma$.

b) *Assume that iA generates an exponentially bounded k-times integrated group on X. Then there exist constants $M, \omega \geq 0$ such that*

$$\|T(z)\| \leq \frac{Me^{\omega|z|}}{(\operatorname{Re} z)^k} \qquad (\operatorname{Re} z > 0).$$

Proof. a) Let $x \in X, k \in \mathbb{N}$ and let $\lambda_0 > \omega$. Then

$$R(\lambda_0, A)^k x = \int_0^\infty e^{-\lambda_0 u} \frac{u^{k-1}}{(k-1)!} T(u)x\, du.$$

Hence,

$$R(\lambda_0, A)^k T(z)x = \int_0^\infty e^{-\lambda_0 u} \frac{u^{k-1}}{(k-1)!} T(u+z)x\, du \quad (\operatorname{Re} z > 0).$$

Setting $z = t + is$ the assumption implies that

$$\begin{aligned} \|R(\lambda_0, A)^k T(z)\| &\leq \frac{M}{(k-1)!} \int_0^\infty e^{-\lambda_0 u} u^{k-1} \frac{e^{\omega|u+t+is|}}{(u+t)^\gamma}\, du \\ &\leq \frac{M}{(k-1)!} \int_0^\infty e^{-\lambda_0 u} u^{k-1-\gamma} e^{\omega|u+t+is|}\, du < \infty, \end{aligned} \tag{3.67}$$

provided $k > \gamma$. Since $z \mapsto R(\lambda_0, A)^k T(z)$ is holomorphic and bounded in a rectangle of the form $\{t + is : 0 < t < 1, -R < s < R\}$ for some $R > 0$ it follows by dominated convergence that

$$\lim_{t \downarrow 0} R(\lambda_0, A)^k T(t + is)x$$

exists for all $s \in \mathbb{R}$ and all $x \in X$. For $s \in \mathbb{R}$ and $x \in X$ set

$$S(s)x := \begin{cases} i^{-k} \lim_{t\downarrow 0} R(\lambda_0, A)^k T(t+is)x & (s \geq 0), \\ (-i)^{-k} \lim_{t\downarrow 0} R(\lambda_0, A)^k T(t+is)x & (s < 0). \end{cases}$$

In order to show that iA generates a k-times integrated semigroup it suffices by Proposition 3.2.7 to verify that $R(\cdot, iA)R(i\lambda_0, iA)^k$ is a Laplace transform. Note first that by the estimate (3.67) there exists a constant $C \geq 0$ such that $\|S(s)\| \leq Ce^{\omega|s|}$ for $s \in \mathbb{R}_+$. We claim that

$$\hat{S}(\lambda) = R(\lambda, iA)R(i\lambda_0, iA)^k \quad (\lambda > \omega).$$

In fact, by Fubini's theorem, the representation formula (3.46) for holomorphic

C_0-semigroups and Cauchy's theorem, we have

$$\begin{aligned}
i^k \int_0^\infty e^{-\lambda s} S(s)\, ds &= \lim_{t\downarrow 0} \int_0^\infty e^{-\lambda s} R(\lambda_0, A)^k T(t+is)\, ds \\
&= \lim_{t\downarrow 0} \int_0^\infty e^{-\lambda s} R(\lambda_0, A)^k \int_\Gamma e^{\mu(t+is)} R(\mu, A)\, d\mu\, ds \\
&= \lim_{t\downarrow 0} \int_\Gamma \int_0^\infty e^{-\lambda s} e^{\mu i s}\, ds\, e^{\mu t} R(\lambda_0, A)^k R(\mu, A)\, d\mu \\
&= \lim_{t\downarrow 0} \int_\Gamma \frac{1}{\lambda - i\mu} e^{\mu t} R(\lambda_0, A)^k R(\mu, A)\, d\mu \\
&= \lim_{t\downarrow 0} R(\lambda, iA) R(\lambda_0, A)^k T(t) \\
&= R(\lambda, iA) R(\lambda_0, A)^k \qquad (\lambda > \omega),
\end{aligned}$$

where Γ denotes the path defined in (3.46). Thus we have $\int_0^\infty e^{-\lambda s} S(s)\, ds = R(\lambda, iA) R(i\lambda_0, iA)^k$ for $\lambda > \omega$. The corresponding result for $-iA$ is proved in exactly the same way.

b) By rescaling, we may assume that $\sigma(A) \subset \{\lambda \in \mathbb{C} : \operatorname{Re}\lambda \le -1\}$ (see Proposition 3.1.9 i) and Proposition 3.2.6). We subdivide the proof into two steps.

Step 1: By assumption, A generates a holomorphic C_0-semigroup T of angle $\pi/2$. Hence, the function $S : z \mapsto S(z) := \int_0^z \frac{(z-\xi)}{(k-1)!}^{k-1} T(\xi)\, d\xi$ is holomorphic in the open right half-plane. We claim that there exist constants $M, \omega \ge 0$ such that

$$\|S(z)\| \le M e^{\omega |z|} \quad (\operatorname{Re} z > 0). \tag{3.68}$$

Integrating by parts, using $\frac{d^n}{d\xi^n} T(\xi) A^{-n} = T(\xi)$, gives

$$S(z)x = T(z)A^{-k}x - \frac{z^{k-1}}{(k-1)!} A^{-1}x - \cdots - A^{-k}x \quad (x \in X).$$

Hence, in order to prove (3.68) it suffices to show that

$$\|T(z)A^{-k}\| \le M e^{\omega |z|} \quad (\operatorname{Re} z > 0), \tag{3.69}$$

for suitable constants $M, \omega \ge 0$. Obviously $(S(t))_{t\ge 0}$ is the k-times integrated semigroup generated by A. For $z \in \{\mu \in \mathbb{C} : \operatorname{Re}\mu > 0\}$ set $z = t + is$ for $t > 0$ and $s \in \mathbb{R}$. Let $(R(s))_{s\in\mathbb{R}}$ be the k-times integrated group generated by iA. For $x \in D(A^k)$ set

$$T(is)x := \frac{d^k}{ds^k} R(s)x = R(s)(iA)^k x + \sum_{n=1}^k \frac{s^n}{n!} (iA)^{n-1} x \qquad (s \in \mathbb{R}).$$

Then

$$T(t+is)x = T(t)T(is)x = T(is)T(t)x \quad (t > 0, s \in \mathbb{R}), \tag{3.70}$$

because the function v given by $v(s) := T(is)T(t)x - T(t+is)x$ is the unique mild solution of the problem $u'(s) = iAu(s)$, $u(0) = 0$, by Lemma 3.2.9 and Lemma 3.2.10.

We consider first the case where $\operatorname{Re} z \geq 1$. By (3.70) we have for $z = t + is$

$$\|T(z)A^{-k}\| \leq \|T(t-1/2)A^{-k}\| \, \|T(is)A^{-k}\| \, \|A^k T(1/2)\|.$$

Since $t \mapsto \|T(t-1/2)A^{-k}\|$ and $s \mapsto \|T(is)A^{-k}\|$ are exponentially bounded we obtain (3.69) in the case where $\operatorname{Re} z \geq 1$.

Next, we consider the case where $0 < \operatorname{Re} z \leq 1$. For $x \in X$ we have by (3.70): $T(z)A^{-2k}x = T(t)A^{-k}T(is)A^{-k}x$. It then follows that there exist constants $M_1, \omega_1 \geq 0$ such that

$$\|T(z)A^{-2k}\| \leq M_1 e^{\omega_1 |z|} \quad \text{and} \quad \|T(is)A^{-k}\| \leq M_1 e^{\omega_1 |s|} \tag{3.71}$$

for $s \in \mathbb{R}$ and $z \in \mathbb{C}$ satisfying $\operatorname{Re} z > 0$. For $z \in \Omega := \{\mu \in \mathbb{C} : 0 \leq \operatorname{Re} \mu \leq 1\}$ and $x \in X$ set

$$f(z) := (\cos z)^{-2\omega_1} T(z)A^{-2k}x.$$

Then f is holomorphic in the interior of Ω and continuous on $\overline{\Omega}$, by (3.70). For $|s| \geq 1$ we have

$$\|f(z)\| \leq 2^{2\omega_1} \|T(z)A^{-2k}x\| \, |e^s - e^{-s}|^{-2\omega_1},$$

which by (3.71) implies that $z \to \|f(z)\|$ is bounded for $z \in \Omega$. For $z = is$ and $z = 1 + is$ we have

$$\begin{aligned} \|f(is)\| &\leq 2^{2\omega_1} \frac{\|T(is)A^{-k}\| \, \|A^{-k}x\|}{|e^s + e^{-s}|^{2\omega_1}} \leq M_2 \|A^{-k}x\| \\ \|f(1+is)\| &\leq 2^{2\omega_1} \frac{\|T(is)A^{-k}\| \, \|T(1)A^{-k}x\|}{|e^s + e^{-2i}e^{-s}|^{2\omega_1}} \leq M_2 \|A^{-k}x\| \end{aligned}$$

for a suitable constant $M_2 \geq 0$. The three-lines lemma [Con73, Theorem VI.3.7] now implies that there exists a constant $M \geq 0$ such that

$$\|T(z)A^{-2k}x\| \leq M e^{2\omega_1 |z|} \|A^{-k}x\| \quad (\operatorname{Re} z > 0).$$

Since $D(A^k)$ is dense in X, the claim follows.

Step 2: Set

$$S_{k+1}(z) := \int_0^z S(\xi)\, d\xi \quad (\operatorname{Re} z > 0).$$

Cauchy's integral formula implies that

$$T(z) = \frac{(k+1)!}{2\pi i} \int_{\gamma_z} \frac{S_{k+1}(\xi)}{(\xi - z)^{k+2}} \, d\xi \quad (\operatorname{Re} z > 0),$$

where γ_z denotes the path defined by the circle with centre z and radius $r = \frac{\operatorname{Re} z}{2}$. It follows that

$$\begin{aligned} T(z) &= \frac{(k+1)!}{2\pi i} \int_{\gamma_z} \frac{S_{k+1}(\xi) - S_{k+1}(z)}{(\xi - z)^{k+2}} \, d\xi \\ &= \frac{(k+1)!}{2\pi r^{k+1}} \int_0^{2\pi} \left(S_{k+1}(z + re^{i\varphi}) - S_{k+1}(z)\right) e^{-i\varphi(k+1)} \, d\varphi. \end{aligned}$$

It follows from the definition of S_{k+1} and (3.68) that

$$\|S_{k+1}(z + re^{i\varphi}) - S_{k+1}(z)\| \le M_3 r e^{\omega_3 |z|} \quad (\operatorname{Re} z > 0)$$

for suitable $M_3, \omega_3 \ge 0$. Inserting this in the above representation of $T(z)$, it follows that

$$\|T(z)\| \le M_4 \frac{e^{\omega_4 |z|}}{r^k} = M_4 \frac{2^k e^{\omega_4 |z|}}{(\operatorname{Re} z)^k} \quad (\operatorname{Re} z > 0)$$

for suitable constants $M_4, \omega_4 \ge 0$. The proof is complete. □

As an application of Theorem 3.9.13 we consider once again "boundary values" of the Gaussian semigroup. More precisely, let $1 \le p < \infty$ and let the operator Δ_p be defined as in (3.60). Then the following corollary holds true.

Corollary 3.9.14. *Let $1 \le p < \infty$ and let $k \in \mathbb{N}_0$. Then $i\Delta_p$ generates an exponentially bounded k-times integrated group on $L^p(\mathbb{R}^n)$ if $k > n|\frac{1}{2} - \frac{1}{p}|$. Moreover, the order of integration is optimal in the sense that $i\Delta_p$ does not generate a k-times integrated semigroup on $L^p(\mathbb{R}^n)$ if $k < n|\frac{1}{2} - \frac{1}{p}|$.*

Proof. It follows from (3.61) and by duality that the assumption of Theorem 3.9.13 a) is satisfied for $\gamma = n|1/2 - 1/p|$. Hence, the first assertion above follows from this theorem. Conversely, suppose that $i\Delta_p$ generates an exponentially bounded k-times integrated semigroup S on $L^p(\mathbb{R}^n)$ for some $1 \le p < \infty$ and $k < n|1/2 - 1/p|$. Let J be the conjugation on the complex space $L^p(\mathbb{R}^n)$; $Jf = \bar{f}$. Then $-i\Delta_p = J(i\Delta_p)J$, which generates the k-times integrated semigroup $JS(\cdot)J$, so $i\Delta_p$ generates an exponentially bounded k-times integrated group. By Theorem 3.9.13 b), there exists a constant $M > 0$ such that $\|T_p(z)\| \le M/(\operatorname{Re} z)^k$ for $z \in \Sigma_{\pi/2}$ with $|z| = 1$, where T_p is the Gaussian semigroup as in Example 3.7.6. However, by (3.62) we have

$$\|T_p(z)\| \ge 2^{-n/2p} (\operatorname{Re} z)^{-n|1/2 - 1/p|} \quad (z \in \Sigma_{\pi/2}, \, |z| = 1),$$

which yields a contradiction. □

More general results for differential and pseudo-differential operators will be given in Section 8.3.

3.10 Intermediate Spaces

It turns out that k-times integrated semigroups are the same as C_0-semigroups up to the choice of the underlying Banach space. This will be made precise in this section.

Throughout this section, Z, X and Y are Banach spaces. We write $Z \hookrightarrow X$ if $Z \subset X$ and there is a constant c such that $\|x\|_X \leq c\|x\|_Z$ for all $x \in Z$. If in addition Z is dense in X we write $Z \overset{d}{\hookrightarrow} X$.

The following lemma is a consequence of the closed graph theorem.

Lemma 3.10.1. *If $Z \hookrightarrow Y$ and $Z \subset X \hookrightarrow Y$, then $Z \hookrightarrow X$.*

Let A be an operator on X. If $Z \hookrightarrow X$ we denote by A_Z the part of A in Z; i.e., $D(A_Z) := \{x \in D(A) \cap Z : Ax \in Z\}$, $A_Z x := Ax$. If A is closed, then A_Z is closed. The following is easy to prove (see also Proposition B.8).

Lemma 3.10.2. *Let A be an operator on X, $Z \hookrightarrow X$. Let $\mu \in \rho(A)$ such that $R(\mu, A)Z \subset Z$. Let B be an operator on Z. Then $B = A_Z$ if and only if $\mu \in \rho(B)$ and $R(\mu, B) = R(\mu, A)|_Z$.*

Let A be a closed operator on X and $k \in \mathbb{N}$. Then $D(A^k)$ is a Banach space for the norm $\|x\|_{A^k} := \|x\| + \|Ax\| + \ldots + \|A^k x\|$. Moreover, $D(A^k) \hookrightarrow X$. We denote by A_k the part of A in $D(A^k)$; i.e., A_k is the operator on the Banach space $D(A^k)$ given by $A_k x = Ax$, $D(A_k) = D(A^{k+1})$. If $\rho(A) \neq \emptyset$, then A_k and A are similar (see Section 3.5 for the definition). In fact, $A_k = U^{-1}AU$ where U may be taken as $U = (\mu - A)^k$ for any $\mu \in \rho(A)$. In particular, $\sigma(A_k) = \sigma(A)$ and A_k generates an (exponentially bounded) m-times integrated semigroup on $D(A^k)$ if and only if A generates an (exponentially bounded) m-times integrated semigroup on X.

The following result on the spectrum of intermediate operators is of general interest.

Proposition 3.10.3. *Let A be an operator on X, $Z \hookrightarrow X$. Assume that $R(\mu, A)Z \subset Z$ for some $\mu \in \rho(A)$ and that $D(A^k) \subset Z$ for some $k \in \mathbb{N}$. Then $\sigma(A_Z) = \sigma(A)$ and $R(\lambda, A_Z) = R(\lambda, A)|_Z$ for all $\lambda \in \rho(A)$.*

Proof. a) Let $B = A_Z$ and $\lambda \in \rho(A)$. Iterating the resolvent equation

$$R(\lambda, A) = R(\mu, A) + (\mu - \lambda)R(\mu, A)R(\lambda, A)$$

gives

$$R(\lambda, A) = \sum_{j=1}^{k} (\mu - \lambda)^{j-1} R(\mu, A)^j + (\mu - \lambda)^k R(\mu, A)^k R(\lambda, A). \tag{3.72}$$

Since $R(\mu, A)Z \subset Z$ and $R(\mu, A)^k X = D(A^k) \subset Z$, it follows that $R(\lambda, A)Z \subset Z$. Hence by Lemma 3.10.2, $\lambda \in \rho(B)$ and $R(\lambda, B) = R(\lambda, A)|_Z$.

b) In order to prove the converse, we observe that $\rho(A) = \rho(A_k)$ since A and A_k are similar operators. Let $Y := D(A^k)$ with the graph norm. Then $Y \hookrightarrow Z$ by Lemma 3.10.1, $D(B^k) \subset Y$ and $A_k = B_Y$. Moreover, $R(\mu, B)Y = R(\mu, A)Y = D(A^{k+1}) \subset Y$. It follows from a) (applied to B instead of A) that $\rho(B) \subset \rho(A_k) = \rho(A)$. $\square$

Our aim is to prove the following result.

Theorem 3.10.4 (Sandwich Theorem). *Let A be an operator on X and let $k \in \mathbb{N}$. The following assertions are equivalent:*

(i) *The operator A generates a k-times integrated semigroup S such that $\omega(S) < \infty$.*

(ii) *There exists a Banach space Y and the generator B of a C_0-semigroup V on Y such that*

  a) *$D(B^k) \subset X \hookrightarrow Y$,*
  b) *$R(\lambda, B)X \subset X$ for some $\lambda \in \rho(B)$, and*
  c) *$A = B_X$.*

(iii) *There exists a Banach space Z such that*

  a) *$D(A^k) \subset Z \hookrightarrow X$,*
  b) *$R(\lambda, A)Z \subset Z$ for some $\lambda \in \rho(A)$, and*
  c) *A_Z generates a C_0-semigroup U on Z.*

Proof. (i) $\Rightarrow$ (ii): Assume that A generates a k-times integrated semigroup S satisfying $\|S(t)\| \leq Me^{\omega t}$ $(t \geq 0)$ where $M, \omega > 0$. For $x \in D(A^k)$, let

$$T(t)x := S(t)A^k x + \frac{t^{k-1}}{(k-1)!}A^{k-1}x + \ldots + tAx + x \quad (t \geq 0). \tag{3.73}$$

By Lemma 3.2.2 (see also Lemma 3.2.10), $v(t) := T(t)x$ is a mild solution of (ACP_0). Hence, $s \mapsto v(t+s)$ is a mild solution of (ACP_0) with initial value $T(t)x$. By Theorem 3.1.3,

$$R(\lambda, A)T(t)x = \int_0^\infty e^{-\lambda s}T(t+s)x\, ds \quad (t \geq 0,\ \lambda > \omega) \tag{3.74}$$

for all $x \in D(A^k)$.

Fix $\mu_0 > b > \omega$, and define a norm $\|\cdot\|_Y$ on X by

$$\|x\|_Y := \sup_{t \geq 0} \|e^{-bt}T(t)R(\mu_0, A)^k x\|_X, \tag{3.75}$$

and let Y be the completion of $(X,\ \|\cdot\|_Y)$. We claim that

$$\|(\lambda - b)R(\lambda, A)x\|_Y \leq \|x\|_Y \quad (\lambda > b,\ x \in X). \tag{3.76}$$

In fact,

$$\begin{aligned}
&\|e^{-bt}T(t)R(\mu_0,A)^kR(\lambda,A)x\|_X \\
&= \left\|e^{-bt}\int_0^\infty e^{-\lambda s}T(s+t)R(\mu_0,A)^kx\,ds\right\|_X \\
&= \left\|\int_0^\infty e^{-(\lambda-b)s}e^{-b(t+s)}T(s+t)R(\mu_0,A)^kx\,ds\right\|_X \\
&\leq \|x\|_Y\int_0^\infty e^{-(\lambda-b)s}\,ds = \frac{\|x\|_Y}{\lambda-b}.
\end{aligned}$$

Hence $R(\lambda,A)$ has a unique extension $R(\lambda)\in\mathcal{L}(Y)$ and $\|(\lambda-b)R(\lambda)\|_{\mathcal{L}(Y)}\leq 1$ $(\lambda>b)$. Then $(R(\lambda))_{\lambda>b}$ is a pseudo-resolvent on Y.

Next we show that

$$\lim_{\lambda\to\infty}\|\lambda R(\lambda)y-y\|_Y = 0 \tag{3.77}$$

for all $y\in Y$. Since $\limsup_{\lambda\to\infty}\|\lambda R(\lambda)\|_{\mathcal{L}(Y)}<\infty$, it suffices to prove (3.77) for $y\in X$, X being dense in Y. Let $x:=R(\mu_0,A)^ky$. Let $\varepsilon>0$. There exists M' such that $\|T(t)x\|_X\leq M'e^{\omega t}$ $(t\geq 0)$. Then

$$\begin{aligned}
&\left\|e^{-bt}\int_0^\infty \lambda e^{-\lambda s}(T(t+s)x-T(t)x)\,ds\right\|_X \\
&\leq e^{-bt}\int_0^\infty \lambda e^{-\lambda s}M'\left(e^{\omega(t+s)}+e^{\omega t}\right)ds \\
&= M'e^{-(b-\omega)t}\left(\frac{\lambda}{\lambda-\omega}+1\right).
\end{aligned}$$

Hence, there exists t_0 such that

$$\sup_{\lambda>2\omega}\sup_{t>t_0}\left\|e^{-bt}\int_0^\infty \lambda e^{-\lambda s}(T(t+s)x-T(t)x)\,ds\right\|_X<\varepsilon. \tag{3.78}$$

Since $t\mapsto T(t)x$ is uniformly continuous on $[0,t_0+1]$, there exists $\tau>0$ such that $\|T(t+s)x-T(t)x\|_X<\varepsilon$ whenever $s\in[0,\tau]$, $t\in[0,t_0]$.

For $\lambda>2\omega$ and $0\leq t\leq t_0$,

$$\begin{aligned}
&\left\|e^{-bt}\int_0^\infty \lambda e^{-\lambda s}(T(t+s)x-T(t)x)\,ds\right\|_X \\
&\leq e^{-bt}\int_0^\tau \lambda e^{-\lambda s}\varepsilon\,ds + e^{-bt}\int_\tau^\infty \lambda e^{-\lambda s}M'(e^{\omega(t+s)}+e^{\omega t})\,ds \\
&\leq \varepsilon + M'\left(\frac{\lambda}{\lambda-\omega}e^{-(\lambda-\omega)\tau}+e^{-\lambda\tau}\right).
\end{aligned} \tag{3.79}$$

By (3.74), (3.75), (3.78) and (3.79),

$$\begin{aligned}
&\limsup_{\lambda\to\infty} \|\lambda R(\lambda, A)y - y\|_Y \\
&\quad= \limsup_{\lambda\to\infty} \sup_{t\geq 0} \left\| e^{-bt} \left(\lambda R(\lambda, A)T(t)x - T(t)x\right)\right\|_X \\
&\quad= \limsup_{\lambda\to\infty} \sup_{t\geq 0} \left\| e^{-bt} \int_0^\infty \lambda e^{-\lambda s} \left(T(t+s)x - T(t)x\right) ds \right\|_X \\
&\quad\leq \varepsilon.
\end{aligned}$$

Since $\varepsilon > 0$ is arbitrary, the claim is proved.

It follows from (3.77) and Proposition B.6 that there exists a densely defined operator B on Y such that $(b, \infty) \subset \rho(B)$ and $R(\lambda, B) = R(\lambda)$ $(\lambda > b)$. By the Hille-Yosida theorem, B generates a C_0-semigroup V on Y satisfying $\|V(t)\|_{\mathcal{L}(Y)} \leq e^{bt}$ $(t \geq 0)$. It follows from Lemma 3.10.2 that $A = B_X$. By definition, $\|R(\mu_0, A)^k y\|_X \leq \|y\|_Y$ $(y \in X)$. Hence, $R(\mu_0, B)^k Y \subset X$; i.e., $D(B^k) \subset X$ and the proof of (ii) is complete.

(ii) $\Rightarrow$ (iii): Let $Z := \left(D(B^k), \|\cdot\|_{B^k}\right)$. Then $A_Z = B_k$ which is similar to B, so A_Z generates a C_0-semigroup U on Z. By Proposition 3.10.3, $\rho(A) = \rho(B)$ and $R(\lambda, A) = R(\lambda, B)|_X$ for all $\lambda \in \rho(B)$. It follows that $R(\lambda, A)Z \subset Z$ for all $\lambda \in \rho(A)$.

(iii) $\Rightarrow$ (i): Note that $D(A^k) \hookrightarrow Z$ by Lemma 3.10.1, and $R(\lambda, A_Z) = R(\lambda, A)|_Z$ for $\lambda \in \rho(A) = \rho(A_Z)$, by Proposition 3.10.3. Let $\mu \in \rho(A)$. Then $\lambda \mapsto R(\lambda, A)R(\mu, A)^k$ is the Laplace transform of $t \mapsto U(t)R(\mu, A)^k$. Now, (i) follows from Proposition 3.2.7. □

Corollary 3.10.5. *Let A be the generator of an exponentially bounded k-times integrated semigroup T on X and let $B \in \mathcal{L}(X, D(A^k))$. Then $A + B$ generates a k-times integrated semigroup S on X satisfying $\omega(S) < \infty$.*

Proof. We use the notation of Theorem 3.10.4 (iii). The operator $B|_Z$ is bounded. By Corollary 3.5.6, $A_Z + B|_Z$ generates a C_0-semigroup on Z. It is clear that $(A+B)_Z = A_Z + B|_Z$. So it will follow from Theorem 3.10.4 that $A+B$ generates an exponentially bounded k-times integrated semigroup on X once we have proved that $A + B$ satisfies conditions a) and b) of Theorem 3.10.4 (iii). It is easy to see that $D((A + B)^k) \subset D(A^k)$, so a) is satisfied.

In order to show b), take $\mu \in \rho(A)$. Then $C := (\mu - A)^k B \in \mathcal{L}(X)$ and by (3.73)

$$\begin{aligned}
R(\lambda, A)B &= R(\lambda, A)R(\mu, A)^k C \\
&= \frac{R(\lambda, A)C}{(\mu - \lambda)^k} - \sum_{j=1}^{k} (\mu - \lambda)^{j-k-1} R(\mu, A)^j C.
\end{aligned}$$

Since $\lambda \mapsto \lambda^{-k} R(\lambda, A)$ is the Laplace transform of an exponentially bounded function, $\limsup_{\lambda\to\infty} \|\lambda^{1-k} R(\lambda, A)\| < \infty$. This implies that $\lim_{\lambda\to\infty} \|R(\lambda, A)B\| = 0$.

Consequently, $(I - R(\lambda, A)B)$ is invertible for large λ. Then by Lemma 3.5.8, $(I - BR(\lambda, A))$ is also invertible for large λ. Hence, $(\lambda - (A+B)) = (I - BR(\lambda, A))(\lambda - A)$ is invertible for large λ. Hence, there exists $\lambda \in \rho(A+B) \cap \rho(A_Z + B|_Z) \cap \rho(A)$. Let $y \in Z$, $x = R(\lambda, A+B)y$. Then $(\lambda - A)x = y + Bx \in Z$. Hence $x = R(\lambda, A)(y + Bx) \in Z$ by Proposition 3.10.3. □

3.11 Resolvent Positive Operators

In this section we assume that X is an ordered Banach space with normal cone X_+ (see Appendix C).

Definition 3.11.1. *An operator A on X is called* resolvent positive *if there exists $\omega \in \mathbb{R}$ such that $(\omega, \infty) \subset \rho(A)$ and $R(\lambda, A) \geq 0$ for all $\lambda > \omega$.*

If A generates a C_0-semigroup T, then A is resolvent positive if and only if T is positive (i.e. $T(t)X_+ \subset X_+$ for all $t \geq 0$). In fact, if T is positive, then

$$R(\lambda, A) = \int_0^\infty e^{-\lambda t} T(t)\, dt \geq 0$$

for all $\lambda > \omega(T)$. The converse follows from Euler's formula

$$T(t)x = \lim_{n \to \infty} \left(I - \tfrac{t}{n} A\right)^{-n} x$$

(see Corollary 3.3.6). But there are interesting examples of resolvent positive operators which do not generate C_0-semigroups; see Section 6.1 for an example.

Let A be a resolvent positive operator. Then for $\lambda > \omega$ one has

$$(-1)^n R(\lambda, A)^{(n)} = n! R(\lambda, A)^{n+1} \geq 0 \tag{3.80}$$

for all $n \in \mathbb{N}$ (see Appendix B). Thus, the function $R(\cdot, A)$ is completely monotonic (cf. Section 2.7).

We first use Bernstein's theorem for real-valued functions to prove some general properties of resolvent positive operators.

Proposition 3.11.2. *Let A be a resolvent positive operator. Denote by*

$$s(A) := \sup\{\operatorname{Re} \lambda : \ \lambda \in \sigma(A)\}$$

the spectral bound *of A. Then $s(A) < \infty$ and*

$$R(\lambda, A) \geq R(\mu, A) \geq 0$$

whenever $s(A) < \lambda < \mu$. Moreover, if $\lambda \in \mathbb{R} \cap \rho(A)$ such that $R(\lambda, A) \geq 0$, then $\lambda > s(A)$. Finally, $s(A) \in \sigma(A)$ if $s(A) > -\infty$.

Proof. Let

$$s := \inf\{\omega : \ (\omega,\infty) \subset \rho(A) \text{ and } R(\lambda, A) \geq 0 \text{ for all } \lambda > \omega\}.$$

By assumption, $s < \infty$. Replacing A by $A - \omega$, we may assume that $s \leq 0$.

a) Let $s < \lambda < \mu$. Then

$$R(\lambda, A) - R(\mu, A) = (\mu - \lambda)R(\lambda, A)R(\mu, A) \geq 0.$$

Thus, $R(\cdot, A)$ is a decreasing function on (s,∞).

b) Assume that $s > -\infty$. Then $s \in \sigma(A)$. In fact, if $s \in \rho(A)$ then $R(s, A) \geq 0$. Moreover, for $\mu < s$ sufficiently close to μ one has

$$R(\mu, A) = \sum_{n=0}^{\infty}(s - \mu)^n R(s, A)^{n+1} \geq 0.$$

This contradicts the definition of s.

c) Let $H_s := \{\lambda \in \mathbb{C} : \ \operatorname{Re}\lambda > s\}$. We claim that $H_s \subset \rho(A)$; this and b) establish that $s = s(A)$. Denote by Ω_0 the connected component of $H_s \cap \rho(A)$ containing (s,∞). If $H_s \not\subset \rho(A)$, there exist $\mu_n \in \Omega_0$ such that $\mu := \lim_{n\to\infty} \mu_n \in H_s \setminus \rho(A)$. Then by Corollary B.3,

$$\sup_{n\in\mathbb{N}} \|R(\mu_n, A)\| = \infty.$$

By the uniform boundedness principle, there exist $x \in X$, $x^* \in X^*$ such that $\sup_{n\in\mathbb{N}} |\langle R(\mu_n, A)x, x^*\rangle| = \infty$. Since $X = \operatorname{span} X_+$ and $X^* = \operatorname{span} X^*_+$ (Proposition C.2), we can assume that $x \in X_+$, $x^* \in X^*_+$. By Bernstein's Theorem 2.7.7, there exists an increasing function $\alpha : \mathbb{R}_+ \to \mathbb{R}$ such that $\alpha(0) = 0$ and

$$\langle R(\lambda, A)x, x^*\rangle = \int_0^\infty e^{-\lambda t}\, d\alpha(t) = \widehat{d\alpha}(\lambda)$$

for all $\lambda > s$. It follows from uniqueness of holomorphic extensions that

$$\langle R(\mu_n, A)x, x^*\rangle = \widehat{d\alpha}(\mu_n)$$

for all $n \in \mathbb{N}$. Consequently,

$$|\langle R(\mu_n, A)x, x^*\rangle| \leq \widehat{d\alpha}(\operatorname{Re}\mu_n) = \langle R(\operatorname{Re}\mu_n, A)x, x^*\rangle \leq \langle R(\lambda, A)x, x^*\rangle,$$

where $\lambda := \inf_{n\in\mathbb{N}} \operatorname{Re}\mu_n > s$ (since $\operatorname{Re}\mu = \lim_{n\to\infty} \operatorname{Re}\mu_n > s$). This is a contradiction.

d) In order to prove the remaining assertion, assume that there exists $\lambda \in \rho(A)$ such that $\lambda < s(A)$ and $R(\lambda, A) \geq 0$. Let $\mu_n \downarrow s(A)$. Since $s(A) \in \sigma(A)$,

$$\|R(\mu_n, A)\| \to \infty \quad \text{as } n \to \infty.$$

As in a), we have

$$R(\lambda, A) \geq R(\mu_n, A) \geq 0 \quad (n \in \mathbb{N}).$$

This is impossible. □

From Theorem 3.11.2 and its proof, we note the following.

Corollary 3.11.3. *Let A be a resolvent positive operator. Then*

$$|\langle R(\lambda, A)x, x^* \rangle| \leq \langle R(\omega, A)x, x^* \rangle \tag{3.81}$$

whenever $\operatorname{Re}\lambda \geq \omega > s(A)$, $x \in X_+$ *and* $x^* \in X_+^*$. *In particular, for each* $\omega > s(A)$,

$$\sup_{\operatorname{Re}\lambda \geq \omega} \|R(\lambda, A)\| < \infty.$$

We need the following identity.

Lemma 3.11.4. *Let A be an operator and $\lambda \in \rho(A)$. Then for all $m \in \mathbb{N}$*

$$(-1)^m \lambda^{m+1} \left(R(\lambda, A)/\lambda\right)^{(m)} /m! = \sum_{k=0}^{m} \lambda^k R(\lambda, A)^{k+1}. \tag{3.82}$$

Proof. This is immediate from Leibniz's rule, since

$$(-1)^k R(\lambda, A)^{(k)}/k! = R(\lambda, A)^{k+1}.$$

□

Now we can prove the following generation theorem.

Theorem 3.11.5. *Let A be a resolvent positive operator. Then A generates a twice integrated semigroup which is Lipschitz continuous on bounded intervals. If $D(A)$ is dense, then A generates a once integrated semigroup.*

Proof. Considering $A - \omega$ instead of A, we can assume that $s(A) < 0$ (see Proposition 3.2.6). Let $m \in \mathbb{N}$, $\lambda \geq 0$. Then

$$\sum_{k=0}^{m-1} \lambda^k R(\lambda, A)^{k+1} = R(0, A) - \lambda^m R(\lambda, A)^m R(0, A). \tag{3.83}$$

For $m = 1$, this is just the resolvent equation. Then (3.83) follows by induction. Consequently,

$$0 \leq \sum_{k=0}^{m-1} \lambda^k R(\lambda, A)^{k+1} \leq R(0, A)$$

for $m \in \mathbb{N}$, $\lambda \geq 0$. It follows from Lemma 3.11.4 that

$$\sup_{\lambda > 0, m \in \mathbb{N}_0} \|\lambda^{m+1} (R(\lambda, A)/\lambda)^{(m)}/m!\| < \infty.$$

Now the claim follows from Theorem 3.3.1 and Theorem 3.3.2. □

More generally, if A is resolvent positive, it follows from Corollary 3.3.13 that the part of A in $Y = \overline{D(A)}$ generates a once integrated semigroup. The following example shows that a resolvent positive operator does not generate a once integrated semigroup in general.

Example 3.11.6. Let $X = C[-1,0] \times \mathbb{C}$ and A be given by

$$D(A) := C^1[-1,0] \times \{0\}, \quad A(f,0) := (f', -f(0)).$$

Then $\rho(A) = \mathbb{C}$ and $R(\lambda, A)(f,c) = (g,0)$ with

$$g(x) := e^{\lambda x} \left(\int_x^0 e^{-\lambda y} f(y)\, dy + c \right)$$

for all $\lambda \in \mathbb{C}$. Thus, A is resolvent positive. Let $e_\lambda \in C[-1,0]$ be given by

$$e_\lambda(x) := e^{\lambda x} \quad (\lambda > 0,\ x \in [-1,0]).$$

Then $(e_\lambda, 0) = R(\lambda, A)(0,1)$. One has $e_\lambda = \int_0^\infty \lambda^2 e^{-\lambda t} k_t\, dt$ where $k_t \in C[-1,0]$ is given by $k_t(x) := 0$ if $x + t \leq 0$ and $k_t(x) := x + t$ otherwise. If A were the generator of a once integrated semigroup, then $\lambda \mapsto e_\lambda/\lambda$ would be a Laplace transform. Hence, $k : \mathbb{R}_+ \to C[-1,0]$ would be differentiable. But $\frac{d}{dt} k_t(x)$ does not exist at $x = -t$ if $t \in (0,1)$. □

The following result shows that the situation is different if X has order continuous norm.

Theorem 3.11.7. *Let A be a resolvent positive operator and assume that X has order continuous norm. Then A generates a once integrated semigroup.*

Proof. Replacing A by $A - \omega$, we may assume that $s(A) \leq 0$. Let $x \in X_+$. Then by Theorem 2.7.18, there exists a unique normalized increasing function $F_x : \mathbb{R}_+ \to X$ such that $F_x(0) = 0$ and

$$R(\lambda, A)x = \int_0^\infty e^{-\lambda t}\, dF_x(t) \quad (\lambda > s(A)).$$

It follows from uniqueness that $F_x(t)$ is additive and positive homogeneous in x for all $t \geq 0$. Hence, there exists a positive linear operator $S_1(t) \in \mathcal{L}(X)$ such that $F_x(t) = S_1(t)x$ for all $t \geq 0$, $x \in X$. Since $S_1(\cdot)$ is increasing, we can define the Riemann integral $S_2(t)x := \int_0^t S_1(s)x\, ds$ for all $x \in X$ (see Corollary 1.9.6). Then

$$R(\lambda, A)x = \lambda \int_0^\infty e^{-\lambda t} S_1(t)x\, dt = \lambda^2 \int_0^\infty e^{-\lambda t} S_2(t)x\, dt$$

for all $x \in X$, $\lambda > \max\{0, s(A)\}$, where the first integral is understood as an improper Riemann integral (see (1.22) and Proposition 1.10.2). We have to show that S_1 is strongly continuous on $\mathbb{R}_+$.

Since $S_2(\cdot)x$ is continuous, S_2 is the twice integrated semigroup generated by A. Thus for $x \in D(A)$, $S_2(\cdot)x$ is continuously differentiable by Lemma 3.2.2. Hence $S_1(\cdot)x$ is continuous if $x \in D(A)$.

Let $x \in X_+$ and $\mu > s(A)$. Then $R(\mu, A)S_1(\cdot)x$ is a normalized increasing function on $\mathbb{R}_+$ and

$$\begin{aligned}\int_0^\infty e^{-\lambda t}\, d(R(\mu,A)S_1(t)x) &= R(\mu,A)\int_0^\infty e^{-\lambda t}\, d(S_1(t)x)\\ &= R(\mu,A)R(\lambda,A)x\\ &= R(\lambda,A)R(\mu,A)x\\ &= \int_0^\infty e^{-\lambda t}\, d(S_1(t)R(\mu,A)x)\end{aligned}$$

for all $\lambda > s(A)$. It follows from uniqueness of the representation (see Theorem 2.7.18) that

$$S_1(t)R(\mu,A)x = R(\mu,A)S_1(t)x.$$

Since the norm is order continuous, $y_+ := \lim_{s\downarrow t} S_1(s)x$ exists. Moreover,

$$\begin{aligned}R(\mu,A)y_+ &= \lim_{s\downarrow t} R(\mu,A)S_1(s)x\\ &= \lim_{s\downarrow t} S_1(s)R(\mu,A)x\\ &= S_1(t)R(\mu,A)x\\ &= R(\mu,A)S_1(t)x.\end{aligned}$$

Since $R(\mu, A)$ is injective, it follows that $y_+ = S_1(t)x$. In the same way one shows that $\lim_{s\uparrow t} S_1(s)x = S_1(t)x$ if $t > 0$. We have shown that S_1 is strongly continuous on $\mathbb{R}_+$. □

The closure of the domain of a resolvent positive operator is of a very special nature if the underlying space is a Banach lattice with order continuous norm.

Let X be a complex Banach lattice; i.e., the complexification of a real Banach lattice. A subspace J of X is called an *ideal* if

a) $x \in J$ implies $\operatorname{Re} x \in J$; and

b) if $x, y \in X$ are real, $|y| \le |x|$ and $x \in J$, then it follows that $y \in J$.

In a space $X := L^p(\Omega, \mu)$, where (Ω, μ) is a σ-finite measure space and $1 \le p < \infty$, every closed ideal J is of the form

$$J = \{f \in L^p(\Omega) : \ f|_{\Omega_0} = 0 \text{ a.e.}\},$$

where Ω_0 is a measurable subset of Ω (see [Sch74, p.157]).

Theorem 3.11.8. *Let X be a Banach lattice with order continuous norm. If A is a resolvent positive operator, then $\overline{D(A)}$ is an ideal.*

Proof. a) Note that by definition X is the complexification of a real Banach lattice $X_{\mathbb{R}}$. Since the resolvent leaves $X_{\mathbb{R}}$ invariant we have $\operatorname{Re} x \in \overline{D(A)}$ whenever $x \in \overline{D(A)}$. Now observe that if J is a closed ideal of $X_{\mathbb{R}}$, then $J \oplus iJ$ is a closed ideal of X. These remarks show that we can assume that X is a real Banach lattice, which we do.

b) We can also assume that $s(A) < 0$ (replacing A by $A - \omega$ otherwise).

c) Let $0 \leq y \leq R(0, A)x$ where $x \in X_+$. We claim that $y \in \overline{D(A)}$. For $\lambda > 0$ we have

$$0 \leq \lambda R(\lambda, A)y \leq \lambda R(\lambda, A)R(0, A)x = R(0, A)x - R(\lambda, A)x \leq R(0, A)x.$$

Since the order interval $[0, R(0, A)x]$ is weakly compact [AB85, Theorem 12.9], there exists a weak limit point z of $\lambda R(\lambda, A)y$ as $\lambda \to \infty$. In particular, $z \in \overline{D(A)}$. Then

$$R(0, A)y - R(\lambda, A)y = \lambda R(0, A)R(\lambda, A)y$$

has $R(0, A)z$ as weak limit point. By the inequality above,

$$\lim_{\lambda \to \infty} R(\lambda, A)y = 0,$$

so $R(0, A)y = R(0, A)z$. Since $R(0, A)$ is injective, it follows that $y = z \in \overline{D(A)}$.

d) Let $y \in D(A)$. Then $|y| \in \overline{D(A)}$. In fact, there exists $x \in X$ such that $y = R(0, A)x$. Hence, $|y| \leq R(0, A)|x|$ and the claim follows from c).

e) If $y \in \overline{D(A)}$, then $|y| \in \overline{D(A)}$. This follows from d) since the absolute value is a continuous mapping.

f) Let $0 \leq y \leq x \in \overline{D(A)}$. Let $x_n \in D(A)$ such that $\lim_{n\to\infty} x_n = x$. It follows from e) that $|x_n| \in \overline{D(A)}$. We have

$$y \wedge |x_n| \leq |x_n| = |R(0, A)Ax_n| \leq R(0, A)|Ax_n|.$$

It follows from c) that $y \wedge |x_n| \in \overline{D(A)}$. Hence, $y = \lim_{n\to\infty}(y \wedge |x_n|) \in \overline{D(A)}$.

g) Let $|y| \leq |x|$, where $x \in \overline{D(A)}$. Then $0 \leq y^+ \leq |x|$, so $y^+ \in \overline{D(A)}$, by e) and f). Similarly, $y^- \in \overline{D(A)}$, and therefore $y = y^+ - y^- \in \overline{D(A)}$. □

In some special cases, densely defined resolvent positive operators are automatically generators of C_0-semigroups.

Theorem 3.11.9. *Let $X = C(K)$ where K is a compact space. Let A be a densely defined resolvent positive operator. Then A generates a positive C_0-semigroup.*

Proof. Since $D(A)$ is dense, there exists a strictly positive function $u \in D(A)$; i.e., $u(x) \geq \varepsilon > 0$ for all $x \in K$ and some $\varepsilon > 0$. We can assume that $s(A) < 0$ (replacing A by $A - \omega$ if necessary). There exists $v \in C(K)$ such that $u = R(0, A)v$.

Then for $\lambda > 0, n \in \mathbb{N}$ and $f \in C(K)$ with $\|f\|_\infty \le 1$,

$$\begin{aligned} |(\lambda R(\lambda, A))^n f| &\le (\lambda R(\lambda, A))^n |f| \\ &\le \frac{1}{\varepsilon}(\lambda R(\lambda, A))^n u \\ &\le \frac{1}{\varepsilon}(\lambda R(\lambda, A))^n R(0, A)|v| \\ &= \frac{1}{\varepsilon}\left(R(0, A)|v| - \sum_{k=0}^{n-1} \lambda^k R(\lambda, A)^{k+1}|v|\right) \\ &\le \frac{1}{\varepsilon} R(0, A)|v|, \end{aligned}$$

using (3.83). It follows that

$$\|(\lambda R(\lambda, A))^n\| \le \frac{1}{\varepsilon}\| R(0, A)|v| \|_\infty$$

for all $\lambda > 0$ and $n \in \mathbb{N}$. Now the claim follows from the Hille-Yosida theorem. □

Now we return to the case when X is an arbitrary ordered Banach space with normal cone. We consider the inhomogeneous Cauchy problem

$$(ACP_f) \quad \begin{cases} u'(t) = & Au(t) + f(t) \quad (t \in [0, \tau]), \\ u(0) = & u_0, \end{cases}$$

where A is a closed operator on X, $\tau > 0$ and $f \in C([0, \tau], X)$. Recall from Section 3.1 that a *mild solution* of (ACP_f) is a function $u \in C([0, \tau], X)$ such that

$$\int_0^t u(s)\, ds \in D(A) \quad \text{and} \quad u(t) - u_0 = A\int_0^t u(s)\, ds + \int_0^t f(s)\, ds$$

for all $t \in [0, \tau]$. The function u is called a *classical solution* if in addition $u \in C^1([0, \tau], X)$. In that case, since A is closed, it follows that $u \in C([0, \tau], D(A))$ and (ACP_f) is satisfied. The following result is a special case of a sharper version of Corollary 3.2.11 c) which is valid for generators of Lipschitz continuous integrated semigroups.

Theorem 3.11.10. *Let A be a resolvent positive operator. Let $u_0 \in D(A)$, $f_0 \in X$ such that $Au_0 + f_0 \in \overline{D(A)}$. Let $f(t) = f_0 + \int_0^t f'(s)\, ds$ where $f' \in L^1((0, \tau), X)$. Then (ACP_f) has a unique mild solution.*

Proof. Denote by S the twice integrated semigroup generated by A. Let

$$v(t) = S(t)u_0 + (S * f)(t).$$

By Lemmas 3.2.9 and 3.2.10, there exists a unique solution if and only if $v \in C^2([0, \tau], X)$. By Proposition 1.3.6, one has $S * f \in C^1([0, \tau], X)$ and

$$\frac{d}{dt}(S * f)(t) = (S * f')(t) + S(t)f_0.$$

Now it follows from Lemma 3.2.2 c) that $v \in C^1([0,\tau],X)$ and

$$v'(t) = S(t)(Au_0 + f_0) + tu_0 + (S * f')(t).$$

Since S is Lipschitz continuous on $[0,\tau]$ (by Theorem 3.11.5), it follows from Proposition 1.3.7 that $S * f' \in C^1([0,\tau],X)$. Since $Au_0 + f_0 \in \overline{D(A)}$, it follows from Lemma 3.3.3 that $S(\cdot)(Au_0 + f_0) \in C^1([0,\tau],X)$. The proof is complete. □

Theorem 3.11.10 will be used in Section 6.2 to solve the heat equation with inhomogeneous boundary conditions. The following result shows that mild solutions of the inhomogeneous problem are positive if the initial value and the inhomogeneity are positive. In Section 6.2 this will be used to prove the parabolic maximum principle.

Theorem 3.11.11. *Let A be a resolvent positive operator, $\tau > 0$, $f \in C([0,\tau],X_+)$, and $u_0 \in X_+$. Let u be a mild solution of (ACP_f). Then $u(t) \geq 0$ for all $t \in [0,\tau]$.*

Proof. Denote by S the twice integrated semigroup generated by A. It follows from Theorem 2.7.15 that S is an increasing convex function. Let

$$w(t) := S(t)u_0 + \int_0^t S(t-r)f(r)\,dr.$$

It follows from Lemma 3.2.9 that $w \in C^2([0,\tau],X)$ and $u(t) = w''(t)$. Thus, it suffices to show that w is convex. We know this already for the first term of w. Define $\widetilde{S}(t) := S(t)$ for $t \geq 0$ and $\widetilde{S}(t) = 0$ for $t < 0$. Since S is increasing and convex and $S(0) = 0$, it follows that $\widetilde{S}(t) : \mathbb{R} \to \mathcal{L}(X)$ is also convex. Hence,

$$\int_0^t S(t-r)f(r)\,dr = \int_0^\infty \widetilde{S}(t-r)f(r)\,dr$$

is convex in $t \geq 0$. □

Next we consider a simple perturbation result.

Proposition 3.11.12. *Let A be a resolvent positive operator. Let $B : D(A) \to X$ be linear and positive (i.e., $Bx \geq 0$ if $x \in D(A) \cap X_+$). If the spectral radius $r(BR(\lambda,A)) < 1$ for some $\lambda > s(A)$, then $A + B$ is resolvent positive and $s(A + B) < \lambda$.*

Notice that $BR(\lambda,A)$ is a linear, positive mapping on X and so it is automatically continuous.

Proof. Let $x \in D(A)$. Then

$$(\lambda - (A+B))x = (I - BR(\lambda,A))(\lambda - A)x.$$

Let

$$S_\lambda := (I - BR(\lambda,A))^{-1} = \sum_{n=0}^\infty (BR(\lambda,A))^n.$$

Then S_λ is a bounded, positive operator on X and

$$R(\lambda, A)S_\lambda(\lambda - (A+B))x = x$$

for all $x \in X$ and

$$(\lambda - (A+B))R(\lambda, A)S_\lambda y = y \quad \text{for all } y \in X.$$

Thus, $\lambda \in \rho(A+B)$ and $R(\lambda, A+B) = R(\lambda, A)S_\lambda \geq 0$. If $\mu > \lambda$, then by Proposition 3.11.2, $\mu \in \rho(A)$ and $BR(\mu, A) \leq BR(\lambda, A)$ and so $r(BR(\mu, A)) \leq r(BR(\lambda, A)) < 1$. Replacing λ by μ, it follows that $\mu \in \rho(A+B)$ and $R(\mu, A+B) \geq 0$ for all $\mu \geq \lambda$. □

The following example shows that, in the theorem above, $A+B$ may not be generator of a C_0-semigroup even if A generates a positive C_0-semigroup.

Example 3.11.13. *Let $\alpha \in (0,1)$. Define the operator A by*

$$Af(x) := -f'(x) + \frac{\alpha}{x}f(x) \quad (x \in (0,1])$$

on $C_0(0,1] := \{f \in C[0,1] : f(0) = 0\}$ with domain $D(A) := \{f \in C^1[0,1] : f'(0) = f(0) = 0\}$. Then A is resolvent positive but not the generator of a C_0-semigroup. Moreover, $s(A) = -\infty$.

Proof. Let $A_0 f := -f'$ with domain $D(A_0) = D(A)$. Then A_0 is the generator of the C_0-semigroup $(T(t))_{t\geq 0}$ given by

$$(T(t)f)(x) = \begin{cases} f(x-t) & (x \geq t), \\ 0 & (x < t). \end{cases}$$

Moreover, $\sigma(A_0) = \emptyset$ and

$$(R(\lambda, A_0)f)(x) = e^{-\lambda x}\int_0^x e^{\lambda y} f(y)\,dy \qquad (\lambda \in \mathbb{C},\ f \in C_0(0,1]).$$

Let $B : D(A_0) \to C_0(0,1]$ be given by

$$(Bf)(x) := \frac{\alpha}{x}f(x) \quad (x > 0), \quad (Bf)(0) := 0.$$

Let $g \in C_0(0,1],\ f := R(0, A)g$. Then

$$|(Bf)(x)| = \left|\frac{\alpha}{x}\int_0^x g(y)\,dy\right| \leq \alpha\|g\|_\infty.$$

Thus $\|BR(0, A_0)\| \leq \alpha < 1$. Now Proposition 3.11.12 implies that $A = A_0 + B$ is resolvent positive.

It remains to show that A is not the generator of a C_0-semigroup. One can easily check that for all $\lambda \in \mathbb{C}$ one has $\lambda \in \rho(A)$ and

$$\begin{aligned}(R(\lambda, A)g)(x) &= e^{-\lambda x} x^{\alpha} \int_0^x y^{-\alpha} e^{\lambda y} g(y)\, dy \\ &= \int_0^x x^{\alpha}(x-t)^{-\alpha} g(x-t) e^{-\lambda t}\, dt \qquad (g \in C_0(0,1]).\end{aligned}$$

Suppose that A generates a C_0-semigroup T. Then

$$(R(\lambda, A)g)(x) = \int_0^{\infty} e^{-\lambda t} (T(t)g)(x)\, dt$$

for sufficiently large λ. It follows from the uniqueness theorem (Theorem 1.7.3) that

$$(T(t)g)(x) = \begin{cases} x^{\alpha}(x-t)^{-\alpha} g(x-t) & (x \geq t), \\ 0 & (x < t). \end{cases}$$

This does not define a bounded operator on $C_0(0,1]$. □

3.12 Complex Inversion and UMD-spaces

In this section, we apply the complex inversion formula for Laplace transforms (Theorem 2.3.4) to orbits of C_0-semigroups; i.e., to solutions of well-posed Cauchy problems. In general this produces a representation only of classical solutions in terms of the resolvent of A, but we shall see in Theorem 3.12.2 that there is a class of Banach spaces where the representation holds for mild solutions.

Proposition 3.12.1. *Let T be a C_0-semigroup on a Banach space X with generator A, and let $\omega > \omega(T)$ and $t \geq 0$. Then*

$$T(t)x = \lim_{k \to \infty} \frac{1}{2\pi} \int_{-k}^{k} e^{(\omega + is)t} R(\omega + is, A)x\, ds$$

for all $x \in D(A)$.

Proof. Replacing $T(t)$ by $e^{-\alpha t}T(t)$ where $\omega(T) < \alpha < \omega$, we may assume that $\omega(T) < 0 < \omega$. Let $x \in D(A)$ and define $F(t) = T(t)x - x$. Then F is differentiable with $F'(t) = T(t)Ax$. Since F' is bounded, $F \in \mathrm{Lip}_0(\mathbb{R}_+, X)$. The Laplace-Stieltjes transform of F is

$$\widehat{dF}(\lambda) = R(\lambda, A)Ax.$$

By Theorem 2.3.4,

$$\begin{aligned} T(t)x - x &= \lim_{k\to\infty} \frac{1}{2\pi} \int_{-k}^{k} e^{(\omega+is)t} \frac{R(\omega+is,A)Ax}{\omega+is}\,ds \\ &= \lim_{k\to\infty} \frac{1}{2\pi} \int_{-k}^{k} e^{(\omega+is)t} \left(R(\omega+is,A)x - \frac{x}{\omega+is} \right) ds \\ &= \lim_{k\to\infty} \left(\frac{1}{2\pi} \int_{-k}^{k} e^{(\omega+is)t} R(\omega+is,A)x\,ds \right) - x, \end{aligned}$$

where we have used a standard contour integral. □

For $f \in L^2(\mathbb{R}, X)$ and $0 < \varepsilon < R$, let

$$(H_{\varepsilon R} f)(t) := \frac{1}{\pi} \int_{\varepsilon \le |t-s| \le R} \frac{f(s)}{t-s}\,ds = (\psi_{\varepsilon R} * f)(t) \quad (t \in \mathbb{R}),$$

where

$$\psi_{\varepsilon R}(t) := \begin{cases} \dfrac{1}{\pi t} & \text{if } \varepsilon \le |t| \le R, \\ 0 & \text{otherwise.} \end{cases}$$

Then $H_{\varepsilon R} \in \mathcal{L}(L^2(\mathbb{R}, X))$, since $\psi_{\varepsilon R} \in L^1(\mathbb{R}, X)$ (see Proposition 1.3.2).

The Banach space X is said to be a UMD-*space* if

$$Hf := \lim_{\substack{\varepsilon\downarrow 0 \\ R\to\infty}} H_{\varepsilon R} f$$

exists in $L^2(\mathbb{R}, X)$ for each $f \in L^2(\mathbb{R}, X)$. Then by the Banach-Steinhaus theorem, H is a bounded linear operator, known as the *Hilbert transform*, on $L^2(\mathbb{R}, X)$.

When $f(t) = \chi_{[a,b]}(t)x$, then Hf exists in $L^2(\mathbb{R}, X)$. Since the step functions are dense in $L^2(\mathbb{R}, X)$, it follows that X is a UMD-space if $\sup_{0<\varepsilon<R<\infty} \|H_{\varepsilon R}\| < \infty$. It follows easily from Plancherel's theorem that any Hilbert space is a UMD-space (see also Proposition E.5). Any space of the form $L^p(\Omega, \mu)$ for $1 < p < \infty$ is a UMD-space. If X is any UMD-space, then X is reflexive and X^* is also a UMD-space. See, for example, [Fra86, Section II].

Theorem 3.12.2. *Let T be a C_0-semigroup on a* UMD-*space X with generator A, and let $\omega > \omega(T)$ and $t > 0$. Then*

$$T(t)x = \lim_{k\to\infty} \frac{1}{2\pi} \int_{-k}^{k} e^{(\omega+is)t} R(\omega+is,A)x\,ds$$

for all $x \in X$.

Proof. Replacing $T(t)$ by $e^{-\omega t}T(t)$, we may assume that $\omega(T) < 0 = \omega$. For $k > 0$ and $t \in \mathbb{R}$, define

$$\begin{aligned} T_k(t) &:= \frac{1}{2\pi}\int_{-k}^{k} e^{ist}R(is,A)\,ds, \\ S_k(t) &:= \frac{1}{2\pi}\int_{-k}^{k} e^{ist}R(is,A)^2\,ds. \end{aligned}$$

Since $T_k(t)x \to T(t)x$ for all $x \in D(A)$ (Proposition 3.12.1) and $D(A)$ is dense, it suffices to show that

$$\sup_k \|T_k(t)\| < \infty$$

for each $t > 0$. Integration by parts gives

$$T_k(t) = \frac{1}{2\pi i t}\left(e^{ikt}R(ik,A) - e^{-ikt}R(-ik,A)\right) + \frac{1}{t}S_k(t).$$

Since $\|R(\pm ik, A)\| \le \int_0^\infty \|T(s)\|\,ds < \infty$, it suffices to show that

$$\sup_k \|S_k(t)\| < \infty.$$

Since X is reflexive, T^* is strongly continuous (Corollary 3.3.9 and Proposition 3.3.14). Let $x \in X$, $x^* \in X^*$ and $t \in \mathbb{R}$. Using Fubini's theorem,

$$\begin{aligned} \langle x, T_k(t)^*x^*\rangle &= \frac{1}{2\pi}\int_{-k}^{k}\int_0^\infty e^{ist}e^{-isr}\langle x, T(r)^*x^*\rangle\,dr\,ds \\ &= \frac{1}{2\pi}\int_0^\infty \frac{e^{ik(t-r)} - e^{-ik(t-r)}}{i(t-r)}\langle x, T(r)^*x^*\rangle\,dr \\ &= \lim_{\substack{\varepsilon\downarrow 0\\ R\to\infty}} \frac{1}{2i}\left(e^{ikt}\langle x, H_{\varepsilon R}(f_k)(t)\rangle - e^{-ikt}\langle x, H_{\varepsilon R}(f_{-k})(t)\rangle\right), \end{aligned}$$

where

$$f_a(r) := \begin{cases} e^{-iar}T(r)^*x^* & (r \ge 0), \\ 0 & (r < 0). \end{cases}$$

It follows that

$$T_k(t)^*x^* = \frac{1}{2i}\left(e^{ikt}H(f_k)(t) - e^{-ikt}H(f_{-k})(t)\right) \quad t\text{-a.e.},$$

where H is the Hilbert transform on $L^2(\mathbb{R}, X^*)$. Hence,

$$\begin{aligned} \left(\int_{-\infty}^{\infty} \|T_k(t)^*x^*\|^2\,dt\right)^{1/2} &\le \frac{\|H\|}{2}\left(\|f_k\|_2 + \|f_{-k}\|_2\right) \\ &\le M\|H\|\,\|x^*\|, \end{aligned}$$

where $M := (\int_0^\infty \|T(s)\|^2\, ds)^{1/2} < \infty$.

For $x \in X$ and $x^* \in X^*$, Fubini's theorem gives

$$\begin{aligned}\langle S_k(t)x, x^*\rangle &= \frac{1}{2\pi}\int_{-k}^{k}\int_0^\infty e^{ist}\langle R(is,A)e^{-isr}T(r)x, x^*\rangle\, dr\, ds \\ &= \int_0^\infty \Big\langle \frac{1}{2\pi}\int_{-k}^{k} e^{is(t-r)}R(is,A)T(r)x\, ds, x^*\Big\rangle\, dr \\ &= \int_0^\infty \langle T(r)x, T_k(t-r)^*x^*\rangle\, dr.\end{aligned}$$

Now the Cauchy-Schwarz inequality gives

$$\begin{aligned}|\langle S_k(t)x, x^*\rangle| &\le \left(\int_0^\infty \|T(r)x\|^2\, dr\right)^{1/2}\left(\int_0^\infty \|T_k(t-r)^*x^*\|^2\, dr\right)^{1/2} \\ &\le M^2\|H\|\,\|x\|\,\|x^*\|.\end{aligned}$$

Thus,

$$\|S_k(t)\| \le M^2\|H\|.$$

□

Example 3.12.3. *Theorem 3.12.2 is not valid if the assumption that X is a UMD-space is omitted.*

Let $X := L^1(\mathbb{R})$ and T be the C_0-semigroup of invertible isometries on X defined by

$$(T(t)f)(r) := f(t+r) \qquad (t \ge 0,\ r \in \mathbb{R}).$$

Let

$$T_k(t) := \frac{1}{2\pi}\int_{-k}^{k} e^{(1+is)t}R(1+is,A)\, ds.$$

A routine calculation shows that

$$(T_k(t)f)(r) = \frac{1}{\pi}\int_{-\infty}^{\infty} f(r+s+t)\phi_{k,t}(s)\, ds,$$

where

$$\phi_{k,t}(s) := \begin{cases}\dfrac{e^{-s}\sin(ks)}{s} & (s \ge -t), \\ 0 & (s < -t).\end{cases}$$

Hence,

$$\begin{aligned}\|T_k(t)\| = \|\phi_{k,t}\|_1 &= \int_{-t}^{\infty} e^{-s}\left|\frac{\sin(ks)}{s}\right| ds \\ &\ge e^{-1}\int_0^1 \frac{|\sin(ks)|}{s}\, ds = e^{-1}\int_0^k \frac{|\sin s|}{s}\, ds \to \infty\end{aligned}$$

as $k \to \infty$. Thus, $\{T_k(t) : k \ge 0\}$ is not uniformly bounded and therefore not strongly convergent on X. □

3.13 Norm-continuous Semigroups and Hilbert Spaces

In this section, we shall consider C_0-semigroups $T = (T(t))_{t\geq 0}$ which are norm-continuous for $t > 0$. This class contains all holomorphic C_0-semigroups, including many examples arising from differential operators. In general, there is no known characterization of such semigroups in terms of the generator and resolvent, but there is a simple characterization in the case of Hilbert spaces.

Proposition 3.13.1. *Let T be a C_0-semigroup on a Banach space X with generator A, and suppose that T is norm-continuous for $t > 0$. Let $\omega > \omega(T)$. Then $\|R(\omega + is, A)\| \to 0$ as $|s| \to \infty$.*

Proof. Since $T : (0, \infty) \to \mathcal{L}(X)$ is norm-continuous,

$$R(\omega + is, A) = \int_0^\infty e^{-(\omega+is)t} T(t)\, dt$$

as an $\mathcal{L}(X)$-valued Bochner integral. Let $H(t) := e^{-\omega t}T(t)$ if $t > 0$, and $H(t) := 0$ if $t < 0$. Then $H \in L^1(\mathbb{R}, \mathcal{L}(X))$, and $R(\omega + is, A) = (\mathcal{F}H)(s)$. The result follows from the Riemann-Lebesgue lemma (Theorem 1.8.1). □

It follows from Neumann series expansions (Corollary B.3) that if $\|R(\omega + is, A)\| \to 0$ as $|s| \to \infty$, then for any real a, $\{\lambda \in \sigma(A) : \operatorname{Re}\lambda \geq a\}$ is bounded, and $\|R(\alpha + is, A)\| \to 0$, uniformly for $\alpha > a$, as $|s| \to \infty$.

The converse of Proposition 3.13.1 does not hold in general, but we will now prove that it is true when X is a Hilbert space.

Theorem 3.13.2. *Let T be a C_0-semigroup on a Hilbert space X with generator A. Let $\omega > \omega(T)$, and suppose that $\|R(\omega + is, A)\| \to 0$ as $|s| \to \infty$. Then T is norm-continuous for $t > 0$.*

Proof. Replacing $T(t)$ by $e^{-\omega t}T(t)$, we may assume that $\omega(T) < 0 = \omega$. Let $x \in D(A)$ and $F(t) := t^2 T(t)x$. Then F is differentiable and $F'(t) = 2tT(t)x + t^2 T(t)Ax$. Hence $F \in \operatorname{Lip}_0(\mathbb{R}_+, X)$ with Laplace-Stieltjes transform

$$\begin{aligned}
\widehat{dF}(\lambda) &= -2\frac{d}{d\lambda}(R(\lambda, A)x) + \frac{d^2}{d\lambda^2}(R(\lambda, A)Ax) \\
&= 2R(\lambda, A)^2 x + 2R(\lambda, A)^3 Ax \\
&= 2\lambda R(\lambda, A)^3 x.
\end{aligned}$$

By Theorem 2.3.4,

$$t^2 T(t)x = \lim_{k\to\infty} \frac{1}{\pi} \int_{-k}^{k} e^{ist} R(is, A)^3 x\, ds. \tag{3.84}$$

Let $M := \left(\int_0^\infty \|T(t)\|^2\,dt\right)^{1/2} < \infty$. Given $\varepsilon > 0$, there exists N such that $\|R(is,A)\| < \varepsilon/4M^2$ whenever $|s| > N$. For $k > N$, $t, t_0 \geq 0$ and $y \in X$,

$$\begin{aligned}
&\left|\left(\int_{N\leq|s|\leq k} \left(e^{ist} - e^{ist_0}\right) R(is,A)^3 x\,ds \middle| y\right)_X\right| \\
&\leq 2\int_{N\leq|s|\leq k} \|R(is,A)^2 x\|\,\|R(is,A)^* y\|\,ds \\
&\leq \frac{\varepsilon}{2M^2}\left(\int_{-\infty}^{\infty} \|R(is,A)x\|^2\,ds\right)^{1/2}\left(\int_{-\infty}^{\infty} \|R(is,A)^* y\|^2\,ds\right)^{1/2} \\
&= \frac{\pi\varepsilon}{M^2}\left(\int_0^\infty \|T(r)x\|^2\,dr\right)^{1/2}\left(\int_0^\infty \|T(r)^* y\|^2\,dr\right)^{1/2} \\
&\leq \pi\varepsilon\|x\|\,\|y\|,
\end{aligned}$$

where $(\cdot|\cdot)_X$ denotes the inner product on X and we have used the Cauchy-Schwarz inequality and Plancherel's theorem for Hilbert spaces (Theorem 1.8.2). Hence,

$$\left\|\int_{N\leq|s|\leq k} \left(e^{ist} - e^{ist_0}\right) R(is,A)^3 x\,ds\right\| \leq \pi\varepsilon\|x\|.$$

It follows from (3.84) that

$$\left\|t^2T(t)x - t_0^2T(t_0)x\right\| \leq \varepsilon\|x\| + \frac{2N\|x\|}{\pi}\sup_{|s|\leq N}\left|e^{ist} - e^{ist_0}\right|\sup_{s\in\mathbb{R}}\|R(is,A)\|^3,$$

so

$$\left\|t^2T(t) - t_0^2T(t_0)\right\| \leq \varepsilon + \frac{2N^2}{\pi}|t-t_0|\left(\int_0^\infty \|T(r)\|\,dr\right)^3.$$

This shows that

$$\limsup_{t\to t_0}\left\|t^2T(t) - t_0^2T(t_0)\right\| \leq \varepsilon.$$

Since $\varepsilon > 0$ is arbitrary, $t \mapsto t^2T(t)$ is norm-continuous, and the result is proved. □

3.14 The Second Order Cauchy Problem

Let A be a closed operator on a Banach space X. Given $x, y \in X$ we consider the problem

$$P^2(x,y)\quad\begin{cases} u''(t) &= Au(t)\quad (t\geq 0)\\ u(0) &= x,\\ u'(0) &= y.\end{cases}$$

A *classical solution* of $P^2(x,y)$ is a function $u \in C^2(\mathbb{R}_+, X)$ such that $u(t) \in D(A)$ for all $t \geq 0$ and $P^2(x,y)$ holds. A *mild solution* is a function $u \in C(\mathbb{R}_+, X)$ such that

$$\int_0^t \int_0^s u(r)\, dr\, ds = \int_0^t (t-s)u(s)\, ds \in D(A)$$

and

$$u(t) = x + ty + A\int_0^t (t-s)u(s)\, ds \tag{3.85}$$

for all $t \geq 0$.

If u is a classical solution, then integrating $P^2(x,y)$ twice shows that u is a mild solution. Conversely, if u is a mild solution and $u \in C^2(\mathbb{R}_+, X)$, then u is a classical solution. This follows from (3.85) and the fact that A is closed.

Proposition 3.14.1. *Let* $u \in C(\mathbb{R}_+, X)$ *with* $\operatorname{abs}(u) < \infty$. *Let* $\omega > \max\{\operatorname{abs}(u), 0\}$. *Then* u *is a mild solution of* $P^2(x,y)$ *if and only if*

$$\hat{u}(\lambda) \in D(A) \quad \text{and} \quad \lambda x + y = (\lambda^2 - A)\hat{u}(\lambda) \tag{3.86}$$

for all $\lambda > \omega$.

Proof. There exists $M \geq 0$ such that $\|v(t)\| \leq Me^{\omega t}$ $(t \geq 0)$, where $v(t) := \int_0^t u(s)\, ds$. Taking Laplace transforms, Corollary 1.7.6 shows that (3.85) holds if and only if

$$\frac{\hat{u}(\lambda)}{\lambda^2} = \int_0^\infty e^{-\lambda t} \int_0^t v(s)\, ds\, dt \in D(A)$$

and

$$\hat{u}(\lambda) = \frac{x}{\lambda} + \frac{y}{\lambda^2} + A\frac{\hat{u}(\lambda)}{\lambda^2} \quad \text{for all } \lambda > \omega.$$

□

Now let $u \in C(\mathbb{R}_+, X)$. Assume that $\operatorname{abs}(u) < \infty$, $\omega > \max\{\operatorname{abs}(u), 0\}$ and $(\omega, \infty) \subset \rho(A)$. Then Proposition 3.14.1 shows that u is a mild solution of $P^2(x,y)$ if and only if

$$\hat{u}(\lambda) = \lambda R(\lambda^2, A)x + R(\lambda^2, A)y \qquad (\lambda > \omega). \tag{3.87}$$

This relation will lead us to consider operators A such that $\lambda R(\lambda^2, A)$ is a Laplace transform.

Definition 3.14.2. *A strongly continuous function* $\operatorname{Cos} : \mathbb{R}_+ \to \mathcal{L}(X)$ *is called a* cosine function *if* $\operatorname{Cos}(0) = I$ *and*

$$2\operatorname{Cos}(t)\operatorname{Cos}(s) = \operatorname{Cos}(t+s) + \operatorname{Cos}(t-s) \quad (t \geq s \geq 0). \tag{3.88}$$

Lemma 3.14.3. *Let* Cos *be a cosine function.*

a) $\omega(\operatorname{Cos}) < \infty$.

b) $\operatorname{Cos}(t)\operatorname{Cos}(s) = \operatorname{Cos}(s)\operatorname{Cos}(t)$ *for all* $s, t \geq 0$.

c) *Define* $\operatorname{Cos}(-t) = \operatorname{Cos}(t)$ *for* $t \geq 0$. *Then* Cos *is strongly continuous on* $\mathbb{R}$, *and*

$$2\operatorname{Cos}(t)\operatorname{Cos}(s) = \operatorname{Cos}(t+s) + \operatorname{Cos}(t-s) \quad (s, t \in \mathbb{R}). \tag{3.89}$$

Proof. a) By the uniform boundedness principle,

$$M := \sup_{0 \leq s \leq 2} \|\operatorname{Cos}(s)\| < \infty.$$

Choose $\omega > 0$ such that $2\|\operatorname{Cos}(1)\|e^{-\omega} + e^{-2\omega} \leq 1$. We claim that $\|\operatorname{Cos}(t)\| \leq Me^{\omega t}$ $(t \geq 0)$. This is certainly true for $t \in [0, 2]$. Assume that it holds for $t \in [0, n]$, where $n \in \mathbb{N}$, $n \geq 2$. We claim that it holds for $t \in [0, n+1]$. Let $t \in (n-1, n]$. Then

$$\begin{aligned} \|\operatorname{Cos}(t+1)\| &= \|2\operatorname{Cos}(t)\operatorname{Cos}(1) - \operatorname{Cos}(t-1)\| \\ &\leq 2\|\operatorname{Cos}(1)\|Me^{\omega t} + Me^{\omega(t-1)} \\ &= \left(2\|\operatorname{Cos}(1)\|e^{-\omega} + e^{-2\omega}\right) Me^{\omega(t+1)} \\ &\leq Me^{\omega(t+1)}. \end{aligned}$$

b) Let $t \geq 0$. Replacing s and t by $t/2$ in (3.88) gives $\operatorname{Cos}(t) = 2\operatorname{Cos}(t/2)^2 - I$. A simple induction shows that, for each $n \geq 1$, $\operatorname{Cos}(t)$ is a polynomial in $\operatorname{Cos}(t/2^n)$, so they commute.

Assume that $\operatorname{Cos}(t)$ commutes with $\operatorname{Cos}(rt/2^n)$ for $r = 1, \ldots, k$, for some $k \geq 1$. Since

$$\operatorname{Cos}\left(\frac{(k+1)t}{2^n}\right) = 2\operatorname{Cos}\left(\frac{kt}{2^n}\right)\operatorname{Cos}\left(\frac{t}{2^n}\right) - \operatorname{Cos}\left(\frac{(k-1)t}{2^n}\right),$$

it follows that $\operatorname{Cos}(t)$ commutes with $\operatorname{Cos}((k+1)t/2^n)$.

By induction, $\operatorname{Cos}(t)$ commutes with $\operatorname{Cos}(rt)/2^n$ for all integers $r, n \geq 1$. Now b) holds by strong continuity of Cos.

c) This follows easily from b). □

We will frequently consider a cosine function to be extended to $\mathbb{R}$ as in Lemma 3.14.3 c) without further notice.

In what follows, the Laplace integrals of operator-valued functions are interpreted as in Sections 1.4 and 1.5.

Proposition 3.14.4. *Let* $\operatorname{Cos} : \mathbb{R}_+ \to \mathcal{L}(X)$ *be strongly continuous. The following assertions are equivalent:*

(i) Cos *is a cosine function.*

(ii) *One has* $\operatorname{abs}(\operatorname{Cos}) < \infty$, *and there exist* $\omega > \max\{\operatorname{abs}(\operatorname{Cos}), 0\}$ *and an operator* A *such that* $(\omega^2, \infty) \subset \rho(A)$ *and*

$$\lambda R(\lambda^2, A) = \int_0^\infty e^{-\lambda t} \operatorname{Cos}(t)\, dt \qquad (\lambda > \omega). \tag{3.90}$$

In that case, we call A the generator *of the cosine function* Cos.

Proof. We extend Cos to an even function on $\mathbb{R}$. Let $\omega > \max\{0, \mathrm{abs}(\mathrm{Cos})\}$. Then for $\lambda, \mu > \omega$ with $\lambda \neq \mu$, one has

$$\int_0^\infty \int_0^\infty e^{-\lambda t} e^{-\mu s} (\mathrm{Cos}(t+s) + \mathrm{Cos}(t-s))\, ds\, dt = \frac{2}{\mu^2 - \lambda^2} (\mu Q(\lambda) - \lambda Q(\mu)), \tag{3.91}$$

where $Q(\lambda) := \int_0^\infty e^{-\lambda t} \mathrm{Cos}(t)\, dt$. In fact,

$$\begin{aligned}
&\int_0^\infty \int_0^\infty e^{-\lambda t} e^{-\mu s} (\mathrm{Cos}(t+s) + \mathrm{Cos}(t-s))\, ds\, dt \\
&= \int_0^\infty e^{-\lambda t} \left\{ \int_t^\infty e^{-\mu(r-t)} \mathrm{Cos}(r)\, dr + \int_{-\infty}^t e^{-\mu(t-r)} \mathrm{Cos}(r)\, dr \right\} dt \\
&= \int_0^\infty e^{-\mu r} \int_0^r e^{(\mu-\lambda)t}\, dt\ \mathrm{Cos}(r)\, dr + \int_0^\infty e^{-\lambda t} \int_{-\infty}^0 e^{-\mu(t-r)} \mathrm{Cos}(r)\, dr\, dt \\
&\quad + \int_0^\infty e^{-\lambda t} \int_0^t e^{-\mu(t-r)} \mathrm{Cos}(r)\, dr\, dt \\
&= \int_0^\infty e^{-\mu r} (\mu - \lambda)^{-1} (e^{(\mu-\lambda)r} - 1) \mathrm{Cos}(r)\, dr \\
&\quad + \int_{-\infty}^0 \int_0^\infty e^{-(\lambda+\mu)t}\, dt\, e^{\mu r} \mathrm{Cos}(r)\, dr + \int_0^\infty \int_r^\infty e^{-(\lambda+\mu)t}\, dt\, e^{\mu r} \mathrm{Cos}(r)\, dr \\
&= \frac{1}{\mu - \lambda} (Q(\lambda) - Q(\mu)) + \frac{1}{\lambda + \mu} \int_0^\infty e^{-\mu r} \mathrm{Cos}(r)\, dr \\
&\quad + \frac{1}{\lambda + \mu} \int_0^\infty e^{-(\lambda+\mu)r} e^{\mu r} \mathrm{Cos}(r)\, dr \\
&= \frac{1}{\mu - \lambda} (Q(\lambda) - Q(\mu)) + \frac{1}{\lambda + \mu} (Q(\mu) + Q(\lambda)) \\
&= \frac{2}{\mu^2 - \lambda^2} (\mu Q(\lambda) - \lambda Q(\mu)).
\end{aligned}$$

Now assume that $Q(\lambda) = \lambda R(\lambda^2, A)$ $(\lambda > \omega)$. Then

$$\begin{aligned}
\frac{2}{\mu^2 - \lambda^2} (\mu Q(\lambda) - \lambda Q(\mu)) &= 2\lambda\mu \frac{R(\lambda^2, A) - R(\mu^2, A)}{\mu^2 - \lambda^2} \\
&= 2\lambda\mu R(\lambda^2, A) R(\mu^2, A) \\
&= 2 \int_0^\infty e^{-\lambda t} \mathrm{Cos}(t)\, dt \int_0^\infty e^{-\mu s} \mathrm{Cos}(s)\, ds.
\end{aligned}$$

Now (3.91) implies by the Uniqueness Theorem 1.7.3 that

$$\mathrm{Cos}(t+s) + \mathrm{Cos}(t-s) = 2\,\mathrm{Cos}(t)\,\mathrm{Cos}(s) \quad \text{for all } s, t \in \mathbb{R}_+. \tag{3.92}$$

Conversely, assume that Cos is a cosine function. By Lemma 3.14.3, (3.92) holds, and by (3.91),

$$\begin{aligned}\frac{1}{\mu^2-\lambda^2}(\mu Q(\lambda)-\lambda Q(\mu)) &= \int_0^\infty\int_0^\infty e^{-\lambda t}e^{-\mu s}\operatorname{Cos}(t)\operatorname{Cos}(s)\,dt\,ds\\ &= Q(\lambda)Q(\mu).\end{aligned}$$

Let $R(\lambda) := \frac{1}{\sqrt{\lambda}}Q(\sqrt{\lambda})$ for $\lambda > \omega^2$. Then

$$\begin{aligned}R(\lambda)R(\mu) &= \frac{1}{\sqrt{\lambda}\sqrt{\mu}}Q(\sqrt{\lambda})Q(\sqrt{\mu})\\ &= \frac{1}{\sqrt{\lambda}\sqrt{\mu}}\frac{1}{\mu-\lambda}\left(\sqrt{\mu}\,Q(\sqrt{\lambda})-\sqrt{\lambda}\,Q(\sqrt{\mu})\right)\\ &= \frac{1}{\mu-\lambda}(R(\lambda)-R(\mu)) \qquad (\lambda,\mu>\omega^2).\end{aligned}$$

Thus, $\{R(\lambda) : \lambda > \omega^2\}$ is a pseudo-resolvent. If $R(\lambda)x = 0$ for all $\lambda > \omega^2$, then by the Uniqueness Theorem, $\operatorname{Cos}(t)x = 0$ for all $t \in \mathbb{R}_+$. Since $\operatorname{Cos}(0) = I$, this implies that $x = 0$. By Proposition B.6, there exists an operator A such that $(\omega^2, \infty) \subset \rho(A)$ and

$$\lambda R(\lambda^2, A) = \lambda R(\lambda^2) = Q(\lambda) = \int_0^\infty e^{-\lambda t}\operatorname{Cos}(t)\,dt$$

for all $\lambda > \omega$. □

Let Cos be a cosine function on X with generator A. The *sine function* $\operatorname{Sin} : \mathbb{R} \to \mathcal{L}(X)$ associated with Cos is defined by

$$\operatorname{Sin}(t) := \int_0^t \operatorname{Cos}(s)\,ds \qquad (t \in \mathbb{R}). \tag{3.93}$$

This means that $\operatorname{Sin}(t)x = \int_0^t \operatorname{Cos}(s)x\,ds$ ($x \in X$), where the integral is a Bochner integral. Then Sin is an odd function satisfying the functional equation

$$2\operatorname{Sin}(t)\operatorname{Sin}(s) = \int_{t-s}^{t+s} \operatorname{Sin}(r)\,dr \qquad (t, s \in \mathbb{R}). \tag{3.94}$$

This follows from integrating (3.89) twice.

The following functional equation is also useful:

$$\operatorname{Sin}(t+s) = \operatorname{Cos}(s)\operatorname{Sin}(t) + \operatorname{Sin}(s)\operatorname{Cos}(t) \qquad (t, s \in \mathbb{R}). \tag{3.95}$$

To see this, differentiate (3.94) to obtain

$$2\operatorname{Sin}(t)\operatorname{Cos}(s) = \operatorname{Sin}(t+s) + \operatorname{Sin}(t-s).$$

Interchanging s and t gives

$$2\operatorname{Sin}(s)\operatorname{Cos}(t) = \operatorname{Sin}(s+t) + \operatorname{Sin}(s-t).$$

Since Sin is odd, adding these two equations gives (3.95).

Moreover, we deduce from (3.90) that

$$R(\lambda^2, A) = \int_0^\infty e^{-\lambda t}\operatorname{Sin}(t)\,dt \qquad (\lambda > \max\{\omega(\operatorname{Cos}), 0\}).$$

We collect some further properties of a cosine function Cos and its relations with the generator A, and the associated sine function Sin.

Proposition 3.14.5. *The following assertions hold:*

a) $\int_0^t (t-s)\operatorname{Cos}(s)x\,ds \in D(A)$ *and* $A\int_0^t (t-s)\operatorname{Cos}(s)x\,ds = \operatorname{Cos}(t)x - x$ *for all* $x \in X$, $t \in \mathbb{R}$.

b) *If* $x \in D(A)$, *then* $\operatorname{Cos}(t)x, \operatorname{Sin}(t)x \in D(A)$ *and* $A\operatorname{Cos}(t)x = \operatorname{Cos}(t)Ax$, $A\operatorname{Sin}(t)x = \operatorname{Sin}(t)Ax$ *for all* $t \in \mathbb{R}$.

c) *Let* $x, y \in X$. *Then* $x \in D(A)$ *and* $Ax = y$ *if and only if* $\int_0^t (t-s)\operatorname{Cos}(s)y\,ds = \operatorname{Cos}(t)x - x$ *for all* $t \in \mathbb{R}$.

d) $D(A) = \{x \in X : \lim_{t\downarrow 0}\frac{2}{t^2}(\operatorname{Cos}(t)x - x)$ *exists*$\}$ *and* $Ax = \lim_{t\downarrow 0}\frac{2}{t^2}(\operatorname{Cos}(t)x - x)$.

e) *A is densely defined.*

f) *For all* $x \in X$, $s, t \in \mathbb{R}$, *one has* $\operatorname{Sin}(t)\operatorname{Sin}(s)x \in D(A)$ *and*

$$A\operatorname{Sin}(t)\operatorname{Sin}(s)x = \tfrac{1}{2}\left(\operatorname{Cos}(t+s)x - \operatorname{Cos}(t-s)x\right).$$

Proof. a) It follows from Proposition 3.14.1 that $\operatorname{Cos}(\cdot)x$ is a mild solution of $P^2(x, 0)$. This implies a) for $t \geq 0$. It follows for $t < 0$ since Cos is even.

b) Let $\mu \in \rho(A)$. It follows from Proposition 3.1.5 that $R(\mu, A)\operatorname{Cos}(t) = \operatorname{Cos}(t)R(\mu, A)$. This clearly implies b) (see Proposition B.7).

c) Assume that $\int_0^t (t-s)\operatorname{Cos}(s)y\,ds = \operatorname{Cos}(t)x - x \;\; (t \geq 0)$. Taking Laplace transforms, we obtain

$$\frac{1}{\lambda}R(\lambda^2, A)y = \lambda R(\lambda^2, A)x - \frac{x}{\lambda} \qquad (\lambda > \max\{\omega(\operatorname{Cos}), 0\}).$$

Hence, $x \in D(A)$ and $Ax = y$. The converse follows from a) and b).

d) Let $x \in D(A)$, $Ax = y$. It follows from c) that

$$\begin{aligned}\frac{2}{t^2}(\operatorname{Cos}(t)x - x) - y &= \frac{2}{t^2}\int_0^t (t-s)\operatorname{Cos}(s)y\,ds - y \\ &= \frac{2}{t^2}\int_0^t (t-s)\big(\operatorname{Cos}(s)y - y\big)\,ds \to 0 \quad \text{as } t \downarrow 0,\end{aligned}$$

since $\operatorname{Cos}(t)y \to y$ as $t \downarrow 0$. Conversely, let $x, y \in X$ with $\lim_{t\downarrow 0} \frac{2}{t^2}(\operatorname{Cos}(t)x - x) = y$. Then by a),

$$A\frac{2}{t^2}\int_0^t (t-s)\operatorname{Cos}(s)x\,ds = \frac{2}{t^2}(\operatorname{Cos}(t)x - x) \to y \quad \text{as } t \downarrow 0.$$

Since $\frac{2}{t^2}\int_0^t (t-s)\operatorname{Cos}(s)x\,ds \to x$ as $t \downarrow 0$ and since A is closed, it follows that $x \in D(A)$ and $Ax = y$.

e) Since $x = \lim_{t\downarrow 0} \frac{2}{t^2}\int_0^t (t-s)\operatorname{Cos}(s)x\,ds$ for all $x \in X$, it follows from a) that $\overline{D(A)} = X$.

f) This follows from (3.94) and a). □

Using the functional equation (3.88), one sees that a cosine function necessarily has non-negative exponential type. More precisely, the following holds.

Proposition 3.14.6. a) *Let* Cos *be a bounded cosine function and* $x \in X$. *If* $\lim_{t\to\infty} \operatorname{Cos}(t)x = 0$, *then* $x = 0$.

b) *In particular,* $\omega(\operatorname{Cos}) \geq 0$ *for each cosine function* Cos.

Proof. a) Since Cos is bounded, it follows from the assumption that

$$x = \lim_{t\to\infty}(\operatorname{Cos}(2t)x + x) = \lim_{t\to\infty} 2(\operatorname{Cos}(t))^2 x = 0. \qquad \square$$

It is natural to transform the second order problem into a first order system. Let A be an operator on X. Consider the operator $\mathcal{A}$ on $X \times X$ given by

$$\begin{aligned} D(\mathcal{A}) &:= D(A) \times X, \\ \mathcal{A}\begin{pmatrix} x \\ y \end{pmatrix} &:= \begin{pmatrix} 0 & I \\ A & 0 \end{pmatrix}\begin{pmatrix} x \\ y \end{pmatrix} = \begin{pmatrix} y \\ Ax \end{pmatrix}. \end{aligned} \tag{3.96}$$

Here, $X \times X$ is considered with the norm $\|(x,y)\|_{X\times X} = \|x\|_X + \|y\|_X$. Let $\lambda \in \mathbb{C}$ such that $\lambda^2 \in \rho(A)$. Then one easily verifies that $\lambda \in \rho(\mathcal{A})$ and $\lambda \in \rho(-\mathcal{A})$ and

$$R(\lambda, \mathcal{A}) = \begin{pmatrix} \lambda R(\lambda^2, A) & R(\lambda^2, A) \\ AR(\lambda^2, A) & \lambda R(\lambda^2, A) \end{pmatrix}, \tag{3.97}$$

$$R(\lambda, -\mathcal{A}) = \begin{pmatrix} \lambda R(\lambda^2, A) & -R(\lambda^2, A) \\ -AR(\lambda^2, A) & \lambda R(\lambda^2, A) \end{pmatrix}. \tag{3.98}$$

Using this we can prove the following theorem.

Theorem 3.14.7. *The operator A generates a cosine function* Cos *on X if and only if $\mathcal{A}$ generates a once integrated semigroup $\mathcal{S}$ on $X \times X$. In that case, $\mathcal{S}$ is given by*

$$\mathcal{S}(t) = \begin{pmatrix} \operatorname{Sin}(t) & \int_0^t \operatorname{Sin}(s)\,ds \\ \operatorname{Cos}(t) - I & \operatorname{Sin}(t) \end{pmatrix}, \tag{3.99}$$

where $\operatorname{Sin}(t) = \int_0^t \operatorname{Cos}(s)\,ds$.

Proof. Assume that A generates a cosine function Cos. Then

$$AR(\lambda^2, A)/\lambda = \lambda R(\lambda^2, A) - I/\lambda = \int_0^\infty e^{-\lambda t}(\mathrm{Cos}(t) - I)\, dt.$$

It follows from (3.97) that

$$\int_0^\infty e^{-\lambda t}\mathcal{S}(t)\, dt = R(\lambda, \mathcal{A})/\lambda \qquad (\lambda > \omega(\mathrm{Cos})).$$

Thus, $\mathcal{S}$ is a once integrated semigroup and $\mathcal{A}$ is its generator.

Conversely, assume that $\mathcal{A}$ generates a once integrated semigroup. Then $R(\lambda, \mathcal{A})/\lambda$ and hence $AR(\lambda^2, A)/\lambda = \lambda R(\lambda^2, A) - I/\lambda$ is a Laplace transform. Since $1/\lambda = \int_0^\infty e^{-\lambda t}\, dt$, $\lambda R(\lambda^2, A)$ is also a Laplace transform; i.e., A is the generator of a cosine function. □

Corollary 3.14.8. *Let A be the generator of a cosine function* Cos *and let* Sin *be the associated sine function. Let $x, y \in X$ and*

$$u(t) := \mathrm{Cos}(t)x + \mathrm{Sin}(t)y \qquad (t \geq 0).$$

Then u is the unique mild solution of $P^2(x, y)$.

Proof. It follows from (3.87) that u is a mild solution. Let v be another one and define

$$w(t) := \int_0^t (t-s)(u(s) - v(s))\, ds.$$

Then w is a classical solution of $P^2(0,0)$. Consider the function $\phi(t) := \begin{pmatrix} w(t) \\ w'(t) \end{pmatrix}$. Then ϕ is a classical solution of the homogeneous Cauchy problem associated with $\mathcal{A}$ with initial value $\phi(0) = 0$. It follows from Lemma 3.2.9 that $\phi \equiv 0$. □

Let A be a bounded operator. Then

$$\mathrm{Cos}(t) := \sum_{n=0}^\infty \frac{t^{2n}}{(2n)!} A^n$$

defines a continuous function from $\mathbb{R}_+$ into $\mathcal{L}(X)$, and for $\lambda > \sqrt{\|A\|}$ one has

$$\begin{aligned}
\int_0^\infty e^{-\lambda t}\,\mathrm{Cos}(t)\, dt &= \sum_{n=0}^\infty \int_0^\infty e^{-\lambda t} \frac{t^{2n}}{(2n)!}\, dt\ A^n \\
&= \sum_{n=0}^\infty \frac{A^n}{\lambda^{2n+1}} \\
&= \frac{1}{\lambda}\left(I - \frac{A}{\lambda^2}\right)^{-1} \\
&= \lambda R(\lambda^2, A).
\end{aligned}$$

Thus, Cos is a cosine function and A is its generator. As a consequence of Theorem 3.14.7, we may characterize cosine functions with bounded generators as follows.

Corollary 3.14.9. *The following assertions are equivalent:*

(i) *The operator $\mathcal{A}$ generates a C_0-semigroup on $X \times X$.*

(ii) *A generates a cosine function* Cos *such that* $\lim_{t\downarrow 0} \|\operatorname{Cos}(t) - I\| = 0$.

(iii) *A is bounded.*

Proof. (i) $\Rightarrow$ (ii): If $\mathcal{A}$ generates a C_0-semigroup, then $\lim_{t\downarrow 0} \|\mathcal{S}(t)\| = 0$. Now (ii) follows from (3.99).

(ii) $\Rightarrow$ (iii): The Abelian Theorem 4.1.2 implies that $\lim_{\lambda\downarrow 0} \|\lambda^2 R(\lambda^2, A) - I\| = 0$. Hence, $\lambda^2 R(\lambda^2, A)$ is invertible for large λ. Consequently, $D(A) = X$.

(iii) $\Rightarrow$ (i): If A is bounded, then $\mathcal{A}$ is also bounded. $\square$

We now deduce a simple perturbation result which will be considerably improved in Corollary 3.14.13.

Corollary 3.14.10. *Let A be the generator of a cosine function and let $B \in \mathcal{L}(X)$. Then $A + B$ generates a cosine function.*

Proof. Consider the operator

$$\mathcal{B} := \begin{pmatrix} 0 & 0 \\ B & 0 \end{pmatrix}$$

on $X \times X$. Then $\mathcal{B}(X \times X) \subset \{0\} \times X \subset D(\mathcal{A})$. It follows from Corollary 3.10.5 that $\mathcal{A} + \mathcal{B} = \begin{pmatrix} 0 & I \\ A + B & 0 \end{pmatrix}$ generates a once integrated semigroup. Now the claim follows from Theorem 3.14.7. $\square$

We have seen that $\mathcal{A}$ does not generate a C_0-semigroup on $X \times X$ unless A is bounded. However, a semigroup exists on a natural "phase space".

Theorem 3.14.11. *The following assertions are equivalent:*

(i) *The operator A generates a cosine function.*

(ii) *There exists a Banach space V such that $D(A) \hookrightarrow V \hookrightarrow X$ and such that the part $\mathcal{B}$ of $\mathcal{A}$ in $V \times X$ generates a C_0-semigroup.*

In that case, the Banach space V is uniquely determined by (ii). We call $V \times X$ the phase space *associated with A. Moreover, one has* $\operatorname{Sin}(\cdot)y \in C(\mathbb{R}, V)$ *for all $y \in X$,* $\operatorname{Cos}(\cdot)x \in C^1(\mathbb{R}, X) \cap C(\mathbb{R}, V)$ *for all $x \in V$,* $\operatorname{Sin}(\cdot)x \in C(\mathbb{R}, D(A))$ *for all $x \in V$, and $\mathcal{B}$ generates a C_0-group $\mathcal{J}$ on $V \times X$ given by*

$$\mathcal{J}(t) = \begin{pmatrix} \operatorname{Cos}(t) & \operatorname{Sin}(t) \\ \operatorname{Cos}'(t) & \operatorname{Cos}(t) \end{pmatrix} = \begin{pmatrix} \operatorname{Cos}(t) & \operatorname{Sin}(t) \\ A\operatorname{Sin}(t) & \operatorname{Cos}(t) \end{pmatrix} \quad (t \in \mathbb{R}), \tag{3.100}$$

where Sin *is the sine function associated with* Cos, *and* $\operatorname{Cos}'(t)x := \frac{d}{dt}\operatorname{Cos}(t)x = A\operatorname{Sin}(t)x \ (x \in V)$.

Here, $V \times X$ is a Banach space for the norm $\|(x,y)\|_{V\times X} := \|x\|_V + \|y\|_X$. Note that the operator $\mathcal{B}$ on $V \times X$ is defined as follows:

$$\begin{aligned} D(\mathcal{B}) &= D(A) \times V, \\ \mathcal{B}\begin{pmatrix} x \\ y \end{pmatrix} &= \begin{pmatrix} 0 & I \\ A & 0 \end{pmatrix}\begin{pmatrix} x \\ y \end{pmatrix} = \begin{pmatrix} y \\ Ax \end{pmatrix}. \end{aligned}$$

Proof of Theorem 3.14.11. (i) $\Rightarrow$ (ii): Assume that A generates a cosine function. Then $\mathcal{A}$ generates a once integrated semigroup on $X \times X$ (Theorem 3.14.7). By the Sandwich Theorem 3.10.4, there exists a Banach space Z such that $D(\mathcal{A}) \hookrightarrow Z \hookrightarrow X \times X$ and such that the part $\mathcal{B}$ of $\mathcal{A}$ in Z generates a C_0-semigroup $\mathcal{J}$. Define $V := \{x \in X : (x,0) \in Z\}$ with norm $\|x\|_V := \|(x,0)\|_Z$. Then $Z = V \times X$. In fact, let $(x,y) \in Z$. Since $(0,y) \in D(\mathcal{A}) \subset Z$, it follows that $(x,0) = (x,y) - (0,y) \in Z$. Hence, $x \in V$ and $(x,y) \in V \times X$. The converse inclusion is obvious.

Since Z is complete and $Z \hookrightarrow X \times X$, it follows that V is complete. It follows from the closed graph theorem that the embedding $y \mapsto (0,y)$ from X into Z is continuous. So there exists $\beta > 0$ such that $\|(0,y)\|_Z \le \beta \|y\|_X \quad (y \in X)$. Thus,

$$\|(x,y)\|_Z \le \|(x,0)\|_Z + \|(0,y)\|_Z \le \|x\|_V + \beta\|y\|_X.$$

Now it follows from the closed graph theorem that $\|(x,y)\|_{V\times X} := \|x\|_V + \|y\|_X$ defines a norm on $V \times X$ which is equivalent to $\|\cdot\|_Z$. Thus, (ii) is proved.

(ii) $\Rightarrow$ (i): Suppose that V is a Banach space such that $D(A) \hookrightarrow V \hookrightarrow X$ and such that $\mathcal{B}$ generates a C_0-semigroup $\mathcal{J}$ on $V \times X$. It follows from the Sandwich Theorem 3.10.4 that $\mathcal{A}$ generates a once integrated semigroup $\mathcal{S}$. Moreover,

$$\mathcal{S}(t)z = \int_0^t \mathcal{J}(s)z\,ds \quad \text{for all} \quad z \in V \times X,\ t \ge 0.$$

In particular, $\mathcal{S}(\cdot)z \in C(\mathbb{R}_+, D(A) \times V) \cap C^1(\mathbb{R}_+, V \times X)$ for all $z \in V \times X$. By Theorem 3.14.7, the operator A generates a cosine function Cos and $\mathcal{S}$ is given by (3.99). Since Sin is odd and Cos is even, this implies that $\mathrm{Sin}(\cdot)y \in C(\mathbb{R}, V)$ for all $y \in X$, $\mathrm{Cos}(\cdot)x \in C^1(\mathbb{R}, X) \cap C(\mathbb{R}, V)$ and $\mathrm{Sin}(\cdot)x \in C(\mathbb{R}, D(A))$ for all $x \in V$ and that $\mathcal{J}$ is given by the first matrix in (3.100) for $t \ge 0$. In order to prove uniqueness of V, we show that V is equal to the space

$$\widetilde{V} := \{x \in X : \mathrm{Sin}(\cdot)x \in C([0,1], D(A))\}.$$

Note that $\widetilde{V}$ is a Banach space for the norm

$$\|x\|_{\widetilde{V}} = \|x\|_X + \sup_{0\le s\le 1} \|A\,\mathrm{Sin}(s)x\|_X,$$

and we show that this norm is equivalent to $\|\cdot\|_V$. Since $\widetilde{V}$ is completely described by the operator A, this proves uniqueness.

As we noted above, one has $V \subset \widetilde{V}$. It follows from the closed graph theorem that the injection is continuous.

Next, we show that $\|x\|_V \le c\|x\|_{\widetilde{V}}$ for all $x \in V$ and some constant $c \ge 0$. First, since $\operatorname{Sin}(\cdot)y \in C(\mathbb{R}, V)$ we note that, by the closed graph theorem again,

$$\| \operatorname{Sin}(t)y\|_V \le c_1\|y\|_X \tag{3.101}$$

for all $y \in X$, $0 \le t \le 2$ and some constant $c_1 \ge 0$. Let $x \in V$. Then by (3.101) and Proposition 3.14.5 f),

$$\begin{aligned}
\|x\|_V &= \left\| \frac{1}{2}\int_0^2 (x - \operatorname{Cos}(t)x)\,dt + \frac{1}{2}\operatorname{Sin}(2)x \right\|_V \\
&\le \frac{1}{2}\|\operatorname{Sin}(2)x\|_V + \frac{1}{2}\left\| \int_0^2 (x - \operatorname{Cos}(t)x)\,dt \right\|_V \\
&\le c_1\|x\|_X + \sup_{0\le t\le 2} \|x - \operatorname{Cos}(t)x\|_V \\
&\le c_1\|x\|_X + \sup_{\substack{0\le t\le 1\\ 0\le s\le 1}} \|\operatorname{Cos}(t+s)x - \operatorname{Cos}(t-s)x\|_V \\
&= c_1\|x\|_X + 2\sup_{\substack{0\le t\le 1\\ 0\le s\le 1}} \|A\operatorname{Sin}(t)\operatorname{Sin}(s)x\|_V \\
&\le c_1\|x\|_X + 2c_1 \sup_{0\le s\le 1} \|A\operatorname{Sin}(s)x\|_X \\
&\le 2c_1\|x\|_{\widetilde{V}}.
\end{aligned}$$

We have shown that the norms $\|\cdot\|_V$ and $\|\cdot\|_{\widetilde{V}}$ are equivalent on V. It remains to show that $\widetilde{V} \subset V$. Let $x \in \widetilde{V}$. Proposition 3.14.5 b) implies that $\operatorname{Cos}(1)x \in \widetilde{V}$. Since $2\operatorname{Sin}(t)\operatorname{Cos}(1)x = \operatorname{Sin}(1+t)x - \operatorname{Sin}(1-t)x$ (by differentiation of (3.94)), it follows that $\operatorname{Sin}(\cdot)x \in C([0,2], D(A))$. Since the function is odd, one has $\operatorname{Sin}(\cdot)x \in C([-2,2], D(A))$. We have to show that $x \in V$. Since $\operatorname{Sin}(t)x \in V$, it suffices to show that $\lim_{t\downarrow 0} \|\frac{1}{t}\operatorname{Sin}(t)x - x\|_{\widetilde{V}} = 0$. By (3.94), we have

$$\begin{aligned}
&\left\| A\operatorname{Sin}(s)\left(\frac{1}{t}\operatorname{Sin}(t)x - x\right)\right\|_X \\
&\quad = \left\| \frac{1}{2t}\int_{s-t}^{s+t} (A\operatorname{Sin}(r)x - A\operatorname{Sin}(s)x)\,dr \right\|_X \to 0
\end{aligned}$$

as $t \downarrow 0$, uniformly for $s \in [0,1]$. This completes the proof of uniqueness of V.

Finally, we show that $\mathcal{J}$ extends to a C_0-group. We have seen that the semigroup $\mathcal{J}$ is given by (3.100), and we can extend $\mathcal{J}$ to $\mathbb{R}$ by the same formulae. For λ sufficiently large,

$$R(\lambda, \mathcal{A})z = \int_0^\infty e^{-\lambda t}\mathcal{J}(t)z\,dt$$

for $z \in V \times X$. Since Sin is odd and Cos is even, it follows from (3.98) that

$$R(\lambda, -\mathcal{A})z = \int_0^\infty e^{-\lambda t} \mathcal{J}(-t)z\, dt.$$

This shows that $\mathcal{J}$ defined by (3.100) is a C_0-group on $V \times X$ and its generator is the part $\mathcal{B}$ of $\mathcal{A}$ in $V \times X$. □

The phase space can be computed in many concrete cases (see Section 7.2 and Examples 3.14.15 and 3.14.16). The phase space is also important in order to obtain classical solutions of the Cauchy problem of second order.

Corollary 3.14.12. *Let A be the generator of a cosine function* Cos *on X. Denote the associated phase space by $V \times X$, and the corresponding sine function by* Sin. *Let $x \in D(A)$, $y \in V$. Then*

$$u(t) := \mathrm{Cos}(t)x + \mathrm{Sin}(t)y$$

defines a classical solution of $P^2(x, y)$.

Proof. Since $y \in V$, one has $\mathrm{Cos}(\cdot)y \in C^1(\mathbb{R}, X)$, and hence $\mathrm{Sin}(\cdot)y \in C^2(\mathbb{R}, X)$. Since $x \in D(A)$, it follows from Proposition 3.14.5 c) that $\mathrm{Cos}(\cdot)x \in C^2(\mathbb{R}, X)$. Thus, $u \in C^2(\mathbb{R}, X)$. Since u is a mild solution and A is closed, it follows that u is a classical solution. □

The following perturbation result improves Corollary 3.14.10. It will be most useful for applications to elliptic operators given in Chapter 7.

Corollary 3.14.13. *Let A be the generator of a cosine function with phase space $V \times X$. Let $B \in \mathcal{L}(V, X)$. Then $A + B$ generates a cosine function with the same phase space.*

Proof. The operator $\mathcal{A} = \begin{pmatrix} 0 & I \\ A & 0 \end{pmatrix}$ on $V \times X$ with domain $D(A) \times V$ generates a C_0-semigroup on $V \times X$. Since $\mathcal{B} = \begin{pmatrix} 0 & 0 \\ B & 0 \end{pmatrix} \in \mathcal{L}(V \times X)$, it follows from Corollary 3.5.6 that $\begin{pmatrix} 0 & I \\ A + B & 0 \end{pmatrix} = \mathcal{A} + \mathcal{B}$ generates a C_0-semigroup on $V \times X$. It follows from Theorem 3.14.11 that $A + B$ generates a cosine function. □

The following corollary will have useful applications to hyperbolic equations in Chapter 7.

Corollary 3.14.14. *Let A be the generator of a cosine function on X with phase space $V \times X$. Then the part A_V of A in V generates a cosine function on V with phase space $D(A) \times V$.*

Proof. The operator $\mathcal{A} = \begin{pmatrix} 0 & I \\ A & 0 \end{pmatrix}$ on $V \times X$ with domain $D(\mathcal{A}) = D(A) \times V$ generates a C_0-semigroup on $V \times X$. Hence the part $\mathcal{G}$ of $\mathcal{A}$ in $D(\mathcal{A})$ also generates a C_0-semigroup on $D(A) \times V$ (see the remarks preceding Proposition 3.10.3). Note that

$$D(\mathcal{G}) = \{(x, y) \in D(A) \times V : \ (y, Ax) \in D(A) \times V\} = D(A) \times D(A_V).$$

Moreover, we have the continuous embeddings $D(A_V) \hookrightarrow D(A) \hookrightarrow V$. Now replacing X by V and V by $D(A)$ in Theorem 3.14.11 we conclude that A_V generates a cosine function on V with phase space $D(A) \times V$. □

Next, we give two examples where the phase space can be determined easily.

Example 3.14.15. *Let B be the generator of a C_0-group U on X. Then $A := B^2$ generates a cosine function* Cos *on X given by*

$$\operatorname{Cos}(t) := \frac{1}{2}(U(t) + U(-t)) \qquad (t \in \mathbb{R}).$$

The phase space is given by $D(B) \times X$, where $D(B)$ carries the graph norm.

Proof. Let $M, \omega \in \mathbb{R}$ such that $\|U(t)\| \leq Me^{\omega|t|}$ $(t \in \mathbb{R})$. Let $\lambda > \omega$. Then $\lambda \in \rho(B) \cap \rho(-B)$ and $(\lambda^2 - A) = (\lambda - B)(\lambda + B)$. Hence, $\lambda^2 \in \rho(A)$ and

$$\begin{aligned} \lambda R(\lambda^2, A) &= -\lambda R(\lambda, B) R(-\lambda, B) \\ &= \frac{1}{2}(R(\lambda, B) - R(-\lambda, B)) \\ &= \frac{1}{2}\int_0^\infty e^{-\lambda t}(U(t) + U(-t))\, dt \\ &= \int_0^\infty e^{-\lambda t} \operatorname{Cos}(t)\, dt. \end{aligned}$$

This shows that Cos is a cosine function with generator A.

For $x \in D(B)$ one has $\operatorname{Cos}'(t)x = \frac{1}{2}(BU(t)x - BU(-t)x)$. Thus, (3.100) defines a strongly continuous, exponentially bounded function $\mathcal{J} : \mathbb{R}_+ \to \mathcal{L}(D(B) \times X)$ whose Laplace transform is the restriction of $R(\lambda, \mathcal{A})$ to $D(B) \times X$, by (3.97). Thus, $\mathcal{J}$ is a C_0-semigroup whose generator is the part of $\mathcal{A}$ in $D(B) \times X$. So $D(B) \times X$ is the phase space, by Theorem 3.14.11. □

Example 3.14.16. *Let A be a selfadjoint operator on a Hilbert space H. Assume that A is* bounded above*; i.e., $(Ax|x)_H \leq \omega\|x\|_H^2$ for all $x \in D(A)$ and some $\omega \in \mathbb{R}$. Then A generates a cosine function.*

Proof. By the Spectral Theorem B.13, we can assume that $H = L^2(\Omega, \mu)$, $Af = mf$, $D(A) = \{f \in H : \ mf \in H\}$, where $m : \Omega \to \mathbb{R}$ is a measurable function. Moreover, the boundedness assumption implies that $m(y) \leq \omega$ for μ-almost all $y \in \Omega$.

First case: Assume that $m \le 0$ a.e. Let $q(y) := i\sqrt{-m(y)}$, $Bf := qf$, $D(B) := \{f \in H : qf \in H\}$. Then B generates the C_0-group U given by $U(t)f := e^{tq}f$. Since $B^2 = A$, it follows from the preceding example that A generates a cosine function Cos. Moreover,

$$\begin{aligned}(\operatorname{Cos}(t)f)(y) &= \frac{1}{2}\left((U(t)+U(-t))\,f\right)(y)\\ &= \frac{1}{2}\left(e^{it\sqrt{-m(y)}}+e^{-it\sqrt{-m(y)}}\right)f(y)\\ &= \cos(t\sqrt{-m(y)})f(y) \qquad (t\in\mathbb{R},\ f\in H,\ y\in\Omega).\end{aligned}$$

Second case: When $m \le \omega$ a.e., we can write $m = m_1 + m_2$ where $m_1 \le 0$, $0 \le m_2 \le \omega$. Then $A = A_1 + B$ where $A_1 f := m_1 f$, $D(A_1) = D(A)$ and $Bf := m_2 f$ defines a bounded operator. It follows from Corollary 3.14.10 and the first case that A generates a cosine function. □

Next, we show that there is always a simple way to go from the second order equation to the first order equation.

Theorem 3.14.17. *Let A be the generator of a cosine function* Cos. *Then A generates a holomorphic C_0-semigroup T of angle $\pi/2$.*

We give two different proofs of this theorem. The first uses the characterization theorem for holomorphic semigroups. However, it does not give the best possible angle.

First proof. Let $M, \omega \ge 0$ such that $\|\operatorname{Cos}(t)\| \le Me^{\omega t}$ $(t \ge 0)$. Then by holomorphic continuation (Proposition B.5) for $\operatorname{Re}\lambda > \omega$ one has $\lambda^2 \in \rho(A)$ and

$$\|\lambda R(\lambda^2, A)\| = \left\|\int_0^\infty e^{-\lambda t}\operatorname{Cos}(t)\,dt\right\| \le \frac{M}{\operatorname{Re}\lambda - \omega}.$$

Let $\omega_1 := 2\omega^2$. Let $\mu \in \mathbb{C}$ such that $\operatorname{Re}\mu > \omega_1$. Write $\mu = re^{i\theta}$, where $-\pi/2 < \theta < \pi/2$, and let $\lambda := \sqrt{r}e^{i\theta/2}$. Then

$$\operatorname{Re}\lambda = \sqrt{r}\cos\frac{\theta}{2} = \sqrt{r}\sqrt{\frac{\cos\theta+1}{2}} \ge \sqrt{\frac{r\cos\theta}{2}} \ge \sqrt{\frac{\omega_1}{2}} = \omega.$$

Hence, $\mu = \lambda^2 \in \rho(A)$ and

$$\begin{aligned}\|\mu R(\mu, A)\| \le |\lambda|\frac{M}{\operatorname{Re}\lambda - \omega} = \frac{M\sqrt{r}}{\sqrt{r}\cos(\theta/2) - \omega} &= \frac{M}{\cos(\theta/2) - \omega/\sqrt{r}}\\ &\le \frac{M}{\frac{1}{\sqrt{2}} - \frac{\omega}{\sqrt{r}}} \le M2\sqrt{2}\end{aligned}$$

if $\frac{1}{\sqrt{2}} - \frac{\omega}{\sqrt{r}} \ge \frac{1}{2\sqrt{2}}$; i.e., if $r \ge 8\omega^2$.

We have shown that $\mu \in \rho(A)$ and $\|\mu R(\mu, A)\| \le M 2\sqrt{2}$ whenever $\operatorname{Re}\mu > \omega_1$ and $|\mu| \ge 8\omega^2$. By Corollary 3.7.17, this implies that A generates a holomorphic semigroup T. Since $D(A)$ is dense, T is a C_0-semigroup. □

The second proof has the advantage of giving a formula which allows one to compute the semigroup T from the cosine function Cos. In fact, we will prove the *Weierstrass formula*

$$T(t)x = \int_0^\infty \frac{e^{-s^2/4t}}{\sqrt{\pi t}} \operatorname{Cos}(s)x\, ds \qquad (t > 0). \tag{3.102}$$

In the context of Example 3.14.15, this formula was established in Corollary 3.7.15.

Second proof of Theorem 3.14.17. Define T by (3.102). Then $T(t) \in \mathcal{L}(X)$ and $T(\cdot)x \in C^\infty((0,\infty), X)$ for all $x \in X$. Let $x \in X$. We show that $\lim_{t\downarrow 0} T(t)x = x$. In fact, putting $s = r\sqrt{t}$ gives

$$T(t)x = \int_0^\infty \frac{e^{-r^2/4}}{\sqrt{\pi}} \operatorname{Cos}(r\sqrt{t})x\, dr \to x \quad \text{as } t \downarrow 0$$

by the dominated convergence theorem, since $\operatorname{Cos}(r\sqrt{t})x \to x$, $\|\operatorname{Cos}(r\sqrt{t})\| \le Me^{\omega r}$ for all $0 < t < 1$, for some M and ω, and $\frac{1}{\sqrt{\pi}} \int_0^\infty e^{-r^2/4}\, dr = 1$.

Let $\lambda > \omega$. Then by Lemma 1.6.7,

$$\begin{aligned} \int_0^\infty e^{-\lambda t} T(t)x\, dt &= \int_0^\infty \int_0^\infty e^{-\lambda t} \frac{e^{-s^2/4t}}{\sqrt{\pi t}}\, dt\, \operatorname{Cos}(s)x\, ds \\ &= \int_0^\infty \frac{1}{\sqrt{\lambda}} e^{-\sqrt{\lambda}s} \operatorname{Cos}(s)x\, ds \\ &= R(\lambda, A)x. \end{aligned}$$

It follows from Definition 3.1.8 that T is a C_0-semigroup and A is its generator.

In order to show that T is holomorphic of angle $\pi/2$, we define

$$T(z)x = \frac{1}{\sqrt{\pi z}} \int_0^\infty e^{-s^2/4z} \operatorname{Cos}(s)x\, ds \qquad (\operatorname{Re} z > 0).$$

Then $T : \mathbb{C}_+ \to \mathcal{L}(X)$ is holomorphic. Let $\theta \in (0, \frac{\pi}{2})$. According to Definition 3.7.1, it remains to show that

$$\sup_{\substack{z\in\Sigma_\theta \\ |z|\le 1}} \|T(z)\| < \infty.$$

Let $z \in \Sigma_\theta$, $|z| \le 1$. Let $u := \operatorname{Re} z/|z|^2$. Then

$$\left\|e^{-s^2/4z} \operatorname{Cos}(s)\right\| \le Me^{-us^2/4} e^{\omega s} = Me^{-u(s-2\omega/u)^2/4} e^{\omega^2/u}.$$

Hence,

$$\begin{aligned}
\|T(z)\| &= \left\| \frac{1}{2\sqrt{\pi z}} \int_{\mathbb{R}} e^{-s^2/4z} \operatorname{Cos}(s)\, ds \right\| \\
&\le \frac{1}{\sqrt{|z|}} \frac{M}{2\sqrt{\pi}} \int_{\mathbb{R}} e^{-u(s-2\omega/u)^2/4}\, ds\, e^{\omega^2/u} \\
&= \frac{1}{\sqrt{|z|}} \frac{M}{\sqrt{\pi}} \int_{\mathbb{R}} e^{-s^2}\, ds\, \frac{1}{\sqrt{u}} e^{\omega^2/u} \\
&= \frac{M}{\sqrt{|z|}} \frac{|z|}{\sqrt{\operatorname{Re} z}} e^{\omega^2 |z|^2/\operatorname{Re} z} \\
&\le \frac{M}{\sqrt{\cos\theta}} e^{\omega^2|z|/\cos\theta} \le \frac{M}{\sqrt{\cos\theta}} e^{\omega^2/\cos\theta}.
\end{aligned}$$

This proves the claim. □

The converse of Theorem 3.14.17 is not true: a generator of a holomorphic C_0-semigroup does not necessarily generate a cosine function. Indeed, generators of cosine functions satisfy a very restrictive spectral condition.

Proposition 3.14.18. *Let A be the generator of a cosine function* Cos. *Then there exists $\omega > 0$ such that the spectrum $\sigma(A)$ of A is contained in the parabola $\{\xi + i\eta : \eta \in \mathbb{R},\ \xi \le \omega^2 - \eta^2/4\omega^2\}$.*

Proof. There exist $\omega, M > 0$ such that

$$\|\operatorname{Cos}(t)\| \le Me^{\omega t} \qquad (t \ge 0).$$

Since $\lambda R(\lambda^2, A) = \int_0^\infty e^{-\lambda t} \operatorname{Cos}(t)\, dt$ for $\lambda > \omega$, it follows from holomorphic continuation (Proposition B.5) that $\lambda^2 \in \rho(A)$ whenever $\operatorname{Re}\lambda > \omega$. It is easy to see that

$$\{\lambda^2 : \lambda \in \mathbb{C},\ \operatorname{Re}\lambda > \omega\} \supset \left\{\xi + i\eta : \eta \in \mathbb{R},\ \xi > \omega^2 - \frac{\eta^2}{4\omega^2}\right\}.$$

□

Example 3.14.19. Let $H := L^2(\mathbb{R})$, $m(s) := -|s| + is$ $(s \in \mathbb{R})$,

$$(Af)(s) := m(s)f(s), \quad D(A) := \{f \in H : mf \in H\}.$$

Then A generates a holomorphic C_0-semigroup on H. Since $\sigma(A) = \{-s \pm is : s \in \mathbb{R}_+\}$, the spectrum of A is not contained in any parabola as described in Proposition 3.14.18. Thus, A does not generate a cosine function. □

3.15 Sine Functions and Real Characterization

In this section, we consider sine functions. These include the integrals of cosine functions as in Section 3.14, and elementary properties of sine functions will be used to prove the generation theorem for cosine functions which is analogous to the Hille-Yosida theorem. We shall also establish some useful perturbation results. Examples of sine functions occurring in applications will be given in the Notes of Chapter 7 and in Chapter 8.

Definition 3.15.1. *An operator A on X* generates a sine function *if there exist $\omega, M \geq 0$ and a strongly continuous function* $\mathrm{Sin} : \mathbb{R}_+ \to \mathcal{L}(X)$ *such that the following properties are satisfied:*

a) $\left\|\int_0^t \mathrm{Sin}(s)\, ds\right\| \leq Me^{\omega t}$ $(t \geq 0)$.

b) $\lambda^2 \in \rho(A)$ *whenever* $\lambda > \omega$.

c) $R(\lambda^2, A) = \int_0^\infty e^{-\lambda t}\, \mathrm{Sin}(t)\, dt$ $(\lambda > \omega)$.

We then call Sin *the* sine function *generated by A.*

If A generates a cosine function Cos, then

$$R(\lambda^2, A) = \frac{1}{\lambda}\int_0^\infty e^{-\lambda t}\, \mathrm{Cos}(t)\, dt = \int_0^\infty e^{-\lambda t} \int_0^t \mathrm{Cos}(s)\, ds\, dt$$

for $\lambda > \omega(\mathrm{Cos})$. Thus A generates the sine function Sin given by $\mathrm{Sin}(t)x := \int_0^t \mathrm{Cos}(s)x\, ds$. In other words, the definition is consistent with the previous notion of Section 3.14.

Next, we establish some relations between a sine function and its generator.

Proposition 3.15.2. *Let* Sin *be a sine function and A be its generator. Then the following hold:*

a) $\int_0^t (t-s)\, \mathrm{Sin}(s)x\, ds \in D(A)$ *and*

$$A\int_0^t (t-s)\, \mathrm{Sin}(s)x\, ds = \mathrm{Sin}(t)x - tx \tag{3.103}$$

for all $x \in X$.

b) *If $x \in D(A)$, then* $\mathrm{Sin}(t)x \in D(A)$ *and* $A\,\mathrm{Sin}(t)x = \mathrm{Sin}(t)Ax$ *for all $t \geq 0$.*

c) *Let $x, y \in X$. Then $x \in D(A)$ and $Ax = y$ if and only if*

$$\int_0^t (t-s)\, \mathrm{Sin}(s)y\, ds = \mathrm{Sin}(t)x - tx \qquad (t \geq 0). \tag{3.104}$$

Proof. a) It follows from (3.87) that $S(\cdot)x$ is a mild solution of $P^2(0,x)$. This is precisely the claim.

b) This follows from Proposition 3.1.5 and Proposition B.7.

c) Let $x, y \in X$ such that (3.104) holds. Taking Laplace transforms on both sides gives

$$\frac{1}{\lambda^2} R(\lambda^2, A)y = R(\lambda^2, A)x - \frac{x}{\lambda^2}.$$

Hence, $x \in D(A)$ and $y = \lambda^2 x - (\lambda^2 - A)x = Ax$. The converse implication follows from a) and b). $\square$

We now prove the characterization theorem for generators of cosine functions. It is mainly of theoretical interest since the condition (3.105) seems to be difficult to verify in concrete cases.

Theorem 3.15.3. *Let A be a densely defined operator on a Banach space X. The following assertions are equivalent:*

(i) *A is the generator of a cosine function.*

(ii) *There exist $\omega, M \geq 0$ such that $(\omega^2, \infty) \subset \rho(A)$ and*

$$\frac{1}{k!} \left\| (\lambda - \omega)^{k+1} \left(\lambda R(\lambda^2, A)\right)^{(k)} \right\| \leq M \tag{3.105}$$

for all $\lambda > \omega$ and $k \in \mathbb{N}_0$.

Proof. (i) $\Rightarrow$ (ii): Assume that A generates a cosine function Cos. There exist $M \geq 0$, $\omega \geq 0$ such that $\|\operatorname{Cos}(t)\| \leq Me^{\omega t}$. Since $\lambda R(\lambda^2, A)x = \int_0^\infty e^{-\lambda t} \operatorname{Cos}(t)x \, dt$ for all $\lambda > \omega$ and $x \in X$, the claim follows from Theorem 2.4.1.

(ii) $\Rightarrow$ (i): Assume that (ii) is satisfied. By Theorem 2.4.1, there exists a function $S : \mathbb{R}_+ \to \mathcal{L}(X)$ satisfying

$$\|S(t+h) - S(t)\| \leq M \int_t^{t+h} e^{\omega s} \, ds \qquad (t, h \geq 0) \tag{3.106}$$

such that

$$R(\lambda^2, A) = \int_0^\infty e^{-\lambda t} S(t) \, dt \qquad (\lambda > \omega).$$

Thus, S is a sine function and A is its generator. Let $x \in D(A)$. Then by Proposition 3.15.2, one has

$$S(t)x - tx = \int_0^t (t-s)S(s)Ax \, ds \qquad (t \geq 0).$$

It follows that $S(\cdot)x \in C^1(\mathbb{R}_+, X)$. Since $D(A)$ is dense in X, it follows from Lemma 3.3.3 that $S(\cdot)x \in C^1(\mathbb{R}_+, X)$ for all $x \in X$. Let $C(t)x := \frac{d}{dt} S(t)x$ ($x \in X, t \geq 0$). It follows that $C(t)$ is linear and by (3.106) that

$$\|C(t)\| \leq Me^{\omega t} \qquad (t \geq 0).$$

Integration by parts shows that

$$\lambda R(\lambda^2, A)x = \int_0^\infty e^{-\lambda t} C(t)x\, dt \qquad (\lambda > \omega)$$

for all $x \in X$. Thus, C is a cosine function and A is its generator. $\square$

Now we resume our investigation of sine functions. The following result parallels Example 3.14.15.

Proposition 3.15.4. *Let A be an operator such that A and $-A$ generate once integrated semigroups. Then A^2 generates a sine function. Moreover, the sine function is exponentially bounded if both integrated semigroups are.*

Proof. There exists $\omega \geq 0$ such that $(\omega, \infty) \subset \rho(A) \cap \rho(-A)$ and $\frac{1}{\lambda}(\lambda - A)^{-1}$ and $\frac{1}{\lambda}(\lambda + A)^{-1}$ are Laplace transforms (in the sense of Definition 3.1.4). Thus,

$$(\lambda^2 - A^2)^{-1} = (\lambda - A)^{-1}(\lambda + A)^{-1}$$

exists for $\lambda > \omega$, and by the resolvent equation we have

$$(\lambda^2 - A^2)^{-1} = \frac{1}{2\lambda}\left((\lambda + A)^{-1} + (\lambda - A)^{-1}\right).$$

Thus, $(\lambda^2 - A^2)^{-1}$ is a Laplace transform. This shows that A^2 generates a sine function Sin given by $\mathrm{Sin}(t) = \frac{1}{2}(S_+(t) + S_-(t))$, where S_+ and S_- are the once integrated semigroups generated by A and $-A$, respectively. $\square$

The next example shows that the generator of a sine function is not necessarily densely defined.

Example 3.15.5. Let B be the generator of the translation group U on $L^1(\mathbb{R})$, given by $(U(t)f)(x) = f(x-t)$. Then B^* and $-B^*$ generate once integrated semigroups on $L^\infty(\mathbb{R}) = L^1(\mathbb{R})^*$ (see Corollary 3.3.7 and Example 3.3.10). By Proposition 3.15.4, $(B^*)^2$ generates a sine function. However, $D\left((B^*)^2\right) \subset D(B^*) = W^{1,\infty}(\mathbb{R})$, which is not dense in $L^\infty(\mathbb{R})$ (cf. also Theorem 4.3.18). $\square$

Next, we establish a perturbation result for sine functions. Our proof is completely different from the corresponding results on cosine functions (Corollary 3.14.10 and Corollary 3.14.13).

Theorem 3.15.6. *Let A be the generator of an exponentially bounded sine function and let $B \in \mathcal{L}(\overline{D(A)}, X)$. Then $A + B$ generates an exponentially bounded sine function.*

Proof. Denote by Sin the sine function generated by A and let $M, \omega \geq 0$ such that $\|\mathrm{Sin}(t)\| \leq Me^{\omega t}$ $(t \geq 0)$. The idea of the proof is to solve the integral equation

$$S^B(t) = \mathrm{Sin}(t) + \int_0^t S^B(s) B\,\mathrm{Sin}(t-s)\, ds. \tag{3.107}$$

First, we remark that $\operatorname{Sin}(t)x \in \overline{D(A)}$ for all $t \geq 0$, $x \in X$. In fact by Proposition 3.15.2, $\int_0^t (t-s)\operatorname{Sin}(t)x\,ds \in D(A)$ for all $t > 0$. Differentiating twice we obtain that $\operatorname{Sin}(t)x \in \overline{D(A)}$. Let $\alpha > 0$. Multiplying (3.107) by $e^{-(\omega+\alpha)t}$, we obtain the equivalent integral equation

$$U(t) = e^{-(\omega+\alpha)t}\operatorname{Sin}(t) + \int_0^t U(s)Be^{-(\omega+\alpha)(t-s)}\operatorname{Sin}(t-s)\,ds. \tag{3.108}$$

Consider now the Banach space

$$\mathcal{C} := \{V : \mathbb{R}_+ \to \mathcal{L}(X) : \ V \text{ is strongly continuous and bounded}\}$$

with norm $\|V\| := \sup_{t\geq 0}\|V(t)\|$. Suppose from now on that $\alpha > M\|B\|$. Then

$$(JV)(t) := \int_0^t V(s)Be^{-(\omega+\alpha)(t-s)}\operatorname{Sin}(t-s)\,ds$$

defines an operator J on $\mathcal{C}$ with norm $\|J\|_{\mathcal{L}(\mathcal{C})} \leq \frac{M\|B\|}{\alpha} < 1$. The integral equation (3.108) can now be written in the form

$$(I-J)U = W$$

with $W(t) := e^{-(\omega+\alpha)t}\operatorname{Sin}(t)$. Hence, it has a unique solution U (given by $(I-J)^{-1}W$). Therefore $S^B(t) := e^{(\omega+\alpha)t}U(t)$ defines a solution of (3.107), which is exponentially bounded.

For $\lambda > \omega+\alpha$, let $Q(\lambda) := \int_0^\infty e^{-\lambda t}S^B(t)\,dt$. Then

$$\begin{aligned}
Q(\lambda) - R(\lambda^2, A) &= \int_0^\infty e^{-\lambda t}(S^B(t) - \operatorname{Sin}(t))\,dt \\
&= \int_0^\infty e^{-\lambda t}\int_0^t S^B(s)B\operatorname{Sin}(t-s)\,ds\,dt \\
&= \int_0^\infty \int_0^t e^{-\lambda s}S^B(s)Be^{-\lambda(t-s)}\operatorname{Sin}(t-s)\,ds\,dt \\
&= \int_0^\infty \int_s^\infty e^{-\lambda s}S^B(s)Be^{-\lambda(t-s)}\operatorname{Sin}(t-s)\,dt\,ds \\
&= \int_0^\infty e^{-\lambda s}S^B(s)B\int_0^\infty e^{-\lambda t}\operatorname{Sin}(t)\,dt\,ds \\
&= Q(\lambda)BR(\lambda^2, A).
\end{aligned}$$

Hence, $Q(\lambda)(I - BR(\lambda^2, A)) = R(\lambda^2, A)$. The operator $I - BR(\lambda^2, A)$ is invertible since

$$\begin{aligned}
\|BR(\lambda^2, A)\| &\leq \|B\|\left\|\int_0^\infty e^{-\lambda t}\operatorname{Sin}(t)\,dt\right\| \\
&\leq \|B\|M(\lambda-\omega)^{-1} < 1.
\end{aligned}$$

Since $(\lambda^2 - A - B) = (I - BR(\lambda^2, A))(\lambda^2 - A)$, it follows that $\lambda^2 \in \rho(A+B)$ and

$$\begin{aligned} R(\lambda^2, A+B) &= R(\lambda^2, A)(I - BR(\lambda^2, A))^{-1} \\ &= Q(\lambda) \\ &= \int_0^\infty e^{-\lambda t} S^B(t)\, dt \end{aligned}$$

for all $\lambda > \omega + \alpha$. This shows that S^B is a sine function and $A+B$ is its generator. $\square$

3.16 Square Root Reduction for Cosine Functions

Let B be the generator of a C_0-group and let $\omega \in \mathbb{R}$. Then by Corollary 3.14.10 and Example 3.14.15, the operator $A := B^2 + \omega$ generates a cosine function. The aim of this section is to establish the following remarkable converse result: If A is the generator of a cosine function on a UMD-space X, then there exist a generator B of a C_0-group and a number $\omega \geq 0$ such that $A = B^2 + \omega$ (see Corollary 3.16.8). By the results of Section 3.8, there is a square root B of $A - \omega$ for ω sufficiently large. We will show that B and $-B$ always generate integrated semigroups, but the UMD-property is needed in order to obtain a C_0-group. We recall the following lemma which is easy to prove.

Lemma 3.16.1. *Let B be a closed operator on X, and $A := B^2$. Let $\lambda \in \mathbb{C}$. If $\lambda^2 \in \rho(A)$, then $\lambda \in \rho(\pm B)$ and*

$$\begin{aligned} R(\lambda, B) &= (\lambda + B) R(\lambda^2, A), \\ R(\lambda, -B) &= (\lambda - B) R(\lambda^2, A). \end{aligned}$$

Proposition 3.16.2. *Let B be a closed operator. Assume that $A := B^2$ generates a cosine function. Then B and $-B$ generate exponentially bounded once integrated semigroups.*

Proof. Denote by Cos and Sin the cosine and the sine functions associated with A. Recall from Proposition 3.14.5 a) that $\int_0^t \mathrm{Sin}(s)x\, ds \in D(A)$ and

$$A \int_0^t \mathrm{Sin}(s)x\, ds = \mathrm{Cos}(t)x - x$$

for all $x \in X$, $t \geq 0$. Define

$$V_+(t)x := \mathrm{Sin}(t)x + B \int_0^t \mathrm{Sin}(s)x\, ds \qquad (t \geq 0). \tag{3.109}$$

Let $\lambda \in \rho(A)$. Then

$$\begin{aligned} B\int_0^t \operatorname{Sin}(s)x\,ds &= BR(\lambda,A)(\lambda - A)\int_0^t \operatorname{Sin}(s)x\,ds \\ &= \lambda BR(\lambda,A)\int_0^t \operatorname{Sin}(s)x\,ds - BR(\lambda,A)(\operatorname{Cos}(t)x - x) \end{aligned}$$

for all $x \in X$. Since $BR(\lambda, A)$ is bounded, it follows that V_+ is strongly continuous and exponentially bounded. Moreover, for large $\lambda > 0$,

$$\begin{aligned} \int_0^\infty e^{-\lambda t} V_+(t)x\,dt &= R(\lambda^2, A)x + \frac{1}{\lambda}BR(\lambda^2,A)x \\ &= \frac{1}{\lambda}(\lambda + B)R(\lambda^2,A)x = \frac{1}{\lambda}R(\lambda,B)x, \end{aligned} \tag{3.110}$$

by the preceding lemma. We have shown that V_+ is a once integrated semigroup and B is its generator. Replacing B by $-B$ shows that $-B$ also generates a once integrated semigroup. □

Now we assume that B is a closed operator such that B^2 generates a cosine function. We want to investigate conditions under which this implies that B generates a C_0-group. Here is a characterization in terms of the behaviour of the sine function on $D(B)$, which is considered as a Banach space for the graph norm.

Proposition 3.16.3. *Let A be the generator of a cosine function* Cos *with associated sine function* Sin. *Assume that B is a closed operator such that $B^2 = A$. Then the following are equivalent:*

(i) *B generates a C_0-group U.*

(ii) *For all $x \in X$, $\operatorname{Sin}(t)x \in D(B)$ for almost all $t \in (0,\infty)$.*

(iii) *$\operatorname{Sin}(\cdot)x \in C(\mathbb{R}, D(B))$ for all $x \in X$.*

(iv) *The phase space of* Cos *is $D(B) \times X$.*

In that case, the group U is given by

$$U(t)x = \operatorname{Cos}(t)x + B\operatorname{Sin}(t)x \tag{3.111}$$

for all $t \in \mathbb{R}$, $x \in X$.

For the proof we need the following results which are of independent interest.

Lemma 3.16.4. *Let $U : \mathbb{R} \to \mathcal{L}(X)$ be a mapping such that*

a) *$U(t+s) = U(t)U(s)$ $(t, s \in \mathbb{R})$,*

b) *$U(0) = I$, and*

c) *$U(\cdot)x$ is measurable for all $x \in X$.*

Then U is strongly continuous.

Proof. a) We show that $M := \sup_{s\in[1,2]} \|U(s)\| < \infty$. Otherwise, by the uniform boundedness principle, there exist $x \in X$, $s_n \in [1,2]$ such that $\|U(s_n)x\| \geq n$ $(n \in \mathbb{N})$. Considering a subsequence if necessary, we can and do assume that $\lim_{n\to\infty} s_n =: \gamma$ exists. Since $\|U(\cdot)x\|$ is measurable, there exists a constant $M_1 \geq 0$ and a measurable set $F \subset [0,\gamma]$ with Lebesgue measure $m(F) > \gamma/2$ such that $\sup_{t\in F} \|U(t)x\| \leq M_1$. Let

$$E_n := \big\{(s_n - t) : \; t \in F \cap [0, s_n]\big\}.$$

Then $m(E_n) \geq \gamma/2$ for large $n \in \mathbb{N}$. Now for $t \in F \cap [0, s_n]$ we have

$$\begin{aligned} n \leq \|U(s_n)x\| &\leq \|U(s_n - t)\| \, \|U(t)x\| \\ &\leq \|U(s_n - t)\| M_1. \end{aligned}$$

Hence, $\|U(s)\| \geq n/M_1$ for all $s \in E_n$. Let $E = \bigcap_{k\in\mathbb{N}} \bigcup_{n\geq k} E_n$. Then $m(E) \geq \gamma/2$ and $\|U(s)\| = \infty$ for all $s \in E$. This is a contradiction.

b) By the group property and a), one has for each $t \in \mathbb{R}$,

$$\sup_{s\in[t+1,t+2]} \|U(s)\| = \sup_{s\in[t+1,t+2]} \|U(s-t)U(t)\| \leq M\|U(t)\| \qquad (t \in \mathbb{R}).$$

Thus U is locally bounded. It follows that for each $x \in X$, $U(\cdot)x$ is locally Bochner integrable.

c) Let $x \in X$, $t_0 \in \mathbb{R}$. Then

$$U(t_0)x = U(t)U(t_0 - t)x$$

and

$$U(t_0 + h)x = U(t)U(t_0 + h - t)x \qquad (t \in [1,2], \; h \in \mathbb{R}).$$

Hence,

$$\begin{aligned} \|U(t_0)x - U(t_0+h)x\| &= \left\| \int_1^2 U(t)\big(U(t_0 - t)x - U(t_0 + h - t)x\big)\, dt \right\| \\ &\leq M \int_1^2 \|U(t_0 - t)x - U(t_0 + h - t)x\| \, dt \\ &\to 0 \qquad \text{as } h \to 0, \end{aligned}$$

using b) and the continuity of shifts on $L^1(\mathbb{R}, X)$. □

Lemma 3.16.5. *Let $F \subset \mathbb{R}$ be a Lebesgue measurable set such that $F + F \subset F$. If $\mathbb{R} \setminus F$ is a null set, then $F = \mathbb{R}$.*

Proof. Assume that $N := \mathbb{R} \setminus F \neq \emptyset$. Let $a \in N$. Then for $x \in F$, the assumption implies that $a - x \in N$ (otherwise, $a = x + (a - x) \in F + F \subset F$). We have shown that $F \subset a - N$, which is impossible since $a - N$ is a null set. □

Proof of Proposition 3.16.3. Assume that (i) holds. One has

$$\operatorname{Cos}(t)x = \frac{1}{2}(U(t)x + U(-t)x)$$

and

$$\operatorname{Sin}(t)x = \frac{1}{2}\int_0^t (U(s) + U(-s))x\,ds.$$

It follows from Proposition 3.1.9 e) that $\operatorname{Sin}(t)x \in D(B)$ and

$$\begin{aligned} B\operatorname{Sin}(t)x &= \frac{1}{2}\big(U(t)x - x - (U(-t)x - x)\big) \\ &= \frac{1}{2}(U(t)x - U(-t)x). \end{aligned}$$

This proves (3.111). We have seen in Example 3.14.15 that (i) implies (iv). Theorem 3.14.11 shows that (iv) implies (iii). The implication (iii) ⇒ (ii) is trivial.

It remains to show that (ii) ⇒ (i). It is obvious that for $\mu \in \rho(B)$ one has

$$R(\mu, B)R(\lambda^2, A) = R(\lambda^2, A)R(\mu, B) \qquad (\lambda > \omega(\operatorname{Cos})).$$

Thus, by Proposition 3.1.5 and Proposition B.7, $x \in D(B)$ implies that $\operatorname{Sin}(t)x$, $\operatorname{Cos}(t)x \in D(B)$ and

$$B\operatorname{Sin}(t)x = \operatorname{Sin}(t)Bx, \quad B\operatorname{Cos}(t)x = \operatorname{Cos}(t)Bx.$$

Recall from (3.95) that

$$\operatorname{Sin}(t+s) = \operatorname{Cos}(s)\operatorname{Sin}(t) + \operatorname{Cos}(t)\operatorname{Sin}(s) \qquad (t, s \in \mathbb{R}). \tag{3.112}$$

Let $x \in X$ and $F := \{t \in \mathbb{R} : \ \operatorname{Sin}(t)x \in D(B)\}$. By hypothesis, $\mathbb{R} \setminus F$ has Lebesgue measure zero. It follows from (3.112) that $F + F \subset F$. Now we conclude from Lemma 3.16.5 that $F = \mathbb{R}$. Define $U(t)x := \operatorname{Cos}(t)x + B\operatorname{Sin}(t)x$ for all $x \in X$, $t \in \mathbb{R}$. Since B is closed, $U(t)$ is also closed. Hence, $U(t) \in \mathcal{L}(X)$. If $x \in D(B)$, then $U(\cdot)x$ is continuous. Since $D(B) \supset D(A)$ which is dense, it follows that $U(\cdot)x$ is measurable for all $x \in X$ (in fact, let $x \in X$, $x_n \in D(B)$ such that $\lim_{n\to\infty} x_n = x$; then $U(t)x = \lim_{n\to\infty} U(t)x_n$). It remains to show the group property. Recall from Proposition 3.14.5 f) that $\operatorname{Sin}(t)\operatorname{Sin}(s)x \in D(A)$ and

$$A\operatorname{Sin}(t)\operatorname{Sin}(s)x = \frac{1}{2}\big(\operatorname{Cos}(t+s)x - \operatorname{Cos}(t-s)x\big) \tag{3.113}$$

for all $t, s \in \mathbb{R}$, $x \in X$. Hence by (3.112),

$$\begin{aligned} & U(t)U(s) \\ = \;& \operatorname{Cos}(t)\operatorname{Cos}(s) + \operatorname{Cos}(t)B\operatorname{Sin}(s) + B\operatorname{Sin}(t)\operatorname{Cos}(s) + B\operatorname{Sin}(t)B\operatorname{Sin}(s) \\ = \;& \operatorname{Cos}(t)\operatorname{Cos}(s) + B\operatorname{Sin}(t+s) + A\operatorname{Sin}(t)\operatorname{Sin}(s) \\ = \;& \frac{1}{2}(\operatorname{Cos}(t+s) + \operatorname{Cos}(t-s)) + B\operatorname{Sin}(t+s) + \frac{1}{2}(\operatorname{Cos}(t+s) - \operatorname{Cos}(t-s)) \\ = \;& \operatorname{Cos}(t+s) + B\operatorname{Sin}(t+s) \\ = \;& U(t+s). \end{aligned}$$

It follows from Lemma 3.16.4 that U is a C_0-group. Since $V_+(t) = \int_0^t U(s)\,ds$ for $t \geq 0$, where V_+ is given by (3.109), we have by (3.110),

$$\int_0^\infty e^{-\lambda t} U(t)x\,dt = \lambda \int_0^\infty e^{-\lambda t} V_+(t)x\,dt = R(\lambda, B)x$$

for large λ. This shows that B is the generator of U. □

We will show that condition (ii) of Proposition 3.16.3 is automatically satisfied if X is a UMD-space (as defined in Section 3.12), $-A$ satisfies the hypotheses of Proposition 3.8.2, and $B = i(-A)^{1/2}$ (as defined in Section 3.8). For this we need some preparation concerning the Hilbert transform.

For $\omega \geq 0$ we define

$$L^\infty_\omega(\mathbb{R}, X) := \left\{ f \in L^1_{loc}(\mathbb{R}, X) : \; \|f\|_{\omega,\infty} := \operatorname*{ess\,sup}_{t\in\mathbb{R}} \|e^{-\omega|t|} f(t)\| < \infty \right\}.$$

Then $L^\infty_\omega(\mathbb{R}, X)$ is a Banach space for the norm $\|\cdot\|_{\omega,\infty}$.

Lemma 3.16.6. *Let X be a* UMD*-space and let $0 \leq \omega < c$. For $f \in L^\infty_\omega(\mathbb{R}, X)$ and $\varepsilon > 0$ define the continuous function $H^c_\varepsilon f : \mathbb{R} \to X$ by*

$$(H^c_\varepsilon f)(t) = \int_{|s|\geq\varepsilon} \frac{e^{-c|s|}}{s} f(t-s)\,ds.$$

Then for each $\tau > 0$, $\lim_{\varepsilon\downarrow 0} H^c_\varepsilon f =: H^c f$ exists in $L^2((-\tau,\tau), X)$. Hence, H^c is a bounded operator from $L^\infty_\omega(\mathbb{R}, X)$ into $L^2((-\tau,\tau), X)$ for each $\tau > 0$.

Proof. Let $\tau \geq 1$, $0 < \varepsilon \leq 1$, $|t| \leq \tau$. Denote by χ the characteristic function of $[-\tau - 1, \tau + 1]$. Then

$$\begin{aligned} (H^c_\varepsilon f)(t) &= \int_{|s-t|\geq\varepsilon} \frac{e^{-c|t-s|}}{t-s} f(s)\,ds \\ &= h_{1\varepsilon}(t) + h_{2\varepsilon}(t) + h_{3\varepsilon}(t), \end{aligned}$$

where

$$\begin{aligned} h_{1\varepsilon}(t) &:= \int_{|s-t|\geq\varepsilon} \frac{1}{t-s}\chi(s)f(s)\,ds, \\ h_{2\varepsilon}(t) &:= \int_{|s-t|\geq\varepsilon} \frac{e^{-c|t-s|}-1}{t-s}\chi(s)f(s)\,ds, \\ h_3(t) &:= \int_{|s|\geq\tau+1} \frac{e^{-c|t-s|}}{t-s}f(s)\,ds. \end{aligned}$$

It is clear that $\|h_3(t)\| \leq c_1\|f\|_{\omega,\infty}$ for $|t| \leq \tau$, where $c_1 \geq 0$ is a constant. Next, observe that $\chi f \in L^2(\mathbb{R}, X)$. If $\psi(t) := \frac{1}{t}(e^{-c|t|}-1)$, then $\psi \in L^2(\mathbb{R})$ and $h_{2\varepsilon}$ is the convolution $((1-\chi_{(-\varepsilon,\varepsilon)})\psi) * (\chi f)$. It follows from Proposition 1.3.2 that $h_{2\varepsilon}$ converges uniformly on $[-\tau,\tau]$ to $\psi * (\chi f)$ as $\varepsilon \downarrow 0$. Finally, let H be the Hilbert transform on $L^2(\mathbb{R}, X)$ and $H_{\varepsilon R}$ be as in Section 3.12. Then $h_{1\varepsilon} = \pi H_{\varepsilon R}(\chi f)|_{[-\tau,\tau]}$ for $R > 2\tau + 1$. Hence, $h_{1\varepsilon} \to \pi H(\chi f)|_{[-\tau,\tau]}$ in $L^2((-\tau,\tau), X)$ as $\varepsilon \downarrow 0$. Thus the first assertion is proved. The second now follows from the Banach-Steinhaus theorem. □

Now we are able to prove the main result.

Theorem 3.16.7 (Fattorini). *Let A be the generator of a cosine function on a* UMD-*space X. Assume that* $(0,\infty) \subset \rho(A)$ *and* $\sup_{\lambda>0}\|\lambda R(\lambda,A)\| < \infty$. *Define* $(-A)^{1/2}$ *as in Proposition* 3.8.2 *and let* $B = i(-A)^{1/2}$. *Then B generates a C_0-group and* $B^2 = A$.

Proof. It is immediate from Proposition 3.8.2 that $B^2 = A$. We want to show condition (ii) of Proposition 3.16.3. Let $c > \omega > \omega(\mathrm{Cos})$. For $t > 0$ and $y \in D(A)$, Proposition 3.8.2 gives

$$\begin{aligned} B\,\mathrm{Sin}(t)y &= \frac{1}{\pi i}\int_0^\infty \lambda^{-1/2}R(\lambda,A)\,\mathrm{Sin}(t)Ay\,d\lambda \\ &= I_1(t)y + I_2(t)y, \end{aligned}$$

where

$$\begin{aligned} I_1(t)y &:= \frac{1}{\pi i}\int_0^{c^2} \lambda^{-1/2}R(\lambda,A)\,\mathrm{Sin}(t)Ay\,d\lambda \\ &= \frac{1}{\pi i}\int_0^{c^2} \lambda^{-1/2}(\lambda R(\lambda,A)\,\mathrm{Sin}(t)y - \mathrm{Sin}(t)y)\,d\lambda \end{aligned}$$

and

$$I_2(t)y := \frac{1}{\pi i}\int_{c^2}^\infty \lambda^{-1/2}R(\lambda,A)\,\mathrm{Sin}(t)Ay\,d\lambda.$$

Since $\omega > \omega(\mathrm{Cos})$, there exists $M \geq 0$ such that $\|\mathrm{Cos}(s)\| \leq Me^{\omega s}$ $(s \geq 0)$. Consequently, $\|\mathrm{Sin}(s)\| \leq Mse^{\omega s}$ $(s \geq 0)$ and

$$\left\|\int_\varepsilon^\infty e^{-\sqrt{\lambda}s}\,\mathrm{Sin}(s)\,\mathrm{Sin}(t)Ay\,ds\right\| \leq c_1\int_0^\infty e^{-\sqrt{\lambda}s}se^{\omega s}\,ds = c_1\frac{1}{(\sqrt{\lambda}-\omega)^2}$$

for all $\varepsilon > 0$, where $c_1 := M\|\operatorname{Sin}(t)Ay\|$. Since the function $\lambda \mapsto \lambda^{-1/2}(\sqrt{\lambda}-\omega)^{-2}$ is in $L^1(c^2, \infty)$, and since

$$R(\lambda, A)\operatorname{Sin}(t)Ay = \int_0^\infty e^{-\sqrt{\lambda}s}\operatorname{Sin}(s)\operatorname{Sin}(t)Ay\,ds,$$

we obtain from the dominated convergence theorem,

$$I_2(t)y = \lim_{\varepsilon\downarrow 0}\frac{1}{\pi i}\int_{c^2}^\infty \lambda^{-1/2}\int_\varepsilon^\infty e^{-\sqrt{\lambda}s}\operatorname{Sin}(s)\operatorname{Sin}(t)Ay\,ds\,d\lambda.$$

Now we apply Fubini's theorem and obtain

$$I_2(t)y = \lim_{\varepsilon\downarrow 0}\frac{2}{\pi i}\int_\varepsilon^\infty \frac{e^{-cs}}{s}\operatorname{Sin}(s)\operatorname{Sin}(t)Ay\,ds.$$

By (3.113), this gives

$$\begin{aligned} I_2(t)y &= \lim_{\varepsilon\downarrow 0}\frac{1}{\pi i}\int_\varepsilon^\infty \frac{e^{-cs}}{s}(\operatorname{Cos}(t+s)-\operatorname{Cos}(t-s))y\,ds \\ &= \lim_{\varepsilon\downarrow 0}\frac{i}{\pi}\int_{|s|\geq\varepsilon}\frac{e^{-c|s|}}{s}\operatorname{Cos}(t-s)y\,ds. \end{aligned}$$

At this point we use that X is a UMD-space, so that we can apply Lemma 3.16.6. We have shown that $I_2(\cdot)y = H_\varepsilon^c(\operatorname{Cos}(\cdot)y)$ for $y \in D(A)$. Let $x \in X$, and let $y_n \in D(A)$ such that $\lim_{n\to\infty} y_n = x$. Then $\operatorname{Cos}(\cdot)y_n$ converges to $\operatorname{Cos}(\cdot)x$ in $L_\omega^\infty(\mathbb{R}, X)$. It follows from Lemma 3.16.6 that $I_2(\cdot)y_n$ converges in $L^2((-\tau,\tau), X)$ as $n \to \infty$ for all $\tau > 0$. Considering a subsequence if necessary, we obtain that $I_2(t)y_n$ converges a.e. in X as $n \to \infty$. Since $I_1(t)$ is a bounded operator it follows that there exists a measurable subset F of $\mathbb{R}$ such that $\mathbb{R}\setminus F$ has Lebesgue measure 0 and $B\operatorname{Sin}(t)y_n$ converges in X as $n \to \infty$ for all $t \in F$. Since B is closed, this implies that $\operatorname{Sin}(t)x \in D(B)$ for all $t \in F$. This finishes the proof. □

Now we obtain the following interesting characterization of generators of cosine functions on UMD-spaces.

Corollary 3.16.8. *Let A be an operator on a* UMD-*space. The following assertions are equivalent:*

(i) *A generates a cosine function.*

(ii) *There exist a generator B of a C_0-group and $\omega \geq 0$ such that $A = B^2 + \omega$.*

Proof. (ii) ⇒ (i): This follows from Example 3.14.15 and Corollary 3.14.10.

(i) ⇒ (ii): Assume that A generates a cosine function Cos. Then A generates a C_0-semigroup by Theorem 3.14.17. Hence, there exists $\omega \geq 0$ such that $(\omega, \infty) \subset \rho(A)$ and $\sup_{\lambda>0}\|\lambda R(\lambda, A-\omega)\| < \infty$. The operator $A-\omega$ is also the generator of a cosine function by Corollary 3.14.10. It follows from Theorem 3.16.7 that there exists a C_0-group with generator B such that $A - \omega = B^2$. □

We conclude this section with some remarks. First, we mention that, given an operator A, there may be infinitely many generators of groups whose square is A. We give an example.

Example 3.16.9. Let $X = l^2$, and define A by

$$\begin{aligned} D(A) &:= \left\{x = (x_n)_{n\in\mathbb{N}} \in l^2 : \ (n^2 x_n)_{n\in\mathbb{N}} \in l^2\right\}, \\ (Ax)_n &:= -n^2 x_n. \end{aligned}$$

Let $(\epsilon_n)_{n\in\mathbb{N}}$ be an arbitrary sequence in $\{-1, 1\}$. Define B on X by $Bx := (i\epsilon_n n x_n)_{n\in\mathbb{N}}$ with domain $D(B) := \{x : \ (nx_n)_{n\in\mathbb{N}} \in l^2\}$. Then B generates the C_0-group U given by

$$U(t)x := (e^{i\epsilon_n nt} x_n)_{n\in\mathbb{N}}$$

and $B^2 = A$. □

It should be mentioned that even if the generator A of a cosine function has a square root which generates a C_0-group there may be other square roots which do not generate C_0-groups. We give an example.

Example 3.16.10. Let A be the generator of a cosine function on a Banach space X. Assume that A is unbounded. Then we know from Theorem 3.14.7 that the operator $\mathcal{A}$ on $X \times X$ given by

$$\begin{aligned} D(\mathcal{A}) &= D(A) \times X, \\ \mathcal{A} &= \begin{pmatrix} 0 & I \\ A & 0 \end{pmatrix}, \end{aligned}$$

does not generate a C_0-semigroup. However, the operator $\mathcal{A}^2$ is given by

$$\begin{aligned} D(\mathcal{A}^2) &= D(A) \times D(A); \\ \mathcal{A}^2 &= \begin{pmatrix} A & 0 \\ 0 & A \end{pmatrix}. \end{aligned}$$

Thus, $\mathcal{A}^2$ generates a cosine function. If A has a square root which generates a C_0-group, then so does $\mathcal{A}^2$, but this square root is different from $\mathcal{A}$. □

There exists an example of a generator A of a cosine function on a Banach space X (which can even be chosen to be reflexive) such that

$$A - \omega \neq B^2$$

for each $\omega \geq 0$ and each generator B of a C_0-group. Of course in that case, X is not a UMD-space. We refer to [Fat85, Section III.8] for such examples.

It is interesting to observe that the question whether a square root reduction exists for a cosine function is equivalent to the existence of a boundary group for a certain holomorphic semigroup.

To be more precise, let A be the generator of a cosine function. Then by Theorem 3.14.17, A generates a holomorphic C_0-semigroup T of angle $\pi/2$. Let $\omega > \omega(T)$. Then $-(\omega - A)^{1/2}$ generates a C_0-semigroup V (see Theorem 3.8.3). Now it follows from Corollary 3.9.10 that $B := i(\omega - A)^{1/2}$ generates a C_0-group if and only if V is a holomorphic semigroup of angle $\pi/2$ and V has a boundary group (in the sense of Proposition 3.9.1). This is not always the case (if X fails to have the UMD property) as we pointed out above. However, we know from Proposition 3.16.2 that V always has an integrated boundary group (in the sense explained in Theorem 3.9.13).

3.17 Notes

Section 3.1

The characterization of mild solutions in terms of Laplace transforms given in Theorem 3.1.3 appeared in [Neu94]. It is at the heart of the theory and it may have been known for a long time. Another way to define mild solutions by approximation schemes ("bonnes solutions") has been given by Bénilan, Crandall and Pazy [BCP88]; see also [BCP90]. This way is most important in non-linear theory where Laplace transforms have not yet been used effectively.

The idea of defining semigroups and their generators directly by the property that the Laplace transform is a resolvent, is from [Are87b], but Theorem 3.1.7 is already contained in [DS59, Corollary VIII.1.16]. We should also mention the approach to semigroups by Laplace transforms given by Kisyński [Kis76].

The characterization of generators in terms of well-posedness of the abstract Cauchy problem for classical solutions is contained in [Kre71]. Related results are contained in the lecture notes of van Casteren [Cas85]. Example 3.1.14 is taken from [Nag86, A-II, Example 1.4].

One may modify Theorem 3.1.12 by requiring a less restrictive existence assumption but a stronger uniqueness assumption by considering the Cauchy problem on a bounded interval $[0, \tau]$:

$$C(\tau) \qquad \begin{cases} u'(t) = & Au(t) \quad (t \in [0,\tau]), \\ u(0) = & x. \end{cases}$$

As before, we call $u \in C([0,\tau], X)$ a mild solution if $\int_0^t u(s)\,ds \in D(A)$ and $u(t) - x = A\int_0^t u(s)\,ds$ for all $t \in [0,\tau]$. Then the following holds.

Theorem 3.17.1. *Let A be a closed operator, $\tau > 0$. The following assertions are equivalent:*

(i) *For all $x \in X$ there exists a unique mild solution of $C(\tau)$.*

(ii) *The operator A generates a C_0-semigroup.*

Other results on the local problem were obtained by Lyubich [Lyu66], Sova [Sov68], Oharu [Oha71] and Sanekata [San75].

Section 3.2

Integrated semigroups were introduced by Arendt in [Are84], [Are87a]. The systematic treatment based on techniques of Laplace transforms was developed by Arendt [Are87b], Neubrander [Neu88] and Kellermann [Kel86]. Theorem 3.2.13 is due to Arendt, Neubrander and Schlotterbeck [ANS92] and Lemma 3.2.14 to Arendt, El-Mennaoui and Keyantuo [AEK94]. The theory of integrated semigroups has been developed in many directions and we refer to the monographs of deLaubenfels [deL94] and Xiao and Liang [XL98] for further information. We do not aim to give an account of the many contributions, but we mention here a few extensions of the theory. First, one may consider α-times integrated semigroups for some $\alpha \in \mathbb{R}_+$; i.e., the power k in Definition 3.2.1 is replaced by some $\alpha \geq 0$ (see [Hie91a]). This leads to sharper regularity results for the solution of the associated Cauchy problem than those which we have described. Moreover, a modification of the Real Representation Theorem 2.2.1 shows that if a C^∞-function $r : (0, \infty) \to X$ satisfies

$$\sup\left\{\|\lambda^{\alpha+1} r^{(n)}(\lambda)/n!\| : \lambda > 0, n \in \mathbb{N} \cup \{0\}\right\} < \infty$$

for some $\alpha > 0$, then there exists a Hölder continuous function F of exponent α satisfying $F(0) = 0$ such that

$$r(\lambda) = \lambda^\alpha \int_0^\infty e^{-\lambda t} F(t)\, dt \qquad (\lambda > 0)$$

(see [Hie91b]). Integrated Volterra equations were considered by Arendt and Kellermann [AK89] and deLaubenfels [deL90]. Second order problems in integrated form have been developed by Kellermann and Hieber [KH89], Arendt and Kellermann [AK89] and Neubrander [Neu89a].

For integrated solutions of implicit differential equations, see the papers of Arendt and Favini [AF93] and Knuckles and Neubrander [KN94]. Concerning the general theory of degenerate differential equations in Banach spaces we refer to the monograph by Favini and Yagi [FY99].

Many years before integrated semigroups were studied, Lions [Lio60] introduced *distribution semigroups.* They have been further developed by Chazarain [Cha71] and Beals [Bea72] who also give applications to partial differential equations. Further information can be found for example in the monograph of Fattorini [Fat83].

In terms of integrated semigroups, Lions's concept may now be described as follows: Let S be a k-times integrated semigroup on X. Denote by $\mathcal{D}(\mathbb{R}_+)$ the test functions on $\mathbb{R}$ with support in $\mathbb{R}_+$. For $\varphi \in \mathcal{D}(\mathbb{R}_+)$ define the operator $T \in \mathcal{L}(X)$ by

$$T(\varphi) := (-1)^k \int_0^\infty \varphi^{(k)}(t) S(t)\, dt.$$

Then $T : \mathcal{D}(\mathbb{R}_+) \to \mathcal{L}(X)$ is a mapping satisfying the semigroup property

$$T(\varphi * \psi) = T(\varphi) T(\psi).$$

It turns out that T defined as above is a *distribution semigroup of exponential type* and that all distribution semigroups of exponential type as defined by Lions in [Lio60] are of this form (see [AK89] for a proof).

More generally, arbitrary distribution semigroups are defined as mappings U : $\mathcal{D}(\mathbb{R}_+) \to \mathcal{L}(X)$ such that

$$U(\varphi * \psi) = U(\varphi) U(\psi)$$

and several other properties are satisfied. To each distribution semigroup one can associate a generator A. These distribution semigroups are equivalent to *local integrated semigroups*. More precisely, a closed operator A generates a distribution semigroup U if and only if there exists $k \in \mathbb{N}$ such that the integrated Cauchy problem $(ACP)_{k+1}$ (see (3.20)) is well-posed on a bounded interval; i.e., A generates a local integrated semigroup of order k. The generators of local integrated semigroups can be completely characterized by spectral properties. See papers of Arendt, El-Mennaoui and Keyantuo [AEK94] or Okazawa and Tanaka [TanO90] for these and related results. These concepts have been extended further in a series of papers by Lumer [Lum90], [Lum92], [Lum94], [Lum97], Cioranescu and Lumer [CL94] and Lumer and Neubrander [LN99], [LN97], and in the monograph of Melnikova and Filinkov [MF01].

A further concept, namely *regularized semigroups*, had been developed by Da Prato [DaP66], and was rediscovered by Davies and Pang [DP87]. Regularized semigroups were extensively studied by deLaubenfels [deL94] and Miyadera and Tanaka [TM92], among many others and are now called C-semigroups, in general. Given a bounded operator C one says, in the language of Laplace transforms, that an operator A generates a *C-semigroup* if $CR(\lambda, A)$ is a Laplace transform (see [HHN92]). Thus, by Proposition 3.2.7, the generator of a k-times integrated semigroup is the same as a $R(\mu, A)^k$-semigroup, where $\mu \in \rho(A)$. Note, however, that in contrast to the situation of integrated semigroups, the generator of a C-semigroup may have an empty resolvent set. Operators with empty resolvent set occur in particular in the context of *Petrovskii correct systems* of partial differential equations. Systems of this type can be treated by the theory of C-semigroups in a very efficient manner (see [HHN92]). Again, we refer to the monographs of deLaubenfels [deL94] and Xiao and Liang [XL98] for further information.

Another concept, which leads to a generalization of generators of bounded semigroups and groups is the following.

A k-times integrated semigroup S is called *tempered* if $\|S(t)\| \leq ct^k$ $(t \geq 0)$ for some $c \geq 0$. Arendt and Kellermann [AK89] showed that a densely defined operator A generates a k-times integrated tempered semigroup if and only if A generates a *smooth distribution semigroup of order* k as introduced by Balabane and Emami-Rad [BE79], [BE85]. Smooth distribution groups allow a spectral calculus similar to the one known for bounded groups, as shown by Balabane, Emami-Rad and Jazar [BEJ93] and Jazar [Jaz95].

It was established in [KH89] that perturbation theory for integrated semigroups is more complicated than for semigroups. We saw in Proposition 3.2.6 that perturbation by a scalar yields a quite complicated formula for the integrated semigroup. In general, generators of integrated semigroups are not even stable under bounded perturbations (in contrast to generators of C_0-semigoups, by Corollary 3.5.6). For instance, let A be the unbounded generator of a cosine function (see Section 3.14), $\mathcal{A}$ be the operator given by (3.96) and $\mathcal{B}$ be the bounded operator on $X \times X$ given by $\mathcal{B}\begin{pmatrix} x \\ y \end{pmatrix} = \begin{pmatrix} -y \\ 0 \end{pmatrix}$. Then $\mathcal{A}$ generates a once integrated semigroup, by Theorem 3.14.7. However, the range of $\mu - (\mathcal{A} + \mathcal{B})$ is contained in $D(A) \times X$ for any $\mu \in \mathbb{C}$, so $\rho(\mathcal{A} + \mathcal{B})$ is empty and $\mathcal{A} + \mathcal{B}$ does not generate a k-times integrated semigroup for any $k \in \mathbb{N}$.

One perturbation theorem for integrated semigroups can be found in Corollary 3.10.5, and we mention another one here. Using Theorem 3.2.8 and an argument similar to Proposition 3.11.12, Kaiser and Weis [KW03] have shown that $A + B$ generates a

$(k+2)$-times integrated semigroup if A generates an exponentially bounded k-times integrated semigroup S and $B : D(A) \to X$ satisfies $\sup\{\|BR(\lambda, A)\| : \operatorname{Re}\lambda = \omega\} < 1$ for some $\omega > \omega(S)$.

Finally, we mention that, more generally, instead of Cauchy problems involving a function and its derivatives at an instant of time, problems with memory may be considered. They lead to the theory of Volterra equations. Their solutions are governed by one-parameter families of operators, more general than semigroups or cosine functions. Vector-valued Laplace transforms also play a decisive role in this theory. We refer to the monograph of Prüss [Prü93] for a comprehensive presentation of Volterra equations. A study of regularized solutions of Volterra equations was initiated in [Liz00].

Section 3.3

The Hille-Yosida theorem (in its general form due to Hille, Yosida, Feller, Miyadera and Phillips) was the starting point of the subsequent development of the theory of semigroups. The classical approach of Yosida using what is now called the *Yosida approximation* and the classical method of Hille based on the convergence of the exponential formula $T(t)x = \lim_{n\to\infty}(I - \frac{t}{n}A)^{-n}x$ can be found in many textbooks on semigroup theory (see, for example, those of Clément et al. [CHA87], Davies [Dav80], [Dav07], Engel and Nagel [EN00], [EN06], Fattorini [Fat83], Goldstein [Gol85], Hille and Phillips [HP57], Kato [Kat66], Nagel et al. [Nag86], Pazy [Paz83] and Yosida [Yos80]). Hille and Phillips already mentioned explicitly on page 364 of [HP57] the problem of how to use Widder's theorem in the proof of the Hille-Yosida theorem. The approach presented here based on the real representation theorem for Laplace transforms solves this problem. Our presentation follows the lines of Arendt [Are87b].

A different condition which is sufficient for an operator to generate a C_0-semigroup has been given by Gomilko [Gom99] and Shi and Feng [SF00]. This condition has the advantage of involving only the square of the resolvent and not higher powers.

Theorem 3.17.2. *Let A be a densely defined operator on a Banach space X and suppose that there exist K and ω such that $\sigma(A) \subset \{\lambda \in \mathbb{C} : \operatorname{Re}\lambda \le \omega\}$ and*

$$\int_{-\infty}^{\infty} \left|\langle R(a+is, A)^2 x, x^*\rangle\right| \, ds \le \frac{K\|x\|\,\|x^*\|}{a-\omega}$$

whenever $a > \omega$, $x \in X$ and $x^* \in X^*$. Then A generates a C_0-semigroup T satisfying $\|T(t)\| \le Me^{\omega t}$ $(t \ge 0)$ for some M.

This theorem can be proved by using the Poisson representation of functions in a Hardy space on a half-plane to establish the conditions of the Hille-Yosida theorem. Alternatively, a complex inversion formula can be used to define $tT(t)x$ when $x \in D(A^2)$ and then the operators $T(t)$ extend by continuity to a C_0-semigroup. The converse of the theorem is true in Hilbert spaces. However, the derivative operator generates the C_0-group of translations on $L^p(\mathbb{R})$, but it does not satisfy the hypotheses of Theorem 3.17.2 when $p \ne 2$.

Proposition 3.3.8 appeared first in a paper of Kato [Kat59]. We mention here that Proposition 3.3.8 is not true if merely the Radon-Nikodym property is assumed instead of reflexivity (see [Are87a]). However, we have the following result [Are87b].

Proposition 3.17.3. *Let X be a Banach space with the Radon-Nikodym property. Let A be an operator on X satisfying the Hille-Yosida condition* (3.23). *Then A generates a semigroup in the sense of Definition* 3.2.5.

The sun-dual $T^{\odot}$ of a C_0-semigroup T was already introduced by Hille and Phillips in [HP57]. Dual semigroups have been investigated systematically by Clément, Dieckmann, Gyllenberg, Thieme and van Neerven. For a comprehensive treatment and precise references, see the monograph by van Neerven [Nee92].

Section 3.4

Dissipative operators and the Lumer-Phillips theorem are classical objects in semigroup theory. Many further results in this direction can be found in the textbooks listed in the Notes of the previous section. It is also possible to characterize generators of positive C_0-semigroups of contractions by a similar notion, namely dispersiveness and a range condition. Even more generally, the norm may be replaced by a "half-norm" as studied by Arendt, Chernoff and Kato [ACK82], Batty and Robinson [BR84] and Nagel et al. [Nag86, Section A-II.2]. If A is a densely defined dissipative operator and $x \in D(A)$, then $\mathrm{Re}\langle Ax, x^* \rangle \leq 0$ for all $x^* \in \mathrm{dN}(x)$ (instead of merely some $x^* \in \mathrm{dN}(x)$ as required by the definition). This is easy to see when A generates a contraction semigroup. For densely defined operators, it was proved in [Bat78]; see also [ACK82, Theorem 2.5].

A very interesting class of contraction semigroups is the class of Ornstein-Uhlenbeck semigroups which arises naturally in the study of stochastic processes. For a thorough study of this class of contraction semigroups, we refer to the book by Lorenzi and Bertoldi [LB07].

The Laplacian with Dirichlet boundary conditions is an easy example of an elliptic operator in divergence form. Generalizations of the arguments used in Example 3.4.7 lead to the theory of quadratic forms. For more information on this topic we refer to the books of Kato [Kat66], Dautray and Lions [DL90] and Davies [Dav80], [Dav95], and the survey article [Are04].

Section 3.5

Hille-Yosida operators were studied by Kato already in [Kat59]. Later, these operators were investigated systematically by Sinestrari [Sin85] and Da Prato and Sinestrari [DS87], where Theorem 3.5.2 was first proved. A proof of the Da Prato-Sinestrari Theorem 3.5.2 based on the theory of integrated semigroups was given by Kellermann and Hieber [KH89]. Theorem 3.5.2 may also be proved via the theory of non-linear semigroups, as shown by Bénilan, Crandall and Pazy [BCP88]. In fact, operators which are not densely defined are natural in the framework of the Crandall-Liggett-Bénilan theorem which corresponds to the Hille-Yosida theorem for non-linear operators. For a third, very different, approach using "abstract Sobolev towers", see the work of Nagel and Sinestrari [NS94]. The renorming Lemma 3.5.4 can be found in Pazy's book [Paz83] and the bounded perturbation Theorem 3.5.5 for Hille-Yosida operators is due to Kellermann and Hieber [KH89].

Section 3.6

The Trotter-Kato theorem is a classical result in semigroup theory. It is frequently proved with the help of the Hille-Yosida theorem (see, for example, [Paz83]). The proof which we give here is based on the approximation theorem for Laplace transforms from Section 1.7, and is due to Jun Xiao and Liang [XL00] who also proved a general Trotter-Kato approximation theorem for integrated semigroups after previous work by Lizama [Liz94] and Nicaise [Nic93]. For other variants of the Trotter-Kato theorem, we refer to Bobrowski's survey article [Bob97b].

Section 3.7

The monographs by Amann [Ama95] and Lunardi [Lun95] are specialized texts on holomorphic semigroups and parabolic problems. Much further information, in particular on maximal regularity, interpolation and extrapolation scales of Banach spaces as well as many applications to non-linear problems can be found there. Our proof of the characterization Theorem 3.7.11 uses only Laplace transform theory. Corollary 3.7.14 is taken from the lecture notes of Nagel et al. [Nag86]. The perturbation Theorem 3.7.25 is due to Desch and Schappacher [DS88] with a different proof. If A generates a C_0-semigroup on a reflexive space and $B : D(A) \to X$ is compact, then the estimate (3.49) holds (see [DS88]) or [EN00, p.179]). However, this is no longer true if X is not reflexive (see [Hes70] for an example). It was also shown in [DS88] (see also [AB06], [AB07], [DSS09]) that if $A + B$ generates a C_0-semigroup for every $B \in \mathcal{L}(D(A), X)$ of rank 1, then A generates a holomorphic C_0-semigroup.

Section 3.8

Fractional powers of sectorial operators are classical objects in semigroup theory, and they now have a very extensive theory (see [MS01] and [Haa06]). Theorem 3.8.1 and many further results in this direction can be found in the books of Amann [Ama95], Haase [Haa06] and others listed in the Notes on Section 3.3, and in the original papers of Balakrishnan [Bal60], Komatsu [Kom66] and others. The assumption that A is a generator is not essential in Theorem 3.8.3. If A is densely defined and $B = -A$ satisfies the assumptions of Proposition 3.8.2, then $-(-A)^{1/2}$ generates a bounded holomorphic C_0-semigroup; see for example, the books of Fattorini [Fat83] or Martínez and Sanz [MS01]. For a sectorial operator B and $\operatorname{Re} z > 0$, one may define B^z to be the algebraic inverse of B^{-z}. If A generates a bounded C_0-semigroup and $0 < \alpha < 1$, then $-(-A)^{\alpha}$ generates a bounded holomorphic C_0-semigroup.

Section 3.9

Boundary values of holomorphic C_0-semigroups already appeared in the book of Hille and Phillips [HP57]. The method which we use here is based on Laplace transform techniques. Theorem 3.9.4 was proved by Hörmander [Hör60], but our approach follows the lines of Arendt, El-Mennaoui and Hieber [AEH97]. Basic information on the Riemann-Liouville semigroup and on fractional integration is in [HP57]; the idea of using the transference principle in this context is in [AEH97]. Results similar to Theorem 3.9.13 were first obtained by Boyadzhiev and deLaubenfels [BdL93] and later improved by El-Mennaoui in [Elm92]. The proof of the implication (ii) $\Rightarrow$ (i) of Theorem 3.9.13 follows the lines of [Elm92]. Corollary 3.9.14 is due to Hieber [Hie91a] with a different proof using techniques from Fourier multipliers; see also [Lan68] and [Sjo70]. Further results in this direction such as behaviour of the critical exponent $k = n|1/2 - 1/p|$ in $L^p(\mathbb{R}^n)$ and $L^1(\mathbb{R}^n)$, results on more general types of operators such as pseudo-differential operators with symbol a of the form $a(\xi) = |\xi|^{\alpha}$ for some $\alpha > 0$ can be found in [Sjo70], [Hie91a] and in Chapter 8. For example, if $1 < p < \infty$, then $i\Delta_p$ generates a k-times integrated semigroup on $L^p(\mathbb{R}^n)$ if and only if $k \geq n|1/2 - 1/p|$. Moreover, $i\Delta_1$ and $i\Delta_\infty$ generate a k-times integrated semigroup on $L^1(\mathbb{R}^n)$ and $L^\infty(\mathbb{R}^n)$, respectively, if and only if $k > n/2$. Further applications of boundaries of holomorphic semigroups are given by El-Mennaoui and Keyantuo, to Schrödinger operators in [EK96a] and to the wave equation in [EK96b]. In particular, they show the remarkable result that the Schrödinger operator $i\Delta$ on $L^p(\Omega)$, where $\Omega = (-\pi, \pi)^n$, with Dirichlet or Neumann boundary conditions, generates a k-times integrated group for $k > \frac{n}{2}|\frac{1}{2} - \frac{1}{p}|$, and that this constant is optimal.

Section 3.10
In this section we follow Arendt, Neubrander and Schlotterbeck [ANS92]. Extensions to fractional integrated semigroups have been given by Keyantuo [Key95a]. The perturbation result Corollary 3.10.5 is due to Kellermann and Hieber [KH89]. Applications of the sandwich theorem to differential operators have been given in [Are91].

Section 3.11
Integrated semigroups were actually introduced first in the context of resolvent positive operators in [Are84]. Theorem 3.11.7 and Proposition 3.11.12 as well as Example 3.11.13 are from [Are87a]. Theorem 3.11.5 and Example 3.11.6 are taken from [Are87b]. Theorem 3.11.9 was proved by Arendt, Chernoff and Kato [ACK82]; it remains valid if X is an ordered Banach space with normal cone X_+ which has non-empty interior. Theorem 3.11.10 is from [Are00], but our proof is different from the original one which is based on the construction of an intermediate Banach lattice where Theorem 3.5.2 can be applied. Theorem 3.11.8 is due to Arendt and Bénilan [AB92a]. It is no longer true on arbitrary Banach lattices: Grabosch and Nagel [GN89] showed that there exists a generator A of a positive C_0-semigroup such that $\overline{D(A^*)}$ is not a sublattice of X^*. Example 3.11.13 also works on $L^p(0,1)$ if $1 < p < \infty$, but not on $L^1(0,1)$. In fact, the following result is due to W. Desch (unpublished). It can also be seen in the framework of Miyadera–Voigt perturbation theory (see [Voi89]).

Theorem 3.17.4. *Let A be the generator of a positive C_0-semigroup on a space X of the form $L^1(\Omega,\mu)$. Let $B : D(A) \to X$ be linear and positive. If $A + B$ is resolvent positive, then $A + B$ generates a C_0-semigroup.*

The analogous result also holds if X is an arbitrary Banach lattice, but A generates a holomorphic, positive C_0-semigroup, as shown by Arendt and Rhandi [AR91].

The theory of resolvent positive operators has been developed further by Thieme [Thi98a] and [Thi98b], where in particular spectral theory and perturbation theory are studied. One may also study asymptotic behaviour. Assume that A is a resolvent positive operator generating a once integrated semigroup S. Then

$$\limsup_{\lambda\to\infty} \|\lambda R(\lambda, A)\| < \infty$$

if and only if

$$\limsup_{t\downarrow 0} \frac{1}{t}\|S(t)\| < \infty.$$

Moreover, $\lim_{t\downarrow 0} \frac{1}{t}S(t) = I$ strongly if and only if $\lim_{\lambda\to\infty} \lambda R(\lambda, A) = I$ strongly (we refer to [Are87a, Proposition 6.9]). Of course, an important class of resolvent positive operators are generators of positive C_0-semigroups. In the spirit of this book, we have concentrated in this section on those results which are related to Laplace transform techniques. We refer to the book [Nag86] edited by Nagel for the general theory of positive semigroups.

Finally, we should mention that there exist natural examples of operators which are not resolvent positive even though the resolvent exists and is positive on some interval $[\lambda_1, \lambda_2]$ where $-\infty < \lambda_1 < \lambda_2 < \infty$. Examples have been given by Greiner, Voigt and Wolff [GVW81] and Ulm [Ulm99].

Section 3.12
Theorem 3.12.2 and Example 3.12.3 are due to Driouich and El-Mennaoui [DE99]. Extensions to resolvent families associated with Volterra equations are given in [CL03] and [Haa08].

The acronym UMD-space stands for "unconditional martingale differences", reflecting the original probabilistic definition of this class of Banach spaces. Burkholder [Bur81] showed that the probabilistic property implies that the Hilbert transform is bounded on $L^p(\mathbb{R}, X)$ for $1 < p < \infty$ and Bourgain [Bou83] established the converse. Every UMD-space is superreflexive (i.e., there is an equivalent norm which is uniformly convex), but there are uniformly convex spaces which are not UMD-spaces. These and other properties of UMD-spaces are discussed in the survey article [Fra86] by Rubio de Francia.

The UMD-property has proved to be very important for the study of maximal regularity of solutions of inhomogeneous Cauchy problems. See, for example, [KW04], [Are04], [AB04].

Section 3.13

A C_0-semigroup T is said to be *eventually norm-continuous* if T is norm-continuous on (τ, ∞) for some $\tau \geq 0$. It was shown in the book of Hille and Phillips [HP57] that if T is eventually norm-continuous, then for any real a, $\{\lambda \in \sigma(A) : \operatorname{Re} \lambda \geq a\}$ is compact.

Theorem 3.13.2 was first proved by You [You92], but the simple proof given here is due to El-Mennaoui and Engel [EE94]. With minor modifications, the proof shows that if X is a Hilbert space and $\|T(\tau)R(\omega + is, A)^n\| \to 0$ as $|s| \to \infty$ for some $n \geq 1$ and $\tau \geq 0$ then T is norm-continuous for $t > \tau$. Other characterizations of immediately and eventually norm-continuous semigroups on Hilbert space have been given by Blasco and Martinez [BM96] and Blake [Bla01]. The latter characterization does not explicitly involve any decay of the resolvent of A (see also [Ile07]).

Goersmeyer and Weis [GW99] have proved that if T is a positive semigroup on an L^p-space and $\|R(\omega + is, A)\| \to 0$ as $|s| \to \infty$, then T is norm-continuous for $t > 0$. On the other hand, Mátrai [Mat08], and Chill and Tomilov [CT09], have constructed examples showing that Theorem 3.13.2 does not extend to all Banach spaces.

Section 3.14

For further literature on cosine functions we refer to the monographs of Fattorini [Fat85] and Goldstein [Gol85] and the survey article of Bobrowski [Bob97b], and the references given there. Kellermann and Hieber [KH89] proved the relation between cosine functions and integrated semigroups (Theorem 3.14.7). Uniqueness of the phase space (in Theorem 3.14.11) is due to Kisyński [Kis72]. Further information about the phase space may be found in the Appendix of [Haa07a].

Corollary 3.14.13 seems to be a natural limit for perturbation results of that type. It is shown in [AB06, Theorem 2.2] (and stated in [AB07, Theorem 5.9]) that if $A + B$ generates a cosine function for every bounded operator $B : D((\omega - A)^\gamma) \to X$ of rank 1, where $\gamma > 1/2$, then A is bounded on X.

More general versions of well-posedness of the Cauchy problem of second order (k-times integrated cosine functions) are considered in [AK89] and, using a different approach, by Takenaka and Okazawa [TakO90]. Higher order problems are covered in [XL98].

Let X be a Hilbert space with inner product $(\cdot|\cdot)$. The following result characterizes generators A of cosine functions X, up to choice of inner product, by a condition on the numerical range $W(A) := \{(Ax|x) : x \in X, \|x\| = 1\}$ related to the spectral condition of Proposition 3.14.18. Here, part a) is due to Crouzeix [Cro04] who extended it to a remarkable piece of work on numerical range and functional calculus for matrices and operators on Hilbert space [Cro07], [Cro08]. Part b) is due to Haase [Haa06, Corollary 7.4.6].

Theorem 3.17.5. *Let A be a densely defined operator on a Hilbert space X.*

a) *If $W(A)$ is contained in the parabola $\Pi_\omega := \{\xi + i\eta : \eta \in \mathbb{R},\ \xi \leq \omega^2 - \eta^2/4\omega^2\}$ for some $\omega > 0$ and there is at least one point in $\rho(A) \cap (\mathbb{C} \setminus \Pi_\omega)$, then A generates a cosine function.*

b) *If A generates a cosine function, then there is an equivalent scalar product on X with respect to which the numerical range of A is contained in Π_ω for some $\omega > 0$.*

Section 3.15
Sine functions were introduced by Arendt and Kellermann [AK89] where in particular the perturbation result Theorem 3.15.6 is proved. Moreover, the following holds.

Theorem 3.17.6. *Let A be a densely defined operator which generates a sine function* Sin *satisfying* $\limsup_{t\downarrow 0} \frac{1}{t}\|\operatorname{Sin}(t)\| < \infty$. *Then A generates a C_0-semigroup T. Moreover, T has a holomorphic extension to the right half-plane.*

The generation theorem for cosine functions (Theorem 3.15.3) is independently due to Da Prato and Giusti [DG67] and Sova [Sov66].

Section 3.16
Theorem 3.16.7 on square root reduction is due to Fattorini [Fat69]. At that time UMD-spaces had not been investigated and Fattorini formulated the result for L^p-spaces ($1 < p < \infty$). But he clearly pointed out that boundedness of the Hilbert transform is the crucial condition. Haase has given a different approach to Fattorini's theorem via functional calculus [Haa07a], and he has shown that $i(-A)^{1/2}$ generates a bounded C_0-group if A generates a bounded cosine function on a UMD-space [Haa09].

Part II

Tauberian Theorems and Cauchy Problems

Again we consider the guide-line of this book, the characteristic equation

$$\hat{u}(\lambda) = R(\lambda, A)x,$$

where u is a solution of the Cauchy problem

$$(CP) \quad \begin{cases} u'(t) = & Au(t) \quad (t \geq 0), \\ u(0) = & x, \end{cases}$$

and $R(\lambda, A)$ is the resolvent of the operator A. The aim of this part is to study the asymptotic behaviour of the solution $u(t)$ as $t \to \infty$.

In applications, one typically has some information about the spectral behaviour of A. This means that, in many cases, we know the Laplace transform $\hat{u}(\lambda)$ for $\operatorname{Re} \lambda > 0$. This part is devoted to the problem of determining the relation between the asymptotic behaviour of $u(t)$ as $t \to \infty$ and that of $\hat{u}(\lambda)$ as $\lambda \downarrow 0$. It contains two chapters. The first, Chapter 4, is devoted to the investigation of arbitrary functions; in Chapter 5 the asymptotic behaviour of solutions of (CP) is investigated.

The Abelian theorems given in Chapter 4 show that convergence of $u(t)$ as $t \to \infty$ implies convergence of $\lambda\hat{u}(\lambda)$ as $\lambda \downarrow 0$. Much more interesting for the applications which we have in mind, are Tauberian theorems, which give statements about the converse implication. However, they do need additional hypotheses, so-called *Tauberian conditions*.

Of particular importance are complex Tauberian theorems where the assumptions involve the behaviour of $\hat{u}(\lambda)$ for λ close to the imaginary axis. Indeed, assumptions of this kind are directly related to spectral properties of the operator via the characteristic equation. We not only consider convergence of functions but we also investigate their periodic behaviour. In fact, Chapter 4 contains an introduction to the theory of almost periodic and asymptotically almost periodic functions. The main results show that under suitable assumptions, countability of the spectrum implies asymptotic almost periodicity.

In the second chapter of this part, Chapter 5, the results of Chapter 4 are applied to solutions of the Cauchy problem, and also a variety of new and special results are obtained. The most important framework is given by generators of C_0-semigroups, but we also consider individual solutions in more general situations. This will be useful, for example, for the investigation of the heat equation with inhomogeneous boundary conditions which we give in Chapter 6.

In the first three sections of Chapter 5 the diverse abscissas of Laplace transform of the semigroup are related to spectral properties. Indeed, it turns out that the relation is not simple, in general. However there are very satisfying results in special cases; for example on Hilbert spaces (Section 5.2) or for positive semigroups (Section 5.3); both are important for applications.

The complex Tauberian theorems from Chapter 4 are applied to solutions of (CP) in many different situations. Asymptotically almost periodic solutions lead to splitting theorems which are considered in Section 5.4. Countability of the spectrum is a criterion for asymptotic periodicity. This is shown in Section 5.5 by exploiting the results of Chapter 4.

Typically, one has to assume boundedness of the solution for these applications. This may be a difficult assumption to verify when the inhomogeneous Cauchy problem is considered, but we show that this is automatic if the underlying semigroup is holomorphic and a certain spectral condition of non-resonance is satisfied. This last topic of Chapter 5 concludes our investigation of the interesting interplay of spectrum and asymptotics.

Chapter 4

Asymptotics of Laplace Transforms

Frequently, convergence of a function $f : \mathbb{R}_+ \to X$ for $t \to \infty$ implies convergence of an average of this function. Assertions of this type are called *Abelian theorems*. A theorem is called *Tauberian* if, conversely, convergence of the function is deduced from the convergence of an average.

The Abelian theorems which we present in Section 4.1 are quite easy to prove. However, the Tauberian theorems corresponding to their converse versions are much more delicate. They need additional hypotheses, so-called *Tauberian conditions*. Section 4.2 is devoted to Tauberian conditions of real type (for example, boundedness or positivity of f).

Interesting applications of these Abelian and real Tauberian theorems to semigroups are given in Section 4.3 where mean ergodicity is discussed. This interrupts the general theme of this chapter, but the results will be useful in the subsequent sections where the notion of mean ergodicity is needed.

In Section 4.4 a complex Tauberian theorem is proved with the help of an elegant contour argument. Here we make assumptions on holomorphic extensions of $\hat{f}$ on the imaginary axis. We restrict ourselves to the case of one singularity in order to keep the ideas more transparent, but this case is already of special interest. For example, an immediate consequence is Gelfand's theorem, saying that a bounded C_0-group is trivial (i.e., the identity) if and only if the spectrum of its generator is reduced to $\{0\}$.

One interesting type of asymptotic behaviour for large time is almost periodicity. The concept is introduced in Section 4.5 where elementary properties are proved for functions on $\mathbb{R}$. In Section 4.6, Loomis's theorem and its vector-valued version are proved by an elegant quotient method which allows one to apply Gelfand's theorem. The basic notion is the Carleman spectrum for a bounded measurable function defined on the line, and Loomis's theorem states that any bounded uniformly continuous function $f : \mathbb{R} \to \mathbb{C}$ with countable spectrum is almost periodic. There is one vector-valued version of Loomis's theorem, which is valid for every Banach space and involves an ergodicity condition. The other vector-valued version holds without further assumptions on the function but a geometric condition on the Banach space is needed.

Functions on the half-line are considered in Section 4.7. The naturally associated "half-line" spectrum is discussed and the main theorem is a complex Tauberian theorem for functions with countable spectrum, which is proved by the same technique as we proved Loomis's theorem.

In Section 4.8 we come back to functions defined on the line showing that the Carleman spectrum (defined by holomorphy) and the Beurling spectrum (defined by the Fourier transform) coincide. This allows us to prove a very general complex Tauberian theorem for functions on the half-line in Section 4.9. Here we use Fourier transform methods which allow one to reduce the problem to an application of Loomis's theorem (in the scalar case).

The structure of this chapter needs some explanation in view of our main purpose, namely the proof of a complex Tauberian theorem on the half-line. We present three different methods by which we prove the result in increasing generality (Theorems 4.4.8, 4.7.7 and 4.9.7); namely, the contour method, the quotient method and the Fourier method. If Loomis's theorem (Corollary 4.6.4) is accepted, the Fourier method of Section 4.9 is the most general. The contour method, presented as the first approach in Section 4.4, is the most elementary. It gives us Gelfand's theorem (Corollary 4.4.12) and other interesting consequences. The quotient method uses Gelfand's theorem. It gives us Loomis's theorem on the line (Section 4.6), and in an elegant way the fairly general complex Tauberian Theorem 4.7.7 on the half-line.

4.1 Abelian Theorems

Throughout this section, f denotes a function in $L^1_{loc}(\mathbb{R}_+, X)$, where X is a Banach space and $\mathbb{R}_+ := [0, \infty)$ is the right half-line.

We consider the following three types of averages.

Definition 4.1.1. *Let $f_\infty \in X$. We say that*

a) $f(t)$ converges to f_∞ in the sense of Abel as $t \to \infty$ *if* $\operatorname{abs}(f) \le 0$ *and* $\text{A-}\lim_{t\to\infty} f(t) := \lim_{\lambda\downarrow 0} \lambda \hat{f}(\lambda) = f_\infty$;

b) $f(t)$ is B-convergent to f_∞ as $t \to \infty$, *or simply write* $\text{B-}\lim_{t\to\infty} f(t) = f_\infty$, *if for every* $\delta > 0$, $\lim_{t\to\infty} \frac{1}{\delta} \int_t^{t+\delta} f(s)\, ds = f_\infty$;

c) $f(t)$ converges to f_∞ in the sense of Cesàro as $t \to \infty$ *if*

$$\text{C-}\lim_{t\to\infty} f(t) := \lim_{t\to\infty} \frac{1}{t} \int_0^t f(s)\, ds = f_\infty.$$

It will be convenient to consider the antiderivative of a function f given by $F(t) := \int_0^t f(s)\,ds$. Then the following Abelian theorem holds.

Theorem 4.1.2. *Let $f_\infty, F_\infty \in X$.*

a) *If* $\lim_{t\to\infty} f(t) = f_\infty$, *then* $\text{B-}\lim_{t\to\infty} f(t) = f_\infty$.

b) *If* $\text{B-}\lim_{t\to\infty} f(t) = f_\infty$, *then* $\text{C-}\lim_{t\to\infty} f(t) = f_\infty$.

c) *If* $\text{C-}\lim_{t\to\infty} f(t) = f_\infty$, *then* $\text{A-}\lim_{t\to\infty} f(t) = f_\infty$.

d) *If* $\lim_{t\to\infty} F(t) = F_\infty$, *then* $\lim_{\lambda\downarrow 0} \hat{f}(\lambda) = F_\infty$.

Proof. a) is obvious.

b) Replacing f by $f - f_\infty$ we can assume that $f_\infty = 0$. Then (taking $\delta = 1$) we have by assumption that $\lim_{t\to\infty} \int_t^{t+1} f(s)\,ds = 0$. Let $\varepsilon > 0$. There exists $t_0 \geq 0$ such that $\left\| \int_t^{t+1} f(s)\,ds \right\| \leq \frac{\varepsilon}{2}$ for all $t \geq t_0$. Moreover, there exists $t_1 \geq t_0$ such that $\frac{1}{t}\int_0^{t_0+1} \| f(s) \|\,ds \leq \frac{\varepsilon}{2}$ for all $t \geq t_1$. Let $t \geq t_1$. Take $n \in \mathbb{N}$ such that $t - n \leq t_0 \leq t - n + 1$. Then

$$\begin{aligned}
\left\| \frac{1}{t}\int_0^t f(s)\,ds \right\| &\leq \left\| \frac{1}{t}\int_0^{t-n+1} f(s)\,ds \right\| + \left\| \frac{1}{t}\int_{t-n+1}^t f(s)\,ds \right\| \\
&\leq \frac{1}{t}\int_0^{t_0+1} \| f(s) \|\,ds + \sum_{k=1}^{n-1} \left\| \frac{1}{t}\int_{t-n+k}^{t-n+k+1} f(s)\,ds \right\| \\
&\leq \frac{\varepsilon}{2} + \frac{n-1}{t}\frac{\varepsilon}{2} \leq \varepsilon.
\end{aligned}$$

This proves the claim.

c) Assume that $\text{C-}\lim_{t\to\infty} f(t) = f_\infty$. Then $F(t) = O(t)$ as $t \to \infty$, so $\hat{f}(\lambda)$ exists for $\operatorname{Re}\lambda > 0$ by Theorem 1.4.3.

Let $\varepsilon > 0$. There exists $\tau > 0$ such that $\| \frac{1}{t}F(t) - f_\infty \| \leq \varepsilon$ $(t \geq \tau)$. Hence by integration by parts,

$$\begin{aligned}
&\limsup_{\lambda\downarrow 0} \| \lambda\hat{f}(\lambda) - f_\infty \| \\
&= \limsup_{\lambda\downarrow 0} \left\| \lambda \int_0^\infty e^{-\lambda t}\,(f(t) - f_\infty)\,dt \right\| \\
&= \limsup_{\lambda\downarrow 0} \left\| \lambda^2 \int_0^\infty e^{-\lambda t}\,(F(t) - t f_\infty)\,dt \right\| \\
&= \limsup_{\lambda\downarrow 0} \left\| \lambda^2 \int_0^\infty t e^{-\lambda t} \left(\frac{1}{t}F(t) - f_\infty \right) dt \right\| \\
&\leq \limsup_{\lambda\downarrow 0} \lambda^2 \int_0^\tau t e^{-\lambda t} \left\| \frac{1}{t}F(t) - f_\infty \right\| dt + \lim_{\lambda\downarrow 0} \varepsilon\lambda^2 \int_\tau^\infty t e^{-\lambda t} dt \\
&= \lim_{\lambda\downarrow 0} \varepsilon\lambda^2 \int_0^\infty t e^{-\lambda t} dt = \varepsilon.
\end{aligned}$$

d) Since $\hat{f}(\lambda) = \widehat{F'}(\lambda) = \lambda\hat{F}(\lambda)$ $(\lambda > 0)$ if $\operatorname{abs}(f) \leq 0$, this follows from the previous parts since convergence implies A-convergence. □

Concerning the behaviour for $t \to 0$, we note:

Proposition 4.1.3. *Let $x \in X$. Assume that* $\operatorname{abs}(f) < \infty$. *Then*

$$\limsup_{\lambda\to\infty} \| \lambda\hat{f}(\lambda) - x \| \leq \limsup_{t\downarrow 0} \| f(t) - x \|.$$

In particular, $\lim_{t\downarrow 0} f(t) = x$ *implies* $\lim_{\lambda\to\infty} \lambda\hat{f}(\lambda) = x$.

Proof. Since $\operatorname{abs}(f) < \infty$, there exist $M, \omega > 0$ such that $\| F(t) \| \leq Me^{\omega t}$ $(t \geq 0)$. Let $b > 0, \lambda > \omega$. Then

$$\begin{aligned}\left\| \int_b^\infty \lambda e^{-\lambda t} f(t)\, dt \right\| &= \left\| -\lambda e^{-\lambda b} F(b) + \lambda^2 \int_b^\infty e^{-\lambda t} F(t)\, dt \right\| \\ &\leq \lambda e^{-\lambda b} \| F(b) \| + \lambda^2 \frac{M}{\lambda - \omega} e^{-(\lambda-\omega)b}.\end{aligned}$$

Hence,

$$\begin{aligned}&\limsup_{\lambda\to\infty} \| \lambda\hat{f}(\lambda) - x \| \\ &= \limsup_{\lambda\to\infty} \left\| \int_0^\infty \lambda e^{-\lambda t} \left(f(t) - x \right) dt \right\| \\ &\leq \lim_{\lambda\to\infty} \left\{ \sup_{0\leq t\leq b} \| f(t) - x \| + \left\| \int_b^\infty \lambda e^{-\lambda t} f(t)\, dt \right\| + \int_b^\infty \lambda e^{-\lambda t}\, dt \, \| x \| \right\} \\ &= \sup_{0\leq t\leq b} \| f(t) - x \|.\end{aligned}$$

□

Here, we used throughout that $\lambda \int_0^\infty e^{-\lambda t}\, dt = 1$ $(\lambda > 0)$. More generally, for $\alpha > -1$ one has

$$\frac{\lambda^{\alpha+1}}{\Gamma(\alpha+1)} \int_0^\infty e^{-\lambda t} t^\alpha\, dt = 1 \quad (\lambda > 0).$$

One shows the following in a similar way as above.

Proposition 4.1.4. *Let $f_\infty \in X, f_0 \in X, \alpha > -1$.*

a) *If* C-$\lim_{t\to\infty} f(t) = f_\infty$, *then* $\lim_{\lambda\downarrow 0} \frac{\lambda^{\alpha+1}}{\Gamma(\alpha+1)} \int_0^\infty e^{-\lambda t} t^\alpha f(t)\, dt = f_\infty$.

b) *If* $\lim_{t\downarrow 0} f(t) = f_0$, *then* $\lim_{\lambda\to\infty} \frac{\lambda^{\alpha+1}}{\Gamma(\alpha+1)} \int_0^\infty e^{-\lambda t} t^\alpha f(t)\, dt = f_0$.

This allows one to compute the derivative of f at 0 from its Laplace transform:

Corollary 4.1.5. *Let $f \in L^1_{loc}(\mathbb{R}_+, X)$ such that* $\operatorname{abs}(f) < \infty$. *Assume that $f \in C^1([0, \tau], X)$ for some $\tau > 0$. Then*

$$\lim_{\lambda\to\infty} \left(\lambda^2 \hat{f}(\lambda) - \lambda f(0) \right) = f'(0).$$

Proof. Let $\alpha = 1$. Applying Proposition 4.1.4 b) to $g(t) := \frac{1}{t}(f(t)-f(0))$, we obtain that $\lim_{\lambda\to\infty}(\lambda^2\hat{f}(\lambda)-\lambda f(0)) = \lim_{\lambda\to\infty}\lambda^2\int_0^\infty e^{-\lambda t}t\frac{1}{t}(f(t)-f(0))\,dt = f'(0)$. □

The following theorem gave the name to this type of result, and to this section. We deduce it from the corresponding result for Laplace transforms. Usually, one uses Abel's "summation by parts" for power series.

Theorem 4.1.6 (Abel's Continuity Theorem). *Let $p(z) := \sum_{n=0}^\infty a_n z^n$ be a power series, where $a_n \in X$. If $\sum_{n=0}^\infty a_n = b$, then $\lim_{z\uparrow 1} p(z) = b$.*

Proof. Let $f(t) = a_n$ if $t \in [n, n+1)$. Then $\lim_{t\to\infty} F(t) = b$. Hence, $\operatorname{abs}(f) \le 0$ and $\hat{f}(\lambda) = \frac{1-e^{-\lambda}}{\lambda}\sum_{n=0}^\infty a_n(e^{-\lambda})^n$ for all $\lambda > 0$. By Theorem 4.1.2 d),

$$\begin{aligned}\lim_{z\uparrow 1} p(z) &= \lim_{z\uparrow 1}\sum_{n=0}^\infty a_n z^n = \lim_{\lambda\downarrow 0}\sum_{n=0}^\infty a_n(e^{-\lambda})^n \\ &= \lim_{\lambda\downarrow 0}\frac{\lambda}{1-e^{-\lambda}}\hat{f}(\lambda) = \lim_{\lambda\downarrow 0}\hat{f}(\lambda) \\ &= \lim_{t\to\infty} F(t) = b.\end{aligned}$$

□

4.2 Real Tauberian Theorems

Let $f \in L^1_{loc}(\mathbb{R}_+, X)$. In the preceding section we established four notions of convergence for $f(t)$ (as $t \to \infty$) of increasing generality: convergence in the usual sense, B-convergence, convergence in the sense of Cesàro, and convergence in the sense of Abel. None of the implications established in Theorem 4.1.2 is reversible. An additional condition, which allows one to reverse one of the implications is called a *Tauberian condition*, and the corresponding result a *Tauberian theorem.* Of particular interest are those results which allow one to deduce convergence in the usual sense from convergence of a mean. In this section we consider assumptions on $f(t)$, i.e. *real Tauberian conditions*, in contrast to conditions on $\hat{f}(\lambda)$, known as *complex Tauberian conditions*, which will be the subject of the next section.

We first consider the reverse implication of Theorem 4.1.2 a).

A. Conditions under which B-$\lim_{t\to\infty} f(t) = f_\infty$ **implies** $\lim_{t\to\infty} f(t) = f_\infty$.

Let $f \in L^1_{loc}(\mathbb{R}_+, X)$. In general, if B-$\lim_{t\to\infty} f(t) = f_\infty$, one cannot conclude that $\lim_{t\to\infty} f(t) = f_\infty$ even if f is bounded. A simple example is the function $f = \sum_{n=1}^\infty \chi_{[n,n+1/n]}$ which has B-limit 0 (as $t \to \infty$) but does not converge.

Definition 4.2.1. *A function $g : \mathbb{R}_+ \to X$ is called* slowly oscillating *if for all $\varepsilon > 0$ there exist $t_0 \ge 0$, $\delta > 0$ such that $\| g(t) - g(s) \| \le \varepsilon$ whenever $s, t \ge t_0$, $|t-s| \le \delta$.*

A bounded continuous function is slowly oscillating if and only if it is uniformly continuous. Indeed, one may think of Definition 4.2.1 as expressing that the function g is uniformly continuous at infinity. This is made more precise by the following characterization.

Proposition 4.2.2. *A function $g : \mathbb{R}_+ \to X$ is slowly oscillating if and only if $g = g_0 + g_1$ where $g_1 : \mathbb{R}_+ \to X$ is uniformly continuous and $g_0 : \mathbb{R} \to X$ satisfies $\lim_{t\to\infty} g_0(t) = 0$.*

Proof. It is clear that the condition is sufficient.

In order to prove the converse, let g be slowly oscillating. Then we find a decreasing sequence $(\delta_n)_{n\in\mathbb{N}} \subset (0,\infty)$ such that $\lim_{n\to\infty} \delta_n = 0$ and an increasing sequence $(t_n)_{n\in\mathbb{N}} \subset [0,\infty)$ such that $\lim_{n\to\infty} t_n = \infty$ and $\| f(t+\delta) - f(t) \| \le \frac{1}{n}$ whenever $t \ge t_n$, $0 < \delta \le \delta_n$. For each $n \in \mathbb{N}$, choose a partition of the interval $[t_n, t_{n+1}]$ so that every subinterval has length smaller than δ_n. Define g_1 to be linear on each subinterval and equal to f at the endpoints of the subintervals. Extend g_1 continuously to $[0, t_1]$. Then g_1 is uniformly continuous and $\lim_{t\to\infty} (f(t) - g_1(t)) = 0$. We put $g_0 := f - g_1$. □

The following Tauberian theorem is easy to prove.

Theorem 4.2.3. *Let $f \in L^1_{loc}(\mathbb{R}_+, X)$ be slowly oscillating and $f_\infty \in X$. If*

$$\text{B-}\lim_{t\to\infty} f(t) = f_\infty, \text{ then } \lim_{t\to\infty} f(t) = f_\infty.$$

Proof. For $\delta > 0$ one has

$$\begin{aligned} \limsup_{t\to\infty} \| f(t) - f_\infty \| &\le \limsup_{t\to\infty} \left\| f(t) - \frac{1}{\delta} \int_t^{t+\delta} f(s)\, ds \right\| \\ &= \limsup_{t\to\infty} \left\| \frac{1}{\delta} \int_t^{t+\delta} (f(t) - f(s))\, ds \right\| \\ &\le \limsup_{t\to\infty} \left(\sup_{t\le s\le t+\delta} \| f(t) - f(s) \| \right) \end{aligned}$$

The last expression tends to 0 as $\delta \downarrow 0$ since f is slowly oscillating. □

Next, we consider:

B. Conditions under which C-$\lim_{t\to\infty} f(t) = f_\infty$ implies $\lim_{t\to\infty} f(t) = f_\infty$.

First, we notice that, in general, convergence in the sense of Cesàro does not imply B-convergence even if f is bounded. For example, let $f(t) := \sin t$. Then C-$\lim_{t\to\infty} f(t) = \lim_{t\to\infty} \frac{1}{t} \int_0^t \sin s\, ds = \lim_{t\to\infty} \frac{1-\cos t}{t} = 0$. But $\frac{1}{\delta} \int_t^{t+\delta} \sin s\, ds = \frac{1}{\delta} (\cos(t+\delta) - \cos t)$, and the B-limit does not exist.

A function $f : \mathbb{R}_+ \to X$ is called *feebly oscillating* (when $t \to \infty$) if

$$\lim_{\substack{t,s\to\infty \\ t/s\to 1}} \| f(t) - f(s) \| = 0.$$

It is clear that every feebly oscillating function is slowly oscillating.

Example 4.2.4. Assume that there exist $\tau > 0, M \geq 0$ such that $\|tf(t)\| \leq M$ $(t \geq \tau)$. Then $F(t) := \int_0^t f(s)\,ds$ is feebly oscillating. In fact, $\| F(t) - F(s) \| \leq M \int_s^t \frac{dr}{r} = M \log\left(\frac{t}{s}\right)$ for $t \geq s \geq \tau$. □

Theorem 4.2.5. *Let $f \in L^1_{loc}(\mathbb{R}_+, X)$ and $f_\infty \in X$. Assume that f is feebly oscillating. If* C-$\lim_{t\to\infty} f(t) = f_\infty$, *then* $\lim_{t\to\infty} f(t) = f_\infty$.

Proof. Let $\varepsilon > 0$. There exist $\delta > 0, t_0 > 0$ such that $\| f(s) - f(t) \| < \varepsilon$ whenever $t \geq t_0,\ s \in [t - \delta t, t + \delta t]$. Hence,

$$\left\| f(t) - \frac{1}{2\delta t} \int_{t(1-\delta)}^{t(1+\delta)} f(s)\,ds \right\| = \left\| \frac{1}{2\delta t} \int_{t(1-\delta)}^{t(1+\delta)} (f(t) - f(s))\,ds \right\| \leq \varepsilon \text{ if } t \geq t_0.$$

The assumption implies that

$$\begin{aligned} &\frac{1}{2\delta t} \int_{t(1-\delta)}^{t(1+\delta)} f(s)\,ds \\ &= \frac{1+\delta}{2\delta} \frac{1}{t(1+\delta)} \int_0^{t(1+\delta)} f(s)\,ds - \frac{1-\delta}{2\delta} \frac{1}{t(1-\delta)} \int_0^{t(1-\delta)} f(s)\,ds \\ &\to f_\infty \quad \text{as } t \to \infty. \end{aligned}$$

It follows that $\lim_{t\to\infty} f(t) = f_\infty$. □

In the next theorem the Tauberian condition is of order-theoretic nature.

Theorem 4.2.6. *Let X be an ordered Banach space with normal cone. Furthermore let $f : \mathbb{R}_+ \to X_+$ be a function such that for some $k \in \mathbb{N}, t \mapsto t^k f(t)$ is increasing. Then*

$$\text{C-}\lim_{t\to\infty} f(t) = f_\infty \text{ implies } \lim_{t\to\infty} f(t) = f_\infty.$$

Proof. Since the cone is normal, there exists $c \geq 0$ such that $u \leq x \leq v$ implies $\| x \| \leq c(\| u \| + \| v \|)$ for $u, x, v \in X$ (see Appendix C). Let $\rho > 1$. The assumption implies that

$$\begin{aligned} \frac{1}{(\rho-1)t} \int_t^{\rho t} f(s)\,ds &= \frac{\rho}{\rho-1} \frac{1}{\rho t} \int_0^{\rho t} f(s)\,ds - \frac{1}{\rho-1} \frac{1}{t} \int_0^t f(s)\,ds \\ &\to f_\infty \quad (t \to \infty) \end{aligned}$$

and, similarly, $\frac{\rho}{(\rho-1)t} \int_{t/\rho}^t f(s)\,ds \to f_\infty \quad (t \to \infty)$.

Since $t^k f(t)$ is increasing, one has

$$\begin{aligned}\int_t^{\rho t} f(s)\,ds &= \int_t^{\rho t} s^k f(s) s^{-k} ds \\ &\geq t^k f(t)(\rho t)^{-k}(\rho t - t) = f(t)t\rho^{-k}(\rho - 1)\end{aligned}$$

and $\int_{t/\rho}^{t} f(s)\,ds \leq f(t)t\rho^{k-1}(\rho-1)$. Thus,

$$\rho^{-k}\frac{\rho}{(\rho-1)t}\int_{t/\rho}^{t} f(s)\,ds - f_\infty \leq f(t) - f_\infty \leq \rho^k \frac{1}{(\rho-1)t}\int_t^{\rho t} f(s)\,ds - f_\infty.$$

Consequently,

$$\begin{aligned}\limsup_{t\to\infty} \| f(t) - f_\infty \| &\leq c \limsup_{t\to\infty} \left\{ \left\| \rho^{-k}\frac{\rho}{(\rho-1)t}\int_{t/\rho}^{t} f(s)\,ds - f_\infty \right\| \right. \\ &\qquad \left. + \left\| \rho^k \frac{1}{(\rho-1)t}\int_t^{\rho t} f(s)\,ds - f_\infty \right\| \right\} \\ &\leq c\,\|f_\infty\|\,(1 - \rho^{-k} + \rho^k - 1).\end{aligned}$$

Letting $\rho \downarrow 1$ yields the claim. □

C. Conditions under which A- $\lim_{t\to\infty} f(t) = f_\infty$ **implies** C- $\lim_{t\to\infty} f(t) = f_\infty$.

In general, Abel convergence does not imply Cesàro convergence. For example, let $f(t) := t \sin t$ $(t \geq 0)$. Then it is easy to see that

$$\text{A-}\lim_{t\to\infty} f(t) = \lim_{\lambda\downarrow 0} 2\lambda^2(1+\lambda^2)^{-2} = 0$$

but $\frac{1}{t}\int_0^t f(s)\,ds = \frac{\sin t}{t} - \cos t$ does not converge. Thus additional hypotheses are needed.

First, we consider a boundedness condition on f.

Theorem 4.2.7. *Let $f_\infty \in X$. Assume that $\sup_{t\geq\tau} \|f(t)\| < \infty$ for some $\tau \geq 0$. If* A- $\lim_{t\to\infty} f(t) = f_\infty$*, then* C- $\lim_\infty f(t) = f_\infty$.

Proof. a) We first assume that $\tau = 0$. For $\beta > 0$ let $e_\beta(t) := e^{-\beta t}$ $(t > 0)$. Then $\text{span}\{e_\beta : \beta > 0\}$ is dense in $L^1(\mathbb{R}_+)$ (by Lemma 1.7.1). By hypothesis,

$$\begin{aligned}\lim_{\alpha\to\infty}\int_0^\infty e^{-s} f(\alpha s)\,ds &= \lim_{\lambda\downarrow 0}\int_0^\infty e^{-s} f\left(\frac{s}{\lambda}\right) ds \\ &= \lim_{\lambda\downarrow 0}\lambda\int_0^\infty e^{-\lambda s} f(s)\,ds = f_\infty.\end{aligned}$$

Hence,

$$\begin{aligned}
\lim_{\alpha\to\infty} \langle \beta e_\beta, f(\alpha\,\cdot)\rangle &= \lim_{\alpha\to\infty} \beta \int_0^\infty e^{-\beta s} f(\alpha s)\,ds \\
&= \lim_{\alpha\to\infty} \int_0^\infty e^{-s} f\left(\frac{\alpha}{\beta}s\right) ds \\
&= f_\infty = \langle \beta e_\beta, f_\infty\rangle \quad \text{for all } \beta > 0.
\end{aligned}$$

It follows that $\lim_{\alpha\to\infty}\langle h, f(\alpha\,\cdot)\rangle = f_\infty \int_0^\infty h(t)\,dt$ for all $h \in L^1(\mathbb{R}_+)$. Letting $h = \chi_{(0,1]}$, we obtain

$$\lim_{\alpha\to\infty} \frac{1}{\alpha}\int_0^\alpha f(s)\,ds = \lim_{\alpha\to\infty}\int_0^1 f(\alpha s)\,ds = \lim_{\alpha\to\infty}\langle h, f(\alpha\,\cdot)\rangle = f_\infty.$$

b) If $\tau > 0$ the result follows by applying a) to $g(t) = f(t+\tau)$. □

Next, we show that positivity is a Tauberian condition.

Theorem 4.2.8 (Karamata). *Let X be an ordered Banach space with normal cone and let $\beta > -1$. Let $f_\infty \in X$ and $f \in L^1_{loc}(\mathbb{R}_+, X)$ such that $f(t) \ge 0$ $(t \in \mathbb{R}_+)$. Suppose that*

a) *$\int_0^\infty e^{-\lambda t} t^\beta f(t)\,dt$ exists for all $\lambda > 0$; and*

b) *$\lim_{\lambda\downarrow 0} \frac{\lambda^{\beta+1}}{\Gamma(\beta+1)} \int_0^\infty e^{-\lambda t} t^\beta f(t)\,dt = f_\infty$.*

Then C-$\lim_{t\to\infty} f(t) = f_\infty$.

This is a converse of Proposition 4.1.4 a). In particular, it follows from Theorem 4.2.8 that for positive $f \in L^1_{loc}(\mathbb{R}_+, X)$,

$$\text{A-}\lim_{t\to\infty} f(t) = f_\infty \quad\text{implies}\quad \text{C-}\lim_{t\to\infty} f(t) = f_\infty.$$

Proof. It follows from the assumption that for $n \in \mathbb{N}_0$,

$$\begin{aligned}
&\lim_{\lambda\downarrow 0} \frac{\lambda^{\beta+1}}{\Gamma(\beta+1)} \int_0^\infty e^{-\lambda t}(e^{-\lambda t})^n f(t) t^\beta dt \\
&= \frac{1}{(n+1)^{\beta+1}} \lim_{\lambda\downarrow 0} \frac{(\lambda(n+1))^{\beta+1}}{\Gamma(\beta+1)} \int_0^\infty e^{-\lambda(n+1)t} f(t) t^\beta dt \\
&= \frac{1}{(n+1)^{\beta+1}} f_\infty = f_\infty \frac{1}{\Gamma(\beta+1)} \int_0^\infty t^\beta e^{-t}(e^{-t})^n dt.
\end{aligned}$$

Consequently, for every polynomial p,

$$\lim_{\lambda\downarrow 0} \frac{\lambda^{\beta+1}}{\Gamma(\beta+1)} \int_0^\infty e^{-\lambda t} p(e^{-\lambda t}) t^\beta f(t)\,dt = f_\infty \frac{1}{\Gamma(\beta+1)} \int_0^\infty t^\beta e^{-t} p(e^{-t})\,dt.$$

Let $q : [0,1] \to \mathbb{R}_+$ be given by $q(x) := 0$ if $x < e^{-1}, q(x) := x^{-1}$ if $x \ge e^{-1}$. Let $\delta > 0$ and let q_1, q_2 be continuous functions such that $0 \le q_1 \le q \le q_2$ on $[0,1]$ and

a) $q_1(x) = q(x)$ if $x \leq e^{-1}$ or $x > e^{-1} + \delta$,

b) $q_2(x) = q(x)$ if $x \leq e^{-1} - \delta$ or $x > e^{-1}$,

c) $\sup_{0\leq x\leq 1}(q_2(x) - q_1(x)) \leq e$.

Now, let $\varepsilon > 0$. Choose $\delta > 0$ such that $\frac{e}{\Gamma(\beta+1)} \int_{t_1}^{t_2} t^\beta e^{-t} dt < \varepsilon$, where $e^{-t_2} = e^{-1} - \delta$ and $e^{-t_1} = e^{-1} + \delta$. By the Stone-Weierstrass theorem, there exist polynomials p_1 and p_2 such that

$$q_1 - \varepsilon \leq p_1 \leq q_1 \leq q \leq q_2 \leq p_2 \leq q_2 + \varepsilon$$

on $[0, 1]$. Define $k_j : \mathbb{R}_+ \to X$ $(j = 1, 2)$ and $h : (0, \infty) \to X$ by:

$$\begin{aligned} k_j(\lambda) &:= \frac{\lambda^{\beta+1}}{\Gamma(\beta+1)} \int_0^\infty t^\beta e^{-\lambda t} p_j(e^{-\lambda t}) f(t)\, dt \quad (\lambda > 0), \\ k_j(0) &:= \frac{f_\infty}{\Gamma(\beta+1)} \int_0^\infty e^{-t} p_j(e^{-t}) t^\beta dt, \\ h(\lambda) &:= \frac{\lambda^{\beta+1}}{\Gamma(\beta+1)} \int_0^\infty e^{-\lambda t} q(e^{-\lambda t}) t^\beta f(t)\, dt \quad (\lambda > 0). \end{aligned}$$

Since $0 \leq p_2 - p_1 \leq q_2 - q_1 + 2\varepsilon$, one has

$$\begin{aligned} 0 \leq k_2(0) - k_1(0) &\leq \frac{f_\infty}{\Gamma(\beta+1)} e \int_{t_1}^{t_2} e^{-t} t^\beta dt + 2\varepsilon \frac{f_\infty}{\Gamma(\beta+1)} \int_0^\infty e^{-t} t^\beta dt \\ &\leq 3\varepsilon f_\infty. \end{aligned}$$

The first part of the proof shows that $\lim_{\lambda\downarrow 0} k_j(\lambda) = k_j(0)$. Let $\lambda_0 > 0$ such that $\| k_j(\lambda) - k_j(0) \| < \varepsilon$ for $0 < \lambda \leq \lambda_0$, $j = 1, 2$. Let $0 < \lambda, \hat{\lambda} \leq \lambda_0$. Then

$$k_1(\lambda) - k_2(\hat{\lambda}) \leq h(\lambda) - h(\hat{\lambda}) \leq k_2(\lambda) - k_1(\hat{\lambda}).$$

Hence, for a fixed constant $c > 0$,

$$\begin{aligned} \| h(\lambda) - h(\hat{\lambda}) \| &\leq c\Big\{ \| k_2(\lambda) - k_1(\hat{\lambda}) \| + \| k_1(\lambda) - k_2(\hat{\lambda}) \| \Big\} \\ &\leq c\Big\{ \| k_2(\lambda) - k_2(0) \| + \| k_2(0) - k_1(0) \| + \| k_1(0) - k_1(\hat{\lambda}) \| \\ &\qquad + \| k_1(\lambda) - k_1(0) \| + \| k_1(0) - k_2(0) \| + \| k_2(0) - k_2(\hat{\lambda}) \| \Big\} \\ &\leq 4c\varepsilon + 2 \cdot 3c\varepsilon \| f_\infty \| \ \text{ whenever } \ 0 < \lambda, \hat{\lambda} \leq \lambda_0. \end{aligned}$$

This shows that $h(\lambda)$ converges as $\lambda \downarrow 0$. Since $h(\lambda) = \frac{\lambda^{\beta+1}}{\Gamma(\beta+1)} \int_0^{1/\lambda} s^\beta f(s)\, ds$, it follows that $\lim_{t\to\infty} t^{-(\beta+1)} \int_0^t s^\beta f(s)\,ds =: g$ exists. Hence,

$$\begin{aligned} f_\infty &= \lim_{\lambda\downarrow 0} \frac{\lambda^{\beta+1}}{\Gamma(\beta+1)} \int_0^\infty e^{-\lambda t} f(t) t^\beta dt \\ &= \lim_{\lambda\downarrow 0} (\beta+1) \frac{\lambda^{\beta+2}}{\Gamma(\beta+2)} \int_0^\infty e^{-\lambda t} t^{\beta+1} \frac{1}{t^{\beta+1}} \int_0^t f(s) s^\beta\, ds\, dt \\ &= (\beta+1) g, \end{aligned}$$

by the Abelian theorem mentioned above (Proposition 4.1.4 a)). Thus, $g = \frac{f_\infty}{\beta+1}$. We have proved that $t^{-(\beta+1)} \int_1^t s^\beta f(s)\,ds \to \frac{f_\infty}{\beta+1}$ as $t \to \infty$. Since convergence implies Cesàro-convergence, we also have

$$\lim_{t\to\infty} \frac{1}{t} \int_1^t \frac{1}{s^{\beta+1}} \int_1^s r^\beta f(r)\,dr\,ds = \frac{f_\infty}{\beta+1}.$$

Hence, integration by parts yields

$$\begin{aligned} \frac{1}{t} \int_1^t f(s)\,ds &= \frac{1}{t} \int_1^t \frac{1}{s^\beta} \frac{d}{ds} \int_1^s r^\beta f(r)\,dr\,ds \\ &= \frac{1}{t^{\beta+1}} \int_1^t r^\beta f(r)\,dr + \frac{\beta}{t} \int_1^t s^{-(\beta+1)} \int_1^s r^\beta f(r)\,dr\,ds \\ &\to \frac{f_\infty}{\beta+1} + \beta \frac{f_\infty}{\beta+1} = f_\infty. \end{aligned}$$

□

D. Conditions under which $\lim_{\lambda\downarrow 0} \hat{f}(\lambda) = F_\infty$ implies $\lim_{t\to\infty} F(t) = F_\infty$.

The following theorem is due to Hardy and Littlewood in the scalar-valued case. Let $f \in L^1_{loc}(\mathbb{R}_+, X)$ and let $F(t) := \int_0^t f(s)\,ds$ $(t \geq 0)$.

Theorem 4.2.9 (Hardy-Littlewood). *Assume that $M := \sup_{t\geq\tau} t\, \| f(t) \| < \infty$ for some $\tau \geq 0$ and let $F_\infty \in X$. If $\lim_{\lambda\downarrow 0} \hat{f}(\lambda) = F_\infty$, then $\lim_{t\to\infty} F(t) = F_\infty$.*

Proof. Replacing f by $f(\cdot + \tau)$ we can assume that $\tau = 0$. For $t > 0$ we have

$$\begin{aligned} \left\| F(t) - \hat{f}\left(\tfrac{1}{t}\right) \right\| &= \left\| \int_0^t f(s)(1 - e^{-s/t})\,ds - \int_t^\infty f(s) e^{-s/t}\,ds \right\| \\ &\leq M \left(\sup_{0<s<t} t \frac{1-e^{-s/t}}{s} + \int_1^\infty e^{-r} \frac{dr}{r} \right) \\ &\leq M \left(\sup_{0<r\leq 1} \frac{1-e^{-r}}{r} + \int_1^\infty e^{-r} \frac{dr}{r} \right) < \infty. \end{aligned}$$

Since $\lim_{t\to\infty} \hat{f}(\frac{1}{t}) = F_\infty$, it follows that F is bounded. But $\text{A-}\lim_{t\to\infty} F(t) = \lim_{\lambda\downarrow 0} \hat{f}(\lambda) = F_\infty$. It follows from Theorem 4.2.7 that $\text{C-}\lim_{t\to\infty} F(t) = F_\infty$. By

Example 4.2.4, F is feebly oscillating. Now Theorem 4.2.5 implies that $\lim_{t\to\infty} F(t) = F_\infty$. □

The following classical example is a beautiful application of the Hardy-Littlewood theorem.

Example 4.2.10. Let $f(t) := \frac{\sin t}{t}$. Then $\hat{f}(\lambda) = -\arctan\lambda + \frac{\pi}{2}$. In fact,

$$\begin{aligned} \frac{d}{d\lambda}\hat{f}(\lambda) &= \int_0^\infty e^{-\lambda t}(-t)f(t)\,dt = -\int_0^\infty e^{-\lambda t}\sin t\,dt \\ &= -\int_0^\infty e^{-\lambda t}\frac{1}{2i}(e^{it} - e^{-it})\,dt \\ &= -\frac{1}{2i}\left(\frac{1}{\lambda - i} - \frac{1}{\lambda + i}\right) = \frac{-1}{1+\lambda^2} \\ &= -\frac{d}{d\lambda}\arctan\lambda. \end{aligned}$$

Since $\lim_{\lambda\to\infty}\hat{f}(\lambda) = 0$, the claim follows. Now by Theorem 4.2.9, it follows that $\int_0^\infty \frac{\sin t}{t}\,dt := \lim_{t\to\infty}\int_0^t \frac{\sin s}{s}\,ds = \lim_{\lambda\downarrow 0}\hat{f}(\lambda) = \frac{\pi}{2}$. □

If f is positive, then $\hat{f}(\lambda)$ is decreasing. So in the real-valued case it is clear that

$$\begin{aligned} \lim_{\lambda\downarrow 0}\hat{f}(\lambda) &= \sup_{\lambda>0}\hat{f}(\lambda) = \sup_{\lambda>0}\sup_{t>0}\int_0^t e^{-\lambda s}f(s)\,ds \\ &= \sup_{t>0}\sup_{\lambda>0}\int_0^t e^{-\lambda s}f(s)\,ds = \sup_{t>0}F(t) = \lim_{t\to\infty}F(t). \end{aligned}$$

If X is an ordered Banach space we cannot argue like this (unless the norm is order continuous). However, using the preceding results we obtain:

Theorem 4.2.11. *Let X be an ordered Banach space with normal cone. Let $f \in L^1_{loc}(\mathbb{R}_+, X)$ and assume that $f(t) \geq 0$ $(t \geq 0)$ and* $\operatorname{abs}(f) \leq 0$. *If* $\lim_{\lambda\downarrow 0}\hat{f}(\lambda) = F_\infty$ *then* $\lim_{t\to\infty} F(t) = F_\infty$.

Proof. We have $\text{A-}\lim_{t\to\infty} F(t) = \lim_{\lambda\downarrow 0}\lambda\hat{F}(\lambda) = \lim_{\lambda\downarrow 0}\hat{f}(\lambda) = F_\infty$. It follows from Karamata's theorem (Theorem 4.2.8) that $\text{C-}\lim_{t\to\infty} F(t) = F_\infty$. Since F is increasing, the claim follows from Theorem 4.2.6. □

E. Conditions under which $\text{A-}\lim_{t\to\infty} f(t) = f_\infty$ implies $\lim_{t\to\infty} f(t) = f_\infty$.

Let $f \in L^1_{loc}(\mathbb{R}_+, X)$ and $F(t) = \int_0^t f(s)\,ds$ $(t \geq 0)$. Assume that $\operatorname{abs}(f) \leq 0$. Since $\lambda\hat{F}(\lambda) = \hat{f}(\lambda)$, every Tauberian theorem of type E yields one of type D. In order to go the other way around, we apply Tauberian theorems of type D to the function f_δ defined by

$$f_\delta(t) := \frac{1}{\delta}\left(f(t+\delta) - f(t)\right) \quad (t \geq 0).$$

Lemma 4.2.12. *Let $f_\infty \in X, \delta > 0$. Consider the following assertions:*

(i) $\lim_{t\to\infty} f(t) = f_\infty$.

(ii) $\lim_{t\to\infty} \frac{1}{\delta}\int_t^{t+\delta} f(s)\,ds = f_\infty$.

(iii) $\lim_{t\to\infty} \int_0^t f_\delta(s)\,ds = f_\infty - \frac{1}{\delta}\int_0^\delta f(s)\,ds$.

(iv) $\lim_{\lambda\downarrow 0} \widehat{f_\delta}(\lambda) = f_\infty - \frac{1}{\delta}\int_0^\delta f(s)\,ds$.

(v) $\text{A-}\lim_{t\to\infty} f(t) = f_\infty$.

Then (i) $\Rightarrow$ (ii) $\Leftrightarrow$ (iii) $\Rightarrow$ (iv) $\Leftrightarrow$ (v).

Proof. The implication (i) $\Rightarrow$ (ii) is obvious, and (ii) is equivalent to (iii) since

$$\int_0^t f_\delta(s)\,ds = \frac{1}{\delta}\int_t^{t+\delta} f(s)\,ds - \frac{1}{\delta}\int_0^\delta f(s)\,ds.$$

Since

$$\widehat{f_\delta}(\lambda) = \frac{1}{\lambda\delta}(e^{\lambda\delta}-1)\lambda\hat{f}(\lambda) - \frac{e^{\lambda\delta}}{\delta}\int_0^\delta e^{-\lambda s} f(s)\,ds, \tag{4.1}$$

(iv) is equivalent to (v). By Theorem 4.1.2, (iii) implies (iv). ☐

Remark 4.2.13. It is easy to see that f is B-convergent if (ii) or (iii) of Lemma 4.2.12 holds for all $\delta \in (0, \delta_0]$ for some $\delta_0 > 0$. See also the Notes for more general results.

Now we can obtain the following Tauberian theorem by applying the Hardy-Littlewood Tauberian theorem to f_δ.

Theorem 4.2.14. *Let $f \in L^1_{loc}(\mathbb{R}_+, X), f_\infty \in X, \delta_0 > 0, \tau \geq 0, M \geq 0$. Assume that*

$$t\|\, f(t) - f(s)\,\| \leq M \tag{4.2}$$

whenever $t \geq \tau, |s-t| \leq \delta_0$. If $\text{A-}\lim_{t\to\infty} f(t) = f_\infty$, then $\lim_{t\to\infty} f(t) = f_\infty$.

Proof. By the assumption, $\limsup_{t\to\infty} t\|\, f_\delta(t)\,\| < \infty$ for all $\delta \in (0, \delta_0)$. Moreover, f is slowly oscillating. Since $\text{A-}\lim_{t\to\infty} f(t) = f_\infty$, Lemma 4.2.12 gives us $\lim_{\lambda\downarrow 0}\widehat{f_\delta}(\lambda) = f_\infty - \frac{1}{\delta}\int_0^\delta f(s)\,ds$. Theorem 4.2.9 implies that $\lim_{t\to\infty}\int_0^t f_\delta(s)\,ds = f_\infty - \frac{1}{\delta}\int_0^\delta f(s)\,ds$ for all $\delta \in (0,\delta_0)$. Hence, $\text{B-}\lim_{t\to\infty} f(t) = f_\infty$ by Lemma 4.2.12 and Remark 4.2.13. Now the claim follows from Theorem 4.2.3. ☐

Now we can actually use Theorem 4.2.14 to prove a slight improvement of the Hardy-Littlewood theorem; i.e., we deduce a Tauberian theorem of type D from a result of type E as indicated above.

Theorem 4.2.15. *Let $f \in L^1_{loc}(\mathbb{R}_+, X)$, $F_\infty \in X$. Assume that for some $\delta > 0$,*

$$\limsup_{t\to\infty} \int_t^{t+\delta} r\| f(r) \| \, dr < \infty.$$

If $\lim_{\lambda\downarrow 0} \hat{f}(\lambda) = F_\infty$, then $\lim_{t\to\infty} F(t) = F_\infty$, where $F(t) := \int_0^t f(s)\, ds$.

Proof. We have

$$\begin{aligned} \limsup_{t\to\infty} \sup_{t\le s\le t+\delta} t\| F(t) - F(s) \| &\le \limsup_{t\to\infty} t \int_t^{t+\delta} \| f(s) \| \, ds \\ &\le \limsup_{t\to\infty} \int_t^{t+\delta} s\| f(s) \| \, ds < \infty. \end{aligned}$$

Thus, F satisfies (4.2). Since $\text{A-}\lim_{t\to\infty} F(t) = \lim_{\lambda\downarrow 0} \hat{f}(\lambda) = F_\infty$, it follows from Theorem 4.2.14 that $\lim_{t\to\infty} F(t) = F_\infty$. □

Next, we consider an order condition.

Theorem 4.2.16. *Let X be an ordered Banach space with normal cone, let $f \in L^1_{loc}(\mathbb{R}_+, X)$ such that $t^k f(t)$ is positive and increasing on $[\tau, \infty)$ for some $\tau \ge 0$ and some $k \in \mathbb{N}$. If $\text{A-}\lim_{t\to\infty} f(t) = f_\infty$, then $\lim_{t\to\infty} f(t) = f_\infty$.*

Proof. Replacing $f(t)$ by $f(t+\tau)$ we can assume that $\tau = 0$. Since f is positive, it follows from Karamata's Theorem 4.2.8 that $\text{C-}\lim_{t\to\infty} f(t) = f_\infty$. Now Theorem 4.2.6 implies that $\lim_{t\to\infty} f(t) = f_\infty$. □

F. Power series.

Let $p(z) := \sum_{n=0}^\infty a_n z^n$ be a power series which converges for $|z| < 1$, where $a_n \in X$. Defining $f \in L^1_{loc}(\mathbb{R}_+, X)$ by

$$f(t) = a_n \quad \text{if} \quad t \in [n, n+1), \tag{4.3}$$

the preceding results yield Tauberian theorems for p. In fact, one has $\text{abs}(f) \le 0$ and

$$\hat{f}(\lambda) = \left(\frac{1-e^{-\lambda}}{\lambda}\right) \sum_{n=0}^\infty a_n e^{-\lambda n} \quad (\operatorname{Re}\lambda > 0). \tag{4.4}$$

Thus, from Theorem 4.2.9 we obtain the following Tauberian counterpart of Theorem 4.1.6.

Theorem 4.2.17 (Hardy). *Assume that $\sup_{n\in\mathbb{N}_0} n\| a_n \| < \infty$, and let $b_\infty \in X$. If $\lim_{z\uparrow 1} p(z) = b_\infty$, then $\sum_{n=0}^\infty a_n = b_\infty$.*

The special case where $\lim_{n\to\infty} na_n = 0$ had been proved by Tauber in 1897 and was the starting point of Tauberian theory.

In the case of power series, theorems of types D and E are equivalent. In fact, let $b_n := \sum_{k=0}^{n} a_k$, or equivalently, $a_0 = b_0, a_n = b_n - b_{n-1}$ $(n \in \mathbb{N})$. Then $q(z) := \sum_{n=0}^{\infty} b_n z^n$ also converges for $|z| < 1$. Moreover,

$$\sum_{k=0}^{\infty} a_k z^k = (1-z) \sum_{k=0}^{\infty} b_k z^k \quad (|z| < 1).$$

Thus, $\text{A-}\lim_{n\to\infty} b_n := \lim_{z\uparrow 1}(1-z)\sum_{k=0}^{\infty} b_k z^k = \lim_{z\uparrow 1}\sum_{k=0}^{\infty} a_k z^k$ whenever one of the limits exists. So we obtain the following

Corollary 4.2.18. *Let $b_n \in X$ such that $\sup_{n\in\mathbb{N}} n\| b_n - b_{n-1} \| < \infty$. If $\text{A-}\lim_{n\to\infty} b_n = b_\infty$, then $\lim_{n\to\infty} b_n = b_\infty$.*

G. Fourier series.

Let Y be a Banach space. Let $f : \mathbb{R} \to Y$ be a continuous 2π-periodic function. By $c_k := \frac{1}{2\pi}\int_{-\pi}^{\pi} f(x)e^{-ikx}dx \in Y$ $(k \in \mathbb{Z})$, we denote the *Fourier coefficients* of f and by

$$s_m(x) := \sum_{k=-m}^{m} c_k e^{ikx} \quad (x \in \mathbb{R})$$

the *Fourier sums*. It is well known that, in general, $s_m(x)$ does not converge as $m \to \infty$. However, it converges in the sense of Cesàro.

Theorem 4.2.19 (Fejér). *One has*

$$\lim_{n\to\infty} \frac{1}{n+1} \sum_{m=0}^{n} s_m(x) = f(x)$$

uniformly in $x \in \mathbb{R}$.

Proof. a) We show that for $m \in \mathbb{N}$, $y \in \mathbb{R}$, $\sum_{k=-m}^{m} e^{iky} = D_m(y)$, where

$$D_m(y) := \begin{cases} \dfrac{\cos my - \cos(m+1)y}{1-\cos y} & (y \notin 2\pi\mathbb{Z}) \\ 2m+1 & (y \in 2\pi\mathbb{Z}) \end{cases}$$

(the so-called *Dirichlet kernel*). In fact, $\sum_{k=-m}^{m} e^{iky} = 1 + \sum_{k=1}^{m}(e^{iky} + e^{-iky})$ is real. Thus,

$$\begin{aligned} (1-\cos y)\sum_{k=-m}^{m} e^{iky} &= \text{Re}\left((1-e^{iy})\sum_{k=-m}^{m} e^{iky}\right) \\ &= \text{Re}(e^{-imy} - e^{i(m+1)y}) \\ &= \cos my - \cos(m+1)y. \end{aligned}$$

b) We obtain from a) and periodicity,

$$\begin{aligned} s_m(x) &= \sum_{k=-m}^{m} \frac{1}{2\pi}\int_{-\pi}^{\pi} f(t)e^{-ikt}dt\, e^{ikx} = \frac{1}{2\pi}\int_{-\pi}^{\pi} f(t)\sum_{k=-m}^{m} e^{ik(x-t)}dt \\ &= \frac{1}{2\pi}\int_{-\pi}^{\pi} f(t)D_m(x-t)\,dt = \frac{1}{2\pi}\int_{-\pi}^{\pi} f(x-t)D_m(t)\,dt. \end{aligned}$$

c) Let

$$\begin{aligned} \sigma_n(x) &:= \frac{1}{n}\sum_{m=0}^{n-1} s_m(x) \\ &= \frac{1}{2\pi n}\int_{-\pi}^{\pi} f(x-t)\frac{1}{1-\cos t}\sum_{m=0}^{n-1}(\cos mt - \cos(m+1)t)\,dt \\ &= \frac{1}{2\pi n}\int_{-\pi}^{\pi} f(x-t)\frac{1-\cos nt}{1-\cos t}\,dt \\ &= \frac{1}{2\pi}\int_{-\pi}^{\pi} f(x-t)F_n(t)\,dt, \end{aligned}$$

where

$$F_n(t) = \begin{cases} \dfrac{1}{n}\left(\dfrac{\sin\frac{nt}{2}}{\sin\frac{t}{2}}\right)^2 & (t \notin 2\pi\mathbb{Z}) \\ n & (t \in 2\pi\mathbb{Z}) \end{cases}$$

(the so-called *Fejér kernel*). Here, we used that $\cos a = 1 - 2(\sin\frac{a}{2})^2$ $(a \in \mathbb{R})$.

d) We have to show that $\sigma_n(x)$ converges uniformly to $f(x)$. If $f \equiv 1$, then clearly $\sigma_n(x) = 1$. It follows from c) that $\frac{1}{2\pi}\int_{-\pi}^{\pi} F_n(t)\,dt = 1$ $(n \in \mathbb{N})$. Moreover, $F_n(t) \geq 0$ $(t \in (-\pi,\pi))$. Since $F_n(t) = F_n(-t)$, we have

$$\begin{aligned} \sigma_n(x) &= \frac{1}{2\pi}\int_{-\pi}^{\pi} f(x+t)F_n(t)\,dt \\ &= \frac{1}{2\pi}\int_{-\pi}^{\pi} \frac{f(x+t)+f(x-t)}{2}F_n(t)\,dt. \end{aligned}$$

Thus, $\sigma_n(x) - f(x) = \frac{1}{2\pi}\int_{-\pi}^{\pi} F_n(t)g(t)\,dt$, where $g(t) := \frac{1}{2}(f(x+t)+f(x-t)) - f(x)$. Let $\varepsilon > 0$. Since f is uniformly continuous, there exists $\delta > 0$ such that $\| g(t) \| \leq \varepsilon$ whenever $|t| \leq \delta$. Thus,

$$\begin{aligned} \left\| \int_{-\delta}^{\delta} F_n(t)g(t)\,dt \right\| &\leq \int_{-\delta}^{\delta} F_n(t)\,\| g(t) \|\,dt \\ &\leq \varepsilon\int_{-\delta}^{\delta} F_n(t)\,dt \leq \varepsilon\cdot 2\pi \end{aligned}$$

and

$$\left\| \int_\delta^\pi F_n(t)g(t)\,dt \right\| \le \|g\|_\infty (\pi-\delta)\frac{1}{n}\frac{1}{(\sin\delta/2)^2} < \|g\|_\infty \frac{\pi}{n}(\sin\delta/2)^{-2}.$$

Similarly, $\| \int_{-\pi}^{-\delta} F_n(t)g(t)\,dt \| \le \| g \|_\infty \frac{\pi}{n}(\sin\delta/2)^{-2}$. Hence, $\| \sigma_n(x) - f(x) \| \le \varepsilon + n^{-1}\| g \|_\infty (\sin\delta/2)^{-2}$. Consequently, $\limsup_{n\to\infty} \| \sigma_n - f \| \le \varepsilon$. ☐

Now we deduce the following from the Tauberian theorem Corollary 4.2.18.

Theorem 4.2.20. *Assume that $f : \mathbb{R} \to Y$ is a continuous 2π-periodic function which is of bounded semivariation on $[-\pi,\pi]$. Then $\lim_{n\to\infty} s_n(x) = f(x)$ uniformly in $x \in \mathbb{R}$.*

Proof. Let $k \in \mathbb{Z}, k \ne 0$. Then $c_k = \frac{1}{2\pi i k}\int_{-\pi}^{\pi} e^{-ikx}df(x)$ by integration by parts. Hence, there exists a constant $c \ge 0$ such that $|k|\,\| c_k \| \le c$ $(k \in \mathbb{Z})$. Consequently, $m\|s_m - s_{m-1}\|_\infty \le m(\| c_m \| + \| c_{-m} \|) \le 2c$. Now the claim follows from Theorem 4.2.9 and Corollary 4.2.18 by choosing as X the Banach space of all continuous 2π-periodic functions on $\mathbb{R}$ with values in Y with the uniform norm $\| \cdot \|_\infty$. ☐

H. Inversion of Laplace transforms.

Now we give a result for Laplace transforms which corresponds to Theorem 4.2.19 and which leads to a proof of the Complex Inversion Theorem 2.3.4. Here X is any Banach space.

Theorem 4.2.21. *Let $f \in L^1_{loc}(\mathbb{R}_+, X)$ such that* $\mathrm{abs}(\|f\|) < \infty$.

a) *If $\omega > \mathrm{abs}(\|f\|)$, then*

$$\lim_{R\to\infty} \int_0^a \left\| \frac{1}{R}\int_0^R \frac{1}{2\pi i}\int_{\omega-ir}^{\omega+ir} e^{\lambda t}\hat{f}(\lambda)\,d\lambda - f(t) \right\| dt = 0$$

for all $a > 0$.

b) *If f is continuous and exponentially bounded, $f(0) = 0$, and $\omega > \omega(f)$, then*

$$\text{C-}\lim_{r\to\infty} \frac{1}{2\pi i}\int_{\omega-ir}^{\omega+ir} e^{\lambda t}\hat{f}(\lambda)\,d\lambda = f(t)$$

uniformly for $t \in [0,a]$, for all $a \ge 0$.

Proof. Let $\Phi(t) := \frac{1}{2\pi}\left(\frac{\sin(t/2)}{t/2}\right)^2$ $(t \in \mathbb{R})$. Then $\Phi \in L^1(\mathbb{R})$, and the inverse Fourier transform of Φ is given by

$$(\mathcal{F}^{-1}\Phi)(s) = \begin{cases} \frac{1}{2\pi}(1-|s|) & (|s| \le 1), \\ 0 & (|s| > 1). \end{cases}$$

Let $\Phi_R(t) := R\Phi(Rt)$, so $(\mathcal{F}^{-1}\Phi_R)(s) = (\mathcal{F}^{-1}\Phi)(s/R)$, and $\{\Phi_R\}$ acts an approximate unit on $L^1(\mathbb{R}, X)$ and $C_0(\mathbb{R}, X)$ as $R \to \infty$ (see Lemma 1.3.3).

Consider f as a function on $\mathbb{R}$ with $f(s) = 0 \ (s < 0)$. Let $\omega > \operatorname{abs}(\|f\|)$, $t \geq 0$. Put

$$g(s) := e^{-\omega(s+t)} f(s+t) \qquad (s \in \mathbb{R}).$$

Then $g \in L^1(\mathbb{R}, X)$ and

$$(\mathcal{F}g)(s) = e^{ist} \hat{f}(\omega + is).$$

By Fubini's theorem and Theorem 1.8.1 b),

$$\begin{aligned}
&\frac{1}{R}\int_0^R \frac{1}{2\pi i}\int_{\omega-ir}^{\omega+ir} e^{\lambda t}\hat{f}(\lambda)\,d\lambda\,dr \\
&= \frac{1}{2\pi}e^{\omega t}\frac{1}{R}\int_{-R}^{R}\int_{|s|}^{R} e^{ist}\hat{f}(\omega+is)\,dr\,ds \\
&= e^{\omega t}\int_{-R}^{R}\frac{1}{2\pi}\left(1-\frac{|s|}{R}\right)(\mathcal{F}g)(s)\,ds \\
&= e^{\omega t}\int_{-\infty}^{\infty}\Phi_R(s)g(s)\,ds \\
&= e^{\omega t}(\Phi_R * f_\omega)(t),
\end{aligned}$$

where $f_\omega(s) := e^{-\omega s}f(s)$ (note that $\Phi_R(s) = \Phi_R(-s)$). Now a) and b) follow, since $\|\Phi_R * f_\omega - f_\omega\|_1 \to 0$ if $\omega > \operatorname{abs}(\|f\|)$ and $\|\Phi_R * f_\omega - f_\omega\|_\infty \to 0$ if $f_\omega \in C_0(\mathbb{R}, X)$. □

Now we are able to give the following proof of Theorem 2.3.4.

Proof of Theorem 2.3.4. Let $F \in \operatorname{Lip}_0(\mathbb{R}_+, X)$. For $\operatorname{Re}\lambda = \omega > 0$, we have $\hat{F}(\lambda) = \widehat{dF}(\lambda)/\lambda$ and

$$\|\widehat{dF}(\lambda)\| \leq \int_0^\infty e^{-\omega t}\|F\|_{\operatorname{Lip}_0(\mathbb{R}_+,X)}\,dt = \|F\|_{\operatorname{Lip}_0(\mathbb{R}_+,X)}/\omega.$$

For $0 \leq t \leq a$ and $S > R > 0$,

$$\begin{aligned}
&\left\|\frac{1}{2\pi i}\int_{\omega-iR}^{\omega+iR} e^{\lambda t}\hat{F}(\lambda)\,d\lambda - \frac{1}{2\pi i}\int_{\omega-iS}^{\omega+iS} e^{\lambda t}\hat{F}(\lambda)\,d\lambda\right\| \\
&\leq \frac{e^{\omega a}}{\pi}\int_R^S \frac{\|F\|_{\operatorname{Lip}_0(\mathbb{R}_+,X)}}{\omega|\omega+is|}\,ds \\
&\leq \frac{e^{\omega a}}{\pi\omega}\|F\|_{\operatorname{Lip}_0(\mathbb{R}_+,X)}\log(S/R).
\end{aligned}$$

Hence, $R \mapsto \frac{1}{2\pi i}\int_{\omega-iR}^{\omega+iR} e^{\lambda\cdot}\hat{F}(\lambda)\,d\lambda$ is feebly oscillating as a function from $\mathbb{R}_+$ to $C([0,a], X)$. It follows from Theorem 4.2.21 and Theorem 4.2.5 that

$$\frac{1}{2\pi i}\int_{\omega-iR}^{\omega+iR} e^{\lambda t}\hat{F}(\lambda)\,d\lambda \to F(t)$$

uniformly for $t \in [0, a]$. □

4.3 Ergodic Semigroups

This section interrupts the general theme of this chapter: we consider convergence in mean of semigroups. This is an interesting illustration of some of the results in Sections 4.1 and 4.2. Moreover, the notions introduced here will be useful in the context of almost periodic functions which form the subject of Section 4.5. We also prove a striking result due to Lotz: every C_0-semigroup on an L^∞-space has a bounded generator (Corollary 4.3.19).

Let A be an operator on a Banach space X such that $(0, \lambda_0) \subset \rho(A)$ for some $\lambda_0 > 0$ and

$$M := \sup_{0<\lambda<\lambda_0} \|\lambda R(\lambda, A)\| < \infty. \tag{4.5}$$

Note that (4.5) is satisfied if A generates a bounded C_0-semigroup T. In fact, then $R(\lambda, A) = \int_0^\infty e^{-\lambda t} T(t)\, dt$ ($\lambda > 0$) and (4.5) holds for $M = \sup_{t\geq 0} \|T(t)\|$. Denote by $\operatorname{Ker} A := \{x \in D(A) : \ Ax = 0\}$ the *kernel* of A and by $\operatorname{Ran} A := \{Ax : \ x \in D(A)\}$ the *range* of A. Let $x \in X$. Since

$$AR(\lambda, A)x = \lambda R(\lambda, A)x - x \quad (0 < \lambda < \lambda_0),$$

it follows from (4.5) that $x \in \overline{\operatorname{Ran} A}$ if and only if

$$\lim_{\lambda \downarrow 0} \lambda R(\lambda, A)x = 0. \tag{4.6}$$

Moreover,

$$\lambda R(\lambda, A)x = x \ (0 < \lambda < \lambda_0) \quad \text{if and only if} \quad x \in \operatorname{Ker} A. \tag{4.7}$$

In particular,

$$\operatorname{Ker} A \cap \overline{\operatorname{Ran} A} = \{0\}. \tag{4.8}$$

In what follows, as elsewhere, limits in Banach spaces are norm-limits unless specified otherwise.

Proposition 4.3.1. *Let A be an operator satisfying* (4.5) *and let $x \in X$.*

a) *The following assertions are equivalent:*

   (i) *There exist $\lambda_n \downarrow 0$ such that $\lambda_n R(\lambda_n, A)x$ converges weakly as $n \to \infty$.*

   (ii) *$x_0 := \lim_{\lambda\downarrow 0} \lambda R(\lambda, A)x$ exists.*

   (iii) *$x \in \operatorname{Ker} A + \overline{\operatorname{Ran} A}$.*

   In that case, $x_0 \in \operatorname{Ker} A$ and $x - x_0 \in \overline{\operatorname{Ran} A}$.

b) *If A generates a bounded C_0-semigroup, then* (i)–(iii) *are equivalent to*

   (iv) *$x_1 := \lim_{t\to\infty} \frac{1}{t} \int_0^t T(s)x\, ds$ exists.*

In that case, $x_1 = x_0$.

Proof. a) (i) $\Rightarrow$ (iii): Assume that $\lambda_n R(\lambda_n, A)x$ converges weakly to y as $n \to \infty$. By the resolvent equation, we have

$$\mu R(\mu, A)\lambda_n R(\lambda_n, A)x = \frac{\mu}{\mu - \lambda_n}(\lambda_n R(\lambda_n, A)x) - \frac{\mu\lambda_n}{\mu - \lambda_n} R(\mu, A)x.$$

for all $\mu \in (0, \lambda_0)$. Taking weak limits gives $\mu R(\mu, A)y = y$. It follows from (4.7) that $y \in \operatorname{Ker} A$. Since $\lambda_n R(\lambda_n, A)x - x = AR(\lambda_n, A)x \in \operatorname{Ran} A$, it follows that $y - x$ is in the weak closure of $\operatorname{Ran} A$, which coincides with the norm closure.

(iii) $\Rightarrow$ (ii): This follows from (4.6) and (4.7).

(ii) $\Rightarrow$ (i): This is trivial.

b) Since the Laplace transform of $u(t) := T(t)x$ is given by $\hat{u}(\lambda) = R(\lambda, A)x$, it follows from the Abelian Theorem 4.1.2 that (iv) implies (ii), and from the Tauberian Theorem 4.2.7 that (ii) implies (iv). □

Corollary 4.3.2. *Let A be an operator satisfying* (4.5).

a) *The following assertions are equivalent:*

(i) $Px := \lim_{\lambda \downarrow 0} \lambda R(\lambda, A)x$ *exists for all* $x \in X$.

(ii) $X = \operatorname{Ker} A \oplus \overline{\operatorname{Ran} A}$.

In that case, P is the projection onto $\operatorname{Ker} A$ *along* $\overline{\operatorname{Ran} A}$.

b) *If $D(A)$ is dense, then* (i) *and* (ii) *are equivalent to*

(iii) $\operatorname{Ker} A$ *separates* $\operatorname{Ker} A^*$.

Note that (iii) means, by definition, that for all $x^* \in \operatorname{Ker} A^*$ such that $x^* \neq 0$ there exists $x \in \operatorname{Ker} A$ such that $\langle x, x^* \rangle \neq 0$.

Proof. a) follows directly from Proposition 4.3.1 and (4.8).

b) Assume that $D(A)$ is dense.

(ii) $\Rightarrow$ (iii): Let $x^* \in \operatorname{Ker} A^*$, $x^* \neq 0$. Then $\langle Ax, x^* \rangle = 0$ for all $x \in D(A)$. Hence $\langle y, x^* \rangle = 0$ for all $y \in \overline{\operatorname{Ran} A}$. Since $X = \operatorname{Ker} A \oplus \overline{\operatorname{Ran} A}$, it follows that $\langle x, x^* \rangle \neq 0$ for some $x \in \operatorname{Ker} A$.

(iii) $\Rightarrow$ (i): We first show that $\operatorname{Ker} A + \operatorname{Ran} A$ is dense in X. In fact, otherwise there exists $x^* \in X^* \setminus \{0\}$ vanishing on $\operatorname{Ker} A + \operatorname{Ran} A$. Since x^* vanishes on $\operatorname{Ran} A$, one has $x^* \in \operatorname{Ker} A^*$. Thus, condition (iii) is violated, and the claim is proved. It follows from (4.6) and (4.7) that $(\lambda R(\lambda, A)x)$ converges as $\lambda \downarrow 0$ for $x \in \operatorname{Ker} A + \operatorname{Ran} A$. This implies convergence for all $x \in X$ by density. □

Definition 4.3.3. *A C_0-semigroup T on X is called* Cesàro-ergodic (*or* mean-ergodic) *if*

$$Qx := \lim_{t \to \infty} \frac{1}{t} \int_0^t T(s)x\, ds \tag{4.9}$$

exists for all $x \in X$. *The semigroup is called* Abel-ergodic *if its generator* A *satisfies* (4.5) *and the equivalent conditions* (i), (ii), (iii) *of Corollary* 4.3.2 *are satisfied.*

Proposition 4.3.4. *Let* T *be a* C_0*-semigroup with generator* A.

a) *If* T *is Cesàro-ergodic, then* T *is Abel-ergodic and* Q *given by* (4.9) *is the projection onto* $\operatorname{Ker} A$ *along* $\overline{\operatorname{Ran} A}$.

b) *Assume that* T *is bounded and Abel-ergodic. Then* T *is Cesàro-ergodic.*

Proof. a) Assume that T is Cesàro-ergodic. It follows from the uniform boundedness principle that there exists $M \geq 0$ such that $\frac{1}{t}\left\|\int_0^t T(s)x\,ds\right\| \leq M\|x\|$ for all $t > 0$, $x \in X$. It follows from Theorem 3.1.7 that $(0,\infty) \subset \rho(A)$ and $R(\lambda, A)x = \int_0^\infty e^{-\lambda t}T(t)x\,dt$ for all $\lambda > 0$, $x \in X$. Integration by parts yields

$$\begin{aligned}
\|\lambda R(\lambda, A)x\| &= \left\|\int_0^\infty \lambda^2 e^{-\lambda t} \int_0^t T(s)x\,ds\,dt\right\| \\
&\leq M\int_0^\infty \lambda^2 e^{-\lambda t} t\,dt\,\|x\| \\
&= M\|x\| \quad (\lambda > 0).
\end{aligned}$$

Thus, condition (4.5) is satisfied. Since $u(t) := T(t)x$ has Laplace tranform $\hat{u}(\lambda) = R(\lambda, A)x$, the claim now follows from Theorem 4.1.2 and Proposition 4.3.1 a).

b) If T is bounded by M, then condition (4.5) is satisfied. So the claim follows from Proposition 4.3.1. ☐

Corollary 4.3.5. *Every bounded* C_0*-semigroup on a reflexive Banach space* X *is Cesàro-ergodic.*

Proof. Let $x \in X$. It follows from reflexivity that condition (i) of Proposition 4.3.1 is satisfied. Thus, T is Abel-ergodic and the claim follows from Proposition 4.3.4. ☐

Condition (iii) of Corollary 4.3.2 is frequently the most convenient way to verify Cesàro-ergodicity of a bounded C_0-semigroup. Note that the dual assertion is always true: the space $\operatorname{Ker} A^*$ separates $\operatorname{Ker} A$ whenever A generates a bounded C_0-semigroup. More generally, the following holds.

Proposition 4.3.6. *Let* A *be a densely defined operator satisfying* (4.5). *Let* $0 \neq x \in \operatorname{Ker} A$. *Then there exists* $x^* \in \operatorname{Ker} A^*$ *such that* $\langle x, x^*\rangle \neq 0$.

Proof. Let $y^* \in X^*$ such that $\langle x, y^*\rangle = 1$. Let x^* be a weak* limit point of $\lambda R(\lambda, A)^* y^*$ as $\lambda \downarrow 0$. Since $\lambda R(\lambda, A)x = x$, it follows that $\langle x, x^*\rangle = 1$. Let $y \in D(A)$. Then

$$\lim_{\lambda\downarrow 0} \lambda R(\lambda, A)Ay = \lim_{\lambda\downarrow 0} \lambda(\lambda R(\lambda, A)y - y) = 0.$$

Hence,

$$\langle Ay, x^* \rangle = \lim_{\lambda \downarrow 0} \langle \lambda R(\lambda, A) Ay, y^* \rangle = 0.$$

Thus, $x^* \in \operatorname{Ker} A^*$. □

We have shown that for bounded C_0-semigroups Abel- and Cesàro-ergodicity are equivalent. This is not always the case for arbitrary C_0-semigroups as the following easy example shows: Let $X := \mathbb{C}^2$, $T(t) := e^{it} \begin{pmatrix} 1 & t \\ 0 & 1 \end{pmatrix}$. Then T is Abel-ergodic but not Cesàro-ergodic. However, by Karamata's theorem (Theorem 4.2.8) both notions do coincide for positive semigroups.

Theorem 4.3.7. *Let T be a positive C_0-semigroup on an ordered Banach space with normal cone. Then T is Abel-ergodic if and only if T is Cesàro-ergodic.*

Next, we extend the results on Cesàro-ergodicity to Lipschitz continuous integrated semigroups.

Let A be an operator on X satisfying the Hille-Yosida condition

$$(0, \infty) \subset \rho(A) \text{ and } \|(\lambda R(\lambda, A))^n\| \leq M \tag{4.10}$$

for all $\lambda > 0$, $n \in \mathbb{N}_0$ and some $M \geq 0$. Then A generates a Lipschitz continuous once integrated semigroup S. More precisely, there exists $S : \mathbb{R}_+ \to \mathcal{L}(X)$ satisfying

$$S(0) = 0, \quad \|S(t) - S(s)\| \leq M|t - s| \quad (s, t > 0)$$

and

$$R(\lambda, A) = \lambda \int_0^\infty e^{-\lambda t} S(t)\, dt \quad (\lambda > 0);$$

see Theorem 3.3.1 and Section 3.5.

If $D(A)$ is dense, then A generates a bounded C_0-semigroup T and $\frac{1}{t}S(t) = \frac{1}{t}\int_0^t T(s)\, ds$. If $D(A)$ is not dense, then T does not exist on X, but the Cesàro means $\frac{1}{t}S(t)$ still make sense. Thus, the following theorem generalizes Proposition 4.3.4, but is more complicated to prove in this more general context.

Theorem 4.3.8. *Let A be the generator of a Lipschitz continuous once integrated semigroup S. The following conditions are equivalent:*

(i) $X = \operatorname{Ker} A \oplus \overline{\operatorname{Ran} A}$.

(ii) $P := \lim_{t \to \infty} \frac{1}{t} S(t)$ *exists in the strong operator topology.*

In that case, P is the projection onto $\operatorname{Ker} A$ *along* $\overline{\operatorname{Ran} A}$.

Proof. (i) ⇒ (ii): If $x \in \operatorname{Ker} A$, then $S(t)x = tx$ $(t > 0)$. Next, let $x = Ay \in \operatorname{Ran} A$. We show that

$$\lim_{t \to \infty} \frac{1}{t} S(t)x = 0.$$

Note that $\|\frac{1}{t}S(t)\| \leq M$. Since by Lemma 3.2.2,

$$S(t)y = ty + \int_0^t S(s)Ay\, ds,$$

one has

$$\lim_{t\to\infty} \frac{1}{t^2} \int_0^t S(s)Ay\, ds = 0.$$

By the Abelian Theorem 4.1.2, it follows that

$$\lim_{t\to\infty} \frac{1}{t} \int_0^t \frac{1}{s^2} \int_0^s S(r)Ay\, dr\, ds = \text{C-}\lim_{t\to\infty} \frac{1}{t^2} \int_0^t S(s)Ay\, ds = 0.$$

Integration by parts yields

$$\begin{aligned}
\text{C-}\lim_{t\to\infty} \frac{1}{t} S(t)Ay &= \lim_{t\to\infty} \frac{1}{t} \int_0^t \frac{1}{s} S(s)Ay\, ds \\
&= \lim_{t\to\infty} \frac{1}{t} \int_0^t \frac{1}{s} \frac{d}{ds} \int_0^s S(r)Ay\, dr\, ds \\
&= \lim_{t\to\infty} \left\{ \frac{1}{t^2} \int_0^t S(r)Ay\, dr + \frac{1}{t} \int_0^t \frac{1}{s^2} \int_0^s S(r)Ay\, dr\, ds \right\} \\
&= 0.
\end{aligned}$$

Now observe that

$$\begin{aligned}
\left\| \frac{1}{t} S(t)x - \frac{1}{s} S(s)x \right\| &\leq \left\| \frac{1}{t}(S(t)x - S(s)x) \right\| + \left| \frac{1}{t} - \frac{1}{s} \right| \|S(s)x\| \\
&\leq \frac{1}{t} |t-s| 2M \|x\|.
\end{aligned}$$

Thus, the function $\frac{1}{t}S(t)Ay$ is feebly oscillating. It follows from the Tauberian Theorem 4.2.5 that

$$\lim_{t\to\infty} \frac{1}{t} S(t)Ay = \text{C-}\lim_{t\to\infty} \frac{1}{t} S(t)Ay = 0.$$

We have shown that $\frac{1}{t}S(t)x$ converges to Px as $t \to \infty$ for $x \in \operatorname{Ker} A + \operatorname{Ran} A$, where P is the projection onto $\operatorname{Ker} A$ along $\overline{\operatorname{Ran} A}$. Since $\|\frac{1}{t}S(t)\| \leq M$, it follows that $\lim_{t\to\infty} \frac{1}{t}S(t)x = Px$ for all $x \in X$.

(ii) $\Rightarrow$ (i): This is an Abelian theorem whose proof is analogous to Theorem 4.1.2 c): Let $x \in X$. By Corollary 4.3.2, we have to show that

$$\lim_{\lambda\downarrow 0} \lambda R(\lambda, A)x = \lim_{\lambda\downarrow 0} \lambda^2 \int_0^\infty e^{-\lambda t} S(t)x\, dt = Px.$$

Let $\varepsilon > 0$. There exists t_0 such that $\|\frac{1}{t}S(t)x - Px\| \le \varepsilon$ for all $t \ge t_0$. Hence

$$\begin{aligned}\limsup_{\lambda\downarrow 0} \|\lambda R(\lambda, A)x - Px\| &= \limsup_{\lambda\downarrow 0} \left\| \int_0^\infty \lambda^2 e^{-\lambda t} t \left(\frac{1}{t} S(t)x - Px \right) dt \right\| \\ &\le \limsup_{\lambda\downarrow 0} \int_0^{t_0} \lambda^2 e^{-\lambda t} t \left\| \frac{1}{t} S(t)x - Px \right\| dt + \varepsilon \\ &= \varepsilon.\end{aligned}$$

□

Example 4.3.9. Consider the operator A on $X := C[0,1]$ given by

$$D(A) := \{ f \in C^2[0,1] : f(0) = f(1), f''(0) = 0 \}, \quad Af := f''.$$

It is not difficult to see that A is dissipative and $(I - A)D(A) = X$. By Corollary 3.4.6, A generates a once integrated semigroup S satisfying

$$\|S(t) - S(s)\| \le |t - s|.$$

Let $f \in X$. Then the constant function $f(0) \in \operatorname{Ker} A$ and $f - f(0) = Ag \in \operatorname{Ran} A$, where

$$g(x) := \int_0^x \int_0^t (f(s) - f(0))\, ds\, dt - x \int_0^1 \int_0^t (f(t) - f(0))\, ds\, dt.$$

It follows from Theorem 4.3.8 that

$$\lim_{t\to\infty} \frac{1}{t} (S(t)f)(x) = f(0)$$

uniformly for all $x \in [0,1]$, for all $f \in X$. □

Next, we fix a bounded C_0-semigroup T with generator A. Recall that the operator $A - i\eta$ generates the C_0-semigroup $(e^{-i\eta t}T(t))_{t\ge 0}$, and we may consider ergodicity for the rescaled semigroup.

Definition 4.3.10. a) *Let $\eta \in \mathbb{R}$. A vector $x \in X$ is called* ergodic *at η (with respect to T) if the mean*

$$M_\eta x := \lim_{t\to\infty} \frac{1}{t} \int_0^t e^{-i\eta s} T(s)x\, ds$$

converges in norm.

b) *A vector x is called* totally ergodic *(with respect to T) if x is ergodic at η for all $\eta \in \mathbb{R}$.*

c) *The semigroup T is called* totally ergodic *if each vector x is totally ergodic; i.e., if the semigroup $(e^{-i\eta t}T(t))_{t\ge 0}$ is Cesàro-ergodic for all $\eta \in \mathbb{R}$.*

By X_e we denote the space of all totally ergodic vectors in X and by X_{e0} the space of all vectors $x \in X_e$ such that $M_\eta x = 0$ for all $\eta \in \mathbb{R}$. Both spaces X_e and X_{e0} are closed and invariant under the semigroup.

It follows from Proposition 4.3.1 that a vector $x \in X$ is totally ergodic if and only if

$$x_\eta := \lim_{\alpha \downarrow 0} \alpha R(\alpha + i\eta, A)x$$

exists for all $\eta \in \mathbb{R}$. In that case, $x_\eta = M_\eta x$. Moreover, $M_\eta x \in \operatorname{Ker}(A - i\eta)$ and $x - M_\eta x \in \overline{\operatorname{Ran}(A - i\eta)}$. In particular,

$$\begin{aligned} T(t)M_\eta x &= e^{i\eta t} M_\eta x \quad (t \geq 0), \\ R(\lambda, A)M_\eta x &= \frac{1}{\lambda - i\eta} M_\eta x \quad (\lambda \in \rho(A),\ \lambda \neq i\eta). \end{aligned}$$

If $x \in X$ is totally ergodic we denote by

$$\operatorname{Freq}(x) := \{\eta \in \mathbb{R} :\ M_\eta x \neq 0\}$$

the set of all *frequencies* of X.

Proposition 4.3.11. *Let x be totally ergodic (with respect to T). Then* Freq(x) *is countable.*

Proof. We can assume that X is separable, as we may replace X by $\overline{\operatorname{span}}\{T(t)x : t \geq 0\}$. Let $M := \sup_{t \geq 0} \|T(t)\|$. Then

$$\|\alpha R(\alpha + i\eta, A)\| \leq M \quad (\alpha > 0, \eta \in \mathbb{R}).$$

Let $x_\eta := M_\eta x = \lim_{t \to \infty} \frac{1}{t} \int_0^t e^{-i\eta s} T(s)x\, ds$ $(\eta \in \mathbb{R})$. For $\eta \in \operatorname{Freq}(x)$, let $y_\eta := x_\eta / \|x_\eta\|$. Then $\|y_\eta\| = 1$ and $\alpha R(\alpha + i\eta, A)y_\eta = y_\eta$ $(\alpha > 0)$ and for $\mu \in \operatorname{Freq}(x)$, $\mu \neq \eta$,

$$\alpha R(\alpha + i\eta, A)y_\mu = \frac{\alpha}{\alpha + i\eta - i\mu} y_\mu \quad (\alpha > 0).$$

Let $\mu, \eta \in \operatorname{Freq}(x)$ such that $\eta > \mu$. Then for $\alpha > 0$,

$$\begin{aligned} M\|y_\eta - y_\mu\| &\geq \|\alpha R(\alpha + i\eta, A)(y_\eta - y_\mu)\| \\ &= \left\| y_\eta - \frac{\alpha}{\alpha + i\eta - i\mu} y_\mu \right\| \\ &\geq 1 - \left| \frac{\alpha}{\alpha + i\eta - i\mu} \right| \\ &= 1 - \left| \frac{1}{1 + i(\eta - \mu)/\alpha} \right|. \end{aligned}$$

Choosing $\alpha = \eta - \mu$, we obtain

$$M\|y_\eta - y_\mu\| \geq 1 - \left| \frac{1}{1+i} \right| = 1 - \left| \frac{1-i}{2} \right| = 1 - \frac{1}{\sqrt{2}} > 0.$$

If Freq(x) is uncountable this contradicts the fact that X is separable. $\square$

It follows from Corollary 4.3.5 that every bounded C_0-semigroup on a reflexive Banach space is totally ergodic. This can be generalized as follows.

Proposition 4.3.12. *Assume that the vector $x \in X$ has relatively weakly compact orbit $\{T(t)x : t \geq 0\}$. Then x is totally ergodic with respect to T.*

Proof. It follows from the assumption that $\{e^{-i\eta t}T(t)x : \ t \geq 0\}$ is relatively weakly compact for all $\eta \in \mathbb{R}$. By Krein's theorem [Meg98, Theorem 2.8.14], the closed convex hull $K := \overline{\text{co}}\{e^{-i\eta t}T(t)x : t \geq 0\}$ is weakly compact. Since $\alpha R(\alpha + i\eta, A)x = \int_0^\infty \alpha e^{-\alpha t} e^{-i\eta t} T(t)x\, dt \in K$ for all $\alpha > 0$, condition (i) of Proposition 4.3.1 is satisfied and so $\frac{1}{t}\int_0^t e^{-i\eta t}T(t)x\, dt$ converges as $t \to \infty$, by Proposition 4.3.1. $\square$

Next we consider positive C_0-semigroups on L^1-spaces.

Proposition 4.3.13. *Let $X = L^1(\Omega, \mu)$ where (Ω, μ) is a σ-finite measure space. Let T be a bounded positive C_0-semigroup on X. If $f \in X_+$ is ergodic at 0, then f is totally ergodic.*

Proof. Since $\alpha R(\alpha, A)f$ converges as $\alpha \downarrow 0$, the set $K := \{\alpha R(\alpha, A)f : 0 < \alpha \leq 1\}$ is relatively compact. It follows that the solid hull $\text{so}(K) := \{g \in X : |g| \leq k \text{ for some } k \in K\}$ is relatively weakly compact (by [AB85, Theorem 13.8]). Let $\eta \in \mathbb{R}$. Then

$$\begin{aligned} |\alpha R(\alpha + i\eta, A)f| &= \left| \int_0^\infty \alpha e^{-t(\alpha + i\eta)} T(t) f\, dt \right| \\ &\leq \int_0^\infty \alpha e^{-\alpha t} T(t) f\, dt = \alpha R(\alpha, A) f. \end{aligned}$$

Thus, $\alpha R(\alpha + i\eta, A)f \in \text{so}(K)$ for $0 < \alpha \leq 1$. Consequently, condition (i) of Proposition 4.3.1 is satisfied. $\square$

In particular, if T is a positive bounded C_0-semigroup on $L^1(\Omega, \mu)$, then T is totally ergodic whenever T is Cesàro-ergodic.

Here is a criterion which is sometimes convenient for verifying total ergodicity.

Proposition 4.3.14. *Let $X := L^1(\Omega, \mu)$, where (Ω, μ) is a σ-finite measure space. Let T be a positive bounded C_0-semigroup on X. Assume that there exists a function $u \in X$ such that $u > 0$ μ-a.e. and $T(t)u \leq u$ for all $t \geq 0$. Then T is totally ergodic.*

Proof. Suppose that $f \in X$ and $0 \leq f \leq u$. For $\alpha > 0$,

$$0 \leq \alpha R(\alpha, A)f \leq \alpha R(\alpha, A)u \leq u.$$

Since $\{g \in X : \ 0 \leq g \leq u\}$ is weakly compact (see [Sch74, Theorem II.5.10 and Proposition II.8.3]), it follows from Proposition 4.3.1 that $\lim_{\alpha \downarrow 0} \alpha R(\alpha, A)f$ exists.

Since $\{f \in X : 0 \le f \le u\}$ is total in X, it follows that T is Abel-ergodic, and hence totally ergodic by Proposition 4.3.4 and Proposition 4.3.13. □

In Proposition 4.3.14, the condition of a subinvariant strictly positive function cannot be omitted. For example, the shift semigroup T on $L^1(\mathbb{R}_+)$ given by

$$(T(t)f)(x) = f(x+t)$$

is not Cesàro-ergodic since $\operatorname{Ker} A = \{0\}$, but $\operatorname{Ker} A^* = \mathbb{C} \cdot 1$ (where 1 denotes the constant 1 function).

Next, we return to general Banach spaces, but we consider the special case where the Abel means in Proposition 4.3.1 converge with respect to the operator norm.

Let A be an operator on X. We say that 0 is a *simple pole* of the resolvent if $\{\lambda \in \mathbb{C} : 0 < |\lambda| < \varepsilon\} \subset \rho(A)$ for some $\varepsilon > 0$ and there exists $0 \neq P \in \mathcal{L}(X)$ such that

$$R(\lambda, A) - \frac{P}{\lambda} \tag{4.11}$$

has a holomorphic extension to the disc $B(0,\varepsilon) := \{\lambda \in \mathbb{C} : |\lambda| < \varepsilon\}$.

For example, if A generates a bounded C_0-semigroup T, $0 \in \sigma(A)$ and A has compact resolvent, then 0 is a simple pole. Indeed, if P is the spectral projection of A associated with $\{0\}$, then $A^m P = 0$ for some $m \in \mathbb{N}$ (see Proposition B.9 and the subsequent remarks). It follows that

$$T(t)P = \sum_{n=0}^{m-1} \frac{t^n A^n P}{n!} \qquad (t \ge 0).$$

Since T is bounded, $AP = 0$. Hence,

$$R(\lambda, A) = R(\lambda, A_Z)(I - P) + P/\lambda,$$

where $Z = (I - P)(X)$ and A_Z is the part of A in Z. Since $0 \notin \sigma(A_Z)$, the claim follows.

Proposition 4.3.15. *Let A be an operator on X. The following assertions are equivalent:*

(i) *There exists $\lambda_0 > 0$ such that $(0, \lambda_0) \subset \rho(A)$ and $P := \lim_{\lambda \downarrow 0} \lambda R(\lambda, A)$ converges in the operator norm.*

(ii) *0 is a simple pole of the resolvent, or $0 \in \rho(A)$.*

(iii) *The range $\operatorname{Ran} A$ is closed, $(0, \lambda_0) \subset \rho(A)$ for some $\lambda_0 > 0$ and $\lambda R(\lambda, A)$ converges in the strong operator topology as $\lambda \downarrow 0$.*

Proof. (i) $\Rightarrow$ (ii): It follows from Proposition 4.3.2 that $X = \operatorname{Ker} A \oplus \overline{\operatorname{Ran} A}$ and that P is the projection onto $\operatorname{Ker} A$ along $Y := \overline{\operatorname{Ran} A}$. Since P commutes with $R(\lambda, A)$, we may consider the part A_Y of A in Y. Then $(0, \lambda_0) \subset \rho(A_Y)$ and $R(\lambda, A_Y) = R(\lambda, A)|_Y$ for $\lambda \in (0, \lambda_0)$ (see Proposition B.8). Since $P = 0$ on Y, it follows that $\lim_{\lambda \downarrow 0} \|\lambda R(\lambda, A_Y)\| = 0$. So there exists $\lambda > 0$ such that $\|\lambda R(\lambda, A_Y)\| \leq \frac{1}{2}$. Hence, $\operatorname{dist}(\lambda, \sigma(A_Y)) \geq \|R(\lambda, A_Y)\|^{-1} \geq 2\lambda$ (see Corollary B.3). Thus $0 \in \rho(A_Y)$, and so $R(\lambda, A)|_Y$ has a holomorphic extension to $B(0, \varepsilon)$ for some $\varepsilon > 0$. Since $R(\lambda, A) = P/\lambda + R(\lambda, A_Y)(I - P)$ ($\lambda \in \rho(A_Y)$, $\lambda \neq 0$), this implies (ii). It also shows that $\operatorname{Ran} A = \operatorname{Ran}(A_Y) = Y$ is closed.

(ii) $\Rightarrow$ (i): It follows from (4.11) that $\lim_{\lambda \downarrow 0} \lambda R(\lambda, A) = P$ in $\mathcal{L}(X)$.

(i) $\Rightarrow$ (iii): Strong convergence is trivial from (i). It has been shown above that $\operatorname{Ran} A$ is closed.

(iii) $\Rightarrow$ (ii): By assumption, $Y := \operatorname{Ran} A$ is closed and $X = \operatorname{Ker} A \oplus Y$ by Corollary 4.3.2. Thus, the part A_Y in Y is invertible. Let P be the projection onto $\operatorname{Ker} A$ along Y. Since $R(\lambda, A) = P/\lambda + R(\lambda, A_Y)(I - P)$ ($\lambda \in \rho(A_Y)$, $\lambda \neq 0$), the result follows. □

Next, we show a very peculiar phenomenon on the space $X := L^\infty(\Omega, \mu)$, where (Ω, μ) is a measure space. Such a space has two remarkable properties concerning convergence of sequences, namely

$$(DP) \quad \begin{cases} x_n \to 0 \text{ weakly in } X \text{ and } x_n^* \to 0 \text{ weakly in } X^* \\ \text{implies } \langle x_n, x_n^* \rangle \to 0 \quad \text{as } n \to \infty; \end{cases}$$

and

$$(G) \quad \begin{cases} x_n^* \to 0 \text{ weak* in } X^* \text{ implies} \\ x_n^* \to 0 \text{ weakly in } X^* \text{ as } n \to \infty. \end{cases}$$

The first property is called the *Dunford-Pettis property.* A space having the second property is called a *Grothendieck space.* It is obvious that the properties (DP) and (G) are inherited by complemented subspaces. We refer to [Sch74, Theorems II.9.7 and II.10.4] for a proof that $L^\infty(\Omega, \mu)$ has these properties. The key argument involving these two properties is expressed in the following lemma.

Lemma 4.3.16. *Let X be a Banach space such that (G) and (DP) are satisfied. Suppose that $T_n \in \mathcal{L}(X)$ $(n \in \mathbb{N})$ such that*

$$\lim_{n \to \infty} \|T_n x\| = 0 \text{ for all } x \in X$$

and

$$\lim_{n \to \infty} \|T_n^* x^*\| = 0 \text{ for all } x^* \in X^*.$$

Then

$$\lim_{n \to \infty} \|T_n^2\| = 0.$$

Proof. Assume that $\|T_n^2\|$ does not converge to 0. Then there exist $\varepsilon > 0$ and a subsequence such that $\|T_{n_k}^2\| \geq 2\varepsilon$. By the Hahn-Banach theorem, we find $x_k \in X, x_k^* \in X^*$ such that $\|x_k\| = \|x_k^*\| = 1$ but

$$\big|\langle T_{n_k} x_k, T_{n_k}^* x_k^*\rangle\big| = \big|\langle T_{n_k}^2 x_k, x_k^*\rangle\big| \geq \varepsilon. \tag{4.12}$$

Let $y_k := T_{n_k} x_k$, $y_k^* := T_{n_k}^* x_k^*$. Then for $x^* \in X^*$,

$$|\langle y_k, x^*\rangle| = |\langle x_k, T_{n_k}^* x^*\rangle| \leq \|T_{n_k}^* x^*\| \to 0 \quad \text{as } k \to \infty.$$

Thus, $y_k \to 0$ weakly.

Let $x \in X$. Then

$$|\langle x, y_k^*\rangle| = |\langle T_{n_k} x, x_k^*\rangle| \leq \|T_{n_k} x\| \to 0 \quad \text{as } k \to \infty.$$

Thus, $y_k^* \to 0$ weak*. It follows from (G) that $y_k^* \to 0$ weakly. Now, (DP) implies that $\langle y_k, y_k^*\rangle \to 0$ as $k \to \infty$. This contradicts (4.12). □

The following surprising result holds in particular on a space $X := L^\infty(\Omega, \mu)$ for any measure space (Ω, μ).

Theorem 4.3.17. *Let X be a Banach space satisfying (DP) and (G). Let A be a densely defined operator on X such that $(0, \lambda_0) \subset \rho(A)$ for some $\lambda_0 > 0$. If $\lambda R(\lambda, A)x$ converges weakly as $\lambda \downarrow 0$ for all $x \in X$, then $\lambda R(\lambda, A)$ converges in $\mathcal{L}(X)$ as $\lambda \downarrow 0$.*

Proof. It follows from Corollary 4.3.2 that $Px := \lim_{\lambda\downarrow 0} \lambda R(\lambda, A)x$ converges in norm for all $x \in X$, and that $X = \operatorname{Ker} A \oplus \overline{\operatorname{Ran} A}$. Replacing A by its part in $\overline{\operatorname{Ran} A}$, we can assume that $\operatorname{Ker} A = 0$ and so $P = 0$.

Let $x^* \in X^*$. Then for $x \in X$ we have

$$\langle x, \lambda R(\lambda, A)^* x^*\rangle = \langle \lambda R(\lambda, A)x, x^*\rangle \to 0 \quad \text{as } \lambda \downarrow 0.$$

Hence, $\lambda R(\lambda, A)^* x^* \to 0$ weak* for all $x^* \in X^*$. It follows from (G) that $\lambda_n R(\lambda_n, A)^* x^* \to 0$ weakly whenever $\lambda_n \downarrow 0$, for all $x^* \in X^*$. Now Corollary 4.3.2 implies that $\|\lambda R(\lambda, A)^* x^*\| \to 0$ as $\lambda \downarrow 0$, for all $x^* \in X^*$.

We have shown that $T_n := \lambda_n R(\lambda_n, A)$ satisfies the hypotheses of Lemma 4.3.16 whenever $\lambda_n \downarrow 0$. It follows that $\|(\lambda R(\lambda, A))^2\| \to 0$ as $\lambda \downarrow 0$. Since $(\lambda R(\lambda, A))^2 = (I + AR(\lambda, A))^2$, it follows that

$$r(I + AR(\lambda, A)) \leq \|(I + AR(\lambda, A))^2\|^{1/2} < 1$$

for $\lambda > 0$ small (where $r(I + AR(\lambda, A))$ denotes the spectral radius). This implies that $AR(\lambda, A)$ is invertible. In particular, $\operatorname{Ran} A = X$. Now the claim follows from Proposition 4.3.15. □

It is also interesting to consider convergence of $\lambda R(\lambda, A)$ for $\lambda \to \infty$. Again, the following result holds in particular on a space $X = L^\infty(\Omega, \mu)$, where (Ω, μ) is a measure space.

Theorem 4.3.18. *Let X be a Banach space satisfying (DP) and (G). Let A be a densely defined operator on X such that $(\lambda_0, \infty) \subset \rho(A)$ and $\sup_{\lambda>\lambda_0} \|\lambda R(\lambda, A)\| < \infty$ for some λ_0. Then A is bounded.*

Proof. By Lemma 3.3.12, $\lim_{n\to\infty}(nR(n, A)x - x) = 0$ for all $x \in X$. Let $x^* \in X^*$. Then it follows that

$$\langle x, nR(n, A)^*x^* - x^*\rangle = \langle nR(n, A)x - x, x^*\rangle \to 0 \qquad \text{as } n \to \infty,$$

for all $x \in X$. Thus, $(nR(n, A)^* - I)x^* \to 0$ weak* as $n \to \infty$. It follows from (G) that $(nR(n, A)^* - I)x^* \to 0$ weakly as $n \to \infty$, for all $x^* \in X^*$. But this implies that x^* is in the weak closure of $D(A^*) = R(\lambda, A^*)X^*$ $(\lambda \geq \lambda_0)$ for all $x^* \in X^*$. Since the weak and norm closures coincide, A^* is densely defined. It follows as above that $\|(nR(n, A)^* - I)x^*\| \to 0$ as $n \to \infty$, for all $x^* \in X^*$. Now Lemma 4.3.16 implies that $\|(nR(n, A) - I)^2\| \to 0$ as $n \to \infty$. In particular,

$$r((nR(n, A) - I)) \leq \|(nR(n, A) - I)^2\|^{1/2} < 1$$

for n sufficiently large. This implies that $nR(n, A)$ is invertible. Hence, $D(A) = X$, so A is bounded by the closed graph theorem. □

We deduce from Theorem 4.3.18 the following surprising and important result.

Corollary 4.3.19 (Lotz). *Let T be a C_0-semigroup on the Banach space $L^\infty(\Omega, \mu)$, where (Ω, μ) is a measure space. Then T has a bounded generator.*

In other words, if a semigroup T defined on $X := L^\infty(\Omega, \mu)$ converges strongly to the identity as $t \downarrow 0$, it converges already in the operator norm.

4.4 Complex Tauberian Theorems: the Contour Method

Let $f \in L^\infty(\mathbb{R}_+, X)$ and let $\hat{f}(\lambda) = \int_0^\infty e^{-\lambda t} f(t)\, dt$ $(\operatorname{Re}\lambda > 0)$ be its Laplace transform. We define the *half-line spectrum* $\operatorname{sp}(f)$ of f by

$$\operatorname{sp}(f) := \Big\{\eta \in \mathbb{R} : \hat{f} \text{ does not have a holomorphic extension to an open neighbourhood of } i\eta \text{ in } \mathbb{C}\Big\}.$$

It turns out that countability of the spectrum with certain growth conditions is a Tauberian hypothesis. Here we prove a Tauberian theorem of type D by completely elementary contour arguments. For simplicity, we consider first the simplest case where the spectrum is empty, giving a qualitative result (Theorem 4.4.1) and then a quantified version (Theorem 4.4.6). Then we adapt the argument to the case when the spectrum consists of one point (Theorem 4.4.8). This suffices for many interesting applications (see Corollaries 4.4.12 and 4.4.13 and Theorems 4.4.14 and 4.4.16) but we shall prove some more general results in Sections 4.7 and 4.9 by other means. In Theorem 4.4.18 we give a related result under different assumptions on f and $\hat{f}$, and using a different contour method.

Theorem 4.4.1. *Let $f \in L^\infty(\mathbb{R}_+, X)$, and assume that* $\operatorname{sp}(f)$ *is empty. Then*

$$\lim_{t\to\infty} \int_0^t f(s)\,ds = \hat{f}(0).$$

Here, $\hat{f}$ denotes the holomorphic extension of $\hat{f}$ to a neighbourhood of 0. As usual, we will denote the open right half-plane $\{\lambda \in \mathbb{C} : \operatorname{Re}\lambda > 0\}$ by $\mathbb{C}_+$.

Proof. We can assume that $\hat{f}(0) = 0$. Otherwise, we replace f by $f - \chi_{(0,1)}\hat{f}(0)$. There is a simply connected open set Ω containing $\overline{\mathbb{C}_+}$ such that $\hat{f}$ has a holomorphic extension (also denoted by $\hat{f}$) to Ω.

Let $g_t(\lambda) := \int_0^t e^{-\lambda s} f(s)\,ds$ ($\lambda \in \mathbb{C}, t \geq 0$). Let $R > 1$, and γ be a contour consisting of the semi-circle $\{\lambda \in \mathbb{C} : |\lambda| = R, \operatorname{Re}\lambda > 0\}$ and a path γ' connecting iR with $-iR$ and lying entirely in $\Omega \cap \{\operatorname{Re}\lambda < 0\}$ (except at the endpoints). Let

$$h(\lambda) := 1 + \frac{\lambda^2}{R^2}.$$

Since $h(0) = 1$, it follows from Cauchy's theorem that

$$\begin{aligned} 0 = -\hat{f}(0) &= -\frac{1}{2\pi i}\int_\gamma e^{\lambda t}\hat{f}(\lambda)h(\lambda)\frac{d\lambda}{\lambda}\,; \\ \int_0^t f(s)\,ds = g_t(0) &= \frac{1}{2\pi i}\int_{|\lambda|=R} e^{\lambda t} g_t(\lambda)h(\lambda)\frac{d\lambda}{\lambda}\,. \end{aligned}$$

Adding up, we have

$$\begin{aligned} \int_0^t f(s)\,ds &= \frac{1}{2\pi i}\int_{\substack{|\lambda|=R \\ \operatorname{Re}\lambda>0}} e^{\lambda t}(g_t(\lambda) - \hat{f}(\lambda))h(\lambda)\frac{d\lambda}{\lambda} \\ &\quad - \frac{1}{2\pi i}\int_{\gamma'} e^{\lambda t}\hat{f}(\lambda)h(\lambda)\frac{d\lambda}{\lambda} \\ &\quad + \frac{1}{2\pi i}\int_{\substack{|\lambda|=R \\ \operatorname{Re}\lambda<0}} e^{\lambda t} g_t(\lambda)h(\lambda)\frac{d\lambda}{\lambda} \\ &=: I_1(t) + I_2(t) + I_3(t). \end{aligned} \tag{4.13}$$

We estimate these three integrals separately.
$I_1(t)$: Let $\lambda := Re^{i\theta}$, $\theta \in (-\frac{\pi}{2}, \frac{\pi}{2})$. Then

$$\begin{aligned} \| e^{\lambda t}(g_t(\lambda) - \hat{f}(\lambda)) \| &= \left\| \int_t^\infty e^{\lambda(t-s)} f(s)\,ds \right\| \\ &= \left\| \int_0^\infty e^{-\lambda s} f(s+t)\,ds \right\| \\ &\leq \frac{\|f\|_\infty}{\operatorname{Re}\lambda} = \frac{\|f\|_\infty}{R\cos\theta}\,; \\ \left|\frac{h(\lambda)}{\lambda}\right| &\leq |1 + e^{2i\theta}|\frac{1}{R} = \frac{2\cos\theta}{R}\,. \end{aligned}$$

Thus,

$$\|I_1(t)\| \le \frac{1}{2\pi}\frac{\|f\|_\infty}{R}\frac{2}{R}\pi R = \frac{\|f\|_\infty}{R} \qquad (t \ge 0).$$

$I_2(t)$: Note that $\lim_{t\to\infty} \| I_2(t) \| = 0$ by the dominated convergence theorem.
$I_3(t)$: Let $\lambda := Re^{i\theta}$, $\theta \in (\frac{\pi}{2}, \frac{3\pi}{2})$. Then

$$\| e^{\lambda t} g_t(\lambda) \| = \left\| \int_0^t e^{\lambda(t-s)} f(s)\, ds \right\| \le \|f\|_\infty \int_0^t e^{Rs\cos\theta} ds \le \frac{\|f\|_\infty}{R|\cos\theta|};$$

$$\left|\frac{h(\lambda)}{\lambda}\right| \le \frac{2|\cos\theta|}{R}.$$

Hence,

$$\| I_3(t) \| \le \frac{1}{2\pi}\frac{\|f\|_\infty}{R}\frac{2}{R}\pi R = \frac{\|f\|_\infty}{R}.$$

These estimates give

$$\limsup_{t\to\infty}\left\| \int_0^t f(s)\, ds \right\| \le \frac{2\|f\|_\infty}{R}. \tag{4.14}$$

Letting $R \to \infty$ proves the claim. □

From the proof above we deduce the following.

Proposition 4.4.2. *Let $f \in L^\infty(\mathbb{R}_+, X)$. If $R > 0$ such that $\operatorname{sp}(f) \cap [-R, R] = \emptyset$, then*

$$\limsup_{t\to\infty}\left\| \int_0^t f(s)\, ds - \hat{f}(0) \right\| \le \frac{2}{R}\limsup_{t\to\infty} \| f(t) \|.$$

Proof. Let $c > \limsup_{t\to\infty} \|f(t)\|$. Choose $\tau \ge 0$ such that $\|f(t+\tau)\| \le c$ for all $t \ge 0$ and apply the estimate (4.14) to $f(\cdot + \tau) - \chi_{(0,1)}\left(\hat{f}(0) - \int_0^\tau f(s)\, ds\right)$. □

The next two corollaries follow from Proposition 4.4.2 by replacing $f(t)$ by $e^{-i\eta t} f(t)$.

Corollary 4.4.3. *Let $f \in L^\infty(\mathbb{R}_+, X)$. If $\eta \notin \operatorname{sp}(f)$, then*

$$\sup_{t\ge 0}\left\| \int_0^t e^{-i\eta s} f(s)\, ds \right\| < \infty.$$

Corollary 4.4.4. *Let $f \in L^\infty(\mathbb{R}_+, X)$ such that $\lim_{t\to\infty} \| f(t) \| = 0$. Then*

$$\hat{f}(i\eta) = \lim_{t\to\infty} \int_0^t e^{-i\eta s} f(s)\, ds$$

for all $\eta \in \mathbb{R}\setminus\operatorname{sp}(f)$.

Next we give a quantified version of Theorem 4.4.1. When $\operatorname{sp}(f)$ is empty, there is a continuous increasing function $M : \mathbb{R}_+ \to (0,\infty)$ such that $\hat{f}$ has a holomorphic extension (also denoted by $\hat{f}$) to

$$\Omega_M := \left\{ \lambda \in \mathbb{C} : \operatorname{Re}\lambda > -\frac{1}{M(|\operatorname{Im}\lambda|)} \right\}$$

and $\|\hat{f}(\lambda)\| \le M(|\operatorname{Im}\lambda|)$ for all $\lambda \in \Omega_M$. Define

$$M_{\log}(s) = M(s)\left[\log(1+M(s)) + \log(1+s)\right]$$

for $s \in \mathbb{R}_+$. The inverse $M_{\log}^{-1}$ of $M_{\log}$ is an increasing function of $(M_{\log}(0),\infty)$ onto $(0,\infty)$. Although we do not require M to be strictly increasing, we shall let M^{-1} denote any increasing function from the range of M to $\mathbb{R}_+$ such that $M(M^{-1}(t)) = t$ for all t in the range of M.

Example 4.4.5. a) If M is bounded, then $M_{\log}^{-1}(t) \sim Ce^t$ as $t \to \infty$, for some constant C.

b) If $M(s) = \beta(1+s)^\alpha$ where $\alpha, \beta > 0$, then

$$M_{\log}^{-1}(t) \sim C_{\alpha,\beta}\left(\frac{t}{\log t}\right)^{1/\alpha} \qquad \text{as } t \to \infty,$$

for some constant $C_{\alpha,\beta}$.

c) If $M(s) = \beta e^{\alpha s}$ where $\alpha, \beta > 0$, then $M_{\log}^{-1}(t) \sim \frac{1}{\alpha}\log t$ as $t \to \infty$. □

Theorem 4.4.6. *Let $f \in L^\infty(\mathbb{R}_+, X)$, and assume that* $\operatorname{sp}(f)$ *is empty. Let M and $M_{\log}$ be as above, and $c \in (0,1)$. Then there exist constants C and t_0, depending only on $\|f\|_\infty$, M and c, such that*

$$\left\| \int_0^t f(s)\,ds - \hat{f}(0) \right\| \le \frac{C}{M_{\log}^{-1}(ct)} \qquad \text{for all } t \ge t_0. \tag{4.15}$$

Remark 4.4.7. In (4.15), the function $M_{\log}^{-1}$ cannot be replaced by M^{-1} in general. For example, if $X = \mathbb{C}$ and $f(t) = e^{-t}$, then the assumptions of Theorem 4.4.6 are satisfied for any continuous increasing function M with $M(0) = 1$. So M^{-1} may increase arbitrarily fast, but

$$\int_0^t f(s)\,ds - \hat{f}(0) = -e^{-t}.$$

On the other hand, M^{-1} and $M_{\log}^{-1}$ behave similarly if M increases rapidly and consistently (see Example 4.4.5). We shall see in Theorem 4.4.14 that, in the context of C_0-semigroups, $M_{\log}^{-1}(ct)$ can never be replaced in (4.15) by a function increasing faster than $M^{-1}(Ct)$ if M is chosen in the optimal way. Example 4.4.15 will show that it may not be possible to replace $M_{\log}^{-1}$ by M^{-1} in (4.15) when M increases arbitrarily rapidly but inconsistently.

Proof of Theorem 4.4.6. We use the same method and notation as in the proof of Theorem 4.4.1, taking $\Omega = \Omega_M$ and choosing γ' and R as follows.

Let $a = c^{-1/2} > 1$ and take γ' to be the union of γ_0, γ_+ and γ_-, where

$$\begin{aligned} \gamma_0(s) &= -\frac{1}{aM(|s|)} \qquad (-R \le s \le R)\,, \\ \gamma_\pm(\tau) &= \tau \pm iR \qquad (-(aM(R))^{-1} \le \tau \le 0)\,. \end{aligned}$$

On $\gamma_\pm$, $|h(\lambda)| \le C/R$, so we can estimate the norm of the integral $I_2(t)$ over $\gamma_\pm$ by

$$\left\| \int_{\gamma_\pm} e^{\lambda t} \hat{f}(\lambda) h(\lambda) \frac{d\lambda}{\lambda} \right\| \le C \int_0^{(aM(R))^{-1}} \frac{1}{R} M(R) \frac{e^{-t\tau}}{R}\, d\tau \le \frac{CM(R)}{R^2 t}\,.$$

Here and subsequently, C denotes a constant which may depend only on M, c and $\|f\|_\infty$, although it may vary from place to place. On γ_0, $h(\lambda)$ is bounded independently of R, and $|e^{t\lambda}| \le e^{-t/aM(R)}$. Hence, we can estimate the norm of the integral over γ_0 by

$$\left\| \int_{\gamma_0} e^{\lambda t} \hat{f}(\lambda) h(\lambda) \frac{d\lambda}{\lambda} \right\| \le CM(R) e^{-t/aM(R)} \int_{\gamma_0} \left| \frac{d\lambda}{\lambda} \right| \le CM(R)(1+\log R) e^{-t/aM(R)}.$$

With these estimates for $I_2(t)$ and those in the proof of Theorem 4.4.1 for $I_1(t)$ and $I_3(t)$, we obtain

$$\left\| \int_0^t f(s)\, ds \right\| \le \frac{2\|f\|_\infty}{R} + C \left(\frac{M(R)}{R^2 t} + M(R)(1+\log R) e^{-t/aM(R)} \right).$$

Given $t > c^{-1} M_{\log}(1)$, choose $R = M_{\log}^{-1}(ct) > 1$. Then

$$\begin{aligned} \frac{M(R)}{R^2 t} &= \frac{c}{2R^2 \log\left((1+M(R))(1+R)\right)} \le \frac{C}{R}\,, \\ M(R)(1+\log R) &\le C \frac{(1+M(R))^a (1+R)^a}{R} = \frac{C}{R} e^{t/aM(R)}\,. \end{aligned}$$

Hence

$$\left\| \int_0^t f(s)\, ds \right\| \le \frac{C}{R} = \frac{C}{M_{\log}^{-1}(ct)}\,.$$

□

Now we consider the case when $\mathrm{sp}(f)$ is a single point, stating only the qualitative result.

Theorem 4.4.8. *Let $f \in L^\infty(\mathbb{R}_+, X)$, and assume that $\mathrm{sp}(f) = \{\eta\}$ and*

$$\sup_{t \ge 0} \left\| \int_0^t e^{-i\eta s} f(s)\, ds \right\| < \infty, \tag{4.16}$$

where $\eta \in \mathbb{R} \backslash \{0\}$. Then $\lim_{t\to\infty} \int_0^t f(s)\, ds = \hat{f}(0)$.

Remark 4.4.9. The growth condition (4.16) in Theorem 4.4.8 cannot be omitted. For example, let $f(t) := e^{it}$. Then $\hat{f}(\lambda) = (\lambda - i)^{-1}$, so $\mathrm{sp}(f) = \{1\}$. However, $\int_0^t f(s)\,ds = \frac{1}{i}(e^{it} - 1)$ does not converge as $t \to \infty$. □

Proof of Theorem 4.4.8. The proof is similar to Theorem 4.4.1, with the following changes.

a) We take $R > |\eta|$ and $0 < \varepsilon < \min\{|\eta|, R - |\eta|\}$. The path γ' now consists of two separate paths connecting iR with $i(\eta + \varepsilon)$ and connecting $i(\eta - \varepsilon)$ with $-iR$. It lies entirely in a simply connected domain where $\hat{f}$ is defined holomorphically, and also in $\{\mathrm{Re}\,\lambda < 0\}$ (except at the four endpoints). The contour γ includes also the semi-circle $\{\lambda \in \mathbb{C} : |\lambda - i\eta| = \varepsilon,\ \mathrm{Re}\,\lambda > 0\}$.

b) Let

$$h(\lambda) := \left(1 + \frac{\varepsilon^2}{(\lambda - i\eta)^2}\right) \frac{\eta^2}{(\eta^2 - \varepsilon^2)} \left(1 + \frac{\lambda^2}{R^2}\right).$$

For $\lambda = Re^{i\theta}$, we have

$$\begin{aligned} \left|\frac{h(\lambda)}{\lambda}\right| &\leq \left(1 + \frac{\varepsilon^2}{(R - |\eta|)^2}\right) \frac{\eta^2}{\eta^2 - \varepsilon^2} |1 + e^{2i\theta}| \frac{1}{R} \\ &= \left(1 + \frac{\varepsilon^2}{(R - |\eta|)^2}\right) \frac{\eta^2}{\eta^2 - \varepsilon^2} \frac{2|\cos\theta|}{R}. \end{aligned} \tag{4.17}$$

c) In (4.13) we have two additional integrals:

$$\begin{aligned} I_4(t) &:= \frac{1}{2\pi i} \int_{\substack{|\lambda - i\eta| = \varepsilon \\ \mathrm{Re}\,\lambda > 0}} e^{\lambda t}(g_t(\lambda) - \hat{f}(\lambda)) h(\lambda) \frac{d\lambda}{\lambda}\,; \\ I_5(t) &:= -\frac{1}{2\pi i} \int_{\substack{|\lambda - i\eta| = \varepsilon \\ \mathrm{Re}\,\lambda > 0}} e^{\lambda t} g_t(\lambda) h(\lambda) \frac{d\lambda}{\lambda} = \frac{1}{2\pi i} \int_{\substack{|\lambda - i\eta| = \varepsilon \\ \mathrm{Re}\,\lambda < 0}} e^{\lambda t} g_t(\lambda) h(\lambda) \frac{d\lambda}{\lambda}. \end{aligned}$$

We estimate these integrals as follows:
$I_4(t)$: Let $\lambda := i\eta + \varepsilon e^{i\theta}$, $\theta \in (-\frac{\pi}{2}, \frac{\pi}{2})$. Let $F_1(t) := \int_0^t e^{-i\eta s} f(s)\,ds$. Then by assumption (4.16), $K := \sup_{t \geq 0} \|F_1(t)\| < \infty$. We have

$$\begin{aligned} \|\, e^{\lambda t}(g_t(\lambda) - \hat{f}(\lambda))\,\| &= \left\| e^{\lambda t} \int_t^\infty e^{-\lambda s} f(s)\,ds \right\| \\ &= \left\| e^{\lambda t} \int_t^\infty e^{-\varepsilon s e^{i\theta}} \frac{d}{ds} F_1(s)\,ds \right\| \\ &= \left\| -e^{\lambda t} e^{-\varepsilon t e^{i\theta}} F_1(t) + e^{\lambda t} \int_t^\infty \varepsilon e^{i\theta} e^{-\varepsilon s e^{i\theta}} F_1(s)\,ds \right\| \\ &\leq K + \varepsilon \left\| \int_t^\infty e^{-\varepsilon e^{i\theta}(s-t)} F_1(s)\,ds \right\| \\ &\leq K + \varepsilon \frac{K}{\varepsilon \cos\theta} = K\left(1 + \frac{1}{\cos\theta}\right)\,; \end{aligned}$$

$$\left|\frac{h(\lambda)}{\lambda}\right| \leq |1+e^{-2i\theta}|\frac{\eta^2}{\eta^2-\varepsilon^2}\,2\,\frac{1}{|\eta|-\varepsilon} = 2\cos\theta\frac{\eta^2}{\eta^2-\varepsilon^2}\,2\,\frac{1}{|\eta|-\varepsilon}\,.$$

Consequently,

$$\|\, I_4(t)\,\| \leq \frac{1}{2\pi}K2\cdot 2\,\frac{\eta^2}{\eta^2-\varepsilon^2}\,2\,\frac{1}{|\eta|-\varepsilon}\varepsilon\pi = 4K\varepsilon\frac{\eta^2}{\eta^2-\varepsilon^2}\frac{1}{|\eta|-\varepsilon}.$$

$I_5(t)$: Let $\lambda := i\eta+\varepsilon e^{i\theta}$, $\theta\in(\frac{\pi}{2},\frac{3\pi}{2})$. Then

$$\begin{aligned}
\|\, e^{\lambda t}g_t(\lambda)\,\| &= \left\|\int_0^t e^{\lambda t}e^{-\lambda s}f(s)\,ds\right\| \\
&= \left\|\int_0^t e^{\lambda t}e^{-\varepsilon s e^{i\theta}}\frac{d}{ds}F_1(s)\,ds\right\| \\
&= \left\|e^{\lambda t}e^{-\varepsilon t e^{i\theta}}F_1(t)+\varepsilon e^{i\theta}\int_0^t e^{\lambda t}e^{-\varepsilon s e^{i\theta}}F_1(s)\,ds\right\| \\
&\leq K+\varepsilon K\int_0^t e^{-\varepsilon(s-t)\cos\theta}ds \\
&\leq K\left(1+\frac{1}{|\cos\theta|}\right)\,.
\end{aligned}$$

Thus, as for $I_4(t)$ one obtains

$$\|\, I_5(t)\,\| \leq 4K\varepsilon\frac{\eta^2}{\eta^2-\varepsilon^2}\,\frac{1}{|\eta|-\varepsilon}.$$

d) The estimates for $I_1(t)$ and $I_3(t)$ now include additional factors appearing in (4.17). The dominated convergence theorem shows that $\lim_{t\to\infty}\|I_2(t)\| = 0$. Thus, we obtain that

$$\begin{aligned}
&\limsup_{t\to\infty}\left\|\int_0^t f(s)\,ds\right\| \\
&\quad\leq \frac{2\|f\|_\infty}{R}\left(1+\frac{\varepsilon^2}{(R-|\eta|)^2}\right)\frac{\eta^2}{\eta^2-\varepsilon^2}+8K\varepsilon\frac{\eta^2}{\eta^2-\varepsilon^2}\,\frac{1}{|\eta|-\varepsilon}\,.
\end{aligned}$$

Letting $\varepsilon\downarrow 0$, we obtain

$$\limsup_{t\to\infty}\left\|\int_0^t f(s)\,ds\right\| \leq \frac{2\|f\|_\infty}{R}.$$

Finally, letting $R\to\infty$ proves the claim. □

We derive from Theorem 4.4.8 a Tauberian theorem of type E for slowly oscillating functions.

Corollary 4.4.10. *Let $f \in L^\infty(\mathbb{R}_+, X)$ be slowly oscillating. Assume that*

a) $\operatorname{sp}(f) = \emptyset$, *or*

b) $\operatorname{sp}(f) = \{\eta\}$ *and* $\sup_{t\geq 0} \| \int_0^t e^{-i\eta s} f(s)\, ds \| < \infty$, *where* $\eta \in \mathbb{R}$.

Then $\lim_{t\to\infty} f(t) = 0$.

Proof. We give the proof in the case when $\operatorname{sp}(f) = \{\eta\}$.

a) Assume that $\eta \neq 0$. For $\delta > 0$, consider $f_\delta(t) := \frac{1}{\delta}(f(\delta + t) - f(t))$ as before. It follows from (4.1) that $\operatorname{sp}(f_\delta) \subset \{\eta\}$. Moreover,

$$\int_0^t e^{-i\eta s} f_\delta(s)\, ds = \delta^{-1} \left(\int_\delta^{\delta+t} e^{i\eta(\delta - s)} f(s)\, ds - \int_0^t e^{-i\eta s} f(s)\, ds \right),$$

which is bounded. It follows from Theorem 4.4.8 that $\lim_{t\to\infty} \int_0^t f_\delta(s)\, ds = \widehat{f_\delta}(0) = -\frac{1}{\delta} \int_0^\delta f(s)\, ds$ (by (4.1)). Hence, B-$\lim_{t\to\infty} f(t) = 0$ by Lemma 4.2.12. Theorem 4.2.3 implies that $\lim_{t\to\infty} f(t) = 0$.

b) Assume that $\eta = 0$. Let $g(t) := e^{it} f(t)$. Then g is bounded and slowly oscillating. Moreover, $\operatorname{sp}(g) \subset \{i\}$ and $\sup_{t\geq 0} \left\| \int_0^t e^{-is} g(s)\, ds \right\| < \infty$. It follows from a) that $\lim_{t\to\infty} g(t) = 0$. Hence, $\lim_{t\to\infty} f(t) = 0$. ☐

Corollary 4.4.11. *Let $f : \mathbb{R}_+ \to X$ be Lipschitz continuous such that*

$$\sup_{t\geq 0} \left\| \int_0^t f(s)\, ds \right\| < \infty. \tag{4.18}$$

Assume that $\operatorname{sp}(f) \subset \{0\}$. *Then* $\lim_{t\to\infty} f(t) = 0$.

Note that (4.18) implies that C-$\lim_{t\to\infty} f(t) = 0$. Thus, Corollary 4.4.11 is a Tauberian theorem of type A.

Proof of Corollary 4.4.11. We show that f is bounded. Then we can apply Corollary 4.4.10. Let $F(t) := \int_0^t f(s)\, ds$. By assumption, $M := \sup_{t\geq 0} \|F(t)\| < \infty$. Moreover, there exists $L \geq 0$ such that

$$\|f(t) - f(s)\| \leq L|t - s| \quad (t, s \geq 0).$$

Let $x^* \in X^*$ such that $\|x^*\| \leq 1$. Then $x^* \circ f$ is differentiable a.e. and

$$\left| \frac{d}{dt} (x^* \circ f)(t) \right| \leq L \quad (t \geq 0).$$

The Taylor expansion for $x^* \circ F$ yields for $s \geq 0$,

$$\langle F(s+1), x^* \rangle = \langle F(s), x^* \rangle + \langle f(s), x^* \rangle + \int_s^{s+1} (s + 1 - t) \frac{d}{dt} \langle f(t), x^* \rangle\, dt.$$

Hence,

$$\begin{aligned} |\langle f(s), x^*\rangle| &\leq 2M + \int_s^{s+1} (s+1-t)\cdot L\,dt \\ &= 2M + \frac{L}{2}. \end{aligned}$$

The Hahn-Banach theorem implies that $\|f(s)\| \leq 2M + \frac{1}{2}L$ $(s \geq 0)$. □

Corollary 4.4.10 and 4.4.11 are first versions of a complex Tauberian theorem where the main hypothesis says that the spectrum is sufficiently small. Generalizations will be given in Section 4.7 (Corollary 4.7.8) and in Section 4.9 (Theorem 4.9.7). The proof of both generalizations will make use of Gelfand's theorem which we obtain now as another immediate consequence of the contour method. In fact, applying Corollary 4.4.11 to bounded groups we obtain the following result.

Corollary 4.4.12 (Gelfand). *Let A be the generator of a bounded C_0-group $U = (U(t))_{t\in\mathbb{R}}$ on X. If $\sigma(A) \subset \{0\}$, then $U(t) = I$ for all $t \in \mathbb{R}$. In particular, if $\sigma(A)$ is empty, then $X = \{0\}$.*

Proof. Let $x \in D(A^2)$, $f(t) := AU(t)x$. Then $\int_0^t f(s)\,ds = U(t)x - x$, which is bounded. Since $\hat{f}(\lambda) = R(\lambda, A)Ax$, we have $\operatorname{sp}(f) \subset \{0\}$. Since $\frac{d}{dt}f(t) = U(t)A^2x$, the function f is Lipschitz continuous. It follows from Corollary 4.4.11 that $\lim_{t\to\infty} f(t) = 0$. Let $M := \sup_{t\in\mathbb{R}} \|U(t)\|$. Since $\|Ax\| = \|U(-t)U(t)Ax\| \leq M\|U(t)Ax\|$, it follows that $Ax = 0$. Hence, $U(t)x = x$ for all $x \in D(A^2)$, $t \in \mathbb{R}$. Since $R(1, A)$ has dense range $D(A)$, the range $D(A^2)$ of $R(1, A)^2$ is also dense. It follows that $U(t) = I$ $(t \in \mathbb{R})$. When $\sigma(A)$ is empty, one can also apply the previous case to $A - i\eta$ for any $\eta \in \mathbb{R}$. □

Corollary 4.4.13. *Let A be the generator of a bounded C_0-group U. Then each isolated point in $\sigma(A)$ is an eigenvalue.*

Proof. Let $i\eta \in \sigma(A)$ be isolated. Consider the spectral projection P associated with $i\eta$, and let $Y := PX \neq \{0\}$ (see Proposition B.9). Then the group leaves the space Y invariant. Consider the restricted group. The spectrum of its generator is reduced to $\{i\eta\}$. Applying Corollary 4.4.12 to $e^{-i\eta t}U(t)|_Y$, one obtains that $e^{-i\eta t}U(t)y = y$ for all $y \in Y$. □

Next, we prove a result on the asymptotic behaviour of a bounded C_0-semigroup T with generator A, specifically the possible decay of $\|T(t)R(\mu, A)\|$ for large t, where $\mu \in \rho(A)$ is fixed. Note that the rate of any such decay is independent of the choice of μ up to multiplicative constants, since $(\mu' - A)R(\mu, A)$ is always a bounded operator.

Let $m : \mathbb{R}_+ \to (0, \infty)$ be a continuous decreasing function such that $m(t) \to 0$ as $t \to \infty$. We let m^{-1} denote any decreasing function from $(0, \infty)$ to $\mathbb{R}_+$ such that $m(m^{-1}(t)) = t$ for all t in the range of m.

Theorem 4.4.14. *Let A be the generator of a bounded C_0-semigroup T, and let $\mu \in \rho(A)$. The following are equivalent:*

(i) $\sigma(A) \cap i\mathbb{R}$ *is empty.*

(ii) $\lim_{t\to\infty} \|T(t)R(\mu, A)\| = 0$.

More precisely:

a) *Assume that $\sigma(A) \cap i\mathbb{R}$ is empty, and let $M : \mathbb{R}_+ \to (0, \infty)$ be a continuous increasing function such that $\|R(is, A)\| \le M(|s|)$ for all $s \in \mathbb{R}$, and let $c \in (0, 1)$. Then there exist C and t_0 such that*

$$\|T(t)R(\mu, A)\| \le \frac{C}{M_{\log}^{-1}(ct)} \qquad \textit{for all } t \ge t_0. \tag{4.19}$$

b) *Assume that $\lim_{t\to\infty} \|T(t)R(\mu, A)\| = 0$, and let $m : \mathbb{R}_+ \to (0, \infty)$ be a continuous decreasing function such that $\|T(t)R(\mu, A)\| \le m(t)$ for $t \ge 0$ and $\lim_{t\to\infty} m(t) = 0$. Then $\sigma(A) \cap i\mathbb{R}$ is empty, and for each $c \in (0, 1)$ there exist C and s_0 such that*

$$\|R(is, A)\| \le Cm^{-1}\left(\frac{c}{|s|}\right) \qquad \textit{whenever } |s| \ge s_0.$$

Proof. The implication from (1) to (2) can be seen from Theorem 4.4.1 with $f(t) = T(t)$. Although that function is not measurable from $\mathbb{R}_+$ to $\mathcal{L}(X)$, its Laplace transform $\hat{f}(\lambda) = R(\lambda, A)$ is holomorphic, and the proof of Theorem 4.4.1 remains valid. Alternatively, we give the following more precise argument for a). We let $K = \sup_{t\ge 0} \|T(t)\|$.

a) First, by Corollary B.3, for any $a > 1$, $\Omega_{aM} \subset \rho(A)$ and $\|R(\lambda, A)\| \le \frac{a}{a-1} M(|\operatorname{Im}\lambda|)$ for all $\lambda \in \Omega_{aM}$.

Let $x \in X$ with $\|x\| \le (a-1)/a$, and let $f(t) = T(t)x$ $(t \ge 0)$. Then $\|f\|_\infty \le K$ and $\hat{f}(\lambda) = R(\lambda, A)x$, so $\|\hat{f}(\lambda)\| \le M(|\operatorname{Im}\lambda|)$ for $\lambda \in \Omega_{aM}$. By Proposition 3.1.9 e),

$$\begin{aligned} T(t)R(\mu, A)x &= R(\mu, A)x + AR(\mu, A) \int_0^t f(s)\, ds \\ &= AR(\mu, A) \left(\int_0^t f(s)\, ds - \hat{f}(0) \right). \end{aligned}$$

Since $AR(\mu, A)$ is a bounded operator, the result follows from Theorem 4.4.8 and its proof, noting that C and t_0 are independent of x for $\|x\| \le (a-1)/a$.

b) Take $s \in \mathbb{R}$, and consider $x^* \in D(A^*)$ with $A^*x^* = isx^*$. Let $x \in D(A)$, and put $g(t) = e^{-ist}\langle T(t)x, x^*\rangle$ $(t \ge 0)$. Then

$$g'(t) = e^{-ist}\langle AT(t)x - isT(t)x, x^*\rangle = 0,$$

so g is constant. By the assumption (ii), $\lim_{t\to\infty} \|T(t)x\| = 0$, so $\lim_{t\to\infty} g(t) = 0$. Hence $0 = g(0) = \langle x, x^* \rangle$. Since $D(A)$ is dense, it follows that $x^* = 0$. By the Hahn-Banach theorem, $A - is$ has dense range.

Let $c' \in (c, 1)$, and $x \in D(A)$. By Proposition 3.1.9 f) applied to the rescaled semigroup,

$$\int_0^t e^{-is\tau} T(\tau)(A - is)x \, d\tau = e^{-ist} T(t)x - x$$

for any $t \geq 0$. Hence,

$$\begin{aligned} Kt\|(A - is)x\| &\geq \left\| \int_0^t e^{-is\tau} T(\tau)(A - is)x \, d\tau \right\| \\ &\geq \|x\| - \|T(t)x\| \\ &\geq \|x\| - m(t)\|(A - \mu)x\| \\ &\geq \|x\| - m(t)\big(\|(A - is)x\| + (|\mu| + |s|)\|x\|\big). \end{aligned}$$

Thus,

$$\|(A - is)x\| \geq \frac{1 - m(t)(|\mu| + |s|)}{Kt + m(t)} \|x\|.$$

Let $t = m^{-1}\left(c'/(|\mu| + |s|)\right)$, so $m(t) \leq c'/(|\mu| + |s|)$. Thus,

$$\|(A - is)x\| \geq \frac{(1 - c')\|x\|}{Km^{-1}(c'/(|\mu| + |s|)) + c'/(|\mu| + |s|)}.$$

This implies that $A - is$ has closed range, and hence it is invertible. Moreover,

$$\|R(is, A)\| \leq \frac{K}{1 - c'} m^{-1}\left(\frac{c'}{|\mu| + |s|}\right) + \frac{c'}{(1 - c')(|\mu| + |s|)} \leq Cm^{-1}\left(\frac{c}{|s|}\right)$$

whenever $|s|$ is sufficiently large. □

Part a) of Theorem 4.4.14 gives the sharpest result when M is chosen as small as possible, that is,

$$M(s) = \sup\{\|R(is', A)\| : |s'| \leq s\}.$$

Then part b) provides the lower bound

$$\|T(t)R(\mu, A)\| \geq \frac{c}{M^{-1}(Ct)}$$

for t sufficiently large and some constants $c > 0$ and C, and the sharpest estimate is given by taking

$$M^{-1}(t) = \inf\{|s| : \|R(is, A)\| \geq t\}.$$

Then the gap between this lower bound and the upper bound in (4.19) is small if M grows rapidly. To understand this more explicitly, we consider diagonal semigroups on Hilbert space.

Example 4.4.15. Let $X = \ell^2$, $(\alpha_n)_{n\in\mathbb{N}}$ be a strictly decreasing sequence in $(0,1]$, and $\lambda_n = -\alpha_n + in$. A C_0-semigroup of contractions on ℓ^2 is defined by

$$T(t)x = \left(e^{\lambda_n t}x_n\right) \qquad (x = (x_n) \in \ell^2).$$

The generator A is given by

$$\begin{aligned} D(A) &= \left\{x \in \ell^2 : (\lambda_n x_n) \in \ell^2\right\} = \left\{x \in \ell^2 : (nx_n) \in \ell^2\right\}, \\ Ax &= (\lambda_n x_n). \end{aligned}$$

Moreover,

$$\begin{aligned} \sigma(A) &= \{\lambda_n : n \in \mathbb{N}\}, \\ \|R(is,A)\| &= \sup\left\{|\lambda_n - is|^{-1} : n \in \mathbb{N}\right\}. \end{aligned}$$

Take the optimal definitions of M and M^{-1} as above. Then $M(k) = \|R(ik,A)\| = 1/\alpha_k$ $(k \in \mathbb{N})$. Let $t_k = 1/\alpha_k$. Then

$$\|T(t_k)A^{-1}\| = \sup_{n\in\mathbb{N}}\left|\frac{e^{\lambda_n t_k}}{\lambda_n}\right| \geq \frac{e^{-1}}{|\lambda_k|} \sim \frac{e^{-1}}{k} = \frac{e^{-1}}{M^{-1}(t_k)} \qquad \text{as } k \to \infty.$$

When (α_n) decays regularly and not too slowly, we may have

$$\|T(t)A^{-1}\| \sim \frac{c}{M^{-1}(t)} \qquad \text{as } t \to \infty. \tag{4.20}$$

For example, when $\alpha_n = n^{-\gamma}$ where $\gamma > 0$,

$$\|T(t)A^{-1}\| \sim \frac{c_\gamma}{t^{1/\gamma}} \sim \frac{c_\gamma}{M^{-1}(t)}.$$

On the other hand, (4.20) may fail if (α_n) has relatively long periods of slow decay. Suppose for example that (α_n) is strictly decreasing and

$$\alpha_{(k+1)!} \geq \frac{\alpha_{k!+1}}{2} \qquad (k \geq 1).$$

Given $c > 0$, let $t_k = 1/(c\alpha_{(k+1)!})$. Then

$$\begin{aligned} M^{-1}(ct_k) &= (k+1)!, \\ \|T(t_k)A^{-1}\| \geq \left|\frac{e^{\lambda_{k!+1}t_k}}{\lambda_{k!+1}}\right| &\geq \frac{e^{-2/c}}{|\lambda_{k!+1}|} \sim \frac{ke^{-2/c}}{M^{-1}(ct_k)}. \end{aligned}$$

Thus there is no upper bound of the form $\|T(t)A^{-1}\| \leq C/M^{-1}(ct)$. By arranging that the subsequence $(\alpha_{k!})$ decreases arbitrarily fast, this phenomenon can be found even when M increases arbitrarily fast. □

The next result on asymptotic behaviour of C_0-semigroups is analogous to the Katznelson-Tzafriri theorem for contractions (see the Notes of this chapter). A C_0-semigroup $T = (T(t))_{t\geq 0}$ is called *eventually differentiable* if there exists $\tau \geq 0$ such that $T(\tau)X \subset D(A)$. Semigroup properties imply that $T(t)X \subset D(A)$ for all $t \geq \tau$ and $T(t)X \subset D(A^2)$ for all $t \geq 2\tau$. By the closed graph theorem, $AT(t) \in \mathcal{L}(X)$ for $t \geq \tau$ and $A^2T(t) \in \mathcal{L}(X)$ for $t \geq 2\tau$.

Theorem 4.4.16. *Let A be the generator of an eventually differentiable, bounded C_0-semigroup. The following are equivalent:*

(i) $\sigma(A) \cap i\mathbb{R} \subset \{0\}$.

(ii) $\lim_{t\to\infty} \| AT(t) \| = 0$.

Proof. Let $K := \sup_{t\geq 0} \| T(t) \|$. Assume that (i) holds. Let $f : \mathbb{R}_+ \to \mathcal{L}(X)$ be defined by $f(t) := AT(t + 2\tau)$. For $x \in X$,

$$\begin{aligned} \|f(t)x - f(s)x\| &= \left\| \int_t^s \frac{d}{dr}(f(r)x)\, dr \right\| \\ &= \left\| \int_t^s A^2T(r + 2\tau)x\, dr \right\| \\ &= \left\| \int_t^s T(r)A^2T(2\tau)x\, dr \right\| \\ &\leq K\|A^2T(2\tau)\|\, |t - s|\, \|x\| \qquad (s, t \geq 0). \end{aligned}$$

Thus f is Lipschitz continuous. Moreover, $\hat{f}(\lambda) = R(\lambda, A)AT(2\tau)$. Hence, we have $\mathrm{sp}(f) \subset \{0\}$. Finally, $\| \int_0^t f(s)\, ds \| = \| T(t+2\tau) - T(2\tau) \| \leq 2K$ $(t \geq 0)$. It follows from Corollary 4.4.11 that $\lim_{t\to\infty} \| f(t) \| = 0$.

Conversely, assume that (ii) holds. Let $i\eta \in \sigma(A)$. By Proposition B.2 d), there exist $x_n \in D(A)$ with $\| x_n \| = 1$ such that $\lim_{n\to\infty} \| (A - i\eta)x_n \| = 0$. Then

$$\begin{aligned} \left(i\eta e^{i\eta t} - AT(t)\right) x_n &= i\eta \left(e^{i\eta t} - T(t)\right) x_n + T(t)\,(i\eta - A)\, x_n \\ &= i\eta e^{i\eta t} \int_0^t e^{-i\eta s} T(s)(i\eta - A)x_n\, ds + T(t)(i\eta - A)x_n \\ &\to 0 \quad \text{as } n \to \infty. \end{aligned}$$

Thus, $i\eta e^{i\eta t} \in \sigma(AT(t))$. Hence, $|\eta| = |i\eta e^{i\eta t}| \leq \| AT(t) \|$ for all $t \geq \tau$. Since $\lim_{t\to\infty} \| AT(t) \| = 0$, we conclude that $\eta = 0$. □

Theorem 4.4.1 shows that $\lim_{t\to\infty} \int_0^t f(s)\, ds = \hat{f}(0)$ if $f \in L^\infty(\mathbb{R}_+, X)$ and $\mathrm{sp}(f)$ is empty. This is not true if the assumption that f is bounded is omitted (see Example 1.5.2). In the final part of this section we shall show that it is true if f is exponentially bounded and $\hat{f}$ has a bounded holomorphic extension to a half-plane $\{\lambda : \operatorname{Re}\lambda > -\varepsilon\}$ for some $\varepsilon > 0$. Indeed, we shall show that

$$\mathrm{abs}(f) \leq \mathrm{hol}_0(\hat{f})$$

whenever f is exponentially bounded, where

$$\mathrm{hol}_0(f) := \inf\{\omega \in \mathbb{R} : \hat{f} \text{ has a bounded holomorphic extension for } \operatorname{Re}\lambda > \omega\}$$

is the *abscissa of boundedness* of $\hat{f}$. We begin with a general estimate.

Proposition 4.4.17. *Let $f \in L^1_{loc}(\mathbb{R}_+, X)$ be exponentially bounded and suppose that $\mathrm{hol}(\hat{f}) \le 0$ and $\hat{f}$ is bounded on $\mathbb{C}_+$. Then there is a constant C such that*

$$\left\| \int_0^\infty \phi(t) f(t)\, dt \right\| \le C \|\mathcal{F}\phi\|_1$$

for all functions $\phi \in L^1(\mathbb{R}_+)$ such that $\mathcal{F}\phi \in L^1(\mathbb{R})$ and $\phi f \in L^1(\mathbb{R}_+, X)$.

Proof. First, we assume that ϕ has compact support and $\mathcal{F}\phi \in L^1(\mathbb{R})$. The Laplace transform $\hat{\phi}$ is defined on $\mathbb{C}$, and $(\mathcal{F}\phi)(s) = \hat{\phi}(is) = (\overline{\mathcal{F}}\phi)(-is)$. Take $\omega > \max(0, \omega(f))$ and $0 < \alpha < \omega$. The function $t \mapsto e^{-\omega t} f(t)$ belongs to $L^1(\mathbb{R}_+, X)$ and its Fourier transform is $s \mapsto \hat{f}(\omega + is)$. Let $\psi \in C_c^\infty(\mathbb{R})$ with $\psi(t) = e^{(\omega-\alpha)t}$ whenever $t \in \operatorname{supp}\phi$. Then $(\overline{\mathcal{F}}\psi) * (\overline{\mathcal{F}}\phi) \in L^1(\mathbb{R})$. By the Fourier Inversion Theorem 1.8.1 d), $\mathcal{F}((\overline{\mathcal{F}}\psi) * (\overline{\mathcal{F}}\phi))(t) = 4\pi^2 \psi(t)\phi(t) = 4\pi^2 e^{(\omega-\alpha)t}\phi(t)$ for all t, and $((\overline{\mathcal{F}}\psi) * (\overline{\mathcal{F}}\phi))(s) = 2\pi\widehat{\phi}(\alpha - \omega - is)$ for all s. By Theorem 1.8.1 b),

$$\begin{aligned}\int_0^\infty e^{-\alpha t} f(t)\phi(t)\, dt &= \int_0^\infty e^{-\omega t} f(t) e^{(\omega-\alpha)t}\phi(t)\, dt \\ &= \frac{1}{2\pi}\int_{-\infty}^\infty \hat{f}(\omega + is)\hat{\phi}(\alpha - \omega - is)\, ds. \qquad (4.21)\end{aligned}$$

Now consider the contour integral

$$\int \hat{f}(z)\hat{\phi}(\alpha - z)\, dz$$

around the rectangle with vertices $\alpha \pm ir$, $\omega \pm ir$, where $r > 0$. The integral along the bottom edge is

$$\int_\alpha^\omega \hat{f}(\xi - ir)\hat{\phi}(\alpha - \xi + ir)\, d\xi.$$

For $\alpha < \xi < \omega$,

$$\hat{\phi}(\alpha - \xi + ir) = \int_0^\infty e^{-(\alpha-\xi)t}\phi(t) e^{-irt}\, dt \to 0$$

as $r \to \infty$, by the Riemann-Lebesgue lemma. Moreover,

$$\begin{aligned}\left|\hat{\phi}(\alpha - \xi + ir)\right| &\le \frac{1}{\sqrt{2\alpha}}\left(\int_0^\infty e^{2\omega t}|\phi(t)|^2\, dt\right)^{1/2}, \\ \left\|\hat{f}(\xi - ir)\right\| &\le \sup_{z \in \mathbb{C}_+} \|\hat{f}(z)\|,\end{aligned}$$

whenever $r > 0$, $\alpha < \xi < \omega$. By the dominated convergence theorem,

$$\lim_{r\to\infty} \int_\alpha^\omega \hat{f}(\xi - ir)\hat{\phi}(\alpha - \xi + ir)\, d\xi = 0.$$

A similar argument shows that the integral along the top edge of the rectangle tends to 0 as $r \to \infty$. By Cauchy's theorem,

$$\lim_{r\to\infty} \left\{ \int_{-r}^{r} \hat{f}(\omega + is)\hat{\phi}(\alpha - \omega - is)\, ds - \int_{-r}^{r} \hat{f}(\alpha + is)\hat{\phi}(-is)\, ds \right\} = 0.$$

From (4.21),

$$\int_0^\infty e^{-\alpha t} f(t)\phi(t)\, dt = \lim_{r\to\infty} \frac{1}{2\pi} \int_{-r}^{r} \hat{f}(\alpha + is)(\mathcal{F}\phi)(-s)\, ds,$$

so

$$\left\| \int_0^\infty e^{-\alpha t}\phi(t)f(t)\, dt \right\| \le \frac{1}{2\pi} \sup_{z\in\mathbb{C}_+} \left\| \hat{f}(z) \right\| \|\mathcal{F}\phi\|_1.$$

Letting $\alpha \downarrow 0$ gives

$$\left\| \int_0^\infty \phi(t)f(t)\, dt \right\| \le \left(\frac{1}{2\pi} \sup_{z\in\mathbb{C}_+} \|\hat{f}(z)\| \right) \|\mathcal{F}\phi\|_1.$$

Now, consider the case when ϕ is any function in $L^1(\mathbb{R}_+)$ such that $\mathcal{F}\phi \in L^1(\mathbb{R})$ and $\int_0^\infty |\phi(t)|\, \|f(t)\|\, dt < \infty$. Let $\psi \in C_c^\infty(\mathbb{R})$ be any function satisfying $0 \le \psi \le 1$, $\psi(0) = 1$ and $\int_{-\infty}^\infty \psi(t)\, dt = 1$. Let $\psi_n(t) = \psi(t/n)$ $(t \in \mathbb{R})$ and $\phi_n(t) = \phi(t)\psi_n(t)$. Then $(2\pi)^{-1}(\mathcal{F}\psi_n)(s) = (2\pi)^{-1}n(\mathcal{F}\psi)(ns)$, which forms a mollifier. Hence, $\mathcal{F}\phi_n = (2\pi)^{-1}\mathcal{F}\phi * \mathcal{F}\psi_n \to \mathcal{F}\phi$ in $L^1(\mathbb{R})$ (see Lemma 1.3.3). Applying the previous result to the functions ϕ_n and taking the limit provides the result. □

Next, we show that the antiderivative can grow at most linearly if $\hat{f}$ is bounded on $\mathbb{C}_+$ and f is exponentially bounded. We shall apply this result to C_0-semigroups in Theorem 5.1.8.

Theorem 4.4.18. *Let $f \in L^1_{loc}(\mathbb{R}_+, X)$ be exponentially bounded and suppose that* $\mathrm{hol}(\hat{f}) \le 0$ *and $\hat{f}$ is bounded on $\mathbb{C}_+$. Then there is a constant c such that*

$$\left\| \int_0^t f(s)\, ds \right\| \le c(1+t) \quad \text{for all } t \ge 0.$$

Proof. Let C be as in Proposition 4.4.17 and let $\omega > \max(0, \omega(f))$, so there exists M such that $\|f(s)\| \le Me^{\omega s}$ for all $s \ge 0$. Take $t > 0$ and let $\alpha := \frac{t}{2}e^{-\omega t}$ and

$\phi := \frac{1}{\alpha}\chi_{(0,\alpha)} * \chi_{(0,t)}$, so

$$\phi(s) = \begin{cases} s/\alpha & (0 \le s < \alpha), \\ 1 & (\alpha \le s < t), \\ \dfrac{t+\alpha-s}{\alpha} & (t \le s < t+\alpha), \\ 0 & (t+\alpha \le s). \end{cases}$$

Then $\alpha \le 1/(2\omega e)$ and

$$(\mathcal{F}\phi)(s) = \frac{1}{\alpha}\left(\frac{1-e^{-i\alpha s}}{is}\right)\left(\frac{1-e^{-ist}}{is}\right) = \frac{4e^{-is\alpha/2}e^{-ist/2}}{\alpha s^2}\sin\frac{\alpha s}{2}\sin\frac{st}{2}.$$

Hence,

$$\begin{aligned} \|\mathcal{F}\phi\|_1 &= \frac{4}{\alpha}\int_0^\infty \frac{|\sin\frac{\alpha s}{2}\sin\frac{st}{2}|}{s^2}\,ds \\ &\le 2\int_0^{1/\alpha}\frac{|\sin\frac{st}{2}|}{s}\,ds + \frac{4}{\alpha}\int_{1/\alpha}^\infty \frac{ds}{s^2} \\ &= 2\int_0^{t/2\alpha}\frac{|\sin u|}{u}\,du + 4 \\ &\le \int_1^{t/2\alpha}\frac{du}{u} + 6 \\ &= \log(t/2\alpha) + 6 \\ &= \omega t + 6. \end{aligned}$$

Also,

$$\int_0^t f(s)\,ds = \int_0^\infty \phi(s)f(s)\,ds + \int_0^\alpha \left(1-\frac{s}{\alpha}\right)f(s)\,ds - \int_t^{t+\alpha}\frac{t+\alpha-s}{\alpha}f(s)\,ds,$$

so

$$\begin{aligned} \left\|\int_0^t f(s)\,ds\right\| &\le \left\|\int_0^\infty \phi(s)f(s)\,ds\right\| + \int_0^\alpha \|f(s)\|\,ds + \int_t^{t+\alpha}\|f(s)\|\,ds \\ &\le C(\omega t+6) + M\alpha e^{\omega\alpha}(1+e^{\omega t}) \\ &\le C(\omega t+6) + M\frac{t}{2}e^{-\omega t}e^{1/2e}2e^{\omega t} \\ &= 6C + \left(C\omega + Me^{1/2e}\right)t. \end{aligned}$$

□

Finally, we prove the result comparing $\mathrm{abs}(f)$ and $\mathrm{hol}_0(\hat{f})$.

Theorem 4.4.19. *Let* $f \in L^1_{loc}(\mathbb{R}_+, X)$ *be exponentially bounded. Then* $\mathrm{abs}(f) \le \mathrm{hol}_0(\hat{f})$.

Proof. Let $\omega > \mathrm{hol}_0(\hat{f})$ and let $f_\omega(t) := e^{-\omega t} f(t)$. Then $\widehat{f_\omega}(\lambda) = \hat{f}(\lambda + \omega)$, so $\mathrm{hol}(\widehat{f_\omega}) < 0$ and $\widehat{f_\omega}$ is bounded on $\mathbb{C}_+$. By Theorem 4.4.18 and Theorem 1.4.3, $\mathrm{abs}(f_\omega) \leq 0$. Hence, $\mathrm{abs}(f) = \mathrm{abs}(f_\omega) + \omega \leq \omega$ whenever $\omega > \mathrm{hol}_0(\hat{f})$. □

Corollary 4.4.20. *Let $f \in L^1_{loc}(\mathbb{R}_+, X)$ be exponentially bounded. Assume that $\hat{f}$ has a bounded holomorphic extension to a half-plane $\{\lambda \in \mathbb{C} : \operatorname{Re}\lambda > -\varepsilon\}$ for some $\varepsilon > 0$. Then*

$$\lim_{t\to\infty} \int_0^t f(s)\, ds = \hat{f}(0).$$

4.5 Almost Periodic Functions

This section is divided into two parts: in the first we describe relatively compact orbits of C_0-groups. The abstract results which we obtain are applied in the second part to characterize almost periodic functions on the real line.

A subset Q of $\mathbb{R}$ is called *relatively dense* if there exists a length $l > 0$ such that

$$[a, a + l] \cap Q \neq \emptyset$$

for all $a \in \mathbb{R}$.

Let $U = (U(t))_{t\in\mathbb{R}}$ be a C_0-group of isometries on a Banach space Y. Our aim is to prove the following.

Theorem 4.5.1. *Let $y \in Y$. The following assertions are equivalent:*

(i) *The set $\{U(t)y : t \in \mathbb{R}\}$ is relatively compact in Y.*

(ii) *The set $\{U(t)y : t \geq 0\}$ is relatively compact in Y.*

(iii) *For all $\varepsilon > 0$, the set $Q_\varepsilon := \{\tau \in \mathbb{R} : \| U(\tau)y - y \| \leq \varepsilon\}$ is relatively dense in $\mathbb{R}$.*

(iv) $y \in \overline{\operatorname{span}}\{x \in Y :$ *there exists* $\eta \in \mathbb{R}$ *such that* $U(t)x = e^{i\eta t}x$ *for all* $t \in \mathbb{R}\}$.

If these four equivalent conditions are satisfied we say that y *has a relatively compact orbit.* Since Y is complete, a subset K of Y is relatively compact if and only if K is precompact (i.e., for every $\varepsilon > 0$ it can be covered by a finite number of ε-balls). We shall use this equivalence frequently without comment, and we shall use the terminology "relatively compact" except in some proofs where precompactness is more relevant.

The equivalence of the conditions (i), (ii) and (iii) will be proved by simple direct arguments. In order to show that they imply (iv) we need the following basic result of harmonic analysis (see [Rud62, Section 1.5.2]).

Proposition 4.5.2. *Let G be a compact abelian group with Haar measure dx. Let $f : G \to \mathbb{C}$ be continuous such that*

$$\int_G f(x)\overline{\gamma(x)}\, dx = 0$$

for every character γ (*i.e., every continuous homomorphism* $\gamma : G \to \mathbb{C} \setminus \{0\}$). *Then* $f(x) = 0$ *for all* $x \in G$.

Proof of Theorem 4.5.1. (i) $\Rightarrow$ (ii): This is trivial.

(ii) $\Rightarrow$ (iii). Let $\varepsilon > 0$. By assumption, there exist $t_1, \dots, t_m \geq 0$ such that for all $t \geq 0$ there exists $j \in \{1, \dots, m\}$ such that $\| U(t)y - U(t_j)y \| \leq \varepsilon$. Let $l := \max_{j=1,\dots,m} t_j$. We show that $[t, t+l] \cap Q_\varepsilon \neq \emptyset$ for all $t \in \mathbb{R}$. Let $t \geq 0$. Choose $j \in \{1, \dots, m\}$ such that $\|U(t)y - U(t_j)y\| \leq \varepsilon$. Let $\tau := t - t_j$. Then $\| U(\tau)y - y \| = \| U(-t_j)(U(t)y - U(t_j)y) \| = \| U(t)y - U(t_j)y \| \leq \varepsilon$. Thus, $\tau \in Q_\varepsilon \cap [t - l, t]$.

Let $t < 0$. Then there exists $j \in \{1, \dots, m\}$ such that $\| U(-t)y - U(t_j)y \| \leq \varepsilon$. Let $\tau := t + t_j$. Then $\| U(\tau)y - y \| = \| U(t)(U(t_j)y - U(-t)y) \| \leq \varepsilon$. Thus, $\tau \in Q_\varepsilon$ and $\tau \in [t, t+l]$. We have shown that $[t, t+l] \cap Q_\varepsilon \neq \emptyset$ for all $t \in \mathbb{R}$.

(iii) $\Rightarrow$ (i): Let $\varepsilon > 0$. By assumption, there exists $l > 0$ such that for all $n \in \mathbb{Z}$, $[ln, ln + l] \cap Q_\varepsilon \neq \emptyset$. Since $\{U(t)y : t \in [0, 2l]\}$ is compact, there exist $t_1, \dots, t_m \in [0, 2l]$ such that for all $s \in [0, 2l]$ there exists $j \in \{1, \dots, m\}$ such that $\| U(s)y - U(t_j)y \| \leq \varepsilon$. Let $t \in \mathbb{R}$. Take $n \in \mathbb{Z}$ such that $t \in [ln, ln + l]$ and choose $\tau \in Q_\varepsilon \cap [-ln, -ln + l]$. Then $t + \tau \in [0, 2l]$. There exists $j \in \{1, \dots, m\}$ such that $\| U(t+\tau)y - U(t_j)y \| \leq \varepsilon$. Thus,

$$\| U(t)y - U(t_j)y \| \leq \| U(t)y - U(t)U(\tau)y \| + \| U(t+\tau)y - U(t_j)y \| \leq 2\varepsilon.$$

We have shown that the orbit $\{U(t)y : t \in \mathbb{R}\}$ is covered by balls $B(U(t_j)y, 2\varepsilon) = \{z \in Y : \| U(t_j)y - z \| \leq 2\varepsilon\}$ $(j = 1, \dots, m)$. Thus, (i) holds.

(i) $\Rightarrow$ (iv): Replacing Y by $\overline{\operatorname{span}}\{U(t)y : t \in \mathbb{R}\}$, we can assume that every orbit $\{U(t)x : t \in \mathbb{R}\}$ $(x \in Y)$ is relatively compact. Then U is bounded, by the uniform boundedness principle. Denote by G the closure of $\{U(t) : t \in \mathbb{R}\}$ in $\mathcal{L}_s(Y)$, the space $\mathcal{L}(Y)$ with respect to the strong operator topology. Then $Gx \subset K(x) := \{U(t)x : t \in \mathbb{R}\}^-$ for all $x \in Y$. By hypothesis, $K(x)$ is compact. We consider the strong operator topology on G, and all limits in this proof will be in that topology. Then G can be identified with a closed subset of $\prod_{x \in Y} K(x)$ via $S \in G \mapsto \{Sx : x \in Y\}$. Hence, G is compact by Tychonov's theorem. Since multiplication is jointly continuous on bounded subsets of $\mathcal{L}_s(X)$, $S, T \in G$ implies $ST \in G$. Let $T \in G$. There exists a net $(t_i)_{i \in I}$ in $\mathbb{R}$ such that $T = \lim_{i \in I} U(t_i)$. Considering a subnet if necessary, we can assume that $S := \lim_{i \in I} U(-t_i)$ exists as well. Then $TS = ST = I$. We show that inversion is continuous. In fact, let $\lim_{i \in I} T_i = T$ in G. It follows from compactness that every subnet of $(T_i)_{i \in I}$ has a subnet $(T_j)_{j \in J}$ such that $\lim_j T_j^{-1}$ exists and joint continuity of multiplication implies that $\lim_j T_j^{-1} = T^{-1}$. This implies that $\lim_i T_i^{-1} = T^{-1}$. Thus, G is a compact abelian group with continuous multiplication.

Denote by $\hat{G}$ the dual group of G and by dS the Haar measure on G. For $\gamma \in \hat{G}$, $x \in Y$ define $P_\gamma x := \int_G \overline{\gamma(S)} Sx \, dS \in Y$. Then $P_\gamma \in \mathcal{L}(Y)$ and $TP_\gamma x = \int_G \overline{\gamma(S)} TSx \, dS = \gamma(T) \int_G \overline{\gamma(TS)} TSx \, dS = \gamma(T) P_\gamma x$ $(T \in G)$. The mapping $\phi(t) := \gamma(U(t))$ is a continuous character on $\mathbb{R}$. Hence, for each $\gamma \in \hat{G}$, there exists $\eta \in \mathbb{R}$ such that $\gamma(U(t)) = e^{i\eta t}$ $(t \in \mathbb{R})$. Thus,

$$\begin{aligned} F &:= \{P_\gamma x : \gamma \in \hat{G}, x \in Y\} \\ &\subset \left\{z \in Y : \text{there exists } \eta \in \mathbb{R} \text{ such that } U(t)z = e^{i\eta t} y \text{ for all } t \in \mathbb{R}\right\}. \end{aligned}$$

It remains to show that span F is dense in Y. Let $\psi \in Y^*$ such that $\langle z, \psi \rangle = 0$ for all $z \in F$. Let $x \in Y$. Then $\int_G \overline{\gamma(S)} \langle Sx, \psi \rangle \, dS = 0$ for all $\gamma \in \hat{G}$. Since $S \mapsto \langle Sx, \psi \rangle$ is a continuous mapping, it follows from Proposition 4.5.2 that $\langle Sx, \psi \rangle = 0$ for all $S \in G$. In particular, $\langle x, \psi \rangle = 0$. Thus, $\psi = 0$. It follows from the Hahn-Banach theorem that $\overline{\text{span}} F = Y$.

(iv) $\Rightarrow$ (i): It is obvious that the set

$$Y_1 := \{x \in Y : \ x \text{ has precompact orbit}\}$$

is a closed subspace of Y. Since $x \in Y_1$ whenever $U(t)x = e^{i\eta t}x$ for all $t \in \mathbb{R}$, the implication follows. □

We recall the following facts from Section 4.3. An element $x \in Y$ is called *totally ergodic* if

$$M_\eta x := \lim_{t \to \infty} \frac{1}{t} \int_0^t e^{-i\eta s} U(s) x \, ds$$

converges for all $\eta \in \mathbb{R}$. In that case,

$$U(t) M_\eta x = e^{i\eta t} M_\eta x \quad (t \in \mathbb{R}).$$

Moreover, the set of all frequencies

$$\text{Freq}(x) := \{\eta \in \mathbb{R} : \ M_\eta x \neq 0\}$$

is countable (by Proposition 4.3.11).

It is obvious that the set Y_e of all totally ergodic vectors is a closed subspace of Y. We introduce the space Y_{ap} of all *almost periodic vectors* (with respect to U) defined by

$$\begin{aligned} Y_{ap} &:= \{x \in Y : \ x \text{ has relatively compact orbit}\} \\ &= \overline{\text{span}}\{x \in Y : \ \text{there exists } \eta \in \mathbb{R} \text{ such that } U(t)x = e^{i\eta t}x \text{ for all } t \in \mathbb{R}\}. \end{aligned}$$

Then Y_{ap} is a closed subspace of Y which is invariant under U. If $x \in Y$ is a *periodic vector*, i.e.,

$$U(t)x = e^{i\xi t} x \quad (t \in \mathbb{R})$$

for some $\xi \in \mathbb{R}$, then $\text{Freq}(x) \subset \{\xi\}$ and

$$M_\eta x = \begin{cases} x & \text{if } \eta = \xi, \\ 0 & \text{if } \eta \neq \xi. \end{cases} \tag{4.22}$$

It follows that all almost periodic vectors are totally ergodic.

The following approximation result shows that every $x \in Y_{ap}$ can be approximated by linear combinations of eigenvectors associated with frequencies of x.

Proposition 4.5.3 (Spectral synthesis). *Let $x \in Y_{ap}$. Then $x \in \overline{\text{span}}\{y \in Y :$ there exists $\eta \in \text{Freq}(x)$ such that $U(t)y = e^{i\eta t}y$ for all $t \in \mathbb{R}\}$.*

Proof. The space $Z := \{y \in Y_{ap} : M_\eta y = 0$ for all $\eta \in \mathbb{R}\backslash\text{Freq}(x)\}$ is closed and invariant under the group. If $\eta \in \mathbb{R}\backslash\text{Freq}(x)$ and $y \in Z$ such that $U(t)y = e^{i\eta t}y$ ($t \in \mathbb{R}$), then $y = M_\eta y = 0$. Now the claim follows from Theorem 4.5.1, applied to the restriction of U to Z. □

Corollary 4.5.4. *Let $x \in Y_{ap}$. Then*

a) $x = 0$ *if and only if* $\text{Freq}(x) = \emptyset$.

b) $\text{Freq}(x) \subset \{\eta\}$ *if and only if* $U(t)x = e^{i\eta t}x$ ($t \in \mathbb{R}$).

c) $\text{Freq}(x) \subset \{\eta_1, \dots, \eta_m\}$ *if and only if* $x = \sum_{j=1}^m x_j$ *with* $U(t)x_j = e^{i\eta_j t}x_j$ *for all* $t \in \mathbb{R}$.

d) *Let* $\tau > 0$. *Then* $U(t+\tau)x = U(t)x$ *for all* $t \in \mathbb{R}$ *if and only if* $\text{Freq}(x) \subset \frac{2\pi}{\tau}\mathbb{Z}$.

Proof. a) and b) follow directly from Proposition 4.5.3. If $\text{Freq}(x) \subset \{\eta_1, \dots, \eta_m\}$, then by Proposition 4.5.3, $x = \lim_{n\to\infty} x_n$, where $x_n = \sum_{j=1}^m x_{nj}$ for some $x_{nj} \in X$ such that $U(t)x_{nj} = e^{i\eta_j t}x_{nj}$ ($j = 1, \dots, m$). We can assume that $\eta_j \neq \eta_k$ for $k \neq j$. Then $M_{\eta_j}x = \lim_{n\to\infty} M_{\eta_j}x_n = \lim_{n\to\infty} x_{nj}$ by (4.22). It follows that $x = \sum_{j=1}^m M_{\eta_j}x$. This proves one implication of c). The other follows from (4.22).

We prove d). Assume that $U(t+\tau)x = U(t)x$ ($t \in \mathbb{R}$). Recall that for $z \in \mathbb{C}, |z| = 1$, one has

$$\lim_{n\to\infty} \frac{1}{n} \sum_{k=0}^{n-1} z^k = \begin{cases} 1 & \text{if } z = 1, \\ 0 & \text{if } z \neq 1. \end{cases}$$

Let $\eta \in \mathbb{R}$. Then

$$\begin{aligned} M_\eta x &= \lim_{n\to\infty} \frac{1}{\tau n} \int_0^{\tau n} U(s)xe^{-i\eta s}\,ds \\ &= \lim_{n\to\infty} \frac{1}{\tau n} \sum_{k=0}^{n-1} \int_{k\tau}^{(k+1)\tau} U(s)xe^{-i\eta s}\,ds \\ &= \lim_{n\to\infty} \frac{1}{\tau n} \sum_{k=0}^{n-1} \int_0^{\tau} U(s)xe^{-i\eta s}\,ds\; e^{-ik\eta\tau} \\ &= \begin{cases} 1/\tau \int_0^\tau U(s)xe^{-i\eta s}\,ds & \text{if } \eta\tau \in 2\pi\mathbb{Z}, \\ 0 & \text{if } \eta\tau \notin 2\pi\mathbb{Z}. \end{cases} \end{aligned}$$

This proves one implication. The other follows directly from Proposition 4.5.3. □

Now we come to the second part of this section where we consider a special group of operators. Let X be a Banach space. By $\text{BUC}(\mathbb{R}, X)$ we denote the space of all bounded uniformly continuous functions on $\mathbb{R}$ with values in X. It is a Banach space for the uniform norm

$$\| f \|_\infty = \sup_{t\in\mathbb{R}} \| f(t) \|.$$

The shift group on $\mathrm{BUC}(\mathbb{R}, X)$ defined by

$$(S(t)f)(r) := f(r+t) \quad (r \in \mathbb{R}, t \in \mathbb{R})$$

is a C_0-group whose generator is denoted by B.

Lemma 4.5.5. *For $f \in \mathrm{BUC}(\mathbb{R}, X)$ one has $f \in D(B)$ if and only if f is differentiable and $f' \in \mathrm{BUC}(\mathbb{R}, X)$. In that case, $Bf = f'$.*

Proof. Let $f \in D(B)$, $g := Bf$. Then $S(t)f - f = \int_0^t S(s)g\,ds$ for all $t \geq 0$. In particular, $f(t) = (S(t)f)(0) = f(0) + \int_0^t g(s)\,ds$. This shows one implication of the claim. Conversely, assume that f is differentiable and $g := f' \in \mathrm{BUC}(\mathbb{R}, X)$. Then $(S(t)f - f)(r) = f(r+t) - f(r) = \int_r^{r+t} g(s)\,ds = \int_0^t (S(s)g)(r)\,ds$. Thus, $S(t)f - f = \int_0^t S(s)g\,ds$. Hence, $f \in D(B)$ and $Bf = g$. □

Let $\eta \in \mathbb{R}, x \in X$. By $e_{i\eta} \otimes x$ we denote the periodic function given by

$$(e_{i\eta} \otimes x)(t) = e^{i\eta t}x \quad (t \in \mathbb{R}).$$

Linear combinations of functions of the form $e_{i\eta} \otimes x$ with $\eta \in \mathbb{R}$, $x \in X$ are called *trigonometric polynomials.*

Definition 4.5.6. *A function $f : \mathbb{R} \to X$ is called* almost periodic *if it can be approximated uniformly on $\mathbb{R}$ by trigonometric polynomials. By*

$$\mathrm{AP}(\mathbb{R}, X) := \overline{\mathrm{span}}\{e_{i\eta} \otimes x : \eta \in \mathbb{R},\, x \in X\}$$

we denote the space of all almost periodic functions on $\mathbb{R}$ with values in X.

Let $\eta \in \mathbb{R}$. A function $f \in \mathrm{BUC}(\mathbb{R}, X)$ satisfies

$$S(t)f = e^{i\eta t}f \text{ for all } t \in \mathbb{R}$$

if and only if $f = e_{i\eta} \otimes f(0)$. Thus, considering the group S on $Y := \mathrm{BUC}(\mathbb{R}, X)$, we have $Y_{ap} = \mathrm{AP}(\mathbb{R}, X)$. Now we can reformulate the results of the first part of this section for this special case.

Let $f \in \mathrm{BUC}(\mathbb{R}, X)$. Let $\varepsilon > 0$. A real number $\tau > 0$ is called an *ε-period* of f if $\|f(\tau + s) - f(s)\| \leq \varepsilon$ for all $s \in \mathbb{R}$.

Theorem 4.5.7. *Let $f \in \mathrm{BUC}(\mathbb{R}, X)$. The following are equivalent:*

(i) *f is almost periodic.*

(ii) *For every $\varepsilon > 0$ the set of all ε-periods is relatively dense in* $\mathbb{R}$.

(iii) *The orbit* $\{S(t)f : t \in \mathbb{R}\}$ *is relatively compact in* $\mathrm{BUC}(\mathbb{R}, X)$.

This is an immediate consequence of Theorem 4.5.1. It follows in particular that for every $f \in \mathrm{AP}(\mathbb{R}, X)$ there exist $t_n \in \mathbb{R}$ such that $\lim_{n\to\infty} t_n = \infty$ and

$$\| f(t_n + s) - f(s) \| \leq \frac{1}{n} \quad \text{for all } s \in \mathbb{R}. \tag{4.23}$$

In particular, if $f \in \mathrm{AP}(\mathbb{R}, X)$, then

$$\| f \|_\infty = \sup_{t \geq \tau} \| f(t) \| \tag{4.24}$$

for all $\tau \in \mathbb{R}$.

For $f \in \mathrm{AP}(\mathbb{R}, X)$, $\eta \in \mathbb{R}$, we define the mean

$$\begin{aligned} M_\eta f \quad &:= \quad \lim_{t\to\infty} \frac{1}{t} \int_0^t e^{-i\eta s} S(s) f \, ds \\ &= \quad \lim_{\alpha \downarrow 0} \alpha R(\alpha + i\eta, B) f. \end{aligned}$$

These limits exist in $\mathrm{BUC}(\mathbb{R}, X)$ (i.e., with respect to the uniform norm) by the remarks preceding Proposition 4.5.3 and by Proposition 4.3.1. Moreover, $S(t)M_\eta f = e^{i\eta t} M_\eta f$ for all $t \in \mathbb{R}$. Evaluating at 0, we deduce that

$$M_\eta f = e_{i\eta} \otimes (M_\eta f)(0). \tag{4.25}$$

As before in the abstract setting, we let

$$\mathrm{Freq}(f) := \{\eta \in \mathbb{R} : M_\eta f \neq 0\}$$

be the set of all *frequencies* of f. By Proposition 4.3.11, this is a countable set. Moreover the following property of spectral synthesis holds.

Proposition 4.5.8 (Spectral synthesis). *Let* $f \in \mathrm{AP}(\mathbb{R}, X)$. *Then*

$$f \in \overline{\mathrm{span}}\{e_{i\eta} \otimes x : \eta \in \mathrm{Freq}(f), x \in f(\mathbb{R}), \}$$

where the closure is taken in $\mathrm{BUC}(\mathbb{R}, X)$.

Proof. Let $X_0 := \overline{\mathrm{span}}\{f(t) : t \in \mathbb{R}\} \subset X$. Then by Theorem 4.5.7, f is also almost periodic when it is considered as a function with values in X_0. So we can assume that $X_0 = X$. Now the claim follows from Proposition 4.5.3. □

From Corollary 4.5.4 we see the following.

Corollary 4.5.9. *Let $f \in \mathrm{AP}(\mathbb{R}, X)$. Then*

a) $f \equiv 0$ *if and only if* $\mathrm{Freq}(f) = \emptyset$.

b) $\mathrm{Freq}(f) \subset \{\eta\}$ *if and only if* $f = e_{i\eta} \otimes f(0)$.

c) f *is a trigonometric polynomial if and only if* $\mathrm{Freq}(f)$ *is finite.*

d) f *is τ-periodic if and only if* $\mathrm{Freq}(f) \subset \frac{2\pi}{\tau}\mathbb{Z}$ *(where $\tau > 0$).*

We call assertion a) the *uniqueness theorem for almost periodic functions.* We also obtain from Theorem 4.5.7 that almost periodic functions have relatively compact range.

Corollary 4.5.10. *Let $f \in \mathrm{AP}(\mathbb{R}, X)$. Then the set $\{f(t) : t \in \mathbb{R}\}^-$ is compact.*

Proof. By hypothesis, the set $\{S(t)f : t \in \mathbb{R}\}$ is relatively compact in $\mathrm{BUC}(\mathbb{R}, X)$. Since evaluation at 0 is a continuous operator from $\mathrm{BUC}(\mathbb{R}, X)$ into X, it follows that the set $\{f(t) : t \in \mathbb{R}\} = \{(S(t)f)(0) : t \in \mathbb{R}\}$ is relatively compact in X. □

A function $f \in \mathrm{BUC}(\mathbb{R}, X)$ is called *weakly almost periodic* if $x^* \circ f$ is almost periodic for all $x^* \in X^*$.

Remark 4.5.11. There are various different definitions of weak almost periodicity in the literature. In particular, a function $f \in \mathrm{BUC}(\mathbb{R}, X)$ is called *weakly almost periodic in the sense of Eberlein* if the set

$$\{S(t)f : t \in \mathbb{R}\}$$

is relatively weakly compact in the Banach space $\mathrm{BUC}(\mathbb{R}, X)$. This property is independent of weak almost periodicity, in general. We refer to the Notes. □

Proposition 4.5.12. *Assume that $f \in \mathrm{BUC}(\mathbb{R}, X)$ has relatively compact range. If f is weakly almost periodic, then f is almost periodic.*

Proof. Since f is separably valued, we can assume that X is separable. Then we can consider X as a closed subspace of $C[0, 1]$ (see [Woj91, p.36]). Let $(P_n)_{n\in\mathbb{N}}$ be a bounded sequence of finite rank operators on $C[0, 1]$ such that $\lim_{n\to\infty} P_n g = g$ for all $g \in C[0, 1]$ (see [Woj91, pp.37,40]). Then $P_n \circ f$ is weakly almost periodic, hence almost periodic since it has values in a finite dimensional space. We have $\lim_{n\to\infty} P_n(f(t)) = f(t)$ for all $t \in \mathbb{R}$. Since f has relatively compact range, this convergence is uniform in $t \in \mathbb{R}$ (see Proposition B.15). Thus, $P_n \circ f$ converges to f in $\mathrm{BUC}(\mathbb{R}, C[0, 1])$. Consequently, f is almost periodic. □

Remark 4.5.13 (Almost periodic orbits). Let U be a bounded C_0-group on a Banach space Y and let $x \in Y$. Then the following are equivalent:

(i) $x \in Y_{ap}$ (i.e., $\{U(t)x : t \in \mathbb{R}\}$ is relatively compact in Y).

(ii) $U(\cdot)x \in \mathrm{AP}(\mathbb{R}, Y)$.

In fact, if $U(\cdot)x \in \mathrm{AP}(\mathbb{R}, Y)$, then $\{U(t)x : t \in \mathbb{R}\}$ is relatively compact, by Corollary 4.5.10. Conversely, let $x \in Y_{ap}$. Let $t_n \in \mathbb{R}$. Then there exists a subsequence such that $\lim_{k\to\infty} U(t_{n_k})x := y$ exists. This implies that $U(t_{n_k} + \cdot)x$ converges to $U(\cdot)y$ in $\mathrm{BUC}(\mathbb{R}, Y)$. Thus, $\{U(t + \cdot)y : t \in \mathbb{R}\}$ is relatively compact in $\mathrm{BUC}(\mathbb{R}, Y)$. □

4.6 Countable Spectrum and Almost Periodicity

In this section we define the spectrum of a function $f \in \mathrm{BUC}(\mathbb{R}, X)$ with the help of the Laplace transform (or more precisely, the Carleman transform). The main result of this section says that, under suitable conditions on the space X, a function $f \in \mathrm{BUC}(\mathbb{R}, X)$ with countable spectrum is almost periodic.

Let $f \in \mathrm{BUC}(\mathbb{R}, X)$. The *Carleman transform* $\hat{f}$ of f is defined by

$$\hat{f}(\lambda) := \begin{cases} \int_0^\infty e^{-\lambda t} f(t)\,dt & (\operatorname{Re}\lambda > 0), \\ -\int_0^\infty e^{\lambda t} f(-t)\,dt & (\operatorname{Re}\lambda < 0). \end{cases}$$

Thus, $\hat{f}$ is a holomorphic function defined on $\mathbb{C}\backslash i\mathbb{R}$.

Remark 4.6.1. Let $f_+ \in \mathrm{BUC}(\mathbb{R}_+, X)$ be the restriction of f to $\mathbb{R}_+$ and $f_- \in \mathrm{BUC}(\mathbb{R}_+, X)$ be given by $f_-(t) = f(-t)$ $(t \in \mathbb{R}_+)$. Then

$$\hat{f}(\lambda) = \begin{cases} \hat{f}_+(\lambda) & (\operatorname{Re}\lambda > 0), \\ -\hat{f}_-(-\lambda) & (\operatorname{Re}\lambda < 0), \end{cases}$$

where $\hat{f}_+$ and $\hat{f}_-$ are the Laplace transforms of f_+ and f_-, respectively. □

We use the same symbol for the Carleman transform and the Laplace transform. This will not lead to confusion.

A point $i\eta \in i\mathbb{R}$ is called *regular* for $\hat{f}$ if $\hat{f}$ has a holomorphic extension to a neighbourhood of $i\eta$ (i.e., $i\eta$ is regular if there exists an open neighbourhood V of $i\eta$ and a holomorphic function $h : V \to X$ such that $h(\lambda) = \hat{f}(\lambda)$ for all $\lambda \in V\backslash i\mathbb{R}$).

The *Carleman spectrum* $\mathrm{sp_C}(f)$ is defined by

$$\mathrm{sp_C}(f) = \{\eta \in \mathbb{R} : \ i\eta \text{ is not regular for } \hat{f}\}. \tag{4.26}$$

The following remark explains why this notion of spectrum is well adapted to spectral theory of C_0-groups.

Remark 4.6.2 (Carleman spectrum and C_0-groups). Let Y be a Banach space and U be a bounded C_0-group on Y with generator A. Then $\sigma(A) \subset i\mathbb{R}$, since A and $-A$ generate bounded C_0-semigroups. Let $x \in Y$ and $f(t) := U(t)x$. Then the Carleman transform $\hat{f}$ of f is given by

$$\hat{f}(\lambda) = R(\lambda, A)x \quad (\lambda \in \mathbb{C}\backslash i\mathbb{R}). \tag{4.27}$$

In particular,

$$i\,\mathrm{sp}_{\mathrm{C}}(f) \subset \sigma(A). \tag{4.28}$$

In fact, (4.27) is clear for $\operatorname{Re}\lambda > 0$. If $\operatorname{Re}\lambda < 0$, then $\hat{f}(\lambda) = -\hat{f}_{-}(-\lambda) = -\int_0^\infty e^{\lambda t}U(-t)x\,dt = -R(-\lambda,-A)x = R(\lambda,A)x$. □

As in Section 4.5, we denote by S the shift group on $\mathrm{BUC}(\mathbb{R},X)$ and by B its generator. For $f \in \mathrm{BUC}(\mathbb{R},X)$ and $s \in \mathbb{R}$, let $f_s := S(s)f$. An easy calculation (see Remark 4.6.2) shows that

$$(R(\lambda,B)f)(s) = \widehat{f_s}(\lambda) = e^{\lambda s}\left(\hat{f}(\lambda) - \int_0^s e^{-\lambda t} f(t)\,dt\right)$$

for $\lambda \in \mathbb{C} \setminus i\mathbb{R}$. We shall see in Lemma 4.6.8 that the singularities of $R(\cdot,B)$ coincide with the singularities of $\hat{f}$; i.e., that the Carleman spectra of $S(\cdot)f$ and $f(\cdot)$ coincide.

Let $\eta \in \mathbb{R}$. A function $f \in \mathrm{BUC}(\mathbb{R},X)$ is called *uniformly ergodic at* η if it is ergodic at η with respect to S; i.e., if

$$M_\eta f := \lim_{t\to\infty} \frac{1}{t}\int_0^t e^{-i\eta s} S(s)f\,ds \tag{4.29}$$

exists in $\mathrm{BUC}(\mathbb{R},X)$. By Section 4.3, this is equivalent to saying that

$$M_\eta f = \lim_{\alpha\downarrow 0} \alpha R(\alpha + i\eta, B)f \tag{4.30}$$

exists in $\mathrm{BUC}(\mathbb{R},X)$. Since $(R(\alpha+i\eta,B)f)(s) = \widehat{f_s}(\alpha+i\eta)$, this can be reformulated by saying that $\alpha\widehat{f_s}(\alpha+i\eta)$ converges as $\alpha\downarrow 0$ uniformly in $s \in \mathbb{R}$.

We say that f is *totally ergodic* if f is uniformly ergodic at each $\eta \in \mathbb{R}$; i.e., if f is totally ergodic with respect to S in the sense of Sections 4.3 and 4.5.

Next, we introduce a geometric condition of Banach spaces. Let c_0 be the Banach space of all complex sequences converging to 0 with the supremum norm (as in Example 1.1.5 b)). We say that X *contains* c_0, and write briefly $c_0 \subset X$, if there exists a closed subspace of X which is isomorphic to c_0. Since closed subspaces of reflexive spaces are reflexive, no reflexive Banach space contains c_0. Also, any space of the form $L^1(\Omega,\mu)$ does not contain c_0. See Appendix D for further information.

Now we can formulate the main theorem of this section.

Theorem 4.6.3. *Let $f \in \mathrm{BUC}(\mathbb{R},X)$ have countable Carleman spectrum. Assume that one of the following conditions is satisfied:*

a) *f is totally ergodic, or*

b) *f has relatively compact range, or*

c) $c_0 \not\subset X$.

Then f is almost periodic.

As a corollary, we note the important scalar case which is due to Loomis. Here, we let $\mathrm{BUC}(\mathbb{R}) := \mathrm{BUC}(\mathbb{R}, \mathbb{C})$ and $\mathrm{AP}(\mathbb{R}) := \mathrm{AP}(\mathbb{R}, \mathbb{C})$.

Corollary 4.6.4 (Loomis's Theorem). *Let $f \in \mathrm{BUC}(\mathbb{R})$ have countable Carleman spectrum. Then f is almost periodic.*

We have seen in the preceding section that every almost periodic function is totally ergodic and has relatively compact range. Thus, conditions a) and b) in Theorem 4.6.3 are necessary for the conclusion to hold. If the Banach space X contains c_0, countability of the Carleman spectrum alone does not imply almost periodicity. This is shown by the following example.

Example 4.6.5. Let $X := c$, the space of all convergent complex sequences $x = (x_n)_{n\in\mathbb{N}}$ with the supremum norm $\|x\| = \sup_n |x_n|$. Then X is isomorphic to c_0. Let

$$f(t) := \left(e^{it/n}\right)_{n\in\mathbb{N}} =: (f_n(t))_{n\in\mathbb{N}}\,.$$

Since $f_n'(t) = \frac{i}{n}e^{it/n}$, the function f is Lipschitz continuous and thus $f \in \mathrm{BUC}(\mathbb{R}, c)$. The Carleman transform $\hat{f}$ of f is given by

$$\hat{f}(\lambda) = \left(\frac{1}{\lambda - i/n}\right)_{n\in\mathbb{N}}$$

for all $\lambda \in \mathbb{C} \setminus i\mathbb{R}$. Consequently, $\mathrm{sp_C}(f) = \{1/n : n \in \mathbb{N}\} \cup \{0\}$, which is countable. However, $f \notin \mathrm{AP}(\mathbb{R}, c)$. In fact, f does not have relatively compact range. To see this, consider $\phi_n \in c^*$ given by $\langle x, \phi_n\rangle := x_n$ for $x = (x_k)_{k\in\mathbb{N}} \in c$, and $\phi_\infty \in c^*$ given by $\langle x, \phi_\infty\rangle := \lim_{k\to\infty} x_k$. Then $\|\phi_n\| \le 1$ $(n \in \mathbb{N} \cup \{\infty\})$ and $\lim_{n\to\infty}\langle x, \phi_n\rangle = \langle x, \phi_\infty\rangle$ for all $x \in c$. Suppose that $K := \{f(t) : t \in \mathbb{R}\}$ is relatively compact in c. Then $\langle x, \phi_n\rangle$ converges to $\langle x, \phi_\infty\rangle$ uniformly on K (see Proposition B.15). In particular, $\lim_{n\to\infty} \mathrm{Im}\langle f(t), \phi_n\rangle = \lim_{n\to\infty} \sin\frac{t}{n} = 0$ uniformly in $t \in \mathbb{R}$, which is absurd. □

We need several auxiliary results for the proof of Theorem 4.6.3. The following is a special kind of maximum principle for holomorphic functions.

Lemma 4.6.6. *Let V be an open neighbourhood of $i\eta$ such that V contains the closed disc $\overline{B}(i\eta, 2r) = \{z \in \mathbb{C} : |z - i\eta| \le 2r\}$. Let $h : V \to X$ be holomorphic and $c \ge 0, k \in \mathbb{N}_0$ such that*

$$\|h(z)\| \le \frac{c}{|\operatorname{Re} z|^k} \qquad \textit{if } |z - i\eta| = 2r,\ \operatorname{Re} z \neq 0.$$

Then $\|h(z)\| \le (4/3)^k\, c r^{-k}$ for all $z \in \overline{B}(i\eta, r)$.

Proof. We can assume that $\eta = 0$ (replacing $h(z)$ by $h(z + i\eta)$ otherwise). Define $g : V \to X$ by $g(z) := \left(1 + \frac{z^2}{(2r)^2}\right)^k h(z)$. Let $|z| = 2r$, $z = 2re^{i\theta}$. Then

$$\begin{aligned} \| g(z) \| &= |1 + e^{i2\theta}|^k \, \| h(z) \| = |e^{i\theta}(e^{-i\theta} + e^{i\theta})|^k \, \| h(z) \| \\ &= 2^k \, |\cos\theta|^k \, \| h(z) \| \leq \frac{c}{r^k} \, . \end{aligned}$$

It follows from the maximum principle that $\| g(z) \| \leq cr^{-k}$ if $|z| \leq 2r$. Let $|z| \leq r$. Then $\left| \left(1 + \frac{z^2}{(2r)^2}\right)^{-1} \right| = \left| \frac{(2r)^2}{(2r)^2 + z^2} \right| \leq \frac{4r^2}{4r^2 - r^2} = \frac{4}{3}$. Hence,

$$\| h(z) \| = \left| \left(1 + \frac{z^2}{(2r)^2}\right)^{-k} \right| \, \| g(z) \| \leq (4/3)^k c r^{-k} \quad (|z| \leq r). \qquad \square$$

We now describe the local spectrum associated with a bounded C_0-group and an individual vector. Let U be an isometric C_0-group on a Banach space Y with generator A. For $x \in Y$ we define the space $Y_x := \overline{\operatorname{span}}\{U(t)x : t \in \mathbb{R}\}$. It is invariant under the group U. We denote by A_x the generator of this restriction group. Then $\sigma(A_x) \subset \sigma(A) \subset i\mathbb{R}$, and $R(\lambda, A_x) = R(\lambda, A)|_{Y_x}$ $(\lambda \in \rho(A))$. As before, $\mathbb{C}_+ := \{z \in \mathbb{C} : \operatorname{Re} z > 0\}$ denotes the right half plane.

Lemma 4.6.7. *Let $x \in Y, \eta \in \mathbb{R}$. The following assertions are equivalent:*

(i) $i\eta \in \rho(A_x)$.

(ii) *There exists an open neighbourhood V of $i\eta$ and a holomorphic function $h : V \to X$ such that $h(\lambda) = R(\lambda, A)x$ for all $\lambda \in \mathbb{C}_+ \cap V$.*

In that case, $h(\lambda) = R(\lambda, A)x$ for all $\lambda \in V \backslash i\mathbb{R}$.

Proof. (i) $\Rightarrow$ (ii): Let $V := \rho(A_x)$, $h(\lambda) := R(\lambda, A_x)x$.

(ii) $\Rightarrow$ (i): Assume (ii) and assume that V is connected. We first show that $h(\lambda) = R(\lambda, A)x$ for all $\lambda \in V \backslash i\mathbb{R}$. Let $\mu \in \rho(A)$. Then $k(\lambda) := (\lambda - A)R(\mu, A)h(\lambda)$ defines a holomorphic function on V such that $k(\lambda) = R(\mu, A)x$ for all $\lambda \in \mathbb{C}_+ \cap V$. It follows from the uniqueness theorem that $(\lambda - A)R(\mu, A)h(\lambda) = k(\lambda) = R(\mu, A)x$ for all $\lambda \in V$. This implies that $R(\mu, A)h(\lambda) = R(\lambda, A)R(\mu, A)x = R(\mu, A)R(\lambda, A)x$ for all $\lambda \in V \backslash i\mathbb{R}$. Since $R(\mu, A)$ is injective, the claim follows.

We now show that $i\eta \in \rho(A_x)$. We have $\| R(\lambda, A) \| = \left\| \int_0^\infty e^{-\lambda t} U(t) \, dt \right\| \leq 1/(\operatorname{Re}\lambda)$ $(\operatorname{Re}\lambda > 0)$, and similarly, $\| R(\lambda, A) \| = \| R(-\lambda, -A) \| \leq -1/(\operatorname{Re}\lambda)$ for $\operatorname{Re}\lambda < 0$. Thus $\| R(\lambda, A) \| \leq 1/|\operatorname{Re}\lambda|$ for all $\lambda \in V \backslash i\mathbb{R}$. Choose $r > 0$ such that $\overline{B}(i\eta, 2r) \subset V$. It follows from the assumption that $R(\cdot, A)z$ has a holomorphic extension to V for all $z \in \operatorname{span}\{U(t)x : t \in \mathbb{R}\}$. Now it follows from Lemma 4.6.6 that

$$\| R(\lambda, A)z \| \leq \frac{4}{3r} \| z \| \text{ for all } \lambda \in B(i\eta, r) \backslash i\mathbb{R}.$$

Hence, $\| R(\lambda, A_x) \| \leq \frac{4}{3r}$ for all $\lambda \in B(i\eta, r)\backslash i\mathbb{R}$. This implies that $i\eta \in \rho(A_x)$. □

We define the *local spectrum* of A at x by

$$\sigma(A, x) := \sigma(A_x) = \mathbb{C}\backslash\rho(A_x).$$

Thus, $i\eta \in \sigma(A, x)$ if and only if the equivalent conditions (i) and (ii) of Lemma 4.6.7 are not satisfied.

Now we consider the shift group S on $\mathrm{BUC}(\mathbb{R}, X)$ with generator B, as before. We show that the Carleman spectrum coincides with the local spectrum with respect to B.

Lemma 4.6.8. *Let $f \in \mathrm{BUC}(\mathbb{R}, X)$. Then*

$$i\,\mathrm{sp_C}(f) = \sigma(B, f).$$

Proof. a) Let $i\eta \in i\mathbb{R} \setminus \sigma(B, f)$. There exists an open neighbourhood V of $i\eta$ and a holomorphic function $h : V \to \mathrm{BUC}(\mathbb{R}, X)$ such that $h(\lambda) = R(\lambda, B)f$ $(\lambda \in V\backslash i\mathbb{R})$. Let $k(\lambda) := h(\lambda)(0)$. Then $k : V \to X$ is holomorphic, $k(\lambda) = (R(\lambda, B)f)(0) = \int_0^\infty e^{-\lambda t}(S(t)f)(0)\,dt = \hat{f}(\lambda)$ for $\lambda \in V \cap \mathbb{C}_+$, and $k(\lambda) = (R(\lambda, B)f)(0) = -(R(-\lambda, -B)f)(0) = -\int_0^\infty e^{\lambda t}(S(-t)f)(0)\,dt = -\int_0^\infty e^{\lambda t} f(-t)\,dt$ for $\lambda \in V$, $\mathrm{Re}\,\lambda < 0$. Thus, $\eta \notin \mathrm{sp_C}(f)$.

b) Let $\eta \in \mathbb{R}\backslash\mathrm{sp_C}(f)$. By assumption, there exists an open neighbourhood V of $i\eta$ and a holomorphic function $h : V \to X$ such that

$$h(\lambda) = \begin{cases} \int_0^\infty e^{-\lambda t} f(t)\,dt & (\lambda \in V \cap \mathbb{C}_+) \\ -\int_0^\infty e^{\lambda t} f(-t)\,dt & (\lambda \in V,\ \mathrm{Re}\,\lambda < 0). \end{cases}$$

For $\lambda \in V$, $s \in \mathbb{R}$, define

$$H(\lambda, s) := e^{\lambda s}\left(h(\lambda) - \int_0^s e^{-\lambda t} f(t)\,dt\right).$$

Then for $\lambda \in V$, $\mathrm{Re}\,\lambda > 0$,

$$(R(\lambda, B)f)(s) = \int_0^\infty e^{-\lambda t} f(t+s)\,dt = H(\lambda, s) \quad (s \in \mathbb{R})$$

and for $\lambda \in V$, $\mathrm{Re}\,\lambda < 0$,

$$(R(\lambda, B)f)(s) = -(R(-\lambda, -B)f)(s) = -\int_0^\infty e^{\lambda t} f(s-t)\,dt = H(\lambda, s).$$

Thus, $\| H(\lambda, s) \| \leq \| f \|_\infty / |\mathrm{Re}\,\lambda|$ for $\lambda \in V \backslash i\mathbb{R}$. By Lemma 4.6.6, this implies that for $r > 0$ such that $\overline{B}(i\eta, 2r) \subset V$, one has $\| H(\lambda, s) \| \leq 4\| f \|_\infty / 3r$ for all $\lambda \in \overline{B}(i\eta, r)$, $s \in \mathbb{R}$. We know that $H(\lambda, \cdot) = R(\lambda, B)f \in \mathrm{BUC}(\mathbb{R}, X)$ if $\mathrm{Re}\,\lambda > 0$. Since $H(\cdot, s)$ is holomorphic on $B(i\eta, r)$ and bounded uniformly in s, it follows from Corollary A.4 that $\lambda \mapsto H(\lambda, \cdot)$ is holomorphic on $B(i\eta, r)$ with values in $\mathrm{BUC}(\mathbb{R}, X)$. Here, we choose linear functionals on $\mathrm{BUC}(\mathbb{R}, X)$ of the form $g \mapsto \langle g(s), x^* \rangle$ for $s \in \mathbb{R}$, $x^* \in X^*$, $\|x^*\| \leq 1$. Now, $i\eta \notin \sigma(B, f)$ by Lemma 4.6.7. □

We deduce from Lemma 4.6.8 the following interesting observation.

Proposition 4.6.9. *Let $f \in \mathrm{BUC}(\mathbb{R}, X)$ and let $\eta \in \mathbb{R} \setminus \mathrm{sp}_{\mathrm{C}}(f)$. Then f is uniformly ergodic at η and $M_\eta f = 0$.*

Proof. It was shown in the proof of Lemma 4.6.8 that $\lambda \mapsto R(\lambda, B)f$ has a holomorphic extension to a neighbourhood of $i\eta$. It follows in particular that $\lim_{\alpha \downarrow 0} \alpha R(\alpha + i\eta, B)f = 0$ in $\mathrm{BUC}(\mathbb{R}_+, X)$. □

Proposition 4.6.10. *Let $f \in \mathrm{AP}(\mathbb{R}, X)$. Then* $\mathrm{sp}_{\mathrm{C}}(f) = \overline{\mathrm{Freq}(f)}$.

Proof. The inclusion $\mathrm{Freq}(f) \subset \mathrm{sp}_{\mathrm{C}}(f)$ is immediate from Proposition 4.6.9, although in this case the fact that f is uniformly ergodic at each point of $\mathbb{R}$ is already known (see Section 4.5). Since $\mathrm{sp}_{\mathrm{C}}(f)$ is closed, it follows that $\overline{\mathrm{Freq}(f)} \subset \mathrm{sp}_{\mathrm{C}}(f)$.

For the converse, let

$$Y_0 = \mathrm{span}\,\{e_{i\eta} \otimes x : \eta \in \mathrm{Freq}(f),\ x \in X\}$$

and $Y := \overline{Y_0}$. By Proposition 4.5.8, $f \in Y$. For $g \in Y_0$, $\hat{g}$ has a holomorphic extension to $\mathbb{C} \setminus i\,\overline{\mathrm{Freq}(f)}$, and $\|\hat{g}(\lambda)\| \leq \|g\|_\infty / |\mathrm{Re}\,\lambda|$. By Lemma 4.6.6, $\|\hat{g}(i\eta)\| \leq 8\|g\|_\infty / \big(3\,\mathrm{dist}(\eta, \mathrm{Freq}(f))\big)$ if $\eta \in \mathbb{R} \setminus \overline{\mathrm{Freq}(f)}$. If $(g_n)_{n \in \mathbb{N}}$ is a sequence in Y_0 such that $\|g_n - f\|_\infty \to 0$, it follows that $\lim_{n\to\infty} \widehat{g_n}(\lambda)$ exists for $\lambda \in \mathbb{C} \setminus i\,\overline{\mathrm{Freq}(f)}$ and by Vitali's Theorem A.5, this gives a holomorphic extension of $\hat{f}$ to $\mathbb{C} \setminus i\,\overline{\mathrm{Freq}(f)}$. Hence, $\mathrm{sp}_{\mathrm{C}}(f) \subset \overline{\mathrm{Freq}(f)}$. □

The condition that $c_0 \not\subset X$ will enter the proof of Theorem 4.6.3 in form of the following theorem due to Kadets.

Theorem 4.6.11 (Kadets). *Assume that $c_0 \not\subset X$ and let $f \in \mathrm{BUC}(\mathbb{R}, X)$. If furthermore $S(t)f - f \in \mathrm{AP}(\mathbb{R}, X)$ for all $t \in \mathbb{R}$, then $f \in \mathrm{AP}(\mathbb{R}, X)$.*

We will give the proof of Kadets's theorem at the end of this section. For $X = \mathbb{C}$, Theorem 4.6.11 is due to H. Bohr. We will give a separate direct proof for this easier case. For Theorem 4.6.3 we use the following spectral-theoretic reformulation of Kadets's theorem.

Denote by $\pi : \mathrm{BUC}(\mathbb{R}, X) \to \mathrm{BUC}(\mathbb{R}, X)/\,\mathrm{AP}(\mathbb{R}, X)$ the quotient map. Recall that the quotient space is a Banach space for the norm

$$\|\pi(f)\| := \inf\{\|f + g\| : g \in \mathrm{AP}(\mathbb{R}, X)\}$$

and π is contractive for this norm. Since $S(t)$ leaves the space $\mathrm{AP}(\mathbb{R}, X)$ invariant, there exists a C_0-group $\widetilde{S}$ on the quotient $\mathrm{BUC}(\mathbb{R}, X)/\,\mathrm{AP}(\mathbb{R}, X)$ given by

$$\widetilde{S}(t)\pi(f) := \pi(S(t)f).$$

We call its generator $\widetilde{B}$. Now Kadets's theorem can be rephrased in the following form.

Corollary 4.6.12. *The following assertions are equivalent:*

(i) $c_0 \not\subset X$.

(ii) $\widetilde{B}$ *has empty point spectrum.*

Proof. (i) $\Rightarrow$ (ii): Theorem 4.6.11 says that $0 \notin \sigma_p(\widetilde{B})$. So it suffices to show that $\sigma_p(\widetilde{B}) = \sigma_p(\widetilde{B}) + i\mathbb{R}$. Let $\eta \in \mathbb{R}$, $e_{i\eta}(r) := e^{i\eta r}$. Then $Q_\eta f := e_{i\eta} \cdot f$ defines an isomorphism Q_η on $\mathrm{BUC}(\mathbb{R}, X)$ which leaves $\mathrm{AP}(\mathbb{R}, X)$ invariant. Moreover, $Q_{-\eta} S(t) Q_\eta = e^{i\eta t} S(t)$. Define $\widetilde{Q}_\eta$ on $\mathrm{BUC}(\mathbb{R}, X)/\mathrm{AP}(\mathbb{R}, X)$ by $\widetilde{Q}_\eta(\pi(f)) := \pi(Q_\eta f)$. Then $\widetilde{Q}_{-\eta} \widetilde{S}(t) \widetilde{Q}_\eta = e^{i\eta t} \widetilde{S}(t)$. This implies the claim.

(ii) $\Rightarrow$ (i): We have to show that if $c_0 \subset X$ then there exists $f \in \mathrm{BUC}(\mathbb{R}, X)$ such that $f \notin \mathrm{AP}(\mathbb{R}, X)$ but $S(t)f - f \in \mathrm{AP}(\mathbb{R}, X)$ for all $t \geq 0$. Since the space c_0 is isomorphic to the space c, it suffices to consider $X = c$. Consider the function $f \in \mathrm{BUC}(\mathbb{R}, c)$ given by $f(t) := \left(e^{it/n}\right)_{n \in \mathbb{N}}$. Then $f \notin \mathrm{AP}(\mathbb{R}, c)$ by Example 4.6.5. Let $t \in \mathbb{R}$, $g := S(t)f - f$. Then $g(s) = \left((e^{it/n} - 1)e^{is/n}\right)_{n \in \mathbb{N}} =: (g_n(s))_{n \in \mathbb{N}}$. Let $h_n(s) := (g_1(s), \ldots, g_n(s), 0, 0, \ldots)$. Since $\lim_{n \to \infty}(e^{it/n} - 1) = 0$, we have $\lim_{n \to \infty} h_n = g$ in $\mathrm{BUC}(\mathbb{R}, c)$. But $h_n \in \mathrm{AP}(\mathbb{R}, c)$ for all $n \in \mathbb{N}$. Thus, $g \in \mathrm{AP}(\mathbb{R}, c)$. □

Proof of Theorem 4.6.3. Let $f \in \mathrm{BUC}(\mathbb{R}, X)$ such that $\mathrm{sp}_\mathrm{C}(f)$ is countable.

Case c): We assume that $c_0 \not\subset X$. Let $Y := \mathrm{BUC}(\mathbb{R}, X)/\mathrm{AP}(\mathbb{R}, X)$ and consider the shift group $\widetilde{S}$ with generator $\widetilde{B}$ on Y. Assume that $f \notin \mathrm{AP}(\mathbb{R}, X)$; i.e., $\widetilde{f} := \pi(f) \neq 0$. We consider the space $Y_{\widetilde{f}} := \overline{\mathrm{span}}\{\widetilde{S}(t)\widetilde{f} : t \in \mathbb{R}\}$ and the induced operator $\widetilde{B}_{\widetilde{f}}$ (see the discussion before Lemma 4.6.7). Let $\eta \in \mathbb{R} \backslash \mathrm{sp}_\mathrm{C}(f)$. Then by Lemma 4.6.8, the function $\lambda \mapsto R(\lambda, B)f$ has a holomorphic extension $h : V \to \mathrm{BUC}(\mathbb{R}, X)$ where V is an open neighbourhood of $i\eta$. Then $\pi \circ h$ is a holomorphic extension of $\lambda \mapsto R(\lambda, \widetilde{B})\widetilde{f} = \pi(R(\lambda, B)f)$. It follows from Lemma 4.6.7 that $i\eta \notin \sigma(\widetilde{B}_{\widetilde{f}})$. We have proved that $\sigma(\widetilde{B}_{\widetilde{f}}) \subset \mathrm{sp}_\mathrm{C}(f)$. Thus, $\sigma(\widetilde{B}_{\widetilde{f}})$ is countable, closed and non-empty (by Corollary 4.4.12), so it contains an isolated point. It follows from Corollary 4.4.13 that $\widetilde{B}_{\widetilde{f}}$ has non-empty point spectrum. Hence, $\widetilde{B}$ also has non-empty point spectrum. This contradicts Corollary 4.6.12. The proof is finished in this case.

Case a): Now assume that f is totally ergodic. Consider the space $\mathcal{E} := \{g \in \mathrm{BUC}(\mathbb{R}, X) : g \text{ is totally ergodic}\}$ which is closed and invariant under the shift group. We denote the shift group on $\mathcal{E}$ also by S and its generator by B. Consider the quotient space $\widetilde{\mathcal{E}} := \mathcal{E}/\mathrm{AP}(\mathbb{R}, X)$, the quotient map $\pi : \mathcal{E} \to \widetilde{\mathcal{E}}$ and the induced group $\widetilde{S}$ on $\widetilde{\mathcal{E}}$ given by $\widetilde{S}(t)\pi(g) := \pi(S(t)g)$ with generator $\widetilde{B}$. It suffices to show that $\widetilde{B}$ has empty point spectrum. Then the proof given in Case a) carries over.

Assume that $g \in \mathcal{E}$, $\eta \in \mathbb{R}$ such that $\widetilde{S}(t)\pi(g) = e^{i\eta t}\pi(g)$ $(t \in \mathbb{R})$. Then $e^{-i\eta t}S(t)g - g \in \mathrm{AP}(\mathbb{R}, X)$ $(t \in \mathbb{R})$. It follows that

$$M_\eta g - g = \lim_{t \to \infty} \frac{1}{t} \int_0^t \left(e^{-i\eta s}S(s)g - g\right) ds \in \mathrm{AP}(\mathbb{R}, X).$$

But $M_\eta g = e_{i\eta} \otimes (M_\eta g)(0) \in \mathrm{AP}(\mathbb{R}, X)$. Thus $g \in \mathrm{AP}(\mathbb{R}, X)$.

Case b): Assume that f has relatively compact range. It follows from Case c) applied in the scalar case that f is weakly almost periodic. Hence, $f \in \mathrm{AP}(\mathbb{R}, X)$ by Proposition 4.5.12. □

Now Theorem 4.6.3 is completely proved in the case a) where f is assumed to be uniformly ergodic at each $\eta \in \mathrm{sp_C}(f)$. The proof in the other cases is complete admitting Kadets's theorem. Note however that for the case b), where f is assumed to have relatively compact range, we merely need Kadets's theorem in the scalar case. The scalar case is much easier to prove and will be particularly important for the general complex Tauberian theorem presented in Section 4.9. For this reason, we first give a direct proof of Kadets's theorem in the scalar case.

The following reformulation will be useful.

Lemma 4.6.13. *Let X be a Banach space. The following assertions are equivalent:*

(i) *If $f \in \mathrm{AP}(\mathbb{R}, X)$ such that $F(t) := \int_0^t f(s)\,ds$ $(t \in \mathbb{R})$ is bounded, then $F \in \mathrm{AP}(\mathbb{R}, X)$.*

(ii) *If $f \in \mathrm{BUC}(\mathbb{R}, X)$ such that $S(t)f - f \in \mathrm{AP}(\mathbb{R}, X)$ for all $t \in \mathbb{R}$, then $f \in \mathrm{AP}(\mathbb{R}, X)$.*

Proof. (ii) ⇒ (i): Let $f \in \mathrm{AP}(\mathbb{R}, X)$ such that $F(t) = \int_0^t f(s)\,ds$ is bounded. Then $F \in \mathrm{BUC}(\mathbb{R}, X)$ and $S(t)F - F = \int_0^t S(s)f\,ds \in \mathrm{AP}(\mathbb{R}, X)$ for all $t \in \mathbb{R}$. Hence, $F \in \mathrm{AP}(\mathbb{R}, X)$ by (ii).

(i) ⇒ (ii): Let $f \in \mathrm{BUC}(\mathbb{R}, X)$ such that $S(t)f - f \in \mathrm{AP}(\mathbb{R}, X)$ for all $t \in \mathbb{R}$. Let $\lambda > 0$ and $g = R(\lambda, B)f$. Then $S(t)g - g \in \mathrm{AP}(\mathbb{R}, X)$ and $g \in D(B)$. It follows that

$$g' = \lim_{t \downarrow 0} 1/t\,(S(t)g - g) \in \mathrm{AP}(\mathbb{R}, X).$$

Thus by (i), $g(t) - g(0) = \int_0^t g'(s)\,ds$ defines an almost periodic function. We have shown that $R(\lambda, B)f \in \mathrm{AP}(\mathbb{R}, X)$ for all $\lambda > 0$. Since $f = \lim_{\lambda\to\infty} \lambda R(\lambda, B)f$ in $\mathrm{BUC}(\mathbb{R}, X)$ (see Proposition 3.1.9 a)), it follows that $f \in \mathrm{AP}(\mathbb{R}, X)$. □

Proof of Theorem 4.6.11 in the scalar case. Let $f \in \mathrm{AP}(\mathbb{R})$ such that $F(t) := \int_0^t f(s)\,ds$ is bounded. We show that $F \in \mathrm{AP}(\mathbb{R})$. Then the result follows from Lemma 4.6.13. It follows easily from Definition 4.5.6 that we can assume that f is real-valued. It is clear that $F \in \mathrm{BUC}(\mathbb{R})$. Let $s_k \in \mathbb{R}$ $(k \in \mathbb{N})$. There exists a subsequence $t_m := s_{k_m}$ such that $g := \lim_{m\to\infty} S(t_m)f$ exists in $\mathrm{BUC}(\mathbb{R})$. We have to show that $S(t_m)F$ has a convergent subsequence. Then $F \in \mathrm{AP}(\mathbb{R})$ by Theorem 4.5.7.

a) Since $(S(t_m)F)(t) = \int_0^{t_m+t} f(s)\,ds = \int_0^t S(t_m)f(s)\,ds + F(t_m)$, and $(F(t_m))_{m\in\mathbb{N}}$ is bounded, it follows that $S(t_m)F$ has a limit point G for compact convergence (by which we mean the topology of uniform convergence on bounded intervals), where $G(t) = \int_0^t g(s)\,ds + d$ for some $d \in \mathbb{R}$. Clearly, $G \in \mathrm{BUC}(\mathbb{R})$ and $\sup G := \sup_{t\in\mathbb{R}} G(t) \le \sup F$, $\inf G \ge \inf F$. Since $S(-t_m)g \to f$ as $m \to \infty$, for the same reason $S(-t_m)G$ has a limit point H for compact convergence, H is differentiable, $H' = f$, $\sup H \leq \sup G$, $\inf H \geq \inf G$. Hence, $H(t) = F(t) + c$ for some $c \in \mathbb{R}$ and $c + \sup F \leq \sup G \leq \sup F$ and $\inf F + c \geq \inf G \geq \inf F$. Hence, $c = 0$. Thus, $\sup G = \sup F$ and $\inf G = \inf F$. This determines d uniquely. So the limit point G is unique. Hence, $S(t_m)F$ converges to $G \in \mathrm{BUC}(\mathbb{R})$ uniformly on bounded intervals.

b) Now it remains to show that $S(t_m)F$ converges uniformly to G. If not, passing to a subsequence we can assume that there exist $\varepsilon > 0$ and $s_m \in \mathbb{R}$ such that $|F(t_m+s_m) - G(s_m)| \geq \varepsilon > 0$ $(m \in \mathbb{N})$. Taking subsequences, we can assume that $h := \lim_{m\to\infty} S(t_m+s_m)f$ and $h_1 := \lim_{m\to\infty} S(s_m)g = h_1$ exist in $\mathrm{BUC}(\mathbb{R})$. Then $h = h_1$, since

$$\begin{aligned} |h(t) - h_1(t)| &\leq |h(t) - f(t+t_m+s_m)| + |f(t+t_m+s_m) - g(s_m+t)| \\ &\quad + |g(s_m+t) - h_1(t)| \\ &\to 0 \qquad \text{as } m \to \infty. \end{aligned}$$

Let $H(t) := \int_0^t h(s)\,ds$. By a), there exist constants c_1, c_2 such that

$$\lim_{m\to\infty} \left(S(t_m+s_m)F\right)(t) = H(t) + c_1 \text{ and } \lim_{m\to\infty} \left(S(s_m)G\right)(t) = H(t) + c_2$$

uniformly on bounded intervals, and $\inf F = \inf H + c_1$, and $\inf G = \inf H + c_2$. Since $\inf F = \inf G$, it follows that $c_1 = c_2$. Thus $\varepsilon \leq \lim_{m\to\infty} |F(t_m+s_m) - G(s_m)| = |c_1 - c_2| = 0$, a contradiction. □

Now we have proved Kadets's theorem in the scalar case, and hence also the proof of Theorem 4.6.3 in the case b), and in particular of Loomis's theorem, is complete.

Before proving Kadets's theorem in full generality, we consider another special case which can be deduced from the scalar case by a short but clever argument.

Recall that a Banach space X is called *uniformly convex* if for all $\varepsilon > 0$ there exists $\delta > 0$ such that for $\|x\| \leq 1, \|y\| \leq 1$,

$$\|x - y\| \geq \varepsilon \Rightarrow \left\| \frac{x+y}{2} \right\| \leq 1 - \delta. \tag{4.31}$$

Every uniformly convex space is reflexive, and L^p-spaces are uniformly convex for $1 < p < \infty$.

Proof of Theorem 4.6.11 *for uniformly convex spaces.*
Let $f \in \mathrm{AP}(\mathbb{R}, X)$ such that $F(t) = \int_0^t f(s)\,ds$ is bounded. We claim that F is almost periodic. We can assume that

$$\|F\|_\infty := \sup_{t\in\mathbb{R}} \|F(t)\| = 1.$$

Since by the scalar case, F is weakly almost periodic, in view of Proposition 4.5.12 it suffices to show that F has precompact range.

Assume that this is false. Then there exist $\varepsilon > 0$, $t_n \in \mathbb{R}$ such that

$$\|F(t_n) - F(t_m)\| \geq \varepsilon \quad (n \neq m).$$

Choose $\delta > 0$ according to (4.31). Let $x^* \in X^*$ such that $\|x^*\| = 1$ and $\|x^* \circ F\|_\infty > 1 - \frac{\delta}{4}$. Replacing $(t_n)_{n\in\mathbb{N}}$ by a subsequence if necessary, we can assume that

$$\begin{aligned} S(t_n)f &\to \widetilde{f} \text{ in } \mathrm{BUC}(\mathbb{R}, X) \\ S(t_n)(x^* \circ F) &\to g \text{ in } \mathrm{BUC}(\mathbb{R}). \end{aligned}$$

Then $\|g\|_\infty \geq 1 - \frac{\delta}{4}$, since $S(-t_n)g \to x^* \circ F$ in $\mathrm{BUC}(\mathbb{R})$. Let $t \in \mathbb{R}$ such that $|g(t)| \geq 1 - \frac{\delta}{2}$. Then

$$\begin{aligned} &\limsup_{\substack{n,m\to\infty \\ n\neq m}} \|F(t_n + t) - F(t_m + t)\| \\ &= \limsup_{\substack{n,m\to\infty \\ n\neq m}} \left\| F(t_n) - F(t_m) + \int_0^t (f(t_n + s) - f(t_m + s))\, ds \right\| \\ &= \limsup_{\substack{n,m\to\infty \\ n\neq m}} \|F(t_n) - F(t_m)\| \geq \varepsilon. \end{aligned}$$

Hence by (4.31),

$$\begin{aligned} 1 - \frac{\delta}{2} \leq |g(t)| &= \lim_{\substack{n,m\to\infty \\ n\neq m}} \frac{|\langle F(t + t_n) + F(t + t_m), x^* \rangle|}{2} \\ &\leq \limsup_{\substack{n,m\to\infty \\ n\neq m}} \frac{\|F(t + t_n) + F(t + t_m)\|}{2} \\ &\leq 1 - \delta. \end{aligned}$$

This is a contradiction. □

For the proof of Kadets's theorem in the general case we use the following characterization which is proved in Appendix D.

Theorem 4.6.14. *Let X be a Banach space. The following are equivalent:*

(i) $c_0 \subset X$.

(ii) *there is a divergent series $\sum_{n=1}^\infty x_n$ in X which is* unconditionally bounded*; i.e., there exists $M \geq 0$ such that*

$$\left\| \sum_{j=1}^m x_{n_j} \right\| \leq M$$

whenever $n_j \in \mathbb{N}$ $(j = 1, 2, \ldots, m)$ such that $n_1 < n_2 < \ldots < n_m$.

Proof of Kadets's Theorem 4.6.11.
Assume that there exists $f \in \mathrm{AP}(\mathbb{R}, X)$ such that $F(t) = \int_0^t f(s)\,ds$ is bounded, but $F \notin \mathrm{AP}(\mathbb{R}, X)$. We show that this implies that $c_0 \subset X$. In view of Lemma 4.6.13, this proves Kadets's theorem.

By the scalar case already proved, F is weakly almost periodic.

a) We show that there exists $\varepsilon > 0$ such that the set

$$U(\varepsilon) := \{t \in \mathbb{R} : \|F(t)\| < \varepsilon\}$$

is not relatively dense. Assume that, on the contrary, $U(\varepsilon)$ is relatively dense for all $\varepsilon > 0$.

Let $\varepsilon > 0$. We show that $\mathbb{R} = \bigcup_{j=0}^n (U(\varepsilon) + s_j)$ for suitable $s_0, \ldots, s_n \in \mathbb{R}$. In fact, there exists $l > 0$ such that $U(\varepsilon/2) \cap [a - l, a] \neq \emptyset$ for all $a \in \mathbb{R}$. Since F is Lipschitz continuous, there exists $\delta > 0$ such that $\|F(z)\| \le \varepsilon/2$ implies $\|F(r)\| \le \varepsilon$ for all $r \in [z, z + \delta]$. Let $s_0 := 0, s_1 := \delta, s_2 := 2\delta, \ldots, s_n := n\delta$, where $n\delta > l$. Let $t \in \mathbb{R}$. Take $s \in [t - l, t] \cap U(\varepsilon/2)$. Then

$$\begin{aligned} [s, s+\delta] &\subset U(\varepsilon) = U(\varepsilon) + s_0, \\ [s+\delta, s+2\delta] = [s, s+\delta] + \delta &\subset U(\varepsilon) + s_1, \\ &\vdots \\ [s + (n-1)\delta, s + n\delta] &\subset U(\varepsilon) + s_n. \end{aligned}$$

Thus, $t \in [s, s + l] \subset \bigcup_{j=0}^n (U(\varepsilon) + s_j)$.

Since $f \in \mathrm{AP}(\mathbb{R}, X)$, the function $S(t)F - F = \int_0^t S(r)f\,dr \in \mathrm{AP}(\mathbb{R}, X)$ for all $t \in \mathbb{R}$. In particular, $S(s_j)F - F$ has precompact range K_j. Now let $t \in \mathbb{R}$. Then $t = s + s_j$ for some $s \in U(\varepsilon)$, $j \in \{0, 1, 2, \ldots, n\}$. Thus, $F(t) = F(s + s_j) = (S(s_j)F - F)(s) + F(s) \in K_j + B(0, \varepsilon)$. Thus, the range of F is contained in $\bigcup_{j=0}^n K_j + B(0, \varepsilon)$. Since $\varepsilon > 0$ is arbitrary, this implies that F has precompact range. This is impossible by Proposition 4.5.12.

b) For $\gamma \in \mathbb{R}, \delta > 0$, let

$$V_\gamma(\delta) := \big\{t \in \mathbb{R} : \|F(t+\gamma) - F(t) - F(\gamma)\| < \delta\big\}.$$

We show that $\bigcap_{j=1}^n V_{\gamma_j}(\delta)$ is relatively dense in $\mathbb{R}$ for all $\delta > 0, \gamma_1, \gamma_2, \ldots, \gamma_n \in \mathbb{R}$. In fact, let $\gamma := \max_{j=1,\ldots,n} |\gamma_j|$. Then

$$\begin{aligned} \|F(t+\gamma_j) - F(t) - F(\gamma_j)\| &= \left\| \int_0^{\gamma_j} (f(t+s) - f(s))\,ds \right\| \\ &\le \gamma . \sup_{0 \le s \le \gamma} \|f(t+s) - f(s)\|. \end{aligned}$$

Thus,

$$\bigcap_{j=1}^n V_{\gamma_j}(\delta) \supset \left\{ t \in \mathbb{R} : \|f(t+s) - f(s)\| \le \frac{\delta}{\gamma} \text{ for all } s \in [0, \gamma] \right\}.$$

The last set is relatively dense since $f \in \mathrm{AP}(\mathbb{R}, X)$.

c) By a) and b), there exists $\varepsilon > 0$ such that the following holds:

$$\text{If } \delta > 0, \gamma_1, \gamma_2, \ldots, \gamma_r \in \mathbb{R}, \text{ then } U(\varepsilon) \not\supset \cap_{j=1}^r V_{\gamma_j}(\delta). \tag{4.32}$$

We construct $t_n \in \mathbb{R}$ such that $\|F(t_n)\| \geq \varepsilon$ $(n \in \mathbb{N})$ and

$$\|F(t_n + t_{j_1} + \ldots + t_{j_r}) - F(t_n) - F(t_{j_1} + \ldots + t_{j_r})\| \leq 2^{-n} \tag{4.33}$$

whenever $j_1, j_2, \ldots, j_r \in \mathbb{N}$, $1 \leq j_1 < j_2 < \ldots < j_r \leq n-1$. Choose $t_1 \in \mathbb{R} \setminus U(\varepsilon)$. Assume that $t_1, \ldots, t_{n-1}$ are constructed. By (4.32), there exists $t_n \in \mathbb{R} \setminus U(\varepsilon)$ such that $t_n \in V_\gamma(2^{-n})$ whenever $\gamma = t_{j_1} + \ldots + t_{j_r}$ with $1 \leq j_1 < j_2 < \ldots < j_r \leq n-1$. This means that (4.33) holds.

d) Let $x_n := F(t_n)$. Then the series $\sum_{n=1}^\infty x_n$ diverges. We show that

$$\left\| \sum_{k=1}^m x_{i_k} \right\| \leq \|F\|_\infty + 1 \tag{4.34}$$

whenever $i_1 < i_2 < \ldots < i_m$. In fact,

$$\begin{aligned} \left\| F\left(\sum_{k=1}^m t_{i_k}\right) - \sum_{k=1}^m F(t_{i_k}) \right\| &\leq \left\| F\left(\sum_{k=1}^m t_{i_k}\right) - F(t_{i_m}) - F\left(\sum_{k=1}^{m-1} t_{i_k}\right) \right\| \\ &\quad + \left\| F\left(\sum_{k=1}^{m-1} t_{i_k}\right) - F(t_{i_{m-1}}) - F\left(\sum_{k=1}^{m-2} t_{i_k}\right) \right\| \\ &\quad + \ldots + \left\| F(t_{i_2} + t_{i_1}) - F(t_{i_2}) - F(t_{i_1}) \right\| \\ &\leq 2^{-i_m} + \ldots + 2^{-i_2} \leq 1. \end{aligned}$$

Thus,

$$\left\| \sum_{k=1}^m x_{i_k} \right\| = \left\| \sum_{k=1}^m F(t_{i_k}) \right\| \leq 1 + \|F\|_\infty.$$

Now it follows from Theorem 4.6.14 that $c_0 \subset X$. □

4.7 Asymptotically Almost Periodic Functions

In this section we study bounded uniformly continuous functions on the half-line. The main result is a Tauberian theorem (Theorem 4.7.7) which says that, under additional assumptions, such a function with countable spectrum is asymptotically almost periodic. Here we again use a quotient method similar to the preceding section. A similar result, in the case when the spectrum has at most one point, was given in Theorem 4.4.8 with a proof by elementary contour integrals.

Let X be a Banach space and denote by $\mathrm{BUC}(\mathbb{R}_+, X)$ the space of all bounded uniformly continuous functions defined on $\mathbb{R}_+$ with values in X. It is a Banach space for the uniform norm

$$\| f \|_\infty = \sup_{t \geq 0} \| f(t) \| \quad (f \in \mathrm{BUC}(\mathbb{R}_+, X)).$$

By $C_0(\mathbb{R}_+, X)$ we denote the closed subspace consisting of all $f \in \mathrm{BUC}(\mathbb{R}_+, X)$ such that $\lim_{t\to\infty} \| f(t) \| = 0$.

For $x \in X$, $\eta \in \mathbb{R}$, we now consider the function $e_{i\eta} \otimes x : t \mapsto e^{i\eta t}x$ on $\mathbb{R}_+$ and denote by

$$\mathrm{AP}(\mathbb{R}_+, X) := \overline{\mathrm{span}}\{e_{i\eta} \otimes x : \eta \in \mathbb{R}, x \in X\}$$

the space of all *almost periodic functions on the half-line* (the closure being taken in $\mathrm{BUC}(\mathbb{R}_+, X)$).

Proposition 4.7.1. *Every* $f \in \mathrm{AP}(\mathbb{R}_+, X)$ *has a unique extension* $\widetilde{f} \in \mathrm{AP}(\mathbb{R}, X)$ *and* $\| f \|_\infty = \| \widetilde{f} \|_\infty$. *Moreover,*

$$\| f \|_\infty = \sup_{t \geq \tau} \| f(t) \| \quad \textit{for all}\, \tau \geq 0. \tag{4.35}$$

Proof. Let $f \in \mathrm{AP}(\mathbb{R}_+, X)$. There exist trigonometric polynomials $f_n \in \mathrm{span}\{e_{i\eta} \otimes x : \eta \in \mathbb{R}, x \in X\}$ such that $f_n \to f$ in $\mathrm{BUC}(\mathbb{R}_+, X)$ as $n \to \infty$. It follows from (4.24) that $(f_n)_{n\in\mathbb{N}}$ is a Cauchy sequence in $\mathrm{BUC}(\mathbb{R}, X)$. Let $\widetilde{f}$ be the limit of $(f_n)_{n\in\mathbb{N}}$ in $\mathrm{BUC}(\mathbb{R}, X)$. Then $\widetilde{f} \in \mathrm{AP}(\mathbb{R}, X)$ and (4.35) follows from (4.24). □

It follows from (4.35) that

$$\mathrm{AP}(\mathbb{R}_+, X) \cap C_0(\mathbb{R}_+, X) = \{0\}. \tag{4.36}$$

By

$$\mathrm{AAP}(\mathbb{R}_+, X) := C_0(\mathbb{R}_+, X) \oplus \mathrm{AP}(\mathbb{R}_+, X),$$

we denote the space of all *asymptotically almost periodic functions* on the half-line. It follows from (4.35) that for $f = f_0 + f_1$ with $f_0 \in C_0(\mathbb{R}_+, X)$, $f_1 \in \mathrm{AP}(\mathbb{R}_+, X)$, one has $\|f_1\|_\infty \leq \|f\|_\infty$. Thus, $\mathrm{AAP}(\mathbb{R}_+, X)$ is a closed subspace of $\mathrm{BUC}(\mathbb{R}_+, X)$.

The following observation is useful for later purposes. By (4.23), there exist $t_n \in \mathbb{R}_+$ such that $\lim_{n\to\infty} t_n = \infty$ and

$$\|f(t_n + s) - f_1(s)\| \leq \frac{1}{n} \quad \text{for all } s \in \mathbb{R}_+. \tag{4.37}$$

In the following, we consider the shift semigroup S on $\mathrm{BUC}(\mathbb{R}_+, X)$ given by

$$(S(t)f)(s) = f(t + s) \quad (s, t \geq 0,\ f \in \mathrm{BUC}(\mathbb{R}_+, X)).$$

It is a C_0-semigroup of contractions, and its generator will be denoted by B.

Similarly to Section 4.6, for $f \in \mathrm{BUC}(\mathbb{R}_+, X)$, $\operatorname{Re}\lambda > 0$ and $s \geq 0$,

$$(R(\lambda, B)f)(s) = \widehat{f_s}(\lambda) = e^{\lambda s}\left(\hat{f}(\lambda) - \int_0^s e^{-\lambda t} f(t)\, dt\right),$$

where $f_s := S(s)f$. We shall see in Lemma 4.7.9 that the singularities of $R(\cdot, B)f$ and $\hat{f}(\cdot)$ in $i\mathbb{R}$ coincide, and we shall exploit this to prove an analogue of part of Theorem 4.6.3 (Theorem 4.7.7).

Let $f \in \mathrm{BUC}(\mathbb{R}_+, X)$. Let $\eta \in \mathbb{R}$. We say that f is *uniformly ergodic at* η if f is ergodic at η with respect to S; i.e., if

$$M_\eta f := \lim_{t\to\infty} \frac{1}{t} \int_0^t e^{-i\eta s} S(s) f\, ds$$

converges in $\mathrm{BUC}(\mathbb{R}_+, X)$. This is equivalent to the convergence of

$$M_\eta f = \lim_{\alpha\downarrow 0} \alpha R(\alpha + i\eta, B) f.$$

In that case,

$$M_\eta f = e_{i\eta} \otimes (M_\eta f)(0). \tag{4.38}$$

A function $f \in \mathrm{BUC}(\mathbb{R}_+, X)$ is called *totally ergodic* if f is ergodic at all $\eta \in \mathbb{R}$. From now on, we denote by $\mathcal{E}(\mathbb{R}_+, X)$ the set of all totally ergodic functions in $\mathrm{BUC}(\mathbb{R}_+, X)$. This is a closed subspace of $\mathrm{BUC}(\mathbb{R}_+, X)$ containing $\mathrm{AAP}(\mathbb{R}_+, X)$.

In the following we will consider the quotient space

$$\widetilde{\mathcal{E}} := \mathcal{E}(\mathbb{R}_+, X) / \mathrm{AAP}(\mathbb{R}_+, X)$$

with quotient map $\pi : \mathcal{E}(\mathbb{R}_+, X) \to \widetilde{\mathcal{E}}$. Then $\widetilde{\mathcal{E}}$ is a Banach space for the norm

$$\|\, \pi(f) \,\| := \inf\{\|f - g\|_\infty : g \in \mathrm{AAP}(\mathbb{R}_+, X)\} \quad (f \in \mathcal{E}(\mathbb{R}_+, X)).$$

Since $\mathcal{E}(\mathbb{R}_+, X)$ and $\mathrm{AAP}(\mathbb{R}_+, X)$ are invariant under the shift semigroup, we can define a C_0-semigroup $\widetilde{S}$ on $\widetilde{\mathcal{E}}$ by

$$\widetilde{S}(t)\pi(f) := \pi\left(S(t)f\right) \quad (t \geq 0,\ f \in \mathcal{E}(\mathbb{R}_+, X)).$$

We denote by $\widetilde{B}$ the generator of $\widetilde{S}$.

The interesting fact about this construction is the following.

Proposition 4.7.2. *Each $\widetilde{S}(t)$ is isometric and surjective. Thus, $\widetilde{S}$ extends to an isometric C_0-group on $\widetilde{\mathcal{E}}$. Moreover, $\widetilde{B}$ has empty point spectrum.*

Proof. It is immediate that $\|\widetilde{S}(t)\| \leq \|S(t)\| = 1$ for all $t \geq 0$. We show that $\widetilde{S}(t)$ is isometric. Let $f \in \mathcal{E}(\mathbb{R}_+, X)$, $t \geq 0$ and $g \in \mathrm{AAP}(\mathbb{R}_+, X)$. Define $h : \mathbb{R}_+ \to X$ by

$$h(s) := \begin{cases} g(s-t) & (s \geq t) \\ g(0) + f(s) - f(t) & (s < t). \end{cases}$$

Then $h \in \mathrm{AAP}(\mathbb{R}_+, X)$ and

$$\begin{aligned} \|f - h\|_\infty &= \sup_{s \geq t} \|f(s) - g(s-t)\| = \sup_{s \geq 0} \|f(s+t) - g(s)\| \\ &= \|S(t)f - g\|_\infty. \end{aligned}$$

Hence, $\|\pi(f)\| \leq \|\pi(S(t)f)\| = \|\widetilde{S}(t)\pi(f)\|$.

Since $S(t)$ maps $\mathcal{E}(\mathbb{R}_+, X)$ onto $\mathcal{E}(\mathbb{R}_+, X)$, it follows that $\widetilde{S}(t)$ is surjective. By Proposition 3.1.23, $\widetilde{S}$ extends to a C_0-group on $\widetilde{\mathcal{E}}$.

It remains to show that the point spectrum of $\widetilde{B}$ is empty. Let $f \in \mathcal{E}(\mathbb{R}_+, X)$ and $\eta \in \mathbb{R}$ such that $\widetilde{S}(t)\pi(f) = e^{i\eta t}\pi(f)$ $(t \geq 0)$. Then $\frac{1}{t}\int_0^t e^{-i\eta s}\widetilde{S}(s)\pi(f)\,ds = \pi(f)$ $(t \geq 0)$. On the other hand, $\lim_{t\to\infty} \frac{1}{t}\int_0^t e^{-i\eta s}S(s)f\,ds = M_\eta f \in \mathrm{AP}(\mathbb{R}_+, X)$. Applying π on both sides, we conclude that $\lim_{t\to\infty}\frac{1}{t}\int_0^t e^{-i\eta s}\widetilde{S}(s)\pi(f)\,ds = \pi(M_\eta f) = 0$. Hence, $\pi(f) = 0$. □

The following simple result is essentially a reformulation of part of Proposition 4.7.2 (see Lemma 4.6.13).

Proposition 4.7.3. *Let $f \in \mathrm{AAP}(\mathbb{R}_+, X)$ and $F(t) := \int_0^t f(s)\,ds$. Suppose that F is bounded and uniformly ergodic at* 0. *Then $F \in \mathrm{AAP}(\mathbb{R}_+, X)$.*

Proof. Note first that

$$S(s)F - F = \int_0^s S(r)f\,dr \ \in \mathrm{AAP}(\mathbb{R}_+, X).$$

Hence,

$$M_0F - F = \lim_{t\to\infty}\frac{1}{t}\int_0^t (S(s)F - F)\,ds \ \in \mathrm{AAP}(\mathbb{R}_+, X).$$

Since M_0F is a constant function, it follows that $F \in \mathrm{AAP}(\mathbb{R}_+, X)$. □

Next, we characterize asymptotically almost periodic functions by relative compactness of the orbits under S. This is analogous to the characterization of almost periodic functions on the line proved in Section 4.5 (see Theorem 4.5.7).

Theorem 4.7.4. *Let $f \in \mathrm{BUC}(\mathbb{R}_+, X)$. The following are equivalent:*

(i) $f \in \mathrm{AAP}(\mathbb{R}_+, X)$.

(ii) *The orbit $\{S(t)f : t \geq 0\}$ is relatively compact in* $\mathrm{BUC}(\mathbb{R}_+, X)$.

Proof. (i) ⇒ (ii): If $h \in C_0(\mathbb{R}_+, X)$, then $\lim_{t\to\infty} S(t)h = 0$. Thus, $\{S(t)h : t \geq 0\}$ is precompact. If $f = e_{i\eta}\otimes x$, then $S(t)f = e^{i\eta t}f$, so f has precompact orbit. It follows that every $f \in \overline{\operatorname{span}}\,(\{e_{i\eta}\otimes x : \eta \in \mathbb{R}, x \in X\} \cup C_0(\mathbb{R}_+, X)) = \mathrm{AAP}(\mathbb{R}_+, X)$ has precompact orbit.

(ii) ⇒ (i): Assume that $O_f := \{S(t)f : t \geq 0\}$ is relatively compact in $\mathrm{BUC}(\mathbb{R}_+, X)$. Then $f \in \mathcal{E}(\mathbb{R}_+, X)$ by Proposition 4.3.12, and $O_{\pi(f)} := \pi(O_f) =$

$\{\widetilde{S}(t)\pi(f) : t \geq 0\}$ is relatively compact in $\widetilde{\mathcal{E}}$. It follows from Theorem 4.5.1 that $\pi(f) \in \overline{\operatorname{span}}\{\widetilde{g} \in \widetilde{\mathcal{E}} : \widetilde{S}(t)\widetilde{g} = e^{i\eta t}\widetilde{g}\ (t \in \mathbb{R}) \text{ for some } \eta \in \mathbb{R}\}$. By Proposition 4.7.2, this set is reduced to $\{0\}$. Thus, $\pi(f) = 0$; i.e., $f \in \mathrm{AAP}(\mathbb{R}_+, X)$. □

Similarly to the results on the line in Section 4.5, one can also characterize asymptotically almost periodic functions by "the relative density of asymptotic ε-periods", but the definition on the half-line is slightly different.

Let $f \in \mathrm{BUC}(\mathbb{R}_+, X)$. For $\varepsilon > 0, \lambda > 0$, we let

$$Q_{\varepsilon,\lambda}(f) := \{\tau \in \mathbb{R}_+ : \ \|f(t+\tau) - f(t)\| \leq \varepsilon \text{ whenever } t \geq \lambda\}.$$

We say that a subset Q of $\mathbb{R}_+$ is *relatively dense in* $\mathbb{R}_+$ if there exists a length $l > 0$ such that

$$[a, a+l] \cap Q \neq \emptyset \quad \text{ for all } a \in \mathbb{R}_+.$$

Theorem 4.7.5. *Let $f \in \mathrm{BUC}(\mathbb{R}_+, X)$. The following assertions are equivalent:*

(i) $f \in \mathrm{AAP}(\mathbb{R}_+, X)$.

(ii) *For all $\varepsilon > 0$ there exists a $\lambda > 0$ such that $Q_{\varepsilon,\lambda}(f)$ is relatively dense in $\mathbb{R}_+$.*

Proof. (ii) ⇒ (i): By Theorem 4.7.4, we have to show that the set $\{S(t)f : t \geq 0\}$ is precompact in $\mathrm{BUC}(\mathbb{R}_+, X)$. Let $\varepsilon > 0$. By assumption, there exists $\lambda > 0$ such that $Q_{\varepsilon,\lambda}(f)$ is relatively dense in $\mathbb{R}_+$. Choose a length l such that $[a, a+l] \cap Q_{\varepsilon,\lambda}(f) \neq \emptyset$ for all $a \geq 0$. Since $\{S(t)f : 0 \leq t \leq 2\lambda + l\}$ is precompact, it suffices to cover $\{S(t)f : \ t \geq 2\lambda + l\}$ by finitely many balls of radius 2ε. There exist $t_1, \ldots, t_m \in [\lambda, \lambda + l]$ such that for all $t \in [\lambda, \lambda + l]$ there exists $j \in \{1, \ldots, m\}$ such that

$$\|S(t)f - S(t_j)f\|_\infty \leq \varepsilon.$$

Now let $t \geq 2\lambda + l$. There exists $\tau \in Q_{\varepsilon,\lambda}(f) \cap [t - \lambda - l, t - \lambda]$. Then $t - \tau \in [\lambda, \lambda + l]$. There exists $j \in \{1, \ldots, m\}$ such that

$$\|S(t-\tau)f - S(t_j)f\|_\infty \leq \varepsilon.$$

Hence,

$$\|S(t)f - S(t_j)f\|_\infty \leq \|S(t)f - S(t-\tau)f\|_\infty + \|S(t-\tau)f - S(t_j)f\|_\infty \leq 2\varepsilon$$

since $\tau \in Q_{\varepsilon,\lambda}(f)$ and $t - \tau \geq \lambda$.

(i) ⇒ (ii): Let $f \in \mathrm{AAP}(\mathbb{R}_+, X)$. Let $\varepsilon > 0$. There exist $\lambda \geq 0$ and $h \in \mathrm{AP}(\mathbb{R}, X)$ such that $\|f(t) - h(t)\| \leq \varepsilon/3$ for all $t \geq \lambda$. Let $\tau > 0$ be an $\varepsilon/3$-period for h. Then $\tau \in Q_{\varepsilon,\lambda}(f)$. Since $\varepsilon/3$-periods for h are relatively dense in $\mathbb{R}$, $Q_{\varepsilon,\lambda}(f)$ is relatively dense in $\mathbb{R}_+$. □

Recall from Section 4.4 that the *half-line spectrum* $\mathrm{sp}(f)$ of a function $f \in \mathrm{BUC}(\mathbb{R}_+, X)$ is defined in the following way: call $i\eta \in i\mathbb{R}$ *regular* for $\hat{f}$ if $\hat{f}$ has a holomorphic extension to a neighbourhood of $i\eta$. Then we set

$$\mathrm{sp}(f) = \{\eta \in \mathbb{R} : \quad i\eta \text{ is not regular for } \hat{f}\} \tag{4.39}$$

The half-line spectrum should not be confused with the Carleman spectrum introduced in the preceding section. Indeed, if f is the restriction of a function $g \in \mathrm{BUC}(\mathbb{R}, X)$, then the half-line spectrum of f is much smaller than the Carleman spectrum of g in general. We make this more precise in the following remark.

Remark 4.7.6 (Comparison of half-line spectrum and Carleman spectrum).
a) Let $f \in \mathrm{BUC}(\mathbb{R}_+, X)$. Assume that there exists $\eta \in \mathbb{R} \setminus \mathrm{sp}(f)$. Then f has at most one extension $g \in \mathrm{BUC}(\mathbb{R}, X)$ such that $\eta \notin \mathrm{sp}_C(g)$.

In fact, assume that there are two extensions $g, h \in \mathrm{BUC}(\mathbb{R}, X)$ such that $\eta \notin \mathrm{sp}_C(g) \cup \mathrm{sp}_C(h)$. Consider $k := g - h$. Then $k|_{\mathbb{R}_+} = 0$. It follows that the Carleman transform $\hat{k}$ of k is 0 on the right half-plane. Since $\hat{k}$ is a holomorphic function on $\mathbb{C} \setminus i\mathbb{R}$ with a holomorphic extension to a neighbourhood of $i\eta$, it follows from the uniqueness theorem for holomorphic functions that $\hat{k}(\lambda) = 0$ also on the left half-plane. This implies that $g - h = 0$ by the uniqueness theorem for Laplace transforms.

b) Let $f \in \mathrm{BUC}(\mathbb{R}_+, X)$ be exponentially decreasing, but $f \neq 0$. Then $\mathrm{sp}_C(g) = \mathbb{R}$ for all extensions $g \in \mathrm{BUC}(\mathbb{R}, X)$ of f, but $\mathrm{sp}(f) = \emptyset$.

In fact, there exist $M \geq 0$, $\varepsilon > 0$ such that $\|f(t)\| \leq Me^{-\varepsilon t}$. Thus the Laplace transform $\hat{f}(\lambda)$ of f has a holomorphic extension to the region $\{\mathrm{Re}\,\lambda > -\varepsilon\}$ and $\hat{f}(\lambda)$ is bounded for $\mathrm{Re}\,\lambda \geq -\varepsilon/2$. Now assume that $g \in \mathrm{BUC}(\mathbb{R}, X)$ is an extension of f such that $\mathrm{sp}_C(g) \neq \mathbb{R}$. Then the Carleman transform $\hat{g}(\lambda)$ of g agrees with $\hat{f}(\lambda)$ for $\mathrm{Re}\,\lambda > 0$, and hence for $\mathrm{Re}\,\lambda > -\varepsilon$ by the uniqueness theorem for holomorphic functions. Since $\hat{g}(\lambda)$ is bounded for $\mathrm{Re}\,\lambda \leq -\varepsilon/2$, it follows that $\hat{g}$ extends to a bounded entire function. So $\hat{g}$ is constant by Liouville's theorem. Since $\lim_{\lambda \to \infty} \hat{f}(\lambda) = 0$, it follows that $\hat{f} = 0$. Hence, $f = 0$. □

It is clear from the preceding remark that the assumption that the half-line spectrum be countable is much less restrictive than countability of the Carleman spectrum. In addition, we have the following observation about countability of a part of the spectrum.

Let $f \in \mathcal{E}(\mathbb{R}_+, X)$. As in Section 4.3, we define the set of all *frequencies* of f by

$$\mathrm{Freq}(f) := \{\eta \in \mathbb{R} : M_\eta f \neq 0\}.$$

Since $(M_\eta f)(0) = \lim_{\alpha \downarrow 0} \alpha \hat{f}(\alpha + i\eta)$, it follows from (4.38) that $\mathrm{Freq}(f) \subset \mathrm{sp}(f)$. It follows from Proposition 4.3.11 that $\mathrm{Freq}(f)$ is countable.

Now we can formulate the main result of this section which is a complex Tauberian theorem. With the help of Proposition 4.7.2, we are able to reduce the proof to an application of Gelfand's theorem (Corollary 4.4.12).

Theorem 4.7.7. *Let f be a totally ergodic function in $\mathrm{BUC}(\mathbb{R}_+, X)$ with countable half-line spectrum $\mathrm{sp}(f)$. Then f is asymptotically almost periodic.*

The following corollary illustrates particularly well the Tauberian character of this theorem.

Corollary 4.7.8. *Let $f \in \mathcal{E}(\mathbb{R}_+, X)$ with countable half-line spectrum.*

a) *If* $\mathrm{Freq}(f) = \emptyset$*, then* $f \in C_0(\mathbb{R}_+, X)$.

b) *If* $\mathrm{Freq}(f) = \{0\}$*, then* $\lim_{t\to\infty} f(t)$ *exists.*

c) *If* $\mathrm{Freq}(f) \subset \frac{2\pi}{\tau}\mathbb{Z}$*, where* $\tau > 0$*, then there exists a* τ*-periodic, continuous function* $g : \mathbb{R}_+ \to X$ *such that* $\lim_{t\to\infty} \| f(t) - g(t) \| = 0$.

Proof. By Theorem 4.7.7, one has $f = g + h \in \mathrm{AP}(\mathbb{R}_+, X) + C_0(\mathbb{R}_+, X)$. Since $\mathrm{Freq}(g) = \mathrm{Freq}(f)$, the claim follows from Corollary 4.5.9. □

For the proof of Theorem 4.7.7 we need the following lemma which is analogous to Lemma 4.6.8.

Lemma 4.7.9. *Let $f \in \mathrm{BUC}(\mathbb{R}_+, X)$. Assume that the Laplace transform $\hat{f}$ of f has a holomorphic extension to a neighbourhood of $i\eta$, where $\eta \in \mathbb{R}$. Then the function $R(\cdot, B)f : \mathbb{C}_+ \to \mathrm{BUC}(\mathbb{R}_+, X)$ has a holomorphic extension to a neighbourhood of $i\eta$.*

Proof. We can assume that $\eta = 0$. Let V be a connected neighbourhood of 0 and $g : V \to X$ be holomorphic such that $g(\lambda) = \hat{f}(\lambda)$ for $\lambda \in V \cap \mathbb{C}_+$. For $\lambda \in V$ let

$$G(\lambda, s) := e^{\lambda s} \left(g(\lambda) - \int_0^s e^{-\lambda t} f(t)\, dt \right) \quad (s \in \mathbb{R}_+).$$

Then for $\mathrm{Re}\,\lambda > 0$,

$$G(\lambda, s) = \int_0^\infty e^{-\lambda t} f(t+s)\, dt = (R(\lambda, B)f)(s).$$

It is clear that $G(\cdot, s) : V \to X$ is holomorphic for all $s \in \mathbb{R}_+$. Choose $r > 0$ such that $\overline{B}(0, 2r) \subset V$. We show that

$$\sup_{s\in\mathbb{R}_+,\, |\lambda|\le r} \| G(\lambda, s) \| < \infty.$$

In fact, let $M := \sup_{|\lambda|\le 2r} \| g(\lambda) \|$. If $\mathrm{Re}\,\lambda > 0$, then

$$\| G(\lambda, s) \| = \| (R(\lambda, B)f)(s) \| \le \|R(\lambda, B)f\|_\infty \le \frac{\|f\|_\infty}{\mathrm{Re}\,\lambda}.$$

If $\mathrm{Re}\,\lambda < 0$, $|\lambda| = 2r$, then $|e^{\lambda s}| \le 1$ and so

$$\begin{aligned} \| G(\lambda, s) \| &\le M + \int_0^s e^{-\mathrm{Re}\,\lambda\,(t-s)}\, dt\, \|f\|_\infty \le M + \frac{\|f\|_\infty}{|\mathrm{Re}\,\lambda|} \\ &\le \frac{2rM + \| f \|_\infty}{|\mathrm{Re}\,\lambda|}. \end{aligned}$$

It follows from Lemma 4.6.6 that $\| G(\lambda, s) \| \leq \frac{4}{3}\left(\frac{2rM+\| f \|_\infty}{r}\right) =: c$ for all $\lambda \in \overline{B}(0,r)$, $s \in \mathbb{R}_+$.

Now the claim follows from Corollary A.4 if we choose linear functionals ψ on $\mathrm{BUC}(\mathbb{R}_+, X)$ which are of the form $\langle f, \psi\rangle := \langle f(s), x^*\rangle$ where $s \in \mathbb{R}_+$, $x^* \in X^*$, $\| x^* \| \leq 1$. □

Proof of Theorem 4.7.7. We keep the notation introduced before Proposition 4.7.2. Let $f \in \mathcal{E}(\mathbb{R}_+, X)$ have countable spectrum. Assume that $\widetilde{f} := \pi(f) \neq 0$. Then

$$\widetilde{\mathcal{E}}_{\widetilde{f}} := \overline{\operatorname{span}}\{\widetilde{S}(t)\widetilde{f} : t \in \mathbb{R}\} \neq \{0\}.$$

Denote by $\widetilde{B}_{\widetilde{f}}$ the generator of the group $\widetilde{S}$ restricted to $\widetilde{\mathcal{E}}_{\widetilde{f}}$. Let $\eta \in \mathbb{R}\setminus \mathrm{sp}(f)$. By Lemma 4.7.9, there exists an open neighbourhood V of $i\eta$ and a holomorphic function $H : V \to \mathrm{BUC}(\mathbb{R}_+, X)$ such that $H(\lambda) = R(\lambda, B)f$ for $\lambda \in V$, $\mathrm{Re}\,\lambda > 0$. Since $H(\lambda) \in \mathcal{E}(\mathbb{R}_+, X)$ for $\mathrm{Re}\,\lambda > 0$, it follows from the identity theorem (Proposition A.2) that $H(\lambda) \in \mathcal{E}(\mathbb{R}_+, X)$ for all $\lambda \in V$. The function $\pi \circ H : V \to \widetilde{\mathcal{E}}$ is holomorphic and for $\lambda \in V \cap \mathbb{C}_+$, $(\pi \circ H)(\lambda) = R(\lambda, \widetilde{B})\pi(f)$. It follows from Lemma 4.6.7 that $i\eta \notin \sigma(\widetilde{B}_{\widetilde{f}})$. Thus, $\sigma(\widetilde{B}_{\widetilde{f}})$ is countable. But then $\sigma(\widetilde{B}_{\widetilde{f}})$ contains an isolated point which is an eigenvalue of $\widetilde{B}_{\widetilde{f}}$ by Corollary 4.4.13. This contradicts Proposition 4.7.2. Thus, $\pi(f) = 0$; i.e., $f \in \mathrm{AAP}(\mathbb{R}_+, X)$. □

An even more general version of Theorem 4.7.7 will be given in Section 4.9. However, before that some further preparation concerning harmonic analysis is given in the following section.

We note an immediate corollary of Lemma 4.7.9.

Corollary 4.7.10. *Let $f \in \mathrm{BUC}(\mathbb{R}_+, X)$. If $\eta \in \mathbb{R} \setminus \mathrm{sp}(f)$, then f is uniformly ergodic at η and $M_\eta f = 0$.*

We now give an example which shows that the condition of total ergodicity is crucial in Theorem 4.7.7, even in the scalar case. This contrasts dramatically with the situation of the entire line where total ergodicity is a consequence of countability of the Carleman spectrum if $c_0 \not\subset X$ (Theorem 4.6.3.

The example also shows that ergodicity has to be uniform with respect to translates of the function. Recall that total ergodicity means by definition that the function is uniformly ergodic at each $\eta \in \mathbb{R}$.

Example 4.7.11. *The function $f(t) := \sin\sqrt{t}$ has half-line spectrum $\mathrm{sp}(f) = \{0\}$ and $0 \notin \mathrm{Freq}(f)$. Since $f(t)$ does not converge to 0 as $t \to \infty$, it follows from Theorem 4.7.7 that f is not uniformly ergodic at 0. However, $\lim_{t\to\infty} \frac{1}{t}\int_0^t f(s)\,ds = 0$.*

Proof. a) We show that

$$\hat{f}(\lambda) = \frac{\sqrt{\pi}e^{-1/4\lambda}}{2\lambda^{3/2}} \quad (\mathrm{Re}\,\lambda > 0).$$

In fact, $f(t) = \sin\sqrt{t} = \sum_{n=0}^{\infty} \frac{(-1)^n t^{n+1/2}}{(2n+1)!}$. Thus, for $\lambda > 0$, integrating term by term, we obtain

$$\begin{aligned}
\hat{f}(\lambda) &= \int_0^\infty \sin\sqrt{t}\, e^{-\lambda t}\, dt \\
&= \sum_{n=0}^{\infty} \frac{(-1)^n}{(2n+1)!} \int_0^\infty t^{n+1/2} e^{-\lambda t}\, dt \\
&= \sum_{n=0}^{\infty} \frac{(-1)^n}{(2n+1)!} \frac{1}{\lambda^{n+3/2}} \int_0^\infty u^{n+1/2} e^{-u}\, du \\
&= \sum_{n=0}^{\infty} \frac{(-1)^n}{(2n+1)!} \lambda^{-n-3/2} \Gamma(n+3/2).
\end{aligned}$$

Standard formulas for the Gamma function give

$$\begin{aligned}
\Gamma(n+3/2) &= (n+1/2)(n-1/2)\ldots\frac{1}{2}\Gamma(1/2) \\
&= \frac{(2n+1)(2n-1)\ldots 1}{2^{n+1}}\sqrt{\pi} = \frac{(2n+1)!}{n!2^{2n+1}}\sqrt{\pi}.
\end{aligned}$$

Thus,

$$\begin{aligned}
\hat{f}(\lambda) &= \sum_{n=0}^{\infty} \lambda^{-n-3/2} \frac{(-1)^n}{n!2^{2n+1}}\sqrt{\pi} \\
&= \frac{\sqrt{\pi}}{2\lambda^{3/2}} \sum_{n=0}^{\infty} \frac{1}{n!}\frac{(-1)^n}{(4\lambda)^n} = \frac{\sqrt{\pi}}{2\lambda^{3/2}} e^{-1/4\lambda}.
\end{aligned}$$

By uniqueness of holomorphic extensions, this formula is true for $\operatorname{Re}\lambda > 0$. It follows that $\operatorname{sp}(f) = \{0\}$.

b) For $t > 0$,

$$\begin{aligned}
\frac{1}{t}\int_0^t f(s)\, ds &= \frac{1}{t}\int_0^{\sqrt{t}} 2u \sin u\, du \\
&= \frac{1}{t}\left[-2u\cos u\right]_0^{\sqrt{t}} + \frac{2}{t}\int_0^{\sqrt{t}} \cos u\, du \\
&= -\frac{2\cos\sqrt{t}}{\sqrt{t}} + \frac{2}{t}\sin\sqrt{t} \to 0
\end{aligned}$$

as $t \to \infty$.

c) One can directly see that f is not uniformly ergodic at 0. In fact, given $t > 0$ we may choose s such that

$$(2n+1/4)\pi < \sqrt{s} < \sqrt{s+t} < (2n+3/4)\pi$$

for some integer n (since $\sqrt{s+t}-\sqrt{s}\to 0$ as $s\to\infty$). Then $f(u)>\frac{1}{\sqrt{2}}$ whenever $s\le u\le s+t$, so $\frac{1}{t}\int_s^{s+t} f(u)\,du>\frac{1}{\sqrt{2}}$. Thus, f is not uniformly ergodic at 0. □

We want to extend Theorem 4.7.7 to bounded measurable functions. Then we can only obtain assertions for the means of f, for example B-convergence, but this is sufficient for interesting applications to power series.

Let $f\in L^\infty(\mathbb{R}_+,X)$. We define the *half-line spectrum* $\operatorname{sp}(f)$ as in Section 4.4; viz.,

$$\operatorname{sp}(f) := \Big\{\eta\in\mathbb{R} : \hat{f}(\lambda) \text{ does not have a holomorphic extension to a neighbourhood of } i\eta\Big\}.$$

Moreover, we say that f is *totally ergodic* if for each $\eta\in\mathbb{R}$,

$$(M_\eta f)(t) := \lim_{\alpha\downarrow 0}\alpha\int_0^\infty e^{-(\alpha+i\eta)s} f(t+s)\,ds$$

converges uniformly in $t\in\mathbb{R}_+$. In that case, it follows as in (4.25) that

$$M_\eta f = e_{i\eta}\otimes x,$$

where $e_{i\eta}(t)=e^{i\eta t}$, $x=(M_\eta f)(0)\in X$. We set

$$\operatorname{Freq}(f) := \{\eta\in\mathbb{R} : M_\eta f\neq 0\}.$$

The following is a Tauberian theorem where B-convergence is deduced.

Theorem 4.7.12. *Let $f\in L^\infty(\mathbb{R}_+,X)$ be totally ergodic. Assume that* $\operatorname{sp}(f)$ *is countable and* $\operatorname{Freq}(f)\subset\{0\}$. *Then* $\text{B-}\lim_{t\to\infty} f(t)=(M_0 f)(0)$.

Proof. Let $\delta>0$, $g(t):=\frac{1}{\delta}\int_t^{t+\delta} f(s)\,ds$. Then $g\in BUC(\mathbb{R}_+,X)$. Integrating by parts we obtain for $\operatorname{Re}\lambda>0$,

$$\begin{aligned}
\hat{g}(\lambda) &= -\frac{1}{\delta\lambda}\int_0^\infty \frac{d}{dt}\left(e^{-\lambda t}\right)\int_t^{t+\delta} f(s)\,ds\,dt\\
&= \frac{1}{\delta\lambda}\left(\int_0^\delta f(s)\,ds+\int_0^\infty e^{-\lambda t}(f(t+\delta)-f(t))\,dt\right)\\
&= \frac{1}{\delta\lambda}\left(\int_0^\delta f(s)\,ds+e^{\lambda\delta}\int_\delta^\infty e^{-\lambda t}f(t)\,dt-\hat{f}(\lambda)\right)\\
&= \frac{1}{\delta\lambda}(e^{\lambda\delta}-1)\hat{f}(\lambda)+\frac{1}{\delta\lambda}\left(\int_0^\delta f(s)\,ds-e^{\lambda\delta}\int_0^\delta e^{-\lambda s}f(s)\,ds\right)\\
&= \frac{1}{\delta\lambda}(e^{\lambda\delta}-1)\hat{f}(\lambda)+\left(\frac{1-e^{\lambda\delta}}{\lambda\delta}\int_0^\delta f(s)\,ds+e^{\lambda\delta}\int_0^\delta f(s)\frac{1-e^{-\lambda s}}{\lambda\delta}\,ds\right).
\end{aligned}$$

The right-hand summand defines an entire function of λ. Since $\frac{1}{\delta\lambda}(e^{\lambda\delta}-1)$ is entire, it follows that $\mathrm{sp}(g) \subset \mathrm{sp}(f)$.

Next, we show that g is totally ergodic. Let $r \in \mathbb{R}$, $g_r(s) := g(s+r)$, $f_r(s) := f(s+r)$. Then $g_r(t) = \frac{1}{\delta}\int_t^{t+\delta} f_r(s)\,ds$. By the computation above, we have for $\mathrm{Re}\,\lambda > 0$,

$$\widehat{g_r}(\lambda) \;=\; \frac{1}{\delta\lambda}(e^{\lambda\delta}-1)\widehat{f_r}(\lambda) + \frac{1-e^{\lambda\delta}}{\lambda\delta}\int_0^\delta f_r(s)\,ds + e^{\lambda\delta}\int_0^\delta f_r(s)\frac{1-e^{-\lambda\delta}}{\lambda\delta}\,ds.$$

Let $\lambda = \alpha + i\eta$ $(\alpha > 0)$. Then $\alpha\widehat{f_r}(\alpha+i\eta)$ converges uniformly in $r \in \mathbb{R}_+$ as $\alpha \downarrow 0$. Consequently, so does $\alpha\widehat{g_r}(\alpha+i\eta)$. We have shown that g is totally ergodic. Moreover, $(M_0 g)(0) = \lim_{\alpha\downarrow 0}\alpha\hat{g}(\alpha) = \lim_{\alpha\downarrow 0}\alpha\hat{f}(\alpha) = (M_0 f)(0)$, whereas for $\eta \neq 0$, $(M_\eta g)(0) = 0$. It follows from Corollary 4.7.8 that $g(t)$ converges as $t \to \infty$; i.e., $f(t)$ is B-convergent as $t \to \infty$. It follows from Theorem 4.1.2 that

$$\text{B-}\lim_{t\to\infty} f(t) = \lim_{t\to\infty} g(t) = (M_0 g)(0) = (M_0 f)(0). \qquad \square$$

Lemma 4.7.13. *Let $f \in L^\infty(\mathbb{R}_+, X)$. Let $\eta \in \mathbb{R}$ such that*

$$\sup_{t\geq 0}\left\|\int_0^t e^{-i\eta s} f(s)\,ds\right\| < \infty.$$

Then f is uniformly ergodic at η and $M_\eta f = 0$.

Proof. Let M be such that $\left\|\int_0^t e^{-i\eta s} f(s)\,ds\right\| \leq M$ $(t \geq 0)$. Let $r \geq 0$. Then

$$\left\|\int_0^t e^{-i\eta s} f(s+r)\,ds\right\| = \left\|e^{i\eta r}\int_t^{t+r} e^{-i\eta s} f(s)\,ds\right\| \leq 2M.$$

$\square$

As an application of Theorem 4.7.12 we obtain the following result on power series. We let $\mathbb{T} := \{z \in \mathbb{C} : |z| = 1\}$.

Proposition 4.7.14. *Let $a_n \in X$ such that $M := \sup_{m\in\mathbb{N}}\|\sum_{n=0}^m a_n\| < \infty$. Consider the power series*

$$p(z) := \sum_{n=0}^\infty a_n z^n \quad (|z| < 1).$$

Assume that for each $z \in \mathbb{T}\setminus\{1\}$ the function p has a holomorphic extension to a neighbourhood of z. Then $\lim_{n\to\infty} a_n = 0$.

Proof. Let $f(t) := a_n$ for $t \in [n, n+1)$, $n \in \mathbb{N}_0$. Then $f \in L^\infty(\mathbb{R}_+, X)$ and $\|f\|_\infty \leq 2M$. For $\mathrm{Re}\,\lambda > 0$, one has

$$\begin{aligned}\hat{f}(\lambda) &= \sum_{m=0}^\infty a_m \int_m^{m+1} e^{-\lambda t}\,dt = \sum_{m=0}^\infty a_m e^{-\lambda m}\int_0^1 e^{-\lambda t}\,dt \\ &= \frac{1-e^{-\lambda}}{\lambda}\sum_{m=0}^\infty a_m e^{-\lambda m}.\end{aligned}$$

Thus, $\mathrm{sp}(f) \subset 2\pi\mathbb{Z}$. We will show that $\sup_{t\geq 0}\left\|\int_0^t e^{-i\eta s}f(s)\,ds\right\| < \infty$ for all $\eta \in 2\pi\mathbb{Z}$. Then Lemma 4.7.13 and Corollary 4.7.10 imply that f is totally ergodic and $M_\eta f = 0$ for all $\eta \in \mathbb{R}$. Finally, Theorem 4.7.12 implies that

$$\lim_{n\to\infty} a_n = \lim_{n\to\infty}\int_n^{n+1} f(s)\,ds = \text{B-}\lim_{n\to\infty} f(t) = 0.$$

Let $\eta = 0$. For $t \geq 0$, let $n \in \mathbb{N}$ such that $t \in [n, n+1)$. Then

$$\int_0^t f(s)\,ds = \sum_{m=0}^{n-1} a_m + a_n(t-n).$$

Thus, $\left\|\int_0^t f(s)\,ds\right\| \leq 3M$.

Let $\eta = 2\pi k,\ k \in \mathbb{Z}\setminus\{0\}$. Let $t \geq 0,\ t \in [n, n+1)$. Then

$$\begin{aligned}\int_0^t e^{-i\eta s}f(s)\,ds &= \sum_{m=0}^{n-1} a_m \int_m^{m+1} e^{-i2\pi kt}\,dt + a_n\int_n^t e^{-i2\pi ks}\,ds\\ &= a_n\int_n^t e^{-i2\pi ks}\,ds.\end{aligned}$$

Hence, $\left\|\int_0^t e^{-i\eta s}f(s)\,ds\right\| \leq 2M$. □

Corollary 4.7.15 (Katznelson-Tzafriri). *Let $T \in \mathcal{L}(X)$ be an operator such that $M := \sup_{n\in\mathbb{N}}\|T^n\| < \infty$. Then $\sigma(T)\cap\mathbb{T} \subset \{1\}$ if and only if $\lim_{n\to\infty}\|T^n(I-T)\| = 0$.*

Proof. a) Assume that $\sigma(T)\cap\mathbb{T}\subset\{1\}$. Let

$$p(z) := \sum_{n=0}^{\infty}(T-I)T^n z^n = (T-I)(I-zT)^{-1} \quad \text{for } |z|<1.$$

Then

$$\left\|\sum_{n=0}^{m}(T-I)T^n\right\| = \|T^{m+1}-I\| \leq M+1.$$

Thus the claim follows from Proposition 4.7.14.

b) Suppose that $\lim_{n\to\infty}\|T^n(I-T)\| = 0$ and $\mu \in \sigma(T)\cap\mathbb{T}$. By the spectral mapping theorem, $\mu^n(1-\mu) \in \sigma(T^n(I-T))$. Then

$$|1-\mu| = |\mu^n(1-\mu)| \leq \|T^n(I-T)\| \to 0.$$

Hence, $\mu = 1$. □

4.8 Carleman Spectrum and Fourier Transform

For functions on the line, the Carleman spectrum is the natural notion if we are interested in the properties of the Laplace transform (or more precisely, the Carleman transform). However, the Fourier transform is also a powerful tool. In order to make it available we show in this section that the Carleman spectrum coincides with the support of the Fourier transform of the given function and also with the Beurling spectrum.

As applications we give several results which allow us to deduce asymptotic properties of a function from the nature of its spectrum. For example, we show that a function $f \in \mathrm{BUC}(\mathbb{R}, X)$ is almost periodic whenever it has discrete spectrum. A function $f \in L^\infty(\mathbb{R}, X)$ is τ-periodic if and only if its spectrum is contained in $\frac{2\pi}{\tau}\mathbb{Z}$.

Let $f \in L^1(\mathbb{R})$. As in Section 1.8 and Appendix E, we denote by $\mathcal{F}f \in C_0(\mathbb{R})$ the Fourier transform given by

$$\mathcal{F}f(t) := \int_{\mathbb{R}} e^{-ist} f(s)\, ds \quad (t \in \mathbb{R}) \tag{4.40}$$

and we let

$$\overline{\mathcal{F}}f(t) := \int_{\mathbb{R}} e^{its} f(s)\, ds \quad (t \in \mathbb{R}). \tag{4.41}$$

By $\mathcal{S}(\mathbb{R})$ we denote the *Schwartz space*; i.e., the space of all infinitely differentiable functions $f : \mathbb{R} \to \mathbb{C}$ such that $\sup_{t\in\mathbb{R}} |f^{(m)}(t)|(1 + |t|)^k < \infty$ for all $m, k \in \mathbb{N}_0$. Then $\mathcal{F}$ is a bijective mapping from $\mathcal{S}(\mathbb{R})$ into $\mathcal{S}(\mathbb{R})$ with inverse $(2\pi)^{-1}\overline{\mathcal{F}}$. See Appendix E for further information.

Let X be a Banach space. By $L^1(\mathbb{R}, (1 + |t|)^{-k}\, dt; X)$ we denote the space of all functions $f \in L^1_{loc}(\mathbb{R}, X)$ such that $\int_{\mathbb{R}} \| f(t) \| (1 + |t|)^{-k}\, dt < \infty$, where $k \in \mathbb{N}_0$. Note that $L^1(\mathbb{R}, X) \subset L^1(\mathbb{R}, (1 + |t|)^{-k}\, dt; X)$ for all $k \in \mathbb{N}_0$.

Let $f \in L^1(\mathbb{R}, (1 + |t|)^{-k}\, dt; X)$, where $k \in \mathbb{N}_0$. Then we define $\mathcal{F}f$ as a linear mapping from $\mathcal{S}(\mathbb{R})$ into X by

$$\langle \varphi, \mathcal{F}f \rangle \; = \int_{\mathbb{R}} f(t)(\mathcal{F}\varphi)(t)\, dt \quad (\varphi \in \mathcal{S}(\mathbb{R})).$$

The *support* of $\mathcal{F}f$ is defined by

$$\operatorname{supp} \mathcal{F}f := \Big\{ \eta \in \mathbb{R} : \text{for all } \varepsilon > 0 \text{ there exists } \varphi \in \mathcal{S}(\mathbb{R}) \text{ such that } \operatorname{supp}\varphi \subset (\eta - \varepsilon, \eta + \varepsilon) \text{ and } \langle \varphi, \mathcal{F}f \rangle, \neq 0 \Big\}. \tag{4.42}$$

In the particular case when $k = 0$, i.e. $f \in L^1(\mathbb{R}, X)$, Fubini's theorem shows that $\langle \varphi, \mathcal{F}f \rangle \; = \int_{\mathbb{R}} \int_{\mathbb{R}} f(t)e^{-ist}\, dt\, \varphi(s)\, ds$. Hence,

$$\operatorname{supp} \mathcal{F}f = \{s \in \mathbb{R} : (\mathcal{F}f)(s) \neq 0\}^-, \tag{4.43}$$

and our notation is consistent with (4.40) and the identification of absolutely regular functions with distributions when $X = \mathbb{C}$ (see Theorem 1.8.1 b) and Appendix E).

If $f \in L^1(\mathbb{R}, (1+|t|)^{-k}\,dt; X)$, then one can define the Carleman transform $\hat{f} : \mathbb{C} \setminus (i\mathbb{R}) \to X$ and the Carleman spectrum $\mathrm{sp}_{\mathrm{C}}(f)$ exactly as in Section 4.6. We are going to show that $\mathrm{supp}\, \mathcal{F}f$ and $\mathrm{sp}_{\mathrm{C}}(f)$ coincide.

We recall the notion of *mollifier* $(\rho_n)_{n\in\mathbb{N}}$ from Section 1.3, but here we shall assume that $\rho_1 \in \mathcal{S}(\mathbb{R})$. The function $\rho_1 \in \mathcal{S}(\mathbb{R})$ satisfies $\int_{\mathbb{R}} \rho_1(t)\,dt = 1$. Then $\rho_n \in \mathcal{S}(\mathbb{R})$ is given by $\rho_n(t) = n\rho_1(nt)$ $(t \in \mathbb{R}, n \in \mathbb{N})$. Thus,

$$\mathcal{F}\rho_n \in \mathcal{S}(\mathbb{R}) \quad (n \in \mathbb{N}), \tag{4.44}$$

and $\mathcal{F}\rho_n(t) = \mathcal{F}\rho_1(\frac{t}{n})$, so

$$\lim_{n\to\infty} (\mathcal{F}\rho_n)(t) = 1 \quad (t \in \mathbb{R}); \tag{4.45}$$

$$|(\mathcal{F}\rho_n)(t)| \le \int_{\mathbb{R}} |\rho_1(s)|\,ds \quad (t \in \mathbb{R},\ n \in \mathbb{N}). \tag{4.46}$$

Moreover, if $\mathrm{supp}\,\rho_1 \subset (-1,1)$, then

$$\mathrm{supp}\,\rho_n \subset \left(-\tfrac{1}{n}, \tfrac{1}{n}\right) \quad (n \in \mathbb{N}); \tag{4.47}$$

and if $\mathrm{supp}\,\mathcal{F}\rho_1 \subset (-1,1)$, then

$$\mathrm{supp}\,\mathcal{F}\rho_n \subset (-n,n) \quad (n \in \mathbb{N}). \tag{4.48}$$

Theorem 4.8.1. *Let $k \in \mathbb{N}_0$, $f \in L^1(\mathbb{R}, (1+|t|)^{-k}\,dt; X)$. Then* $\mathrm{sp}_{\mathrm{C}}(f) = \mathrm{supp}\,\mathcal{F}f$.

Proof. Let $\eta \in \mathbb{R}$. Then by the definition of the Carleman transform, for $\alpha > 0$, one has

$$\begin{aligned} \hat{f}(\alpha + i\eta) &= \int_0^\infty e^{-\alpha t} e^{-i\eta t} f(t)\,dt \quad \text{and} \\ \hat{f}(-\alpha + i\eta) &= -\int_{-\infty}^0 e^{\alpha t} e^{-i\eta t} f(t)\,dt. \end{aligned}$$

Hence,

$$\hat{f}(\alpha + i\eta) - \hat{f}(-\alpha + i\eta) = \int_{-\infty}^\infty e^{-\alpha|t|} e^{-i\eta t} f(t)\,dt. \tag{4.49}$$

Let $\varphi \in \mathcal{S}(\mathbb{R})$. Then by Fubini's theorem,

$$\begin{aligned} \langle \varphi, \mathcal{F}f \rangle &= \int_{\mathbb{R}} f(t)(\mathcal{F}\varphi)(t)\,dt \\ &= \lim_{\alpha\downarrow 0} \int_{\mathbb{R}} e^{-\alpha|t|} f(t)(\mathcal{F}\varphi)(t)\,dt \\ &= \lim_{\alpha\downarrow 0} \int_{\mathbb{R}} e^{-\alpha|t|} f(t) \int_{\mathbb{R}} e^{-ist}\varphi(s)\,ds\,dt \\ &= \lim_{\alpha\downarrow 0} \int_{\mathbb{R}} \int_{\mathbb{R}} e^{-\alpha|t|} e^{-ist} f(t)\,dt\,\varphi(s)\,ds. \end{aligned}$$

Thus by (4.49), we obtain

$$\langle \varphi, \mathcal{F}f \rangle = \lim_{\alpha \downarrow 0} \int_{\mathbb{R}} \left(\hat{f}(\alpha + is) - \hat{f}(-\alpha + is) \right) \varphi(s)\, ds \tag{4.50}$$

for all $\varphi \in \mathcal{S}(\mathbb{R})$.

From this, the inclusion $\operatorname{supp} \mathcal{F}f \subset \operatorname{sp}_{\mathbb{C}}(f)$ follows immediately: Assume that $\eta \notin \operatorname{sp}_{\mathbb{C}}(f)$. Then there exists $\varepsilon > 0$ such that $\hat{f}$ has a bounded holomorphic extension to $B(i\eta, \varepsilon)$. This implies that $\lim_{\alpha \downarrow 0} \left(\hat{f}(\alpha + is) - \hat{f}(-\alpha + is)\right) = 0$ whenever $s \in (\eta - \varepsilon, \eta + \varepsilon)$. Thus, by (4.50) and the dominated convergence theorem, $\langle \varphi, \mathcal{F}f \rangle = 0$ whenever $\varphi \in \mathcal{S}(\mathbb{R})$ and $\operatorname{supp} \varphi \subset (\eta - \varepsilon, \eta + \varepsilon)$.

Next, we prove the other inclusion: $\operatorname{sp}_{\mathbb{C}}(f) \subset \operatorname{supp} \mathcal{F}f$.

a) We first deal with the special case where $k = 0$; i.e., $f \in L^1(\mathbb{R})$. Then $\mathcal{F}f$ is a continuous function. Let $\eta \in \mathbb{R} \setminus \operatorname{supp} \mathcal{F}f$. Then there exists $\varepsilon > 0$ such that $\mathcal{F}f$ vanishes on $(\eta - \varepsilon, \eta + \varepsilon)$. Putting

$$\hat{f}(i\beta) := \int_0^{\infty} e^{-i\beta t} f(t)\, dt = -\int_{-\infty}^{0} e^{-i\beta t} f(t)\, dt$$

for $\beta \in (\eta - \varepsilon, \eta + \varepsilon)$, $\hat{f}$ has a continuous extension to a neighbourhood of $i\eta$. It follows from Morera's theorem that this extension is holomorphic. Thus, $\eta \notin \operatorname{sp}_{\mathbb{C}}(f)$.

b) Now let $k \in \mathbb{N}_0$ be arbitrary. We first give an estimate of $\hat{f}$. Let $M := \int_{\mathbb{R}} (1 + |t|)^{-k} \| f(t) \|\, dt$. Then

$$\| \hat{f}(\lambda) \| \le M k^k e^{1-k} |\operatorname{Re} \lambda|^{-k} \tag{4.51}$$

whenever $0 < |\operatorname{Re} \lambda| < 1$. In fact, if $0 < \operatorname{Re} \lambda < 1$, then

$$\begin{aligned} \| \hat{f}(\lambda) \| &\le \int_0^{\infty} e^{-\operatorname{Re} \lambda t} \| f(t) \|\, dt \\ &\le M \sup_{t > 0} \left((1 + t)^k e^{-\operatorname{Re} \lambda t} \right) \\ &\le M k^k e^{1-k} (\operatorname{Re} \lambda)^{-k}. \end{aligned}$$

The estimation is similar for $-1 < \operatorname{Re} \lambda < 0$.

Now let $\eta \in \mathbb{R} \setminus \operatorname{supp} \mathcal{F}f$. Then there exists $\varepsilon > 0$ such that $\langle \varphi, \mathcal{F}f \rangle = 0$ for all $\varphi \in \mathcal{S}(\mathbb{R})$ with $\operatorname{supp} \varphi \subset (\eta - 2\varepsilon, \eta + 2\varepsilon)$. Let $(\rho_n)_{n \in \mathbb{N}}$ be a mollifier in $\mathcal{S}(\mathbb{R})$, such that $\operatorname{supp} \rho_n \subset (-\frac{1}{n}, \frac{1}{n})$. Let $f_n := (\mathcal{F}\rho_n) \cdot f$. Since $\mathcal{F}\rho_n \in \mathcal{S}(\mathbb{R})$, we have $f_n \in L^1(\mathbb{R}, X)$.

For $n > 1/\varepsilon$, we have $\operatorname{supp} \mathcal{F}f_n \cap (\eta - \varepsilon, \eta + \varepsilon) = \emptyset$. In fact, let $\varphi \in \mathcal{S}(\mathbb{R})$ such that $\operatorname{supp} \varphi \subset (\eta - \varepsilon, \eta + \varepsilon)$. Then $\langle \mathcal{F}f_n, \varphi \rangle = \int_{\mathbb{R}} f_n \mathcal{F}\varphi = \int_{\mathbb{R}} f \mathcal{F}\rho_n \cdot \mathcal{F}\varphi = \int f \mathcal{F}(\rho_n * \varphi) = 0$ since $\rho_n * \varphi \in \mathcal{S}(\mathbb{R})$ and $\operatorname{supp} \rho_n * \varphi \subset \operatorname{supp} \rho_n + \operatorname{supp} \varphi \subset (-\frac{1}{n}, \frac{1}{n}) + (\eta - \varepsilon, \eta + \varepsilon) \subset (\eta - 2\varepsilon, \eta + 2\varepsilon)$.

From part a) of the proof we conclude that $\hat{f}_n$ has a holomorphic extension to $B(i\eta, \varepsilon)$. Since $\| f_n(t) \| \le \|f(t)\| \, \|\rho_1\|_1$, we obtain from (4.51) (replacing f by f_n) that $\| \hat{f}_n(\lambda) \| \le M\|\rho\|_1 k^k e^{1-k} |\operatorname{Re}\lambda|^{-k}$ whenever $0 < |\operatorname{Re}\lambda| < 1$. Now it follows from Lemma 4.6.6 that

$$\| \hat{f}_n(\lambda) \| \le c \qquad (\lambda \in B(i\eta, \varepsilon/2)) \tag{4.52}$$

for all $n \in \mathbb{N}$, where c is a constant independent of $n \in \mathbb{N}$. Since $f_n(t) \to f(t)$ as $n \to \infty$ for all $t \in \mathbb{R}$ (see Lemma 1.3.3), it follows from the dominated convergence theorem that $\lim_{n\to\infty} \hat{f}_n(\lambda) = \hat{f}(\lambda)$ if $\operatorname{Re}\lambda > 0$. It follows from Vitali's theorem (Theorem A.5) that $\hat{f}$ has a holomorphic extension to $B(i\eta, \varepsilon/2)$. Thus, $\eta \notin \operatorname{sp_C}(f)$. □

The following are consequences of Theorem 4.8.1, and extensions of parts of Corollary 4.5.9 (see Proposition 4.6.10).

Theorem 4.8.2. *Let $f \in L^1(\mathbb{R}, (1+|t|)^{-k}\, dt; X)$ for some $k \in \mathbb{N}_0$.*

a) $\operatorname{sp_C}(f) = \emptyset$ *if and only if* $f(t) = 0$ *a.e.*

b) $\operatorname{sp_C}(f) = \{0\}$ *if and only if f is a polynomial.*

c) $\operatorname{sp_C}(f) \subset \{\eta_1 \dots \eta_m\}$ *if and only if there exist polynomials $p_1, \dots, p_m$ such that* $f(t) = \sum_{j=1}^m p_j(t) e^{i\eta_j t}$ $(t \in \mathbb{R})$.

Proof. a) Assume that $\operatorname{sp_C}(f) = \emptyset$. Then it follows from the proof of Theorem 4.8.1 that $\int_{\mathbb{R}} f(t)\mathcal{F}\varphi(t)\, dt = 0$ for all $\varphi \in C_c^\infty(\mathbb{R})$. Since $C_c^\infty(\mathbb{R})$ is dense in $\mathcal{S}(\mathbb{R})$ for the canonical topology of $\mathcal{S}(\mathbb{R})$ and since $\mathcal{F}$ is continuous (see Appendix E), it follows that $\int_{\mathbb{R}} f(t)(\mathcal{F}\varphi)(t)\, dt = 0$ for all $\varphi \in \mathcal{S}(\mathbb{R})$. Hence, $\int_{\mathbb{R}} f(t)\psi(t)\, dt = 0$ for all $\psi \in \mathcal{S}(\mathbb{R})$. This implies that $f(t) = 0$ a.e.

b) Assume that $\operatorname{sp_C}(f) = \{0\}$. Then the Laurent series

$$\hat{f}(\lambda) = \sum_{n=-\infty}^{\infty} a_n \lambda^n$$

converges for $\lambda \in \mathbb{C} \setminus \{0\}$, where

$$a_n := \frac{1}{2\pi i} \int_{|z|=r} f(z) z^{-n-1}\, dz \quad (n \in \mathbb{Z})$$

independently of $r > 0$.

From (4.51) we obtain a constant $c > 0$ such that

$$\| \hat{f}(\lambda) \| \le c |\operatorname{Re}\lambda|^{-k} \tag{4.53}$$

for $0 < |\operatorname{Re}\lambda| < 1$. This implies that

$$\left\| \frac{1}{2\pi i} \int_{|z|=r} z^{n+k-1} \Big(1 + \frac{z^2}{r^2}\Big)^k \hat{f}(z)\, dz \right\| \le c\, 2^k r^n \qquad (n \in \mathbb{Z},\ r \in (0,1))$$

(observe that for $z = re^{i\theta}$, $\| (1 + \frac{z^2}{r^2})^k \hat{f}(z) \| \le |1 + e^{i2\theta}|^k c\, |r \cos\theta|^{-k} = 2^k c r^{-k}$). Hence,

$$\begin{aligned} c\,2^k r^n &\ge \left\| \sum_{j=0}^{k} \binom{k}{j} r^{-2j} \cdot \frac{1}{2\pi i} \int_{|z|=r} z^{n+k-1+2j} \hat{f}(z)\, dz \right\| \\ &= \left\| \sum_{j=0}^{k} \binom{k}{j} r^{-2j} a_{-n-k-2j} \right\| \end{aligned}$$

for all $r \in (0,1)$. Multiplying by r^{2k}, we obtain

$$c2^k r^{n+2k} \ge \left\| \sum_{j=0}^{k} \binom{k}{j} r^{2k-2j} a_{-n-k-2j} \right\|.$$

Now let $n \ge 1-2k$. Then the left-hand term, as well as all terms on the right except for $j = k$, converge to 0 as $r \downarrow 0$. Consequently, $a_{-n-3k} = 0$ whenever $n \ge 1 - 2k$, i.e., $a_m = 0$ if $m < -k$, so $\hat{f}(\lambda) = \sum_{n=-k}^{\infty} a_n \lambda^n$. Let $f_0(t) := \sum_{j=0}^{k} a_{-j-1} t^j / j!$. Then $\hat{f}_0(\lambda) = \sum_{n=-k}^{-1} a_n \lambda^n$. Thus $\hat{f} - \hat{f}_0$ is entire and so $\operatorname{sp_C}(f - f_0) = \emptyset$. It follows from a) that $f(t) = f_0(t)$ a.e.

c) We prove the assertion by induction. Let $m = 1$. Then $\operatorname{sp_C}(f) = \{\eta_1\}$. Let $g(t) := e^{-i\eta_1 t} f(t)$. Then $\operatorname{sp_C}(g) = 0$. It follows from b) that g is a polynomial.

Now assume that the assertion holds for m and that $\operatorname{sp_C}(f) = \{\eta_1, \dots, \eta_{m+1}\}$, where $\eta_j \ne \eta_l$ if $j \ne l$. Replacing $f(t)$ by $e^{-i\eta_{m+1} t} f(t)$ if necessary, we can assume that $\eta_{m+1} = 0$. It follows from the proof of b) that 0 is a pole of $\hat{f}$. Moreover, there exists a polynomial p_{m+1} such that $(f - p_{m+1})\widehat{\ }$ has a holomorphic extension to a neighbourhood of 0. Thus, $\operatorname{sp_C}(f - p_{m+1}) \subset \{\eta_1, \dots, \eta_m\}$. Now it follows from the inductive assumption that $(f - p_{m+1})(t) = \sum_{n=1}^{m} p_n(t) e^{i\eta_n t}$ $(t \in \mathbb{R})$. ☐

Corollary 4.8.3. *Let $f \in L^\infty(\mathbb{R}, X)$.*

a) *If $\operatorname{sp_C}(f) = \{0\}$, then f is constant.*

b) *If $\operatorname{sp_C}(f)$ is finite, then f is a trigonometric polynomial.*

If $f \in L^\infty(\mathbb{R}, X)$, there is an alternative way to describe the spectrum of f. We define the *Beurling spectrum* by

$$\operatorname{sp_B}(f) := \Big\{ \eta \in \mathbb{R} : \text{ for all } \varepsilon > 0 \text{ there exists } g \in L^1(\mathbb{R}) \text{ such that } \operatorname{supp} \mathcal{F}g \subset (\eta - \varepsilon, \eta + \varepsilon) \text{ and } f * g \ne 0 \Big\}.$$

Proposition 4.8.4. *Let $f \in L^\infty(\mathbb{R}, X)$. Then $\operatorname{sp_C}(f) = \operatorname{sp_B}(f)$.*

Proof. By Theorem 4.8.1, we have

$$\begin{aligned}
\mathrm{sp_C}(f) &= \operatorname{supp} \mathcal{F}f \\
&= \Big\{\eta \in \mathbb{R} : \text{for all } \varepsilon > 0 \text{ there exists } \varphi \in \mathcal{S}(\mathbb{R}) \text{ such that} \\
&\qquad\qquad \operatorname{supp} \varphi \subset (\eta - \varepsilon, \eta + \varepsilon) \text{ and } \int_{\mathbb{R}} f(t)(\mathcal{F}\varphi)(t)\,dt \neq 0\Big\} \\
&= \Big\{\eta \in \mathbb{R} : \text{for all } \varepsilon > 0 \text{ there exists } h \in \mathcal{S}(\mathbb{R}) \text{ such that} \\
&\qquad\qquad \operatorname{supp} \overline{\mathcal{F}}h \subset (\eta - \varepsilon, \eta + \varepsilon) \text{ and } \int_{\mathbb{R}} f(t)h(t)\,dt \neq 0\Big\} \\
&= \Big\{\eta \in \mathbb{R} : \text{for all } \varepsilon > 0 \text{ there exists } h \in \mathcal{S}(\mathbb{R}) \text{ such that} \\
&\qquad\qquad \operatorname{supp} \mathcal{F}h \subset (\eta - \varepsilon, \eta + \varepsilon) \text{ and } \int_{\mathbb{R}} f(t)h(-t)\,dt \neq 0\Big\} \\
&= \Big\{\eta \in \mathbb{R} : \text{for all } \varepsilon > 0 \text{ there exists } h \in \mathcal{S}(\mathbb{R}) \text{ such that} \\
&\qquad\qquad \operatorname{supp} \mathcal{F}h \subset (\eta - \varepsilon, \eta + \varepsilon) \text{ and } (f * h)(0) \neq 0\Big\}.
\end{aligned}$$

This shows that $\mathrm{sp_C}(f) \subset \mathrm{sp_B}(f)$.

In order to show the converse, let $\eta \in \mathrm{sp_B}(f)$. Let $\varepsilon > 0$. Then there exists $g \in L^1(\mathbb{R})$ such that $\operatorname{supp} \mathcal{F}g \subset (\eta - \varepsilon, \eta + \varepsilon)$ and $f * g \neq 0$. Replacing g by $g(\cdot - t)$ if necessary, we can assume that $(f * g)(0) \neq 0$.

Take a mollifier $(\rho_n)_{n\in\mathbb{N}}$ in $\mathcal{S}(\mathbb{R})$ such that $\operatorname{supp} \rho_n \subset (-\frac{1}{n}, \frac{1}{n})$, and define $g_n := \overline{\mathcal{F}}\rho_n \cdot g$. By (4.45), (4.46) and the dominated convergence theorem, $\lim_{n\to\infty} g_n = g$ in $L^1(\mathbb{R})$. It follows from Proposition 1.3.2 that $\lim_{n\to\infty}(g_n * f)(0) = (g * f)(0) \neq 0$.

We have $\rho_n * \mathcal{F}g \in C^\infty(\mathbb{R})$ (see Proposition 1.3.6) and $\operatorname{supp}(\rho_n * \mathcal{F}g) \subset (-\frac{1}{n}, \frac{1}{n}) + (\eta - \varepsilon, \eta + \varepsilon) \subset (\eta - 2\varepsilon, \eta + 2\varepsilon)$ if $n > 1/\varepsilon$. Moreover, $(\rho_n * \mathcal{F}g)(s) = \int \rho_n(s - r)\mathcal{F}g(r)\,dr = \int \rho_n(s - r) \int g(t)e^{-irt}dt\,dr = \int \rho_n(r) \int g(t)e^{-i(s-r)t}dt\,dr = \mathcal{F}((\overline{\mathcal{F}}\rho_n)\cdot g)(s)$. Hence, $\mathcal{F}g_n = \mathcal{F}((\overline{\mathcal{F}}\rho_n)\cdot g) \in C_c^\infty(\mathbb{R}) \subset \mathcal{S}(\mathbb{R})$. Since $\mathcal{F}$ is a bijection from $\mathcal{S}(\mathbb{R})$ onto $\mathcal{S}(\mathbb{R})$, it follows that $g_n \in \mathcal{S}(\mathbb{R})$. We have shown that there exists $n > 1/\varepsilon$ such that $g_n \in \mathcal{S}(\mathbb{R})$, $\operatorname{supp} \mathcal{F}g_n \subset (\eta - 2\varepsilon, \eta + 2\varepsilon)$ and $(f * g_n)(0) \neq 0$. Since $\varepsilon > 0$ is arbitrary, it follows that $\eta \in \operatorname{supp} \mathcal{F}f = \mathrm{sp_C}(f)$. □

So far, we have established the identity of three different notions of spectrum of a function $f \in L^\infty(\mathbb{R}, X)$: the Carleman spectrum, the support of its Fourier transform, and the Beurling spectrum. Next, we consider two situations in which a special form of the spectrum of a function tells us a lot about the nature of the function itself (Theorems 4.8.7 and 4.8.8).

We need the following auxiliary result.

Lemma 4.8.5. *Let $f \in L^\infty(\mathbb{R}, X)$ and $g \in L^1(\mathbb{R})$. Then $\mathrm{sp_C}(f * g) \subset \mathrm{sp_C}(f) \cap \operatorname{supp} \mathcal{F}g$.*

Proof. a) Let $\eta \in \mathbb{R} \setminus \mathrm{sp_C}(f)$. By Proposition 4.8.4, there exists $\varepsilon > 0$ such that

$f * h = 0$ whenever $h \in L^1(\mathbb{R})$, $\operatorname{supp} \mathcal{F}h \subset (\eta - \varepsilon, \eta + \varepsilon)$. Hence, $(f * g) * h = (f * h) * g = 0$. Thus, $\eta \notin \operatorname{sp_B}(f * g) = \operatorname{sp_C}(f * g)$.

b) Let $\eta \notin \operatorname{supp} \mathcal{F}g$. We show that $\eta \notin \operatorname{sp_B}(f * g)$. There exists $\varepsilon > 0$ such that $(\mathcal{F}g)(r) = 0$ for all $r \in (\eta - \varepsilon, \eta + \varepsilon)$. Let $h \in L^1(\mathbb{R})$ such that $\operatorname{supp} \mathcal{F}h \subset (\eta - \varepsilon, \eta + \varepsilon)$. Then $\mathcal{F}(g * h) = \mathcal{F}g \cdot \mathcal{F}h = 0$. Thus, $g * h = 0$. Hence, $(f * g) * h = 0$. This proves the claim. □

Remark 4.8.6. The same property is true if $f \in L^1(\mathbb{R}, (1 + |t|)^{-k}\, dt; X)$ for some $k \in \mathbb{N}_0$ and $g \in \mathcal{S}(\mathbb{R})$. □

Now we can prove that a function $f \in \mathrm{BUC}(\mathbb{R}, X)$ is almost periodic whenever it has discrete Carleman spectrum. Thus, the geometric condition "$c_0 \not\subset X$", which appeared in Theorem 4.6.3, is not needed if the spectrum does not have any accumulation point.

Theorem 4.8.7. *Let $f \in \mathrm{BUC}(\mathbb{R}, X)$. If the spectrum $\operatorname{sp_C}(f)$ of f is discrete, then f is almost periodic.*

Proof. Let $(\rho_n)_{n \in \mathbb{N}}$ be a mollifier in $\mathcal{S}(\mathbb{R})$ such that $\operatorname{supp} \mathcal{F}\rho_1 \subset (-1, 1)$. By Proposition 1.3.2 and Lemma 4.8.5, $f * \rho_n \in \mathrm{BUC}(\mathbb{R}, X)$ and $\operatorname{sp_C}(f * \rho_n) \subset \operatorname{sp_C}(f) \cap \operatorname{supp} \mathcal{F}\rho_n \subset \operatorname{sp_C}(f) \cap (-n, n)$. Thus, $f * \rho_n$ has finite spectrum, and it follows from Corollary 4.8.3 that $f * \rho_n \in \mathrm{AP}(\mathbb{R}, X)$. By Lemma 1.3.3, it follows that $f = \lim_{n \to \infty} f * \rho_n \in \mathrm{AP}(\mathbb{R}, X)$. □

Next, we give a spectral characterization of periodic functions extending part of Corollary 4.5.9. If g is a function defined on $\mathbb{R}$ we let $\check{g}(t) = g(-t)$ $(t \in \mathbb{R})$.

Theorem 4.8.8. *Let $f \in L^\infty(\mathbb{R}, X)$ and let $\tau > 0$. Then f is τ-periodic (i.e., $f(t + \tau) = f(t)$ t-a.e.) if and only if $\operatorname{sp_C}(f) \subset \frac{2\pi}{\tau}\mathbb{Z}$.*

Proof. a) Let f be τ-periodic. Then

$$\hat{f}(\lambda) = (1 - e^{-\lambda\tau})^{-1} \int_0^\tau e^{-\lambda t} f(t)\, dt$$

for $\lambda \in \mathbb{C} \setminus i\mathbb{R}$. Thus, $\hat{f}$ extends to a meromorphic function with at most simple poles at $\lambda_n := 2\pi i n/\tau$ $(n \in \mathbb{Z})$. The residue at λ_n is given by

$$c_n := \frac{1}{\tau} \int_0^\tau e^{-2\pi i n t/\tau} f(t)\, dt \quad (n \in \mathbb{Z}).$$

Therefore we have

$$\operatorname{sp_C}(f) = \{2\pi n/\tau : n \in \mathbb{Z}, c_n \neq 0\}.$$

b) Assume that $\operatorname{sp_C}(f) \subset \frac{2\pi}{\tau}\mathbb{Z}$. Let $(\rho_n)_{n \in \mathbb{N}}$ be a mollifier in $\mathcal{S}(\mathbb{R})$ such that $\operatorname{supp} \mathcal{F}\rho_n \subset (-n, n)$ and $\check{\rho}_n = \rho_n$ $(n \in \mathbb{N})$. Then $\rho_n * f \in \mathrm{BUC}(\mathbb{R}, X)$ and $\operatorname{sp_C}(\rho_n * f) \subset \operatorname{supp} \mathcal{F}\rho_n \cap \operatorname{sp_C}(f) \subset (-n, n) \cap \operatorname{sp_C}(f) \subset (-n, n) \cap \frac{2\pi}{\tau}\mathbb{Z}$. It follows from Corollary 4.8.3 and Proposition 4.6.10 that $\rho_n * f$ is a τ-periodic trigonometric polynomial.

Let $\varphi \in C_c(\mathbb{R})$. Then $\lim_{n\to\infty} \rho_n * \varphi = \varphi$ in $L^1(\mathbb{R})$. Thus, $\int f(t+\tau)\,\varphi(t)\,dt = \lim_{n\to\infty} \int f(t+\tau)\,(\rho_n * \varphi)(t)\,dt = \lim_{n\to\infty} \int (f * \rho_n)(t+\tau)\,\varphi(t)\,dt = \lim_{n\to\infty} \int (f * \rho_n)(t)\,\varphi(t)\,dt = \lim_{n\to\infty} \int f(t)\,(\rho_n * \varphi)(t)\,dt = \int f(t)\,\varphi(t)\,dt$. Since $\varphi \in C_c(\mathbb{R})$ is arbitrary, it follows that $f(t+\tau) = f(t)$ t-a.e. □

4.9 Complex Tauberian Theorems: the Fourier Method

In this section we present an approach via Fourier transforms to complex Tauberian theorems for Laplace transforms. This method was already used in a restricted form by Ingham in 1935 (see [Ing35]). It will eventually lead to the most general complex Tauberian theorem presented in this book (Theorem 4.9.7).

It is not such a great difference to consider bounded measurable functions which are slowly oscillating, instead of uniformly continuous functions as we did before, but the point is that the notion of spectrum is changed. Instead of considering a point $i\eta$ as regular if the Laplace transform has a holomorphic extension in a neighbourhood of $i\eta$, we merely assume that a locally integrable extension to the imaginary axis exists. This leads to a smaller spectrum which we call the *weak half-line spectrum.* Thus, asking that this small spectrum be countable is a weaker hypothesis. It turns out that this weaker hypothesis is more natural or easier to verify in some applications (see the Notes of this section).

In the case where the weak half-line spectrum is empty, the proof is completely elementary and leads to a slight generalization of Ingham's Tauberian theorem. We consider this case first (Theorem 4.9.5). The general case will be proved by a Hahn-Banach argument which allows us to apply Loomis's theorem in a quite tricky way.

Let $f \in L^\infty(\mathbb{R}_+, X)$. We define the weak half-line spectrum $\mathrm{sp}_w(f)$ of f as follows.

Definition 4.9.1. *Let $\eta \in \mathbb{R}$. We say that $i\eta$ is a* weakly regular point *for $\hat{f}$ if there exist $\varepsilon > 0$ and $h \in L^1((\eta-\varepsilon, \eta+\varepsilon), X)$ such that*

$$\hat{f}(\alpha + i\cdot) \to h \text{ in the distributional sense on } (\eta-\varepsilon, \eta+\varepsilon) \text{ as } \alpha \downarrow 0. \tag{4.54}$$

Then the weak half-line spectrum $\mathrm{sp}_w(f)$ *of f is defined as the set of all real numbers which are not weakly regular for $\hat{f}$.*

Of course, (4.54) means by definition that

$$\lim_{\alpha\downarrow 0} \int_{\mathbb{R}} \hat{f}(\alpha + is)\varphi(s)\,ds = \int_{\eta-\varepsilon}^{\eta+\varepsilon} h(s)\varphi(s)\,ds \tag{4.55}$$

for all test functions $\varphi \in \mathcal{D}(\eta-\varepsilon, \eta-\varepsilon) = C_c^\infty(\eta-\varepsilon, \eta-\varepsilon)$.

It is clear that $\mathrm{sp}_w(f)$ is a closed subset of $\mathbb{R}$. Moreover, if $\hat{f}(\lambda)$ has a continuous extension to $\mathbb{C}_+ \cup (i(\eta-\varepsilon), i(\eta+\varepsilon))$, where $\eta \in \mathbb{R}$, $\varepsilon > 0$, then clearly $\eta \notin \mathrm{sp}_w(f)$. In particular, we have $\mathrm{sp}_w(f) \subset \mathrm{sp}(f)$. The inclusion is strict in general as the following example shows.

Example 4.9.2. Let $X := l^2$,

$$f(t) := \left(n^{-1}e^{-t/n}\right)_{n\in\mathbb{N}}.$$

Then $\hat{f}$ has a continuous extension to $\mathbb{C}_+ \cup i\mathbb{R}$, but not a holomorphic extension to a neighbourhood of 0. □

Lemma 4.9.3. *Let $f \in L^\infty(\mathbb{R}_+, X)$. Then there exists $h \in L^1_{loc}(\mathbb{R}\setminus \mathrm{sp}_w(f), X)$ such that*

$$\hat{f}(\alpha + i\cdot) \to h \text{ in the distributional sense on } \mathbb{R}\setminus\mathrm{sp}_w(f) \text{ as } \alpha \downarrow 0. \tag{4.56}$$

We set $\hat{f}(is) := h(s)$ $(s \in \mathbb{R}\setminus \mathrm{sp}_w(f))$.

Proof. For all $x \in \mathbb{R}\setminus\mathrm{sp}_w(f)$, we find an open neighbourhood U_x of x and $h_x \in L^1(U_x, X)$ such that $\hat{f}(\alpha+i\cdot) \to h_x$ in the distributional sense on U_x as $\alpha \downarrow 0$. Clearly, $h_x(t) = h_y(t)$ almost everywhere on $U_x \cap U_y$ whenever $x, y \in \mathbb{R}\setminus\mathrm{sp}_w(f)$. Hence, there exists a function $h \in L^1_{loc}(\mathbb{R}\setminus\mathrm{sp}_w(f), X)$ such that

$$h|_{U_x} = h_x \quad (x \in \mathbb{R}\setminus\mathrm{sp}_w(f)).$$

It remains to show (4.56). Let $\varphi \in \mathcal{D}(\mathbb{R}\setminus\mathrm{sp}_w(f))$. Let $K := \mathrm{supp}\,\varphi$. There exist $x_1, x_2, \ldots, x_n$ such that the sets $U_j := U_{x_j}$ $(j = 1, 2, \ldots, n)$ cover K. Let $\varphi_j \in \mathcal{D}(\mathbb{R}\setminus\mathrm{sp}_w(f))$ $(j = 1, 2, \ldots, n)$ be a partition of unity subordinate to this covering; i.e., $0 \le \varphi_j \le 1, \mathrm{supp}\,\varphi_j \subset U_j$ $(j = 1, 2, \ldots, n), \sum_{j=1}^n \varphi_j(x) = 1$ for all $x \in K$. Then $\varphi = \sum_{j=1}^n \psi_j$, where $\psi_j := \varphi\varphi_j$. Since $\mathrm{supp}\,\psi_j \subset U_j$,

$$\lim_{\alpha\downarrow 0}\int_{\mathbb{R}} \hat{f}(\alpha+is)\psi_j(s)\,ds = \int_{U_j} h(s)\psi_j(s)\,ds$$

for all $j = 1, 2, \ldots, n$. Hence,

$$\lim_{\alpha\downarrow 0}\int_{\mathbb{R}} \hat{f}(\alpha+is)\varphi(s)\,ds = \int_{\mathbb{R}\setminus\mathrm{sp}_w(f)} h(s)\varphi(s)\,ds. \tag{4.57}$$

Since $\varphi \in \mathcal{D}(\mathbb{R}\setminus\mathrm{sp}_w(f))$ is arbitrary, this is precisely the meaning of (4.56). □

Lemma 4.9.4. *Let $f \in L^\infty(\mathbb{R}_+, X)$. Then*

$$\varphi * f \in C_0(\mathbb{R}_+, X)$$

for all $\varphi \in \mathcal{S}(\mathbb{R})$ such that $\mathcal{F}\varphi \in C_c^\infty(\mathbb{R})$ and $\mathrm{supp}\,\mathcal{F}\varphi \cap \mathrm{sp}_w(f) = \emptyset$.

Here as elsewhere, we identify a function defined on $\mathbb{R}_+$ with its extension by 0 on $\mathbb{R}$. In particular,

$$(\varphi * f)(t) = \int_0^\infty f(s)\varphi(t-s)\,ds \quad (t \in \mathbb{R}).$$

Proof of Lemma 4.9.4. Let $\varphi \in \mathcal{S}(\mathbb{R})$ such that $\mathcal{F}\varphi \in C_c^\infty(\mathbb{R})$ and $\operatorname{supp}\mathcal{F}\varphi \cap \operatorname{sp}_w(f) = \emptyset$. Then $\mathcal{F}\varphi \cdot \hat{f}(i\cdot) \in L^1(\mathbb{R}, X)$, where $\hat{f}(i\cdot)$ was defined in Lemma 4.9.3. Let $t \geq 0$. Then the inverse Fourier transform of the function $s \mapsto \varphi(t-s)$ is the function $\eta \mapsto (2\pi)^{-1}e^{i\eta t}\mathcal{F}\varphi(\eta)$. Thus by Theorem 1.8.1 b),

$$\begin{aligned}
(\varphi * f)(t) &= \int_0^\infty f(s)\varphi(t-s)\,ds \\
&= \lim_{\alpha\downarrow 0} \int_0^\infty e^{-\alpha s} f(s)\varphi(t-s)\,ds \\
&= \lim_{\alpha\downarrow 0} \quad (2\pi)^{-1} \int_{\mathbb{R}} \hat{f}(\alpha + i\eta)e^{int}\mathcal{F}\varphi(\eta)\,d\eta \\
&= (2\pi)^{-1} \int_{\mathbb{R}} \hat{f}(i\eta)\mathcal{F}\varphi(\eta)e^{i\eta t}\,d\eta.
\end{aligned}$$

It follows from the Riemann-Lebesgue lemma (Theorem 1.8.1) that $\lim_{t\to\infty}(\varphi * f)(t) = 0$. □

Now we prove the complex Tauberian theorem in the case where the weak half-line spectrum is empty.

Theorem 4.9.5 (Ingham). *Let $f \in L^\infty(\mathbb{R}_+, X)$ be slowly oscillating. If $\operatorname{sp}_w(f) = \emptyset$, then*

$$\lim_{t\to\infty} f(t) = 0.$$

Proof. a) We show that $g * f \in C_0(\mathbb{R}_+, X)$ for all $g \in L^1(\mathbb{R})$. By Proposition 1.3.2, $Tg := (g * f)|_{\mathbb{R}_+}$ defines a bounded linear operator from $L^1(\mathbb{R})$ into $\mathrm{BUC}(\mathbb{R}_+, X)$. Let ϕ be a continuous linear functional on $\mathrm{BUC}(\mathbb{R}_+, X)$ vanishing on $C_0(\mathbb{R}_+, X)$. By the Hahn-Banach theorem, it suffices to show that $\langle Tg, \phi\rangle = 0$ for all $g \in L^1(\mathbb{R})$. Let $h := T^*\phi \in L^\infty(\mathbb{R})$. Then

$$\langle g * f, \phi\rangle = \int_{\mathbb{R}} hg\,dt$$

for all $g \in L^1(\mathbb{R})$. It follows from Lemma 4.9.4 that $\int_{\mathbb{R}} h\varphi\,dt = 0$ if $\varphi \in \mathcal{S}(\mathbb{R})$ and $\mathcal{F}\varphi \in C_c^\infty(\mathbb{R})$. Hence, $\operatorname{supp}\mathcal{F}h = \emptyset$ (see (4.42)). Thus, $\operatorname{sp}_C(h) = \emptyset$ by Theorem 4.8.1 which implies that $h = 0$ by Theorem 4.8.2. This proves the claim. We have shown that $g * f \in C_0(\mathbb{R}_+, X)$ for all $g \in L^1(\mathbb{R})$.

b) Taking in particular $g := \frac{1}{\delta}\chi_{[0,\delta]}$, it follows from a) that $\text{B-}\lim_{t\to\infty} f(t) = 0$. Now it follows from Theorem 4.2.3 that $\lim_{t\to\infty} f(t) = 0$. □

We will see in Example 5.5.7 that in general the hypothesis that g be slowly oscillating cannot be omitted.

Next, we consider the case of countable weak spectrum and we extend Theorem 4.7.12. We need the notion of uniform ergodicity on a subset of $\mathbb{R}$, for bounded measurable functions, similar to the total ergodicity of Section 4.7.

Definition 4.9.6. *Let $f \in L^\infty(\mathbb{R}_+, X)$ and let E be a subset of $\mathbb{R}$. We say that f is* uniformly ergodic on E *if for all $\eta \in E$ the limit*

$$(M_\eta f)(t) = \lim_{\alpha \downarrow 0} \alpha \int_0^\infty e^{-(\alpha+i\eta)s} f(t+s)\, ds$$

exists uniformly in $t \in \mathbb{R}_+$.

As in (4.25) it follows that

$$M_\eta f = e_{i\eta} \otimes x$$

where $e_{i\eta}(t) = e^{i\eta t}$, $x = (M_\eta f)(0) \in X$.

The following complex Tauberian theorem is the main result of this section. Its proof is based on the same idea as the one of Theorem 4.9.5, but it is less elementary since Loomis's theorem is used.

Theorem 4.9.7. *Let $f \in L^\infty(\mathbb{R}_+, X)$ be slowly oscillating. Assume that $\mathrm{sp}_w(f)$ is countable and that f is uniformly ergodic on $\mathrm{sp}_w(f)$. Then*

$$f = f_0 + f_1,$$

where $f_1 \in \mathrm{AP}(\mathbb{R}_+, X)$ and $\lim_{t\to\infty} f_0(t) = 0$. In particular, if $f \in \mathrm{BUC}(\mathbb{R}_+, X)$ then $f \in \mathrm{AAP}(\mathbb{R}_+, X)$.

Proof. a) We show that $g * f \in \mathrm{AAP}(\mathbb{R}_+, X)$ for all $g \in L^1(\mathbb{R})$ (and for this we do not need the assumption that f is slowly oscillating). As in the proof of Theorem 4.9.5, we consider the operator $T : L^1(\mathbb{R}) \to \mathrm{BUC}(\mathbb{R}_+, X)$ given by $Tg := (g * f)|_{\mathbb{R}_+}$. Let $\phi \in \mathrm{BUC}(\mathbb{R}_+, X)^*$ be a continuous linear functional on $\mathrm{BUC}(\mathbb{R}_+, X)$ which vanishes on $\mathrm{AAP}(\mathbb{R}_+, X)$. We have to show that $\langle g * f, \phi \rangle = 0$ for all $g \in L^1(\mathbb{R})$. Then the claim follows from the Hahn-Banach theorem.

Let $h := T^*\phi$. As in the proof of Theorem 4.9.5 it follows from Lemma 4.9.4 that $\mathrm{sp}_C(h) \subset \mathrm{sp}_w(f)$. Hence, $\mathrm{sp}_C(h)$ is countable. Let $\varphi \in L^1(\mathbb{R})$. Since $\mathrm{sp}_C(\varphi * h) \subset \mathrm{sp}_C(h)$ (by Lemma 4.8.5), $\mathrm{sp}_C(\varphi * h)$ is also countable. Moreover, $\varphi * h \in \mathrm{BUC}(\mathbb{R})$ by Proposition 1.3.2. It follows from Loomis's theorem (Corollary 4.6.4) that $\varphi * h \in \mathrm{AP}(\mathbb{R})$ for all $\varphi \in L^1(\mathbb{R})$.

Next, let $\varphi \in C_c^\infty(0, \infty)$. We show that $M_\eta(\varphi * h) = 0$ for all $\eta \in \mathbb{R}$ in order to conclude that $\varphi * h = 0$ by Corollary 4.5.9.

First, since $\mathrm{Freq}(\varphi * h) \subset \mathrm{sp}_C(\varphi * h) \subset \mathrm{sp}_w(f)$, it is clear that $M_\eta(\varphi * h) = 0$ for all $\eta \in \mathbb{R} \setminus \mathrm{sp}_w(f)$.

Now, let $\eta \in \mathrm{sp}_w(f)$. Let $c_\eta := M_\eta(\varphi * h)(0)$. Then

$$M_\eta(\varphi * h)(t) = c_\eta e^{i\eta t} \quad \text{(see (4.25))}.$$

We have to show that $c_\eta = 0$.

For $\lambda \in \mathbb{R}$, we define $e_\lambda(t) := e^{\lambda t}$ $(t \in \mathbb{R})$. Note that

$$\begin{aligned} c_\eta &= \lim_{\alpha \downarrow 0} \alpha \int_0^\infty e^{-(\alpha+i\eta)t} (\varphi * h)(t)\, dt \\ &= \lim_{\alpha \downarrow 0} \alpha \langle \chi_{\mathbb{R}_+} e_{-(\alpha+i\eta)}, \varphi * h\rangle \\ &= \lim_{\alpha \downarrow 0} \alpha \langle h, \chi_{\mathbb{R}_-} e_{\alpha+i\eta} * \varphi\rangle \\ &= \lim_{\alpha \downarrow 0} \alpha \langle (f * \chi_{\mathbb{R}_-} e_{\alpha+i\eta} * \varphi)_{|\mathbb{R}_+}, \phi\rangle. \end{aligned}$$

However,

$$k_\alpha(t) := \alpha \int_{-\infty}^0 f(t-s) e^{i(\alpha+i\eta)s}\, ds = \alpha \int_0^\infty f(t+s) e^{-i(\alpha+i\eta)s}\, ds$$

converges to the function $e_{i\eta} \otimes x$ as $\alpha \downarrow 0$ uniformly in $t \geq 0$. Since $\varphi \in C_c^\infty(0,\infty)$, this implies that $(k_\alpha * \varphi)(t)$ converges to

$$e^{i\eta t} \int_0^\infty e^{-i\eta s} \varphi(s)\, ds\, x =: k(t)$$

uniformly in $t \geq 0$ as $\alpha \downarrow 0$. Thus, $c_\eta = \langle k, \phi\rangle$. Since $k \in \mathrm{AP}(\mathbb{R}_+, X)$, it follows that $c_\eta = 0$.

We have shown that $h * \varphi = 0$ for all $\varphi \in C_c^\infty(0,\infty)$. Choosing a mollifier $(\rho_n)_{n\in\mathbb{N}} \subset C_c^\infty(-\infty, 0)$, we deduce that for all $k \in L^1(\mathbb{R})$,

$$\langle k, h\rangle = \lim_{n\to\infty} \langle k * \rho_n, h\rangle = \lim_{n\to\infty} \langle k, \check{\rho}_n * h\rangle = 0.$$

Thus, $T^*\phi = h = 0$. Here, for $v : \mathbb{R} \to X$ we let $\check{v}(t) = v(-t)$. It follows that $\langle g * f, \phi\rangle = 0$ for all $\phi \in \mathrm{BUC}(\mathbb{R}_+, X)^*$ vanishing on $\mathrm{AAP}(\mathbb{R}_+, X)$. Hence, $g * f \in \mathrm{AAP}(\mathbb{R}_+, X)$.

b) Since f is slowly oscillating, $f = g_0 + g_1$ where $g_1 \in \mathrm{BUC}(\mathbb{R}_+, X)$ and $\lim_{t\to\infty} g_0(t) = 0$, by Proposition 4.2.2. Let $(\rho_n)_{n\in\mathbb{N}} \subset C_c^\infty(0,\infty)$ be a mollifier with $\operatorname{supp} \rho_n \subset (0, 1/n)$. Then $\rho_n * g_0 \in C_0(\mathbb{R}_+, X)$, and $\rho_n * f \in \mathrm{AAP}(\mathbb{R}_+, X)$ by a). Hence, $\rho_n * g_1 = \rho_n * f - \rho_n * g_0 \in \mathrm{AAP}(\mathbb{R}_+, X)$. Since $g_1 \in \mathrm{BUC}(\mathbb{R}_+, X)$, $\lim_{n\to\infty} \rho_n * g_1 = g_1$ in $\mathrm{BUC}(\mathbb{R}_+, X)$. Thus, $g_1 \in \mathrm{AAP}(\mathbb{R}_+, X)$. Now, $g_1 = f_1 + f_2$ with $f_1 \in \mathrm{AP}(\mathbb{R}_+, X)$, $f_2 \in C_0(\mathbb{R}_+, X)$. Thus, $f = f_1 + (f_2 + g_0)$. □

4.10 Notes

Sections 4.1 and 4.2

The prototype for Abelian theorems was Abel's classical continuity theorem (Theorem 4.1.6) which was proved by N. Abel [Abe26] in 1826. Tauberian theorems form their counterpart and the first result of this type is Tauber's classical theorem from 1897

[Tau97] which is mentioned in connection with Hardy's Theorem 4.2.17. There is an enormous amount of literature on Tauberian theorems. We refer to the monographs by Widder [Wid71] and by Doetsch [Doe50, Volume I] for the classical results. The monographs of Pitt [Pit58] and van de Lune [Lun86], and Chatterji's historical account [Cha84], cover some of the subsequent developments, and Korevaar's book [Kor04] is a wide-ranging account which provides much historical information up to recent times.

The notion of B-limit was introduced by Arendt and Prüss [AP92] following ideas of Batty [Bat90] showing how to pass from theorem of type D to results of type E. Theorem 4.2.15 illustrates in a surprising way how this strategy can be used. The presentation given here is close to [AP92].

With the help of Wiener's Tauberian theorem the following clarification of B-convergence was given by Arendt and Batty [AB00].

Theorem 4.10.1. *Let $u \in L^\infty(\mathbb{R}, X), u_\infty \in X$, and suppose that*

$$\lim_{t\to\infty} \frac{1}{\delta} \int_t^{t+\delta} u(s)\, ds = u_\infty$$

holds for $\delta = \delta_1$ and $\delta = \delta_2$, where δ_1 and δ_2 are rationally independent. Then

$$\lim_{t\to\infty} (\rho * u)(t) = \left(\int_{\mathbb{R}} \rho\right) u_\infty$$

for all $\rho \in L^1(\mathbb{R})$. In particular, $\text{B-lim}_{t\to\infty} u(t) = u_\infty$.

Karamata's theorem has important applications to the study of eigenvalue distributions (see e.g. [Sim79, Theorem 10.6]). An elegant short proof of Karamata's theorem in the scalar case has been given by König [Kön60]. Further Abelian and Tauberian theorems for positive vector-valued functions were given by El-Mennaoui [Elm94].

Another direction of Tauberian theorems for Laplace transforms occurs in the framework of limitation theory. A final result in this direction has been obtained by Stadtmüller and Trautner [ST81] and by Kratz and Stadtmüller [KS90] in the discrete case.

Section 4.3

The results at the beginning of the section are quite standard.

A systematic study of mean convergence of integrated semigroups was carried out by El-Mennaoui [Elm92]. In particular, Theorem 4.3.7, saying that for positive C_0-semigroups Cesàro-ergodicity and Abel-ergodicity are equivalent, can be extended to integrated semigroups. We state a special case explicitly as follows:

Let E be a Banach lattice with order continuous norm. Let A be a resolvent positive operator, so A generates a once integrated semigroup S (by Theorem 3.11.7). Then

$$X = \operatorname{Ker} A \oplus \overline{\operatorname{Ran} A}$$

if and only if

$$Px = \lim_{t\to\infty} \frac{1}{t} S(t)x$$

exists for all $x \in E$. In that case, P is the projection onto $\operatorname{Ker} A$ along $\overline{\operatorname{Ran} A}$. See [Elm92, Theorem 4.1].

Theorem 4.3.7 and Theorem 4.3.8 are explicitly proved in Arendt and Prüss [AP92]. More generally, El-Mennaoui [Elm92] studied strong convergence of $\frac{1}{t^k}S(t)$ (as $t \to \infty$), where S is a k-times integrated semigroup and $k \in \mathbb{N}$. This is a very natural problem. In fact, if A is the generator of S, then for $x \in D(A^{k+1})$, $S(\cdot)x \in C^{k+1}((0,\infty), X)$ and $u(t) := S^{(k)}(t)x$ defines a classical solution of the Cauchy problem

$$\begin{cases} u'(t) = & Au(t) \quad (t \geq 0), \\ u(0) = & x. \end{cases}$$

Thus, $\frac{1}{t^k}S(t)x = \frac{1}{t^k}\int_0^t \frac{(t-s)^{k-1}}{(k-1)!}u(s)\, ds$ is the kth Cesàro mean of the solution u and describes its asymptotic behaviour for $t \to \infty$.

Proposition 4.3.13 is due to Arendt and Batty [AB92a] who also showed that the result does not hold on every Banach lattice. In fact, an example is given in [AB92a] which shows that a positive C_0-semigroup on a space $C(K)$, where K is compact, may well be Cesàro-ergodic without being totally ergodic. However, in Proposition 4.3.13 the L^1-space may be replaced by any Banach lattice with order continuous norm.

The striking properties concerning C_0-semigroups on L^∞-spaces are due to Lotz [Lot85]. In fact, it had been proved before by Kishimoto and Robinson [KR81] that every generator of a positive C_0-semigroup on an L^∞-space is bounded, and independently of Lotz, Coulhon [Cou84] had proved Corollary 4.3.19 for contraction semigroups. However, it was Lotz who discovered the interesting interplay of the geometric properties (G) and (DP). He proved in particular Lemma 4.3.16 and also proved ergodic theorems in the discrete case (i.e., for power bounded operators).

Other examples of Banach spaces having both properties (DP) and (G) are the following:

a) $C(K)$, where K is an F-space (cf. the Notes of Section 2.7).

b) $H^\infty(\mathbb{D})$, the space of all bounded holomorphic functions on the unit disc $\mathbb{D} := \{z \in \mathbb{C} : |z| < 1\}$ with the supremum norm.

We refer to [Nag86, Section A-II.3] and the references given there.

Section 4.4.

A version of the complex Tauberian Theorem 4.4.1 was proved by Ingham [Ing35]. The ingenious contour argument proof of a case of Ingham's theorem was given in Korevaar's beautiful article [Kor82] which also gives an elementary proof of the prime number theorem based on Ingham's Tauberian theorem. Korevaar was inspired by Newman [New80] who proved the corresponding result for Dirichlet series; see also the book by Newman [New98]. The quantified version presented in Theorem 4.4.6 is due to Batty and Duyckaerts [BD08]. Remark 4.4.7 shows that the estimate (4.15) is quite sharp when M is bounded or grows very slowly, and Borichev and Tomilov [BT10] have shown that it is optimal for polynomial growth.

Theorem 4.4.8, as stated here, is due to Arendt and Batty [AB88], who also gave other versions allowing a countable number or even a null set of singularities. Here, merely the case of one singularity is presented which allows several interesting applications including those given in this section and later in the chapter. A quantified version is given in [BD08]. Further versions are contained in the work of Arendt and Prüss [AP92]. Cesàro convergence is investigated in [AB95] with the help of contour arguments similar to those of Theorem 4.4.8. For example, it is shown that a function $f \in L^1_{loc}(\mathbb{R}_+, X)$ is Cesáro convergent if $\|f(t)\| = O(t)$ as $t \to \infty$ and every point of $i\mathbb{R}$ is either regular for $\hat{f}$ or a pole of order 1.

Corollary 4.4.12 is contained in Arveson's work on spectral subspaces [Arv82], but it is usually associated with Gelfand's name. More precisely, it is the corresponding result on bounded operators which is due to Gelfand (saying that an isometry whose spectrum is $\{1\}$ is necessarily the identity). Since the weak spectral mapping theorem holds for bounded C_0-groups [Nag86, A-III, Theorem 7.4], Corollary 4.4.12 follows immediately from Gelfand's classical theorem on isometries. Another elegant proof of Gelfand's theorem is due to Allan and Ransford [AR89], see also the survey article of Zemanek [Zem94]. Extensions of Gelfand's theorem have been obtained by Zarrabi [Zar93]. He proved in particular that an invertible contraction T on a Banach space X with countable spectrum is already an isometry if it satisfies the growth condition

$$\lim_{n\to\infty} \frac{\log \|T^{-n}\|}{\sqrt{n}} = 0.$$

Theorem 4.4.14 is taken from [BD08]. Remarkably the implication (ii) $\Rightarrow$ (i) does not seem to have been in the earlier literature of operator semigroups. The implication (i) $\Rightarrow$ (ii) was already implicit in [AB88] (see [Bat94, p.41]). Quantified versions were obtained for polynomial growth of M on arbitrary Banach spaces [BEPS06], and for polynomial or exponential growth of M [Bur98], [Leb96], [LR05], using methods which generally give slightly slower rates of decay than (4.19). For $M(s) = \beta(1+s)^\alpha$, it is shown in [BT10] that the rate in (4.19) is optimal on general Banach spaces, but $M_{\log}^{-1}$ can be replaced by M^{-1} in the case of Hilbert spaces. A version of Theorem 4.4.14 when $\sigma(A) \cap i\mathbb{R}$ is finite is given in [Mar11].

These results have applications to energy decay of damped wave equations of the form

$$\begin{cases} \dfrac{\partial^2 u}{\partial t^2} = \Delta u - 2a(x)\dfrac{\partial u}{\partial t} & (t > 0, x \in \Omega), \\ u(x,t) = 0 & (t > 0, x \in \partial\Omega), \\ u(\cdot, 0) = u_0 \in H_0^1(\Omega), & \dfrac{\partial u}{\partial t}(\cdot, 0) = u_1 \in L^2(\Omega). \end{cases} \tag{4.58}$$

Here, Ω is a smooth or convex open subset of $\mathbb{R}^n$ with boundary $\partial\Omega$, and $a : \Omega \to [0, \infty)$ is a bounded continuous function representing the damping at each point. Let X be the Hilbert space $H_0^1(\Omega) \times L^2(\Omega)$, and let A be the operator on X defined by

$$\begin{aligned} D(A) &:= \left(H^2(\Omega) \cap H_0^1(\Omega)\right) \times H_0^1(\Omega), \\ A &:= \begin{pmatrix} 0 & I \\ \Delta & -2a(x) \end{pmatrix}. \end{aligned}$$

Then A generates a C_0-semigroup of contractions on X, and the energy of the solution u of (4.58) is given by $E(u,t) := \frac{1}{2}\|T(t)(u_0, u_1)\|_X^2$. The following result was proved by Lebeau [Leb96] with a slightly slower rate of decay in b), and the optimal rate given here was established in [Bur98]. Once a) is proved, b) follows directly from Theorem 4.4.14 a) with $M(s) = \beta e^{\alpha s}$.

Theorem 4.10.2. *Consider the damped wave equation* (4.58), *where Ω is a bounded open set and a is not identically zero. We assume that Ω is smooth or convex. Let A and $E(u,t)$ be as above. Then the following hold.*

a) *$\sigma(A) \subset \{\lambda \in \mathbb{C} : -2\|a\|_\infty \le \operatorname{Re}\lambda < 0\}$, and there exist $\alpha, \beta > 0$ such that $\|R(is, A)\| \le \beta e^{\alpha|s|}$ for all $s \in \mathbb{R}$.*

b) *There is a constant C such that $E(u,t) \le \dfrac{C\|(u_0,u_1)\|^2_{H^2\times H^1}}{(\log(2+t))^2}$ for all $u_0 \in H^2(\Omega) \cap H^1_0(\Omega)$, $u_1 \in H^1_0(\Omega)$, and $t \ge 0$.*

When the data is sufficiently smooth that $(u_0, u_1) \in D(A^k)$, the rate of energy decay in b) can readily be improved to $O\left((\log t)^{-k}\right)$. In many cases, the rate is much faster than logarithmic. When the domain of damping $\{x \in \Omega : a(x) > 0\}$ satisfies the "geometric optics condition", the energy decay is exponential (and uniform with respect to $\|(u_0, u_1)\|_X$), i.e., $\omega(T) < 0$ and $\|R(is, A)\|$ is bounded [BLR89]. When Ω is a rectangle in $\mathbb{R}^2$ and the domain of damping is a strip across Ω, the geometric optics condition is not satisfied but $\|R(is, A)\| = O(s^2)$ as $|s| \to \infty$ [LR05, Example 3]. In this case, the optimal rate of decay $E(u,t) = O(t^{-1})$ follows from [BT10, Theorem 2.4].

Laplace transform methods can also be applied to local energy decay of wave equations on exterior domains, a subject which already featured prominently in the 1967 monograph by Lax and Phillips [LP67]. In this context, $\Omega = \mathbb{R}^n \setminus K$ where K is a compact obstacle, and $a = 0$ in (4.58). The quantity of study is now the local energy of the solutions, and one restricts attention to initial data supported in a fixed compact set. Thus one is interested in the asymptotic behaviour of $\|P_1 T(t) R(\mu, A) P_2\|$ where P_1 and P_2 are multiplications by characteristic functions of balls. The semigroup property does not hold for $P_1 T(t) P_2$, but Laplace transform results are still applicable. However $P_1 R(\lambda, A) P_2$ has a singularity at the origin when n is even, so a quantified version of Theorem 4.4.8 is used (see [BD08]). The general result, originally obtained in [Bur98], is that the local energy decays at least logarithmically. For the case when the obstacle is "not trapping", earlier results had shown that the decay occurs at an exponential rate if n is odd and at a polynomial rate if n is even (see [MRS77], [MS78]).

Theorem 4.4.16 is due to Arendt and Prüss [AP92]; it is a continuous analogue of the Katznelson-Tzafriri theorem [KT86] (see Corollary 4.7.15). Most of these results were originally proved by means of harmonic analysis; the link with Korevaar's contour methods for complex Tauberian theorems was first established by Allan, Ransford and O'Farrell [AOR87] whose work inspired [AB88].

Theorem 4.4.18 and Theorem 4.4.19 were discovered by Blake [Bla99] (see also [BB00]) who developed a method used by van Neerven [Nee96b] in the case of orbits of C_0-semigroups (see the Notes on Section 5.1). These results are valid not only when f is exponentially bounded, but even when $e^{-\omega t} f(t) \in L^p(\mathbb{R}_+, X)$ for some ω and some $p > 1$. However, they are not valid in the case $p = 1$. Bloch [Blo49] gave an example where $e^{-\omega t} f(t) \in L^1(\mathbb{R}_+)$ whenever $\omega > 0$, $\operatorname{abs}(f) = 0$ and $\operatorname{hol}_0(\hat{f}) = -\infty$. His example can be adapted to show that the estimate $c(1+t)$ in the conclusion of Theorem 4.4.18 is sharp for the class of exponentially bounded functions f [Bat03].

Section 4.5

The material presented here is standard. Almost periodic functions were introduced by Harald Bohr [Boh25] in 1925, and the first edition of his book [Boh47] was published in 1934. Further textbooks are those of Levitan and Zhikov [LZ82], Fink [Fin74] and Amerio and Prouse [AP71]. Concerning the role of almost periodic functions for dynamical systems governed by partial differential equations, we refer to Haraux [Har91].

Some authors use the terminology "scalarly almost periodic" functions instead of weakly almost periodic functions in order to distinguish from the notion of "weakly almost periodic in the sense of Eberlein". We refer to Milnes [Mil80] and to the Notes of Section 4.7 for more information and comparison of these different notions.

Section 4.6
It was Loomis [Loo60] who proved that bounded uniformly continuous functions with countable spectrum are almost periodic in the scalar case. The extension to the vector-valued case when $c_0 \not\subset X$ was included in the book of Levitan and Zhikov [LZ82] after Kadets had proved his striking result (Theorem 4.6.11) in [Kad69]. For the proof of Kadets's theorem we follow [LZ82].

The condition of total ergodicity was used by Levitan [Lev66] in the context of antiderivatives of almost periodic functions. Theorem 4.6.3 in case a) is due to Ruess and Vũ [RV95], but with a different proof. The condition of total ergodicity turned out to be crucial in later developments on the half-line (see Sections 4.7, 4.9). The proof of Theorem 4.6.3 which we give here, is due to Arendt and Batty [AB97], but Lemma 4.6.6 and Lemma 4.6.7 are taken from Batty, van Neerven and Räbiger [BNR98a]. Several results presented here can be extended to measurable functions. In particular, a version of Theorem 4.6.3 remains true if $f \in L^\infty(\mathbb{R}, X)$ is slowly oscillating at infinity and a priori not continuous. Then the result says that f is equal almost everywhere to an almost periodic function (see the paper of Arendt and Batty [AB00, Corollary 3.3]).

A harmonic analytic approach to countable spectrum on the line and almost periodicity was taken by Baskakov [Bas78], [Bas85] and Basit [Bas95], [Bas97], using the Beurling spectrum instead of the Carleman spectrum (c.f. Section 4.8).

Section 4.7
The main result of this section, Theorem 4.7.7 is due to Batty, van Neerven and Räbiger [BNR98a], [BNR98b] who also proved Lemma 4.7.9. However, their proof is more complicated and based on the countable spectrum theorem (Theorem 5.5.4). The direct proof via the quotient method given here is due to Arendt and Batty [AB99] who proved Proposition 4.7.2 in particular. Example 4.7.11 was considered by Staffans [Sta81, p.608], Ruess and Vũ [RV95, Example 3.12] and Batty, van Neerven and Räbiger [BNR98b]. The proof of the Katznelson-Tzafriri theorem (Corollary 4.7.15) presented here is similar to the one given by Arendt and Prüss [AP92].

Here our emphasis is on (strong) asymptotic almost periodicity, but there are many interesting results on weak versions of these notions. A function $f \in \mathrm{BUC}(\mathbb{R}_+, X)$ is called *weakly asymptotically almost periodic* in the sense of Eberlein (in short, Eberlein-w.a.a.p.) if the set $\{S(t)f : t \geq 0\}$ is relatively weakly compact in $\mathrm{BUC}(\mathbb{R}_+, X)$. Ruess and Summers have investigated this notion in a series of articles [RS86], [RS87], [RS88a], [RS88b], [RS89], [RS90a], [RS90b], [RS92a], [RS92b]; see also papers of Ruess [Rue91], [Rue95] and Rosenblatt, Ruess and Sentilles [RRS91]. They show in a convincing way that this is the right notion for evolution equations. It is different from weak asymptotic almost periodicity (w.a.a.p.) in the sense that $x^* \circ f$ is asymptotically almost periodic for all $x^* \in X^*$. An example of an Eberlein-w.a.a.p. orbit of a bounded C_0-semigroup, which is not w.a.a.p. is given in [RS90b, p.180].

The notion of Eberlein-w.a.a.p. functions is particularly useful in the context of the mean ergodic theorem for non-linear semigroups; see [RS87], [RS88a], [RS90a] and [RS92a]. An Eberlein-w.a.a.p. function f splits, $f = f_0 + f_1$, where f_1 is almost periodic and f_0 is such that 0 is in the weak closure of $\{S(t)f_0 : t \geq 0\}$ in $\mathrm{BUC}(\mathbb{R}_+, X)$ (see

Theorem 5.4.11). This implies that $M_\eta f_0 = 0$ for all $\eta \in \mathbb{R}$, but otherwise the asymptotic behaviour as $t \to \infty$ of the function f_0 is very weak. It can still happen that for some sequence $t_n \to \infty$, $\|S(t_n)f_0 - f_0\|_\infty \to 0$ (see [RRS91, Section 3]).

Section 4.8
In this section we closely follow the book of Prüss [Prü93] where in particular Proposition 4.8.4 is proved with the help of our favourite fudge factor. A different proof is given in Davies's book [Dav80, Chapter 8]. Theorem 4.8.7 on discrete spectrum is contained in a paper of Arendt and Schweiker [AS99] with a slightly different proof. In more abstract contexts the result appeared already in work of Baskakov [Bas85], Beurling [Beu47] and Reiter [Rei52].

Section 4.9
In this section we follow closely Chill's thesis [Chi98a] (see also [Chi98b]). In particular, Theorem 4.9.7 is due to Chill with the proof which we give here. This theorem seems to be a definitive complex Tauberian theorem involving countable spectrum. It is worth mentioning that uniform ergodicity is automatic outside the weak half-line spectrum. More precisely, Chill [Chi98a, Lemma 1.16] proved the following.

Lemma 4.10.3. *Let $f \in L^\infty(\mathbb{R}_+, X)$ and let $\eta \in \mathbb{R} \setminus \mathrm{sp}_w(f)$. Then f is uniformly ergodic at η.*

The more general weak half-line spectrum seems to be more natural in the context of Volterra equations. A first investigation of asymptotic behaviour of the corresponding solution operators (which are called *resolvents* in the theory of Volterra equations) has been carried out by Arendt and Prüss [AP92] (see also [Prü93]). Theorem 4.9.7 can now be more directly applied, as shown by Chill and Prüss [CP01] and Fasangova and Prüss [FP01].

Chapter 5

Asymptotics of Solutions of Cauchy Problems

In this chapter, we give various results concerning the long-time asymptotic behaviour of mild solutions of homogeneous and inhomogeneous Cauchy problems on $\mathbb{R}_+$ (see Section 3.1 for the definitions and basic properties). For the most part, we shall assume that the homogeneous problem is well-posed, so that the operator A generates a C_0-semigroup T, mild solutions of the homogeneous problem (ACP_0) are given by $u(t) = T(t)x =: u_x(t)$ (Theorem 3.1.12), and mild solutions of the inhomogeneous problem (ACP_f) are given by $u(t) = T(t)x + (T * f)(t)$, where $T * f$ is the convolution of T and f (Proposition 3.1.16). In typical applications, the operator A and its spectral properties will be known, but solutions u will not be known explicitly, so the objective is to obtain information about the behaviour of u from the spectral properties of A. To achieve this, we shall apply the results of earlier chapters, making use of the fact that the Laplace transform of u can easily be described in terms of the resolvent of A.

In Section 5.1, we obtain general relations between spectral bounds of A, abscissas associated with the Laplace transform of T, growth bounds of T and its associated integrated semigroup, and the behaviour of convolutions $T * f$ for general f.

In Sections 5.2 and 5.3, we give more precise relations between spectral and growth bounds in the cases of semigroups on Hilbert spaces and positive semigroups on Banach lattices.

In Section 5.4, we use the general theory of ergodicity and asymptotic almost periodicity (Sections 4.3 and 4.7) to obtain splitting theorems (Glicksberg-deLeeuw theorems) for C_0-semigroups with relatively (weakly) compact orbits. In Section 5.5, we apply the complex Tauberian theorem (Theorem 4.7.7 or Theorem 4.9.7) to the case when the imaginary part of the spectrum of A is countable.

In Section 5.6, we consider the asymptotic behaviour of $T * f$, showing in particular that $T * f$ is bounded when T is a bounded holomorphic C_0-semigroup, f is bounded, and T and f are out of phase in a sense described by their Laplace transforms.

5.1 Growth Bounds and Spectral Bounds

Let T be a C_0-semigroup on X with generator A. Recall from Section 3.1 that T is exponentially bounded, and the *exponential growth bound* $\omega(T)$ is defined by:

$$\omega(T) = \inf \left\{ \omega \in \mathbb{R} : \text{there exists } M_\omega \text{ such that } \|T(t)\| \le M_\omega e^{\omega t} \text{ for all } t \ge 0 \right\}.$$

By the uniform boundedness principle applied to $\{e^{-\omega t}T(t) : t \ge 0\}$,

$$\begin{aligned} \omega(T) &= \inf \Big\{ \omega \in \mathbb{R} : \text{for each } x \in X, \text{ there exists } M_{\omega,x} \\ &\qquad\qquad \text{such that } \|T(t)x\| \le M_{\omega,x} e^{\omega t} \text{ for all } t \ge 0 \Big\} \\ &= \sup_{x \in X} \omega(u_x), \end{aligned} \tag{5.1}$$

where $u_x(t) := T(t)x$. This suggests several possible ways of defining other bounds, for example by replacing $\omega(u_x)$ by $\operatorname{hol}(\widehat{u_x})$ or $\operatorname{abs}(u_x)$, and/or taking the supremum in (5.1) not over all $x \in X$, but only over $x \in D(A)$ (i.e., considering classical solutions u_x of the homogeneous Cauchy problem rather than mild solutions). Later in this section, we shall consider such bounds, and also bounds associated with the spectrum and resolvent of the generator A, but first we establish some properties of $\omega(T)$.

The first elementary result exploits the semigroup property of T to obtain some simple properties of the growth bound.

Proposition 5.1.1. *Let T be a C_0-semigroup on X. Then*

a) $\omega(T) = \lim_{t\to\infty} t^{-1} \log \|T(t)\| = \inf_{t>0} t^{-1} \log \|T(t)\|$.

b) *The spectral radius $r(T(t))$ of $T(t)$ is $e^{t\omega(T)}$.*

c) *Let $x \in X$ and $\omega \in \mathbb{R}$, and suppose that $\int_0^\infty e^{-\omega t}\|T(t)x\|\,dt < \infty$. Then $\omega(u_x) \le \omega$.*

Proof. a) For $\omega > \omega(T)$,

$$\frac{\log \|T(t)\|}{t} \le \frac{\log M_\omega}{t} + \omega,$$

so

$$\limsup_{t\to\infty} t^{-1} \log \|T(t)\| \le \omega.$$

Hence,

$$\limsup_{t\to\infty} t^{-1}\log\|T(t)\| \le \omega(T).$$

For the reverse inequality we may assume that $\|T(t)\| > 0$ for all $t \ge 0$. For $\tau > 0$ and $n\tau \le t < (n+1)\tau$,

$$\begin{aligned} \|T(t)\| &= \|T(\tau)^n T(t-n\tau)\| \le C_\tau \|T(\tau)\|^n \\ &\le C'_\tau \|T(\tau)\|^{t/\tau} = C'_\tau \exp\left((\tau^{-1}\log\|T(\tau)\|)t\right), \end{aligned}$$

where

$$C_\tau = \sup_{0\le s\le\tau} \|T(s)\|, \qquad C'_\tau = \begin{cases} C_\tau & \text{if } \|T(\tau)\| \ge 1, \\ \frac{C_\tau}{\|T(\tau)\|} & \text{if } \|T(\tau)\| < 1. \end{cases}$$

Thus,

$$\omega(T) \le \tau^{-1}\log\|T(\tau)\|$$

for all $\tau > 0$. This proves a).

b) By the spectral radius formula and a),

$$\begin{aligned} r(T(t)) = \lim_{n\to\infty} \|T(t)^n\|^{1/n} = \lim_{n\to\infty} \|T(nt)\|^{1/n} &= \lim_{n\to\infty} \exp\left(\frac{t}{nt}\log\|T(nt)\|\right) \\ &= e^{t\omega(T)}. \end{aligned}$$

c) Take $\tau \ge 1$. For $\tau - 1 \le t \le \tau$,

$$\|u_x(\tau)\| = \|T(\tau)x\| = \|T(\tau-t)T(t)x\| \le C_1\|T(t)x\| = C_1\|u_x(t)\|.$$

Hence,

$$e^{-\omega\tau}\|u_x(\tau)\| \le C_1 \int_{\tau-1}^{\tau} e^{-\omega\tau}\|u_x(t)\|\,dt \le C_1 e^{|\omega|}\int_0^\infty e^{-\omega t}\|u_x(t)\|\,dt < \infty.$$

This shows that $\tau \mapsto e^{-\omega\tau}\|u_x(\tau)\|$ is bounded on $[1,\infty)$ and hence on $\mathbb{R}_+$, so $\omega(u_x) \le \omega$. □

It follows from Proposition 5.1.1 c) that

$$\begin{aligned} \omega(T) &= \inf\left\{\omega\in\mathbb{R} : \int_0^\infty e^{-\omega t}\|T(t)x\|\,dt < \infty \text{ for all } x \in X\right\} \qquad (5.2) \\ &= \sup\{\mathrm{abs}(\|u_x\|) : x \in X\} \\ &= \mathrm{abs}(\|T\|). \end{aligned}$$

The following result (which we call Datko's theorem, although there were other contributions; see the Notes) shows that the infimum in (5.2) is never attained. Indeed, when applied to $e^{-\omega t}T(t)$ with $p = 1$, condition b) shows that $\omega(T) < \omega$ if $\int_0^\infty e^{-\omega t}\|T(t)x\|\,dt < \infty$ for all $x \in X$.

Recall from Section 1.3 that $T * f$ denotes the convolution of T with f, so

$$(T * f)(t) := \int_0^t T(t-s)f(s)\,ds = \int_0^t T(s)f(t-s)\,ds$$

when $f \in L^1_{\mathrm{loc}}(\mathbb{R}_+, X)$.

Theorem 5.1.2 (Datko's Theorem). *Let T be a C_0-semigroup on X, and let $1 \le p < \infty$. The following are equivalent:*

(i) $\omega(T) < 0$.

(ii) *For all $x \in X$, $u_x \in L^p(\mathbb{R}_+, X)$.*

(iii) *For all $f \in L^p(\mathbb{R}_+, X)$, $T * f \in L^p(\mathbb{R}_+, X)$.*

(iv) *For all $f \in L^\infty(\mathbb{R}_+, X)$, $T * f \in L^\infty(\mathbb{R}_+, X)$.*

(v) *For all $f \in C_0(\mathbb{R}_+, X)$, $T * f \in C_0(\mathbb{R}_+, X)$.*

(vi) *For all $f \in C_0(\mathbb{R}_+, X)$,*

$$\sup_{t\ge 0} \left\| \int_0^t T(s)f(s)\,ds \right\| < \infty.$$

(vii) *For all $f \in \mathrm{AP}(\mathbb{R}_+, X)$,*

$$\sup_{t\ge 0} \left\| \int_0^t T(s)f(s)\,ds \right\| < \infty.$$

(viii) *There is a constant C such that*

$$\left\| \int_0^t T(s)f(s)\,ds \right\| \le C \sup_{0\le s\le t} \|f(s)\| \tag{5.3}$$

for all $f \in C([0,t], X)$ and all $t \ge 0$.

Proof. First, assume that (i) holds. Then there exist M and $\alpha > 0$ such that $\|T(t)\| \le Me^{-\alpha t}$ for all $t \ge 0$. Since $\|u_x(t)\| \le Me^{-\alpha t}\|x\|$, $u_x \in L^p(\mathbb{R}_+, X)$, so (ii) holds.

Proposition 1.3.5 shows that (iii), (iv) and (v) hold. The proofs of (vi), (vii) and (viii) all follow from the estimate

$$\left\| \int_0^t T(s)f(s)\,ds \right\| \le \int_0^t Me^{-\alpha s} \sup_{0\le r\le t} \|f(r)\|\,ds \le \frac{M}{\alpha} \sup_{0\le s\le t} \|f(s)\|.$$

(ii) ⇒ (i): By hypothesis, $x \mapsto u_x$ maps X into $L^p(\mathbb{R}_+, X)$, and it is easy to check that this map has closed graph. Hence, there is a constant C such that

$$\int_0^\infty \|T(t)x\|^p \, dt \le C\|x\|^p \tag{5.4}$$

for all $x \in X$.

Suppose that $\omega(T) \ge 0$. By Proposition 5.1.1 b), there exists $\lambda \in \sigma(T(1))$ with $|\lambda| = e^{\omega(T)} \ge 1$, and λ is in the topological boundary of $\sigma(T(1))$. Now λ is an approximate eigenvalue of $T(1)$ (Proposition B.2), so there is a sequence (x_k) in X such that $\|x_k\| = 1$ and $\lim_{k\to\infty} \|T(1)x_k - \lambda x_k\| = 0$. Then $\lim_{k\to\infty} \|T(1)^n x_k - \lambda^n x_k\| = 0$ for $n \in \mathbb{N}$. Passing to a subsequence of (x_k), we may assume that

$$\|x_k\| = 1, \qquad \|T(1)^n x_k - \lambda^n x_k\| \le \frac{1}{2} \quad (n = 1, 2, \dots, k).$$

Hence, $\|T(n)x_k\| \ge \frac{1}{2}$ $(n = 1, 2, \dots, k)$. If $n - 1 \le t \le n \le k$, then

$$\frac{1}{2} \le \|T(n)x_k\| = \|T(n-t)T(t)x_k\| \le C_1\|T(t)x_k\|,$$

where $C_1 = \sup_{0\le s\le 1} \|T(s)\|$. Thus, $\|T(t)x_k\| \ge 1/(2C_1)$ whenever $0 \le t \le k$, so

$$\int_0^\infty \|T(t)x_k\|^p \, dt \ge \left(\frac{1}{2C_1}\right)^p k\|x_k\|^p \quad (k = 1, 2, \dots).$$

This contradicts (5.4). It follows that $\omega(T) < 0$.

(iii) ⇒ (ii): Choose $\omega > \max(0, \omega(T))$. Take $x \in X$, and let $f(t) := e^{-\omega t}T(t)x$. Then $f \in L^p(\mathbb{R}_+, X)$, so (iii) implies that $T * f \in L^p(\mathbb{R}_+, X)$. But

$$(T * f)(t) = \int_0^t T(s)\left(e^{-\omega(t-s)}T(t-s)x\right) ds = \left(\frac{1 - e^{-\omega t}}{\omega}\right) u_x(t).$$

Thus, $\|u_x(t)\| \le \left(\frac{\omega}{1 - e^{-\omega}}\right) \|(T * f)(t)\|$ for $t \ge 1$, and u_x is bounded on $[0, 1]$, so $u_x \in L^p(\mathbb{R}_+, X)$.

(iv) or (v) ⇒ (viii): Define $V_t : C_0(\mathbb{R}_+, X) \to X$ by $V_t g := (T * g)(t)$. Either (iv) or (v) implies that $\sup_{t\ge 0} \|V_t g\| < \infty$ for each $g \in C_0(\mathbb{R}_+, X)$. By the uniform boundedness principle, there is a constant C such that

$$\left\|\int_0^t T(s)g(t-s)\, ds\right\| \le C\|g\|_\infty$$

for all $g \in C_0(\mathbb{R}_+, X)$.

Given $t \ge 0$ and $f \in C([0, t], X)$, choose $g \in C_0(\mathbb{R}_+, X)$ such that $g(s) = f(t - s)$ whenever $0 \le s \le t$ and $\|g\|_\infty = \sup_{0\le s\le t} \|f(s)\|$. Then

$$\left\|\int_0^t T(s)f(s)\, ds\right\| = \left\|\int_0^t T(s)g(t-s)\, ds\right\| \le C\|g\|_\infty = C \sup_{0\le s\le t} \|f(s)\|.$$

(vi) or (vii) $\Rightarrow$ (viii): As in the proof of (iv) or (v) $\Rightarrow$ (viii), (vi) (respectively, (vii)) implies that there is a constant C such that

$$\left\| \int_0^t T(s)g(s)\,ds \right\| \le C\|g\|_\infty$$

for all $g \in C_0(\mathbb{R}_+, X)$ (respectively, $g \in \mathrm{AP}(\mathbb{R}_+, X)$). Given $t \ge 0$ and $f \in C([0,t], X)$ there exists an extension $g \in C_0(\mathbb{R}_+, X)$ (respectively, a periodic extension g) such that $\|g\|_\infty = \sup_{0\le s\le t} \|f(s)\|$. It follows that

$$\left\| \int_0^t T(s)f(s)\,ds \right\| \le C \sup_{0\le s\le t} \|f(s)\|.$$

(viii) $\Rightarrow$ (i): Take $\omega > \max(0, \omega(T))$, so there exists M such that $\|T(s)\| \le Me^{\omega s}$ for all $s \ge 0$. For $t \ge 0$ and $x \in X$, let $f(s) := e^{\omega s}T(t-s)x$ $(0 \le s \le t)$. Then (5.3) gives

$$\left(\frac{e^{\omega t} - 1}{\omega}\right) \|T(t)x\| \le CMe^{\omega t}\|x\|.$$

Thus,

$$\|T(t)x\| \le \frac{CM\omega\|x\|}{1 - e^{-\omega}}$$

for all $t \ge 1$. It follows that $M_0 := \sup_{t\ge 0} \|T(t)\| < \infty$. Putting $f(s) := T(t-s)x$ in (5.3) gives

$$t\|T(t)x\| \le CM_0\|x\|$$

for all $t \ge 0$ and all $x \in X$. Thus, $\|T(t)\| \le CM_0/t < 1$ for sufficiently large t, so $\omega(T) = \inf_{t>0} t^{-1} \log \|T(t)\| < 0$. ☐

The following corollary of Datko's theorem can also be proved by a variation of the method of Lemma 3.2.14, without assuming that T is a semigroup.

Corollary 5.1.3. *Let T be a C_0-semigroup on X. Then there exists $x \in X$ such that $\omega(u_x) = \omega(T)$. In particular, if for each $x \in X$ there exist M_x and $\alpha_x > 0$ such that $\|T(t)x\| \le M_x e^{-\alpha_x t}$ for all $t \ge 0$, then there exist M and $\alpha > 0$ such that $\|T(t)\| \le Me^{-\alpha t}$ for all $t \ge 0$.*

Proof. The result is trivial if $\omega(T) = -\infty$, so we assume that $\omega(T) > -\infty$. Replacing $T(t)$ by $e^{-\omega(T)t}T(t)$, we may assume that $\omega(T) = 0$. Then the result follows immediately from Theorem 5.1.2, (ii) $\Rightarrow$ (i). ☐

Recall from Sections 1.4 and 1.5 that

$$\begin{aligned} \mathrm{abs}(T) &:= \sup\{\mathrm{abs}(u_x) : x \in X\} \\ &= \inf\left\{\omega \in \mathbb{R} : \text{for all } x \in X,\ \lim_{\tau\to\infty} \int_0^\tau e^{-\omega t}T(t)x\,dt \text{ exists}\right\}, \end{aligned}$$

$$\begin{aligned}
\hat{T}(\lambda)x &:= \lim_{t\to\infty}\int_0^t e^{-\lambda s}T(s)x\,ds = \widehat{u_x}(\lambda) \quad (\lambda > \operatorname{abs}(T),\ x\in X),\\
\operatorname{hol}(\hat{T}) &:= \inf\Big\{\omega\in\mathbb{R} : \hat{T} \text{ extends to a holomorphic}\\
&\qquad\qquad \text{function from } \{\operatorname{Re}\lambda > \omega\} \text{ to } \mathcal{L}(X)\Big\}\\
&= \sup\{\operatorname{hol}(\widehat{u}_x) : x\in X\}\\
&= \inf\Big\{\omega\in\mathbb{R} : \text{for all } x\in X,\ \widehat{u}_x \text{ has a holomorphic}\\
&\qquad\qquad \text{extension to } \{\lambda\in\mathbb{C} : \operatorname{Re}\lambda > \omega\}\Big\}.
\end{aligned}$$

We now define

$$\begin{aligned}
\omega_1(T) &:= \sup\{\omega(u_x) : x\in D(A)\}\\
&= \inf\Big\{\omega\in\mathbb{R} : \int_0^\infty e^{-\omega t}\|T(t)x\|\,dt < \infty \text{ for all } x\in D(A)\Big\}\\
&= \inf\Big\{\omega\in\mathbb{R} : \text{for all } x\in D(A), \text{ there exists } M_{\omega,x}\\
&\qquad\qquad \text{such that } \|T(t)x\| \le M_{\omega,x}e^{\omega t} \text{ for all } t\ge 0\Big\}\\
&= \inf\Big\{\omega\in\mathbb{R} : \text{there exists } M_\omega \text{ such that}\\
&\qquad\qquad \|T(t)R(\lambda,A)\| \le M_\omega e^{\omega t} \text{ for all } t\ge 0\Big\}.
\end{aligned}$$

Here, λ is any point in $\rho(A)$. The equalities follow from the definition of $\omega(u_x)$, Proposition 5.1.1 c) and the uniform boundedness principle.

It is clear from (1.10), (1.14) and the definitions that $\operatorname{hol}(\widehat{T}) \le \operatorname{abs}(T) \le \omega(T)$ and $\omega_1(T) \le \omega(T)$.

The *spectral bound* $s(A)$ of the generator A of T is defined by

$$s(A) := \sup\{\operatorname{Re}\lambda : \lambda\in\sigma(A)\},$$

with the convention that $s(A) = -\infty$ if $\sigma(A)$ is empty.

The following results make precise the relation between the spectrum and resolvent of A and abscissas associated with the Laplace transform of T. In particular, Proposition 5.1.4 shows that $s(A) < \infty$.

Proposition 5.1.4. *Let T be a C_0-semigroup on X with generator A. Then*

$$s(A) = \operatorname{hol}(\hat{T}). \tag{5.5}$$

Moreover, for $x\in X$,

$$\widehat{u_x}(\lambda) = R(\lambda,A)x \tag{5.6}$$

whenever $\operatorname{Re}\lambda > s(A)$,

$$R(\lambda,A)x = \lim_{\tau\to\infty}\int_0^\tau e^{-\lambda t}T(t)x\,dt \tag{5.7}$$

whenever $\operatorname{Re}\lambda > \operatorname{abs}(T)$, *and*

$$\sup\{\|R(\lambda, A)\| : \operatorname{Re}\lambda > \omega\} < \infty \tag{5.8}$$

whenever $\omega > \omega(T)$.

Proof. The functions $\hat{T}(\lambda)$ and $R(\lambda, A)$ are holomorphic for $\operatorname{Re}\lambda > \operatorname{hol}(\hat{T})$ and for $\lambda \in \rho(A)$, respectively, and they coincide for $\lambda > \omega(T)$. Thus $\hat{T}$ has a holomorphic extension for $\operatorname{Re}\lambda > \min(s(A), \operatorname{hol}(\hat{T}))$ and this implies that $\operatorname{hol}(\hat{T}) \le s(A)$ and (5.6) and (5.7) hold. The equality (5.5) now follows from Theorem 3.1.7.

For $\omega > \omega' > \omega(T)$, there exists M such that $\|T(t)\| \le Me^{\omega' t}$ for all t. For $\operatorname{Re}\lambda > \omega$,

$$\|R(\lambda, A)x\| \le \int_0^\infty e^{-\operatorname{Re}\lambda t}\|T(t)x\|\,dt \le \frac{M\|x\|}{\operatorname{Re}\lambda - \omega'},$$

so

$$\sup\{\|R(\lambda, A)\| : \operatorname{Re}\lambda > \omega\} \le \frac{M}{\omega - \omega'} < \infty. \qquad \square$$

When $\operatorname{Re}\lambda = \operatorname{abs}(T)$, the existence of the limit in (5.7) for all $x \in X$ does not correspond exactly to the existence of $R(\lambda, A)$ (consider, for example, $T(t) = I$ with $\lambda \in i\mathbb{R}$, $\lambda \ne 0$). The following result describes the relation between these two properties and stability of classical solutions of the homogeneous problem.

Proposition 5.1.5. *Let T be a C_0-semigroup on X, and let $\lambda \in \mathbb{C}$. The following are equivalent:*

(i) $\lim_{t\to\infty} \int_0^t e^{-\lambda s}T(s)x\,ds$ *exists for all* $x \in X$.

(ii) $\lambda \in \rho(A)$ *and* $\lim_{t\to\infty} \|e^{-\lambda t}T(t)x\| = 0$ *for all* $x \in D(A)$.

In that case,

$$R(\lambda, A)x = \lim_{t\to\infty} \int_0^t e^{-\lambda s}T(s)x\,ds$$

for all $x \in X$.

Proof. Replacing $T(t)$ by $e^{-\lambda t}T(t)$, we may assume that $\lambda = 0$.

(i) $\Rightarrow$ (ii): Suppose that $Bx := \lim_{t\to\infty} \int_0^t T(s)x\,ds$ exists for all $x \in X$. Then

$$\frac{1}{h}(T(h)Bx - Bx) = -\frac{1}{h}\int_0^h T(s)x\,ds \to -x$$

as $h \downarrow 0$. Thus, $Bx \in D(A)$ and $ABx = -x$ for all $x \in X$.

Now suppose that $x \in D(A)$. Then

$$BAx = \lim_{t\to\infty} \int_0^t T(s)Ax\,ds = \lim_{t\to\infty} T(t)x - x,$$

by Proposition 3.1.9. Thus, $y := \lim_{t\to\infty} T(t)x$ exists. Since $\lim_{t\to\infty} \int_0^t T(s)x\,ds$ also exists, it follows that $y = 0$ and $BAx = -x$ (for all $x \in D(A)$). Thus A has an algebraic inverse $-B$. By Proposition B.1, $0 \in \rho(A)$ and $B = -A^{-1} = R(0, A)$.

(ii) ⇒ (i): Suppose that $0 \in \rho(A)$ and $\lim_{t\to\infty} \|T(t)y\| = 0$ for all $y \in D(A)$. Then, for $x \in X$,

$$\begin{aligned} \int_0^t T(s)x\,ds &= -\int_0^t T(s)AR(0,A)x\,ds \\ &= R(0,A)x - T(t)R(0,A)x \\ &\to R(0,A)x \end{aligned}$$

as $t \to \infty$. □

The following result is the analogue of Theorem 1.4.3 for semigroups.

Proposition 5.1.6. *Let T be a C_0-semigroup on X with generator A, let S be the associated integrated semigroup:*

$$S(t)x := \int_0^t T(s)x\,ds \quad (x \in X),$$

and let

$$\widetilde{S}(t) := \begin{cases} S(t) - R(0,A) & \text{if } 0 \in \rho(A), \\ S(t) & \text{if } 0 \in \sigma(A). \end{cases}$$

Then

$$\operatorname{abs}(T) = \omega_1(T) = \omega(\widetilde{S}).$$

Proof. The fact that $\operatorname{abs}(T) = \omega(\widetilde{S})$ follows from Proposition 1.4.5, Remark 1.4.6 and Proposition 5.1.5. To prove that $\operatorname{abs}(T) = \omega_1(T)$, we may assume that $\omega(T) < 0$ (replacing $T(t)$ by $e^{-\omega t}T(t)$). Then $0 \in \rho(A)$. By Proposition 3.1.9, $S(t)x \in D(A)$ and $AS(t)x = T(t)x - x$ for all $x \in X$. Hence,

$$\widetilde{S}(t)x = S(t)x - R(0,A)x = T(t)A^{-1}x.$$

Thus,

$$\operatorname{abs}(T) = \omega(\widetilde{S}) = \sup\{\omega(u_{A^{-1}x}) : x \in X\} = \omega_1(T).$$ □

It turns out that the spectral bound $s(A)$ is of limited use in the study of asymptotic behaviour—the spectrum of an operator may be unstable under small perturbations. However, such instability can only occur when the norm of the resolvent is large, so it is more useful to consider the *pseudo-spectral bound* defined by

$$s_0(A) := \inf\Big\{\omega > s(A) : \text{there exists } C_\omega \text{ such that } \|R(\lambda,A)\| \le C_\omega \text{ whenever } \operatorname{Re}\lambda > \omega\Big\}.$$

It is clear from (5.8) that

$$s(A) \le s_0(A) \le \omega(T).$$

It follows from (5.5) and the uniform boundedness principle that

$$s_0(A) = \sup \{\mathrm{hol}_0(\widehat{u_x}) : x \in X\},$$

where $\mathrm{hol}_0(\hat{f})$ is the abscissa of boundedness of $\hat{f}$ defined in Sections 1.5 and 4.4. Theorem 4.4.19 provides the following result.

Theorem 5.1.7. *Let T be a C_0-semigroup on X with generator A. Then* $\mathrm{abs}(T) \le s_0(A)$.

Proof. By Theorem 4.4.19, $\mathrm{abs}(u_x) \le \mathrm{hol}_0(\widehat{u_x})$ for each $x \in X$. Hence, $\mathrm{abs}(T) = \sup_x \mathrm{abs}(u_x) \le \sup_x \mathrm{hol}_0(\widehat{u_x}) = s_0(A)$. □

Proceeding as in the proof of Theorem 4.4.14 a), we can deduce from Theorem 4.4.18 the following more precise information about asymptotic behaviour for individual vectors.

Theorem 5.1.8. *Let T be a C_0-semigroup on X with generator A, and let S be the associated integrated semigroup. Let $x \in X$, and suppose that* $\mathrm{hol}(\widehat{u_x}) \le 0$ *and $\widehat{u_x}$ is bounded on $\mathbb{C}_+$. Then there is a constant c (depending on x) such that*

$$\|S(t)x\| \le c(1+t)$$

for all $t \ge 0$. Moreover, for each $\mu \in \rho(A)$, there is a constant c_μ (depending on x and μ) such that

$$\|T(t)R(\mu, A)x\| \le c_\mu(1+t)$$

for all $t \ge 0$.

Proof. The first statement is immediate from Theorem 4.4.18. By Proposition 3.1.9,

$$\begin{aligned} T(t)R(\mu, A)x &= R(\mu, A)x + \int_0^t T(s)AR(\mu, A)x \, ds \\ &= R(\mu, A)x + (\mu R(\mu, A) - I)S(t)x, \end{aligned}$$

and the second statement follows. □

We now summarise the general relations between spectral bounds, abscissas and growth bounds associated with semigroups, obtained in Proposition 5.1.4, Proposition 5.1.6 and Theorem 5.1.7.

Theorem 5.1.9. *Let T be a C_0-semigroup on X with generator A. Then*

$$s(A) = \mathrm{hol}(\hat{T}) \le \omega_1(T) = \mathrm{abs}(T) \le s_0(A) \le \omega(T).$$

Now, we give two examples which show that none of the inequalities in Theorem 5.1.9 can be replaced by an equality, and we shall give a further example in Section 5.3. In Theorem 5.1.12 and the next two sections of this chapter, we shall see that further equalities are valid under various additional assumptions on X and/or T.

Example 5.1.10. *There is a C_0-semigroup T on a Hilbert space X such that $s(A) < \omega_1(T) < s_0(A) = \omega(T)$.*

Let X be the Hilbert space

$$\begin{aligned} X &:= \left\{ x = (x_n)_{n\in\mathbb{N}} : x_n \in \mathbb{C}^n, \sum_{n=1}^{\infty} \|x_n\|^2 < \infty \right\}, \\ \|x\| &:= \left(\sum_{n=1}^{\infty} \|x_n\|^2 \right)^{1/2}, \end{aligned}$$

where the norm on $\mathbb{C}^n$ is the Euclidean norm. Let $B_n := (\beta_{i,j}^{(n)})_{1\le i,j\le n}$ be the $n \times n$ complex matrix with $\beta_{i,i+1}^{(n)} = 1$ for $1 \le i < n$, $\beta_{i,j}^{(n)} = 0$ otherwise, and let $A_n := i2^n I_n + B_n$. Let A be the operator on X defined by

$$\begin{aligned} D(A) &:= \left\{ x \in X : \sum_{n=1}^{\infty} 2^{2n} \|x_n\|^2 < \infty \right\}, \\ Ax &:= (A_n x_n)_{n\in\mathbb{N}}. \end{aligned}$$

Since $\|B_n\| = 1$ and $B_n^n = 0$,

$$\left\| e^{tA_n} \right\| = \left\| e^{tB_n} \right\| \le \sum_{j=0}^{n-1} \frac{t^j}{j!} \le e^t.$$

On the other hand, if $x_n := n^{-1/2}(1,1,1,\dots,1)^T \in \mathbb{C}^n$, then $\|x_n\| = 1$ and

$$\left\| e^{tB_n} x_n - e^t x_n \right\|^2 = \frac{1}{n} \sum_{m=0}^{n-1} \left(\sum_{j=0}^{m} \frac{t^j}{j!} - e^t \right)^2 \to 0$$

as $n \to \infty$. Thus, $\sup_n \left\| e^{tA_n} \right\| = e^t$.

We may define $T(t) : X \to X$ by

$$T(t)x := \left(e^{tA_n} x_n \right)_{n\in\mathbb{N}}.$$

Then $\|T(t)\| = e^t$, and T is a C_0-semigroup with generator A and $\omega(T) = 1$.

For $x \in D(A)$,

$$\begin{aligned}
\|T(t)x\|^2 = \sum_{n=1}^{\infty} \left\|e^{tA_n}x_n\right\|^2 &\leq \sum_{n=1}^{\infty} \left(\sum_{j=0}^{n-1} \frac{t^j}{j!}\right)^2 \|x_n\|^2 \\
&\leq \sum_{n=1}^{\infty} \left(\sum_{j=0}^{n-1} \frac{t^j}{2^j j!}\right)^2 2^{2n}\|x_n\|^2 \leq e^t \sum_{n=1}^{\infty} 2^{2n}\|x_n\|^2.
\end{aligned}$$

Thus, $\omega_1(T) \leq \frac{1}{2}$.

On the other hand, if $0 < \alpha < 1/2$ and $x_n = \alpha^n(1,1,\ldots,1)^T \in \mathbb{C}^n$, then $x = (x_n) \in D(A)$ and

$$\begin{aligned}
\|T(t)x\|^2 &= \sum_{n=1}^{\infty}\sum_{m=0}^{n-1} \left(\sum_{j=0}^{m} \frac{t^j}{j!}\right)^2 \alpha^{2n} \\
&\geq \sum_{n=1}^{\infty}\sum_{r=0}^{2n-2} \left(\sum_{\substack{j+k=r \\ 0\leq j,k\leq n-1}} \frac{r!}{j!k!}\right) \frac{t^r}{r!}\alpha^{2n} \\
&\geq \sum_{n=1}^{\infty}\sum_{r=0}^{2n-2} \frac{2^r}{r+1}\frac{t^r}{r!}\alpha^{2n} \\
&= \sum_{r=0}^{\infty}\sum_{n\geq \frac{r}{2}+1} \alpha^{2n}\frac{2^r t^r}{(r+1)!} \\
&\geq \sum_{r=0}^{\infty} \frac{\alpha^{r+3}}{1-\alpha^2}\frac{2^r t^r}{(r+1)!} = \frac{\alpha^2}{1-\alpha^2}\frac{e^{2\alpha t}-1}{2t}.
\end{aligned}$$

Thus, $\omega(u_x) \geq \alpha$. It follows that $\omega_1(T) = \frac{1}{2}$.

To calculate the spectral bounds, note that $\sigma(A_n) = \{i2^n\}$ and

$$\|R(\lambda, A_n)\| = \|R(\lambda - i2^n, B_n)\| \leq \frac{1}{|\lambda - i2^n| - 1} \tag{5.9}$$

if $|\lambda - i2^n| > 1$. It follows that $\sup_n \|R(\lambda, A_n)\| < \infty$ whenever $\lambda \notin \{i2^n : n \in \mathbb{N}\}$. Hence, $\sigma(A) = \{i2^n : n = 1, 2, \ldots\}$, so $s(A) = 0$.

It also follows from (5.9) that $s_0(A) \leq 1$. On the other hand,

$$\|R(1 + i2^n, A)\| \geq \|R(1, B_n)\| = \left\|\sum_{j=0}^{n-1} B_n^j\right\| \geq n^{1/2}.$$

Hence, $s_0(A) = 1$. □

We shall see in Section 5.2 that the equality $s_0(A) = \omega(T)$ holds for all C_0-semigroups on Hilbert space.

Example 5.1.11. *There is a positive C_0-semigroup on a (reflexive) Banach lattice X such that $s(A) = \omega_1(T) = s_0(A) < \omega(T)$.*

Let $X := L^p(1,\infty) \cap L^q(1,\infty)$, where $1 \le p \le q < \infty$. Then X is a Banach lattice with the natural ordering and norm:

$$\|f\| := \max(\|f\|_p,\ \|f\|_q).$$

Let T_p be the C_0-semigroup on $L^p(1,\infty)$ defined by

$$(T_p(t)g)(s) := g(se^t),$$

and let T be the positive C_0-semigroup on X obtained by restricting T_p to X. Let A and A_p be the generators of T and T_p, respectively.

For $f \in X$,

$$\begin{aligned}
\|T(t)f\| &= \max\left\{\left(\int_1^\infty |f(se^t)|^p\,ds\right)^{1/p},\ \left(\int_1^\infty |f(se^t)|^q\,ds\right)^{1/q}\right\} \\
&= \max\left\{e^{-t/p}\left(\int_{e^t}^\infty |f(r)|^p\,dr\right)^{1/p},\ e^{-t/q}\left(\int_{e^t}^\infty |f(r)|^q\,dr\right)^{1/q}\right\} \\
&\le \max\left\{e^{-t/p}\|f\|_p,\ e^{-t/q}\|f\|_q\right\} \\
&\le e^{-t/q}\|f\|.
\end{aligned}$$

On the other hand, if

$$f(s) := \begin{cases} 1 & (e^t \le s \le e^t + 1), \\ 0 & \text{otherwise}, \end{cases}$$

then $\|f\| = 1$ and $\|T(t)f\| = e^{-t/q}$. Thus, $\|T(t)\| = e^{-t/q}$ and $\omega(T) = -1/q$.

For $\operatorname{Re}\lambda < -1/p$, let $f_\lambda(s) := s^\lambda$. Then $f_\lambda \in X$ and $T(t)f_\lambda = e^{\lambda t}f_\lambda$, so $f_\lambda \in D(A)$ and $Af_\lambda = \lambda f_\lambda$. Hence,

$$\sigma(A) \supseteq \left\{\lambda \in \mathbb{C} : \operatorname{Re}\lambda \le -\tfrac{1}{p}\right\},$$

and $s(A) \ge -1/p$.

In the case $p = q$, we now know that $s(A_p) = \omega(T_p) = -1/p$, and for $\operatorname{Re}\lambda := \alpha > -1/p$ and $f \in L^p(1,\infty)$,

$$(R(\lambda, A_p)f)(s) = \int_0^\infty e^{-\lambda t} f(se^t)\,dt = s^\lambda \int_s^\infty \frac{f(r)}{r^{\lambda+1}}\,dr.$$

For $1 < p < q < \infty$ and p' such that $\frac{1}{p} + \frac{1}{p'} = 1$, we have

$$\begin{aligned}
\int_1^\infty |(R(\lambda, A_p)f)(s)|^q \, ds &= \int_1^\infty s^{\alpha q} \left| \int_s^\infty \frac{f(r)}{r^{\lambda+1}} \, dr \right|^q \, ds \\
&\le \int_1^\infty s^{\alpha q} \|f\|_p^q \left(\int_s^\infty \frac{dr}{r^{(\alpha+1)p'}} \right)^{q/p'} \, ds \\
&= \frac{\|f\|_p^q}{((\alpha+1)p'-1)^{q/p'}} \int_1^\infty s^{-q/p} \, ds \\
&= \frac{\|f\|_p^q p}{((\alpha+1)p'-1)^{q/p'}(q-p)}.
\end{aligned}$$

If $1 = p < q < \infty$, then

$$\int_1^\infty |(R(\lambda, A_p)f)(s)|^q \, ds \le \frac{\|f\|_1^q}{q-1}.$$

In each case, $R(\lambda, A_p)$ maps $L^p(1,\infty)$ into X, so $D(A_p) \subset X$. Since

$$R(\lambda, A) = \hat{T}(\lambda) = \widehat{T_p}(\lambda)|_X = R(\lambda, A_p)|_X$$

for $\lambda > -1/q$, it follows that A is the part of A_p in X, $\sigma(A) \subset \sigma(A_p) \subset \{\operatorname{Re}\lambda \le -1/p\}$, and $R(\lambda, A) = R(\lambda, A_p)|_X$ for $\operatorname{Re}\lambda > -1/p$ (see Proposition B.8). For $f \in X$ and $\operatorname{Re}\lambda = \alpha > -1/p$,

$$\begin{aligned}
\|R(\lambda, A)f\| &\le \max\left\{ \|R(\lambda, A)f\|_p, \frac{\|f\|_p p^{1/q}}{((\alpha+1)p'-1)^{1/p'}(q-p)^{1/q}} \right\} \\
&\le \max\left\{ \frac{p}{\alpha p+1}, \frac{p^{1/q}}{((\alpha+1)p'-1)^{1/p'}(q-p)^{1/q}} \right\} \|f\|
\end{aligned}$$

if $1 < p < q < \infty$;

$$\|R(\lambda, A)f\| \le \max\left\{ \frac{1}{\alpha+1}, \frac{1}{(q-1)^{1/q}} \right\} \|f\|$$

if $1 = p < q < \infty$. Thus, $s_0(A) = -1/p$. It follows from Theorem 5.1.9 that $\omega_1(T) = -1/p$. □

We shall see in Section 5.3 that the equality $s(A) = s_0(A)$ holds for all positive semigroups on Banach lattices, while the equality $s(A) = \omega(T)$ holds for all positive semigroups on L^p-spaces. We conclude this section by showing that $s(A) = \omega(T)$ for all holomorphic semigroups.

Theorem 5.1.12. *Let T be a holomorphic C_0-semigroup on X with generator A. Then $\omega(T) = s(A)$. Moreover, there exists $\lambda \in \sigma(A)$ such that $\operatorname{Re}\lambda = \omega(T)$.*

Proof. For each $x \in X$, u_x has a holomorphic extension to a sector Σ_θ, given by $u_x(z) = T(z)x$. By Theorem 2.6.2,

$$\begin{aligned} \omega(u_x) &= \inf \left\{ \omega \in \mathbb{R} : \lambda \mapsto R(\lambda, A)x \text{ has a holomorphic} \right. \\ &\qquad\qquad \left. \text{extension to } \{\lambda : \operatorname{Re}\lambda > \omega\} \right\} \\ &\le s(A). \end{aligned}$$

Hence,

$$\omega(T) = \sup_{x \in X} \omega(u_x) \le s(A).$$

The final statement follows from the fact that $\{\lambda \in \sigma(A) : \operatorname{Re}\lambda \ge s(A) - 1\}$ is nonempty and compact (see Theorem 3.7.11 and Corollary 3.7.17). □

5.2 Semigroups on Hilbert Spaces

Example 5.1.10 shows that there are C_0-semigroups on Hilbert spaces such that $s(A) < \omega_1(T) < s_0(A)$. In that example, $s_0(A) = \omega(T)$, and we now show that this equality always holds on Hilbert spaces.

Theorem 5.2.1. *Let T be a C_0-semigroup on a Hilbert space X with generator A. Then $s_0(A) = \omega(T)$.*

Proof. Let $x \in X$. For $\omega > \omega(T)$, the function $s \mapsto R(\omega + is, A)x$ on $\mathbb{R}$ is the Fourier transform of the function $t \mapsto e^{-\omega t}T(t)x$ on $\mathbb{R}_+$. By Plancherel's Theorem 1.8.2,

$$\int_0^\infty e^{-2\omega t} \|T(t)x\|^2 \, dt = \frac{1}{2\pi} \int_{-\infty}^\infty \|R(\omega + is, A)x\|^2 \, ds. \tag{5.10}$$

Suppose that $s_0(A) < \omega(T)$ and let $C := \sup_{\operatorname{Re}\lambda > \omega(T)} \|R(\lambda, A)\| < \infty$. For $\omega(T) < \omega_1 < \omega_2$,

$$R(\omega_1 + is, A)x = R(\omega_2 + is, A)x + (\omega_2 - \omega_1)R(\omega_1 + is, A)R(\omega_2 + is, A)x,$$

so

$$\|R(\omega_1 + is, A)x\| \le (1 + C(\omega_2 - \omega_1)) \|R(\omega_2 + is, A)x\|.$$

By (5.10),

$$\int_0^\infty e^{-2\omega_1 t} \|T(t)x\|^2 \, dt \le (1 + C(\omega_2 - \omega_1))^2 \int_0^\infty e^{-2\omega_2 t} \|T(t)x\|^2 \, dt.$$

Letting $\omega_1 \downarrow \omega(T)$ gives

$$\int_0^\infty e^{-2\omega(T)t} \|T(t)x\|^2 \, dt \le (1 + C(\omega_2 - \omega(T)))^2 \int_0^\infty e^{-2\omega_2 t} \|T(t)x\|^2 \, dt < \infty$$

for all $x \in X$. The implication (ii) $\Rightarrow$ (i) of Theorem 5.1.2, with $p = 2$ and $T(t)$ replaced by $e^{-\omega(T)t}T(t)$, gives a contradiction. $\square$

Theorem 5.2.1 is not valid for $X = L^p(0,1)$ $(1 < p < \infty, p \neq 2)$.

Example 5.2.2. Let $2 < q < \infty$. Example 5.1.11 shows that there is a C_0-semigroup T_q on $L^2(1,\infty) \cap L^q(1,\infty)$ whose generator A_q satisfies $s_0(A_q) = -\frac{1}{2} < -\frac{1}{q} = \omega(T_q)$. There is a linear homeomorphism J_q of $L^q(0,1)$ onto $L^2(1,\infty) \cap L^q(1,\infty)$ [LT77, Corollary II.2.e.8]. Let $S_q(t) := J_q^{-1}T_q(t)J_q$. Then S_q is a C_0-semigroup on $L^q(0,1)$ whose generator B_q satisfies $s_0(B_q) = -\frac{1}{2} < -\frac{1}{q} = \omega(S_q)$.

Let $1 < p < 2$, and let q be the conjugate index, so that $\frac{1}{p} + \frac{1}{q} = 1$ and $L^p(0,1) = L^q(0,1)^*$. Let $S_p(t) := S_q(t)^*$. By Corollary 3.3.9, S_p is a C_0-semigroup on $L^p(0,1)$, whose generator $B_p = B_q^*$ satisfies $s_0(B_p) = -\frac{1}{2} < -\frac{1}{q} = \omega(S_p)$. $\square$

We shall see in the next section that $\omega(T) = s_0(A) = s(A)$ for all positive semigroups on L^p-spaces.

The analogue of Theorem 5.2.1 for individual orbits is not true.

Example 5.2.3. *There is a C_0-semigroup T on a Hilbert space X with a vector $x \in X$ and a real number $a < \omega(u_x)$ such that $\widehat{u_x} = R(\cdot, A)x$ has a bounded holomorphic extension to the half-plane $\{\lambda \in \mathbb{C} : \operatorname{Re}\lambda > a\}$.*

Let $X := L^2(1,\infty)$ and $(T(t)f)(s) := f(se^t)$. By Example 5.1.11, $s(A) = \omega(T) = \omega(A) = -1/2$. Let A_1 be the generator of the C_0-semigroup on $L^1(1,\infty) \cap L^2(1,\infty)$ obtained by restricting T, so $s_0(A_1) = -1$, again by Example 5.1.11. For $f \in L^1(1,\infty) \cap L^2(1,\infty)$, $R(\cdot, A_1)f$ has an extension to a bounded holomorphic map of $\{\lambda \in \mathbb{C} : \operatorname{Re}\lambda > a\}$ into $L^1(1,\infty) \cap L^2(1,\infty)$ whenever $a > -1$. But $R(\lambda, A)f = R(\lambda, A_1)f$ when $\operatorname{Re}\lambda$ is large, so $R(\cdot, A)f$ has an extension to a bounded holomorphic map of $\{\lambda \in \mathbb{C} : \operatorname{Re}\lambda > a\}$ into $L^2(1,\infty)$. However, it is possible to choose $f \in L^1(1,\infty) \cap L^2(1,\infty)$ such that $\omega(u_f) = -1/2$ (where $\omega(u_f)$ is calculated in $L^2(1,\infty)$). For example, let

$$f(s) = \begin{cases} 1 & (e^n \leq s \leq e^n + n^{-2};\ n \in \mathbb{N}), \\ 0 & \text{otherwise}. \end{cases}$$

Then $(T(n)f)(s) = 1$ for $1 \leq s \leq 1 + n^{-2}e^{-n}$, so $\|T(n)f\|_2 > n^{-1}e^{-n/2}$. Hence, $\omega(u_f) \geq -1/2$. On the other hand, $\omega(u_f) \leq \omega(T) = -1/2$. $\square$

5.3 Positive Semigroups

Let T be a C_0-semigroup on an ordered Banach space X, with generator A. We recall from Section 3.11 that T is positive if and only if A is resolvent positive.

Example 5.1.11 shows that there are positive C_0-semigroups T on spaces of the form $L^p(\Omega,\mu) \cap L^q(\Omega,\mu)$ $(1 \leq p < q < \infty)$ such that $s(A) < \omega(T)$. On the other hand, we shall show in this section that $s(A) = \operatorname{abs}(T) = s_0(A)$ for all positive semigroups on any ordered Banach space with normal cone, and that $s(A) = \omega(T)$ for all positive semigroups on $L^p(\Omega, \mu)$. Note that Proposition 3.11.2 shows that $s(A) \in \sigma(A)$ if A generates a positive semigroup and $\sigma(A)$ is non-empty. The following result makes this more precise. We give a proof that $\operatorname{abs}(T) \in \sigma(A)$ based on Theorem 1.5.3, whereas Proposition 3.11.2 was proved by means of Bernstein's Theorem 2.7.7.

Theorem 5.3.1. *Let X be an ordered Banach space with normal cone and let T be a positive C_0-semigroup on X with generator A. Then*

$$s(A) = \omega_1(T) = \operatorname{abs}(T) = s_0(A).$$

Moreover, $s(A) \in \sigma(A)$ if $s(A) > -\infty$.

Proof. By Theorem 5.1.9, $s(A) \leq \omega_1(T) = \operatorname{abs}(T) \leq s_0(A)$. It suffices to prove that $\operatorname{abs}(T) \in \sigma(A)$ if $\operatorname{abs}(T) > -\infty$, and that $\sup_{\operatorname{Re}\lambda>\omega} \|R(\lambda, A)\| < \infty$ whenever $\omega > \operatorname{abs}(T)$, so $\operatorname{abs}(T) = s_0(A)$.

Suppose that $\operatorname{abs}(T) > -\infty$ and $\operatorname{abs}(T) \in \rho(A)$. Let $\varepsilon > 0$ such that the ball $B(\operatorname{abs}(T), \varepsilon) \subset \rho(A)$. For $x \in X_+$, the function u_x is positive with Laplace transform $R(\lambda, A)x$ ($\operatorname{Re}\lambda > \omega_1(T)$), which has a holomorphic extension to $B(\operatorname{abs}(T), \varepsilon)$. By Theorem 1.5.3, $\operatorname{abs}(u_x) \leq \operatorname{abs}(T) - \varepsilon$. By linearity, $\operatorname{abs}(u_x) \leq \operatorname{abs}(T) - \varepsilon$ for all $x \in X$. But this contradicts the definition of $\operatorname{abs}(T)$. This proves that $\operatorname{abs}(T) \in \sigma(A)$ if $\operatorname{abs}(T) > -\infty$, and hence $s(A) = \operatorname{abs}(T)$. By Corollary 3.11.3, $\sup_{\operatorname{Re}\lambda>\omega} \|R(\lambda, A)\| < \infty$. □

Example 5.3.2. *There exist a Hilbert space X which is a vector lattice with continuous lattice operations, and a positive C_0-semigroup T on X such that $s(A) = \omega_1(T) < s_0(A) = \omega(T)$.*

Let X be the Sobolev space

$$H^1(1,\infty) := \{f \in L^2(1,\infty) : f' \in L^2(1,\infty)\},$$

(see Appendix E) with

$$\|f\|_{H^1(1,\infty)} := \left(\|f\|_2^2 + \|f'\|_2^2\right)^{1/2}.$$

Then X is a Hilbert space, and it is a vector lattice with the properties that $\| |f| \| = \|f\|$ [DL90, Chapter IV, Section 7, Proposition 6] and lattice operations are continuous (see [BY84, p.219]), but it is not a Banach lattice and the positive cone is not normal.

Let T be the C_0-semigroup on $H^1(1,\infty)$ given by $(T(t)f)(s) := f(se^t)$. The generator A of T is given by

$$\begin{aligned} D(A) &= \left\{f \in H^1(1,\infty) : s \mapsto sf'(s) \in H^1(1,\infty)\right\}, \\ (Af)(s) &= sf'(s). \end{aligned}$$

The semigroup governs the following very natural partial differential equation:

$$\begin{aligned} \frac{\partial u}{\partial t} &= s\frac{\partial u}{\partial s}, && (t > 0,\ s > 1), \\ u(0,s) &= u_0(s), && (s > 1), \end{aligned}$$

where $u(t,s) := (T(t)u_0)(s)$.

For $\alpha < -1/2$, the function $f_\alpha(s) := s^\alpha$ lies in X and $T(t)f_\alpha = e^{\alpha t} f_\alpha$. Hence, $s(A) \geq -1/2$. For $f \in H^1(1,\infty)$,

$$\begin{aligned} \|T(t)f\|^2_{H^1(1,\infty)} &= \int_1^\infty \left|f(se^t)\right|^2 ds + \int_1^\infty e^{2t}\left|f'(se^t)\right|^2 ds \\ &= e^{-t}\int_{e^t}^\infty |f(r)|^2\,dr + e^t\int_{e^t}^\infty \left|f'(r)\right|^2 dr \\ &\leq e^t\|f\|^2_{H^1(1,\infty)}. \end{aligned}$$

Thus, $\omega(T) \leq 1/2$. Choose non-zero $g \in C_c^\infty(\mathbb{R})$ with support in $\mathbb{R}_+$. Given $t \geq 0$, let $f(s) := g(s - e^t)$. Then $\|f\|^2_{H^1(1,\infty)} = \int_0^\infty \left(|g(s)|^2 + |g'(s)|^2\right) ds$ and $\|T(t)f\|^2_{H^1(1,\infty)} \geq e^t \int_1^\infty |g'(r)|^2\,dr$. It follows that $\omega(T) \geq 1/2$, so $s_0(A) = 1/2$ by Theorem 5.2.1. We shall show that $\operatorname{abs}(T) \leq -1/2$. It then follows from Theorem 5.1.9 that $s(A) = \omega_1(T) = -1/2$.

Let S be the corresponding C_0-semigroup on $L^2(1,\infty)$ with generator B, so $\operatorname{abs}(S) = \omega_1(S) = \omega(S) = -1/2$ (Example 5.1.11). Let $\omega > -1/2$ and $f \in H^1(1,\infty)$. For $t > 0$, let

$$\begin{aligned} g_t &:= \int_0^t e^{-\omega r} T(r)f\,dr \in H^1(1,\infty), \\ g &:= R(\omega, B)f \in L^2(1,\infty). \end{aligned}$$

Then $\lim_{t\to\infty} \|g_t - g\|_2 = 0$, since $\omega > \operatorname{abs}(S)$. We have to show that $\lim_{t\to\infty} \|g_t - g\|_{H^1(1,\infty)} = 0$.

By Proposition 3.1.9, $g_t \in D(A-\omega) = D(A) \subset D(B)$, and

$$(A-\omega)g_t = (B-\omega)g_t = e^{-\omega t}S(t)f - f.$$

But $(Ag_t)(s) = sg_t'(s)$, so

$$g_t' = h\cdot\left(\omega g_t - f + e^{-\omega t}S(t)f\right),$$

where $h(s) := s^{-1}$. Since $|h(s)| \leq 1$ for all $s \in (1,\infty)$, it is clear that

$$\lim_{t\to\infty} \|g_t' - h(\omega g - f)\|_2 = 0,$$

so $\lim_{t\to\infty} g_t$ exists in $H^1(1,\infty)$. Thus, $\operatorname{abs}(u_f) \leq -1/2$ for all $f \in H^1(1,\infty)$, so $\operatorname{abs}(T) \leq -1/2$. □

Now we consider positive semigroups on $L^p(\Omega,\mu)$, where we aim to show that $s(A)=\omega(T)$. For $p=2$, this result is immediate from Theorems 5.3.1 and 5.2.1. There is also an easy proof for $p=1$, which we present in Proposition 5.3.7. The general case needs some preliminaries. We work with the product space $\mathbb{R}\times\Omega$, and we use a vector-valued norm on $L^p(\mathbb{R}\times\Omega)$. We begin by defining this norm, and establishing its properties.

Let (Ω,μ) be a σ-finite measure space, and consider $\mathbb{R}\times\Omega$ to be equipped with the product of Lebesgue measure m on $\mathbb{R}$ and the given measure μ on Ω. Let $1\le p<\infty$. We write $L^p(\Omega)$ for $L^p(\Omega,\mu)$ and we identify $L^p(\mathbb{R}\times\Omega)$ with $L^p(\mathbb{R},L^p(\Omega))$, so that the notations $g(t,y)$ and $g(t)(y)$ $(t\in\mathbb{R}, y\in\Omega)$ are interchangeable. We consider the non-linear map $\Phi : L^p(\mathbb{R},L^p(\Omega))\to L^p(\Omega)$ given by

$$\Phi(g) := \left(\int_{-\infty}^{\infty} |g(t)|^p\,dt\right)^{1/p}.$$

This is the composition of three maps:

$$\begin{aligned}
&\Phi_1 : L^p(\mathbb{R}\times\Omega)\to L^1(\mathbb{R}\times\Omega), && \Phi_1(g):=|g|^p,\\
&\Phi_2 : L^1(\mathbb{R}\times\Omega)\to L^1(\Omega), && \Phi_2(h)(y):=\int_{-\infty}^{\infty} h(t,y)\,dt,\\
&\Phi_3 : L^1(\Omega)\to L^p(\Omega), && \Phi_3(k):=|k|^{1/p}.
\end{aligned}$$

This makes it clear that Φ is well defined.

Lemma 5.3.3. *Let $g,h\in L^p(\mathbb{R},L^p(\Omega))$, $f\in L^\infty(\Omega)$, $s\in\mathbb{R}$. Then*

a) $\|\Phi(g)\|_{L^p(\Omega)}=\|g\|_{L^p(\mathbb{R}\times\Omega)}$.

b) $\Phi(g_s)=\Phi(g)$, *where* $g_s(t)=g(t+s)$.

c) $\Phi(f\cdot g)=|f|\Phi(g)$, *where* $(f\cdot g)(t,y)=f(y)g(t,y)$.

d) $\Phi(g+h)\le\Phi(g)+\Phi(h)$ *in* $L^p(\Omega)$.

e) Φ *is continuous.*

Proof. a), b) and c) are all trivial. To prove d), let $G_y(t):=g(t,y)$, $H_y(t):=h(t,y)$ $(t\in\mathbb{R}, y\in\Omega)$. For μ-almost all y, $G_y\in L^p(\mathbb{R})$ and $H_y\in L^p(\mathbb{R})$, so Minkowski's inequality gives

$$\|G_y+H_y\|_{L^p(\mathbb{R})}\le\|G_y\|_{L^p(\mathbb{R})}+\|H_y\|_{L^p(\mathbb{R})}. \tag{5.11}$$

But

$$\|G_y\|_{L^p(\mathbb{R})}=\left(\int_{-\infty}^{\infty}|g(t,y)|^p\,dt\right)^{1/p}=\Phi(g)(y),$$

etc., so (5.11) gives

$$\Phi(g+h)(y)\le\Phi(g)(y)+\Phi(h)(y).$$

This holds μ-a.e., so $\Phi(g+h) \le \Phi(g) + \Phi(h)$ in $L^p(\Omega)$.

Now, e) follows from a) and d). By d),

$$\Phi(g) \le \Phi(g-h) + \Phi(h),$$

so

$$\Phi(g) - \Phi(h) \le \Phi(g-h).$$

Similarly,

$$\Phi(h) - \Phi(g) \le \Phi(h-g) = \Phi(g-h).$$

Since $\Phi(g)$ etc. are real-valued, this shows that

$$|\Phi(g) - \Phi(h)| \le \Phi(g-h).$$

Hence,

$$\|\Phi(g) - \Phi(h)\|_{L^p(\Omega)} \le \|\Phi(g-h)\|_{L^p(\Omega)} = \|g-h\|_{L^p(\mathbb{R},L^p(\Omega))}. \qquad \square$$

Lemma 5.3.3 shows that Φ is a convex, vector-valued function. The next lemma is a vector-valued instance of Jensen's inequality.

Lemma 5.3.4. *Let $G : [a,b] \to L^p(\mathbb{R}, L^p(\Omega))$ be continuous. Then*

$$\Phi\left(\int_a^b G(t)\,dt\right) \le \int_a^b \Phi(G(t))\,dt \quad \textit{in } L^p(\Omega).$$

Proof. By Lemma 5.3.3, c) and d),

$$\Phi\left(\frac{b-a}{2^n}\sum_{r=0}^{2^n-1} G\left(\frac{rb+(2^n-r)a}{2^n}\right)\right) \le \frac{b-a}{2^n}\sum_{r=0}^{2^n-1} \Phi\left(G\left(\frac{rb+(2^n-1)a}{2^n}\right)\right).$$

Letting $n \to \infty$ and using the continuity of Φ (Lemma 5.3.3 e)) gives the result. $\square$

For $g \in L^p(\mathbb{R}, L^p(\Omega))$ and a bounded operator T on $L^p(\Omega)$, we may define $T \circ g \in L^p(\mathbb{R}, L^p(\Omega))$ by

$$(T \circ g)(t) = T(g(t)).$$

Proposition 5.3.5. *Let T be a positive bounded linear operator on $L^p(\Omega)$, and $g \in L^p(\mathbb{R}, L^p(\Omega))$. Then*

$$\Phi(T \circ g) \le T(\Phi(g)). \tag{5.12}$$

Proof. Both sides of (5.12) depend continuously on g, so it suffices to assume that g is a simple function

$$g(t)(y) = \sum_{k=1}^{n} \chi_{A_k}(t) g_k(y),$$

where $A_1, \dots, A_n$ are disjoint Borel subsets of $\mathbb{R}$, and $g_1, \dots, g_n \in L^p(\Omega)$. Let $h_k := m(A_k)^{1/p} g_k$ ($k = 1, \dots, n$). Then

$$\Phi(T \circ g) = \left(\sum_{k=1}^n m(A_k) |Tg_k|^p \right)^{1/p} = \left(\sum_{k=1}^n |Th_k|^p \right)^{1/p},$$

$$T(\Phi(g)) = T \left(\sum_{k=1}^n m(A_k) |g_k|^p \right)^{1/p} = T \left(\sum_{k=1}^n |h_k|^p \right)^{1/p}.$$

Take $\alpha_k \in \mathbb{Q} + i\mathbb{Q}$ with $\sum_{k=1}^n |\alpha_k|^{p'} \le 1$, where p' is the conjugate index of p ($\max_k |\alpha_k| \le 1$ if $p = 1$). By Hölder's inequality,

$$\operatorname{Re} \left(\sum_{k=1}^n \alpha_k h_k \right) \le \left(\sum_{k=1}^n |h_k|^p \right)^{1/p} = \Phi(g).$$

Applying T gives

$$\operatorname{Re} \left(\sum_{k=1}^n \alpha_k T h_k \right) = T \left(\operatorname{Re} \sum_{k=1}^n \alpha_k h_k \right) \le T(\Phi(g)).$$

Now

$$\begin{aligned} &\left(\sum_{k=1}^n |(Th_k)(y)|^p \right)^{1/p} \\ = \; &\sup \left\{ \operatorname{Re} \left(\sum_{k=1}^n \alpha_k (Th_k)(y) \right) : \alpha_k \in \mathbb{Q} + i\mathbb{Q}, \sum_{k=1}^n |\alpha_k|^{p'} \le 1 \right\} \\ \le \; &T(\Phi(g))(y) \quad \mu\text{-a.e.} \end{aligned}$$

Thus, $\Phi(T \circ g) \le T(\Phi(g))$. □

Theorem 5.3.6. *Let (Ω, μ) be a σ-finite measure space, $1 \le p < \infty$, and T be a positive C_0-semigroup on $L^p(\Omega)$, with generator A. Then $s(A) = \omega(T)$.*

Proof. First, assume that $s(A) < 0$. Fix $\alpha > \max(0, \omega(T))$. Take $f \in L^p(\Omega)$ and define $g \in L^p(\mathbb{R}, L^p(\Omega))$ by

$$g(t) := \begin{cases} e^{-\alpha t} T(t) f & (t \ge 0), \\ 0 & (t < 0). \end{cases}$$

Define $G : \mathbb{R}_+ \to L^p(\mathbb{R}, L^p(\Omega))$ by

$$G(s) := T(s) \circ g_{-s},$$

where $g_{-s}(t) := g(t-s)$. Thus,

$$G(s)(t) = \begin{cases} e^{-\alpha(t-s)}T(t)f & (0 \le s \le t), \\ 0 & (-\infty < t < s). \end{cases}$$

Then

$$\begin{aligned} \Phi\left(\int_0^m G(s)\,ds\right) &= \left(\int_0^\infty \left|\int_0^{\min(m,t)} e^{-\alpha(t-s)}T(t)f\,ds\right|^p dt\right)^{1/p} \\ &= \frac{1}{\alpha}\left\{\int_0^\infty \left(e^{-\alpha\max(0,t-m)} - e^{-\alpha t}\right)^p |T(t)f|^p\,dt\right\}^{1/p}. \end{aligned}$$

Thus,

$$\begin{aligned} 0 &\le \frac{1}{\alpha}\left\{\int_0^\infty \left(e^{-\alpha\max(0,t-m)} - e^{-\alpha t}\right)^p |T(t)f|^p\,dt\right\}^{1/p} \\ &= \Phi\left(\int_0^m G(s)\,ds\right) \\ &\le \int_0^m \Phi(G(s))\,ds \\ &= \int_0^m \Phi\left(T(s)\circ g_{-s}\right)\,ds \\ &\le \int_0^m T(s)\left(\Phi(g_{-s})\right)\,ds \\ &= \int_0^m T(s)(\Phi(g))\,ds, \end{aligned}$$

where we have used Lemma 5.3.4, Proposition 5.3.5 and Lemma 5.3.3 b) in the third, fifth and sixth lines respectively. Since $\operatorname{abs}(T) = s(A) < 0$ (Theorem 5.3.1),

$$\int_0^m T(s)(\Phi(g))\,ds \to R(0,A)(\Phi(g)) \quad \text{in } L^p(\Omega),$$

as $m \to \infty$ (Proposition 5.1.4). By the monotone convergence theorem,

$$0 \le \frac{1}{\alpha}\left\{\int_0^\infty (1-e^{-\alpha t})^p\,|T(t)f|^p\,dt\right\}^{1/p} \le R(0,A)(\Phi(g)).$$

Hence,

$$\left(\frac{1-e^{-\alpha}}{\alpha}\right)\left(\int_1^\infty |T(t)f|^p\,dt\right)^{1/p} \le R(0,A)(\Phi(g)).$$

Taking norms in $L^p(\Omega)$ gives

$$\int_\Omega \int_1^\infty |(T(t)f)(y)|^p\,dt\,d\mu(y) \le \left(\frac{\alpha}{1-e^{-\alpha}}\right)^p \|R(0,A)\|^p \|\Phi(g)\|^p_{L^p(\Omega)},$$

so

$$\int_1^\infty \|T(t)f\|_{L^p(\Omega)}^p \, dt < \infty.$$

It follows from Theorem 5.1.2 (ii) $\Rightarrow$ (i), that $\omega(T) < 0$.

In the general case, we may apply the case above to the semigroup $e^{-\omega t}T(t)$ for $\omega > s(A)$, and we deduce that $\omega(T) < \omega$ whenever $\omega > s(A)$. Thus $\omega(T) \leq s(A)$. □

For $p = 1$, there is a much simpler proof of Theorem 5.3.6. A Banach lattice X is said to be an *L-space* if

$$\|x + y\| = \|x\| + \|y\| \quad \text{for all } x, y \in X_+. \tag{5.13}$$

If (Ω, μ) is any measure space, then $L^1(\Omega, \mu)$ is an L-space. If Ω is a locally compact, Hausdorff space, then $C_0(\Omega)^*$ (which can be identified with the space of all regular Borel measures on Ω) is an L-space. On the other hand, any L-space is isomorphic as a Banach lattice to a space of the form $L^1(\Omega, \mu)$ [Sch74, Theorem II.8.5].

Proposition 5.3.7. *Let T be a positive C_0-semigroup on an L-space X, with generator A. Then $s(A) = \omega(T)$.*

Proof. By (5.13), there exists $x^* \in X_+^*$ such that $\langle x, x^* \rangle = \|x\|$ for all $x \in X_+$. Let $\omega > \operatorname{abs}(T) = s(A)$ (Theorem 5.3.1). For $x \in X_+$ and $\tau \geq 0$,

$$\int_0^\tau e^{-\omega t}\|T(t)x\| \, dt = \left\langle \int_0^\tau e^{-\omega t}T(t)x \, dt, x^* \right\rangle \leq \langle R(\omega, A)x, x^* \rangle.$$

Hence,

$$\int_0^\infty e^{-\omega t}\|T(t)x\| \, dt < \infty$$

for all $x \in X_+$, and so for all $x \in X$. It follows from Datko's theorem (Theorem 5.1.2) that $\omega(T) < \omega$ whenever $\omega > s(A)$, and therefore $\omega(T) = s(A)$. □

Theorem 5.3.6 is also true in the case $p = \infty$ (even without the assumption that the semigroup is positive), but for the trivial reason that the semigroup is norm-continuous (Corollary 4.3.19). A more interesting case is that of spaces of the form $C_0(\Omega)$, and this can be deduced from Proposition 5.3.7 by duality.

Theorem 5.3.8. *Let Ω be a locally compact, Hausdorff space, and T be a positive C_0-semigroup on $C_0(\Omega)$, with generator A. Then $s(A) = \omega(T)$.*

Proof. Let $X := C_0(\Omega)^*$, which is an L-space and therefore has order-continuous norm. Let $Y := \overline{D(A^*)}$. By Theorem 3.11.8, Y is a closed ideal in X, so Y is also an L-space. Let $T(t)^\odot := T(t)^*|_Y$. By Proposition 3.3.14, $T^\odot$ is a C_0-semigroup on Y whose generator $A^\odot$ is the part of A^* in Y. Moreover, $\sigma(A^\odot) = \sigma(A)$ (see

Propositions B.8 and B.11). Let $\omega > s(A) = s(A^{\odot})$. By Proposition 5.3.7, there exists M such that $\|T^{\odot}(t)\| \le Me^{\omega t}$ for all $t \ge 0$. For $x \in X$ and $y^* \in Y$,

$$|\langle T(t)x, y^* \rangle| \le Me^{\omega t}\|x\|\,\|y^*\|.$$

For $x^* \in X^*$ and $\lambda > \omega(T)$, $R(\lambda, A)^*x^* \in D(A^*) \subset Y$. Furthermore, $x = \lim_{\lambda\to\infty} \lambda R(\lambda, A)x$ (Proposition 3.1.9), and $c := \limsup_{\lambda\to\infty} \lambda\|R(\lambda, A)\| < \infty$ since T is exponentially bounded. Hence,

$$\begin{aligned} |\langle T(t)x, x^* \rangle| &= \lim_{\lambda\to\infty} |\langle T(t)x, \lambda R(\lambda, A)^*x^* \rangle| \\ &\le \limsup_{\lambda\to\infty} Me^{\omega t}\|x\|\lambda\|R(\lambda, A)^*x^*\| \\ &\le cMe^{\omega t}\|x\|\,\|x^*\|. \end{aligned}$$

It follows that $\|T(t)\| \le cMe^{\omega t}$, so $\omega(T) < \omega$ whenever $\omega > s(A)$. □

5.4 Splitting Theorems

Let T be a bounded C_0-semigroup on X, and let $x \in X$. In this section, we shall apply the theory of asymptotic behaviour of functions on $\mathbb{R}_+$, as developed in Chapter 4, to the special case of the orbit u_x, where

$$u_x(t) := T(t)x.$$

We shall see that ergodicity and (asymptotic) almost periodicity of u_x correspond to natural semigroup properties of x, and also to compactness properties of the orbit. In particular, the main results will be two splitting theorems. We saw already in Proposition 4.3.12 that a vector x is totally ergodic with respect to T if the orbit $\{T(t)x : t \ge 0\}$ is relatively weakly compact. The weak splitting theorem shows that x can be uniquely decomposed as $x = x_0 + x_1$, where x_0 is totally ergodic with all means $M_\eta x_0 = 0$, and x_1 is in the closed linear span of the unimodular eigenvectors of T. The strong splitting theorem states that if the orbit of x is relatively compact in the norm topology, then $\lim_{t\to\infty} \|T(t)x_0\| = 0$.

As in Section 4.3, we let X_e denote the space of all vectors $x \in X$ which are totally ergodic with respect to T. Thus, $x \in X_e$ if and only if

$$M_\eta x := \lim_{t\to\infty} \frac{1}{t} \int_0^t e^{-i\eta s} T(s)x \, ds$$

exists (in the norm topology of X), for each $\eta \in \mathbb{R}$. We also let

$$\begin{aligned} X_{e0} &:= \{x \in X_e : M_\eta x = 0 \text{ for all } \eta \in \mathbb{R}\}, \\ X_0 &:= \{x \in X : \|T(t)x\| \to 0 \text{ as } t \to \infty\}. \end{aligned}$$

Since T is bounded, all these are closed T-invariant subspaces of X, and

$$X_0 \subset X_{e0} \subset X_e.$$

In Example 5.4.3, we shall exhibit these subspaces in a very fundamental example of multiplier semigroups, but we first give two very simple general results.

Proposition 5.4.1. *Let T be a bounded C_0-semigroup on X. Then $u_x \in \mathrm{BUC}(\mathbb{R}_+, X)$ for each $x \in X$. Moreover, the map $x \mapsto u_x$ is bounded and linear from X into $\mathrm{BUC}(\mathbb{R}_+, X)$.*

Proof. Uniform continuity of u_x follows from the strong continuity of T and the estimate

$$\|u_x(t+h) - u_x(t)\| = \|T(t)(T(h)x - x)\| \leq M\|T(h)x - x\|,$$

where $M = \sup_{s\geq 0} \|T(s)\|$. The other properties are immediate. □

Recall from Section 4.7 that a function $f \in \mathrm{BUC}(\mathbb{R}_+, X)$ is *totally ergodic* if it is totally ergodic with respect to the shift semigroup on $\mathrm{BUC}(\mathbb{R}_+, X)$; i.e., for each $\eta \in \mathbb{R}$,

$$(M_\eta f)(t) := \lim_{\tau\to\infty} \frac{1}{\tau} \int_0^\tau e^{-i\eta s} f(t+s)\, ds$$

exists in X, uniformly for $t \geq 0$. The space $\mathcal{E}(\mathbb{R}_+, X)$ of all totally ergodic functions is a closed subspace of $\mathrm{BUC}(\mathbb{R}_+, X)$.

Proposition 5.4.2. *Let T be a bounded C_0-semigroup on X, and let $x \in X$. Then $u_x \in \mathcal{E}(\mathbb{R}_+, X)$ if and only if $x \in X_e$. In that case,*

$$M_\eta(u_x)(t) = T(t)M_\eta x = e^{i\eta t} M_\eta x$$

for all $t \geq 0$.

Proof. Observe that

$$\frac{1}{\tau}\int_0^\tau e^{-i\eta s} u_x(t+s)\, ds = T(t)\left(\frac{1}{\tau}\int_0^\tau e^{-i\eta s} T(s)x\, ds\right).$$

It follows easily that $u_x \in \mathcal{E}(\mathbb{R}_+, X)$ if and only if $x \in X_e$, and that then $M_\eta(u_x)(t) = T(t)M_\eta x$. By (4.38), $M_\eta(u_x)(t) = e^{i\eta t} M_\eta(u_x)(0) = e^{i\eta t} M_\eta x$. □

Let T be a bounded C_0-semigroup on X with generator A. A vector $x \in X$ is a *unimodular eigenvector* of T if there exists $\eta \in \mathbb{R}$ such that $T(t)x = e^{i\eta t}x$ for all $t \geq 0$ (equivalently, $x \in D(A)$ and $Ax = i\eta x$). Let

$$X_{ap} = \overline{\mathrm{span}}\,\{\text{unimodular eigenvectors of } T\}$$

be the space of *almost periodic vectors* of T. Then X_{ap} is a closed T-invariant subspace of X. Proposition 5.4.2 shows that $M_\eta x \in X_{ap}$ whenever $x \in X_e$ and $\eta \in \mathbb{R}$. When T is the restriction of a C_0-group of isometries on X, this definition of X_{ap} is consistent with the definition given in Section 4.5.

Example 5.4.3. Let μ be a Borel measure on $\overline{\mathbb{C}}_- := \{\lambda \in \mathbb{C} : \operatorname{Re}\lambda \le 0\}$, and let $X := L^p(\mu) := L^p(\overline{\mathbb{C}}_-, \mu)$ for some $1 \le p < \infty$. Let T be the multiplier semigroup given by

$$(T(t)f)(\lambda) := e^{\lambda t} f(\lambda) \quad (f \in X, \lambda \in \overline{\mathbb{C}}_-, t \ge 0).$$

The generator A is given by

$$\begin{aligned} D(A) &= \left\{ f \in X : \int_{\overline{\mathbb{C}}_-} |\lambda f(\lambda)|^p \, d\mu(\lambda) < \infty \right\}, \\ (Af)(\lambda) &= \lambda f(\lambda). \end{aligned}$$

Furthermore,

$$\begin{aligned} \sigma(A) &= \operatorname{supp}\mu, \\ \sigma_p(A) = \sigma_p(A^*) &= \{\lambda : \lambda \text{ atom of } \mu\}. \end{aligned}$$

The measure μ can be decomposed as $\mu = \mu_- + \nu = \mu_- + \nu_a + \nu_n$, where μ_- is the restriction of μ to $\mathbb{C}_- := \{\lambda \in \mathbb{C} : \operatorname{Re}\lambda < 0\}$, ν is the restriction of μ to $i\mathbb{R}$, and ν_a and ν_n are the atomic and non-atomic parts of ν, respectively. Since μ_-, ν_a and ν_n are carried by disjoint sets, X splits in a natural way as

$$X = L^p(\mu_-) \oplus L^p(\nu) = L^p(\mu_-) \oplus L^p(\nu_a) \oplus L^p(\nu_n).$$

It is easy to verify that T is totally ergodic (this is automatic for $1 < p < \infty$, since X is then reflexive), and

$$\begin{aligned} X_e &= X, \\ X_{e0} &= L^p(\mu_-) \oplus L^p(\nu_n), \\ X_0 &= L^p(\mu_-), \\ X_{ap} &= L^p(\nu_a). \end{aligned}$$

For $f \in X$, u_f is asymptotically almost periodic if and only if $f \in L^p(\mu_-) \oplus L^p(\nu_a)$. Moreover, there is a bounded C_0-group U on $L^p(\nu)$ such that $U(t)f = T(t)f$ for all $f \in L^p(\nu)$ and all $t \ge 0$. □

Although multiplier semigroups are very special in some ways, we shall see in the results which follow that some of the features of Example 5.4.3 hold very generally.

Proposition 5.4.4. *Let T be a bounded C_0-semigroup on X. Then*

a) $X_{ap} \subset X_e$.

b) $X_{ap} = \{x \in X : u_x \in \mathrm{AP}(\mathbb{R}_+, X)\}$.

c) $X_{e0} \cap X_{ap} = \{0\}$.

d) *There is a bounded C_0-group U on X_{ap} such that $T(t)x = U(t)x$ for all $x \in X_{ap}$ and all $t \geq 0$.*

Proof. Suppose that y is a unimodular eigenvector, with $u_y(t) = T(t)y = e^{i\eta t}y$ for all $t \geq 0$. Then $u_y \in \mathrm{AP}(\mathbb{R}_+, X)$ and $y = T(t)(e^{-i\eta t}y) \in T(t)(X_{ap})$. By linearity and continuity (Proposition 5.4.1), $u_x \in \mathrm{AP}(\mathbb{R}_+, X)$ for all $x \in X_{ap}$, and $T(t)(X_{ap})$ is a dense subspace of X_{ap}.

Let $x \in X_{ap}$. Then $u_x \in \mathcal{E}(\mathbb{R}_+, X)$, so $x \in X_e$ by Proposition 5.4.2. Let $t \geq 0$. By (4.23), there exist $s_n \in \mathbb{R}_+$ such that

$$\lim_{n\to\infty} \|x - T(s_n + t)x\| = \lim_{n\to\infty} \|u_x(0) - u_x(s_n + t)\| = 0.$$

Hence,

$$\|x\| = \lim_{n\to\infty} \|T(s_n)T(t)x\| \leq M\|T(t)x\|,$$

where $M := \sup_{s\geq 0} \|T(s)\|$. Thus,

$$M^{-1}\|x\| \leq \|T(t)x\| \leq M\|x\|, \tag{5.14}$$

for all $x \in X_{ap}$ and all $t \geq 0$. This implies that $T(t)(X_{ap})$ is closed, and we saw above that it is a dense subspace of X_{ap}. It follows from this and (5.14) that $T|_{X_{ap}}$ extends to a bounded C_0-group U on X_{ap}.

If $x \in X_{e0} \cap X_{ap}$, then $M_\eta(u_x)(0) = M_\eta x = 0$ for all η, so $x = 0$ by Proposition 4.7.1 and Corollary 4.5.9.

Finally, suppose that $u_x \in \mathrm{AP}(\mathbb{R}_+, X)$ and let $\pi : X \to X/X_{ap} =: \widetilde{X}$ be the quotient map. Then $\pi \circ u_x \in \mathrm{AP}(\mathbb{R}_+, \widetilde{X})$ and for each $\eta \in \mathbb{R}$,

$$M_\eta(\pi \circ u_x)(0) = \pi(M_\eta(u_x)(0)) = \pi(M_\eta x) = 0,$$

since $M_\eta x$ is a unimodular eigenvector (Proposition 5.4.2). By Proposition 4.7.1 and Corollary 4.5.9, $\pi \circ u_x = 0$, so $u_x(t) \in X_{ap}$ for all $t \geq 0$. In particular, $x = u_x(0) \in X_{ap}$. □

In Proposition 5.4.15, we shall extend part b) of Proposition 5.4.4 to individual solutions of homogeneous Cauchy problems, and in particular to individual orbits of (unbounded) semigroups. On the other hand, the following example shows that the assumption that T is bounded is important for many aspects of Proposition 5.4.4.

Example 5.4.5. *There is an (unbounded) C_0-semigroup T on a Banach space X such that $\omega(T) = 0$, the set of vectors x such that $\lim_{t\to\infty} \|T(t)x\| = 0$ is dense in X, and the span of the unimodular eigenvectors of T is dense in X.*

Let $w : \mathbb{R}_+ \to \mathbb{R}_+$ be a continuous function with the following properties:

a) $w(0) = 1$,

b) w is strictly decreasing,

c) $\lim_{t\to\infty} w(t) = 0$,

d) $w(s+t) \geq w(s)w(t)$ $(s, t \geq 0)$.

Let

$$\begin{aligned} X &:= \left\{ f : \mathbb{R}_+ \to \mathbb{C} : f \text{ continuous}, \lim_{t\to\infty} f(t)w(t) = 0 \right\}, \\ \|f\| &:= \sup_{t\geq 0} |f(t)|w(t), \\ (T(t)f)(s) &:= f(s+t). \end{aligned}$$

Then T is a C_0-semigroup on X, with $\|T(t)\| = 1/w(t)$, and we may choose w such that $\omega(T) = 0$ (for example, $w(t) = (1+t)^{-1}$). For $\eta \in \mathbb{R}$, $e_{i\eta}$ is a unimodular eigenvalue of T, where $e_{i\eta}(t) := e^{i\eta t}$.

Suppose that $f \in X$ has compact support. Then $\lim_{t\to\infty} \|T(t)f\| = 0$. Given $\varepsilon > 0$, choose τ such that $\operatorname{supp} f \subset [0, \tau]$ and $w(\tau) < \varepsilon \left(\sup_{t\geq 0} |f(t)| + \varepsilon\right)^{-1}$. By Fejér's Theorem (Theorem 4.2.19), there is a trigonometric polynomial p of period τ such that $|f(t) - p(t)| < \varepsilon$ whenever $0 \leq t \leq \tau$. For $t > \tau$, $f(t) = 0$ and there exists $t' \in [0, \tau]$ such that $p(t) = p(t')$, so

$$|f(t) - p(t)|w(t) = |p(t')|w(t) \leq (|f(t')| + \varepsilon)w(\tau) < \varepsilon.$$

Thus, $\|f - p\| < \varepsilon$. Since the functions of compact support are dense in X, it follows that the unimodular eigenvectors span a dense subspace of X, and also that the vectors f with $\lim_{t\to\infty} \|T(t)f\| = 0$ are dense. □

The following theorem is one of the main results of this section.

Theorem 5.4.6 (Strong Splitting Theorem). *Let T be a bounded C_0-semigroup on X, and let $x \in X$. The following are equivalent:*

(i) $x \in X_0 \oplus X_{ap}$.

(ii) u_x *is asymptotically almost periodic.*

(iii) $\{T(t)x : t \geq 0\}$ *is relatively compact.*

Moreover, if $x = x_0 + x_1$ where $x_0 \in X_0$ and $x_1 \in X_{ap}$, then $\|x_1\| \leq M\|x\|$, where $M = \sup_{t\geq 0} \|T(t)\|$.

Proof. The implication (i) ⇒ (ii) follows immediately from the fact that $\mathrm{AAP}(\mathbb{R}_+, X) = C_0(\mathbb{R}_+, X) \oplus \mathrm{AP}(\mathbb{R}_+, X)$, together with Proposition 5.4.4 b).

Suppose that u_x is asymptotically almost periodic. Then $u_x = f + g$ for some $f \in C_0(\mathbb{R}_+, X)$ and $g \in \mathrm{AP}(\mathbb{R}_+, X)$. By (4.37), there is a sequence (t_n) in $\mathbb{R}_+$ such that

$$\lim_{n\to\infty} \|u_x(t_n + s) - g(s)\| = 0,$$

uniformly for $s \geq 0$. Let $x_1 := g(0) = \lim_{n\to\infty} u_x(t_n)$. For $s \geq 0$,

$$g(s) = \lim_{n\to\infty} u_x(t_n + s) = \lim_{n\to\infty} T(s)u_x(t_n) = T(s)x_1.$$

Thus, $g = u_{x_1}$. By Proposition 5.4.4 b), $x_1 \in X_{ap}$. Moreover, $u_{x-x_1} = f \in C_0(\mathbb{R}_+, X)$, so $x - x_1 \in X_0$. By Proposition 5.4.4 c), x_1 is unique. Moreover, $\|x_1\| = \lim_{n\to\infty} \|T(t_n)x\| \leq M\|x\|$. This proves the implication (ii) $\Rightarrow$ (i) and the final statement of the theorem.

The implication (ii) $\Rightarrow$ (iii) follows from the fact that any asymptotically almost periodic function has relatively compact range (see Theorem 4.7.4).

Now suppose that $\{T(t)x : t \geq 0\}$ is relatively compact. Let (t_n) be a sequence in $\mathbb{R}_+$. By assumption, there is a subsequence (t_{n_k}) such that $T(t_{n_k})x$ converges to a limit x_1 in X. Then $(S(t_{n_k})u_x)(t) = T(t + t_{n_k})x \to T(t)x_1$ uniformly for $t \geq 0$. Thus, $\{S(t)u_x : t \geq 0\}$ is relatively compact in $\mathrm{BUC}(\mathbb{R}_+, X)$, so Theorem 4.7.4 shows that u_x is asymptotically almost periodic. □

A bounded C_0-semigroup T on X is said to be *asymptotically almost periodic* if the equivalent conditions of Theorem 5.4.6 are satisfied for every $x \in X$. In other words, T is asymptotically almost periodic if and only if $X = X_0 \oplus X_{ap}$ (as a topological direct sum). In the literature, such semigroups are often called "almost periodic", but we will not use this loose terminology.

We shall see in Section 5.5 that a totally ergodic semigroup with generator A is asymptotically almost periodic if $\sigma(A) \cap i\mathbb{R}$ is countable. We can see this immediately in the special case when A has compact resolvent, i.e., $R(\lambda, A)$ is compact for $\lambda \in \rho(A)$ (see Appendix B).

Proposition 5.4.7. *Let T be a bounded C_0-semigroup on X such that the generator A of T has compact resolvent. Then $X = X_0 \oplus X_{ap}$.*

Proof. Choose $\lambda \in \rho(A)$. For $x \in D(A)$,

$$\{T(t)x : t \geq 0\} = \{R(\lambda, A)T(t)(\lambda I - A)x : t \geq 0\},$$

which is relatively compact. It follows by density of $D(A)$ that $\{T(t)x : t \geq 0\}$ is relatively compact for all $x \in X$, so the result follows from Theorem 5.4.6. □

Now we turn towards the weak splitting theorem. First, we require some preliminary results of a very classical nature. Recall from Section 4.5 that a complex-valued *trigonometric polynomial* is a function $p : \mathbb{R} \to \mathbb{C}$ of the form

$$p = \sum_{j=1}^{m} \lambda_j e_{i\eta_j}$$

for some $m \in \mathbb{N}$, $\lambda_j \in \mathbb{C}$ and distinct $\eta_j \in \mathbb{R}$, where $e_{i\eta}(t) = e^{i\eta t}$. Then p is totally ergodic with means

$$M_\eta(p)(0) = \begin{cases} \lambda_j & \text{if } \eta = \eta_j \text{ for some } j, \\ 0 & \text{otherwise.} \end{cases}$$

We shall write $\mu_\eta(p)$ for $M_\eta(p)(0)$.

Proposition 5.4.8. *Let $\eta_1, \eta_2, \dots, \eta_k \in \mathbb{R}$, and $\varepsilon > 0$. There is a trigonometric polynomial $p : \mathbb{R} \to \mathbb{C}$ such that*

a) $p(t) \geq 0$ *for all* t,

b) $\mu_0(p) = 1$,

c) $1 \geq \mu_{\eta_j}(p) > 1 - \varepsilon$ *for* $j = 1, 2, \dots, k$.

Proof. Let $\{\eta_1', \eta_2', \dots, \eta_m'\}$ be a basis over $\mathbb{Q}$ of the $\mathbb{Q}$-linear span of $\{\eta_1, \eta_2, \dots, \eta_k\}$, so that $\eta_j = \sum_{r=1}^m \beta_{jr}\eta_r'$ for some unique $\beta_{jr} \in \mathbb{Q}$. Replacing η_r' by η_r'/n_r and β_{jr} by $\beta_{jr}n_r$ where n_r is a common multiple of the denominators of $\beta_{1r}, \beta_{2r}, \dots, \beta_{mr}$, we may assume that $\beta_{jr} \in \mathbb{Z}$.

For $j = 1, 2, \dots, m$ and $N \in \mathbb{N}$, let F_{jN} be the Fejér kernel corresponding to the frequency η_j' (see the proof of Theorem 4.2.19):

$$F_{jN}(t) := \sum_{n=-N}^{N} \left(1 - \frac{|n|}{N}\right) e^{in\eta_j' t} = \frac{1}{N} \left(\frac{\sin \frac{N\eta_j' t}{2}}{\sin \frac{\eta_j' t}{2}} \right)^2 .$$

Let

$$\begin{aligned} p_N(t) &:= \prod_{j=1}^m F_{jN}(t) \\ &= \sum_{1 \leq n_1, \dots, n_m \leq N} \left(1 - \frac{|n_1|}{N}\right) \dots \left(1 - \frac{|n_m|}{N}\right) \exp\left(i(n_1\eta_1' + \dots + n_m\eta_m')t\right). \end{aligned}$$

Then $p_N(t) \geq 0$ and the $\mathbb{Q}$-independence of $\{\eta_1', \dots, \eta_m'\}$ implies that

$$\begin{aligned} \mu_0(p_N) &= 1, \\ \mu_{\eta_j}(p_N) &= \left(1 - \frac{|\beta_{j1}|}{N}\right) \dots \left(1 - \frac{|\beta_{jm}|}{N}\right) \end{aligned}$$

for $j = 1, 2, \dots, k$. The result follows by choosing N sufficiently large. $\square$

Corollary 5.4.9. *There is a net $(p_\alpha)_{\alpha \in \Lambda}$ of complex-valued trigonometric polynomials such that*

a) $p_\alpha(t) \geq 0$ *for all* $t \in \mathbb{R}$ *and all* $\alpha \in \Lambda$;

b) $\mu_0(p_\alpha) = 1$ *for all* $\alpha \in \Lambda$; *and*

c) $\lim_\alpha \mu_\eta(p_\alpha) = 1$ *for all* $\eta \in \mathbb{R}$.

Proof. Let $\mathcal{P}$ be the set of all trigonometric polynomials p such that $p(t) \geq 0$ for all t and $\mu_0(p) = 1$. For $p \in \mathcal{P}$, define $\nu_p : \mathbb{R} \to \mathbb{C}$ by

$$\nu_p(\eta) := \mu_\eta(p).$$

By Proposition 5.4.8, the constant function 1 is in the closure of $\{\nu_p : p \in \mathcal{P}\}$ in $\mathbb{C}^{\mathbb{R}}$ for the topology of pointwise convergence, and the result follows. □

Proposition 5.4.10. *Let T be a bounded C_0-semigroup on X, and let $x \in X_e$. Then there is a net $(x_\alpha)_{\alpha\in\Lambda}$ in X such that*

a) *For each α, $x_\alpha \in X_{ap}$;*

b) *For each α, x_α is in the closed convex hull of $\{T(t)x : t \geq 0\}$; and*

c) $\lim_\alpha M_\eta(x_\alpha) = M_\eta(x)$ *for all $\eta \in \mathbb{R}$.*

Proof. Given a complex-valued trigonometric polynomial $p(t) = \sum_\eta \mu_\eta(p)e^{i\eta t}$ (where $\mu_\eta(p) = 0$ for all except finitely many η), let

$$\begin{aligned} p \cdot x &:= \sum_\eta \mu_{-\eta}(p) M_\eta(x) \\ &= \lim_{t\to\infty} \frac{1}{t}\int_0^t \sum_\eta \mu_{-\eta}(p) e^{-i\eta s} T(s)x\,ds \\ &= \lim_{t\to\infty} \frac{1}{t}\int_0^t p(s)T(s)x\,ds. \end{aligned}$$

Since $M_\eta(x)$ is a unimodular eigenvector of T, $p \cdot x \in X_{ap}$ and $M_\eta(p \cdot x) = \mu_{-\eta}(p)M_\eta(x)$.

Suppose that $p(t) \geq 0$ for all t and $\mu_0(p) = 1$. Then

$$p \cdot x = \lim_{t\to\infty} \frac{\frac{1}{t}\int_0^t p(s)T(s)x\,ds}{\frac{1}{t}\int_0^t p(s)\,ds}.$$

For each $t > 0$,

$$\frac{\frac{1}{t}\int_0^t p(s)T(s)x\,ds}{\frac{1}{t}\int_0^t p(s)\,ds}$$

is the mean value of u_x with respect to a probability measure on $[0, t]$, and therefore it is in the closed convex hull of the orbit of x. It follows that $p \cdot x$ also lies in this closed set.

The result now follows by taking $(p_\alpha)_{\alpha\in\Lambda}$ as in Corollary 5.4.9 and putting $x_\alpha := p_\alpha \cdot x$. □

Now we can give the second main theorem of this section, which is the analogue of Theorem 5.4.6 for the weak topology.

Theorem 5.4.11 (Weak Splitting Theorem). *Let T be a bounded C_0-semigroup on X, let $x \in X$, and suppose that $\{T(t)x : t \geq 0\}$ is relatively weakly compact. Then $x = x_0 + x_1$ for some unique $x_0 \in X_{e0}$ and $x_1 \in X_{ap}$. Moreover, $\|x_1\| \leq M\|x\|$, where $M := \sup_{t\geq 0} \|T(t)\|$.*

Proof. The uniqueness follows from Proposition 5.4.4. For existence, we may replace X by the closed linear span of the orbit of x, so we may assume that every vector in X is totally ergodic with respect to T. Let $(x_\alpha)_{\alpha\in\Lambda}$ be as in Proposition 5.4.10. The property b) of Proposition 5.4.10 shows that $\|x_\alpha\| \leq M\|x\|$ and $\{x_\alpha : \alpha \in \Lambda\}$ is relatively weakly compact, by Krein's theorem [Meg98, Theorem 2.8.14]. Hence, there is a subnet which is weakly convergent to a limit $x_1 \in X$ with $\|x_1\| \leq M\|x\|$. Since $\{x_\alpha\} \subset X_{ap}$ and X_{ap} is norm closed and hence weakly closed, $x_1 \in X_{ap}$. Let $x_0 := x - x_1$. Since $\lim_\alpha M_\eta(x - x_\alpha) = 0$ and M_η is bounded, hence weakly continuous on X, $M_\eta(x_0) = 0$. □

Corollary 5.4.12. *Let T be a bounded C_0-semigroup on a reflexive space X. Then $X = X_{e0} \oplus X_{ap}$ as a topological direct sum.*

Comparison of Theorems 5.4.11 and 5.4.6 provides the following corollary.

Corollary 5.4.13. *Let T be a bounded C_0-semigroup on X, let $x \in X$, and suppose that x is totally ergodic with respect to T with all means $M_\eta x = 0$ and that the orbit $\{T(t)x : t \geq 0\}$ is relatively compact. Then $\lim_{t\to\infty} \|T(t)x\| = 0$.*

Corollary 5.4.14. *Let T be a bounded C_0-semigroup on X, let $x \in X$, and suppose that x is totally ergodic with respect to T and that the orbit $\{T(t)x : t \geq 0\}$ is relatively compact. If $M_\eta x = 0$ for all $\eta \in \mathbb{R} \setminus \{0\}$, then $\lim_{t\to\infty} T(t)x$ exists.*

Proof. This follows by applying Corollary 5.4.13 with x replaced by $x - M_0 x$. □

Next, we consider mild solutions of Cauchy problems, without assuming the existence of a semigroup. For some purposes, it is not natural to specify an initial value, and we therefore extend our previous terminology by saying that $u \in C(\mathbb{R}_+, X)$ is a *mild solution* of the abstract Cauchy problem

$$(ACP_f) \qquad u'(t) = Au(t) + f(t) \quad (t \geq 0),$$

where $f \in L^1_{loc}(\mathbb{R}_+, X)$, if $\int_0^t u(r)\,dr \in D(A)$ and $u(t) = u(0) + A\int_0^t u(r)\,dr + \int_0^t f(r)\,dr$ for all $t \geq 0$. It is easy to see that this implies that $u(t) = u(s) + A\int_s^t u(r)\,dr + \int_s^t f(r)\,dr$ for $t \geq s \geq 0$.

We begin by describing the almost periodic solutions of the homogeneous Cauchy problem, thereby extending Proposition 5.4.4 c).

Proposition 5.4.15. *Let A be a closed linear operator on X, and let $u \in \mathrm{AP}(\mathbb{R}_+, X)$ be a mild solution of (ACP_0). For each $\varepsilon > 0$, there exist $n \in \mathbb{N}$, $x_1, \ldots, x_n \in D(A)$, and $\eta_1, \ldots, \eta_n \in \mathbb{R}$ such that $Ax_j = i\eta_j x_j$ and*

$$\left\| u(t) - \sum_{j=1}^{n} e^{i\eta_j t} x_j \right\| < \varepsilon \tag{5.15}$$

for all $t \geq 0$.

Proof. Let Z be the set of all $u \in \mathrm{BUC}(\mathbb{R}_+, X)$ which are mild solutions of (ACP_0). Let (u_n) be a sequence in Z, $u \in \mathrm{BUC}(\mathbb{R}_+, X)$ and suppose that $\|u_n - u\|_\infty \to 0$. For $t \geq 0$, $\int_0^t u_n(s)\,ds \in D(A)$ and

$$u_n(t) = u_n(0) + A \int_0^t u_n(s)\,ds.$$

Since A is closed, it follows on letting $n \to \infty$ that

$$u(t) = u(0) + A \int_0^t u(s)\,ds.$$

Thus, Z is a closed subspace of $\mathrm{BUC}(\mathbb{R}_+, X)$. Moreover, if S is the shift semigroup on $\mathrm{BUC}(\mathbb{R}_+, X)$ and $u \in Z$, then $S(t)u \in Z$ for all $t \geq 0$.

Now, suppose that $u \in Z \cap \mathrm{AP}(\mathbb{R}_+, X)$. By Proposition 4.7.1, $u = g|_{\mathbb{R}_+}$ for some $g \in \mathrm{AP}(\mathbb{R}, X)$. Let

$$Y := \overline{\operatorname{span}}\{S_{\mathbb{R}}(t)g : t \in \mathbb{R}\} \subset \mathrm{BUC}(\mathbb{R}, X),$$

where $S_{\mathbb{R}}$ is the shift group on $\mathrm{BUC}(\mathbb{R}, X)$. Applying Theorem 4.5.1 to the restriction of $S_{\mathbb{R}}$ to Y shows that

$$g \in \overline{\operatorname{span}}\,\{e_{i\eta} \otimes x : \eta \in \mathbb{R}, x \in X, e_{i\eta} \otimes x \in Y\}. \tag{5.16}$$

Let $t \in \mathbb{R}$. By (4.23) there exist $\tau_n \to \infty$ such that

$$\lim_{n\to\infty} \|S_{\mathbb{R}}(t + \tau_n)g - S_{\mathbb{R}}(t)g\|_\infty = 0. \tag{5.17}$$

For $t + \tau_n > 0$, $(S_{\mathbb{R}}(t + \tau_n)g)|_{\mathbb{R}_+} = S(t + \tau_n)u$, so it follows from (5.17) and the first paragraph that $(S_{\mathbb{R}}(t)g)|_{\mathbb{R}_+} \in Z$ for every $t \in \mathbb{R}$, and hence that $h|_{\mathbb{R}_+} \in Z$ for every $h \in Y$.

Now suppose that $h := e_{i\eta} \otimes x \in Y$. Take $t > 0$ and let

$$\zeta := \begin{cases} \dfrac{e^{i\eta t} - 1}{i\eta} & \text{if } \eta \neq 0, \\ t & \text{if } \eta = 0. \end{cases}$$

We can choose t in such a way that $\zeta \neq 0$. Then

$$x = \frac{1}{\zeta} \int_0^t h(s)\,ds \in D(A),$$

and

$$e^{i\eta t}x = h(t) = h(0) + A \int_0^t h(s)\,ds = x + \zeta Ax.$$

Hence, $Ax = i\eta x$. The result follows from this and (5.16). □

Now we show that the splitting of $\mathrm{AAP}(\mathbb{R}_+, X)$ as $C_0(\mathbb{R}_+, X) \oplus \mathrm{AP}(\mathbb{R}_+, X)$ respects mild solutions of inhomogeneous Cauchy problems.

Proposition 5.4.16. *Let A be a closed linear operator on X, let $u_0 \in C_0(\mathbb{R}_+, X)$, $u_1 \in \mathrm{AP}(\mathbb{R}_+, X)$, $f_0 \in C_0(\mathbb{R}_+, X)$, $f_1 \in \mathrm{AP}(\mathbb{R}_+, X)$, $u = u_0 + u_1$, and $f = f_0 + f_1$. Suppose that u is a mild solution of (ACP$_f$). Then u_0 and u_1 are mild solutions of (ACP$_{f_0}$) and (ACP$_{f_1}$), respectively.*

Proof. It is easy to verify that the function $t \mapsto (u_1(t), f_1(t))$ is almost periodic (with values in $X \times X$), and hence that $t \mapsto (u(t), f(t))$ is asymptotically almost periodic. By (4.37), there is a sequence $(\tau_n)_{n\geq 1}$ in $\mathbb{R}_+$ such that $\tau_n \to \infty$ and

$$\sup_{t\geq 0} \|u(t+\tau_n) - u_1(t)\| \to 0, \qquad \sup_{t\geq 0} \|f(t+\tau_n) - f_1(t)\| \to 0$$

as $n \to \infty$. Since u is a mild solution of (ACP$_f$),

$$u(t+\tau_n) = A\left(\int_0^t u(s+\tau_n)\,ds\right) + \int_0^t f(s+\tau_n)\,ds + u(\tau_n).$$

Since A is closed, it follows on letting $n \to \infty$ that

$$u_1(t) = A\left(\int_0^t u_1(s)\,ds\right) + \int_0^t f_1(s)\,ds + u_1(0),$$

so u_1 is a mild solution of (ACP$_{f_1}$). By linearity, $u_0 = u - u_1$ is a mild solution of (ACP$_{f_0}$). □

The following corollary is a generalization of part of the strong splitting theorem (Theorem 5.4.6, (i) ⇔ (ii)).

Corollary 5.4.17. *Let T be a C_0-semigroup on X, let $x \in X$, and suppose that u_x is asymptotically almost periodic. Then there exist unique x_0 and x_1 in X such that*

a) $x = x_0 + x_1$;

b) $\lim_{t\to\infty} \|T(t)x_0\| = 0$; *and*

c) *There is a sequence (y_n) in the linear span of the unimodular eigenvectors of T such that $\lim_{n\to\infty} T(t)y_n = T(t)x_1$ uniformly for $t \geq 0$.*

Conversely, if a), b) *and* c) *hold, then u_x is asymptotically almost periodic.*

Proof. Let $u_x = v_0 + v_1$, where $v_0 \in C_0(\mathbb{R}_+, X)$ and $v_1 \in \mathrm{AP}(\mathbb{R}_+, X)$. By Proposition 5.4.16 (with A equal to the generator of T and $f_0 = f_1 = 0$), $v_0 = u_{x_0}$ and $v_1 = u_{x_1}$ for some $x_0, x_1 \in X$. Now, a) and b) are immediate, c) follows from Proposition 5.4.15, and the uniqueness follows from (4.36). The converse statement is immediate, as in Theorem 5.4.6. □

5.5 Countable Spectral Conditions

In this section, we shall give various results showing that solutions of well-posed homogeneous Cauchy problems are asymptotically almost periodic under assumptions including boundedness of the solution and countability of the purely imaginary part of the (local) spectrum of A, using the results and methods of Chapter 4. In Section 5.6, we shall extend some of the results to inhomogeneous Cauchy problems which are not necessarily associated with C_0-semigroups.

Suppose that A generates a C_0-semigroup T on a complex Banach space X. For $x \in X$, we again put $u_x(t) := T(t)x$ $(t \geq 0)$. Then u_x is an exponentially bounded function and its Laplace transform $\widehat{u_x}(\lambda)$ coincides with $R(\lambda, A)x$ for large real λ. We shall assume that u_x is bounded, and our first step is to relate the half-line spectrum $\operatorname{sp}(u_x)$ of u_x in the sense of Section 4.7 to the operator A.

The *imaginary local resolvent set* $\rho_u(A, x)$ of A at x is defined to be the set of all points $i\eta \in i\mathbb{R}$ such that there exist an open set U containing $\mathbb{C}_+ \cup \{i\eta\}$ and a holomorphic function $g : U \to X$ such that $g(\lambda) \in D(A)$ and $(\lambda - A)g(\lambda) = x$ for all $\lambda \in \mathbb{C}_+$. We shall see in Proposition 5.5.1 that $g(\lambda) = R(\lambda, A)x$ whenever $\lambda \in \mathbb{C}_+ \cap \rho(A)$, so g is uniquely determined if U is connected.

The *imaginary local spectrum* $\sigma_u(A, x)$ of A at x is:

$$\sigma_u(A, x) := i\mathbb{R} \setminus \rho_u(A, x).$$

It is clear from the definition that $\rho_u(A, x)$ is open in $i\mathbb{R}$, so $\sigma_u(A, x)$ is closed. When A generates a bounded C_0-group, Lemma 4.6.7 shows that $\sigma_u(A, x)$ coincides with the local spectrum $\sigma(A, x)$ defined in Section 4.6.

The following is a local version of Proposition B.5.

Proposition 5.5.1. *Let T be a C_0-semigroup on X with generator A, and let $x \in X$.*

a) *Let V be a connected open subset of $\mathbb{C}$, let $g : V \to X$ be holomorphic, and suppose that there is a subset U of V, with a limit point in V, such that $g(\lambda) \in D(A)$ and $(\lambda - A)g(\lambda) = x$ whenever $\lambda \in U$. Then $g(\lambda) \in D(A)$ and $(\lambda - A)g(\lambda) = x$ whenever $\lambda \in V$.*

b) *If $\operatorname{Re}\lambda > \operatorname{hol}(\widehat{u_x})$, then $\widehat{u_x}(\lambda) \in D(A)$ and $(\lambda - A)\widehat{u_x}(\lambda) = x$.*

c) *If $\operatorname{Re}\lambda > \operatorname{hol}(\widehat{u_x})$ and $\lambda \in \rho(A)$, then $\widehat{u_x}(\lambda) = R(\lambda, A)x$.*

Proof. a) Fix $\mu \in \rho(A)$. For $\lambda \in U$,

$$(\lambda - A)R(\mu, A)g(\lambda) = R(\mu, A)x.$$

Since $AR(\mu, A) = \mu R(\mu, A) - I$ is a bounded operator on X, the left-hand side is holomorphic on V. By uniqueness of holomorphic extensions, this formula is true whenever $\lambda \in V$. Hence,

$$g(\lambda) = R(\mu, A)x - (\lambda - \mu)R(\mu, A)g(\lambda) \in D(A),$$

and

$$(\lambda - A)g(\lambda) = x,$$

since $R(\mu, A)$ is injective.

Statement b) follows by applying a) with $g := \widehat{u_x}$, and c) is then immediate. □

The following result identifies $\sigma_u(A, x)$ with $\mathrm{sp}(u_x)$ when u_x is bounded.

Proposition 5.5.2. *Let T be a C_0-semigroup on X with generator A. Let $x \in X$, and suppose that u_x is bounded. Then*

a) $\sigma_u(A, x) \subset \sigma(A) \cap i\mathbb{R}$.

b) $\mathrm{sp}(u_x) = \{\eta \in \mathbb{R} : i\eta \in \sigma_u(A, x)\}$.

Proof. a) By Proposition 5.5.1 c), $\widehat{u_x}(\lambda) = R(\lambda, A)x$ whenever $\lambda \in \mathbb{C}_+ \cap \rho(A)$. Thus, we may define a holomorphic function $g : \mathbb{C}_+ \cup \rho(A) \to X$ by

$$g(\lambda) := \begin{cases} \widehat{u_x}(\lambda) & (\lambda \in \mathbb{C}_+), \\ R(\lambda, A)x & (\lambda \in \rho(A)). \end{cases}$$

This shows that $\sigma_u(A, x) \subset \sigma(A) \cap i\mathbb{R}$.

b) Suppose that $i\eta \notin \sigma_u(A, x)$, so there exist an open set U containing $\mathbb{C}_+ \cup \{i\eta\}$ and a holomorphic function $g : U \to X$ such that $g(\lambda) \in D(A)$ and $(\lambda - A)g(\lambda) = x$ for all $\lambda \in \mathbb{C}_+$. Then $g(\lambda) = R(\lambda, A)x = \widehat{u_x}(\lambda)$ whenever $\lambda \in \rho(A) \cap \mathbb{C}_+$. By uniqueness of holomorphic extensions, $g(\lambda) = \widehat{u_x}(\lambda)$ whenever $\lambda \in \mathbb{C}_+$, so $\eta \notin \mathrm{sp}(u_x)$.

Conversely, suppose that $\eta \notin \mathrm{sp}(u_x)$, so there exists a connected open set V containing $i\eta$ and a holomorphic function $g : V \to X$ such that $g(\lambda) = \widehat{u_x}(\lambda)$ for all $\lambda \in V \cap \mathbb{C}_+$. By Proposition 5.5.1 b), $g(\lambda) \in D(A)$ and $(\lambda - A)g(\lambda) = x$ for all $x \in V \cap \mathbb{C}_+$. By Proposition 5.5.1 a), this holds for all $\lambda \in V$. Thus, $i\eta \notin \sigma_u(A, x)$. □

The next four results appear in decreasing order of generality, with the assumptions becoming stronger but less technical.

Theorem 5.5.3. *Let T be a C_0-semigroup on X with generator A. Let $x \in X$, and suppose that the following conditions are satisfied:*

a) *$u_x : t \mapsto T(t)x$ is bounded and uniformly continuous;*

b) *$\sigma_u(A, x)$ is countable; and*

c) *For each $i\eta \in \sigma_u(A, x)$, $\lim_{\alpha\downarrow 0} \alpha T(s)\widehat{u_x}(\alpha + i\eta)$ exists, uniformly for $s \ge 0$.*

Then u_x is asymptotically almost periodic. If all the limits in c) *are zero, then $\|T(t)x\| \to 0$ as $t \to \infty$.*

Proof. By Proposition 5.5.2, condition b) is equivalent to $\mathrm{sp}(u_x)$ being countable. Moreover, c) is equivalent to total ergodicity of u_x. Indeed, if S denotes the shift semigroup on $\mathrm{BUC}(\mathbb{R}_+, X)$, then

$$(S(s)u_x)(t) = T(s)u_x(t), \text{ so } \left(\widehat{S(s)u_x}\right)(\lambda) = T(s)\widehat{u_x}(\lambda).$$

The result follows from Theorem 4.7.7 or 4.9.7. □

Theorem 5.5.4. *Let T be a bounded C_0-semigroup on X, with generator A. Let $x \in X$, and suppose that the following conditions are satisfied:*

a) *$\sigma_u(A, x)$ is countable; and*

b) *$x \in \mathrm{Ker}(A - i\eta) + \overline{\mathrm{Ran}(A - i\eta)}$ for each $i\eta \in \sigma_u(A, x)$.*

Then $x \in X_0 \oplus X_{ap}$. If $x \in \overline{\mathrm{Ran}(A - i\eta)}$ for each $i\eta \in \sigma_u(A, x)$, then $x \in X_0$.

Proof. By Proposition 5.4.1, u_x is uniformly continuous and bounded. Moreover, $\widehat{u_x}(\alpha + i\eta) = R(\alpha + i\eta, A)x$ for $\alpha > 0$, $\lim_{\alpha\downarrow 0} \alpha T(s)\widehat{u_x}(\alpha + i\eta)$ exists, uniformly for $s \geq 0$, by Proposition 4.3.1 and the boundedness of T, and the limit is 0 if $x \in \overline{\mathrm{Ran}(A - i\eta)}$. Now the results follow from Theorem 5.5.3. □

The results of Section 4.3 (applied with $T(t)$ replaced by $e^{-i\eta t}T(t)$) show that condition b) of Theorem 5.5.4 is equivalent to any of the following:

(i) If $x^* \in D(A^*)$, $A^*x^* = i\eta x^*$ for some $i\eta \in \sigma_u(A, x)$ and $\langle y, x^* \rangle = 0$ for all $y \in \mathrm{Ker}(A - i\eta)$, then $\langle x, x^* \rangle = 0$;

(ii) $\lim_{\alpha\downarrow 0} \alpha R(\alpha + i\eta, A)x$ exists in X whenever $i\eta \in \sigma_u(A, x)$;

(iii) $\lim_{t\to\infty} t^{-1} \int_0^t e^{-i\eta s}T(s)x\, ds$ exists whenever $i\eta \in \sigma_u(A, x)$.

Since the Abel limit exists (and equals 0) when $i\eta \in \rho_u(A, x)$, the restriction in each condition that $i\eta \in \sigma_u(A, x)$ is redundant.

Recall from Definition 4.3.10 that a bounded C_0-semigroup T on X is said to be *totally ergodic* if the condition (iii) above is satisfied for all $x \in X$ (and all $\eta \in \mathbb{R}$). By Proposition 5.4.2, T is totally ergodic if and only if each orbit u_x is a totally ergodic function. As in the case of individual η and x discussed above, this property can be characterized in several ways in terms of the generator A.

Recall also from Section 5.4 that T is said to be *asymptotically almost periodic* if each orbit u_x is asymptotically almost periodic (see Theorem 5.4.6 for equivalent properties). Thus, every asymptotically almost periodic semigroup is totally ergodic, but the converse is not true (see Example 5.5.9). The following result shows that the converse is true under a countable spectral condition. Note that $\mathrm{Ran}(A - i\eta)$ is dense in X if and only if $i\eta$ does not belong to the point spectrum $\sigma_p(A^*)$ of A^*, and that $\sigma_p(A) \cap i\mathbb{R} \subset \sigma_p(A^*)$ when A generates a bounded semigroup (see Proposition 4.3.6).

Theorem 5.5.5 (Countable Spectrum). *Let T be a bounded C_0-semigroup on X with generator A, and suppose that $\sigma(A) \cap i\mathbb{R}$ is countable.*

a) *If T is totally ergodic, then T is asymptotically almost periodic.*

b) *If $\sigma_p(A^*) \cap i\mathbb{R}$ is empty, then $\|T(t)x\| \to 0$ as $t \to \infty$, for each $x \in X$.*

Proof. This is immediate from Proposition 5.5.2, Theorem 5.5.4 and the remarks above. □

Theorem 5.5.6. *Let T be a bounded C_0-semigroup on a reflexive space X with generator A, and suppose that $\sigma(A) \cap i\mathbb{R}$ is countable. Then the following hold:*

a) *T is asymptotically almost periodic.*

b) *If $\sigma_p(A) \cap i\mathbb{R} \subset \{0\}$, then there is a projection P such that $\lim_{t\to\infty} T(t) = P$ in the strong operator topology.*

Proof. a) This is immediate from Corollary 4.3.5 and Theorem 5.5.5 a).

b) By a), $X = X_0 \oplus X_{ap}$. Let P be the projection onto X_{ap} along X_0. The assumption on $\sigma_p(A) \cap i\mathbb{R}$ implies that $X_{ap} = \{x \in X : T(t)x = x \ (t \geq 0)\}$, and the result follows. □

Theorem 5.5.6 is also true if T is a positive, bounded, Cesàro-ergodic C_0-semigroup on $L^1(\Omega, \mu)$ and $\sigma(A) \cap i\mathbb{R}$ is countable. This follows from Proposition 4.3.14.

We saw in Proposition 5.4.7 that Theorem 5.5.5 can be proved more directly when A has compact resolvent. There is also a simple proof when T is a bounded holomorphic C_0-semigroup in the sense of Definition 3.7.1. Then $\sigma(A) \cap i\mathbb{R} \subset \{0\}$ by Theorem 3.7.11, and there is a constant c such that $\|AT(t)\| \leq c/t$ for all $t > 0$, by Theorem 3.7.19. Hence, $\lim_{t\to\infty} T(t)Ay = 0$ for all $y \in D(A)$, and $\lim_{t\to\infty} T(t)(x_1 + x_2) = x_1$ whenever $x_1 \in \operatorname{Ker} A$ and $x_2 \in \operatorname{Ran} A$. If A is totally ergodic, then $\operatorname{Ker} A + \operatorname{Ran} A$ is dense in X, so $\lim_{t\to\infty} T(t)x$ exists for all $x \in X$.

The following example shows that the assumption of uniform continuity cannot be omitted from Theorem 5.5.3.

Example 5.5.7. *There is a C_0-semigroup T on a Hilbert space X and a vector $x \in X$ such that $\sigma(A) \cap i\mathbb{R}$ is empty and u_x is bounded, but u_x is not asymptotically almost periodic (and not uniformly continuous, by Theorem 5.5.3).*

Let $X := \ell^2$ and T be the C_0-semigroup defined by:

$$\begin{aligned}
(T(t)x)_{2n-1} &:= e^{\lambda_n t}(x_{2n-1} + t x_{2n}), \\
(T(t)x)_{2n} &:= e^{\lambda_n t} x_{2n},
\end{aligned}$$

where $\lambda_n := in - 1/n$. The generator A is given by:

$$\begin{aligned}
D(A) &= \left\{x \in \ell^2 : (n x_n) \in \ell^2\right\}, \\
(Ax)_{2n-1} &= x_{2n} + \lambda_n x_{2n-1}, \\
(Ax)_{2n} &= \lambda_n x_{2n}, \\
\sigma(A) &= \{\lambda_n : n \geq 1\}.
\end{aligned}$$

Now take $x \in \ell^2$ given by:

$$x_{2n-1} := 0, \qquad x_{2n} := n^{-3/2}.$$

Then

$$\begin{aligned}(T(t)x)_{2n-1} &= \frac{te^{\lambda_n t}}{n^{3/2}},\\ (T(t)x)_{2n} &= \frac{e^{\lambda_n t}}{n^{3/2}}.\end{aligned}$$

A simple argument involving Riemann sums of $s^{-3}e^{-2/s}$ shows that

$$\|T(t)x\|^2 = \sum_{n=1}^{\infty} \frac{1+t^2}{n^3} e^{-2t/n} \to \int_0^\infty s^{-3}e^{-2/s}\,ds = \int_0^\infty ue^{-2u}\,du = \frac{1}{4}$$

as $t \to \infty$. Thus, all the assumptions of Theorem 5.5.3, except uniform continuity, are satisfied, with c) vacuous, and T has no unimodular eigenvectors since A has no imaginary eigenvalues. However, $\lim_{t\to\infty} \|T(t)x\| \neq 0$, so u_x is not asymptotically almost periodic, by Corollary 5.4.17. □

Theorem 5.5.4 was obtained from Theorem 5.5.3 by observing that the convergence in condition c) of Theorem 5.5.3 is automatically uniform when T is uniformly bounded. The next example shows that this may not be valid for individual bounded orbits of unbounded semigroups.

Example 5.5.8. *There is a (norm-continuous, unbounded) C_0-semigroup T on a Hilbert space X and a vector $x \in X$ such that $\sigma(A) \cap i\mathbb{R} = \{0\}$, $0 \notin \sigma_p(A) \cup \sigma_p(A^*)$, and u_x is bounded and uniformly continuous, but u_x is not asymptotically almost periodic (and not totally ergodic).*

Let $X := \ell^2$, and T be the norm-continuous semigroup given by

$$\begin{aligned}(T(t)x)_{2n-1} &:= e^{-t/n}(x_{2n-1} + tx_{2n}),\\ (T(t)x)_{2n} &:= e^{-t/n}x_{2n}.\end{aligned}$$

The generator A is the bounded operator given by:

$$\begin{aligned}(Ax)_{2n-1} &= x_{2n} - \frac{x_{2n-1}}{n},\\ (Ax)_{2n} &= -\frac{x_{2n}}{n},\\ \sigma(A) &= \left\{-\frac{1}{n} : n \geq 1\right\} \cup \{0\}.\end{aligned}$$

Moreover, $0 \notin \sigma_p(A) \cup \sigma_p(A^*)$, so $\overline{\operatorname{Ran} A} = X$, and $X_{ap} = \{0\}$.

Now take

$$x_{2n-1} := 0, \qquad x_{2n} := n^{-3/2}.$$

As in Example 5.5.7, $\|T(t)x\| \to \frac{1}{2}$ as $t \to \infty$. Thus, $u_x : t \mapsto T(t)x$ is bounded. Since A is bounded, it follows that $t \mapsto AT(t)x = u_x'(t)$ is bounded, so u_x is uniformly continuous. Since T has no unimodular eigenvectors and $\lim_{t\to\infty} \|T(t)x\| \neq 0$, it follows from Corollary 5.4.17 that u_x is not asymptotically almost periodic. By Theorem 4.7.7 or 4.9.7 (see also Theorem 5.5.3), u_x is not totally ergodic.

It is not difficult to verify directly that $\lim_{\alpha\downarrow 0} \alpha R(\alpha, A)x$ does not exist, which also shows that u_x is not totally ergodic, and therefore not asymptotically almost periodic. □

The next example shows that the countability condition in Theorem 5.5.5 (and hence in other results of this section) is best possible in a certain sense.

Example 5.5.9. *Let E be any uncountable closed subset of $\mathbb{R}$. There is a C_0-group T of isometries on a Banach space X (even a Hilbert space) such that $\sigma(A) \subset iE$ and $\sigma_p(A^*)$ is empty, but T is not (asymptotically) almost periodic.*

Choose $[a, b]$ so that $E \cap [a, b]$ is uncountable, and let

$$E' := \{x \in E \cap [a, b] : \text{for all } \varepsilon > 0, E \cap [a, b] \cap (x - \varepsilon, x + \varepsilon) \text{ is uncountable}\}.$$

Then $(E \cap [a, b]) \setminus E'$ is countable, and E' is compact with no isolated points. If E contains no interval, then E' is homeomorphic to the Cantor set [Wil70, Theorem 30.3]. If E contains an interval, then E clearly contains a subset homeomorphic to the Cantor set. There is a non-zero non-atomic Borel measure on the Cantor set (for example, the Lebesgue-Stieltjes measure associated with the Lebesgue-Cantor function [Tay73, Sections 2.7,4.5]). Hence, there is a non-zero non-atomic measure μ supported by iE. Let T be the multiplier (semi)group on $L^p(\mu)$, where $1 \le p < \infty$ (see Example 5.4.3). Then

$$\sigma(A) = \operatorname{supp} \mu \subset iE, \qquad \sigma_p(A^*) = \emptyset,$$

but T is not asymptotically almost periodic. □

Suppose that T is a bounded, totally ergodic semigroup on X and there is a dense subspace Y of X such that $\sigma_u(A, y)$ is countable for each $y \in Y$. From Theorem 5.5.4 and the fact that $X_0 \oplus X_{ap}$ is closed, it follows that T is asymptotically almost periodic. The following example shows that the converse does not hold.

Example 5.5.10. *There is a C_0-semigroup T on a Banach space X such that $\lim_{t\to\infty} \|T(t)x\| = 0$ for all $x \in X$, but $\sigma_u(A, x) = i\mathbb{R}$ whenever $x \neq 0$.*

Let $X := L^1(\mathbb{R}_+, w(t)dt)$, where $w : \mathbb{R}_+ \to \mathbb{R}_+$ satisfies:

a) w is non-increasing,

b) $\lim_{t\to\infty} w(t) = 0$,

c) For each $a > 0$, there is a constant $c_a > 0$ such that $w(t) \ge c_a e^{-at}$ for all $t \ge 0$.

Let T be the C_0-semigroup of contractions on X defined by

$$(T(t)f)(s) := \begin{cases} f(s-t) & (s \geq t \geq 0), \\ 0 & (t > s \geq 0), \end{cases}$$

and let A be the generator. For $f \in X$,

$$\|T(t)f\| = \int_t^\infty |f(s-t)| w(s)\, ds = \int_0^\infty |f(s)| w(s+t)\, ds \to 0$$

as $t \to \infty$.

For $\operatorname{Re}\lambda < 0$, let

$$g_\lambda(s) := \frac{e^{\lambda s}}{w(s)} \quad (s \geq 0).$$

Then g_λ is bounded, by c), so $e_\lambda(s) := e^{\lambda s}$ defines an element of X^*. For $f \in X$ and $t \geq 0$,

$$\begin{aligned} \langle T(t)f, e_\lambda \rangle &= \int_t^\infty f(s-t) e^{\lambda s}\, ds = \int_0^\infty f(s) e^{\lambda(s+t)}\, ds \\ &= e^{\lambda t} \int_0^\infty f(s) e^{\lambda s} ds = e^{\lambda t} \langle f, e_\lambda \rangle. \end{aligned}$$

Thus,

$$T^*(t) e_\lambda = e^{\lambda t} e_\lambda \quad (t \geq 0),$$

so $e_\lambda \in D(A^*)$ and $A^* e_\lambda = \lambda e_\lambda$.

Suppose that $f \in X$ and $\lambda \mapsto R(\lambda, A)f$ has a holomorphic extension F to a connected neighbourhood V of some point $i\eta_0$ in $i\mathbb{R}$. We shall show that $f = 0$. For $\lambda \in V \cap \mathbb{C}_+$,

$$R(1, A)f = F(\lambda) + (\lambda - 1) R(1, A) F(\lambda).$$

By uniqueness of holomorphic extensions,

$$R(1, A)f = F(\lambda) + (\lambda - 1) R(1, A) F(\lambda) = (\lambda - A) R(1, A) F(\lambda)$$

for all $\lambda \in V$. Hence, for $\lambda \in V \cap \mathbb{C}_-$,

$$\int_0^\infty e^{\lambda s} (R(1, A)f)(s)\, ds = \langle R(1, A)f, h_\lambda \rangle = \langle R(1, A) F(\lambda), (\lambda - A^*) h_\lambda \rangle = 0.$$

As a function of λ, the left-hand side is holomorphic on $\mathbb{C}_-$ and vanishes on $\mathbb{C}_- \cap V$. Therefore,

$$\int_0^\infty e^{\lambda s} (R(1, A)f)(s)\, ds = 0 \quad (\lambda \in \mathbb{C}_-).$$

By uniqueness of Laplace transforms, this implies that $R(1, A)f = 0$. Hence, $f = 0$, since $R(1, A)$ is injective. □

5.6 Solutions of Inhomogeneous Cauchy Problems

In this section, we consider the asymptotic behaviour of mild solutions of inhomogeneous Cauchy problems. In practice, once one knows that the solution is bounded and uniformly continuous, it often follows that further asymptotic properties are inherited by u from asymptotic properties of a semigroup generated by A (or spectral properties of A) and of the inhomogeneity f. For example, Theorems 5.6.6 and 5.6.8 are results of this type under assumptions of non-resonance and countable spectrum, respectively. Thus, we are interested first in conditions which ensure that a mild solution is bounded and uniformly continuous.

We first assume that A generates a bounded C_0-semigroup T on X, and that $f \in L^1_{loc}(\mathbb{R}_+, X)$ is given. Recall from Proposition 3.1.16 that the unique solution of the inhomogeneous Cauchy problem

$$(ACP_f) \qquad \begin{cases} u'(t) = & Au(t) + f(t), \\ u(0) = & x, \end{cases}$$

is given by

$$u(t) = T(t)x + (T * f)(t),$$

where

$$(T * f)(t) = \int_0^t T(t-s)f(s)\,ds.$$

Thus, we are seeking conditions which ensure that $T * f$ is bounded and uniformly continuous.

We consider first the simple case when $\omega(T) < 0$.

Proposition 5.6.1. *Let T be a semigroup on X with generator A, and suppose that $\omega(T) < 0$.*

a) *If $f \in L^\infty(\mathbb{R}_+, X)$, then $T * f$ is bounded.*

b) *If $f \in \mathrm{BUC}(\mathbb{R}_+, X)$, then $T * f \in \mathrm{BUC}(\mathbb{R}_+, X)$.*

c) *If $f \in \mathrm{AAP}(\mathbb{R}_+, X)$, then $T * f \in \mathrm{AAP}(\mathbb{R}_+, X)$.*

d) *If $f_\infty := \lim_{t\to\infty} f(t)$ exists, then $\lim_{t\to\infty}(T * f)(t) = R(0, A)f_\infty$.*

e) *If $f \in \mathrm{AP}(\mathbb{R}_+, X)$, then there exist unique $x \in X$ and $g \in \mathrm{AP}(\mathbb{R}_+, X)$ such that $(T * f)(t) = T(t)x + g(t)$ for all $t \geq 0$. If f is τ-periodic, then g is τ-periodic.*

Proof. Parts a), b) and the case $f_\infty = 0$ of d) all follow from Proposition 1.3.5. Moreover, the map $f \mapsto T * f$ is bounded on $\mathrm{BUC}(\mathbb{R}_+, X)$.

Let $Y := \{u_x : x \in X\}$, where $u_x(t) = T(t)x$. Since $\omega(T) < 0$, Y is a closed subspace of $C_0(\mathbb{R}_+, X)$. Let $Z := Y \oplus \mathrm{AP}(\mathbb{R}_+, X)$, a closed subspace of

$\mathrm{AAP}(\mathbb{R}_+, X)$. Suppose that $f = e_{i\eta} \otimes y$ for some $\eta \in \mathbb{R}$ and $y \in X$, so $f(t) = e^{i\eta t} y$. Then

$$\begin{aligned}
(T * f)(t) &= e^{i\eta t} \int_0^t e^{-i\eta s} T(s) y \, ds \\
&= e^{i\eta t} \int_0^t \frac{d}{ds} \left(-e^{-i\eta s} T(s) R(i\eta, A) y \right) ds \\
&= T(t) \left(-R(i\eta, A) y \right) + e^{i\eta t} R(i\eta, A) y. \qquad (5.18)
\end{aligned}$$

Thus, $T * f \in Z$, and the almost periodic part of $T * f$ has the same period as f. By linearity and continuity, $T * f \in Z$ for all $f \in \mathrm{AP}(\mathbb{R}_+, X)$, and the almost periodic part of $T * f$ is τ-periodic when f is τ-periodic. This proves e), and d) follows from the case $f_\infty = 0$ and the case $\eta = 0$ of (5.18). Finally, c) follows from d) and e). □

In the context of Proposition 5.6.1 a), we cannot conclude that $T * f$ is uniformly continuous.

Example 5.6.2. *There exist a C_0-semigroup S on a Hilbert space X with $\omega(S) < 0$ and a function $f \in L^\infty(\mathbb{R}_+, X)$ such that $S * f$ is not uniformly continuous.*

Let T, X and x be as in Example 5.5.7. Take $\omega > \omega(T)$, and let $S(t) := e^{-\omega t} T(t)$ and $f(t) := T(t)x$. Then f is bounded and continuous, and

$$(S * f)(t) = \int_0^t e^{-\omega s} T(t) x \, ds = \frac{1 - e^{-\omega t}}{\omega} T(t) x.$$

Since $T(\cdot)x$ is not uniformly continuous, $S * f$ is not uniformly continuous. □

Remark 5.6.3. In part e) of Proposition 5.6.1, $T * f$ is an asymptotically almost periodic mild solution of the inhomogeneous Cauchy problem (ACP_f), with initial value 0. The decompositions $f = 0 + f$ and $T * f = u_x + g$, where $u_x(t) = T(t)x$, correspond to the splitting $\mathrm{AAP}(\mathbb{R}_+, X) = C_0(\mathbb{R}_+, X) \oplus \mathrm{AP}(\mathbb{R}_+, X)$. As predicted by Proposition 5.4.16, u_x is a mild solution of $(ACP)_0$ with initial value x, and g is a mild solution of $(ACP)_f$ with initial value $-x$. When f is τ-periodic, $x = R(1, T(\tau))((T * f)(\tau))$ and it is easy to verify directly that $T * f - u_x$ is τ-periodic. □

Now suppose that T is bounded but $\omega(T) = 0$. Then Datko's theorem (Theorem 5.1.2) shows that there exist bounded f such that $T * f$ is unbounded. The simplest examples in which $T * f$ is unbounded arise from eigenvalues. If $f(t) = T(t)x = e^{i\eta t} x$ for all $t \geq 0$, then $(T * f)(t) = t e^{i\eta t} x$, and this is unbounded if $x \neq 0$. Note that there is resonance between T and f, reflected in the fact that $i\eta \in \sigma(A) \cap i \,\mathrm{sp}(f)$.

In order to obtain positive results showing that $T * f$ is bounded when $\omega(T) = 0$, we have to place some constraints on f and we may also impose assumptions on T. The first possibility (which is somewhat dual to the case considered in Proposition 5.6.1) is to assume that $f \in L^1(\mathbb{R}_+, X)$.

Proposition 5.6.4. *Let T be a bounded C_0-semigroup on X, and let $f \in L^1(\mathbb{R}_+, X)$. Then*

a) $T * f \in \mathrm{BUC}(\mathbb{R}_+, X)$.

b) *If T is asymptotically almost periodic, then $T * f \in \mathrm{AAP}(\mathbb{R}_+, X)$.*

c) *If $\lim_{t\to\infty} T(t)$ exists in the strong operator topology, then $\lim_{t\to\infty}(T * f)(t)$ exists.*

d) *If $\lim_{t\to\infty} T(t) = 0$ in the strong operator topology, then $\lim_{t\to\infty}(T * f)(t) = 0$.*

Proof. First, note that $\|(T * f)(t)\| \le M\|f\|_1$, where $M := \sup_{t\ge 0}\|T(t)\|$. Thus, the map $f \mapsto T * f$ is bounded, and by density it suffices to prove the results when f has support in $[0, \tau]$ for some τ. But then

$$(T * f)(t) = T(t-\tau)((T * f)(\tau))$$

for $t \ge \tau$, and all the results are immediate. □

Another way to obtain results that $T * f$ is bounded is to impose a condition of non-resonance by assuming that $\sigma(A) \cap i\,\mathrm{sp}(f)$ is empty. However, there are examples (see Examples 5.1.10 and 5.1.11 after rescaling) of bounded semigroups where $\sigma(A) \cap i\mathbb{R}$ is empty (so non-resonance occurs for all f), but $\omega(T) = 0$, and then by Datko's theorem there exist bounded f such that $T * f$ is unbounded. Thus, non-resonance is not sufficient on its own to obtain positive results about boundedness of $T * f$; we need to impose further assumptions on T, or f, or both. If T is holomorphic and $\omega(T) = 0$, then $\sigma(A) \cap i\mathbb{R}$ is nonempty (Theorem 5.1.12), and we now establish that non-resonance implies boundedness of $T * f$ when T is holomorphic.

Theorem 5.6.5 (Non-resonance Theorem). *Let T be a holomorphic C_0-semigroup on X with generator A such that $\sup_{t\ge 0}\|T(t)\| < \infty$. Let $f \in L^\infty(\mathbb{R}_+, X)$, and suppose that $\sigma(A) \cap i\,\mathrm{sp}(f)$ is empty. Then $T * f \in \mathrm{BUC}(\mathbb{R}_+, X)$.*

Proof. Note first that T is norm-continuous on $(0, \infty)$, and that $-i\sigma(A) \cap \mathbb{R}$ is compact (by Corollary 3.7.18) and disjoint from $\mathrm{sp}(f)$ by assumption. Let $\psi \in C_c^\infty(\mathbb{R})$ be such that $\psi = 1$ on a neighbourhood of $-i\sigma(A) \cap \mathbb{R}$ in $\mathbb{R}$ and $\psi = 0$ on a neighbourhood of $\mathrm{sp}(f)$. Define $G, H : \mathbb{R} \to \mathcal{L}(X)$ by

$$G(t) := \begin{cases} T(t) - \int_0^\infty (\mathcal{F}^{-1}\psi)(t-s)T(s)\,ds & (t \ge 0), \\ -\int_0^\infty (\mathcal{F}^{-1}\psi)(t-s)T(s)\,ds & (t < 0), \end{cases}$$

$$H(\eta) := \begin{cases} (1-\psi(\eta))R(i\eta, A) & (i\eta \in \rho(A)), \\ 0 & (i\eta \in \sigma(A)). \end{cases}$$

Then $G \in L^\infty(\mathbb{R}, \mathcal{L}(X))$ and G is continuous on $\mathbb{R} \setminus \{0\}$. Also, $H \in C^\infty(\mathbb{R}, \mathcal{L}(X))$ and, for all large $|\eta|$, $H(\eta) = R(i\eta, A)$, so $H''(\eta) = -2R(i\eta, A)^3$. By Corollary 3.7.18, there is a constant C such that $\|H''(\eta)\| \le C|\eta|^{-3}$ for all large $|\eta|$. Hence, $H'' \in L^1(\mathbb{R}, \mathcal{L}(X))$ and $\mathcal{F}^{-1}H'' : \mathbb{R} \to \mathcal{L}(X)$ is bounded. We shall use this to show that $G \in L^1(\mathbb{R}, \mathcal{L}(X))$.

Let $\rho \in C_c^\infty(\mathbb{R})$. By two applications of Theorem 1.8.1 b), the dominated convergence theorem and integration by parts,

$$\begin{aligned}
\int_{-\infty}^{\infty} (\mathcal{F}\rho)(t)(\mathcal{F}^{-1}H'')(t)\,dt &= \int_{-\infty}^{\infty} \rho(\eta)H''(\eta)\,d\eta \\
&= \int_{-\infty}^{\infty} \rho''(\eta)H(\eta)\,d\eta \\
&= \lim_{\xi\downarrow 0} \int_{-\infty}^{\infty} \rho''(\eta)(1-\psi(\eta))R(\xi+i\eta, A)\,d\eta \\
&= \lim_{\xi\downarrow 0} \int_{-\infty}^{\infty} (\mathcal{F}\rho'')(t)G_\xi(t)\,dt \\
&= \lim_{\xi\downarrow 0} \int_{-\infty}^{\infty} (-t^2)(\mathcal{F}\rho)(t)G_\xi(t)\,dt,
\end{aligned}$$

where

$$G_\xi(t) := \begin{cases} e^{-\xi t}T(t) - \int_0^\infty (\mathcal{F}^{-1}\psi)(t-s)e^{-\xi s}T(s)\,ds & (t \ge 0), \\ -\int_0^\infty (\mathcal{F}^{-1}\psi)(t-s)e^{-\xi s}T(s)\,ds & (t < 0). \end{cases}$$

By the dominated convergence theorem, $\|G_\xi(t) - G(t)\| \to 0$ as $\xi \downarrow 0$. Moreover,

$$\|G_\xi(t)\| \le \left(1 + \|\mathcal{F}^{-1}\psi\|_1\right) M$$

for all $\xi > 0$ and $t \in \mathbb{R}$, where $M := \sup_{t\ge 0} \|T(t)\|$. By the dominated convergence theorem again,

$$\int_{-\infty}^{\infty} (\mathcal{F}\rho)(t)(\mathcal{F}^{-1}H'')(t)\,dt = \int_{-\infty}^{\infty} (\mathcal{F}\rho)(t)(-t^2)G(t)\,dt.$$

Since this holds for all $\rho \in C_c^\infty(\mathbb{R})$, it follows that

$$(\mathcal{F}^{-1}H'')(t) = -t^2 G(t)$$

a.e. (in fact, everywhere, since both sides are continuous). Hence, $G \in L^1(\mathbb{R}, \mathcal{L}(X))$.

Now define $g, h : \mathbb{R} \to X$ by

$$\begin{aligned}
g(t) &:= \int_0^\infty (\mathcal{F}^{-1}\psi)(t-s)f(s)\,ds, \\
h(\eta) &:= \begin{cases} \psi(\eta)\hat{f}(i\eta) & (\eta \notin \operatorname{sp}(f)), \\ 0 & (\eta \in \operatorname{sp}(f)). \end{cases}
\end{aligned}$$

Then $g \in \mathrm{BUC}(\mathbb{R}, X)$ (by Proposition 1.3.2 c)) and $h \in C_c^\infty(\mathbb{R}, X)$. A similar argument to the previous paragraph shows that

$$(\mathcal{F}^{-1}h'')(t) = -t^2 g(t),$$

so that $g \in L^1(\mathbb{R}, X)$. For $t \geq 0$,

$$\begin{aligned}(T * f)(t) &= \int_0^\infty G(t-s)f(s)\,ds + \int_0^\infty \int_0^\infty (\mathcal{F}^{-1}\psi)(t-s-r)T(r)f(s)\,dr\,ds \\ &= \int_0^\infty G(t-s)f(s)\,ds + \int_0^\infty T(r)g(t-r)\,dr.\end{aligned}$$

Both of these terms may be regarded as being convolutions of functions on $\mathbb{R}$, where $f(s) = 0$ for $s < 0$ and $T(r) = 0$ for $r < 0$. Since T is bounded and $g \in L^1(\mathbb{R}, X)$, and f is bounded and $G \in L^1(\mathbb{R}, \mathcal{L}(X))$, both terms are bounded and uniformly continuous on $\mathbb{R}$ and hence on $\mathbb{R}_+$ (see Propositions 1.3.2, 1.3.5 and Remark 1.3.8). □

Next, we give a result which shows that asymptotic properties of $T * f$ generally follow from those of T and f when non-resonance holds and $T * f$ is bounded and uniformly continuous.

Theorem 5.6.6. *Let T be a bounded C_0-semigroup on X, and let $f \in \mathrm{BUC}(\mathbb{R}_+, X)$. Suppose that $\sigma(A) \cap i\,\mathrm{sp}(f)$ is empty and $T * f \in \mathrm{BUC}(\mathbb{R}_+, X)$.*

a) *If T is asymptotically almost periodic and $f \in \mathrm{AAP}(\mathbb{R}_+, X)$, then $T * f \in \mathrm{AAP}(\mathbb{R}_+, X)$.*

b) *If $\lim_{t\to\infty} T(t)$ exists in the strong operator topology and $\lim_{t\to\infty} f(t)$ exists, then $\lim_{t\to\infty}(T * f)(t)$ exists.*

c) *If $\lim_{t\to\infty} T(t) = 0$ in the strong operator topology and $\lim_{t\to\infty} f(t) = 0$, then $\lim_{t\to\infty}(T * f)(t) = 0$.*

Proof. First, let Y be any of the spaces $\mathrm{BUC}(\mathbb{R}_+, X)$, $\mathrm{AAP}(\mathbb{R}_+, X)$, $C_0(\mathbb{R}_+, X)$ or the space of continuous functions $f : \mathbb{R}_+ \to X$ such that $\lim_{t\to\infty} f(t)$ exists. We assume that Y contains all orbits u_x of T ($x \in X$), $f \in Y$ and $\sigma(A) \cap i\,\mathrm{sp}(f)$ is empty, and we shall prove that $(T * \varphi * f)|_{\mathbb{R}_+} \in Y$ whenever $\varphi \in L^1(\mathbb{R})$ and $\mathcal{F}\varphi \in C_c^\infty(\mathbb{R})$. Here we are regarding T and f as being defined on $\mathbb{R}$ with $T(t) = 0$ and $f(t) = 0$ for $t < 0$, and the convolutions are defined accordingly (see Section 1.3). We shall also consider the convolution $T * \varphi : \mathbb{R} \to \mathcal{L}(X)$ defined by

$$(T * \varphi)(t)x = \int_0^\infty \varphi(t-s)T(s)x\,ds.$$

Note that $T * \varphi$ is bounded and uniformly norm-continuous (see Proposition 1.3.5 c)).

We proceed in a similar way as in Theorem 5.6.5. Let $\psi \in C_c^\infty(\mathbb{R})$ be such that $\psi = 1$ on a neighbourhood of $-i\sigma(A) \cap \operatorname{supp}(\mathcal{F}\varphi)$ and $\psi = 0$ on a neighbourhood of $\operatorname{sp}(f)$. Define $G, H : \mathbb{R} \to \mathcal{L}(X)$ by:

$$\begin{aligned} G &:= T * \varphi - T * \varphi * \mathcal{F}^{-1}\psi, \\ H(\eta) &:= \begin{cases} (\mathcal{F}\varphi)(\eta)(1-\psi(\eta))R(i\eta, A) & (i\eta \in \rho(A)), \\ 0 & (i\eta \in \sigma(A)). \end{cases} \end{aligned}$$

Then $G \in \mathrm{BUC}(\mathbb{R}, X)$, $H \in C_c^\infty(\mathbb{R}, \mathcal{L}(X))$, and as in the proof of Theorem 5.6.5, $(\mathcal{F}^{-1}H'')(t) = -t^2 G(t)$. Hence, $G \in L^1(\mathbb{R}, \mathcal{L}(X))$. It follows that $(G * f)|_{\mathbb{R}_+} \in Y$, (see Remark 5.6.3 b)).

Let $g := \varphi * \mathcal{F}^{-1}\psi * f$, and let

$$h(\eta) := \begin{cases} (\mathcal{F}\varphi)(\eta)\psi(\eta)\hat{f}(i\eta) & (\eta \notin \operatorname{sp}(f)), \\ 0 & (\eta \in \operatorname{sp}(f)). \end{cases}$$

Then $g \in \mathrm{BUC}(\mathbb{R}, X)$, $h \in C_c^\infty(\mathbb{R}, X)$ and $(\mathcal{F}^{-1}h'')(t) = -t^2 g(t)$. Hence, $g \in L^1(\mathbb{R}, X)$. We claim that this implies that $(T*g)|_{\mathbb{R}_+} \in Y$. Note that $T*(\chi_{(a,b)} \otimes x) = \chi_{(a,b)} * u_x \in Y$ (see Remark 5.6.3 b)). Since the step functions are dense in $L^1(\mathbb{R}, X)$ and convolution with T is a bounded map from $L^1(\mathbb{R}, X)$ into $L^\infty(\mathbb{R}, X)$, it follows that $(T * g)|_{\mathbb{R}_+} \in Y$, as claimed.

Since

$$\begin{aligned} T * \varphi * f &= G * f + T * \varphi * \mathcal{F}^{-1}\psi * f \\ &= G * f + T * g, \end{aligned}$$

it follows that $(T * \varphi * f)|_{\mathbb{R}_+} \in Y$, as claimed.

Now suppose that $T * f \in \mathrm{BUC}(\mathbb{R}_+, X)$. Let (φ_n) be an approximate unit in $L^1(\mathbb{R})$ such that $\mathcal{F}\varphi_n \in C_c^\infty(\mathbb{R})$ for every n (see Lemma 1.3.3). Since $(T*f)(0) = 0$, we can consider $T*f$ as an element of $\mathrm{BUC}(\mathbb{R}, X)$ and we obtain that $T*\varphi_n*f = \varphi_n * (T * f) \to T * f$ uniformly as $n \to \infty$. It follows that $(T * f)|_{\mathbb{R}_+} \in Y$, and this completes the proof. □

Finally in this section, we consider the situation when the assumption of non-resonance is replaced by a countable spectral condition. As in Theorem 5.6.6, we assume that the inhomogeneity f and the given mild solution u of (ACP_f) are bounded and uniformly continuous (and totally ergodic), and we aim to show that asymptotic properties of f are transferred to u. In contrast to Section 5.5 and the earlier part of this section, we consider problems which may not be associated with C_0-semigroups; i.e., A is not assumed to be a generator. Thus we shall extend Theorem 5.5.3 in several directions.

Let $f, u \in \mathrm{BUC}(\mathbb{R}_+, X)$. Recall from Section 5.4 that u is said to be a *mild solution* of the inhomogeneous Cauchy problem

$$(ACP_f) \qquad u'(t) = Au(t) + f(t) \quad (t \geq 0),$$

if $\int_0^t u(s)\,ds \in D(A)$ and

$$u(t) = u(0) + A\int_0^t u(s)\,ds + \int_0^t f(s)\,ds$$

for all $t \geq 0$. As in Section 4.7, we let S be the shift semigroup on $\mathrm{BUC}(\mathbb{R}_+, X)$, $(S(t)f)(s) := f(s+t)$, and B be the generator of S, so

$$(R(\lambda, B)f)(t) = \int_0^\infty e^{-\lambda s} f(s+t)\,ds = \widehat{f_t}(\lambda) \qquad (\lambda \in \mathbb{C}_+, t \geq 0),$$

where $f_t(s) := f(s+t)$. The following proposition relates the Laplace transform, spectrum and ergodicity of u to the properties of f.

Proposition 5.6.7. *Let A be a closed operator on X, let $f \in \mathrm{BUC}(\mathbb{R}_+, X)$, and let u be a bounded, uniformly continuous, mild solution of the inhomogeneous Cauchy problem (ACP_f). Then*

a) *For $t \geq 0$ and $\lambda \in \mathbb{C}_+ \cap \rho(A)$,*

$$(R(\lambda, B)u)(t) = R(\lambda, A)\left((R(\lambda, B)f)(t)\right) + R(\lambda, A)(u(t)).$$

b) $\mathrm{sp}(u) \subset \mathrm{sp}(f) \cup \{\eta \in \mathbb{R} : i\eta \in \sigma(A)\}$.

c) *If f is uniformly ergodic at η where $i\eta \in \rho(A) \cap i\mathbb{R}$, then u is uniformly ergodic at η and $M_\eta u(t) = R(i\eta, A)(M_\eta f(t))$.*

Proof. Note first that

$$u(s+t) = u(t) + A(v_t(s)) + g_t(s), \tag{5.19}$$

where $v_t(s) := \int_t^{s+t} u(r)\,dr$ and $g_t(s) := \int_t^{s+t} f(r)\,dr$. Moreover, for $\mathrm{Re}\,\lambda > 0$,

$$(R(\lambda, B)u)(t) = \int_0^\infty e^{-\lambda s} u(s+t)\,ds = \lambda \widehat{v_t}(\lambda),$$

and similarly

$$(R(\lambda, B)f)(t) = \lambda \widehat{g_t}(\lambda).$$

For $\lambda \in \mathbb{C}_+$, the integral $\int_0^\infty e^{-\lambda s} v_t(s)\,ds$ is absolutely convergent, and it follows from (5.19) that $\int_0^\infty e^{-\lambda s} A(v_t(s))\,ds$ is also absolutely convergent. By Proposition 1.6.3, $\widehat{v_t}(\lambda) \in D(A)$ and

$$A(\widehat{v_t}(\lambda)) = \int_0^\infty e^{-\lambda s} u(s+t)\,ds - \widehat{g_t}(\lambda) - \lambda^{-1} u(t).$$

Hence,

$$(\lambda - A)((R(\lambda, B)u)(t)) = (R(\lambda, B)f)(t) + u(t),$$

and a) follows.

Since $\hat{f}(\lambda) = (R(\lambda, B)f)(0)$ and $\hat{u}(\lambda) = (R(\lambda, B)u)(0)$, it follows from a) that

$$\hat{u}(\lambda) = R(\lambda, A)\big(\hat{f}(\lambda)\big) + R(\lambda, A)(u(0))$$

for $\lambda \in \mathbb{C}_+ \cap \rho(A)$. If $i\eta \in \rho(A)$ and $\eta \notin \operatorname{sp}(f)$, then $\hat{f}$ has a holomorphic extension near $i\eta$, and then $\hat{u}$ also has a holomorphic extension. This proves b).

Since $M_\eta f = \lim_{\alpha\downarrow 0} \alpha R(\alpha + i\eta, B)f$, c) follows from a). □

We now give the extension of Theorem 5.5.3 to inhomogeneous Cauchy problems. Although this case is not a corollary of the Tauberian theorem (Theorem 4.7.7), the idea of the proof is similar.

Theorem 5.6.8. *Let A be a closed operator on X such that $\sigma(A) \cap i\mathbb{R}$ is countable. Let $f : \mathbb{R}_+ \to X$ be asymptotically almost periodic, and u be a bounded, uniformly continuous, mild solution of the inhomogeneous Cauchy problem (ACP_f), and suppose that u is uniformly ergodic at η whenever $i\eta \in \sigma(A) \cap i\mathbb{R}$. Then*

a) *u is asymptotically almost periodic, and*

$$i\,\operatorname{Freq}(u) \cap \rho(A) = i\,\operatorname{Freq}(f) \cap \rho(A).$$

b) *If $\tau > 0$, f is τ-periodic and $i\,\operatorname{Freq}(u) \subset \rho(A) \cup \left(\frac{2\pi i}{\tau}\right)\mathbb{Z}$, then $u = u_0 + u_1$, where u_0 is a mild solution of (ACP_0), $\lim_{t\to\infty} u_0(t) = 0$, and u_1 is a τ-periodic solution of (ACP_f).*

c) *If $i\,\operatorname{Freq}(f) \subset \sigma(A) \cup \{0\}$ and $i\,\operatorname{Freq}(u) \subset \rho(A) \cup \{0\}$, then $\lim_{t\to\infty} u(t) = R(0, A)(M_0 f(0))$.*

d) *If $i\,\operatorname{Freq}(f) \subset \sigma(A)$ and $i\,\operatorname{Freq}(u) \subset \rho(A)$, then $\lim_{t\to\infty} u(t) = 0$.*

Proof. By Proposition 5.6.7 c), u is uniformly ergodic at η whenever $i\eta \in \rho(A) \cap i\mathbb{R}$ so u is totally ergodic. Let B be the generator of the shift semigroup on the space $\mathcal{E}(\mathbb{R}_+, X)$ (see Section 4.7). By Proposition 5.6.7 a),

$$(R(\lambda, B)u)(t) = (R(\lambda, B)(R(\lambda, A) \circ f))(t) + R(\lambda, A)(u(t))$$

for $\lambda \in \mathbb{C}_+ \cap \rho(A)$. The space $\operatorname{AAP}(\mathbb{R}_+, X)$ is invariant under the operation of composition with a fixed member of $\mathcal{L}(X)$ and under $R(\lambda, B)$, so $R(\lambda, B)(R(\lambda, A) \circ f) \in \operatorname{AAP}(\mathbb{R}_+, X)$. Hence,

$$R(\lambda, \widetilde{B})\pi(u) = \pi(R(\lambda, B)u) = \pi(R(\lambda, A) \circ u)$$

whenever $\lambda \in \mathbb{C}_+ \cap \rho(A)$, where

$$\pi : \mathcal{E}(\mathbb{R}_+, X) \to Y := \widetilde{\mathcal{E}}(\mathbb{R}_+, X)/\operatorname{AAP}(\mathbb{R}_+, X)$$

is the quotient map and $\widetilde{B}$ is the generator of the C_0-group $\widetilde{S}$ on $\widetilde{\mathcal{E}}$ induced by the shift semigroup (see Proposition 4.7.2). This shows that the map $\lambda \mapsto R(\lambda, \widetilde{B})\pi(u)$ has a holomorphic extension to a map $h : \rho(A) \to \widetilde{\mathcal{E}}$, given by $h(\lambda) = \pi(R(\lambda, A)\circ u)$. It now follows as in the proof of Theorem 4.7.7 that $\pi(u) = 0$, so u is asymptotically almost periodic.

The fact that $i\,\mathrm{Freq}(u) \cap \rho(A) = i\,\mathrm{Freq}(f) \cap \rho(A)$ follows from Proposition 5.6.7 c), and the remaining statements follow from Corollary 4.7.8 and Proposition 5.4.16. □

The following corollary generalizes several parts of Proposition 5.6.1.

Corollary 5.6.9. *Let A be a closed operator on X such that $\sigma(A) \cap i\mathbb{R}$ is empty. Let $f \in \mathrm{BUC}(\mathbb{R}_+, X)$, and u be a bounded, uniformly continuous, mild solution of the inhomogeneous Cauchy problem (ACP_f).*

a) *If f is totally ergodic, then u is totally ergodic.*

b) *If f is asymptotically almost periodic, then u is asymptotically almost periodic, and* $\mathrm{Freq}(u) = \mathrm{Freq}(f)$.

c) *If f is τ-periodic, then $u = u_0 + u_1$, where u_0 is a mild solution of (ACP_0), $\lim_{t\to\infty} u_0(t) = 0$, and u_1 is a τ-periodic solution of (ACP_f).*

d) *If $f_\infty := \lim_{t\to\infty} f(t)$ exists, then $\lim_{t\to\infty} u(t) = R(0, A)f_\infty$.*

e) *If $\lim_{t\to\infty} f(t) = 0$, then $\lim_{t\to\infty} u(t) = 0$.*

Proof. a) follows from Proposition 5.6.7 c), and the remaining statements from Theorem 5.6.8. □

5.7 Notes

Accounts of the asymptotic behaviour of C_0-semigroups have appeared in the books of Daletskii and Krein [DK74], Levitan and Zhikov [LZ82], Nagel et al. [Nag86], van Neerven [Nee96c], Chicone and Latushkin [CL99] and Eisner [Eis10].

Section 5.1
The growth bound $\omega(T)$ appeared in the book of Hille and Phillips [HP57], while $\omega_1(T)$ arose in papers of D'Jacenko [Jac76] and Zabczyk [Zab79]. It is possible to define higher order and fractional growth bounds in the following way.

For $\mu > \omega(T)$, $\mu - A$ is a sectorial operator, and the fractional powers $(\mu - A)^\alpha$ and $R(\mu, A)^\alpha$ are defined whenever $\alpha \geq 0$ (see the Notes on Section 3.8). Define

$$\begin{aligned}\omega_\alpha(T) &:= \omega(\|T(\cdot)R(\mu, A)^\alpha\|) \\ &= \inf\{\omega(u_x) : x \in D((\mu - A)^\alpha)\}.\end{aligned}$$

This is independent of $\mu > \omega(T)$, since $D((\mu - A)^\alpha)$ is independent of μ (see Proposition 3.8.2). With only minor modifications, the proof of Proposition 5.1.5 shows that

$$\omega_{\alpha+1}(T) = \sup\{\mathrm{abs}(u_x) : x \in D((\mu - A)^\alpha)\}.$$

Moreover, the resolvent identity may be used to show that

$$\mathrm{hol}(\widehat{u}_{R(\mu,A)x}) = \mathrm{hol}(\widehat{u_x}),$$

from which it follows that

$$s(A) = \inf\{\mathrm{hol}(\widehat{u_x}) : x \in D(A^n)\} \le \omega_n(T)$$

for all positive integers n, and hence that $s(A) \le \omega_\alpha(T)$ for all $\alpha \ge 0$.

Theorem 5.1.2 is mostly due to Datko [Dat70], [Dat72], with contributions also from Pazy [Paz72], van Neerven [Nee96a] and Schüler and Vũ [SV98]. The equality of $\omega_1(T)$ and $\mathrm{abs}(T)$ (Proposition 5.1.6) and Proposition 5.1.5 were established by Neubrander [Neu86]. Theorem 5.1.7 was first proved by Weis and Wrobel [WW96], following preliminary results of Slemrod [Sle76] (showing that $\omega_2(T) \le s_0(A)$) and van Neerven, Straub and Weis [NSW95] (showing that $\omega_\alpha(T) \le s_0(A)$ whenever $\alpha > 1$). In [WW96], interpolation theory was used to show that $\omega_\alpha(T)$ is a convex function of α and therefore it is continuous for $\alpha > 0$. Thus, Theorem 5.1.7 followed from the result of [NSW95]. A Banach space X is said to have *Fourier type* p (where $1 \le p \le 2$) if the Fourier transform defines a bounded linear map from $L^p(\mathbb{R}, X)$ into $L^{p'}(\mathbb{R}, X)$, where $1/p + 1/p' = 1$. Every Banach space has Fourier type 1; every superreflexive space has Fourier type p for some $p > 1$; X has Fourier type 2 if and only if X is (isomorphic to) a Hilbert space (see also the Notes on Section 1.8). van Neerven, Straub and Weis [NSW95] showed that $\omega_\alpha(T) \le s_0(A)$ if X has Fourier p and $\alpha > 1/p - 1/p'$, and Weis and Wrobel [WW96] extended this to the case when $\alpha = 1/p - 1/p'$ and $p < 2$ (for the case $p = 2$, see the notes on Section 5.2). van Neerven [Nee09] has given another variation of the result. Assume that X has (Rademacher) type q_1 and cotype q_2, where $1 \le q_1 \le 2$ and $q_2 \ge 2$, and let $\alpha = 1/q_1 - 1/2$; for example, X may be any L^p-space ($1 \le p < \infty$) and then $\alpha = \left|\frac{1}{p} - \frac{1}{2}\right|$. Then $\omega_\alpha(T)$ is bounded above by an abscissa for the resolvent of A at least as large as, but often equal to, $s_0(A)$. For the definitions and properties of type and cotype, see [Woj91] for example.

Trefethen and Embree [TE05] provide many examples of the behaviour of the pseudo-spectrum (sets where the norm of the resolvent of an operator is large) including strong evidence that the pseudo-spectrum is much more stable than the spectrum under perturbations. For $\alpha \ge 0$, there is an associated pseudo-spectral bound defined by

$$s_\alpha(A) = \inf\Big\{\omega > s(A) : \text{there exists } C_\omega \text{ such that } \|R(\lambda, A)\| \le C_\omega(1+|\lambda|)^\alpha \text{ whenever } \operatorname{Re}\lambda > \omega\Big\}.$$

The proofs of Theorem 5.1.7 can be modified to show that $\omega_{\alpha+1}(T) \le s_\alpha(A)$ for any semigroup on any Banach space (earlier, Slemrod [Sle76] proved that $\omega_{n+2}(T) \le s_n(A)$, and Wrobel [Wro89] showed that $\omega_{n+1}(T) \le s_n(A)$ if X has non-trivial Fourier type).

van Neerven [Nee96b] proved Theorem 5.1.8 by means of Laplace inversion along a well-chosen contour. Both conclusions that growth is at most linear are sharp.

Suppose that $x \in X$, and $\widehat{u_x}$ is defined and bounded on $\mathbb{C}_+$, and let $\alpha > 1$. Under certain additional assumptions, $\|T(t)R(\mu, A)^\alpha x\| \to 0$ as $t \to \infty$. This was established by Huang and van Neerven [HN99] when X has the analytic Radon-Nikodym property, and by Batty, Chill and van Neerven [BCN98] when T is sun-reflexive in the sense of [Nee92]; i.e., when $R(\mu, A)$ is weakly compact, by a theorem of de Pagter [Pag89]. In particular, it is true if X is reflexive (the Hilbert space case was first covered by Huang [Hua99]). However, if $X = C_0(\mathbb{R})$, T is the shift group: $(T(t)f)(s) = f(s+t)$, and $g \in C_c(\mathbb{R})$, then $\widehat{u_g}$ exists and is bounded on $\mathbb{C}_+$, but $\|T(t)R(\mu, A)^\alpha g\| = \|R(\mu, A)^\alpha g\|$ for all $t \geq 0$. Blake [Bla99] has obtained further results related to Theorem 5.1.8. A detailed account of many of these refinements of Theorems 5.1.7 and 5.1.8 is given in the book of van Neerven [Nee96c, Chapter 4].

Example 5.1.10 is a modification, due to Wrobel [Wro89], of an example of Zabczyk [Zab75]. Example 5.1.11 is due to Arendt [Are94b], and van Neerven [Nee96d] has analyzed it for many rearrangement-invariant function spaces. The first example of a positive C_0-semigroup with $s(A) < \omega(T)$ was given by Greiner, Voigt and Wolff [GVW81], and the first example of a positive C_0-group by Wolff [Wol81].

Theorem 5.1.12 is a consequence of the spectral mapping theorem:

$$\sigma(T(t)) \setminus \{0\} = \left\{e^{\mu t} : \mu \in \sigma(A)\right\},$$

which is valid for eventually norm-continuous semigroups. It appeared in the book of Hille and Phillips [HP57] and it can be proved either by product space techniques (see [Nag86]) or by Banach algebra methods (see [HP57] or [Dav80]), but we do not know of a proof by Laplace transform methods of either the spectral mapping theorem for holomorphic semigroups, or Theorem 5.1.12 for eventually norm-continuous semigroups. Greiner and Müller [GM93] obtained a spectral mapping theorem for all exponentially bounded integrated semigroups. Martinez and Mazon [MM96], Blake [Bla01] and Nagel and Poland [NP00] have established a version of the spectral mapping theorem for asymptotically norm-continuous semigroups (see the Notes on Section 5.2).

Section 5.2

Theorem 5.2.1 originated in the work of Gearhart [Gea78] who considered the case when T is a contraction semigroup on a Hilbert space X. He showed that $e^{\mu t} \in \rho(T(t))$ if and only $\mu + (2\pi i/t)\mathbb{Z} \subset \rho(A)$ and $\sup_{n\in\mathbb{Z}} \|R(A, \mu + 2\pi i n/t)\| < \infty$. This was extended to the non-contractive case with new and simpler proofs, independently by Herbst [Her83], Howland [How84], Huang [Hua85] and Prüss [Prü84]. A consequence of these results is that if the resolvent of A is bounded on the imaginary axis, then T is "hyperbolic"; i.e., X splits as a topological direct sum of closed, T-invariant subspaces, $X = X_- \oplus X_+$, such that $\omega(T|_{X_-}) < 0$ and there is a C_0-group U on X_+ with $T(t)|_{X_+} = U(-t)$ and $\omega(U) < 0$. A particularly beautiful proof of Theorem 5.2.1 was given by Latushkin and Montgomery-Smith [LM95] which is based on the theory of "evolution semigroups". See also the monograph by Chicone and Latushkin [CL99] where many far-reaching consequences are given.

Our short proof of Theorem 5.2.1 is a simplified version of a proof given by Weiss [Wei88]. For higher order growth and spectral bounds, Weiss [Wei90] and Wrobel [Wro89] showed that $\omega_n(T) = s_n(A)$ for semigroups on Hilbert space. For Banach spaces, Greiner [Gre84] gave a characterisation of $\sigma(T(t))$ based on Fejér's theorem, from which Theorem 5.2.1 can be deduced in the case of Hilbert spaces. Examples 5.2.2 and 5.2.3 are due to Weis [Wei98] and Arendt [Are94b], respectively.

Building on preliminary work of Martinez and Mazon [MM96], Blake [Bla01] proved the following variant of Theorem 5.2.1, showing that the absence of norm-continuity for large t is reflected in a rather precise way in the shape of the spectrum and the growth of the resolvent (i.e., the shape of the pseudo-spectrum).

For a C_0-semigroup T with generator A, let

$$\begin{aligned} s_0^\infty(A) &:= \inf\Big\{\omega \in \mathbb{R} : \text{there exist } b_\omega > 0 \text{ and } C_\omega > 0 \text{ such that } \lambda \in \rho(A) \text{ and} \\ &\qquad \|R(\mu, A)\| \le C_\omega \text{ whenever } \operatorname{Re}\lambda > \omega \text{ and } |\operatorname{Im}\lambda| > b_\omega\Big\}, \\ \delta(T) &:= \inf\Big\{\omega \in \mathbb{R} : \text{there exists } M_\omega > 0 \text{ such that} \\ &\qquad \limsup_{h\to 0} \|T(t+h) - T(t)\| \le M_\omega e^{\omega t} \text{ for all } t \ge 0\Big\}. \end{aligned}$$

We say that T is *asymptotically norm-continuous* if $\delta(T) < \omega(T)$.

Theorem 5.7.1. *Let T be a C_0-semigroup on a Hilbert space, with generator A. Then $s_0^\infty(A) = \delta(T)$.*

Section 5.3
Part of Theorem 5.3.1 was proved by Greiner, Voigt and Wolff [GVW81], and the remainder by Neubrander [Neu86]. Example 5.3.2 is due to Arendt [Are94b]. Theorem 5.3.6 was first proved by Weis [Wei95], answering a question which had been open for some time. The proof in [Wei95] used interpolation theory and the theory of evolution semigroups developed by Latushkin and Montgomery-Smith [LM95]. The simplified proof given here follows a later method of Weis [Wei96]. Proposition 5.3.5 appeared in a proof given by Montgomery-Smith [Mon96] but it can alternatively be established by methods shown in [Haa07b]. Another variant of the proof of Theorem 5.3.6 is given in [Wei98]. The special cases of $p = 1$ (Proposition 5.3.7) and $p = 2$ had been proved by Derndinger [Der80] and Greiner and Nagel [GN83], respectively. Our proof of Proposition 5.3.7 is taken from [Der80]. Theorem 5.3.8 was proved by Derndinger [Der80] for compact Ω and by Batty and Davies [BD82] for locally compact Ω. Note that the proofs of Proposition 5.3.7 and Theorem 5.3.8 do not use the lattice properties, and the equality $s(A) = \omega(T)$ holds for positive semigroups on ordered Banach spaces where either the norm on X_+ or the norm on X_+^* is additive in the sense of (5.13) (see [BD82]). In particular, this is true for C^*-algebras (the result for positive semigroups on unital C^*-algebras was first established by Groh and Neubrander [GN81]). For a positive semigroup on a Banach lattice with (Rademacher) type p and cotype q, one has $\omega_\alpha(T) \le s_0(A)$ for $\alpha = 1/p - 1/q$ [Nee09] (see the Notes on Section 5.1).

There are many aspects of asymptotic behaviour of positive semigroups which are not covered in this book. An early account of the theory was given in [Nag86]. See also the Notes on Section 5.5.

Section 5.4
Splitting theorems for relatively compact orbits of semigroups, such as Theorems 5.4.6 and 5.4.11, are often associated with the names of Glicksberg and de Leeuw. In [GL61], they were the first to obtain such a theorem for general Banach spaces, following special cases due to Jacobs [Jac56, etc]. In those papers, the splitting theory was carried out for very general semigroups of operators, and the methods were algebraic and topological. For one-parameter semigroups, the Glicksberg-deLeeuw theorem is the following variant of Theorem 5.4.11.

Theorem 5.7.2 (Glicksberg-deLeeuw Theorem). *Let T be a bounded C_0-semigroup on X, and suppose that $\{T(t)x : t \ge 0\}$ is weakly relatively compact for each $x \in X$. Then*

$$X = X_{w0} \oplus X_{ap},$$

where

$$X_{w0} := \{x \in X : 0 \text{ is in the weak closure of } \{T(t)x : t \geq 0\}\}.$$

Accounts of the general Glicksberg-deLeeuw theory can be found in the books of Krengel [Kre85, Section 2.4], van Neerven [Nee96c, Section 5.7], Engel and Nagel [EN00, Section V.2] and Eisner [Eis10, Section I.1].

It is easy to deduce the strong splitting theorem (Theorem 5.4.6) from Theorem 5.7.2, but comparison with the weak splitting theorem (Theorem 5.4.11) is more delicate. A priori, it is not clear that the spaces X_{e0} and X_{w0} are contained in each other. However, it is easy to verify that the following properties are equivalent for a totally ergodic semigroup T on X:

(i) There is a closed T-invariant subspace Y of X such that $X = Y \oplus X_{ap}$ as a topological direct sum.

(ii) There is a bounded projection P of X onto X_{ap} such that $PT(t) = T(t)P$ for all $t \geq 0$.

(iii) $X = X_{e0} \oplus X_{ap}$.

When these properties hold, $Y = \operatorname{Ker} P = X_{e0}$. The weak and strong splitting theorems and the Glicksberg-deLeeuw theorem provide important special cases when the properties hold, and they show in particular that $X_{e0} = X_{w0}$ if the orbits of T are weakly relatively compact.

There are some totally ergodic C_0-semigroups for which the properties (i), (ii) and (iii) above do not hold. In particular, Woodward [Woo74] showed that the space $\mathcal{E}(\mathbb{R})$ of all totally ergodic functions in $\mathrm{BUC}(\mathbb{R})$ is strictly larger than $\mathcal{E}_0(\mathbb{R}) \oplus \mathrm{AP}(\mathbb{R})$, where $\mathcal{E}_0(\mathbb{R})$ is the space of totally ergodic functions whose means are all zero, and therefore (iii) does not hold for the C_0-group of shifts on $\mathcal{E}(\mathbb{R})$. The weak splitting theorem shows that there is such a splitting of the space of all Eberlein-w.a.a.p. functions (see the Notes on Section 4.7).

The descriptions of the spaces X_{e0} and X_{w0} are both rather weak, and the strongest result has been obtained by Ruess and Summers [RS90b].

Theorem 5.7.3. *Let T be a bounded C_0-semigroup on X, let $x \in X$, and suppose that $\{T(t)x : t \geq 0\}$ is relatively weakly compact. Then u_x is Eberlein-w.a.a.p. Moreover, $x = x_0 + x_1$, where $x_1 \in X_{ap}$ and u_{x_0} is Eberlein-w.a.a.p. with means $M_\eta u_{x_0} = 0$ for all $\eta \in \mathbb{R}$.*

Ruess and Summers [RS87], [RS88a], [RS90a], [RS92a], [RS92b] have also carried out detailed investigations of orbits, and almost-orbits, of non-linear semigroups and solutions of non-autonomous Cauchy problems. Their results show that Eberlein's notion of weak asymptotic almost periodicity is important even in those cases.

The construction of trigonometric polynomials as in Proposition 5.4.8 by means of Fejér kernels occurred in Bohr's book [Boh47] where it was used in a proof (attributed to De La Vallée Poussin) of a fundamental property of almost periodic functions. Other techniques in the proof of Theorem 5.4.11 appear in the work of Datry and Muraz [DM95], [DM96], who have considered splittings in a very abstract situation.

Section 5.5

There is a very large literature on the subject of the local spectrum for a bounded operator, or a commuting family of bounded operators. The book of Erdelyi and Wang

[EW85] includes an account of the theory for unbounded operators. In the literature, the operators are usually assumed to satisfy the "single-valued extension property", but in the context of Section 5.5, we consider only a peripheral part of the local spectrum, for which we have chosen a definition which makes this property hold automatically (see Proposition 5.5.1) and which is consistent with the notion of spectrum for functions (see Proposition 5.5.2).

An alternative notion of imaginary local spectrum, $\tilde{\sigma}_u(T,x)$, of a C_0-semigroup T has been introduced by Batty and Yeates [BY00], using ideas of Albrecht [Alb81]. When T is bounded, the definition of $\tilde{\sigma}_u(T,x)$ is as follows.

For a bounded semigroup T and $f \in L^1(\mathbb{R}_+)$, define $f(T) \in \mathcal{L}(X)$ by $f(T)x = \int_0^\infty f(t)T(t)x\,dt$. A point $i\eta \in i\mathbb{R}$ is in $\tilde{\rho}_u(T,x)$ if there exist $n \in \mathbb{N}$, $f_1, \ldots, f_n \in L^1(\mathbb{R}_+)$, a neighbourhood V of the point $((\mathcal{F}f_1)(-\eta), \ldots, (\mathcal{F}f_n)(-\eta))$ in $\mathbb{C}^n$ and holomorphic functions $g_i : V \to X$ for each $i = 1, \ldots, n$ such that

$$\sum_{i=1}^{n} (z_i - f_i(T))\, g_i(z) = x \quad (z = (z_1, \ldots, z_n) \in V).$$

Then $\tilde{\sigma}_u(T,x) := i\mathbb{R} \setminus \tilde{\rho}_u(T,x)$.

While it is easy to see that $\tilde{\sigma}_u(T,x)$ is contained in $\sigma_u(A,x)$, it remains open whether equality holds. Theorem 5.5.3 remains valid when $\sigma_u(A,x)$ is replaced by $\tilde{\sigma}_u(T,x)$ (see [BY00]). Part b) of Theorem 5.5.5 was proved independently by Arendt and Batty [AB88] and Lyubich and Vũ [LV88]. The proof in [AB88] was based on the contour integral method of Section 4.4, combined with an unusual argument by transfinite induction. The functional analytic proof in [LV88] is related to the quotient method of Sections 4.7 and 4.8. Part a) of Theorem 5.5.5 was proved by Lyubich and Vũ [LV90a]. Prototypes of Theorem 5.5.5 for norm-continuous semigroups, some discrete semigroups, and the case of empty peripheral spectrum had been obtained by Sklyar and Shirman [SS82], Atzmon [Atz84], and Huang [Hua83] (see also [Hua93a], [Hua93b]), respectively.

A different proof of Theorem 5.5.5 b) was subsequently given by Esterle, Strouse and Zouakia [ESZ92]. We summarize their method in the next few paragraphs.

Let E be a closed subset of $\mathbb{R}$. A function $f \in L^1(\mathbb{R}_+)$ is said to be *of spectral synthesis with respect to* E if there is a sequence (g_n) in $L^1(\mathbb{R})$ such that $\lim_{n\to\infty} \|g_n - f\|_1 = 0$ and, for each n, $\mathcal{F}g_n$ vanishes on a neighbourhood of E. (Here, we are regarding f as a member of $L^1(\mathbb{R})$ with $f(t) = 0$ for $t < 0$.) If f is of spectral synthesis with respect to E, then $\mathcal{F}f$ vanishes on E. If the boundary of E is countable and $\mathcal{F}f$ vanishes on E, then f is of spectral synthesis with respect to E [Kat68, p.230].

For a bounded semigroup T such that $\sigma(A) \cap i\mathbb{R}$ is countable, it was shown by Esterle, Strouse, and Zouakia [ESZ92], using an abstract Mittag-Leffler theorem (a generalization of Baire's category theorem), that the linear span of the union of the ranges of $f(T)$ for all f which are of spectral synthesis with respect to $i\sigma(A) \cap \mathbb{R}$ is dense in X. Then, Theorem 5.5.5 b) follows from the following analogue of the Katznelson-Tzafriri theorem [KT86].

Theorem 5.7.4. *Let T be a bounded C_0-semigroup on X and let $f \in L^1(\mathbb{R}_+)$ be of spectral synthesis with respect to $i\sigma(A) \cap \mathbb{R}$. Then $\|T(t)f(T)\| \to 0$ as $t \to \infty$.*

Theorem 5.7.4 was proved in [ESZ92] using methods of harmonic analysis, and in [Vu92] using a functional analytic method. In the next paragraph, we sketch a proof which uses the ideas of Section 4.9, in particular a version of Ingham's Theorem 4.9.5 for functions on $\mathbb{R}$.

Let $g \in L^1(\mathbb{R})$ be such that $\mathcal{F}g$ vanishes in a neighbourhood of $i\sigma(A) \cap \mathbb{R}$, and let $(\breve{g} * T)(t)x = \int_0^\infty g(s-t)T(s)x\,ds$. Then $\breve{g} * T \in \mathrm{BUC}(\mathbb{R}, X)$, and it is not difficult to show that $\breve{g} * T$ has distributional Fourier transform

$$H(s) = \begin{cases} (\mathcal{F}g)(-s)R(is, A) & (is \in \rho(A)), \\ 0 & (is \in \sigma(A)), \end{cases}$$

in the sense that

$$\int_{-\infty}^{\infty} (\breve{g} * T)(t)(\mathcal{F}\varphi)(t)\,dt = \int_{-\infty}^{\infty} H(s)\varphi(s)\,ds$$

for all $\varphi \in C_c^\infty(\mathbb{R})$. If $\rho \in \mathcal{S}(\mathbb{R})$ is such that $\mathcal{F}\rho$ has compact support, then $(\mathcal{F}\rho) \cdot H \in L^1(\mathbb{R}, \mathcal{L}(X))$ with $\mathcal{F}^{-1}((\mathcal{F}\rho) \cdot H) = \rho * (\breve{g} * T) \in C_0(\mathbb{R}, \mathcal{L}(X))$ by the Riemann-Lebesgue lemma. Using a mollifier (ρ_n), it follows that $\breve{g} * T \in C_0(\mathbb{R}, \mathcal{L}(X))$. Choosing g to approximate f, it follows that $T(t)f(T) = (\check{f} * T)(t) \to 0$ as $t \to \infty$.

Another proof of Theorem 5.5.5 b) using the abstract Mittag-Leffler theorem has been given in [BCT02]. This was one of a series of papers, also including [Tom01], [CT03], [CT04] and [BCT07], in which Chill and Tomilov investigated stability of semigroups in the sense of strong operator convergence to 0. An excellent survey is given in [CT07]. For general Banach spaces the main results are as follows.

Theorem 5.7.5. *Let T be a bounded C_0-semigroup T on X, with generator A. Then $\lim_{t\to\infty} \|T(t)x\| = 0$ for all $x \in X$ if any of the following conditions holds:*

a) $\left\{x \in X : \lim_{\alpha\downarrow 0} \left\|\alpha R(\alpha + is, A)^2 x\right\| = 0 \text{ for all } s \in \mathbb{R}\right\}$ *is dense in X.*

b) *For some $\gamma > 1$,* $\left\{x \in X : \lim_{\alpha\downarrow 0} \int_{\mathbb{R}} \left\|\alpha^{\gamma-1} R(\alpha + is, A)^\gamma x\right\| \,ds = 0\right\}$ *is dense in X.*

c) *For each $s_0 \in \mathbb{R}$, there exists a neighbourhood U of s_0 in $\mathbb{R}$ such that*

$$\left\{x \in X : \lim_{\alpha\downarrow 0} \int_U \left\|\alpha R(\alpha + is, A)^2 x\right\| \,ds = 0\right\}$$

is dense in X.

The assumptions of Theorem 5.5.5 b) imply that $\bigcap_{s\in\mathbb{R}} \mathrm{Ran}(A - is)$ is dense in X, by the abstract Mittag-Leffler theorem. If $x \in \mathrm{Ran}(A - is)$, then $\lim_{\alpha\downarrow 0} \left\|\alpha R(\alpha + is, A)^2 x\right\| = 0$. So Theorem 5.5.5 b) follows from case a) of Theorem 5.7.5, which is proved by means of an edge-of-the-wedge theorem.

Theorem 5.7.5 can be improved when X has non-trivial Fourier type. For example, for a Hilbert space X, density of

$$\left\{x \in X : \lim_{\alpha\downarrow 0} \left\|\sqrt{\alpha} R(\alpha + is, A)x\right\| = 0 \text{ for all } s \in \mathbb{R}\right\}$$

is sufficient for stability. Moreover there is a variant of the integral condition b) of Theorem 5.7.5 which is both necessary and sufficient on Hilbert spaces. We refer the reader to the survey [CT07], or the original articles, for the details.

Greenfield [Gre94] (see also [BBG96]) proved the following quantitative version of Theorem 5.5.5 b).

Theorem 5.7.6. *Let T be a C_0-semigroup of contractions on X, and suppose that $\sigma(A) \cap i\mathbb{R}$ is countable. Let X_1^* be the weak*-closed linear span of the unimodular eigenvectors of the dual semigroup T^* on X^*. Then*

$$\begin{aligned}\lim_{t\to\infty} \|T(t)x\| &= \inf\{\|x-y\| : y \in X_0\} \\ &= \sup\{|\langle x, x^*\rangle| : x^* \in X_1^*, \|x^*\| \le 1\}\end{aligned}$$

for each $x \in X$.

A consequence of Theorem 5.5.6 is that if X is superreflexive (i.e., if there is an equivalent uniformly convex norm on X), and A generates a bounded C_0-semigroup T on X and $\sigma(A) \cap i\mathbb{R}$ is countable, then every ultrapower of T is asymptotically almost periodic. Huang and Räbiger [HR94] proved a converse result, that $\sigma(A) \cap i\mathbb{R}$ is countable if every ultrapower of T is asymptotically almost periodic (the discrete version was given earlier by Nagel and Räbiger [NR93]; see also the work of Räbiger and Wolff [RW95]).

There was a considerable interval before the global Theorem 5.5.5 was improved to the local Theorem 5.5.3. Batty and Vũ [BV90] gave a few results on individual orbits, and an intermediate stage between the global and the local was considered by Huang [Hua93a], [Hua93b] and Batty [Bat96]. Theorem 5.5.3 was first proved by Batty, van Neerven and Räbiger [BNR98b], and the method given here is from [AB99].

Theorems 5.5.5 and 5.5.3, and their discrete analogues, have been extended in various directions: to weighted results (showing that $\|T(t)x\|/w(t) \to 0$ where w is a weight on $\mathbb{R}_+$ and $\|T(t)\| \le w(t)$), by Allan and Ransford [AR89], Vũ [Vu93], Kérchy [Kér97] and Batty and Yeates [BY00]; to representations of subsemigroups of locally compact abelian groups, by Lyubich and Vũ [LV90b], Batty and Vũ [BV92], Batty and Yeates [BY00] and Kérchy [Kér99]; to once integrated semigroups, by El-Mennaoui [Elm94] (using Theorem 5.5.5 and the extrapolation construction of Section 3.10); and to Volterra equations, by Arendt and Prüss [AP92].

Examples 5.5.7–5.5.10 are taken from papers of Arendt and Batty [AB88], Batty and Vu [BV90], and Batty, van Neerven and Räbiger [BNR98b]. Using a direct sum of weighted shifts, van Neerven [Nee00] has extended Example 4.7.11 to give an example of a C_0-semigroup T with a vector x such that u_x is bounded and uniformly continuous, $\sigma_u(A, x) = \{0\}$, and $\lim_{\alpha\downarrow 0} \alpha T(s)\widehat{u_x}(\alpha)$ exists for each $s \ge 0$, but u_x is not asymptotically almost periodic. Thus, the assumption of uniform convergence cannot be omitted from condition c) of Theorem 5.5.3.

Perron-Frobenius theory provides a special interplay between spectral properties of the generator and asymptotic behaviour of a positive semigroup. The lecture notes [Nag86] give a complete account of the state of the art in 1986. They were written before the results of Section 5.5 on countable spectrum were discovered. Consequently some results can now be formulated or proved more easily, and we describe some of them here.

A basic result in Perron-Frobenius theory is the cyclicity of the boundary spectrum. If A generates a bounded positive C_0-semigroup and $is \in \sigma(A)$ for some $s \in \mathbb{R}$, then $ins \in \sigma(A)$ for all $n \in \mathbb{Z}$. If the semigroup is eventually norm-continuous (i.e., norm-continuous on (t_0, ∞) for some $t_0 \ge 0$), then it follows that $\sigma(A) \cap i\mathbb{R} \subset \{0\}$. The following result was proved in [Nag86] by means of the discrete Katznelson-Tzafriri theorem (Corollary 4.7.15), but it now has the same proof as Theorem 5.5.6 b).

Theorem 5.7.7. *Let T be a bounded, positive, eventually norm-continuous, Cesàro-ergodic C_0-semigroup on a Banach lattice. Then there is a projection P such that $\lim_{t\to\infty} T(t) = P$ in the strong operator topology.*

A positive C_0-semigroup on a Banach lattice X is said to be *irreducible* if the only closed T–invariant ideals of X are $\{0\}$ and X (see Appendix C for the definition of ideals). If A generates a bounded, positive, irreducible, C_0-semigroup and A has compact resolvent, then Perron-Frobenius theory establishes that $\sigma(A) \cap i\mathbb{R}$ is either empty or equal to $2\pi i\tau\mathbb{Z}$ for some $\tau \geq 0$. The following is then an easy consequence of Proposition 5.4.7 (see [Are04, Sections 3.4, 3.5]).

Theorem 5.7.8. *Let T be a bounded, positive, irreducible C_0-semigroup on a Banach lattice X, and assume that the generator A has compact resolvent. Then there is a periodic C_0-group U on a closed subspace Y of X such that, for each $x \in X$ there exists $y \in Y$ such that $\lim_{t\to\infty} \|T(t)x - U(t)y\| = 0$.*

Let $X = L^p(\Omega, \mu)$, where (Ω, μ) is a σ-finite measure space and $1 \leq p < \infty$. A bounded operator S on X is a *kernel operator* if there is a measurable function $k : \Omega\times\Omega \to \mathbb{C}$ such that $(Sf)(y) = \int_\Omega k(y, z)\, d\mu(z)$ for μ-almost all $y \in \Omega$ and all $f \in X$. For abstract characterizations of kernel operators, see [Sch74, Proposition IV.9.8] or [Mey91, Theorem 3.3.11]. All bounded operators on ℓ^p are kernel operators. Many elliptic operators generate irreducible C_0-semigroups of kernel operators, and the properties of the kernels have been much studied (see [Dav90], [Ouh05], [Rob91]).

The following result is due to Greiner [Gre82] (see also [Are08]).

Theorem 5.7.9. *Let T be a positive, bounded, irreducible C_0-semigroup on $L^p(\Omega, \mu)$, with generator A. Suppose that $T(t_0)$ is a kernel operator for some $t_0 > 0$ and that $0 \in \sigma_p(A)$. Then there is a projection P of rank one such that $\lim_{t\to\infty} T(t) = P$ in the strong operator topology.*

It is remarkable that any positive, bounded C_0-semigroup of kernel operators on $L^p(\Omega, \mu)$ satisfies $\sigma_p(A) \cap i\mathbb{R} \subset \{0\}$. This is also due to Greiner [Gre82] and it was discussed more recently in [Dav05], [Kei06], [Wol07] and [Are08]. The Gaussian semigroup on $L^1(\mathbb{R}^n)$ is positive, contractive and irreducible, and it consists of kernel operators (see Example 3.7.6). However 0 is not an eigenvalue of its generator, and the semigroup does not converge.

It should be mentioned that for cosine functions countable spectrum also gives the expected asymptotic behaviour. The following result is due to Arendt and Batty [AB97, Proposition 4.9] (see also [Bas85, Theorem 10]).

Theorem 5.7.10. *Let* Cos *be a bounded cosine function on a Banach space X with generator A. Assume that the following conditions hold:*

a) $c_0 \not\subset X$;

b) *$\sigma(A)$ is countable; and*

c) $0 \notin \sigma(A)$.

Then for each $x, y \in X$ the function

$$u(t) := \mathrm{Cos}(t)x + \mathrm{Sin}(t)y \qquad (t \in \mathbb{R})$$

is almost periodic.

Note that u is the unique mild solution of $P^2(x,y)$ as defined in Section 3.14. Here,

$$\mathrm{Sin}(t)y := \int_0^t \mathrm{Cos}(s)y\,ds \qquad (t \in \mathbb{R}).$$

Since Cos is bounded, one knows that $\sigma(A) \subset (-\infty, 0]$. An equivalent formulation of the theorem is to say that, under the conditions a), b) and c), one has

$$X = \overline{\mathrm{span}}\left\{x \in D(A) : \text{ there exists } \eta \in \mathbb{R} \text{ such that } Ax = -\eta^2 x\right\}.$$

Surveys of the topics of this section have previously been written by Batty [Bat94], Vũ [Vu97], van Neerven [Nee96c, Chapter 5], and Chill and Tomilov [CT07].

Section 5.6

A version of Theorem 5.6.5 was first given by Basit [Bas97] under the stronger assumption that f has an extension $g \in \mathrm{BUC}(\mathbb{R}, X)$ such that $\sigma(A) \cap i\,\mathrm{sp}_C(g)$ is empty (see Section 4.6). Basit's method originated in Lyapunov's finite-dimensional theory, and it was developed and applied to infinite-dimensional Cauchy problems on the line by Vũ [Vu91], Ruess and Vũ [RV95] and Schüler and Vũ [SV98]. It involves solving operator equations of the form $AY - YB = C$ (Lyapunov equations) for an operator Y from a subspace of $\mathrm{BUC}(\mathbb{R}, X)$ into X. The proof of Theorem 5.6.5 given here is due to Batty and Chill [BC99] who carried out the basic argument in a more general situation in which T is not necessarily a semigroup. It shows that the conclusion of boundedness in Theorem 5.6.5 is valid, not only for holomorphic semigroups, but also for many eventually norm-continuous semigroups and for asymptotically norm-continuous semigroups on Hilbert space. Blake [Bla99] has further refined the method and extended Theorem 5.6.5 to various classes of asymptotically norm-continuous semigroups on Banach spaces, including eventually differentiable semigroups (for the definition of asymptotically norm-continuous semigroups, see the Notes on Section 5.2 above). Theorem 5.6.6 also appeared in [BC99]. The role of smooth functions in the proof also leads to another result of [BC99] that $T * f$ is bounded and uniformly continuous if T is bounded, $R(i\eta, A)$ exists and is bounded for large $|\eta|$, and f is bounded with bounded derivatives of first and second order.

Proposition 5.6.7 and Theorem 5.6.8 are due to Arendt and Batty [AB99], following results of Ruess and Vũ [RV95] for inhomogeneous Cauchy problems on $\mathbb{R}$. Batty, Hutter and Räbiger [BHR99] obtained a version for periodic Cauchy problems. Applications to inhomogeneous Volterra equations have been given by Arendt and Batty [AB00], Chill and Prüss [CP01] and Fasangova and Prüss [FP01].

Part III

Applications and Examples

In this part of the book we present some applications and examples which illustrate how the theory developed in Parts I and II can be used. There are three chapters which are independent of each other; they all use basic concepts from distribution theory which can be found in Appendix E.

In Chapter 6 the heat equation with inhomogeneous boundary conditions is investigated. The idea of the approach presented here is to work entirely in spaces of continuous functions. We assume that Ω is a bounded open set on which the Dirichlet problem is well-posed. This is a very weak regularity assumption on the boundary, and it is well known from potential theory. Based on this assumption, the results of this chapter rely on the methods developed in Parts I and II and do not use complicated results of partial differential equations. Resolvent positive operators (as developed in Section 3.11) play an important role giving the transition from the elliptic problem to a parabolic problem. Results of Part II will be used to show how the asymptotic behaviour of the given function on the boundary determines the asymptotic behaviour of the solution.

In the approach of Chapter 6 we do not use Hilbert space techniques at all. This is different in Chapter 7 where we prove well-posedness of a fairly general hyperbolic equation in $L^2(\Omega)$. The results are based on the theory of cosine functions as they are presented in Section 3.14. Most important is the role of the phase space introduced there. We need a brief introduction to quadratic form methods which is given in Section 7.1. Here we only consider the most simple case; the spectral theorem for selfadjoint operators (as stated in Appendix B) plays a major role in this introduction. Then the results of Chapter 7 follow from the general theory of cosine functions given in Section 3.14.

In Chapter 8 differential operators with constant coefficients on $\mathbb{R}^n$, and more generally pseudo-differential operators, are considered. With the help of the notion of integrated semigroups, precise results on well-posedness and regularity of the corresponding parabolic problem in L^p-spaces are obtained. In particular, it will be shown that the wave equation is not well-posed in the semigroup sense on $L^p(\mathbb{R}^n)$ for $p \neq 2$. This explains why Chapter 7 is restricted to $L^2(\Omega)$. However, it will be shown that the wave equation as well as some other equations from mathematical physics lead to k-times integrated semigroups on L^p-spaces. An important issue is to determine the best possible value of k depending on p. This tells us something about the regularity properties of the equations which we consider. A principal tool in Chapter 8 is the theory of Fourier multipliers. A resumé of some of the required results, including Mikhlin's theorem about Fourier multipliers on $L^p(\mathbb{R}^n)$ for $1 < p < \infty$, is given in Appendix E without proofs. Some further results on Fourier multipliers are proved in Section 8.2, including a weak form of Mikhlin's theorem which is valid for $L^1(\mathbb{R}^n)$.

Chapter 6

The Heat Equation

In this chapter we consider the Laplacian on spaces of continuous functions. If $\Omega \subset \mathbb{R}^n$ is an open, bounded set with boundary $\partial\Omega$ which is Dirichlet regular, we will show that the Laplacian generates a holomorphic semigroup on the space

$$C_0(\Omega) := \{u \in C(\overline{\Omega}) : \ u|_{\partial\Omega} = 0\}.$$

Furthermore, using the theory of resolvent positive operators developed in Section 3.11 we show that the heat equation with inhomogeneous boundary conditions is well-posed. We use the results of Chapter 5 to study the asymptotic behaviour of its solutions.

6.1 The Laplacian with Dirichlet Boundary Conditions

Let $\Omega \subset \mathbb{R}^N$ be an open, bounded set with boundary $\partial\Omega =: \Gamma$. Given $\varphi \in C(\Gamma)$, we consider the *Dirichlet problem*

$$D(\varphi) \quad \begin{cases} u \in C(\overline{\Omega}), \\ u|_\Gamma = \varphi, \\ \Delta u = 0 \qquad \text{in } \mathcal{D}(\Omega)'. \end{cases}$$

Here and throughout this chapter, $\mathcal{D}(\Omega)'$ is the space of all distributions on Ω, and we identify functions in $C(\overline{\Omega})$ with their restrictions to Ω and locally integrable functions on Ω with distributions on Ω. Thus, $C(\overline{\Omega}) \subset C(\Omega) \subset L^1_{loc}(\Omega) \subset \mathcal{D}(\Omega)'$, and the third line of $D(\varphi)$ says that $\int_\Omega u\Delta\psi \, dx = 0$ for all $\psi \in \mathcal{D}(\Omega)$ (see Appendix E). Although we allow complex-valued functions φ and u, $D(\varphi)$ is essentially a real problem; u is a solution of $D(\varphi)$ if and only if $\operatorname{Re} u$ and $\operatorname{Im} u$ are solutions of $D(\operatorname{Re}\varphi)$ and $D(\operatorname{Im}\varphi)$ respectively. It is well known that a function satisfying $D(\varphi)$ is in $C^\infty(\Omega)$ (see [Rud91, p.220], for example). So we look for harmonic functions in Ω having a continuous extension to the boundary with prescribed boundary values. We will not talk about methods to solve the Dirichlet problem. For us, it serves as a reference problem. In fact, this problem is well studied in potential theory (see [DL90, Chapter II], [Hel69], [GT83], [Kel67], [Lan72]).

Many geometric properties of the boundary are known to be sufficient for well-posedness of the Dirichlet problem.

Definition 6.1.1. *The set Ω is called* Dirichlet regular *if for all $\varphi \in C(\Gamma)$ there exists a solution of $D(\varphi)$.*

Examples 6.1.2. a) If $n = 1$, then each bounded open set is Dirichlet regular (see [DL90, Chapter II, Section 4, Example 6]). On the other hand, if $n \geq 2$ and $\Omega \subset \mathbb{R}^n$ is open, then $\Omega \setminus \{z\}$ is not Dirichlet regular for any $z \in \Omega$ (see [DL90, Chapter II, Section 4, Remark 1]).

b) If the boundary of Ω is C^1, or more generally, Lipschitz continuous, then Ω is Dirichlet regular (see [DL90, Chapter II, Section 4, Proposition 4]).

c) For $n = 2$, any simply connected $\Omega \subset \mathbb{R}^2$ is Dirichlet regular (see [Con73, Chapter X, Corollary 4.18]).

d) For $n = 3$, Lebesgue's cusp is a simply connected set which is not Dirichlet regular (see [Lan72, p.287], [AD08]). □

Next, we establish the elliptic maximum principle. It will be important for us to consider distributional inequalities which are easy to define. Let $f \in \mathcal{D}(\Omega)'$. We write

$$f \geq 0 \;\text{ if }\; \langle \varphi, f\rangle \geq 0 \;\text{ for all }\; \varphi \in \mathcal{D}(\Omega)_+, \tag{6.1}$$

where $\mathcal{D}(\Omega)_+ := \{\varphi \in \mathcal{D}(\Omega) : \varphi(x) \geq 0 \;\text{ for all } x \in \Omega\}$. If $f \in L^1_{loc}(\Omega)$ is identified with a distribution in $\mathcal{D}(\Omega)'$, then $f \geq 0$ as a distribution if and only if $f(x) \geq 0$ a.e. in Ω.

Theorem 6.1.3 (Elliptic maximum principle). *Let $M \geq 0,\ \lambda \geq 0,\ u \in C(\overline{\Omega})$ such that*

a) *$\lambda u - \Delta u \leq 0$ in $\mathcal{D}(\Omega)'$; and*

b) *$u|_\Gamma \leq M$.*

Then $u \leq M$ on $\overline{\Omega}$.

Proof. By considering the real and imaginary parts of u separately, we may assume that u is real-valued. Let $c := \max_{x\in\overline{\Omega}} u(x)$.

First case: We assume that $u \in C^2(\Omega)$. Assume that $c > M$. Let $\gamma > \sqrt{\lambda}$ and $\delta := \sup_{x\in\Omega} e^{\gamma x_1}$ (where $x = (x_1, \ldots, x_n)$). Choose $\varepsilon > 0$ such that $M + \varepsilon\delta < c$. Let $v(x) := u(x) + \varepsilon e^{\gamma x_1}$. Then $v \in C^2(\Omega) \cap C(\overline{\Omega})$ and $v \leq M + \varepsilon\delta < c$ on Γ, but $\max_{x\in\overline{\Omega}} v(x) \geq c$. Thus, there exists $x_0 \in \Omega$ such that $v(x_0) = \max_\Omega v(x)$. It follows that $D_j^2 v(x_0) = \frac{d^2}{dt^2} v(x_0 + te_j) \le 0$ (where $e_j = (0,0,\dots,1,0,\dots,0)$), and so $\Delta v(x_0) \le 0$. Hence,

$$\begin{aligned} 0 \le \lambda v(x_0) - \Delta v(x_0) &= \lambda u(x_0) - \Delta u(x_0) + \varepsilon e^{\gamma x_{01}}(\lambda - \gamma^2) \\ &\le \varepsilon e^{\gamma x_{01}}(\lambda - \gamma^2), \end{aligned}$$

which is a contradiction.

Second case: Now let u be arbitrary and assume that $c > M$. Let

$$K := \{x \in \Omega : \ u(x) = c\}.$$

Then assumption b) implies that K is a non-empty compact subset of Ω. Let $\Omega' \subset \Omega$ be open such that $K \subset \Omega' \subset \overline{\Omega'} \subset \Omega$. Then $c_1 := \sup_{\partial\Omega'} u(x) < c$. Let $c_1 < c_2 < c$, $c_2 > M$. Denote by $(\rho_k)_{k\in\mathbb{N}}$ a mollifier in $C_c^\infty(\mathbb{R}^n)$ with $\rho_k \ge 0$ and $\operatorname{supp} \rho_k \subset \{y \in \mathbb{R}^n : \ |y| < 1/k\}$ (see Section 1.3). Then

$$v_k(x) := (\rho_k * u)(x) = \int_{|y|<1/k} u(x-y)\rho_k(y)\, dy$$

is defined for $x \in \Omega_k := \{y \in \Omega : \ \operatorname{dist}(y,\Gamma) > 1/k\}$ and $v_k \in C^\infty(\Omega_k)$. Moreover, v_k converges to u uniformly on compact subsets of Ω as $k \to \infty$. Hence, there exists $k \in \mathbb{N}$ such that $\overline{\Omega'} \subset \Omega_k$, $\sup_{\Omega'} v_k > c_2$ and $\sup_{\partial\Omega'} v_k < c_2$. But

$$\lambda v_k(x) - \Delta v_k(x) = \langle \rho_k(x - \cdot), \lambda u - \Delta u\rangle \le 0$$

for all $x \in \Omega'$. This is impossible by the first case. □

It follows immediately from Theorem 6.1.3 that $D(\varphi)$ has at most one solution for all each $\varphi \in C(\Gamma)$ and the solution is real if φ is real-valued.

Next, we consider the space $X := C(\overline{\Omega}) \times C(\Gamma)$ which is a Banach lattice for the ordering

$$(u,\varphi) \ge 0 \Longleftrightarrow u \ge 0 \ \text{ and } \ \varphi \ge 0$$

$(u \in C(\overline{\Omega}), \varphi \in C(\Gamma))$ and the norm

$$\|(u,\varphi)\| := \max\left\{\|u\|_{C(\overline{\Omega})}, \ \|\varphi\|_{C(\Gamma)}\right\},$$

with

$$\begin{aligned} \|u\|_{C(\overline{\Omega})} &:= \max_{x\in\overline{\Omega}} |u(x)|, \\ \|\varphi\|_{C(\Gamma)} &:= \max_{z\in\Gamma} |\varphi(z)|. \end{aligned}$$

On $C(\overline{\Omega})$ we consider the Laplacian $\Delta_{\max}$ with maximal domain; i.e.,

$$\begin{aligned} D(\Delta_{\max}) &:= \{u \in C(\overline{\Omega}) : \ \Delta u \in C(\overline{\Omega})\}, \\ \Delta_{\max} u &:= \Delta u \ \text{ in } \ \mathcal{D}(\Omega)'. \end{aligned}$$

It is obvious that $\Delta_{\max}$ is a closed operator.

Remark 6.1.4. It is known that $D(\Delta_{\max}) \not\subset C^2(\Omega)$ whenever Ω is a non-empty open set in $\mathbb{R}^n$ $(n \geq 2)$. However, we will see in Lemma 6.1.5 that $D(\Delta_{\max})$ is contained in $C^1(\Omega)$ (cf. Remark 3.7.7 b)). Thus, there always exist functions $u \in C^1(\Omega)$ such that $\Delta u \in C(\Omega)$ but for some i, j, the distribution $D_i D_j u$ is not a function in $C(\Omega)$. This fact may be considered as unpleasant. However, for our purposes it does not matter. We will see that solutions of the heat equation are always of class C^∞. □

We consider the operator A on X given by

$$\begin{aligned} D(A) &:= D(\Delta_{\max}) \times \{0\}, \\ A(u,0) &:= (\Delta u, -u|_\Gamma). \end{aligned}$$

Thus, for $u \in D(\Delta_{\max})$, $f \in C(\overline{\Omega})$, $\varphi \in C(\Gamma)$, we have $-A(u,0) = (f, \varphi)$ if and only if

$$\begin{cases} -\Delta u &= f \text{ in } \mathcal{D}(\Omega)', \\ u|_\Gamma &= \varphi; \end{cases} \tag{6.2}$$

i.e., if and only if u solves Poisson's equation. For this reason we call A the *Poisson operator*. Since $\Delta_{\max}$ is closed, it follows that A is also closed.

By E_n we denote the *Newtonian potential*; i.e., $E_n : \mathbb{R}^n \setminus \{0\} \to \mathbb{R}$ is given by

$$E_n(x) := \begin{cases} |x|/2 & \text{if } n = 1, \\ \dfrac{\log|x|}{2\pi} & \text{if } n = 2, \\ -\dfrac{1}{n(n-2)\omega_n} \dfrac{1}{|x|^{n-2}} & \text{if } n \geq 3, \end{cases}$$

where $\omega_n := |B(0,1)|$ is the volume of the unit ball in $\mathbb{R}^n$. Then $E_n \in C^\infty(\mathbb{R}^n \setminus \{0\})$ and E_n, $D_j E_n \in L^1_{loc}(\mathbb{R}^n)$ $(j = 1, \ldots, n)$, as is easy to see.

Let $f \in C_c(\mathbb{R}^n)$. Then one has $v := E_n * f \in C^1(\mathbb{R}^n)$. Moreover,

$$\Delta v = f \quad \text{in} \quad \mathcal{D}(\mathbb{R}^n)'. \tag{6.3}$$

We refer to ([DL90, Chapter II, Section 3]) for this standard fact of distribution theory. Frequently, $v = E_n * f$ is called the *Newtonian potential* of f. Note however, that $v \notin C^2(\mathbb{R}^n)$ in general.

We deduce the following regularity result which will be useful. In the proof and elsewhere in this chapter, we do not distinguish notationally between functions on Ω and their restrictions to $\Omega' \subset \Omega$.

Lemma 6.1.5. *Let $u, f \in C(\Omega)$ such that $\Delta u = f$ in $\mathcal{D}(\Omega)'$. Then $u \in C^1(\Omega)$. If $f \in C^k(\Omega)$ for some $k \in \mathbb{N}$, then $u \in C^{k+1}(\Omega)$.*

Proof. a) Let Ω' be open such that $\overline{\Omega'} \subset \Omega$. Let $\rho \in \mathcal{D}(\Omega)$ such that $\rho(x) = 1$ on $\overline{\Omega'}$. Consider $\rho f \in C_c(\mathbb{R}^n)$ and $v := E_n * (\rho f) \in C^1(\mathbb{R}^n)$. Then $\Delta v = \rho f$ in $\mathcal{D}(\mathbb{R}^n)'$ (see (6.3)). Hence,

$$\Delta(u - v) = f - \rho f = 0 \quad \text{in} \quad \mathcal{D}(\Omega')'.$$

Thus, $u-v$ is harmonic and so in $C^\infty(\Omega')$. Consequently, $u = (u-v)+v \in C^1(\Omega')$. Since Ω' is arbitrary, the first assertion is proved.

b) We prove the second assertion. It is true for $k = 0$ by a). Assume that it holds for $k \in \mathbb{N}_0$. Assume that $f \in C^{k+1}(\Omega)$. Then $u \in C^1(\Omega)$ by a) and $\Delta D_j u = D_j \Delta u = D_j f$ in $\mathcal{D}(\Omega)'$. Since $D_j f \in C^k(\Omega)$, it follows that $D_j u \in C^{k+1}(\Omega)$ $(j = 1, \dots, n)$ by the inductive hypothesis. Hence, $u \in C^{k+2}(\Omega)$. □

Theorem 6.1.6. *Assume that Ω is Dirichlet regular. The Poisson operator A is resolvent positive and $s(A) < 0$.*

Proof. a) Let $\lambda \geq 0$ and suppose that $\lambda \in \rho(A)$. Then $R(\lambda, A) \geq 0$. In fact, let $f \in C(\overline{\Omega})$, $\varphi \in C(\Gamma)$, $(u, 0) = R(\lambda, A)(f, \varphi)$. Then $\lambda u - \Delta u = f$ in $\mathcal{D}(\Omega)'$ and $u|_\Gamma := \varphi$. If $f \leq 0$ and $\varphi \leq 0$, it follows from Theorem 6.1.3 that $u \leq 0$ in $\overline{\Omega}$.

b) We show that $0 \in \rho(A)$. Let $f \in C(\overline{\Omega})$ and $\varphi \in C(\Gamma)$. Let $\tilde{f} \in C_c(\mathbb{R}^n)$ be an extension of f. Let $w := -E_n * \tilde{f}$. Then $w \in C(\mathbb{R}^n)$ and $-\Delta w = f$ in $\mathcal{D}(\Omega)'$. Let v be the solution of the Dirichlet problem $D(\varphi - \psi)$, where $\psi = w|_\Gamma$. Then $u = v + w \in C(\overline{\Omega})$, $u|_\Gamma = \varphi$ and

$$\Delta u = \Delta v + \Delta w = \Delta w = -f \text{ in } \mathcal{D}(\Omega)'.$$

We have shown that A is surjective. It follows from Theorem 6.1.3 that A is injective. Hence, $0 \in \rho(A)$ since A is closed.

c) Let $Q := \mathbb{R}_+ \cap \rho(A)$. Then by a), $R(\lambda, A) \geq 0$ for all $\lambda \in Q$ and

$$R(0, A) - R(\lambda, A) = \lambda R(\lambda, A) R(0, A) \geq 0,$$

hence $0 \leq R(\lambda, A) \leq R(0, A)$ for all $\lambda \in Q$. Thus, $\|R(\lambda, A)\| \leq \|R(0, A)\|$ for all $\lambda \in Q$. By Corollary B.3, it follows that $|\lambda - \mu| \geq \|R(\lambda, A)\|^{-1} \geq \|R(0, A)\|^{-1}$ for all $\lambda \in Q$, $\mu \in \sigma(A)$. Since $0 \in Q$, this implies that $Q = \mathbb{R}_+$. It follows from Proposition 3.11.2 that $s(A) < 0$. □

In the remainder of this chapter we assume that Ω is Dirichlet regular.

The Poisson operator is not densely defined and is not a Hille-Yosida operator since

$$\|\lambda R(\lambda, A)\| \geq \lambda \qquad (\lambda > 0).$$

In fact, let $(u, 0) = R(\lambda, A)(0, 1_\Gamma)$. Then $u|_\Gamma = 1_\Gamma$. Hence by Theorem 6.1.3,

$$\|\lambda R(\lambda, A)\| \geq \|\lambda(u, 0)\| = \lambda.$$

Moreover, since the polynomials are dense in $C(\overline{\Omega})$ by the Stone-Weierstrass theorem, it follows that

$$\overline{D(A)} = C(\overline{\Omega}) \times \{0\}. \tag{6.4}$$

If we consider the part A_c of A in $\overline{D(A)} = C(\overline{\Omega}) \times \{0\}$, then we obtain a Hille-Yosida operator as we shall see in the next theorem. The operator A_c is given by

$$\begin{aligned} D(A_c) &= \{(u, 0) : \, u \in D(\Delta_{\max}), \, u|_\Gamma = 0\}, \\ A_c(u, 0) &= (\Delta_{\max} u, 0). \end{aligned}$$

Since $R(\lambda, A)X \subset \overline{D(A)}$, it follows from Proposition B.8 that $(s(A), \infty) \subset \rho(A_c)$ and $R(\lambda, A_c) = R(\lambda, A)|_{\overline{D(A)}}$ for all $\lambda > s(A)$. Thus, A_c is a resolvent positive operator and $s(A_c) \leq s(A) < 0$.

Identifying $C(\overline{\Omega}) \times \{0\}$ with $C(\overline{\Omega})$, A_c is identified with the operator Δ_c on $C(\overline{\Omega})$ given by

$$\begin{aligned} D(\Delta_c) &= C_0(\Omega) \cap D(\Delta_{\max}), \\ \Delta_c u &= \Delta u \text{ in } \mathcal{D}(\Omega)'. \end{aligned}$$

Here, $C_0(\Omega) = \{u \in C(\overline{\Omega}) : u|_\Gamma = 0\}$.

Theorem 6.1.7. *Assume that Ω is Dirichlet regular. Then the operator Δ_c on $C(\overline{\Omega})$ is dissipative and resolvent positive, and $s(\Delta_c) < 0$.*

Proof. We have established above that Δ_c is resolvent positive and $s(\Delta_c) < 0$. It remains to show that Δ_c is dissipative. Let $t > 0$, $u \in D(\Delta_c)$, $u - t\Delta u = f$. We have to show that $\|u\|_{C(\overline{\Omega})} \leq \|f\|_{C(\overline{\Omega})}$. Let $M := \|f\|_{C(\overline{\Omega})}$. Let $\theta \in [0, 2\pi]$ and $v := \operatorname{Re}(e^{i\theta}u)$. Then $(v - M) - t\Delta(v - M) = \operatorname{Re}(e^{i\theta}f) - M \leq 0$ in $\mathcal{D}(\Omega)'$. It follows from Theorem 6.1.3 that $v - M \leq 0$; i.e., $\operatorname{Re}(e^{i\theta}u) \leq M$ for all θ. We deduce that $\|u\|_{C(\overline{\Omega})} \leq M$. □

Note that the operator Δ_c is also not densely defined. So we consider the part Δ_0 of Δ_c in $C_0(\Omega) = \overline{D(\Delta_c)}$. Then Δ_0 is given by

$$\begin{aligned} D(\Delta_0) &= \{u \in C_0(\Omega) : \Delta u \in C_0(\Omega)\}, \\ \Delta_0 u &= \Delta u \text{ in } \mathcal{D}(\Omega)'. \end{aligned}$$

Since $\mathcal{D}(\Omega) \subset D(\Delta_0)$, the operator Δ_0 is densely defined. Moreover, for $\lambda > s(\Delta_c)$, $R(\lambda, \Delta_c)C_0(\Omega) \subset D(\Delta_c) \subset C_0(\Omega)$. Consequently, $(s(\Delta_c), \infty) \subset \rho(\Delta_0)$ and $R(\lambda, \Delta_0) = R(\lambda, \Delta_c)|_{C_0(\Omega)}$ for all $\lambda > s(\Delta_c)$. Hence, $\|\lambda R(\lambda, \Delta_0)\| \leq \|\lambda R(\lambda, \Delta_c)\| \leq 1$ for $\lambda > 0$. Applying the Hille-Yosida theorem we obtain the following result.

Theorem 6.1.8. *Assume that Ω is Dirichlet regular. Then the operator Δ_0 generates a positive contractive C_0-semigroup T_0 on $C_0(\Omega)$.*

Next, we prove holomorphy of T_0. More generally, the following holds.

Theorem 6.1.9. *Assume that Ω is Dirichlet regular. Then Δ_c generates a bounded holomorphic semigroup T_c on $C(\overline{\Omega})$. The operator Δ_0 generates a bounded holomorphic C_0-semigroup on $C_0(\Omega)$.*

Proof. We recall from Example 3.7.6 that the Laplacian generates a bounded holomorphic C_0-semigroup on $C_0(\mathbb{R}^n)$; i.e., defining the operator $L := \Delta_{C_0(\mathbb{R}^n)}$ on $C_0(\mathbb{R}^n)$ by

$$\begin{aligned} D(L) &:= \{f \in C_0(\mathbb{R}^n) : \Delta f \in C_0(\mathbb{R}^n)\}, \\ Lu &:= \Delta u \text{ in } \mathcal{D}(\mathbb{R}^n)', \end{aligned}$$

there exists $M \geq 0$ such that

$$\lambda \in \rho(L) \text{ and } \|\lambda R(\lambda, L)\| \leq M \tag{6.5}$$

whenever $\operatorname{Re}\lambda > 0$.

We show a similar estimate for Δ_c. Let $f \in C(\overline{\Omega})$, $\operatorname{Re}\lambda > 0$, $g := R(\lambda, \Delta_c)f$. Let $\tilde{f} \in C_0(\mathbb{R}^n)$ be an extension of f with $\|\tilde{f}\|_{C_0(\mathbb{R}^n)} = \|f\|_{C(\overline{\Omega})}$ and let $\tilde{g} := R(\lambda, L)\tilde{f}$. Let $\varphi := \tilde{g}|_\Gamma$ and $(w, 0) = R(\lambda, A)(0, \varphi)$, where A is the Poisson operator on $X = C(\overline{\Omega}) \times C(\Gamma)$. Then $w \in C(\overline{\Omega})$, $\lambda w - \Delta w = 0$ in $\mathcal{D}(\Omega)'$ and $w|_\Gamma = \varphi = \tilde{g}|_\Gamma$. Thus, $h := \tilde{g} - w \in C_0(\Omega)$ and $\lambda h - \Delta h = \lambda\tilde{g} - \Delta\tilde{g} = f$ in $\mathcal{D}(\Omega)'$. It follows that $h = g$. Observe that

$$\|R(\lambda, A)\| \leq \|R(0, A)\|$$

by Corollary 3.11.3 (since A is resolvent positive and $s(A) < 0$). Hence,

$$\|w\|_{C(\overline{\Omega})} \leq \|R(0, A)\| \; \|\varphi\|_{C(\Gamma)} \leq \|R(0, A)\| \; \|\tilde{g}\|_{C(\overline{\Omega})}.$$

Hence, setting $c := 1 + \|R(0, A)\|$ we have,

$$\|g\|_{C(\overline{\Omega})} = \|\tilde{g} - w\|_{C(\overline{\Omega})} \leq c\|\tilde{g}\|_{C(\overline{\Omega})} \leq \frac{cM}{|\lambda|}\|\tilde{f}\|_{C_0(\mathbb{R}^n)} = \frac{cM}{|\lambda|}\|f\|_{C(\overline{\Omega})}$$

by (6.5).

We have shown that

$$\|\lambda R(\lambda, \Delta_c)\| \leq cM \qquad (\operatorname{Re}\lambda > 0). \tag{6.6}$$

It follows from Corollary 3.7.12 that Δ_c generates a bounded holomorphic semigroup T_c on $C(\overline{\Omega})$. Since Δ_0 is the part of Δ_c in $\overline{D(\Delta_c)} = C_0(\Omega)$, the second assertion is an immediate consequence (see Remark 3.7.13). □

We should mention that

$$\|T_c(t)\| \leq 1 \qquad \text{for all} \quad t > 0. \tag{6.7}$$

This follows from Proposition 3.7.16. In fact, one can show that Δ_c is dissipative (regularizing as in Theorem 6.1.3). We will not use (6.7) and we omit the proof. Moreover,

$$T_0(t) = T_c(t)|_{C_0(\Omega)} \qquad (t > 0). \tag{6.8}$$

This follows from (3.46). It follows from Theorem 5.1.12 or Theorem 5.3.8 that $s(\Delta_0) = \omega(T_0)$. Since $s(\Delta_0) < 0$, we conclude that $\omega(T_0) < 0$. Since T_c is holomorphic we have $T_c(t)C(\overline{\Omega}) \subset D(\Delta_c) \subset C_0(\Omega)$ for all $t > 0$, hence $T_c(t+s) = T_0(t)T_c(s)$ for all $t > 0$, $s > 0$. We conclude that

$$\|T_c(t)\| \leq Me^{-\varepsilon t} \qquad (t \geq 0) \tag{6.9}$$

for some $M > 0$ and $\varepsilon > 0$.

Finally, as a consequence of Theorem 6.1.9 we note the following regularity result.

Proposition 6.1.10. *Assume that Ω is Dirichlet regular. Let $u_0 \in C_0(\Omega)$, and let $u(t,x) = (T_0(t)u_0)(x)$ $(t \geq 0, x \in \overline{\Omega})$. Then*

$$\begin{cases} u \in C^\infty((0,\infty) \times \Omega) \cap C(\mathbb{R}_+ \times \overline{\Omega}), \\ u_t(t,x) = \Delta u(t,x) \quad (t > 0, x \in \Omega), \\ u(0,x) = u_0(x) \quad (x \in \Omega). \end{cases} \tag{6.10}$$

Proof. Since T_0 is holomorphic, for $f \in C_0(\Omega)$ we have $T_0(\cdot)f \in C^\infty((0,\infty), D(\Delta_0^k))$ for all $k \in \mathbb{N}$. It follows from Lemma 6.1.5 that $D(\Delta_0^k) \subset C^k(\Omega)$. The closed graph theorem implies that for each open set Ω' such that $\overline{\Omega'} \subset \Omega$ the restriction map of $D(\Delta_0^k)$ into $C^k(\overline{\Omega'})$ is continuous where $D(\Delta_0^k)$ carries the graph norm and $C^k(\overline{\Omega'})$ the norm

$$\|u\|_{C^k(\overline{\Omega'})} = \max\{\|D^\alpha u\|_{C(\overline{\Omega'})} : |\alpha| \leq k\},$$

where $\alpha = (\alpha_1, \ldots, \alpha_n)$ is a multi-index, $|\alpha| = \sum_{j=1}^n \alpha_i$, $D^\alpha = D_1^{\alpha_1} \ldots D_n^{\alpha_n}$ and $D^0 u = u$. In particular, $T(\cdot)f \in C^\infty((0,\infty), C^k(\overline{\Omega'}))$ for all $k \in \mathbb{N}$. This implies that the function satisfies (6.10). □

6.2 Inhomogeneous Boundary Conditions

In Section 6.1 we solved the elliptic problem showing that the Poisson operator is resolvent positive. Now we prove well-posedness of an evolutionary problem with time-dependent boundary conditions, by converting it into an inhomogeneous Cauchy problem and using the results of Chapter 3. We keep the notation of Section 6.1.

Let $\tau > 0$. Given $u_0 \in C(\overline{\Omega})$ and $\varphi \in C([0,\tau], C(\Gamma))$, we consider the parabolic problem

$$P_\tau(u_0,\varphi) \quad \begin{cases} u'(t) &= \Delta u(t) \qquad (t \in [0,\tau]), \\ u(t)|_\Gamma &= \varphi(t) \qquad (t \in [0,\tau]), \\ u(0) &= u_0. \end{cases}$$

Remark 6.2.1. Let $n = 3$. Then Ω is a solid body and the solution u of $P_\tau(u_0,\varphi)$ describes the heat flow in Ω. More precisely, for $x \in \Omega$, $u_0(x)$ is the given initial temperature at the point $x \in \Omega$. For $t \in [0,\tau]$, $z \in \Gamma$, the quantity $\varphi(t)(z)$ is the given temperature at z at the time t. We may imagine that the boundary is heated by some resistance surrounding Ω. Then the solution $u(t)(x)$ is the temperature at the point $x \in \Omega$ at the time $t \in [0,\tau]$. □

Definition 6.2.2. *A* mild solution *of $P_\tau(u_0,\varphi)$ is a function $u \in C([0,\tau], C(\overline{\Omega}))$ such that*

$$\Delta \int_0^t u(s)\,ds = u(t) - u_0 \quad \textit{in } \mathcal{D}(\Omega)' \tag{6.11}$$

and

$$u(t)|_\Gamma = \varphi(t) \tag{6.12}$$

for all $t \in [0, \tau]$.

Note that (6.11) implies in particular that $\int_0^t u(s)\,ds \in D(\Delta_{\max})$ for all $t \in [0, \tau]$.

Consider the Poisson operator A on $X = C(\overline{\Omega}) \times C(\Gamma)$ given by $A(u, 0) = (\Delta u, -u|_\Gamma)$ on $D(A) = D(\Delta_{\max}) \times \{0\}$. Given $U_0 \in X$ and $\Phi \in C([0, \tau], X)$, we consider the Cauchy problem

$$\begin{cases} U'(t) = & AU(t) + \Phi(t) \quad (t \in [0, \tau]), \\ U(0) = & U_0. \end{cases} \tag{6.13}$$

Proposition 6.2.3. *Let* $u_0 \in C(\overline{\Omega})$, $U_0 := (u_0, 0)$, *and let* $\varphi \in C([0, \tau], C(\Gamma))$, $\Phi(t) := (0, \varphi(t))$. *Let* $U \in C([0, \tau], X)$. *Then* U *is a mild solution of* (6.13) *if and only if* U *is of the form* $U(t) = (u(t), 0)$ $(t \in [0, \tau])$ *where* u *is a mild solution of* $P_\tau(u_0, \varphi)$.

Proof. If U is a mild solution of (6.13), then

$$U(t) = \frac{d}{dt} \int_0^t U(s)\,ds \in \overline{D(A)} = C(\overline{\Omega}) \times \{0\}$$

for all $t \in [0, \tau]$. Thus, $U(t) = (u(t), 0)$ for some $u \in C([0, \tau], C(\overline{\Omega}))$. Now the claim is immediate from the definition of A and Definition 6.2.2. □

Now let $u_0 \in D(\Delta_{\max})$, $U_0 = (u_0, 0) \in D(A)$. Let $\varphi \in C([0, \tau], C(\Gamma))$, $\Phi(t) = (0, \varphi(t))$. Then

$$AU_0 + \Phi(0) = (\Delta u_0, -u_0|_\Gamma + \varphi(0)).$$

Thus, the consistency condition

$$AU_0 + \Phi(0) \in \overline{D(A)} = C(\overline{\Omega}) \times \{0\} \tag{6.14}$$

from Theorem 3.11.10 becomes

$$u_0|_\Gamma = \varphi(0). \tag{6.15}$$

This is obviously a necessary condition for the existence of a mild solution of $P_\tau(u_0, \varphi)$. Now we obtain the following from Theorem 3.11.10.

Proposition 6.2.4. *Assume that* Ω *is Dirichlet regular. Let* $\varphi_0 \in C(\Gamma)$, $\varphi' \in L^1((0, \tau), C(\Gamma))$, $\varphi(t) := \varphi_0 + \int_0^t \varphi'(s)\,ds$ $(t \in [0, \tau])$. *Let* $u_0 \in D(\Delta_{\max})$. *If condition* (6.15) *is satisfied, then there exists a unique mild solution of* $P_\tau(u_0, \varphi)$.

Next, we obtain the weak parabolic maximum principle as a direct consequence of Theorem 3.11.11. It will serve as an a priori estimate for solutions of $P_\tau(u_0, \varphi)$.

Proposition 6.2.5 (Parabolic maximum principle). *Assume that Ω is Dirichlet regular. Let u be a mild solution of $P_\tau(u_0,\varphi)$, where u_0 and φ are real-valued. Let $c_+,\ c_- \in \mathbb{R}$ be constants such that*

$$\begin{array}{lllll} c_- & \le & u_0 & \le c_+ & \text{and} \\ c_- & \le & \varphi(t) & \le c_+ & (t \in [0,\tau]). \end{array}$$

Then $c_- \le u(t) \le c_+$ $(t \in [0,\tau])$.

Proof. Note that $e(t) := c_+$ defines a mild solution of $P_\tau(c_+, c_+)$. Let $v(t) := c_+ - u(t)$. Then v is a mild solution of $P_\tau(c_+ - u_0, c_+ - \varphi)$. Since $c_+ - u_0 \ge 0$ and $c_+ - \varphi(t) \ge 0$, it follows from Theorem 3.11.11, applied to (6.13) with $U(t) = (v(t), 0)$, $U_0 = (c_+ - u_0, 0)$, $\Phi(t) = (0, c_+ - \varphi(t))$, that $c_+ - u(t) = v(t) \ge 0$ for all $t \in [0,\tau]$. The other inequality is proved in a similar way. □

Let u be a mild solution of $P_\tau(u_0,\varphi)$, where u_0 and φ may be complex-valued. By considering $\mathrm{Re}(e^{i\theta}u)$, which is a mild solution of $P_\tau(\mathrm{Re}(e^{i\theta}u_0), \mathrm{Re}(e^{i\theta}\varphi))$, it follows from Proposition 6.2.5 that

$$\|u\|_{C([0,\tau],C(\overline{\Omega}))} \le \max\left\{\|\varphi\|_{C([0,\tau],C(\Gamma))}\,,\ \|u_0\|_{C(\overline{\Omega})}\right\}. \tag{6.16}$$

Here, we consider $C([0,\tau], C(\overline{\Omega}))$ and $C([0,\tau], C(\Gamma))$ as Banach spaces for the norms

$$\|u\|_{C([0,\tau],C(\overline{\Omega}))} = \sup_{0\le t\le\tau} \|u(t)\|_{C(\overline{\Omega})} \text{ and } \|\varphi\|_{C([0,\tau],C(\Gamma))} = \sup_{0\le t\le\tau} \|\varphi(t)\|_{C(\Gamma)},$$

respectively.

Now we can prove well-posedness of $P_\tau(u_0,\varphi)$.

Theorem 6.2.6. *Assume that Ω is Dirichlet regular. Let $u_0 \in C(\overline{\Omega})$ and $\varphi \in C([0,\tau], C(\Gamma))$ such that $u_0|_\Gamma = \varphi(0)$. Then there exists a unique mild solution of $P_\tau(u_0,\varphi)$.*

Proof. Uniqueness follows from Lemma 3.2.9. For existence, choose $u_{0n} \in D(\Delta_{\max})$ such that $\lim_{n\to\infty} u_{0n} = u_0$ in $C(\overline{\Omega})$. Choose $\varphi_n \in C^1([0,\tau], C(\Gamma))$ such that $\varphi_n(0) = u_{0n}|_\Gamma$ and $\varphi_n \to \varphi$ as $n \to \infty$ in $C([0,\tau], C(\Gamma))$. For example, one may let $\varphi_n(t) := (1 - \lambda(nt))\psi_n(t) + \lambda(nt)u_{0n}|_\Gamma$, where $\lambda(s) := (1 - \min(s,1))^2$, $\psi_n \in C^1([0,\tau], C(\Gamma))$ and $\|\psi_n - \varphi\|_{C([0,\tau],C(\Gamma))} < 1/n$. By Proposition 6.2.4, there exists a unique mild solution u_n of $P_\tau(u_{0n}, \varphi_n)$. By (6.16), we have

$$\|u_n - u_m\|_{C([0,\tau],C(\overline{\Omega}))} \le \max\left\{\|\varphi_n - \varphi_m\|_{C([0,\tau],C(\Gamma))}\,,\ \|u_{0n} - u_{0m}\|_{C(\overline{\Omega})}\right\}.$$

Hence, $(u_n)_{n\in\mathbb{N}}$ is a Cauchy sequence in $C([0,\tau], C(\overline{\Omega}))$. Let $u := \lim_{n\to\infty} u_n$ in $C([0,\tau], C(\overline{\Omega}))$. Then $u(t)|_\Gamma = \lim_{n\to\infty} \varphi_n(t) = \varphi(t)$. Since $\Delta_{\max}\int_0^t u_n(s)\,ds = u_n(t) - u_{0n}$ and $\Delta_{\max}$ is closed, it follows that

$$\int_0^t u(s)\,ds = \lim_{n\to\infty}\int_0^t u_n(s)\,ds \in D(\Delta_{\max}) \text{ and } \Delta_{\max}\int_0^t u(s)\,ds = u(t) - u_0$$

for all $t \in [0, \tau]$. We have shown that u is a mild solution of $P_\tau(u_0, \varphi)$. $\square$

So far, we have seen that for each $u_0 \in C(\overline{\Omega})$ and $\varphi \in C([0, \tau], C(\Gamma))$ satisfying $\varphi(0) = u_0|_\Gamma$ there exists a unique mild solution. We now show that the mild solution u is always of class C^∞ on $(0, \tau] \times \overline{\Omega}$. In fact, we may identify u with a continuous function defined on $[0, \tau] \times \overline{\Omega}$ with values in $\mathbb{R}$ by letting $u(t, x) := u(t)(x)$ $(t \in [0, \tau],\ x \in \overline{\Omega})$. Then the following holds.

Theorem 6.2.7. *Assume that Ω is Dirichlet regular. Let $u_0 \in C(\overline{\Omega})$ and $\varphi \in C([0, \tau], C(\Gamma))$ such that $u_0|_\Gamma = \varphi(0)$. Let u be the mild solution of $P_\tau(u_0, \varphi)$. Then*

$$u \in C^\infty((0, \tau] \times \Omega).$$

Proof. a) Assume that $u_0 = 0$. Let $v(t) := \int_0^t u(s)\, ds$. Then

$$v \in C^1([0, \tau], C(\overline{\Omega})),\ v(t) \in D(\Delta_{\max}) \text{ and } v'(t) = \Delta v(t) \text{ for } t \in [0, \tau].$$

Let $0 < t_0 \le \tau$, $x_0 \in \Omega$. Choose $r > 0$ such that $\overline{B}(x_0, r) \subset \Omega$, and let $C := [0, \tau] \times \overline{B}(x_0, r)$ and $C' := [t_0/2, \tau] \times \overline{B}(x_0, r/2)$. Choose $\xi \in C^\infty([0, \tau] \times \mathbb{R}^n)$ such that $\xi \equiv 1$ on C', $\xi \equiv 0$ on $([0, \tau] \times \mathbb{R}^n) \setminus C$, and $\xi \equiv 0$ on $[0, t_0/4] \times \mathbb{R}^n$. Let $w := \xi \cdot v$ on $[0, \tau] \times \Omega$, $w := 0$ on $[0, \tau] \times (\mathbb{R}^n \setminus \Omega)$. Then $w \in C^1([0, \tau], C_0(\mathbb{R}^n))$, and

$$w'(t) = \xi'(t) v(t) + \xi(t) v'(t) \quad \text{on} \quad [0, \tau] \times \Omega,$$

with $w'(t) = 0$ on $[0, \tau] \times (\mathbb{R}^n \setminus \Omega)$. It follows from Lemma 6.1.5 and the closed graph theorem that $D(\Delta_{\max}) \hookrightarrow C^1(\Omega)$ when $D(\Delta_{\max})$ carries the graph norm and $C^1(\Omega)$ has the natural Fréchet topology. In particular, ∇v is continuous on $[0, \tau] \times C$. We have

$$\Delta w(t) = \xi(t) \Delta v(t) + 2 \nabla \xi(t) \cdot \nabla v(t) + \Delta \xi(t) v(t) \text{ on } [0, \tau] \times \Omega.$$

Let

$$f(t) := \begin{cases} \xi'(t) v(t) - 2 \nabla \xi(t) \cdot \nabla v(t) - \Delta \xi(t) v(t) & \text{on } [0, \tau] \times \Omega, \\ 0 & \text{on } [0, \tau] \times (\mathbb{R}^n \setminus \Omega). \end{cases}$$

Then $f \in C([0, \tau], C_0(\mathbb{R}^n))$ and $w'(t) = \Delta w(t) + f(t)$ on $[0, \tau] \times \mathbb{R}^n$. Denote by G the Gaussian semigroup on $C_0(\mathbb{R}^n)$, i.e.

$$G(t) g := k_t * g, \quad \text{where} \quad k_t(x) := (4 \pi t)^{-n/2} e^{-|x|^2/4t}.$$

Since $w(0) = 0$, it follows from Proposition 3.1.16 that

$$\begin{aligned} w(t) &= \int_0^t G(t-s) f(s)\, ds, \quad \text{i.e.,} \\ w(t, x) &= \int_0^t \int_{\mathbb{R}^n} (4 \pi (t-s))^{-n/2} e^{-|x-y|^2/4(t-s)} f(s, y)\, dy\, ds \end{aligned}$$

for all $0 < t \le t_0$, $x \in \mathbb{R}^n$. Since $f \equiv 0$ on C' and outside C, the integrand has no singularities for (t,x) in the interior of C'. Thus, w is of class C^∞ in a neighbourhood of (t_0,x_0) in $(0,\tau]\times\Omega$. Since $v = w$ in C' and (t_0,x_0) is arbitrary, it follows that v, and hence also u, belong to $C^\infty((0,\tau]\times\Omega)$.

b) Now consider the general case when $u_0|_\Gamma = \varphi(0)$. Let w_0 be the solution of the Dirichlet problem $D(\varphi(0))$. Consider $v(t) := u(t) - w_0$. Then v is a mild solution of

$$v'(t) = \Delta v(t) \ \ (t \in [0,\tau])$$

and $v(0)|_\Gamma = 0$. Denote by T_0 the C_0-semigroup generated by Δ_0 on $C_0(\Omega)$. Let $w(t) := v(t) - T_0(t)v(0)$. Then w is a mild solution of $P_\tau(0,\varphi - \varphi(0))$. Hence, $w \in C^\infty((0,\tau]\times\Omega)$ by a). Since $T_0(\cdot)v(0) \in C^\infty((0,\infty)\times\Omega)$ by Proposition 6.1.10, the proof is complete. □

Now we can reformulate the results. For this, we consider the parabolic domain

$$\Omega_\tau := (0,\tau]\times\Omega$$

with parabolic boundary

$$\Gamma_\tau = (\{0\}\times\overline{\Omega}) \cup ((0,\tau]\times\Gamma),$$

where $\tau > 0$. Thus, Ω_τ is a cylinder and Γ_τ is the topological boundary without the top.

Theorem 6.2.8. *Assume that Ω is Dirichlet regular. Then for every $\psi \in C(\Gamma_\tau)$ there exists a unique function $u \in C(\overline{\Omega}_\tau)\cap C^\infty(\Omega_\tau)$ such that*

$$\begin{cases} u_t - \Delta u &= 0 \ \ \text{in } \Omega_\tau, \ \ \text{and} \\ u|_{\Gamma_\tau} &= \psi. \end{cases} \tag{6.17}$$

Thus, (6.17) is formulated exactly as the Dirichlet problem, the Laplacian being replaced by the parabolic operator $\frac{d}{dt} - \Delta$, Ω by the parabolic domain Ω_τ and Γ by the parabolic boundary Γ_τ.

6.3 Asymptotic Behaviour

We keep the notation of the preceding section. But now we consider the problem on the half-line

$$P_\infty(u_0,\varphi)\ \begin{cases} u'(t) &= \Delta u(t) \qquad (t \ge 0), \\ u(t)|_\Gamma &= \varphi(t) \qquad (t \ge 0), \\ u(0) &= u_0, \end{cases}$$

where $u_0 \in C(\overline{\Omega})$ and $\varphi \in C(\mathbb{R}_+, C(\Gamma))$ are given functions. As before, by a *mild solution* of $P_\infty(u_0, \varphi)$ we understand a function $u \in C(\mathbb{R}_+, C(\overline{\Omega}))$ such that

$$\begin{aligned} \Delta \int_0^t u(s)\, ds &= u(t) - u_0 \quad \text{in } \mathcal{D}(\Omega)' \text{ and} \\ u(t)|_\Gamma &= \varphi(t) \end{aligned}$$

for all $t \in \mathbb{R}_+$.

We assume throughout this section that Ω is Dirichlet regular.

It follows from Theorem 6.2.6 that for each $u_0 \in C(\overline{\Omega})$ and $\varphi \in C(\mathbb{R}_+, C(\Gamma))$ such that $u_0|_\Gamma = \varphi(0)$ there exists a unique mild solution u of $P_\infty(u_0, \varphi)$. In this section we study the asymptotic behaviour of $u(t)$ as $t \to \infty$. The results are analogous to (and in some cases, consequences of) abstract results given in Sections 5.4 and 5.6. We start with Cesàro convergence.

Proposition 6.3.1. *Let $\varphi : \mathbb{R}_+ \to C(\Gamma)$ be continuous and bounded. Assume that*

$$\lim_{t\to\infty} \frac{1}{t} \int_0^t \varphi(s)\, ds = \varphi_\infty \quad \text{in } C(\Gamma).$$

Let $u_0 \in C(\overline{\Omega})$ satisfying $u_0|_\Gamma = \varphi(0)$ and let u be the mild solution of $P_\infty(u_0, \varphi)$. Then

$$\lim_{t\to\infty} \frac{1}{t} \int_0^t u(s)\, ds = u_\infty \quad \text{exists in } C(\overline{\Omega}).$$

Moreover, $\Delta u_\infty = 0$ in $\mathcal{D}(\Omega)'$ and $u_\infty|_\Gamma = \varphi_\infty$.

Proof. By (6.16), u is bounded. Taking Laplace transforms, we have $\lambda\hat{u}(\lambda) - \Delta\hat{u}(\lambda) = u_0$ and $\hat{u}(\lambda)|_\Gamma = \hat{\varphi}(\lambda)$ $(\lambda > 0)$. Denote by $w(\lambda)$ the solution of the Dirichlet problem $D(\hat{\varphi}(\lambda))$. Then $\hat{u}(\lambda) - w(\lambda) \in C_0(\Omega)$ and

$$\lambda(\hat{u}(\lambda) - w(\lambda)) - \Delta(\hat{u}(\lambda) - w(\lambda)) = u_0 - \lambda w(\lambda).$$

Thus, $\hat{u}(\lambda) - w(\lambda) = R(\lambda, \Delta_c)(u_0 - \lambda w(\lambda))$. Let u_∞ be the solution of $D(\varphi_\infty)$. By Theorem 4.1.2, $\lim_{\lambda\downarrow 0} \lambda\hat{\varphi}(\lambda) = \mathrm{C} - \lim_{t\to\infty} \varphi(t) = \varphi_\infty$. It follows from the maximum principle (Theorem 6.1.3) that $\lambda w(\lambda)$ converges to a function u_∞ in $C(\overline{\Omega})$ as $\lambda \downarrow 0$. Clearly, u_∞ solves $D(\varphi_\infty)$. Thus, $\hat{u}(\lambda) - w(\lambda) \to R(0, \Delta_c)(u_0 - u_\infty)$ in $C(\overline{\Omega})$ as $\lambda \downarrow 0$. Consequently, $\lambda(\hat{u}(\lambda) - w(\lambda)) \to 0$ in $C(\overline{\Omega})$ as $\lambda \downarrow 0$. This implies that $\lim_{\lambda\downarrow 0} \lambda\hat{u}(\lambda) = u_\infty$ in $C(\overline{\Omega})$. By Theorem 4.2.7, this implies the claim. □

Next we consider uniform continuity.

Proposition 6.3.2. *Let $\varphi \in \mathrm{BUC}(\mathbb{R}_+, C(\Gamma))$ and $u_0 \in C(\overline{\Omega})$ such that $u_0|_\Gamma = \varphi(0)$. Let u be the mild solution of $P_\infty(u_0, \varphi)$. Then $u \in \mathrm{BUC}(\mathbb{R}_+, C(\overline{\Omega}))$.*

Proof. By (6.16), u is bounded. For $\delta > 0$ let $u_\delta(t) := u(t+\delta) - u(t)$, $\varphi_\delta(t) := \varphi(t+\delta) - \varphi(t)$ $(t \geq 0)$. Then u_δ is the mild solution of $P_\infty(u_\delta(0), \varphi_\delta)$. Since

$\varphi_\delta \to 0$ in $\mathrm{BUC}(\mathbb{R}_+, C(\Gamma))$ and $u_\delta \to 0$ in $C(\overline{\Omega})$ as $\delta \downarrow 0$, it follows from (6.16) that $u_\delta(t) \to 0$ as $\delta \downarrow 0$ uniformly on $\mathbb{R}_+$. This means that u is uniformly continuous. □

Using Propositions 6.2.3 and 6.3.2, the results of Chapter 5 on inhomogeneous Cauchy problems give the following.

Theorem 6.3.3. *Let $\varphi \in \mathrm{AAP}(\mathbb{R}_+, C(\Gamma))$ and $u_0 \in C(\overline{\Omega})$ such that $u_0|_\Gamma = \varphi(0)$. Denote by u the mild solution of $P_\infty(u_0, \varphi)$. Then*

a) *$u \in \mathrm{AAP}(\mathbb{R}_+, C(\overline{\Omega}))$ and $\mathrm{Freq}(u) = \mathrm{Freq}(\varphi)$.*

b) *If $\varphi = \varphi_1 + \varphi_2$ where $\varphi_1 \in \mathrm{AP}(\mathbb{R}_+, C(\Gamma))$, $\varphi_2 \in C_0(\mathbb{R}_+, C(\Gamma))$, and $u = u_1 + u_2$ where $u_1 \in \mathrm{AP}(\mathbb{R}_+, C(\overline{\Omega}))$ and $u_2 \in C_0(\mathbb{R}_+, C(\overline{\Omega}))$, then u_1 is the mild solution of $P_\infty(u_1(0), \varphi_1)$ and u_2 is the mild solution of $P_\infty(u_2(0), \varphi_2)$.*

c) *If $\lim_{t\to\infty} \varphi(t) = \varphi_\infty$ exists in $C(\Gamma)$, then $\lim_{t\to\infty} u(t) = u_\infty$ where u_∞ is the solution of the Dirichlet problem $D(\varphi_\infty)$.*

Proof. a) Let $X := C(\overline{\Omega}) \times C(\Gamma)$. Consider the function $U : \mathbb{R}_+ \to X$ defined by $U(t) := (u(t), 0)$. Then by Proposition 6.3.2, $U \in \mathrm{BUC}(\mathbb{R}_+, X)$. Let $\Phi(t) := (0, \varphi(t))$. Then $\Phi \in \mathrm{AAP}(\mathbb{R}_+, X)$ and Proposition 6.3.2 shows that U is a mild solution of

$$\begin{cases} U'(t) &= AU(t) + \Phi(t) \quad (t \geq 0), \\ U(0) &= (u_0, 0), \end{cases} \tag{6.18}$$

where A is the Poisson operator. Since $s(A) < 0$, it follows from Corollary 5.6.9 that $U \in \mathrm{AAP}(\mathbb{R}_+, X)$, hence $u \in \mathrm{AAP}(\mathbb{R}_+, C(\overline{\Omega}))$. Moreover, it also follows that $\mathrm{Freq}(u) = \mathrm{Freq}(U) = \mathrm{Freq}(\Phi) = \mathrm{Freq}(\varphi)$.

b) This is a direct consequence of Propositions 5.4.16 and 6.2.3.

c) By Corollary 5.6.9, $\lim_{t\to\infty} U(t)$ exists and equals $R(0, A)(0, \varphi_\infty) = (u_\infty, 0)$, so that u_∞ is the solution of $D(\varphi_\infty)$. □

Corollary 6.3.4. *Let $\varphi \in \mathrm{AP}(\mathbb{R}_+, C(\Gamma))$. Then there exists a unique $u_0 \in C(\overline{\Omega})$ satisfying $u_0|_\Gamma = \varphi(0)$ such that the mild solution u of $P_\infty(u_0, \varphi)$ is almost periodic.*

Proof. Existence: Let $v_0 \in C(\overline{\Omega})$ such that $v_0|_\Gamma = \varphi(0)$. Let v be the mild solution of $P_\infty(v_0, \varphi)$. Then $v = v_1 + v_2$ where $v_1 \in \mathrm{AP}(\mathbb{R}_+, C(\overline{\Omega}))$ and $v_2 \in C_0(\mathbb{R}_+, C(\overline{\Omega}))$, by Theorem 6.3.3. Moreover, v_1 is the mild solution of $P_\infty(v_1(0), \varphi)$. So we may choose $u_0 = v_1(0)$.

Uniqueness: Assume that the mild solution $\tilde{u}$ of $P_\infty(\tilde{u}_0, \varphi)$ is almost periodic. Then $v = u - \tilde{u} \in \mathrm{AP}(\mathbb{R}_+, C(\overline{\Omega}))$ and v is the mild solution of $P_\infty(u_0 - \tilde{u}_0, 0)$. It follows from Theorem 6.3.3 c) that $\lim_{t\to\infty} v(t) = 0$. Hence, $v(t) \equiv 0$ by (4.36). □

In the situation of Corollary 6.3.4, if φ is τ-periodic, then u is also τ-periodic. This follows from Corollary 4.5.4, since $\mathrm{Freq}(u) = \mathrm{Freq}(\varphi) \subset \frac{2\pi}{\tau}\mathbb{Z}$ (see also Corollary 5.6.9 c).

Finally, we consider the inhomogeneous heat equation with inhomogeneous boundary conditions.

Given $u_0 \in C(\overline{\Omega})$, $\varphi \in C(\mathbb{R}_+, C(\Gamma))$ such that $u_0|_\Gamma = \varphi(0)$ and $f \in C(\mathbb{R}_+, C(\overline{\Omega}))$, we consider the problem

$$P_\infty(u_0, \varphi, f) \begin{cases} u'(t) = & \Delta u(t) + f(t) \qquad (t \geq 0), \\ u(t)|_\Gamma = & \varphi(t) \qquad (t \geq 0), \\ u(0) = & u_0. \end{cases}$$

A *mild solution* is a continuous function $u : \mathbb{R}_+ \to C(\overline{\Omega})$ such that $u(0) = u_0$, $u(t)|_\Gamma = \varphi(t)$, $\int_0^t u(s)\, ds \in D(\Delta_{\max})$ and

$$\Delta \int_0^t u(s)\, ds + \int_0^t f(s)\, ds = u(t) - u_0 \text{ in } \mathcal{D}(\Omega)' \tag{6.19}$$

for all $t \geq 0$. By Theorem 6.2.6, there is at most one mild solution of $P_\infty(u_0, \varphi, f)$.

By Proposition 3.7.22, the function

$$v(t) := \int_0^t T_c(t-s) f(s)\, ds \qquad (t \geq 0) \tag{6.20}$$

defines a mild solution of $P_\infty(0, 0, f)$. Since Ω is Dirichlet regular, by Theorem 6.2.6, there exists a unique mild solution w of $P_\infty(u_0, \varphi, 0)$. Hence, $u = v + w$ is a mild solution of $P_\infty(u_0, \varphi, f)$. We have shown the following.

Theorem 6.3.5. *Let $f \in C(\mathbb{R}_+, C(\overline{\Omega}))$, $\varphi \in C(\mathbb{R}_+, C(\Gamma))$ and $u_0 \in C(\overline{\Omega})$ such that $u_0|_\Gamma = \varphi(0)$. Then there exists a unique mild solution of $P_\infty(u_0, \varphi, f)$.*

Recall that $\omega(T_c) < \infty$. Thus, $v = T_c * f \in \mathrm{BUC}(\mathbb{R}_+, C(\overline{\Omega}))$ (respectively, $\mathrm{AAP}(\mathbb{R}_+, C(\overline{\Omega}))$) if $f \in \mathrm{BUC}(\mathbb{R}_+, C(\overline{\Omega}))$ (respectively, $f \in \mathrm{AAP}(\mathbb{R}_+, C(\overline{\Omega}))$), by Proposition 5.6.1. So we obtain the following from Theorem 6.3.3.

Theorem 6.3.6. *Let $f \in \mathrm{AAP}(\mathbb{R}_+, C(\overline{\Omega}))$, $\varphi \in \mathrm{AAP}(\mathbb{R}_+, C(\Gamma))$ and $u_0 \in C(\overline{\Omega})$. Assume that $u_0|_\Gamma = \varphi(0)$. Let u be the mild solution of $P_\infty(u_0, \varphi, f)$. Then $u \in \mathrm{AAP}(\mathbb{R}_+, C(\overline{\Omega}))$.*

Corollary 6.3.7. *Let $f : \mathbb{R}_+ \to C(\overline{\Omega})$ and $\varphi : \mathbb{R}_+ \to C(\Gamma)$ be continuous. Assume that $\lim_{t\to\infty} f(t) = f_\infty$ exists in $C(\overline{\Omega})$ and $\lim_{t\to\infty} \varphi(t) = \varphi_\infty$ in $C(\Gamma)$. Let $u_0 \in C(\overline{\Omega})$ such that $u_0|_\Gamma = \varphi(0)$. Let u be the mild solution of $P_\infty(u_0, \varphi, f)$. Then $\lim_{t\to\infty} u(t) = u_\infty$ exists in $C(\overline{\Omega})$ and*

$$\begin{cases} u_\infty|_\Gamma & = \varphi_\infty, \\ -\Delta u_\infty & = f_\infty \text{ in } \mathcal{D}(\Omega)'. \end{cases} \tag{6.21}$$

6.4 Notes

The approach to solving the heat equation with the help of the Poisson operator is taken from [Are00], where general strongly elliptic operators in divergence form with bounded measurable coefficients are also considered. Greiner [Gre87] developed an abstract perturbation theory for boundary conditions. Theorems 6.1.7 and 6.1.8 are proved in [AB98], where it is also shown that Dirichlet regularity is a necessary condition. Lumer and Schnaubelt [LS99] consider also non-cylindrical domains. Lumer gave a proof of the holomorphy of the semigroup generated by Δ_0 which is based on the maximum principle (cf. [LP79]). In Theorem 6.1.9 we use the properties of resolvent positive operators to prove holomorphy.

Theorem 6.2.7 is an adaptation of Evans's proof [Eva98, Section 2.3, Theorem 8] to the solutions defined here. Theorem 6.2.8 was probably first proved by Tychonoff [Tyc38] in 1938 with the help of integral equations. Other proofs were given by Fulks [Ful56], [Ful57] and Babuŝka and Výborný [BV62] (see also the work of Lumer [Lum75]).

Concerning an L^p-approach to boundary value problems via holomorphic semigroups we refer to the monograph by Taira [Tai95].

Chapter 7

The Wave Equation

In this chapter we study the wave equation

$$u_{tt} = \Delta u$$

on an open subset Ω of $\mathbb{R}^n$. We will use the theory of cosine functions and work on $L^2(\Omega)$. We first consider the Laplacian with Dirichlet boundary conditions. This is a selfadjoint operator and well-posedness is a consequence of the spectral theorem. A further aim is to replace the Laplace operator by a general elliptic operator. This will be done by a perturbation theorem for selfadjoint operators which we prove in Section 7.1.

We give a brief introduction to symmetric sesquilinear forms which are the natural tool for proving selfadjointness. However, we restrict ourselves to the minimum needed to show that quite general equations can be solved in a simple way by functional analytical methods. The restriction to Dirichlet boundary conditions and to second order operators is not essential; we choose these in order to present the simplest case.

7.1 Perturbation of Selfadjoint Operators

A selfadjoint operator which is bounded above generates a cosine function (see Example 3.14.16). We will present a perturbation result in terms of the form domain which again leads to a generator of a cosine function with the same phase space. It is important to know this since the phase space yields the natural domain for initial data for classical solutions (Corollary 3.14.12).

We will see in the following section that the abstract setting which we present here is very well adapted to elliptic operators.

Let H be a Hilbert space with scalar product $(\cdot|\cdot)_H$ and norm $\|\cdot\|_H$. We consider another Hilbert space V with scalar product $(\cdot|\cdot)_V$ and norm $\|\cdot\|_V$. Moreover, we assume that V is continuously embedded into H with dense image.

This means, we assume that $V \subset H$, that V is dense in H and that there exists a constant $\omega > 0$ such that

$$\omega \|u\|_H^2 \leq \|u\|_V^2 \tag{7.1}$$

for all $u \in V$. We use the abbreviation

$$V \overset{d}{\hookrightarrow} H$$

for these three properties.

In this situation, we associate to V an operator A_H on H by postulating:

$$\begin{aligned} D(A_H) \quad &:= \quad \Big\{ u \in V : \text{ there exists } f \in H \text{ such that} \\ &\qquad\qquad (u|v)_V = -(f|v)_H \text{ for all } v \in V \Big\}, \\ A_H u \quad &:= \quad f. \end{aligned}$$

Note that f is uniquely determined by u since V is dense in H.

We call A_H *the operator on H associated with V.*

Proposition 7.1.1. *The operator A_H is selfadjoint and bounded above by $-\omega$. In particular, A_H generates a holomorphic C_0-semigroup T on H of angle $\pi/2$ satisfying*

$$\|T(z)\| \leq e^{-\omega \operatorname{Re} z} \qquad (\operatorname{Re} z > 0).$$

Moreover, A_H generates a cosine function.

Proof. a) A_H is symmetric. In fact, let $u, v \in D(A_H)$. Then by the definition of A_H,

$$(A_H u|v)_H = -(u|v)_V = -\overline{(v|u)_V} = \overline{(A_H v|u)_H} = (u|A_H v)_H.$$

b) For $u \in D(A_H)$ one has

$$(A_H u|u)_H = -(u|u)_V \leq -\omega \|u\|_H^2,$$

by (7.1). Thus, A_H is bounded above by $-\omega$.

c) We show that $-A_H$ is surjective. Let $f \in H$. Then $F(v) := (v|f)_H$ defines a continuous linear form on V. By the Riesz-Fréchet lemma, there exists a unique $u \in V$ such that $(v|f)_H = (v|u)_V$ for all $v \in V$. Hence by the definition of A_H, one has $u \in D(A_H)$ and $-A_H u = f$. It follows from Theorem B.14 that A_H is selfadjoint. The remaining two assertions follow from Example 3.7.5 and Example 3.14.16. □

Let us consider multiplication operators as a first example. It is very simple; nevertheless, it is a generic example by the spectral theorem (Theorem B.13).

Example 7.1.2 (Multiplication operators). Let (Y, μ) be a measure space, $\omega > 0$ and $m : Y \to [\omega, \infty)$ be measurable. Let $H := L^2(Y, \mu)$ and let

$$V := L^2(Y, m\, d\mu) := \left\{ u \in H : \int_Y |u|^2 m\, d\mu < \infty \right\}.$$

Then V is a Hilbert space for the scalar product

$$(u|v)_V := \int_Y u(x)\overline{v(x)}m(x)\, d\mu(x)$$

and $V \overset{d}{\hookrightarrow} H$.

It is easy to see that the operator A_H on H associated to V is the multiplication operator given by

$$\begin{aligned} D(A_H) &= \{u \in H : mu \in H\}, \\ A_H u &= -mu. \end{aligned}$$

(cf. Example B.12). The cosine function Cos generated by A_H is given by

$$(\mathrm{Cos}(t)f)(x) = (\cos t\sqrt{m(x)})f(x)$$

(cf. Example 3.14.16). Thus,

$$\mathrm{Cos}(t) = \frac{1}{2}(U(t) + U(-t)),$$

where U is the C_0-group on H given by

$$(U(t)f)(x) = e^{it\sqrt{m(x)}}f(x)$$

(cf. Example 3.14.16). The generator B of U is given by

$$\begin{aligned} D(B) &= \{f \in H : \sqrt{m} \cdot f \in H\}, \\ Bf &= i\sqrt{m} \cdot f. \end{aligned}$$

Thus, $V = D(B)$. It follows from Example 3.14.15 that the phase space of Cos is $V \times H$. □

In Example 7.1.2, the phase space of the cosine function associated with V is $V \times H$. In fact, this is the case whenever $V \overset{d}{\hookrightarrow} H$. The spectral theorem applied to the selfadjoint operator A_H shows that there is a measure space (Y, μ), a measurable function $m : Y \to [\omega, \infty)$ and a unitary equivalence U taking H onto $L^2(Y, \mu)$ and A_H onto the operator appearing in Example 7.1.2. The next result, which is the converse of Proposition 7.1.1, shows that V is uniquely determined by A_H. Hence, it follows from Example 7.1.2 that U takes V onto $L^2(Y, m\, d\mu)$ and that the phase space is $V \times H$.

Proposition 7.1.3. *Let B be a selfadjoint operator on H which is bounded above by $-\omega$, where $\omega > 0$. Then there exists a unique Hilbert space V such that $V \overset{d}{\hookrightarrow} H$ and such that B is the operator associated with V. Moreover, the phase space associated with the cosine function generated by B is $V \times H$.*

Proof. Uniqueness: Assume that B is associated with V where $V \overset{d}{\hookrightarrow} H$. We show that $D(B)$ is dense in V. In fact, let $v \in V$ such that $(u|v)_V = 0$ for all $u \in D(B)$. Then

$$(Bu|v)_H = -(u|v)_V = 0$$

for all $u \in D(B)$. Since B is selfadjoint, it follows that $v \in D(B)$ and $Bv = 0$. Since B is invertible, we conclude that $v = 0$. This proves the claim. Now we observe that

$$\|u\|_V^2 = (u|u)_V = -(Bu|u)_H$$

for all $u \in D(B)$. Thus, V is the completion of $D(B)$ for the norm $\|u\|_V := \sqrt{-(Bu|u)_H}$. Moreover, $(u|v)_V = -(Bu|v)_H$ for all $u, v \in D(B)$. Thus, the scalar product is also determined by B.

Existence: Using the spectral theorem, we may assume that B is a multiplication operator. Then the assertion is proved in Example 7.1.2. □

Propositions 7.1.1 and 7.1.3 establish a bijective correspondence between selfadjoint operators B on H which are bounded above by $-\omega$ (where $\omega > 0$) and Hilbert spaces V such that $V \overset{d}{\hookrightarrow} H$ and (7.1) is satisfied. One frequently calls V the *form domain* of B, and $(\cdot|\cdot)_V$ the *sesquilinear form associated with B*.

From Corollary 3.14.13 we now deduce our first perturbation result for selfadjoint operators.

Corollary 7.1.4. *Let V be a Hilbert space such that $V \overset{d}{\hookrightarrow} H$ and let A_H be the operator associated with V on H. Let $C \in \mathcal{L}(V, H)$. Then $A_H + C$ generates a cosine function on H with phase space $V \times H$.*

Our next aim is to introduce another kind of perturbation. For this we need some preparation.

Let $V \overset{d}{\hookrightarrow} H$. A mapping $\varphi : V \to \mathbb{C}$ is called *antilinear* if

$$\begin{aligned} \varphi(u+v) &= \varphi(u) + \varphi(v) \quad (u, v \in V), \quad \text{and} \\ \varphi(\lambda u) &= \bar{\lambda}\varphi(u) \quad (u \in V, \lambda \in \mathbb{C}). \end{aligned}$$

By

$$V' := \{\varphi :\ V \to \mathbb{C} :\ \varphi \text{ antilinear and continuous}\}$$

we denote the *antidual* of V. It is a Banach space for the norm

$$\|\varphi\|_{V'} := \sup_{\|u\|_V \le 1} |\varphi(u)|.$$

We embed H into V' in the following way. For $f \in H$ we define $\varphi_f \in V'$ by

$$\varphi_f(u) := (f|u)_H.$$

It is clear that the mapping $f \mapsto \varphi_f : H \to V'$ is linear, injective and continuous. This is the desired embedding.

For $\varphi \in V'$ we use the notation

$$(\varphi|u) := \varphi(u) \qquad (u \in V).$$

Thus, if $f \in H$ we have

$$(\varphi_f|u) = (f|u)_H \qquad (u \in V).$$

By the Riesz-Fréchet lemma, the mapping $A : V \to V'$ given by

$$(Au|v) := -(u|v)_V \qquad (u, v \in V)$$

is an isometric isomorphism from V onto V'. In particular, V' is itself a Hilbert space for the scalar product

$$(f|g)_{V'} := (A^{-1}f|A^{-1}g)_V \qquad (f, g \in V').$$

It follows from (7.1) that

$$\omega\|\varphi_f\|_{V'}^2 \leq \|f\|_H^2 \tag{7.2}$$

for all $f \in H$. In fact,

$$\begin{aligned} \omega\|\varphi_f\|_{V'}^2 &= \omega \sup_{\substack{u \in V \\ \|u\|_V \leq 1}} |(f|u)_H|^2 \\ &\leq \sup_{\substack{u \in V \\ \sqrt{\omega}\|u\|_H \leq 1}} |(f|\sqrt{\omega}u)_H|^2 \\ &\leq \|f\|_H^2, \end{aligned}$$

by the Cauchy-Schwarz inequality.

Now we identify H with a subspace of V', and we identify f and φ_f for $f \in H$. In particular, for $f \in H$ we write

$$(f|v) = (f|v)_H \qquad (v \in V).$$

Having this in mind, the proof of the following proposition is easy.

Proposition 7.1.5. *The operator A on V' is selfadjoint with upper bound $-\omega$. Moreover, $H \times V'$ is the phase space associated with the cosine function generated by A on V'. The operator A_H associated with V is the part of A in H.*

Proof. We first prove that

$$(Au|f)_{V'} = -(u|f)_H \tag{7.3}$$

for all $u \in V,\ f \in H$.

In fact, let $w = A^{-1}f \in V$. Then

$$(w|u)_V = -(Aw|u) = -(f|u)_H.$$

Hence,

$$(Au|f)_{V'} = (u|A^{-1}f)_V = (u|w)_V = -(u|f)_H.$$

It follows from (7.3) that for all $u, v \in V$,

$$(Au|v)_{V'} = -(u|v)_H = -\overline{(v|u)_H} = \overline{(Av|u)_{V'}} = (u|Av)_{V'}.$$

Thus, A is symmetric. Moreover,

$$(Au|u)_{V'} = -\|u\|_H^2 \leq -\omega\|u\|_{V'}^2$$

for all $u \in V$ by (7.2). Thus, A is bounded above by $-\omega$. Since A is surjective, it follows from Theorem B.14 that A is selfadjoint. In particular, $V = D(A)$ is dense in V'. Thus, $H \overset{d}{\hookrightarrow} V'$. Let B be the operator on V' associated with H. It follows from (7.3) that B is an extension of A. Since both operators are invertible, they are equal. It follows from Proposition 7.1.3 that the phase space associated with the cosine function generated by B on V' is $H \times V'$. The final assertion is easy to verify. □

We will illustrate Proposition 7.1.5 by considering multiplication operators in the following example. The discussion before Proposition 7.1.3 shows that they describe the most general situation; in fact, the example gives an alternative proof of Proposition 7.1.5.

Example 7.1.6 (Antidual associated with multiplication operator). Let $H := L^2(Y,\mu)$, $V := L^2(Y, m\,d\mu)$ where $m : Y \to [\omega, \infty)$ is measurable, $\omega > 0$. Then we may identify V' with $L^2(Y, \frac{1}{m}\,d\mu)$ by letting

$$(w|v) := \int_Y w\bar{v}\,d\mu$$

for all $w \in L^2(Y, \frac{1}{m}\,d\mu)$, $v \in V$. Since for $f \in H$, one has

$$(\varphi_f|v) = \int_Y f\bar{v}\,d\mu,$$

the embedding $H \hookrightarrow V' : f \mapsto \varphi_f$ corresponds to the identity mapping $L^2(Y,\mu) \hookrightarrow L^2(Y, \frac{1}{m}\,d\mu)$. Now the mapping $A : V \to V'$ is given by $Au = -mu$. In fact, for $u \in V$ one has

$$(Au|v) = -\int_Y u\bar{v}m\,d\mu$$

for all $v \in V$.

Note that A, considered as an operator on V', is associated to the subspace $H = L^2(Y, d\mu)$ of $V' = L^2(Y, \frac{1}{m}\, d\mu)$. Thus, A generates a cosine function on V' and the associated phase space is $H \times V'$. □

Now we are in the position to prove the following general perturbation result.

Theorem 7.1.7. *Let $V \overset{d}{\hookrightarrow} H$ and identify H with a subspace of V' in the canonical way. Denote by $A : V \to V'$ the isomorphism given by the Riesz-Fréchet lemma. Let $C \in \mathcal{L}(V, H)$, $B \in \mathcal{L}(H, V')$. Then the part $(A+B+C)_H$ of $A+B+C$ in H generates a cosine function* Cos *on H whose phase space is $V \times H$.*

Note that

$$D((A+B+C)_H) = \{u \in V : \ Au + Bu + Cu \in H\}.$$

Proof. We consider A as an unbounded operator on V' with domain V. Then we know from Proposition 7.1.5 that A generates a cosine function on V' with associated phase space $H \times V'$. Since $B \in \mathcal{L}(H, V')$, $A+B$ also generates a cosine function on V' with phase space $H \times V'$ (by Corollary 3.14.13). Now we apply Corollary 3.14.14 to deduce that the part $(A+B)_H$ of $A+B$ in H generates a cosine function with associated phase space $D(A+B) \times H = V \times H$. Since $C \in \mathcal{L}(V, H)$, one has $(A+B+C)_H = (A+B)_H + C$. Now the claim follows by another application of Corollary 3.14.13. □

Corollary 7.1.8. *Under the conditions of Theorem* 7.1.7, *the operator $(A+B+C)_H$ generates a holomorphic C_0-semigroup on H of angle $\pi/2$.*

Proof. This follows from Theorem 3.14.17. □

7.2 The Wave Equation in $L^2(\Omega)$

Let $\Omega \subset \mathbb{R}^n$ be an open set. In this section we will consider the wave equation on Ω with Dirichlet boundary conditions. We consider first the Laplacian and then more general elliptic operators.

For this, we recall some distributional notions (see Appendix E). We denote the first Sobolev space in $L^2(\Omega)$ by $H^1(\Omega)$, i.e., $H^1(\Omega) = W^{1,2}(\Omega)$, and we define $H_0^1(\Omega) := \overline{\mathcal{D}(\Omega)}^{H^1(\Omega)}$. This allows us to give a meaning to Dirichlet boundary conditions: for $f \in H^1(\Omega)$ we say that $f|_{\partial\Omega} = 0$ *weakly* if $f \in H_0^1(\Omega)$.

As usual, we consider $L^2(\Omega)$ as a subspace of $\mathcal{D}(\Omega)'$. In particular, if $f \in H^1(\Omega)$, then $\Delta f \in \mathcal{D}(\Omega)'$ is defined by

$$\langle \varphi, \Delta f \rangle = \langle \Delta\varphi, f \rangle = \sum_{j=1}^{n} \int_\Omega f D_j^2 \varphi \, dx = -\sum_{j=1}^{n} \int_\Omega D_j f D_j \varphi \, dx$$

for all $\varphi \in \mathcal{D}(\Omega)$. Thus, to say that $\Delta f \in L^2(\Omega)$ means that there exists a function $g \in L^2(\Omega)$ such that

$$-\sum_{j=1}^{n} \int_{\Omega} D_j f D_j \varphi \, dx = \int_{\Omega} g\varphi \, dx$$

for all $\varphi \in \mathcal{D}(\Omega)$. We then identify g and Δf.

Next, we define the Laplacian with Dirichlet boundary conditions. This example has already been given in Chapter 3 (Example 3.4.7). Here we show how it fits into the setting of the preceding section. After that, it will be easy to investigate more general elliptic operators than the Laplacian.

Example 7.2.1 (The Dirichlet Laplacian). *Define the operator $\Delta_{L^2(\Omega)}$ on $L^2(\Omega)$ by*

$$\begin{aligned} D(\Delta_{L^2(\Omega)}) &:= \left\{ f \in H_0^1(\Omega) : \Delta f \in L^2(\Omega) \right\}, \\ \Delta_{L^2(\Omega)} f &:= \Delta f. \end{aligned}$$

Then $\Delta_{L^2(\Omega)}$ is selfadjoint and bounded above by 0. Moreover, $\Delta_{L^2(\Omega)}$ generates a cosine function with phase space $H_0^1(\Omega) \times L^2(\Omega)$.

Proof. Let $H := L^2(\Omega)$ and $V := H_0^1(\Omega)$ with scalar product

$$(u|v)_V := \int_{\Omega} u\bar{v} \, dx + \sum_{j=1}^{n} \int_{\Omega} D_j u D_j \bar{v} \, dx.$$

Then clearly, $V \overset{d}{\hookrightarrow} H$. Let B be the operator on H which is associated with V. We show that $\Delta_{L^2(\Omega)} = B + I$. In fact, let $u \in D(B)$, $Bu =: f$. Then $u \in H_0^1(\Omega)$ and

$$-\sum_{j=1}^{n} \int_{\Omega} D_j u D_j \varphi \, dx - \int_{\Omega} u\varphi \, dx = -(u|\bar{\varphi})_V = \int_{\Omega} f\varphi \, dx$$

for all $\varphi \in H_0^1(\Omega)$. Taking $\varphi \in \mathcal{D}(\Omega)$, we obtain

$$\begin{aligned} \langle \varphi, \Delta u \rangle = \langle \Delta\varphi, u \rangle &= \sum_{j=1}^{n} \int_{\Omega} u D_j^2 \varphi \, dx \\ &= -\sum_{j=1}^{n} \int_{\Omega} D_j u D_j \varphi \, dx = \int_{\Omega} f\varphi \, dx + \int_{\Omega} u\varphi \, dx. \end{aligned}$$

Hence, $\Delta u = f + u$. This shows that $u \in D(\Delta_{L^2(\Omega)})$ and $Bu = \Delta_{L^2(\Omega)} u - u$. Conversely, let $u \in D(\Delta_{L^2(\Omega)})$. Then $u \in H_0^1(\Omega)$ and for $\varphi \in \mathcal{D}(\Omega)$,

$$-(u|\varphi)_V = -\sum_{j=1}^{n} \int_{\Omega} D_j u D_j \bar{\varphi} \, dx - \int_{\Omega} u\bar{\varphi} \, dx = \langle \bar{\varphi}, \Delta u \rangle - \int_{\Omega} u\bar{\varphi} \, dx.$$

Since $\mathcal{D}(\Omega)$ is dense in $H_0^1(\Omega)$, it follows that $-(u|\varphi)_V = (\Delta u - u \mid \varphi)_{L^2(\Omega)}$ for all $\varphi \in H_0^1(\Omega)$. Thus, $u \in D(B)$ and $Bu = \Delta u - u = \Delta_{L^2(\Omega)}u - u$.

We have shown that $B = \Delta_{L^2(\Omega)} - I$. Hence, $\Delta_{L^2(\Omega)} = B + I$ is also selfadjoint. Since B generates a cosine function with phase space $H_0^1(\Omega) \times L^2(\Omega)$, so does $\Delta_{L^2(\Omega)}$, by Corollary 3.14.13. □

Now we obtain the following well-posedness result for the wave equation.

Theorem 7.2.2 (Wave equation). *Let $f \in H_0^1(\Omega)$ such that $\Delta f \in L^2(\Omega)$. Let $g \in H_0^1(\Omega)$. Then there exists a unique function $u \in C^2(\mathbb{R}_+, L^2(\Omega))$ such that*

a) *$\Delta u(t) \in L^2(\Omega)$ for $t \geq 0$;*

b) *$u(t) \in H_0^1(\Omega)$ for $t \geq 0$;*

c) *$u''(t) = \Delta u(t)$ for $t \geq 0$;*

d) *$u(0) = f,\ u'(0) = g$.*

Proof. Denote by $\Delta_{L^2(\Omega)}$ the Dirichlet Laplacian and by Cos the cosine function generated by $\Delta_{L^2(\Omega)}$ on $L^2(\Omega)$ (see Example 7.2.1). Let Sin be the associated sine function. Then $u(t) = \mathrm{Cos}(t)f + \mathrm{Sin}(t)g$ is a solution of

$$\begin{cases} u \in C^2(\mathbb{R}_+, L^2(\Omega)), & \\ u(t) \in D(\Delta_{L^2(\Omega)}) & (t \geq 0) \\ u''(t) = \Delta_{L^2(\Omega)}u(t) & (t \geq 0) \\ u(0) = f,\ u'(0) = g, & \end{cases}$$

by Corollary 3.14.12. Uniqueness follows from Corollary 3.14.8. □

Next, we consider general uniformly elliptic operators of second order. Let $a_{ij} \in L^\infty(\Omega)$ be complex-valued coefficients such that $a_{ij} = \overline{a_{ji}}$ and

$$\sum_{i,j=1}^{n} a_{ij}(x)\xi_i\overline{\xi_j} \geq \alpha|\xi|^2$$

for all $\xi \in \mathbb{R}^n$, x-a.e. on Ω, where $\alpha > 0$ is fixed. This last condition is called *uniform ellipticity* and is equivalent to saying that the smallest eigenvalue of the hermitian matrix $(a_{ij}(x))$ is at least α for almost all $x \in \Omega$.

Let $b_i, c_i, d \in L^\infty(\Omega)$ be complex-valued functions ($i = 1, 2, \ldots, n$). We consider the formal elliptic second order operator

$$Lu := \sum_{i,j=1}^{n} D_i(a_{ij}D_ju) + \sum_{j=1}^{n}(D_j(b_ju) + c_jD_ju) + du. \tag{7.4}$$

It is possible to give a sense to L by multiplying (7.4) by a test function and integration by parts. More precisely, we define L as follows.

Let $u \in H_0^1(\Omega)$. Define the distribution $Lu \in \mathcal{D}(\Omega)'$ by

$$\begin{aligned}\langle \varphi, Lu\rangle &= -\sum_{i,j=1}^{n} \int_\Omega a_{ij}(x)(D_j u)(x)(D_i\varphi)(x)\,dx \\ &\quad -\sum_{j=1}^{n}\int_\Omega b_j(x)u(x)(D_j\varphi)(x)\,dx + \sum_{j=1}^{n}\int_\Omega c_j(x)(D_j u)(x)\varphi(x)\,dx \\ &\quad + \int_\Omega d(x)u(x)\varphi(x)\,dx\end{aligned}$$

for all $\varphi \in \mathcal{D}(\Omega)$. Then $L : H_0^1(\Omega) \to \mathcal{D}(\Omega)'$ is linear. Consider the part L_H of L in $H := L^2(\Omega)$; i.e., L_H is the operator on H given by

$$\begin{aligned} D(L_H) &= \{u \in H_0^1(\Omega) :\ Lu \in L^2(\Omega)\} \\ L_H u &= Lu.\end{aligned}$$

Then the following holds.

Theorem 7.2.3. *The operator L_H generates a cosine function on $L^2(\Omega)$ with phase space $H_0^1(\Omega) \times L^2(\Omega)$.*

Proof. Let $\alpha > 0$ and consider $V := H_0^1(\Omega)$ with the scalar product

$$(u|v)_V = \int_\Omega \sum_{i,j=1}^{n} a_{ij}(x) D_i u(x) D_j \bar{v}(x)\,dx + \alpha \int_\Omega u(x)\bar{v}(x)\,dx.$$

Then $\|u\|_V = \sqrt{(u|u)_V}$ is equivalent to the given norm on $H_0^1(\Omega)$. Thus, V is a Hilbert space and $V \overset{d}{\hookrightarrow} L^2(\Omega)$. We identify $L^2(\Omega)$ with a subspace of V'. Define $B \in \mathcal{L}(H, V')$ by

$$(Bu|\varphi) := -\int_\Omega \sum_{j=1}^{n} b_j u D_j\varphi\,dx \qquad (\varphi \in V = H_0^1(\Omega))$$

and $C \in \mathcal{L}(V, H)$ by

$$Cu := \sum_{j=1}^{n} c_j D_j u + du - \alpha u.$$

Let $A : V \to V'$ be the isomorphism given by the Riesz-Fréchet lemma. It follows from Theorem 7.1.7 that the operator $(A + B + C)_H$ given by

$$\begin{aligned} D((A+B+C)_H) &:= \left\{u \in H_0^1(\Omega) :\ Au + Bu + Cu \in L^2(\Omega)\right\}, \\ (A+B+C)_H u &:= Au + Bu + Cu,\end{aligned}$$

generates a cosine function on $L^2(\Omega)$ with associated phase space $H_0^1(\Omega) \times L^2(\Omega)$. Since $\mathcal{D}(\Omega)$ is dense in $H_0^1(\Omega)$, the restriction mapping $V' \to \mathcal{D}(\Omega)'$ is injective;

thus, we may identify V' with a subspace of $\mathcal{D}(\Omega)'$. With this identification one has $L_H = (A+B+C)_H$. □

Now we deduce from Corollary 3.14.12 well-posedness of the following hyperbolic problem. As before, $L : H_0^1(\Omega) \to \mathcal{D}(\Omega)'$ denotes the elliptic operator associated with the coefficients a_{ij}, b_i, c_i, d.

Corollary 7.2.4 (Hyperbolic equation). *Let $f \in H_0^1(\Omega)$ such that $Lf \in L^2(\Omega)$ and let $g \in H_0^1(\Omega)$. Then there exists a unique function $u \in C^2(\mathbb{R}_+, L^2(\Omega))$ satisfying*

a) $u(t) \in H_0^1(\Omega), \quad Lu(t) \in L^2(\Omega) \quad (t \geq 0)$;

b) $u''(t) = Lu(t) \quad (t \geq 0)$;

c) $u(0) = f, \quad u'(0) = g$.

Proof. This follows from Theorem 7.2.3 and Corollary 3.14.12 as in the proof of Theorem 7.2.2. □

Since each generator of a cosine function is also the generator of a holomorphic C_0-semigroup (by Theorem 3.14.17), we also obtain well-posedness of the corresponding parabolic equation. The result is an extension of Example 3.7.24.

Corollary 7.2.5 (Parabolic equation). *Let $f \in L^2(\Omega)$. Then there exists a unique function $u \in C^\infty((0,\infty), L^2(\Omega)) \cap C(\mathbb{R}_+, L^2(\Omega))$ such that*

a) $u(t) \in H_0^1(\Omega), \quad Lu(t) \in L^2(\Omega) \quad (t > 0)$;

b) $u'(t) = Lu(t) \quad (t > 0)$;

c) $u(0) = f$.

Proof. Denote by T the holomorphic C_0-semigroup generated by L_H. Then $u(t) = T(t)f$ is the unique solution of a), b) and c), by Theorem 3.7.19. □

7.3 Notes

Section 7.1 merely contains a direct approach to constructing selfadjoint operators by scalar products. We refer to the textbooks [Dav80], [Dav95], [Kat66] and [RS72] for a systematic treatment of quadratic form methods.

Elliptic operators generating cosine functions are described in the monographs by Fattorini [Fat83], [Fat85] and Goldstein [Gol85], but the perturbation arguments leading to Theorem 7.1.7 and Corollary 7.2.4 may be new (in the case $B \neq 0$; i.e., when the coefficients b_j do not vanish).

Section 7.2 gives a fairly general well-posedness result on $L^2(\Omega)$, and the restriction to Dirichlet boundary conditions has been chosen merely for convenience. However, these results are definitely restricted to $L^2(\Omega)$ and no longer valid on $L^p(\Omega)$ $(p \neq 2)$. This will be made precise in Example 8.4.9. On $L^1(\mathbb{R}^n)$ or $C_0(\mathbb{R}^n)$ the Laplacian does not generate a cosine function. However, the following holds for $n = 3$.

Theorem 7.3.1. *Let* $X := L^1(\mathbb{R}^3)$ *or* $C_0(\mathbb{R}^3)$, *and*

$$\begin{cases} D(\Delta_X) & := \{f \in X : \Delta f \in X\}, \\ \Delta_X f & := \Delta f. \end{cases}$$

Then Δ_X *generates a sine function* Sin *on* X *given by*

$$(\mathrm{Sin}(t)f)(x) = \frac{1}{t\sigma_2} \int_{\partial B(x,t)} f(z)\, d\sigma(z)$$

where σ *denotes the surface measure on* $\partial B(x,t) := \{z \in \mathbb{R}^3 : |x - z| = t\}$ *and* $\sigma_2 := 2|B(0,1)|$ *is the surface area of the 2-dimensional sphere. Thus,* $(\mathrm{Sin}(t)f)(x)$ *is* t*-times the mean of* f *over the sphere* $\partial B(x,t)$.

This can be seen by inspecting the proofs given in [Eva98, Section 2.4].

Generalizing the method of spherical means, a systematic treatment of an "abstract Laplacian" (given as the closure of the sum of n generators of commuting cosine functions) is given by Keyantuo [Key95b].

Chapter 8

Translation Invariant Operators on $L^p(\mathbb{R}^n)$

In this chapter we consider differential operators with constant coefficients, and more generally pseudo-differential operators on $L^p(\mathbb{R}^n)$. The realization $\mathrm{Op}_p(a)$ in $L^p(\mathbb{R}^n)$ of such an operator is translation invariant. We assume that the "symbol" a satisfies certain smoothness and growth assumptions. In particular, when a is a polynomial, then $\mathrm{Op}_p(a)$ is a differential operator. In the following sections we investigate the question under which conditions on the symbol a the operator $\mathrm{Op}_p(a)$ generates a C_0-semigroup or an integrated semigroup on $L^p(\mathbb{R}^n)$. This question is closely related to the problem whether e^{ta} or $\int_0^t \frac{(t-s)^{k-1}}{k!} e^{sa}\, ds$ is a Fourier multiplier for $L^p(\mathbb{R}^n)$. In Section 8.2 (see also Appendix E) we consider Fourier multipliers in some detail. Since we are interested in the case $p = 1$ as well as $1 < p < \infty$, we need to include Fourier multipliers on $L^1(\mathbb{R}^n)$. Bernstein's lemma and a partition of unity argument are our main tools. Proposition 8.2.3 gives a simple criterion for a function to belong to the Fourier algebra $\mathcal{F}L^1(\mathbb{R}^n)$ and hence to be a Fourier multiplier on $L^1(\mathbb{R}^n)$. These techniques are essential for our main results.

Assuming a suitable growth condition on the symbol a, which is in particular fulfilled for elliptic and even hypoelliptic polynomials, we prove in Section 8.3 that the operator $\mathrm{Op}_p(a)$ associated to a generates a k-times integrated semigroup on $L^p(\mathbb{R}^n)$ for some $k \in \mathbb{N}$ provided $\rho(\mathrm{Op}_p(a)) \neq \emptyset$ and the range of a lies in a left half-plane. Observe that the result covers the case of the operator $i\Delta$ which has already been considered in Section 3.9 (Theorem 3.9.4 and Corollary 3.9.14). It is interesting to note that the order of integration stated in Theorem 8.3.6 is in fact optimal for homogeneous symbols of the form $a(\xi) = i|\xi|^m$. These results are also closely related to the existence of the boundary group of the Poisson semigroup on $L^p(\mathbb{R}^n)$ (see Corollary 8.3.11) and to Littman's result on the cosine function generated by the Laplacian on $L^p(\mathbb{R}^n)$ (see Theorem 8.3.12).

In the final section of this chapter, we consider systems of differential operators with constant coefficients on L^p-spaces. Brenner's result (Theorem 8.4.3) states that first order symmetric, hyperbolic systems generate C_0-semigroups on $L^p(\mathbb{R}^n)^N$ for $p \neq 2$ if and only if the matrices commute. This means that the solutions of the wave equation, Maxwell's equation or Dirac's equation are not governed by a C_0-semigroup on L^p-spaces if $p \neq 2$. We prove, however, that the solutions are given by integrated semigroups on these spaces.

8.1 Translation Invariant Operators and C_0-semigroups

In this section we consider Cauchy problems

$$\begin{cases} u'(t) = & Au(t) \quad (t \geq 0), \\ u(0) = & u_0, \end{cases}$$

where A is the realization of a pseudo-differential operator in a function space X of the form $X = L^p(\mathbb{R}^n)$ $(1 \leq p < \infty)$ or $C_0(\mathbb{R}^n)$, and $u_0 \in X$. More precisely, let $m > 0$ and $\rho \in [0,1]$. We define $S^m_{\rho,0}$ to be the set of all functions $a \in C^\infty(\mathbb{R}^n)$ such that for each multi-index $\alpha \in \mathbb{N}_0^n$ there exists a constant C_α such that

$$|D^\alpha a(\xi)| \leq C_\alpha (1 + |\xi|)^{m - \rho|\alpha|} \quad (\xi \in \mathbb{R}^n).$$

Obviously, a polynomial of order m belongs to $S^m_{1,0}$. We call $a \in C^\infty(\mathbb{R}^n)$ a *symbol* if $a \in S^m_{\rho,0}$ for some $m > 0$ and some $\rho \in [0,1]$. For a symbol a we define the *pseudo-differential operator* $\mathrm{Op}(a)$ associated to a by

$$\mathrm{Op}(a)u(x) := \int_{\mathbb{R}^n} e^{ix\cdot\xi} a(\xi)\mathcal{F}u(\xi)d\xi \qquad (x \in \mathbb{R}^n,\ u \in \mathcal{S}(\mathbb{R}^n)),$$

where $x \cdot \xi$ is the scalar product of x and ξ, and $\mathcal{F}u$ denotes the Fourier transform of u. The operator $\mathrm{Op}_X(a)$ defined by

$$\begin{aligned} \mathrm{Op}_X(a)f &:= \mathcal{F}^{-1}(a\mathcal{F}f), \\ D(\mathrm{Op}_X(a)) &:= \{f \in X : \mathcal{F}^{-1}(a\mathcal{F}f) \in X\}, \end{aligned} \tag{8.1}$$

is called the *realization* of $\mathrm{Op}(a)$ in X, or the X*-realization* of $\mathrm{Op}(a)$. When $X = L^p(\mathbb{R}^n)$, we may write $\mathrm{Op}_p(a)$ for $\mathrm{Op}_X(a)$. Here, $\mathcal{F}^{-1}(a\mathcal{F}f)$ is interpreted in the sense of distributions (see Appendix E) as follows: as usual, we identify $f \in X$ with $T_f \in \mathcal{S}(\mathbb{R}^n)'$ given by

$$\langle \varphi, T_f \rangle = \int_{\mathbb{R}^n} f\varphi \qquad (\varphi \in \mathcal{S}(\mathbb{R}^n)).$$

The Fourier transform is an isomorphism of $\mathcal{S}(\mathbb{R}^n)'$ (see (E.10)) which implies that $\mathcal{F}f \in \mathcal{S}(\mathbb{R}^n)'$. Since a and all its derivatives are polynomially bounded, $a \cdot \mathcal{F}f$ is a well defined element of $\mathcal{S}(\mathbb{R}^n)'$ (see (E.3)). Hence, $\mathcal{F}^{-1}(a\mathcal{F}f) \in \mathcal{S}(\mathbb{R}^n)'$ since $\mathcal{F}$ is an isomorphism of $\mathcal{S}(\mathbb{R}^n)'$.

It is not difficult to verify that $\mathrm{Op}_X(a)$ is a closed operator in X whenever a is a symbol. In addition, $\mathrm{Op}_X(a)$ is densely defined since $\mathcal{S}(\mathbb{R}^n) \subset D(\mathrm{Op}_X(a))$. Moreover, by (E.12), $\mathrm{Op}_X(a)$ is a differential operator of order m with constant coefficients $a_\alpha \in \mathbb{C}$, i.e.,

$$\mathrm{Op}_X(a)f = \sum_{|\alpha|\le m} a_\alpha D^\alpha f \quad (f \in D(\mathrm{Op}_X(a))), \tag{8.2}$$

when a is the polynomial of order m of the form

$$a(\xi) = \sum_{|\alpha|\le m} a_\alpha (i\xi)^\alpha \quad (\xi \in \mathbb{R}^n). \tag{8.3}$$

A polynomial a of the form (8.3) is called *elliptic* if its *principal part* a_m, defined by

$$a_m(\xi) := \sum_{|\alpha|=m} a_\alpha (i\xi)^\alpha \quad (\xi \in \mathbb{R}^n),$$

vanishes only at $\xi = 0$. We call $\mathrm{Op}_X(a)$ an *elliptic operator* on X if a is an elliptic polynomial. Moreover, a polynomial a is called *hypoelliptic* if

$$\frac{D^\alpha a(\xi)}{a(\xi)} \to 0 \quad \text{as } |\xi| \to \infty$$

whenever $|\alpha| \neq 0$.

Our first lemma in this section shows that for operators $\mathrm{Op}_X(a)$ under consideration there is a close relationship between the resolvent set of $\mathrm{Op}_X(a)$ and Fourier multipliers for X. For the definition of Fourier multipliers and the space $\mathcal{M}_X(\mathbb{R}^n)$ we refer to Appendix E, but we note that a symbol a is a Fourier multiplier for X if and only if $\mathrm{Op}_X(a)$ is a bounded operator on X. In order to simplify our notation, we also write $\mathcal{A}_X$ for $\mathrm{Op}_X(a)$ if no confusion seems likely. Finally, given a symbol a we set

$$a(\mathbb{R}^n) := \{a(\xi) : \xi \in \mathbb{R}^n\}.$$

Lemma 8.1.1. *Let X be one of the spaces $L^p(\mathbb{R}^n)$ $(1 \le p < \infty)$ or $C_0(\mathbb{R}^n)$. Let a be a symbol and let $\lambda \in \mathbb{C}$. Then $\lambda \in \rho(\mathcal{A}_X)$ if and only if $(\lambda - a)$ is nowhere zero and $(\lambda - a)^{-1} \in \mathcal{M}_X(\mathbb{R}^n)$. In particular, $a(\mathbb{R}^n) \subset \sigma(\mathcal{A}_X)$.*

Proof. Assume that $(\lambda - a)^{-1} \in \mathcal{M}_X(\mathbb{R}^n)$. For $f \in X$ set

$$T_{r_\lambda} f := \mathcal{F}^{-1}((\lambda - a)^{-1}\mathcal{F}f).$$

If $f \in D(\mathcal{A}_X)$, then $T_{r_\lambda}(\lambda - \mathcal{A}_X)f = \mathcal{F}^{-1}((\lambda - a)^{-1}(\lambda - a)\mathcal{F}f) = f$. Moreover, if $f \in X$, then $T_{r_\lambda} f \in D(\mathcal{A}_X)$ since $\mathcal{F}^{-1}((\lambda - a)(\lambda - a)^{-1}\mathcal{F}f) = f \in X$. Hence,

$$(\lambda - \mathcal{A}_X)T_{r_\lambda} f = \mathcal{F}^{-1}((\lambda - a)(\lambda - a)^{-1}\mathcal{F}f) = f.$$

We have therefore proved that $(\lambda - \mathcal{A}_X)$ is invertible with inverse operator T_{r_λ}.

Conversely, let $\lambda \in \rho(\mathcal{A}_X)$. If $f \in D(\mathcal{A}_X)$, then $\tau_a f \in D(\mathcal{A}_X)$ and $\mathcal{A}_X \tau_a f = \tau_a \mathcal{A}_X f$, where $(\tau_a f)(x) := f(x-a)$. Hence, $R(\lambda, \mathcal{A}_X)\tau_a = \tau_a R(\lambda, \mathcal{A}_X)$. Proposition E.1 shows that there exists $r_\lambda \in \mathcal{M}_X(\mathbb{R}^n)$ satisfying $R(\lambda, \mathcal{A}_X)f = \mathcal{F}^{-1}(r_\lambda \mathcal{F} f)$ $(f \in X)$. Therefore,

$$f = (\lambda - \mathcal{A}_X)R(\lambda, \mathcal{A}_X)f = \mathcal{F}^{-1}((\lambda - a)r_\lambda \mathcal{F} f) \qquad (f \in X),$$

which implies by the uniqueness theorem for Fourier transforms that $(\lambda - a)r_\lambda = 1$ a.e. Since a is continuous and r_λ is bounded, it follows that $\lambda \notin a(\mathbb{R}^n)$ and $r_\lambda = (\lambda - a)^{-1}$ a.e. □

Remark 8.1.2. The above proof shows that, given $\lambda \in \rho(\mathcal{A}_X)$, we have

$$R(\lambda, \mathcal{A}_X)f = \mathcal{F}^{-1}(r_\lambda \mathcal{F} f) \quad (f \in X),$$

where $r_\lambda = (\lambda - a)^{-1}$. □

Proposition 8.1.3. *Let X be one of the spaces $L^p(\mathbb{R}^n)$ $(1 \le p < \infty)$ or $C_0(\mathbb{R}^n)$. Let a be a symbol. Then the following assertions are equivalent:*

(i) *$e^{ta} \in \mathcal{M}_X(\mathbb{R}^n)$ for all $t \ge 0$ and there exist constants $M, \omega \ge 0$ such that*

$$\|e^{ta}\|_{\mathcal{M}_X(\mathbb{R}^n)} \le M e^{\omega t} \quad (t \ge 0).$$

(ii) *$\mathcal{A}_X$ generates a C_0-semigroup on X.*

Proof. (i) ⇒ (ii): By replacing $a(\xi)$ by $a(\xi) - \omega$, we may assume without loss of generality that $\omega = 0$. It follows from Proposition E.2 and the assumptions that

$$\sup_{\xi \in \mathbb{R}^n} e^{t \operatorname{Re} a(\xi)} = \|e^{ta}\|_{\mathcal{M}_2(\mathbb{R}^n)} \le \|e^{ta}\|_{\mathcal{M}_X(\mathbb{R}^n)} \le M \quad (t \ge 0). \tag{8.4}$$

Hence, $\operatorname{Re} a \le 0$. For $\lambda > 0$ and $f \in X$, the integral $\int_0^\infty e^{-\lambda t}\mathcal{F}^{-1}(e^{ta}\mathcal{F})f\,dt$ converges in X and it is easy to see that it coincides in $\mathcal{S}(\mathbb{R}^n)'$ with $\mathcal{F}^{-1}((\lambda - a)^{-1})\mathcal{F} f$. Hence, $(\lambda - a)^{-1} \in \mathcal{M}_X(\mathbb{R}^n)$ and

$$\|(\lambda - a)^{-1}\|_{\mathcal{M}_X(\mathbb{R}^n)} \le \int_0^\infty e^{-\lambda t}\|e^{ta}\|_{\mathcal{M}_X(\mathbb{R}^n)}\,dt \le M\int_0^\infty e^{-\lambda t}dt \le \frac{M}{\lambda}.$$

By Lemma 8.1.1, we conclude that $(0, \infty) \subset \rho(\mathcal{A}_X)$.

For $f \in X$ set

$$T(t)f := \begin{cases} \mathcal{F}^{-1}(e^{ta}\mathcal{F} f) & (t > 0), \\ f & (t = 0). \end{cases}$$

By assumption, $T(t) \in \mathcal{L}(X)$ for all $t \ge 0$. In order to prove that the mapping $T : \mathbb{R}_+ \to \mathcal{L}(X)$ is strongly continuous, assume, for the time being, that $f \in D(\mathcal{A}_X)$. Since

$$a\int_0^t e^{sa}ds = e^{ta} - 1 \qquad (t \ge 0)$$

it follows easily that

$$\|T(t)f - f\|_X = \left\| \int_0^t \mathcal{F}^{-1}(e^{sa}\mathcal{F})\mathcal{A}_X f\, ds \right\|_X \le Mt\|\mathcal{A}_X f\|_X \qquad (t \ge 0)$$

for $f \in D(\mathcal{A}_X)$. Since $D(\mathcal{A}_X)$ is dense in X and since $\|T(t)\|_{\mathcal{L}(X)} \le M$ by assumption, it follows that T is strongly continuous.

Finally, let $f \in \mathcal{S}(\mathbb{R}^n)$. By Fubini's theorem,

$$\begin{aligned}\int_0^\infty e^{-\lambda t}\mathcal{F}^{-1}(e^{ta}\mathcal{F}f)\, dt &= \mathcal{F}^{-1}\left(\int_0^\infty e^{-\lambda t} e^{ta}\mathcal{F}f\, dt\right) \\ &= \mathcal{F}^{-1}((\lambda - a)^{-1}\mathcal{F}f) \qquad (\lambda > 0).\end{aligned}$$

Since $(\lambda - a)^{-1} \in \mathcal{M}_X(\mathbb{R}^n)$, it follows from Remark 8.1.2 that

$$R(\lambda, \mathcal{A}_X) = \int_0^\infty e^{-\lambda t}T(t)\, dt \quad (\lambda > 0).$$

Thus, the assertion follows from Theorem 3.1.7.

(ii) $\Rightarrow$ (i): Denote by T_X the C_0-semigroup generated by $\mathcal{A}_X$. Since $T_X(t)$ commutes with translations for all $t \ge 0$, it follows from Proposition E.1 that there exists $u_t \in \mathcal{M}_X(\mathbb{R}^n)$ such that

$$T_X(t)f = \mathcal{F}^{-1}(u_t\mathcal{F}f) \qquad (f \in \mathcal{S}(\mathbb{R}^n),\ t \ge 0).$$

By Theorem 3.1.7, we have $\|T_X(t)\|_{\mathcal{L}(X)} \le Me^{\omega t}$ $(t \ge 0)$ for some $M, \omega \ge 0$. Let $f \in \mathcal{S}(\mathbb{R}^n), \varphi \in C_c^\infty(\mathbb{R}^n)$. For λ sufficiently large, we have by Remark 8.1.2, Definition 3.1.6 and Fubini's theorem that

$$\begin{aligned}\int_0^\infty e^{-\lambda t}\langle \varphi, e^{ta}\mathcal{F}f\rangle\, dt &= \langle \varphi, (\lambda - a)^{-1}\mathcal{F}f\rangle \\ &= \langle \mathcal{F}\varphi, \mathcal{F}^{-1}(\lambda - a)^{-1}\mathcal{F}f\rangle \\ &= \langle \mathcal{F}\varphi, R(\lambda, \mathcal{A}_X)f\rangle \\ &= \int_0^\infty e^{-\lambda t}\langle \mathcal{F}\varphi, \mathcal{F}^{-1}u_t\mathcal{F}f\rangle\, dt.\end{aligned}$$

The uniqueness theorem for Laplace transforms implies that

$$\langle \varphi, e^{ta}\mathcal{F}f\rangle = \langle \varphi, u_t\mathcal{F}f\rangle$$

for all $t \ge 0$, $\varphi \in C_c^\infty(\mathbb{R}^n)$ and $f \in \mathcal{S}(\mathbb{R}^n)$. This implies that $u_t = e^{ta}$ a.e. for each $t \ge 0$. Thus, $e^{ta} \in \mathcal{M}_X(\mathbb{R}^n)$ and

$$\|e^{ta}\|_{\mathcal{M}_X(\mathbb{R}^n)} = \|u_t\|_{\mathcal{M}_X(\mathbb{R}^n)} = \|T_X(t)\|_{\mathcal{L}(X)} \le Me^{\omega t} \quad (t \ge 0). \qquad \square$$

Since $\mathcal{M}_2(\mathbb{R}^n) = L^\infty(\mathbb{R}^n)$ (see Proposition E.2 b)), the following corollary is obvious.

Corollary 8.1.4. *Let a be a symbol. Then $\mathcal{A}_{L^2(\mathbb{R}^n)}$ generates a C_0-semigroup on $L^2(\mathbb{R}^n)$ if and only if there exists $\omega \in \mathbb{R}$ such that*

$$\operatorname{Re} a(\xi) \leq \omega \text{ for all } \xi \in \mathbb{R}^n.$$

A necessary condition for e^{ta} to belong to $\mathcal{M}_X(\mathbb{R}^n)$ for $t > 0$ is given in the next lemma.

Lemma 8.1.5. *Let X be one of the spaces $L^p(\mathbb{R}^n)$ $(1 \leq p < \infty)$ or $C_0(\mathbb{R}^n)$. Consider a polynomial a of order m with principal part a_m. Suppose that $e^{ta} \in \mathcal{M}_X(\mathbb{R}^n)$ for all $t \geq 0$ and that $\|e^{ta}\|_{\mathcal{M}_X(\mathbb{R}^n)} \leq Me^{\omega t}$ $(t \geq 0)$ for suitable constants $M, \omega \geq 0$. Then $e^{a_m} \in \mathcal{M}_X(\mathbb{R}^n)$.*

Proof. Let $a = a_m + a_{m-1} + \cdots + a_0$, where each term a_j is homogeneous of degree j $(j = 0, \ldots, m)$. The change of variables $\xi \mapsto t^{-1/m}\xi$ implies by Proposition E.2 e) that u_t, defined by

$$u_t(\xi) := e^{a_m(\xi)} e^{ta_{m-1}(t^{-1/m}\xi)} \ldots e^{ta_0},$$

belongs to $\mathcal{M}_X(\mathbb{R}^n)$ and $\|e^{ta}\|_{\mathcal{M}_X(\mathbb{R}^n)} = \|u_t\|_{\mathcal{M}_X(\mathbb{R}^n)}$ for all $t > 0$. By assumption, there exists $C > 0$ such that $\|u_t\|_{\mathcal{M}_X(\mathbb{R}^n)} \leq C$ for all $t \in (0, 1)$. Since $\lim_{t\downarrow 0} u_t(\xi) = e^{a_m(\xi)}$ for all $\xi \in \mathbb{R}^n$, it follows from Proposition E.2 f) that $e^{a_m} \in \mathcal{M}_X(\mathbb{R}^n)$. □

When $\mathcal{A}_X$ is a first order differential operator of the form

$$\mathcal{A}_X f = \sum_{j=1}^{n} a_j D_j f + a_0 f,$$

where $a_j \in \mathbb{C}$ $(j = 0, 1, \ldots, n)$, then $\mathcal{A}_X$ generates a C_0-semigroup given by

$$(T(t)f)(x) = e^{a_0 t} f(x + ta),$$

where $a = (a_1, \ldots, a_n)$, $X = L^p(\mathbb{R}^n)$ $(1 \leq p < \infty)$ or $X = C_0(\mathbb{R}^n)$. In the following proposition a converse assertion is proved.

Proposition 8.1.6. *Let X be one of the spaces $L^p(\mathbb{R}^n)$ $(1 \leq p < \infty, p \neq 2)$ or $C_0(\mathbb{R}^n)$. Assume that $\mathcal{A}_X$ is a differential operator of the form (8.2) on X such that the symbol of the principal part a_m satisfies* $\operatorname{Re} a_m = 0$. *Then $\mathcal{A}_X$ generates a C_0-semigroup on X if and only if the order m of $\mathcal{A}_X$ is* 1.

Proof. It follows from Proposition 8.1.3 and Lemma 8.1.5 that $e^{a_m} \in \mathcal{M}_p(\mathbb{R}^n)$ for $X = L^p(\mathbb{R}^n)$ and $e^{a_m} \in \mathcal{M}_\infty(\mathbb{R}^n)$ for $X = C_0(\mathbb{R}^n)$. Since $\operatorname{Re} a_m = 0$ and $p \neq 2$ by assumption, it follows from Theorem E.4 a) that $m = 1$.

Conversely, if $m = 1$, then $\|e^{ta}\|_{\mathcal{M}_X(\mathbb{R}^n)} = e^{ta_0}$ for all $t \geq 0$ and the assertion follows from Proposition 8.1.3. □

Note that the special case of the symbol $a(\xi) = -i|\xi|^2$ was already considered in Theorem 3.9.4.

8.2 Fourier Multipliers

In this section on Fourier multipliers we give several sufficient conditions for a function to be a Fourier multiplier for $L^p(\mathbb{R}^n)$. The results presented in the following are the basis of our subsequent analysis of Cauchy problems in $L^p(\mathbb{R}^n)$ corresponding to operators of the form (8.1).

We start with a classical result due to Bernstein. Recall from Appendix E that for $j \in \mathbb{N}_0$ the space $H^j(\mathbb{R}^n)$ is defined to be the space of all functions $f \in L^2(\mathbb{R}^n)$ whose distributional derivatives $D^\alpha f$ belong to $L^2(\mathbb{R}^n)$ for $|\alpha| \leq j$. Plancherel's theorem implies that $f \in H^j(\mathbb{R}^n)$ if and only if $\xi \mapsto \xi^\alpha \mathcal{F}f(\xi)$ belong to $L^2(\mathbb{R}^n)$ for all $|\alpha| \leq j$. It is not hard to verify that there exist constants $C_1, C_2 > 0$ such that

$$C_1(1+|\xi|^2)^j \leq \sum_{|\alpha|\leq j} |\xi^\alpha|^2 \leq C_2(1+|\xi|^2)^j \quad (\xi \in \mathbb{R}^n),$$

from which it follows that $f \in H^j(\mathbb{R}^n)$ if and only if $\xi \mapsto (1+|\xi|^2)^{j/2}\mathcal{F}f(\xi)$ belongs to $L^2(\mathbb{R}^n)$, and that the norm $\left(\sum_{|\alpha|\leq j} \|D^\alpha f\|^2_{L^2(\mathbb{R}^n)}\right)^{1/2}$ is equivalent to $\|(1+|\cdot|^2)^{j/2}\mathcal{F}f(\cdot)\|_{L^2(\mathbb{R}^n)}$. Thus,

$$H^j(\mathbb{R}^n) = \left\{f \in L^2(\mathbb{R}^n) : \mathcal{F}^{-1}((1+|\cdot|^2)^{j/2}\mathcal{F}f(\cdot)) \in L^2(\mathbb{R}^n)\right\}.$$

Lemma 8.2.1 (Bernstein). *Let $u \in H^j(\mathbb{R}^n)$ for some $j > \frac{n}{2}$. Then $\mathcal{F}u \in L^1(\mathbb{R}^n)$ and there exists a constant C (depending only on n and j) such that*

$$\|\mathcal{F}u\|_{L^1(\mathbb{R}^n)} \leq C\|u\|^{1-(n/2j)}_{L^2(\mathbb{R}^n)}\left(\sum_{|\alpha|=j}\|D^\alpha u\|_{L^2(\mathbb{R}^n)}\right)^{n/2j} \qquad (u \in H^j(\mathbb{R}^n)).$$

Proof. For $R > 0$, we obtain by the Cauchy-Schwarz inequality and Plancherel's theorem,

$$\begin{aligned}
\|\mathcal{F}u\|_{L^1(\mathbb{R}^n)} &= \int_{|\xi|\leq R} 1\cdot|\mathcal{F}u(\xi)|\,d\xi + \int_{|\xi|\geq R} |\xi|^{-j}|\xi|^j|\mathcal{F}u(\xi)|\,d\xi \\
&\leq \left(\int_{|\xi|\leq R} 1\,d\xi\right)^{1/2} \|u\|_{L^2(\mathbb{R}^n)} \\
&\quad + \left(\int_{|\xi|\geq R}|\xi|^{-2j}\,d\xi\right)^{1/2}\left(\int_{|\xi|\geq R}|\xi|^{2j}|\mathcal{F}u(\xi)|^2\,d\xi\right)^{1/2} \\
&\leq CR^{n/2}\|u\|_{L^2(\mathbb{R}^n)} + CR^{(n/2)-j}\sum_{|\alpha|=j}\|D^\alpha u\|_{L^2(\mathbb{R}^n)}
\end{aligned}$$

for some constant C (depending on n and j). The assertion follows by choosing $R := \|u\|^{-1/j}_{L^2(\mathbb{R}^n)}\left(\sum_{|\alpha|=j}\|D^\alpha u\|_{L^2(\mathbb{R}^n)}\right)^{1/j}$. □

Our next result on a "partition of unity" will be very useful in the sequel.

Lemma 8.2.2. *There exists $\varphi \in C_c^\infty(\mathbb{R}^n)$ satisfying $\varphi \geq 0$, $\operatorname{supp}\varphi \subset \{\xi \in \mathbb{R}^n : \frac{1}{2} < |\xi| < 2\}$, and*

$$\sum_{k\in\mathbb{Z}} \varphi(2^{-k}\xi) = 1 \quad (\xi \neq 0).$$

Proof. Choose $f \in C_c^\infty(\mathbb{R}^n)$ such that $\operatorname{supp} f \subset \{\xi \in \mathbb{R}^n : \frac{1}{2} < |\xi| < 2\}$, $f \geq 0$ and $f(\xi) > 0$ if $\frac{1}{\sqrt{2}} \leq |\xi| \leq \sqrt{2}$. For $\xi \in \mathbb{R}^n$, set $f_0(\xi) := \sum_{k\in\mathbb{Z}} f(2^{-k}\xi)$. Then $f_0 \in C^\infty(\mathbb{R}^n\backslash\{0\})$, $f_0(\xi) > 0$ for all $\xi \in \mathbb{R}^n \setminus \{0\}$ and $f_0(2^{-k}\xi) = f_0(\xi)$ for all $k \in \mathbb{Z}$ and all $\xi \neq 0$. Hence, the function φ defined by

$$\varphi(\xi) := \begin{cases} 0 & (\xi = 0), \\ \dfrac{f(\xi)}{f_0(\xi)} & (\xi \neq 0), \end{cases}$$

satisfies the desired assertions. □

A very efficient sufficient condition for a function to belong to $\mathcal{M}_p(\mathbb{R}^n)$ is given by Mikhlin's theorem (see Theorem E.3). In fact, let $j := \min\{k \in \mathbb{N} : k > \frac{n}{2}\}$ and define $\mathcal{M}_M$ as

$$\mathcal{M}_M := \left\{m : \mathbb{R}^n \to \mathbb{C} : m \in C^j(\mathbb{R}^n\backslash\{0\}),\, |m|_M < \infty\right\},$$

where the norm $|\cdot|_M$ is defined as

$$|m|_M := \max_{|\alpha|\leq j} \sup_{\xi\in\mathbb{R}^n\backslash\{0\}} |\xi|^{|\alpha|}|D^\alpha m(\xi)|.$$

Mikhlin's theorem then states that $\mathcal{M}_M \hookrightarrow \mathcal{M}_{L^p(\mathbb{R}^n)}$ provided $1 < p < \infty$ (see Theorem E.3). Note that Mikhlin's theorem does not hold for $p = 1$. In the following we give a simple criterion for a function to belong to $\mathcal{M}_{L^1(\mathbb{R}^n)}$. We set $\mathcal{F}L^1(\mathbb{R}^n) := \{\mathcal{F}g : g \in L^1(\mathbb{R}^n)\} = \{f \in C_0(\mathbb{R}^n) : \mathcal{F}f \in L^1(\mathbb{R}^n)\}$. This space is a Banach space for the norm inherited from $L^1(\mathbb{R}^n)$; i.e.,

$$\|f\|_{\mathcal{F}L^1(\mathbb{R}^n)} := \|\mathcal{F}^{-1}f\|_{L^1(\mathbb{R}^n)} = (2\pi)^{-n}\|\mathcal{F}f\|_{L^1(\mathbb{R}^n)}.$$

The convolution theorem for Fourier transforms shows that $\mathcal{F}L^1(\mathbb{R}^n) \subset \mathcal{M}_1(\mathbb{R}^n)$ isometrically; i.e., $\|f\|_{\mathcal{F}L^1(\mathbb{R}^n)} = \|f\|_{\mathcal{M}_1(\mathbb{R}^n)}$ for $f \in \mathcal{F}L^1(\mathbb{R}^n)$. Moreover, by Proposition E.2, $\mathcal{F}L^1(\mathbb{R}^n) \hookrightarrow \mathcal{M}_p(\mathbb{R}^n)$ for $1 \leq p \leq \infty$. Let $\varepsilon > 0$ and put $j := \min\{k \in \mathbb{N} : k > \frac{n}{2}\}$. Define

$$\mathcal{M}_\varepsilon := \{m \in C^j(\mathbb{R}^n) : |m|_{\mathcal{M}_\varepsilon} < \infty\},$$

where

$$|m|_{\mathcal{M}_\varepsilon} := \max_{|\alpha|\leq j} \sup_{\xi\in\mathbb{R}^n} |\xi|^{|\alpha|+\varepsilon}|D^\alpha m(\xi)|.$$

Then $(\mathcal{M}_\varepsilon, |\cdot|_{\mathcal{M}_\varepsilon})$ is a Banach space and the following holds true.

Proposition 8.2.3. *Let* $\varepsilon > 0$. *Then* $\mathcal{M}_\varepsilon \hookrightarrow \mathcal{F}L^1(\mathbb{R}^n)$.

Proof. Let $m \in \mathcal{M}_\varepsilon$. Choose $\psi \in C_c^\infty(\mathbb{R}^n)$ such that $\psi(\xi) = 1$ whenever $|\xi| \le 2$ and write $m = \psi m + (1-\psi)m$. It follows from Bernstein's Lemma 8.2.1 that $\mathcal{F}(\psi m) \in L^1(\mathbb{R}^n)$. We may therefore assume that $m(\xi) = 0$ whenever $|\xi| \le 2$.

Let φ be a function as in Lemma 8.2.2, and for $k \in \mathbb{Z}$ set

$$m_k := m\varphi_k, \quad \varphi_k(\xi) := \varphi(2^{-k}\xi) \qquad (\xi \in \mathbb{R}^n).$$

Then $m = \sum_{k=1}^\infty m_k$. By Leibniz's rule, we obtain for α with $|\alpha| \le j$,

$$\begin{aligned} |D^\alpha m_k(\xi)| &= \left| \sum_{\beta \le \alpha} \binom{\alpha}{\beta} D^{\alpha-\beta} m(\xi) 2^{-k|\beta|} (D^\beta \psi)(2^{-k}\xi) \right| \\ &\le C \sum_{\beta \le \alpha} \binom{\alpha}{\beta} |\xi|^{-(|\alpha-\beta|+\varepsilon)} 2^{-k|\beta|} \|D^\beta \psi\|_\infty \\ &\le C \sum_{\beta \le \alpha} \binom{\alpha}{\beta} \frac{2^{-k|\beta|}}{2^{k(|\alpha-\beta|+\varepsilon)}} \\ &\le C 2^{-k(|\alpha|+\varepsilon)}, \end{aligned}$$

where C denotes a constant (which may differ from line to line). Here, we used the fact that $2^{k-1} < |\xi|$ for $\xi \in \operatorname{supp} m_k$. The L^2-norm of $D^\alpha m_k$ may hence be estimated as follows:

$$\begin{aligned} \|D^\alpha m_k\|_{L^2(\mathbb{R}^n)} &\le C 2^{-k(|\alpha|+\varepsilon)} \left(\int_{2^{k-1}<|\xi|<2^{k+1}} 1 \, d\xi \right)^{1/2} \\ &\le C 2^{-k(|\alpha|+\varepsilon)} \left(2^{kn}\right)^{1/2} \qquad (|\alpha| \le j). \end{aligned}$$

This implies that

$$\|D^\alpha m_k\|_{L^2(\mathbb{R}^n)} \le C 2^{-k(|\alpha|+\varepsilon-(n/2))} \qquad (k \ge 1, |\alpha| \le j).$$

Hence, $m_k \in H^j(\mathbb{R}^n)$ and it follows from Lemma 8.2.1 that

$$\begin{aligned} \|\mathcal{F} m_k\|_{L^1(\mathbb{R}^n)} &\le C \left(2^{-k(\varepsilon-(n/2))}\right)^{1-(n/2j)} \left(2^{-k(j+\varepsilon-(n/2))}\right)^{n/2j} \\ &= C 2^{-k\varepsilon}. \end{aligned}$$

Therefore, $\|\mathcal{F}m\|_{L^1(\mathbb{R}^n)} \le \sum_{k=1}^\infty \|\mathcal{F}m_k\|_{L^1(\mathbb{R}^n)} < \infty$ and it follows that $m \in \mathcal{F}L^1(\mathbb{R}^n)$. The closed graph theorem implies that the embedding $\mathcal{M}_\varepsilon \hookrightarrow \mathcal{F}L^1(\mathbb{R}^n)$ is continuous (in fact, the constants above are proportional to $|m|_{\mathcal{M}_\varepsilon}$). □

Lemma 8.2.4. *Let* $a : \mathbb{R}^n \to \mathbb{C}$ *be continuous such that* $a \in C^j(\mathbb{R}^n \setminus \{0\})$, *where* $j = \min\{k \in \mathbb{N} : k > \frac{n}{2}\}$. *Assume that there exist constants* $m > 0$ *and* $C_\alpha > 0$ *such that*

$$|D^\alpha a(\xi)| \le C_\alpha |\xi|^{m-|\alpha|} \qquad (0 < |\alpha| \le j, \ |\xi| \le 1, \ \xi \ne 0).$$

Let $\psi \in C_c^\infty(\mathbb{R}^n)$ such that $\psi(\xi) = 1$ for all $\xi \in \mathbb{R}^n$ with $|\xi| \le 1$. Then $a\psi \in \mathcal{F}L^1(\mathbb{R}^n)$.

Proof. Replacing $a(\xi)$ by $a(\xi) - a(0)$, we may assume that $a(0) = 0$. Then

$$|a(\xi)| = \left| \int_0^1 \xi \cdot \nabla a(t\xi)\, dt \right| \le C_0 |\xi|^m$$

for some constant C_0.

For $k \in \mathbb{Z}_-$, set $v_k(\xi) := a(\xi)\psi(\xi)\varphi(2^{-k}\xi)$, where φ is a function as in Lemma 8.2.2. The L^2-norm of $D^\alpha v_k$ may be estimated exactly as in the proof of Proposition 8.2.3, giving

$$\|D^\alpha v_k\|_{L^2(\mathbb{R}^n)} \le C 2^{k(m-|\alpha|+(n/2))} \qquad (k \le -1,\ |\alpha| \le j).$$

It follows from Lemma 8.2.1 that

$$\|\mathcal{F}v_k\|_{L^1(\mathbb{R}^n)} \le C 2^{kmn/2j}.$$

Thus,

$$\sum_{k=-\infty}^{-1} \|\mathcal{F}v_k\|_{L^1(\mathbb{R}^n)} \le C \sum_{k=1}^{\infty} 2^{-kmn/2j} < \infty,$$

and it follows that $\sum_{k=-\infty}^{-1} v_k \in \mathcal{F}L^1(\mathbb{R}^n)$. Since $a\psi - \sum_{k=-\infty}^{-1} v_k \in H^j(\mathbb{R}^n)$, Lemma 8.2.1 gives the result. □

Examples 8.2.5. The assumptions of Lemma 8.2.4 are in particular satisfied for the functions $a : \mathbb{R}^n \to \mathbb{C}$ defined by

a) $a(\xi) := c|\xi|^m \quad (m > 0,\ c \in \mathbb{C})$;

b) $a(\xi) := e^{i|\xi|^m} \quad (m > 0)$;

c) $a(\xi) := \displaystyle\int_0^1 \frac{(1-s)^{k-1}}{(k-1)!} e^{is|\xi|^m}\, ds \quad (m > 0,\ k \in \mathbb{N})$.

□

Proposition 8.2.6. *Let $1 \le p \le \infty$ and let $m \in C^j(\mathbb{R}^n)$ for some $j > \frac{n}{2}$. Suppose that $m(\xi) = 0$ whenever $|\xi| \le 1$. Let $\varepsilon > 0$ and $\rho \in (-\infty, 1]$. Assume that there exists a constant $M \ge 1$ such that*

$$\sup_{0<|\alpha|\le j} \left(\sup_{|\xi|\ge 1} |D^\alpha m(\xi)|\, |\xi|^{\varepsilon+\rho|\alpha|} \right)^{1/|\alpha|} \le M,$$

$$\sup_{|\xi|\ge 1} |m(\xi)| |\xi|^\varepsilon \le M.$$

If $\varepsilon > n\left|\frac{1}{2}-\frac{1}{p}\right|(1-\rho)$, then $m \in \mathcal{M}_p(\mathbb{R}^n)$ and there exists a constant C (depending on n, p, ρ and ε but otherwise independent of m and M) such that

$$\|m\|_{\mathcal{M}_p(\mathbb{R}^n)} \leq CM^{1+n\left|\frac{1}{2}-\frac{1}{p}\right|}.$$

Proof. By Proposition E.2 c), we may assume without loss of generality that $1 \leq p \leq 2$. Let φ be a function as in Lemma 8.2.2. For $k \in \mathbb{Z}$, put $m_k := m\varphi_k$, where $\varphi_k(\xi) = \varphi(2^{-k}\xi)$ for $\xi \in \mathbb{R}^n$. We claim that

$$\|m\|_{\mathcal{M}_p(\mathbb{R}^n)} \leq \sum_{k=0}^{\infty} \|m_k\|_{\mathcal{M}_p(\mathbb{R}^n)} < \infty.$$

Observe that the first inequality follows from the assumption that $m(\xi) = 0$ for $|\xi| \leq 1$. In order to estimate $\|m_k\|_{\mathcal{M}_p(\mathbb{R}^n)}$, note that $|\xi| > 2^{k-1}$ for $\xi \in \operatorname{supp} m_k$. By Leibniz's rule, we have

$$\begin{aligned} |D^\alpha m_k(\xi)| &= \left|\sum_{\beta \leq \alpha} \binom{\alpha}{\beta} D^{\alpha-\beta} m(\xi) 2^{-k|\beta|} (D^\beta \varphi)(2^{-k}\xi)\right| \\ &\leq \begin{cases} C_0 M 2^{-k\varepsilon} & (|\alpha| = 0), \\ C_\alpha M^{|\alpha|} 2^{k(-\varepsilon-\rho|\alpha|)} & (|\alpha| \neq 0), \end{cases} \end{aligned}$$

for suitable constants $C_0, C_\alpha > 0$. Consequently, there exist constants $C_{\alpha n}$ such that

$$\|D^\alpha m_k\|_{L^2(\mathbb{R}^n)} \leq \begin{cases} C_{0n} M 2^{-k\varepsilon} 2^{kn/2} & (|\alpha| = 0), \\ C_{\alpha n} M^{|\alpha|} 2^{k(-\varepsilon-\rho|\alpha|)} 2^{kn/2} & (|\alpha| \neq 0). \end{cases}$$

Choosing now $j > \frac{n}{2}$, we conclude by Lemma 8.2.1 that

$$\begin{aligned} \|m_k\|_{\mathcal{M}_1(\mathbb{R}^n)} &= \|m_k\|_{\mathcal{F}L^1(\mathbb{R}^n)} \\ &\leq C\left(M 2^{-k\varepsilon} 2^{kn/2}\right)^{1-(n/2j)} \left(M^j 2^{k(-\varepsilon-\rho j)} 2^{kn/2}\right)^{n/2j} \\ &\leq CM^{1-(n/2j)} M^{n/2} 2^{k(-\varepsilon+n(1-\rho)/2)}, \end{aligned}$$

for a suitable constant $C > 0$. Setting $\theta := 2\left(1-\frac{1}{p}\right)$ for $p \in (1,2)$, it follows from Proposition E.2 d) that

$$\begin{aligned} \|m_k\|_{\mathcal{M}_p(\mathbb{R}^n)} &\leq \|m_k\|^{1-\theta}_{\mathcal{M}_1(\mathbb{R}^n)} \|m_k\|^{\theta}_{\mathcal{M}_2(\mathbb{R}^n)} \\ &\leq CM^{1+n\left|\frac{1}{2}-\frac{1}{p}\right|} 2^{k\left(-\varepsilon+(1-\rho)n\left|\frac{1}{2}-\frac{1}{p}\right|\right)}. \end{aligned}$$

Thus, $\sum_{k=0}^{\infty} \|m_k\|_{\mathcal{M}_p(\mathbb{R}^n)} < \infty$ and the proof is complete. □

For a symbol $a \in S^m_{\rho,0}$ and $r > 0$, we consider the following growth hypothesis:

(H_r): There exist constants $C, L > 0$ such that

$$|a(\xi)| \geq C|\xi|^r$$

for all $\xi \in \mathbb{R}^n$ with $|\xi| \geq L$.

It is clear that if $a \in S^m_{\rho,0}$ satisfies (H_r) then $r \leq m$.

Remark 8.2.7. We note that by the Seidenberg-Tarski theorem (see [Hör83, Theorem 11.1.3]), Hypothesis (H_r) is in particular satisfied for all polynomials a satisfying $|a(\xi)| \to \infty$ as $|\xi| \to \infty$. Hence, assumption (H_r) holds for hypoelliptic polynomials. If a is an elliptic polynomial of order m, then (H_r) is satisfied with $r = m$.

Lemma 8.2.8. *Let $1 \leq p \leq \infty, N \in \mathbb{N}, m \in (0,\infty)$, $\rho \in [0,1]$ and $r > 0$. Suppose that $a \in S^m_{\rho,0}$ satisfies (H_r) and that $0 \notin a(\mathbb{R}^n)$. If $N > n|\frac{1}{2} - \frac{1}{p}|\left(\frac{m-\rho-r+1}{r}\right)$, then $a^{-N} \in \mathcal{M}_p(\mathbb{R}^n)$.*

Proof. Let $\psi \in C_c^\infty(\mathbb{R}^n)$ such that

$$\psi(\xi) := \begin{cases} 1 & (|\xi| \leq \max(L,1)), \\ 0 & (|\xi| \geq L+1), \end{cases}$$

where L is the constant arising in Hypothesis (H_r). Then, writing $a^{-N} = \psi a^{-N} + (1-\psi)a^{-N}$, we conclude by Lemma 8.2.1 that it suffices to prove that $(1-\psi)a^{-N} \in \mathcal{M}_p(\mathbb{R}^n)$. Now, $D^\alpha((1-\psi)a^{-N})(\xi) = D^\alpha(a^{-N})(\xi)$ for $|\xi| \geq L+1$. Using the assumption that $a \in S^m_{\rho,0}$ and a satisfies (H_r), and noting that $r \leq m$, one sees that

$$|D^\alpha(a^{-N})(\xi)| \leq C_\alpha |\xi|^{-rN+(m-r-\rho)|\alpha|} \qquad (|\xi| \geq L+1),$$

for suitable constants C_α. Hence, the assertion follows from Proposition 8.2.6. □

Lemma 8.2.9. *Let $1 \leq p \leq \infty$, $N \in \mathbb{N}$, $m \in (0,\infty)$, $\rho \in [0,1]$, $r > 0$ and let $a \in S^m_{\rho,0}$. Assume that $\sup_{\xi\in\mathbb{R}^n} \operatorname{Re} a(\xi) \leq -1$ and that Hypothesis (H_r) is satisfied. If $N > n|\frac{1}{2} - \frac{1}{p}|\left(\frac{1+m-\rho}{r}\right)$, then $e^{ta}a^{-N} \in \mathcal{M}_p(\mathbb{R}^n)$ and there exists a constant C (depending on N, n, ρ, p, m and r but otherwise independent of a) such that*

$$\|e^{ta}a^{-N}\|_{\mathcal{M}_p(\mathbb{R}^n)} \leq C(1+t)^{n|\frac{1}{2}-\frac{1}{p}|} \qquad (t \geq 0).$$

Proof. By Proposition E.2 c), we may restrict ourselves to the case $1 \leq p \leq 2$. Let $\psi \in C_c^\infty(\mathbb{R}^n)$ such that $0 \leq \psi \leq 1$ and

$$\psi(\xi) := \begin{cases} 1 & (|\xi| \leq L_1), \\ 0 & (|\xi| \geq L_1 + 1), \end{cases}$$

where $L_1 := \max(L, C^{-1/r}, 1)$ and C, L are the constants appearing in (H_r). For $t \geq 0$, we set $u_t := e^{ta}a^{-N}$. By Lemma 8.2.1, we conclude that $\psi u_t \in \mathcal{M}_1(\mathbb{R}^n)$ and that

$$\|\psi u_t\|_{\mathcal{M}_1(\mathbb{R}^n)} \leq C_n(1+t)^{n/2} \qquad (t \geq 0),$$

for some constant C_n. Since

$$\|\psi u_t\|_{\mathcal{M}_2(\mathbb{R}^n)} = \|\psi u_t\|_{L^\infty(\mathbb{R}^n)} \leq 1$$

for all $t \geq 0$, it follows from Proposition E.2 d) that

$$\|\psi u_t\|_{\mathcal{M}_p(\mathbb{R}^n)} \leq C_n(1+t)^{n|\frac{1}{2}-\frac{1}{p}|} \qquad (t \geq 0),$$

for a suitable constant C_n. Writing $u_t = \psi u_t + (1-\psi)u_t$, we conclude that it remains to prove the assertion for $(1-\psi)u_t$ instead of u_t. Now, by Leibniz's rule,

$$D^\alpha u_t = \sum_{\beta+\gamma=\alpha} \frac{\alpha!}{\beta!\gamma!} D^\beta(e^{ta}) D^\gamma(a^{-N}) \qquad (t \geq 0).$$

Since $a \in S^m_{\rho,0}$, we have

$$|(D^\beta e^{ta})(\xi)| \leq C_\beta(1+t)^{|\beta|}|\xi|^{|\beta|(m-\rho)} \qquad (|\xi| \geq L).$$

As in the proof of Lemma 8.2.8,

$$|(D^\gamma a^{-N})(\xi)| \leq C_\gamma|\xi|^{-rN+|\gamma|(m-r-\rho)} \qquad (|\xi| \geq L),$$

and it follows that there exists a constant $C > 1$ such that

$$\sup_{0<|\alpha|\leq j} \sup_{|\xi|\geq 1} \left(|D^\alpha [(1-\psi)(\xi)u_t(\xi)]| \, |\xi|^{rN+|\alpha|(\rho-m)} \right)^{1/|\alpha|} \leq C(1+t) \qquad (t > 0).$$

Since $L \geq C^{-1/r}$, we see that there exists a constant $C > 1$ such that

$$\sup_{|\xi|\geq 1} ((1-\psi)(\xi)\, u_t)(\xi)|\xi|^{rN} \leq C$$

for $t \geq 0$. Hence, the assertion follows from Proposition 8.2.6. □

8.3 L^p-spectra and Integrated Semigroups

For a symbol $a \in S^m_{\rho,0}$ and $r > 0$, consider again the Hypothesis (H_r) introduced in the previous Section 8.2:

(H_r): There exist constants $C, L > 0$ such that

$$|a(\xi)| \geq C|\xi|^r$$

for all $\xi \in \mathbb{R}^n$ with $|\xi| \geq L$.

In order to obtain a precise description of $\sigma(\mathrm{Op}_p(a))$, Lemma 8.1.1 shows that we need to decide whether or not the function $(\lambda - a)^{-1}$ is an L^p-multiplier. In general, this is a difficult matter. However, if the symbol a satisfies Hypothesis (H_r), the situation is much simpler. Indeed, we have the following result. Recall that $a(\mathbb{R}^n)$ was defined as $a(\mathbb{R}^n) = \{a(\xi) : \xi \in \mathbb{R}^n\}$.

Proposition 8.3.1. *Let $1 \le p < \infty$, $m \in (0,\infty)$, $\rho \in [0,1]$ and $r > 0$. Suppose that $a \in S^m_{\rho,0}$ satisfies (H_r). If $\rho(\mathrm{Op}_p(a)) \neq \emptyset$, then $\sigma(\mathrm{Op}_p(a)) = \sigma(\mathrm{Op}_2(a)) = a(\mathbb{R}^n)$.*

The following lemma will be useful in the proof of Proposition 8.3.1.

Lemma 8.3.2. *Let $1 \le p < \infty, m \in (0,\infty)$ and $\rho \in [0,1]$. Suppose that $a \in S^m_{\rho,0}$ and $\rho(\mathrm{Op}_p(a)) \neq \emptyset$. Let q be a polynomial of order k of the form $q(t) = c_k t^k + \ldots + c_0$ ($t \in \mathbb{R}$), with coefficients $c_0, \ldots, c_k \in \mathbb{C}$. Then*

$$\mathrm{Op}_p(q(a)) = q(\mathrm{Op}_p(a)).$$

Proof. It is clear that $\mathrm{Op}_p(q(a))$ is an extension of $q(\mathrm{Op}_p(a))$. Moreover, we have $D(q(\mathrm{Op}_p(a))) = D(\mathrm{Op}_p(a)^k)$. We are claiming that $D(\mathrm{Op}_p(q(a))) = D(\mathrm{Op}_p(a)^k)$, and we shall prove this by induction on k. For $k = 1$, this is trivial. Let $\mu \in \rho(\mathrm{Op}_p(a))$. Then, there exist $d_0 \in \mathbb{C}, d_k \in \mathbb{C}\backslash\{0\}$ and polynomials q_1, q_2 of degree $k-1$ such that

$$q(t) = (\mu - t)q_1(t) + d_0 = d_k t^k + q_2(t) \qquad (t \in \mathbb{R}).$$

For $f \in D(\mathrm{Op}_p(q(a))) \subset L^p(\mathbb{R}^n)$, we have $\mathcal{F}^{-1}((\mu - a)q_1(a)\mathcal{F}f) + d_0 f \in L^p(\mathbb{R}^n)$. Hence, $\mathcal{F}^{-1}((\mu - a)q_1(a)\mathcal{F}f) \in L^p(\mathbb{R}^n)$. Since $\mu \in \rho(\mathrm{Op}_p(a))$, we have $(\mu - a)^{-1} \in \mathcal{M}_p(\mathbb{R}^n)$ by Lemma 8.1.1. Thus, $\mathcal{F}^{-1}(q_1(a)\mathcal{F}f) \in L^p(\mathbb{R}^n)$. Therefore, we have $f \in D(\mathrm{Op}_p(q_1(a))) = D(\mathrm{Op}_p(a)^{k-1}) = D(\mathrm{Op}_p(q_2(a)))$ by the induction hypothesis. Moreover, $d_k\mathcal{F}^{-1}(a^k\mathcal{F}f) = \mathcal{F}^{-1}(q(a)\mathcal{F}f) - \mathcal{F}^{-1}(q_2(a)\mathcal{F}f) \in L^p(\mathbb{R}^n)$. Thus, $\mathrm{Op}_p(a)^{k-1}f \in D(\mathrm{Op}_p(a))$ and $f \in D(\mathrm{Op}_p(a)^k)$ as required. □

Proof of Proposition 8.3.1. Note first that Lemma 8.1.1 together with the fact that $\mathcal{M}_2(\mathbb{R}^n) = L^\infty(\mathbb{R}^n)$ implies that $\sigma(\mathrm{Op}_2(a))$ coincides with $a(\mathbb{R}^n)$. Since $a(\mathbb{R}^n) \subset \sigma(\mathrm{Op}_p(a))$ by Lemma 8.1.1, we only need to prove that $\sigma(\mathrm{Op}_p(a)) \subset a(\mathbb{R}^n)$. Choose $\lambda \in \mathbb{C}\backslash a(\mathbb{R}^n)$. The assumption (H_r) and Lemma 8.2.8 imply that $(\lambda - a)^{-N} \in \mathcal{M}_p(\mathbb{R}^n)$ if N is sufficiently large. Therefore, by Lemma 8.1.1, $0 \in \rho(\mathrm{Op}_p((\lambda - a)^N))$. It follows from Lemma 8.3.2 that

$$\mathrm{Op}_p((\lambda - a)^N) = \left(\mathrm{Op}_p(\lambda - a)\right)^N = \left(\lambda - \mathrm{Op}_p(a)\right)^N. \tag{8.5}$$

This implies that $(\lambda - \mathrm{Op}_p(a))^N$ is invertible. It follows that $\lambda - \mathrm{Op}_p(a)$ is invertible with inverse $(\lambda - \mathrm{Op}_p(a))^{N-1}((\lambda - \mathrm{Op}_p(a))^N)^{-1}$, which is bounded by the closed graph theorem. □

A quantitative version of Proposition 8.3.1 is given in the following theorem.

Theorem 8.3.3. *Let $1 \le p < \infty$, $m \in (0,\infty)$, $\rho \in [0,1]$ and $r > 0$. Suppose that $a \in S^m_{\rho,0}$ satisfies (H_r). Then the following assertions hold true:*

a) *If $n\left|\frac{1}{2} - \frac{1}{p}\right| \left(\frac{m-\rho-r+1}{r}\right) < 1$, then $\sigma(\mathrm{Op}_p(a)) = \sigma(\mathrm{Op}_2(a))$.*

b) *If $\rho \neq 1$, then the bound given in a) is optimal; i.e., if $n\left|\frac{1}{2} - \frac{1}{p}\right| \left(\frac{1-\rho}{m}\right) > 1$, there exists $a \in S^m_{\rho,0}$, satisfying (H_r) with $r = m$, such that $\sigma(\mathrm{Op}_p(a)) \neq \sigma(\mathrm{Op}_2(a))$.*

Proof. The assertion a) follows by combining Proposition 8.3.1 with Lemma 8.2.8 and Lemma 8.1.1. In order to prove assertion b), let $m \in \left(0, \frac{n(1-\rho)}{2}\right)$ and let $a : \mathbb{R}^n \to \mathbb{C}$ be a C^∞-function such that

$$a(\xi) := \begin{cases} |\xi|^m e^{i|\xi|^{1-\rho}} & (|\xi| \ge 2), \\ 1 & (|\xi| \le 1), \end{cases}$$

and $|a(\xi)| \ge 1$ for all $\xi \in \mathbb{R}^n$. Then $a \in S^m_{\rho,0}$ and (H_r) is satisfied with $r = m$. It follows from Theorem E.4 b) that $a^{-1} \in \mathcal{M}_p(\mathbb{R}^n)$ only if $n\left|\frac{1}{2} - \frac{1}{p}\right| \le \frac{m}{1-\rho}$. Therefore, by Lemma 8.1.1, $0 \in \sigma(\mathrm{Op}_p(a))$ if $n\left|\frac{1}{2} - \frac{1}{p}\right|\frac{1-\rho}{m} > 1$. Since $0 \notin a(\mathbb{R}^n) = \sigma(\mathrm{Op}_2(a))$, the assertion follows. □

Observing that for elliptic polynomials of degree m we have $\rho = 1$ and (H_r) is satisfied with $r = m$, we immediately have the following corollary.

Corollary 8.3.4. *Let $1 \le p < \infty$ and let a be an elliptic polynomial. Then*

$$\sigma(\mathrm{Op}_p(a)) = a(\mathbb{R}^n).$$

Remark 8.3.5. It is worthwhile noticing that assertion a) of Theorem 8.3.3 is no longer true if Hypothesis (H_r) is not satisfied. In fact, consider the symbol a given by

$$a(\xi) := -i(\xi_1 + \xi_2^2 + \xi_3^2 - i) \quad (\xi \in \mathbb{R}^3).$$

Then $\sigma(\mathrm{Op}_2(a)) = \{z \in \mathbb{C} : \mathrm{Re}\, z = -1\}$, but, by Theorem E.4 c)(i), we have $a^{-1} \notin \mathcal{M}_p(\mathbb{R}^n)$ if $p \neq 2$. Hence, by Lemma 8.1.1, $0 \in \sigma(\mathrm{Op}_p(a))$ whenever $p \neq 2$. □

We now consider the question whether operators associated to symbols $a \in S^m_{\rho,0}$ satisfying Hypothesis (H_r) are generators of integrated semigroups on L^p-spaces. To this end, let N_p be the smallest integer such that

$$N_p > n\left|\frac{1}{2} - \frac{1}{p}\right| \left(\frac{1+m-\rho}{r}\right) \quad (1 \le p < \infty).$$

We then have the following result (see Corollary 3.9.14 for the special case where $a(\xi) = -i|\xi|^2$).

Theorem 8.3.6. *Let $1 \le p < \infty, m \in (0,\infty)$, $\rho \in [0,1]$ and $r > 0$. Suppose that $a \in S^m_{\rho,0}$ satisfies (H_r). Then the following assertions are equivalent:*

(i) *$\rho(\mathrm{Op}_p(a)) \neq \emptyset$ and $\sup_{\xi\in\mathbb{R}^n} \mathrm{Re}\, a(\xi) \le \omega$ for some $\omega \in \mathbb{R}$.*

(ii) *The operator $\mathrm{Op}_p(a)$ generates an N_p-times integrated semigroup on $L^p(\mathbb{R}^n)$.*

(iii) *$\sigma(\mathrm{Op}_p(a)) \subset \{z \in \mathbb{C} : \mathrm{Re}\, z \le \omega\}$ for some $\omega \in \mathbb{R}$.*

Proof. (i) $\Rightarrow$ (ii): By rescaling we may assume that $\omega = -1$ (see Proposition 3.2.6). It follows from Proposition 8.3.1 that $0 \in \rho(\mathrm{Op}_p(a))$. For $t \ge 0$ and $k \in \mathbb{N}$ define the function $u_t^k : \mathbb{R}^n \to \mathbb{C}$ by

$$u_t^k := \int_0^t \frac{(t-s)^{k-1}}{(k-1)!} e^{sa} ds. \tag{8.6}$$

Integrating by parts we obtain

$$u_t^k = \frac{e^{ta}}{a^k} - \sum_{j=1}^{k} \frac{1}{(k-j)!} \frac{t^{k-j}}{a^j}. \tag{8.7}$$

We conclude from Lemma 8.1.1 and from the fact that $\mathcal{M}_p(\mathbb{R}^n)$ is a Banach algebra that there exists a constant C such that

$$\left\| \sum_{j=1}^{k} \frac{1}{(k-j)!} \frac{t^{k-j}}{a^j} \right\|_{\mathcal{M}_p(\mathbb{R}^n)} \le C(1+t)^{k-1} \qquad (t \ge 0). \tag{8.8}$$

By assumption, the symbol a satisfies (H_r). It thus follows from Lemma 8.2.9 that

$$\left\| \frac{e^{ta}}{a^{N_p}} \right\|_{\mathcal{M}_p(\mathbb{R}^n)} \le C(1+t)^{n|\frac{1}{2}-\frac{1}{p}|} \qquad (t \ge 0), \tag{8.9}$$

for some constant C. Combining (8.8) with (8.9) it follows that $u_t^{N_p} \in \mathcal{M}_p(\mathbb{R}^n)$ for all $t \ge 0$ and that

$$\|u_t^{N_p}\|_{\mathcal{M}_p(\mathbb{R}^n)} \le C(1+t)^{\alpha} \qquad (t \ge 0),$$

for some constants C, α.

For $f \in L^p(\mathbb{R}^n)$ and $t \ge 0$ set

$$S(t)f := \mathcal{F}^{-1}(u_t^{N_p} \mathcal{F} f).$$

Since

$$a \int_0^t u_s^{N_p} ds = u_t^{N_p} - \frac{t^{N_p}}{(N_p)!} \qquad (t \ge 0),$$

it follows that for $f \in \mathcal{S}(\mathbb{R}^n)$ and $r > t \geq 0$ we have

$$\begin{aligned}
&\|S(t)f - S(r)f\|_{L^p(\mathbb{R}^n)} \\
&\qquad \leq \|\operatorname{Op}_p(a)f\|_{L^p(\mathbb{R}^n)} \int_t^r \|u_s^{N_p}\|_{\mathcal{M}_p(\mathbb{R}^n)}\, ds + (r^{N_p} - t^{N_p})\|f\|_{L^p(\mathbb{R}^n)}.
\end{aligned}$$

Thus, $S(\cdot)f : \mathbb{R}_+ \to L^p(\mathbb{R}^n)$ is continuous. Since $\mathcal{S}(\mathbb{R}^n)$ is dense in $L^p(\mathbb{R}^n)$ and S is locally bounded, it follows that $S : \mathbb{R}_+ \to \mathcal{L}(L^p(\mathbb{R}^n))$ is strongly continuous.

It remains to show that the generator of S is $\operatorname{Op}_p(a)$. To this end, fix $\lambda > 0$ and let $f \in \mathcal{S}(\mathbb{R}^n)$. It follows from Fubini's theorem that

$$\begin{aligned}
\int_0^\infty e^{-\lambda t} S(t)f\, dt &= \int_0^\infty e^{-\lambda t} \mathcal{F}^{-1}(u_t^{N_p} \mathcal{F}f)\, dt \\
&= \mathcal{F}^{-1}\left(\int_0^\infty e^{-\lambda t} u_t^{N_p}\, dt\, \mathcal{F}f \right) \\
&= \mathcal{F}^{-1}\left(\frac{1}{\lambda^{N_p}} (\lambda - a)^{-1} \mathcal{F}f \right) = \frac{1}{\lambda^{N_p}} R(\lambda, \operatorname{Op}_p(a))f.
\end{aligned}$$

Since $\mathcal{S}(\mathbb{R}^n)$ is dense in $L^p(\mathbb{R}^n)$, the assertion follows by Definition 3.2.1.

(ii) $\Rightarrow$ (iii): This follows from the definition of an integrated semigroup.

(iii) $\Rightarrow$ (i): This is a consequence of Proposition 8.3.1. □

The following is immediate from Theorem 8.3.3 and Theorem 8.3.6.

Corollary 8.3.7. *Let $1 \leq p < \infty$, $m \in (0, \infty), \rho \in [0, 1]$ and $r > 0$. Assume that $a \in S^m_{\rho,0}$ satisfies (H_r) and $\sup_{\xi \in \mathbb{R}^n} \operatorname{Re} a(\xi) \leq \omega$ for some $\omega \in \mathbb{R}$. For $N \in \mathbb{N}$, there exists a constant $\delta_N > 0$ such that $\operatorname{Op}_p(a)$ generates an N-times integrated semigroup on $L^p(\mathbb{R}^n)$ provided $\left|\frac{1}{2} - \frac{1}{p}\right| < \delta_N$.*

Example 8.3.8. The example of the symbol a given by

$$a(\xi) := (-i)(\xi_1 - \xi_2^2 - \xi_3^2 - i)(\xi_1 + \xi_2^2 + \xi_3^2 + i) \qquad (\xi \in \mathbb{R}^3)$$

shows that $\operatorname{Op}_p(a)$ generates an integrated semigroup on $L^p(\mathbb{R}^3)$ only for certain values of p. Indeed, we verify that $\sup_\xi \operatorname{Re} a(\xi) = 0$ and (H_r) is satisfied with $r = 1$. Hence, by Theorem 8.3.3, we see that $\rho(\operatorname{Op}_p(a)) \neq \emptyset$ provided $\left|\frac{1}{2} - \frac{1}{p}\right| < \frac{1}{9}$. Therefore, $\operatorname{Op}_p(a)$ generates a once integrated semigroup on $L^p(\mathbb{R}^3)$ provided $\left|\frac{1}{2} - \frac{1}{p}\right| < \frac{1}{12}$. However, by Theorem E.4 c)(ii), $\sigma(\operatorname{Op}_p(a)) \neq a(\mathbb{R}^n)$ if $\left|\frac{1}{2} - \frac{1}{p}\right| > \frac{3}{8}$. Proposition 8.3.1 implies that $\sigma(\operatorname{Op}_p(a)) = \mathbb{C}$ if $\left|\frac{1}{2} - \frac{1}{p}\right| > \frac{3}{8}$. Thus, $\operatorname{Op}_p(a)$ does not generate an N-times integrated semigroup on $L^p(\mathbb{R}^3)$ for any N for those values of p. □

We consider now the case where a is no longer a symbol belonging to $S^m_{\rho,0}$ but a is a homogeneous function of the form

$$a(\xi) = i|\xi|^m \quad \text{or} \quad a(\xi) = -i|\xi|^m \qquad (\xi \in \mathbb{R}^n) \tag{8.10}$$

for some $m > 0$. In that case, the realization of the pseudo-differential operator associated to a in function spaces X of the form $L^p(\mathbb{R}^n)$ $(1 \le p < \infty)$ or $C_0(\mathbb{R}^n)$ is defined as follows. For $f \in X$ and a of the form (8.10), define $a\mathcal{F}f \in \mathcal{S}(\mathbb{R}^n)'$ as the mapping

$$\varphi \mapsto \int_{\mathbb{R}^n} f\mathcal{F}(a\varphi)\,dx \quad (\varphi \in \mathcal{S}(\mathbb{R}^n)). \tag{8.11}$$

Notice that $\mathcal{F}(a\varphi) \in L^2(\mathbb{R}^n) \cap L^\infty(\mathbb{R}^n)$ by Plancherel's theorem and the Riemann-Lebesgue lemma. We even have $\mathcal{F}(a\varphi) \in L^1(\mathbb{R}^n)$ by Lemma 8.2.4 and Example 8.2.5. Together with the inequality

$$\|f\|_q \le \|f\|_1^\theta \|f\|_\infty^{1-\theta} \qquad (1 < q < \infty,\ \theta := 1/q),$$

this implies that $\mathcal{F}(a\varphi) \in L^q(\mathbb{R}^n)$ for $1 \le q \le \infty$. It follows that the mapping given in (8.11) is well defined. It is also not difficult to verify that $a\mathcal{F}f \in \mathcal{S}(\mathbb{R}^n)'$. We thus define, for a of the form (8.10),

$$\begin{aligned} \mathrm{Op}_X(a) &:= \mathcal{F}^{-1}(a\mathcal{F}f), \\ D(\mathrm{Op}_X(a)) &:= \{f \in X : \mathcal{F}^{-1}(a\mathcal{F}f) \in X\}. \end{aligned} \tag{8.12}$$

We note that the assertions of Lemma 8.1.1 and Remark 8.1.2 remain true if the symbol a is replaced by a homogeneous function a of the form (8.10).

By the proof of Theorem 8.3.6, the operator $\mathrm{Op}_p(a) := \mathrm{Op}_{L^p(\mathbb{R}^n)}(a)$, defined as in (8.12), generates a k-times integrated semigroup on $L^p(\mathbb{R}^n)$ $(1 \le p < \infty)$ for some $k > 0$ if and only if $u_t^k \in \mathcal{M}_p(\mathbb{R}^n)$, where u_t^k is defined by

$$u_t^k(\xi) := \int_0^t \frac{(t-s)^{k-1}}{(k-1!)} e^{sa(\xi)}\,ds \qquad (\xi \in \mathbb{R}^n).$$

Notice first that

$$u_t^k(\xi) = t^k \int_0^1 \frac{(1-s)^{k-1}}{(k-1)!} e^{sta(\xi)}\,ds \qquad (\xi \in \mathbb{R}^n).$$

The change of variables $\xi \mapsto t^{-1/m}\xi$ implies by Proposition E.2 e) that $u_t^k \in \mathcal{M}_p(\mathbb{R}^n)$ if and only if $u_1^k \in \mathcal{M}_p(\mathbb{R}^n)$, and then

$$\|u_t^k\|_{\mathcal{M}_p(\mathbb{R}^n)} = t^k \|u_1^k\|_{\mathcal{M}_p(\mathbb{R}^n)}.$$

In order to determine whether $u_1^k \in \mathcal{M}_p(\mathbb{R}^n)$, let $\psi \in C_c^\infty(\mathbb{R}^n)$ such that $\psi(\xi) = 1$ for $|\xi| \le 1$. It follows from Lemma 8.2.4 and from Example 8.2.5 that $\psi u_1^k \in \mathcal{F}L^1(\mathbb{R}^n) \subset \mathcal{M}_p(\mathbb{R}^n)$ $(1 \le p < \infty)$. Furthermore,

$$(1-\psi)u_1^k = (1-\psi)\frac{e^a}{a^k} - (1-\psi)\sum_{j=1}^k \frac{1}{(k-j)!}\frac{1}{a^j}.$$

Now, since $\left|D^\alpha((1-\psi)\sum_{j=1}^k \frac{1}{(k-j)!}\frac{1}{a^j})(\xi)\right| \leq \frac{C_\alpha}{|\xi|^{m+|\alpha|}}$ for $|\alpha| \leq l$ with $l > \frac{n}{2}$, it follows from Proposition 8.2.3 that the second term on the right-hand side above belongs to $\mathcal{M}_p(\mathbb{R}^n)$. By Theorem E.4 b), we conclude that $u_1^k \in \mathcal{M}_p(\mathbb{R}^n)$ when $1 < p < \infty$ (respectively, $p = 1$) if and only if $k \geq n\left|\frac{1}{2} - \frac{1}{p}\right|$ (respectively, $k > \frac{n}{2}$) in the case $m \neq 1$; and $u_1^k \in \mathcal{M}_p(\mathbb{R}^n)$ when $1 < p < \infty$ (respectively, $p = 1$) if and only if $k \geq (n-1)\left|\frac{1}{2} - \frac{1}{p}\right|$ (respectively, $k > \frac{n-1}{2}$) for the case $m = 1$.

Consider now the case $k = 0$. It follows from the proof of Proposition 8.1.3 that $\mathrm{Op}_p(a) := \mathrm{Op}_{L^p(\mathbb{R}^n)}(a)$, defined as in (8.12), generates a C_0-semigroup on $L^p(\mathbb{R}^n)$ $(1 \leq p < \infty)$ if and only if $u_t^0 : \xi \mapsto e^{it|\xi|^m} \in \mathcal{M}_p(\mathbb{R}^n)$ and $\|u_t^0\|_{\mathcal{M}_p(\mathbb{R}^n)}$ is exponentially bounded in t. The change of variables $\xi \mapsto t^{-1/m}\xi$ implies by Proposition E.2 e) that $\|u_t^0\|_{\mathcal{M}_p(\mathbb{R}^n)} = \|u_1^0\|_{\mathcal{M}_p(\mathbb{R}^n)}$ for all $t > 0$. Let now $\psi \in C_c^\infty(\mathbb{R}^n)$ such that $\psi(\xi) = 1$ for $\xi \in \mathbb{R}^n$ with $|\xi| \leq 1$, and write $u_1^0 = u_1^0\psi + u_1^0(1-\psi)$. Then $u_1^0\psi \in \mathcal{M}_p(\mathbb{R}^n)$ by Lemma 8.2.4 and Example 8.2.5. It follows from Theorem E.4 b) that $u_1^0 \in \mathcal{M}_p(\mathbb{R}^n)$ when $1 < p < \infty$ (respectively, $p = 1$) if and only if $n\left|\frac{1}{2} - \frac{1}{p}\right| \leq 0$ (respectively, $\frac{n}{2} < 0$) in the case $m \neq 1$; and $u_1^0 \in \mathcal{M}_p(\mathbb{R}^n)$ when $1 < p < \infty$ (respectively, $p = 1$) if and only if $(n-1)\left|\frac{1}{2} - \frac{1}{p}\right| \leq 0$ (respectively, $\frac{n-1}{2} < 0$) for the case $m = 1$. We have therefore proved the following result.

Theorem 8.3.9. *Let $1 \leq p < \infty$, $k \in \mathbb{N}_0$ and $m > 0$. Define $a : \mathbb{R}^n \to \mathbb{C}$ by $a(\xi) := i|\xi|^m$.*

a) *If $m \neq 1$ and $1 < p < \infty$ (respectively, $p = 1$), then $\mathrm{Op}_p(a)$ generates a k-times integrated semigroup on $L^p(\mathbb{R}^n)$ if and only if $k \geq n\left|\frac{1}{2} - \frac{1}{p}\right|$ (respectively, $k > \frac{n}{2}$).*

b) *If $m = 1$ and $1 < p < \infty$ (respectively, $p = 1$), then $\mathrm{Op}_p(a)$ generates a k-times integrated semigroup on $L^p(\mathbb{R}^n)$ if and only if $k \geq (n-1)\left|\frac{1}{2} - \frac{1}{p}\right|$ (respectively, $k > \frac{n-1}{2}$).*

Remark 8.3.10. a) We note that the assertions of Theorem 8.3.9 remain true for the homogeneous function a given by $a(\xi) := -i|\xi|^m$.

b) Let $\mathrm{Op}_{C_0}(a)$ be the operator on $C_0(\mathbb{R}^n)$ defined as in (8.12) with $L^p(\mathbb{R}^n)$ replaced by $C_0(\mathbb{R}^n)$. Then the assertions of Theorem 8.3.9 remain true if $\mathrm{Op}_p(a)$ in Theorem 8.3.9 is replaced by $\mathrm{Op}_{C_0}(a)$ and $1/p$ by 0. □

Theorem 8.3.9 has interesting consequences for boundary values of holomorphic semigroups as discussed in Section 3.9. In fact, consider the Poisson semigroup T defined as in Example 3.7.9. There we proved that T is a bounded holomorphic C_0-semigroup of angle $\pi/2$ on $L^p(\mathbb{R}^n)$ for $1 \leq p < \infty$. Its generator is given by $A_p f = \mathcal{F}^{-1}(-|\cdot|\mathcal{F}f)$ for $f \in D(A_p) = \{f \in L^p(\mathbb{R}^n) : \mathcal{F}^{-1}(-|\cdot|\mathcal{F}f) \in L^p(\mathbb{R}^n)\}$. Theorem 8.3.9 and Remark 8.3.10 imply the following corollary.

Corollary 8.3.11. a) *Let $1 < p < \infty$ and let T_p be the Poisson semigroup on $L^p(\mathbb{R}^n)$. Then T_p admits a boundary group on $L^p(\mathbb{R}^n)$ in the sense of Proposition 3.9.1 if and only if $(n-1)\left|\frac{1}{2} - \frac{1}{p}\right| \leq 0$, i.e., if and only if $n = 1$ or $p = 2$.*

b) *Let T_1 be the Poisson semigroup on $L^1(\mathbb{R}^n)$. Then T_1 does not admit a boundary group on $L^1(\mathbb{R}^n)$ in the sense of Proposition* 3.9.1.

Another interesting consequence of Theorem 8.3.9 concerns the question whether the Laplacian Δ_p generates a cosine function on $L^p(\mathbb{R}^n)$ (see Example 3.7.6, Section 3.9 and Section 3.14). Indeed, let $1 \le p < \infty$ and recall that Δ_p on $L^p(\mathbb{R}^n)$ may be written as

$$\begin{aligned} \Delta_p f &= \mathcal{F}^{-1}(-|\cdot|^2 \mathcal{F} f), \\ D(\Delta_p) &= \{ f \in L^p(\mathbb{R}^n) : \mathcal{F}^{-1}(-|\cdot|^2 \mathcal{F} f) \in L^p(\mathbb{R}^n) \}. \end{aligned}$$

Since Δ_p generates a bounded C_0-semigroup on $L^p(\mathbb{R}^n)$ (see Example 3.7.6), it follows that $(0,\infty) \subset \rho(\Delta_p)$ and $\sup_{\lambda>0} \|\lambda R(\lambda, \Delta_p)\| < \infty$. Moreover, by Remark 8.1.2 we have

$$R(\lambda, \Delta_p) f = \mathcal{F}^{-1}\left(\frac{1}{\lambda + |\cdot|^2} \mathcal{F} f \right) \qquad (\lambda > 0,\ f \in \mathcal{S}(\mathbb{R}^n)).$$

For the time being, let $n = 1$. Then $\Delta_p = A_p^2$, where A_p is the generator of the C_0-group T_p of shifts considered in Example 3.3.10. By Example 3.14.15, Δ_p generates a cosine function Cos on $L^p(\mathbb{R})$ given by

$$\mathrm{Cos}(t) = \frac{1}{2}\left(T_p(t) + T_p(-t)\right) \qquad (t \in \mathbb{R}).$$

For $n > 1$, the situation is different. Indeed, suppose that Δ_p generates a cosine function Cos on $L^p(\mathbb{R}^n)$ for $n > 1$. For the time being, let $1 < p < \infty$. Then $L^p(\mathbb{R}^n)$ is a UMD-space and it follows from Theorem 3.16.7 that $i(-\Delta_p)^{1/2}$, defined as in Proposition 3.8.2, generates a C_0-group U on $L^p(\mathbb{R}^n)$. By Example 3.8.5 or by noting that $|\xi| = \int_0^\infty \lambda^{-\frac{1}{2}} \frac{|\xi|}{\lambda + |\xi|^2}\, d\lambda$ ($\xi \in \mathbb{R}^n$), we see that $-(-\Delta_p)^{1/2}$ coincides with the generator of the Poisson semigroup T on $L^p(\mathbb{R}^n)$; i.e.,

$$-(-\Delta_p)^{1/2} f = \mathcal{F}^{-1}(-|\xi| \mathcal{F} f) \qquad (f \in \mathcal{S}(\mathbb{R}^n)).$$

Since the Poisson semigroup T is a bounded holomorphic C_0-semigroup of angle $\pi/2$ on $L^p(\mathbb{R}^n)$, $i(-\Delta_p)^{1/2}$ generates the boundary semigroup of T in the sense of Proposition 3.9.1. By Corollary 8.3.11, we conclude that this implies that $p = 2$, since we assumed that $n > 1$.

Finally, consider the case $p = 1$ and assume that Δ_1 generates a cosine function Cos on $L^1(\mathbb{R}^n)$ for $n > 1$. Observe that Δ_2 generates a cosine function Cos on $L^2(\mathbb{R}^n)$ given by

$$\mathrm{Cos}(t) f = \mathcal{F}^{-1}(\cos(t|\cdot|) \mathcal{F} f) \qquad (t \in \mathbb{R},\ f \in L^2(\mathbb{R}^n)).$$

The Riesz-Thorin interpolation theorem [Hör83, Theorem 7.1.12] implies that Δ_p generates a cosine function on $L^p(\mathbb{R}^n)$ for $1 < p < 2$. This contradicts the assertion proved above and we have therefore proved the following result.

Theorem 8.3.12. *Let $1 \le p < \infty$ and assume that the Laplacian Δ_p generates a cosine function on $L^p(\mathbb{R}^n)$. Then $n = 1$ or $p = 2$.*

8.4 Systems of Differential Operators on L^p-spaces

In this section we consider initial value problems for systems of the form

$$\begin{cases} \dfrac{\partial u}{\partial t} = & Au \qquad (t \geq 0, x \in \mathbb{R}^n), \\ u(0,x) = & u_0(x) \quad (x \in \mathbb{R}^n), \end{cases}$$

where $u : \mathbb{R}_+ \times \mathbb{R}^n \to \mathbb{C}^N$ and A is an $N \times N$-matrix whose entries $(A_{ij})_{1 \leq i,j \leq N}$ are differential operators with constant coefficients of order m_{ij} in the sense of (8.2). The realization of A in $L^p(\mathbb{R}^n)^N$ $(1 \leq p < \infty)$ is defined as follows: let $a : \mathbb{R}^n \to \mathcal{L}(\mathbb{C}^N)$ be of the form

$$a(\xi) := \begin{pmatrix} a_{11}(\xi) & \dots & a_{1N}(\xi) \\ \vdots & & \vdots \\ a_{N1}(\xi) & \dots & a_{NN}(\xi) \end{pmatrix} \qquad (\xi \in \mathbb{R}^n), \tag{8.13}$$

where $a_{ij}(\xi) := \sum_{|\alpha| \leq m_{ij}} a_{ij\alpha}(i\xi)^\alpha$. Let $m := \max\{m_{i,j} : 1 \leq i,j \leq N\}$. Then $a(\xi) = a_0(\xi) + a_1(\xi) + \cdots + a_m(\xi)$, where each term a_j $(0 \leq j \leq m)$, is homogeneous of degree j. The term a_m is called the *principal part* of a. For $1 \leq p < \infty$ we define

$$\begin{aligned} \mathcal{A}_p f &:= \mathcal{F}^{-1}(a\mathcal{F}f), \\ D(\mathcal{A}_p) &:= \left\{ f \in L^p(\mathbb{R}^n)^N : \mathcal{F}^{-1}(a\mathcal{F}f) \in L^p(\mathbb{R}^n)^N \right\}, \end{aligned} \tag{8.14}$$

where the Fourier transform of vector-valued functions is defined by applying the transform elementwise. The proof of Lemma 8.1.1 and Proposition 8.1.3 imply the following result.

Lemma 8.4.1. *The operator $\mathcal{A}_2$ generates a C_0-semigroup on $L^2(\mathbb{R}^n)^N$ if and only if there exists $\omega \in \mathbb{R}$ such that*

$$\sup \left\{ \left\| e^{t(a(\xi) - \omega I)} \right\| : \xi \in \mathbb{R}^n,\, t \geq 0 \right\} < \infty. \tag{8.15}$$

Matrix-valued symbols a satisfying (8.15) have been completely characterized by Kreiss. The following consequence of his result will be very useful in the sequel.

Proposition 8.4.2. *Suppose that a of the form* (8.13) *satisfies* (8.15). *Assume that $\sigma(a_m(\xi)) \subset i\mathbb{R}$ for all $\xi \in \mathbb{R}^n$. Then there exists $S \in L^\infty(\mathbb{R}^n, \mathcal{L}(\mathbb{C}^N))$ such that $S(\xi)$ is invertible and $S(\xi)^{-1}a(\xi)S(\xi)$ is diagonal for all $\xi \in \mathbb{R}^n$.*

For a proof of Proposition 8.4.2 we refer to [Kre59].

In the following we examine in detail the special case of symmetric hyperbolic systems on $L^p(\mathbb{R}^n)^N$. To this end, let $a : \mathbb{R}^n \to \mathcal{L}(\mathbb{C}^N)$ be of the form

$$a(\xi) := \sum_{j=1}^n M_j(i\xi_j),$$

where $M_1, \ldots, M_n$ are hermitian $N \times N$-matrices. Then the following holds true.

Theorem 8.4.3 (Brenner). *Let $1 \le p < \infty$ such that $p \ne 2$. Let $a : \mathbb{R}^n \to \mathcal{L}(\mathbb{C}^N)$ be given by*

$$a(\xi) = \sum_{j=1}^{n} M_j(i\xi_j) \tag{8.16}$$

where $M_1, \dots, M_n$ are hermitian matrices. Then $\mathcal{A}_p$ generates a C_0-semigroup on $L^p(\mathbb{R}^n)^N$ if and only if the matrices $M_1, \dots, M_n$ commute.

We base the proof of Theorem 8.4.3 on the following two lemmas. Here, $\mathcal{M}_p^N(\mathbb{R}^n)$ is the space of all $(N \times N)$-matrices $m = (m_{ij})$ where $m_{ij} \in \mathcal{M}_p(\mathbb{R}^n)$ for $i, j = 1, 2, \dots, N$ (see Appendix E).

Lemma 8.4.4. *Let $1 \le p < \infty$ such that $p \ne 2$. Let a be of the form (8.16) and set $b = -ia$. Suppose that $e^a \in \mathcal{M}_p^N(\mathbb{R}^n)$. Then the eigenvalues $\lambda_k(\cdot)$ of $b(\cdot)$ can be chosen in such a way that*

$$\lambda_k(\xi) = \sum_{j=1}^{n} \lambda_{kj}\xi_j \qquad (\xi \in \mathbb{R}^n), \tag{8.17}$$

where $\lambda_{kj} \in \mathbb{R}$ for $1 \le j \le n$, $1 \le k \le N$.

Proof. The implicit function theorem implies that there is an open ball $U \subset \mathbb{R}^n$ and C^∞-functions $\lambda_k : U \to \mathbb{R}$ and $u_k : U \to \mathbb{C}^N$ $(k = 1, \dots, N)$ such that, for all $\xi \in U$, $\|u_k(\xi)\| \ge 1$, $b(\xi)u_k(\xi) = \lambda_k(\xi)u_k(\xi)$ and $\{u_k(\xi) : k = 1, \dots, N\}$ is a basis of $\mathbb{C}^N$. For the time being, we fix k and write λ and u for λ_k and u_k.

Let $\xi_0 \in U$ and let $\psi \in C_c^\infty(U)$ such that $\psi(\xi_0) = 1$. Choose $v \in C_c^\infty(\mathbb{R}^n, \mathbb{C}^N)$ such that $u(\xi) \cdot v(\xi) = 1$ $(\xi \in U)$.

For $t > 0$, we have

$$\psi(\xi)e^{it\lambda(\xi)}u(\xi) = \psi(\xi)e^{ta(\xi)}u(\xi) \qquad (\xi \in U),$$

and therefore

$$\psi(\xi)e^{it\lambda(\xi)} = e^{ta(\xi)}\psi(\xi)u(\xi) \cdot v(\xi) \qquad (\xi \in U).$$

Since $\psi u, v \in C_c^\infty(\mathbb{R}^n, \mathbb{C}^N)$, each of their coordinates belongs to $\mathcal{M}_p(\mathbb{R}^n)$. It follows from the homogeneity of a and Proposition E.2 e) that there exists a constant C such that

$$\|\psi e^{it\lambda}\|_{\mathcal{M}_p(\mathbb{R}^n)} \le C\|e^{ta}\|_{\mathcal{M}_p^N(\mathbb{R}^n)} = C\|e^a\|_{\mathcal{M}_p^N(\mathbb{R}^n)}. \tag{8.18}$$

Define now $\mu(\xi) := \lambda(\xi_0 + \xi) - \lambda(\xi_0) - \xi \cdot \nabla\lambda(\xi_0)$ if $\xi_0 + \xi \in U$, and set

$$f_t(\xi) := \begin{cases} \psi(\xi_0 + t^{-1/2}\xi)e^{it\mu(t^{-1/2}\xi)} & \text{if } \xi_0 + t^{-1/2}\xi \in U, \\ 0 & \text{otherwise.} \end{cases}$$

For $g \in \mathcal{M}_p(\mathbb{R}^n)$ and $x \in \mathbb{R}^n$, let $(\tau_x g)(\xi) := g(\xi - x)$ and let $h_x(\xi) := e^{i\xi\cdot x}$. Then $\mathcal{F}^{-1}(\tau_x g \cdot \mathcal{F})\varphi = h_x \cdot \mathcal{F}^{-1}(g\mathcal{F})(h_{-x} \cdot \varphi)$ and $\mathcal{F}^{-1}((h_x \cdot g)\mathcal{F})\varphi = \tau_{-x}(\mathcal{F}^{-1}(g\mathcal{F})\varphi)$ for all $\varphi \in \mathcal{S}(\mathbb{R}^n)$. It follows that $\tau_x g,\ h_x \cdot g \in \mathcal{M}_p(\mathbb{R}^n)$ and

$$\|\tau_x g\|_{\mathcal{M}_p(\mathbb{R}^n)} = \|h_x \cdot g\|_{\mathcal{M}_p(\mathbb{R}^n)} = \|g\|_{\mathcal{M}_p(\mathbb{R}^n)}.$$

Using these relations, (8.18) and Proposition E.2 e), we have that $\|f_t\|_{\mathcal{M}_p(\mathbb{R}^n)} \leq C\|e^a\|_{\mathcal{M}_p^N(\mathbb{R}^n)}$ for all $t > 0$. Moreover, $\lim_{t\to\infty} f_t(\xi) = e^{iP(\xi)}$ uniformly on compact subsets of $\mathbb{R}^n$, where $P(\xi) := \frac{1}{2}\sum_{i,j=1}^n D_iD_j\lambda(\xi_0)\xi_i\xi_j$. Proposition E.2 f) implies that $e^{iP} \in \mathcal{M}_p(\mathbb{R}^n)$. However, since we assumed that $p \neq 2$, Theorem E.4 a) implies that $P \equiv 0$. Hence, all the second derivatives of λ vanish at an arbitrary point $\xi_0 \in U$, which implies that λ is linear on U.

We have shown that

$$\lambda_k(\xi) = \lambda_{k0} + \sum_{j=1}^n \lambda_{kj}\xi_j \qquad (\xi \in U,\ k = 1, \dots, N),$$

where $\lambda_{k0},\ \lambda_{kj} \in \mathbb{R}$ $(k = 1, \dots, N,\ j = 1, \dots, n)$. It follows that

$$\det(zI - b(\xi)) = \prod_{k=1}^N \Big(z - \lambda_{k0} - \sum_{j=1}^n \lambda_{kj}\xi_j\Big) \qquad (\xi \in U,\ z \in \mathbb{C}). \tag{8.19}$$

Since both sides of the equation above are polynomials in ξ, it follows that (8.19) holds for $\xi \in \mathbb{R}^n$ and $z \in \mathbb{C}$. By homogeneity, $\lambda_{k0} = 0$ for all $k = 1, \dots, N$. Thus, we can choose

$$\lambda_k(\xi) = \sum_{j=1}^n \lambda_{kj}\xi_j \qquad (\xi \in \mathbb{R}^n). \qquad \square$$

Lemma 8.4.5. *Let $a : \mathbb{R}^n \to \mathcal{L}(\mathbb{C}^N)$ be of the form* (8.16) *and let $b = -ia$. Assume that the eigenvalues $\lambda(\cdot)$ of $b(\cdot)$ are of the form* (8.17). *Then the matrices $M_1, \dots, M_n$ commute.*

Proof. Let $\lambda_j(\cdot)$, $j = 1, \dots, r$ be the distinct linear functions representing the eigenvalues of $b(\cdot)$ for $\xi \in \mathbb{R}^n$. Denote by V the set where two or more eigenvalues coincide. Then

$$b(\xi) = \sum_{j=1}^r \lambda_j(\xi)P_j(\xi) \qquad (\xi \in \mathbb{R}^n\backslash V),$$

where $P_j(\xi)$ are orthogonal projections given by

$$P_j(\xi) = \frac{\prod_{k\neq j}(b(\xi) - \lambda_k(\xi))}{\prod_{k\neq j}(\lambda_j(\xi) - \lambda_k(\xi))} =: \frac{N_j(\xi)}{D_j(\xi)}.$$

Since $\|P_j(\xi)\| = 1$ for $\xi \in \mathbb{R}^n\backslash V$, we have $\|N_j(\xi)\| = |D_j(\xi)|$ for $\xi \in \mathbb{R}^n \setminus V$, and hence for all $\xi \in \mathbb{R}^n$ by continuity. The entries of $N_j(\xi)$ are polynomials, so this implies that they are divisible by each of the linear factors $\lambda_j(\xi)-\lambda_k(\xi)$ of $D_j(\xi)$. Since $N_j(\xi)$ and $D_j(\xi)$ both have degree $r-1$, it follows that $P_j(\xi)$ is constant for all $\xi \in \mathbb{R}^n \backslash V$. Set $P_j := P_j(\xi)$. Then

$$b(\xi)=\sum_{j=1}^{r} \lambda_j(\xi) P_j \qquad (\xi \in \mathbb{R}^n \backslash V), \tag{8.20}$$

and by continuity (8.20) holds for all $\xi \in \mathbb{R}^n$. Obviously, the projections P_j commute and it thus follows that the matrices $M_1, \ldots, M_n$ also commute. □

Proof of Theorem 8.4.3. The proof of Theorem 8.4.3 may now be completed as follows. Suppose that $\mathcal{A}_p$ generates a C_0-semigroup on $L^p(\mathbb{R}^n)$. Then, by the proof of Proposition 8.1.3, $e^{ta} \in \mathcal{M}_p^N(\mathbb{R}^n)$ and $\|e^{ta}\|_{\mathcal{M}_p^N(\mathbb{R}^n)} = \|e^a\|_{\mathcal{M}_p^N(\mathbb{R}^n)}$ by Proposition E.2 e). Lemma 8.4.4 and Lemma 8.4.5 imply that the matrices commute.

The converse implication is easy to prove. In fact, if $M_1, \ldots, M_n$ commute, then the M_j may be simultaneously diagonalised by a unitary matrix U so that $D_j = UM_jU^*$, where $D_j = \operatorname{diag}(\lambda_{j1}, \ldots, \lambda_{jn})$. Hence,

$$e^{ta(\xi)} = \exp\left(it\sum_{j=1}^{n} U^* D_j U \xi_j\right) = U^* \exp\left(it \sum_{j=1}^{n} D_j \xi_j\right) U$$

and it follows that $e^{ta} \in \mathcal{M}_p^N(\mathbb{R}^n)$ and $\sup_{t\geq 0} \|e^{ta}\|_{\mathcal{M}_p^N(\mathbb{R}^n)} < \infty$. The assertion now follows as in the proof of Proposition 8.1.3. The C_0-semigroup generated by $\mathcal{A}_p$ may be written explicitly in terms of U and N translation semigroups on $L^p(\mathbb{R}^n)$. □

The following result describes the generalisation of Theorem 8.4.3 to the situation of systems of arbitrary order m.

Theorem 8.4.6 (Brenner). *Let $1 \leq p < \infty$ such that $p \neq 2$. Assume that $a : \mathbb{R}^n \to \mathcal{L}(\mathbb{C}^N)$ of the form* (8.13) *satisfies $\sigma(a_m(\xi)) \subset i\mathbb{R}$ for all $\xi \in \mathbb{R}^n$. Then $\mathcal{A}_p$ (defined as in* (8.14)*) generates a C_0-semigroup on $L^p(\mathbb{R}^n)^N$ if and only if there exist commuting diagonalisable matrices $M_1, \ldots, M_n$, with real eigenvalues such that*

$$a_m(\xi) = \sum_{j=1}^{n} M_j (i\xi_j) \qquad (\xi \in \mathbb{R}^n). \tag{8.21}$$

We do not aim to give here a detailed proof of Theorem 8.4.6. For this we refer to [Bre73]. We only notice that Proposition 8.1.5 generalizes to the situation of systems discussed in Theorem 8.4.6. Hence, if $\mathcal{A}_p$ generates a C_0-semigroup on $L^p(\mathbb{R}^n)^N$, then $e^{ta} \in \mathcal{M}_p^N(\mathbb{R}^n)$ which implies that $e^{a_m} \in \mathcal{M}_p^N(\mathbb{R}^n)$, where a_m denotes the principal part of a. One can now show that the order m is necessarily 1 and that a_m is of the form (8.21).

Starting from this situation, we show in the following that the operator $\mathcal{A}_p$ generates a k-times integrated semigroup on $L^p(\mathbb{R}^n)^N$ for suitable $k > 0$ provided

$\mathcal{A}_2$ generates a C_0-semigroup on $L^2(\mathbb{R}^n)^N$ and further additional assumptions on the symbol a are satisfied. More precisely, the following holds true.

Theorem 8.4.7. *Let $1 < p < \infty$. Assume that a of the form (8.13) is homogeneous of degree m for some $m \geq 1$. Suppose that $\sigma(a(\xi)) \subset i\mathbb{R}$ for all $\xi \in \mathbb{R}^n$ and that the number of distinct eigenvalues of $a(\xi)$ is constant, and the rank of $a(\xi)$ is also constant, for $\xi \in \mathbb{R}^n\backslash\{0\}$.*

a) *If $\mathcal{A}_2$ generates a C_0-semigroup on $L^2(\mathbb{R}^n)^N$, then $\mathcal{A}_p$ generates a k-times integrated semigroup on $L^p(\mathbb{R}^n)^N$ provided $k > n|\frac{1}{2} - \frac{1}{p}|$.*

b) *If $\mathcal{A}_2$ generates a C_0-semigroup on $L^2(\mathbb{R}^n)^N$ and in addition $\sigma(a(\xi)) = \{i\alpha_1|\xi|, \ldots, i\alpha_N|\xi|\}$ for all $\xi \in \mathbb{R}^n$, where $\alpha_1, \ldots, \alpha_N \in \mathbb{R}$, then $\mathcal{A}_p$ generates a k-times integrated semigroup on $L^p(\mathbb{R}^n)^N$ provided $k \geq (n-1)|\frac{1}{2} - \frac{1}{p}|$.*

The proof of Theorem 8.4.7 is based on the following lemma.

Lemma 8.4.8. *Let the assumption of Theorem* 8.4.7 a) *be satisfied. Then the function $u_t^k : \mathbb{R}^n \to \mathcal{L}(\mathbb{C}^N)$ defined by*

$$u_t^k(\xi) := \int_0^t \frac{(t-s)^{k-1}}{(k-1)!} e^{sa(\xi)}\, ds \qquad (t \geq 0, \xi \in \mathbb{R}^n)$$

belongs to $\mathcal{M}_p^N(\mathbb{R}^n)$ provided $k > n|\frac{1}{2} - \frac{1}{p}|$. If the assumption b) is also satisfied, then $u_t \in \mathcal{M}_p^N(\mathbb{R}^n)$ provided $k \geq (n-1)|\frac{1}{2} - \frac{1}{p}|$.

Proof. By assumption, a is homogeneous of order m and $\sigma(a(\xi)) \subset i\mathbb{R}$ for all $\xi \in \mathbb{R}^n$. It follows from Lemma 8.4.1 and Proposition 8.4.2 that $a(\xi)$ is diagonalisable for all $\xi \in \mathbb{R}^n$. Therefore and by virtue of our assumptions, the minimal polynomial of the matrix $-ia(\xi)$ has only $(K+1)$ simple roots $\lambda_l(\xi)$ $(l = 0, \ldots, K)$ for some K satisfying $0 \leq K \leq N-1$. Observe next that the eigenvalues $\lambda_l(\cdot)$ are homogeneous functions of degree m since $a(\cdot)$ is homogeneous of degree m.

For $t \geq 0$ and $\xi \in \mathbb{R}^n \setminus \{0\}$, let q be a polynomial of degree K in one variable such that $q(\lambda_l(\xi)) = e^{it\lambda_l(\xi)}$ for $l = 0, \ldots, K$. Then

$$e^{ta(\xi)} = q(a(\xi)).$$

We now examine the form of the coefficients $C_j(t, \xi)$ of q.

Denote by $L(\xi)$ the $(K+1) \times (K+1)$-matrix whose l-th row is given by $(\lambda_l(\xi)^K, \lambda_l(\xi)^{K-1}, \ldots, \lambda_l(\xi), 1)$ $(l = 0, \ldots, K)$. Since

$$\det L(\xi) = \prod_{l<j\leq K} (\lambda_l(\xi) - \lambda_j(\xi)) \neq 0$$

for all $\xi \neq 0$, we have

$$C_j(t, \xi) = (\det L(\xi))^{-1} \sum_{l=0}^{K} (\det C_j^l(\xi)) e^{it\lambda_l(\xi)},$$

where $C_j^l(\xi)$ is the $(K+1)\times(K+1)$-matrix defined by replacing the element $\lambda_l(\xi)^{K-j}$ of $L(\xi)$ by 1 and all other elements of $L(\xi)$ of row l and column j by 0. Therefore,

$$\begin{aligned} e^{ta(\xi)} &= \sum_{j=0}^{K} C_j(\xi)(a(\xi))^{K-j} \\ &= \sum_{l=0}^{K} e^{it\lambda_l(\xi)} \frac{1}{\det L(\xi)} \sum_{j=0}^{K} (\det C_j^l(\xi)) a(\xi)^{K-j} \qquad (t \geq 0,\ \xi \neq 0). \end{aligned}$$

Moreover, since $\xi \mapsto \det L(\xi)$ is homogeneous of degree $h := (K+1)Km/2$ and $\xi \mapsto \det C_j^l(\xi)$ is homogeneous of degree $h-(K-j)m$ for all $l \in \{0,\dots,K\}$, the functions $\Phi_l : \mathbb{R}^n \setminus \{0\} \to \mathcal{L}(\mathbb{C}^N)$ given by

$$\Phi_l(\xi) := \frac{1}{\det L(\xi)} \sum_{j=0}^{K} (\det C_j^l(\xi))(a(\xi))^{K-j} \qquad (l = 0,\dots,K)$$

are homogeneous of degree 0.

Since by assumption the number of distinct eigenvalues of $a(\xi)$ is constant for $\xi \in \mathbb{R}^n\setminus\{0\}$, it follows from [Kat82, Theorem II.5.13a] that the eigenvalues $\lambda_l(\cdot)$ are C^∞-functions on $\mathbb{R}^n\setminus\{0\}$. Hence,

$$e^{ta(\xi)} = \sum_{l=0}^{K} e^{it\lambda_l(\xi)} \Phi_l(\xi), \tag{8.22}$$

where $\Phi_l \in C^\infty(\mathbb{R}^n\setminus\{0\}, \mathcal{L}(\mathbb{C}^N))$ is homogeneous of degree 0 for $l = 0,\dots,K$. It follows from Mikhlin's theorem (see Theorem E.3) that $\Phi_l \in \mathcal{M}_p^N(\mathbb{R}^n)$. By assumption, either λ_l is identically zero on $\mathbb{R}^n \setminus \{0\}$ or it is homogeneous of degree m on $\mathbb{R}^n \setminus \{0\}$ and therefore satisfies (H_r) with $r = m$. It follows from the proof of Theorem 8.3.9, Proposition 8.2.6 and Lemma 8.2.1 that

$$\xi \mapsto \int_0^t \frac{(t-s)^{k-1}}{(k-1)!} e^{is\lambda_l(\xi)} ds$$

belongs to $\mathcal{M}_p(\mathbb{R}^n)$ provided $k > n|\frac{1}{2} - \frac{1}{p}|$. Since $\max_{1\leq i,j\leq N}\{\|a_{ij}\|_{\mathcal{M}_p(\mathbb{R}^n)}\}$ is an equivalent norm to $\|a\|_{\mathcal{M}_p^N(\mathbb{R}^n)}$, we conclude from (8.22) that $u_t^k \in \mathcal{M}_p^N(\mathbb{R}^n)$ if $k > n|\frac{1}{2} - \frac{1}{p}|$.

If $\sigma(a(\xi)) = \{i\alpha_1|\xi|,\dots,i\alpha_N|\xi|\}$ for $\alpha_1,\dots,\alpha_N \in \mathbb{R}$ and all $\xi \in \mathbb{R}^n$, then the representation (8.22) together with Theorem 8.3.9 implies that $u_t^k \in \mathcal{M}_p^N(\mathbb{R}^n)$ if $k \geq (n-1)|\frac{1}{2} - \frac{1}{p}|$. □

Proof of Theorem 8.4.7. Thanks to Lemma 8.4.8 it is now no longer difficult to extend the proof of Theorem 8.3.6 to the present situation of systems and to show that $\mathcal{A}_p$ is the generator of a k-times integrated semigroup S on $L^p(\mathbb{R}^n)^N$ given by

$$S(t)f := \mathcal{F}^{-1}(u_t^k \mathcal{F}f) \qquad (t \geq 0, f \in L^p(\mathbb{R}^n)^N). \qquad \square$$

We finish this section by applying Theorem 8.4.7 to certain systems arising in mathematical physics. We start with the wave equation on $\mathbb{R}^n$.

Example 8.4.9 (Wave equation on $\mathbb{R}^n$). Consider the classical wave equation

$$w_{tt} = \Delta w \quad (t \in \mathbb{R},\ x \in \mathbb{R}^n).$$

Introducing the variable $u := (\nabla w, w_t)^T$, the wave equation can be written as a symmetric, hyperbolic system with

$$a(\xi) = i \begin{pmatrix} 0 & \cdot & \cdot & \cdot & \cdot & \xi_1 \\ \vdots & & & & & \vdots \\ \cdot & & & & & \xi_n \\ \xi_1 & \xi_2 & \cdot & \cdot & \xi_n & 0 \end{pmatrix}_{(n+1)\times(n+1)} \qquad (\xi \in \mathbb{R}^n).$$

It follows from Theorem 8.4.3 that the operator $\mathcal{A}_p$ in $L^p(\mathbb{R}^n)^{n+1}$ associated with a does not generate a C_0-semigroup on $L^p(\mathbb{R}^n)^{n+1}$ if $p \neq 2$ and $n > 1$. This property of the wave equation was observed first by Littman [Lit63]. The eigenvalues of $a(\xi)$ are $\lambda_0(\xi) = 0$ of multiplicity $(n-1)$ and $\lambda_{1,2}(\xi) = \pm i|\xi|$ each of multiplicity 1 ($\xi \in \mathbb{R}^n$). Hence, given $p \in (1, \infty)$, by Theorem 8.4.7, the operator $\mathcal{A}_p$ on $L^p(\mathbb{R}^n)^{n+1}$ associated with a generates a k-times integrated semigroup on $L^p(\mathbb{R}^n)^{n+1}$ if $k \geq (n-1)\left|\frac{1}{2} - \frac{1}{p}\right|$. $\square$

Example 8.4.10 (Maxwell's equations). We consider Maxwell's equations in the case where current and charge densities are zero and units are chosen so that the speed of light is one. Then Maxwell's equations can be written as

$$\frac{\partial}{\partial t}\begin{pmatrix} u \\ v \end{pmatrix} = \begin{pmatrix} 0 & -\operatorname{rot} \\ \operatorname{rot} & 0 \end{pmatrix}\begin{pmatrix} u \\ v \end{pmatrix}, \quad \begin{pmatrix} u(0) \\ v(0) \end{pmatrix} = \begin{pmatrix} u_0 \\ v_0 \end{pmatrix},$$

where $u, v : \mathbb{R}^3 \to \mathbb{C}^3$. Note that Maxwell's equations may be rewritten as a symmetric, hyperbolic system satisfying the assumptions of Theorem 8.4.7 b). In fact, $a : \mathbb{R}^3 \to \mathcal{L}(\mathbb{C}^6)$ given by

$$a(\xi) = i \begin{pmatrix} 0 & 0 & 0 & 0 & -\xi_3 & \xi_2 \\ 0 & 0 & 0 & \xi_3 & 0 & -\xi_1 \\ 0 & 0 & 0 & -\xi_2 & \xi_1 & 0 \\ 0 & \xi_3 & -\xi_2 & 0 & 0 & 0 \\ -\xi_3 & 0 & \xi_1 & 0 & 0 & 0 \\ \xi_2 & -\xi_1 & 0 & 0 & 0 & 0 \end{pmatrix}$$

has eigenvalues $\lambda_0(\xi) = 0, \lambda_{1,2}(\xi) = \pm i|\xi|$ ($\xi \in \mathbb{R}^3$), each of multiplicity 2. Hence, by Theorem 8.4.7 b), the Maxwell operator $\mathcal{A}_p$ associated to a generates a once integrated semigroup on $L^p(\mathbb{R}^3)^6$. Note that, by Theorem 8.4.3, the Maxwell operator $\mathcal{A}_p$ does not generate a C_0-semigroup on $L^p(\mathbb{R}^3)^6$ if $p \neq 2$. □

Example 8.4.11 (Dirac's equation). The relativistic description of the motion of a particle of mass m with spin 1/2 is provided by Dirac's equation

$$\frac{\partial u}{\partial t}(x,t) = c\sum_{j=1}^{3} M_j \frac{\partial u}{\partial x_j}(x,t) - M_4 \frac{mc^2}{ih} u(x,t) + V(x,t) \qquad (x \in \mathbb{R}^3, t \geq 0).$$

Here, u is a function defined on $\mathbb{R}^3 \times \mathbb{R}_+$ which takes values in $\mathbb{C}^4$, c is the speed of light, h is Planck's constant, and M_1, M_2, M_3, M_4 are 4×4 matrices given by

$$M_1 := \begin{pmatrix} 0 & 0 & 0 & 1 \\ 0 & 0 & 1 & 0 \\ 0 & 1 & 0 & 0 \\ 1 & 0 & 0 & 0 \end{pmatrix}, \qquad M_2 := \begin{pmatrix} 0 & 0 & 0 & -i \\ 0 & 0 & i & 0 \\ 0 & -i & 0 & 0 \\ i & 0 & 0 & 0 \end{pmatrix},$$

$$M_3 := \begin{pmatrix} 0 & 0 & 1 & 0 \\ 0 & 0 & 0 & -1 \\ 1 & 0 & 0 & 0 \\ 0 & -1 & 0 & 0 \end{pmatrix}, \qquad M_4 := \begin{pmatrix} 1 & 0 & 0 & 0 \\ 0 & 1 & 0 & 0 \\ 0 & 0 & -1 & 0 \\ 0 & 0 & 0 & -1 \end{pmatrix}.$$

If $V \equiv 0$ and units are chosen so that all constants are equal to 1, then Dirac's equation may be written as a symmetric, hyperbolic system on $X := L^p(\mathbb{R}^3)^4$ $(1 < p < \infty)$ of the form

$$\begin{aligned} \frac{d}{dt}\begin{pmatrix} v \\ w \end{pmatrix} &= \begin{pmatrix} 0 & A_p \\ A_p & 0 \end{pmatrix}\begin{pmatrix} v \\ w \end{pmatrix} + i\begin{pmatrix} I & 0 \\ 0 & -I \end{pmatrix}\begin{pmatrix} v \\ w \end{pmatrix} \qquad (t \geq 0), \\ \begin{pmatrix} v(0) \\ w(0) \end{pmatrix} &= \begin{pmatrix} v_0 \\ w_0 \end{pmatrix}, \end{aligned}$$

where

$$A_p := \begin{pmatrix} \frac{\partial}{\partial x_3} & \frac{\partial}{\partial x_1} - i\frac{\partial}{\partial x_2} \\ \frac{\partial}{\partial x_1} + i\frac{\partial}{\partial x_2} & -\frac{\partial}{\partial x_3} \end{pmatrix}$$

with domain $D(A_p) := \{f \in L^p(\mathbb{R}^3)^2 : A_p f \in L^p(\mathbb{R}^3)^2\}$ in $L^p(\mathbb{R}^3)^2$.

It follows from Theorem 8.4.3 that the Dirac operator

$$\mathcal{D}_p := \begin{pmatrix} 0 & A_p \\ A_p & 0 \end{pmatrix} + i\begin{pmatrix} I & 0 \\ 0 & -I \end{pmatrix},$$

with domain $D(\mathcal{D}_p) := D(A_p) \times D(A_p)$, generates a C_0-semigroup on $L^p(\mathbb{R}^3)^4$ if and only if $p = 2$. In the following, we show that the Dirac operator $\mathcal{D}_p$ generates a twice integrated semigroup on $L^p(\mathbb{R}^3)^4$ if $1 < p < \infty$.

In order to do so, let A be a linear operator on a Banach space Y, let $\mathcal{A}$ on $Y \times Y$ be given by

$$\mathcal{A} := \begin{pmatrix} 0 & A \\ A & 0 \end{pmatrix}, \qquad D(\mathcal{A}) := D(A) \times D(A),$$

and let $\mathcal{B}$ be the bounded operator on $Y \times Y$ defined by $\mathcal{B} := i \begin{pmatrix} I & 0 \\ 0 & -I \end{pmatrix}$. Then the following holds true.

Lemma 8.4.12. *Let $k \in \mathbb{N}_0$. Then, the operator $\mathcal{A}$ generates an (exponentially bounded) k-times integrated semigroup on $Y \times Y$ if and only if A and $-A$ generate (exponentially bounded) k-times integrated semigroups on Y.*

Proof. Define $U \in \mathcal{L}(Y \times Y)$ by $U := \frac{1}{\sqrt{2}} \begin{pmatrix} I & I \\ I & -I \end{pmatrix}$. Then $\mathcal{A} = UDU^{-1}$, where $D := \begin{pmatrix} A & 0 \\ 0 & -A \end{pmatrix}$. Since D generates a k-times integrated semigroup if and only if A and $-A$ both do so, the result follows from the remarks before Theorem 3.5.7. $\square$

Lemma 8.4.13. *Assume that $\mathcal{A}$ generates an exponentially bounded once integrated semigroup on $Y \times Y$. Then $\mathcal{A} + \mathcal{B}$ with domain $D(\mathcal{A})$ generates an exponentially bounded twice integrated semigroup on $Y \times Y$.*

Proof. It follows from Lemma 8.4.12 that A and $-A$ generate once integrated semigroups on Y which are exponentially bounded. Moreover, Proposition 3.15.4 implies that A^2 generates an exponentially bounded sine function $(\mathrm{Sin}(t))_{t\geq 0}$ on Y. Thus,

$$R(\lambda^2, A^2) = \int_0^\infty e^{-\lambda t} \,\mathrm{Sin}(t)\, dt \qquad (\lambda > \mathrm{abs}(\mathrm{Sin})).$$

We conclude that $(\lambda - \mathcal{A} - \mathcal{B})$ is invertible for $\lambda > \mathrm{abs}(\mathrm{Sin})$ and that

$$R(\lambda, \mathcal{A} + \mathcal{B}) = \begin{pmatrix} \lambda + i & A \\ A & \lambda - i \end{pmatrix} R(\lambda^2 + 1, A^2) \qquad (\lambda > \mathrm{abs}(\mathrm{Sin})).$$

It follows from Theorem 3.15.6 that $A^2 - I$ generates a sine function $(\mathrm{Sin}_I(t))_{t\geq 0}$ on Y. Thus,

$$R(\lambda^2 + 1, A^2) = R(\lambda^2, A^2 - I) = \int_0^\infty e^{-\lambda t} \,\mathrm{Sin}_I(t)\, dt,$$

for λ sufficiently large. For $t \geq 0$, we set

$$\begin{aligned}
S_{11}(t) &:= \int_0^t \mathrm{Sin}_I(s)\, ds + i \int_0^t (t-s)\,\mathrm{Sin}_I(s)\, ds, \\
S_{22}(t) &:= \int_0^t \mathrm{Sin}_I(s)\, ds - i \int_0^t (t-s)\,\mathrm{Sin}_I(s)\, ds, \\
S_{12}(t) &:= S_{21}(t) := A \int_0^t (t-s)\,\mathrm{Sin}_I(s)\, ds.
\end{aligned}$$

We note that $S_{12}(t)$ and $S_{21}(t)$ are well defined for $t \geq 0$, since by Proposition 3.15.2, $\int_0^t (t-s)\,\mathrm{Sin}_I(s)x\,ds \in D(A^2)$ for all $x \in Y$ and all $t \geq 0$. We now set

$$S(t) := \begin{pmatrix} S_{11}(t) & S_{12}(t) \\ S_{21}(t) & S_{22}(t) \end{pmatrix} \qquad (t \geq 0).$$

Then $\mathrm{abs}(S) < \infty$ and we verify that

$$R(\lambda, \mathcal{A} + \mathcal{B}) = \lambda^2 \int_0^\infty e^{-\lambda t} S(t)\,dt \qquad (\lambda > \mathrm{abs}(S)).$$

Thus, $\mathcal{A} + \mathcal{B}$ generates a twice integrated semigroup on $Y \times Y$. □

Finally, consider again the situation of the Dirac equation. The eigenvalues of the symbol of A_p may be computed to be $\lambda_{1,2}(\xi) := \pm i|\xi|$. Hence, by Theorem 8.4.7 it follows that A_p and $-A_p$ generate exponentially bounded once integrated semigroups on $L^p(\mathbb{R}^3)^2$. By Lemma 8.4.12, $\mathcal{A}_p := \begin{pmatrix} 0 & A_p \\ A_p & 0 \end{pmatrix}$ generates an exponentially bounded once integrated semigroup on $L^p(\mathbb{R}^3)^2 \times L^p(\mathbb{R}^3)^2$. Furthermore by Lemma 8.4.13, $\mathcal{D}_p = \mathcal{A}_p + \mathcal{B}$ generates a twice integrated semigroup on $L^p(\mathbb{R}^3)^4$. We have thus proved the following result.

Theorem 8.4.14. *Let $1 < p < \infty$. Then the Dirac operator $\mathcal{D}_p = \mathcal{A}_p + \mathcal{B}$ on $L^p(\mathbb{R}^3)^4$, with domain $D(A_p) \times D(A_p)$, generates a twice integrated semigroup on $L^p(\mathbb{R}^3)^4$.*

8.5 Notes

Section 8.1

Most of the content of this section is more or less standard. The results given in Proposition 8.1.3 and Proposition 8.1.6 are based on the well known properties of Fourier multipliers listed in Appendix E.

Section 8.2

An excellent reference for more information on Fourier multipliers is [Ste93]. A proof of Bernstein's result (Lemma 8.2.1) can be found for instance in [Hör83]. Proposition 8.2.3 is due to Hieber [Hie91a]. The remaining part of this section follows the lines of [Hie95].

Section 8.3

The result described in Theorem 8.3.3 on L^p-spectral independence for pseudo-differential operators is due to Hieber [Hie95]. Corollary 8.3.4 was first shown by Iha and Schubert [IS70]. For further results on invariance of the L^p-spectrum of certain classes of pseudo-differential operators, see [Sch71] and [LS97]. Theorem 8.3.6 and Theorem 8.3.9 on integrated semigroups generated by operators associated to symbols $a \in S^m_{\rho,0}$ or to homogeneous symbols a of the form $a(\xi) = |\xi|^m$ are due to Hieber [Hie91a], [Hie95]. For related results see also [Lan68], [Sjo70], [BE85], [deL94]. It seems that Corollary 8.3.11 on the boundary group of the Poisson semigroup does not exist in the literature. It is however strongly related to L^p estimates of the wave equation; see [Per80]. The result described in Theorem 8.3.12 on the cosine function generated by the Laplacian in $L^p(\mathbb{R}^n)$ was first proved by Littman [Lit63] by direct calculations (not using the theory of cosine functions).

Section 8.4

The Cauchy problem for systems of differential operators with constant coefficients of the form described in Theorem 8.4.3 and Theorem 8.4.6 was investigated in detail by Brenner (see [Bre66] and [Bre73]). Theorem 8.4.3 and Theorem 8.4.6 are due to him. Our proof follows essentially the lines of [Bre66]. For the rest of the section we follow closely [Hie91c]. Example 8.4.11 and Theorem 8.4.14 can be found in [Hie91d]. For further information on the systems discussed in Section 8.4, see also Chapter 1 of [Fat83].

Appendix A

Vector-valued Holomorphic Functions

Let X be a Banach space and let $\Omega \subset \mathbb{C}$ be an open set. A function $f : \Omega \to X$ is *holomorphic* if

$$f'(z_0) := \lim_{\substack{h \to 0 \\ h \in \mathbb{C} \setminus \{0\}}} \frac{f(z_0 + h) - f(z_0)}{h} \tag{A.1}$$

exists for all $z_0 \in \Omega$.

If f is holomorphic, then f is continuous and *weakly holomorphic* (i.e. $x^* \circ f$ is holomorphic for all $x^* \in X^*$). If $\Gamma := \{\gamma(t) : t \in [a, b]\}$ is a finite, piecewise smooth contour in Ω, we can form the contour integral $\int_\Gamma f(z)\, dz$. This coincides with the Bochner integral $\int_a^b f(\gamma(t))\gamma'(t)\, dt$ (see Section 1.1). Similarly we can define integrals over infinite contours when the corresponding Bochner integral is absolutely convergent. Since

$$\left\langle \int_\Gamma f(z)\, dz, x^* \right\rangle = \int_\Gamma \langle f(z), x^* \rangle\, dz,$$

many properties of holomorphic functions and contour integrals may be extended from the scalar to the vector-valued case, by applying the Hahn-Banach theorem. For example, Cauchy's theorem is valid, and also Cauchy's integral formula:

$$f(w) = \frac{1}{2\pi i} \int_{|z - z_0| = r} \frac{f(z)}{z - w}\, dz \tag{A.2}$$

whenever f is holomorphic in Ω, the closed ball $\overline{B}(z_0, r)$ is contained in Ω and $w \in B(z_0, r)$. As in the scalar case one deduces Taylor's theorem from this.

Proposition A.1. *Let $f : \Omega \to X$ be holomorphic, where $\Omega \subset \mathbb{C}$ is open. Let $z_0 \in \Omega$, $r > 0$ such that $\overline{B}(z_0, r) \subset \Omega$. Then*

$$f(z) = \sum_{n=0}^{\infty} a_n (z - z_0)^n$$

converges absolutely for $|z - z_0| < r$, *where*

$$a_n := \frac{1}{2\pi i} \int_{|z-z_0|=r} \frac{f(z)}{(z - z_0)^{n+1}}\, dz.$$

We also mention a special form of the identity theorem.

Proposition A.2 (Identity theorem for holomorphic functions). *Let Y be a closed subspace of a Banach space X. Let Ω be a connected open set in $\mathbb{C}$ and $f : \Omega \to X$ be holomorphic. Assume that there exists a convergent sequence $(z_n)_{n\in\mathbb{N}} \subset \Omega$ such that $\lim_{n\to\infty} z_n \in \Omega$ and $f(z_n) \in Y$ for all $n \in \mathbb{N}$. Then $f(z) \in Y$ for all $z \in \Omega$.*

Note that for $Y = \{0\}$, we obtain the usual form of the identity theorem.

Proof. Let $x^* \in Y^0 := \{y^* \in X^* : \langle y, y^* \rangle = 0 \ (y \in Y)\}$. Then $x^* \circ f(z_n) = 0$ for all $n \in \mathbb{N}$. It follows from the scalar identity theorem that $x^* \circ f(z) = 0$ for all $z \in \Omega$. Hence, $f(z) \in Y^{00} = Y$ for all $z \in \Omega$. □

In the following we show that every weakly holomorphic function is holomorphic. Actually, we will prove a slightly more general assertion which turns out to be useful. A subset N of X^* is called *norming* if

$$\|x\|_1 := \sup_{x^* \in N} |\langle x, x^* \rangle|$$

defines an equivalent norm on X. A function $f : \Omega \to X$ is called *locally bounded* if $\sup_K \|f(z)\| < \infty$ for all compact subsets K of Ω.

Proposition A.3. *Let $\Omega \subset \mathbb{C}$ be open and let $f : \Omega \to X$ be locally bounded such that $x^* \circ f$ is holomorphic for all $x^* \in N$, where N is a norming subset of X^*. Then f is holomorphic.*

In particular, if $X = \mathcal{L}(Y, Z)$, where Y, Z are Banach spaces, and if $f : \Omega \to X$ is locally bounded, then the following are equivalent:

(i) *f is holomorphic.*

(ii) *$f(\cdot)y$ is holomorphic for all $y \in Y$.*

(iii) *$\langle f(\cdot)y, z^* \rangle$ is holomorphic for all $y \in Y$, $z^* \in Z^*$.*

Proof. We can assume that $\|x\|_1 = \|x\|$ for all $x \in X$. In order to show holomorphy at $z_0 \in \Omega$ we can assume that $z_0 = 0$, replacing Ω by $\Omega - z_0$ otherwise. For small $h, k \in \mathbb{C}\backslash\{0\}$, let

$$u(h, k) := \frac{f(h) - f(0)}{h} - \frac{f(k) - f(0)}{k}.$$

We have to show that for $\varepsilon > 0$ there exists $\delta > 0$ such that $\| u(h,k) \| \leq \varepsilon$ whenever $|h| \leq \delta$ and $|k| \leq \delta$. Let $r > 0$ such that $\overline{B}(0,2r) \subset \Omega$ and

$$M := \sup_{z \in \overline{B}(0,2r)} \| f(z) \| < \infty.$$

Then by Cauchy's integral formula, for $|z| < r$, $|h| \leq r$, $|k| \leq r$, $h, k \neq 0$, $x^* \in N$,

$$\begin{aligned}\langle u(h,k), x^* \rangle &= \frac{1}{2\pi i} \int_{|z|=2r} \langle f(z), x^* \rangle \left\{ \frac{1}{h} \left(\frac{1}{z-h} - \frac{1}{z} \right) - \frac{1}{k} \left(\frac{1}{z-k} - \frac{1}{z} \right) \right\} dz \\ &= \frac{h-k}{2\pi i} \int_{|z|=2r} \frac{\langle f(z), x^* \rangle}{z(z-h)(z-k)} \, dz.\end{aligned}$$

Hence, $|\langle u(h,k), x^* \rangle| \leq |h-k| M / r^2$. Since N is norming, we deduce that

$$\| u(h,k) \| \leq |h-k| \, \frac{M}{r^2}.$$

This proves the claim. □

Corollary A.4. *Let $\Omega \subset \mathbb{C}$ be a connected open set and $\Omega_0 \subset \Omega$ be open. Let $h : \Omega_0 \to X$ be holomorphic. Assume that there exists a norming subset N of X^* such that for all $x^* \in N$ there exists a holomorphic extension $H_{x^*} : \Omega \to \mathbb{C}$ of $x^* \circ h$. If $\sup_{\substack{x^* \in N \\ z \in \Omega}} |H_{x^*}(z)| < \infty$, then h has a unique holomorphic extension $H : \Omega \to X$.*

Proof. Again we assume that $\| \cdot \|_1 = \| \cdot \|$. Let

$$Y := \Big\{ y = (y_{x^*})_{x^* \in N} \subset \mathbb{C} : \|y\|_\infty := \sup_{x^* \in N} |y_{x^*}| < \infty \Big\},$$

and let $H : \Omega \to Y$ be given by $H(z) := \big(H_{x^*}(z) \big)_{x^* \in N}$. It follows from Proposition A.3 that H is holomorphic. By $x \in X \mapsto (\langle x, x^* \rangle)_{x^* \in N}$, one defines an isometric injection from X into Y. Since $H(z) \in X$ for $z \in \Omega_0$, it follows from the identity theorem (Proposition A.2) that $H(z) \in X$ for all $z \in \Omega$. □

We will extend Proposition A.3 considerably in Theorem A.7. Before that we prove Vitali's theorem.

Theorem A.5 (Vitali). *Let $\Omega \subset \mathbb{C}$ be open and connected. Let $f_n : \Omega \to X$ be holomorphic ($n \in \mathbb{N}$) such that*

$$\sup_{\substack{n \in \mathbb{N} \\ z \in B(z_0,r)}} \|f_n(z)\| < \infty$$

whenever $\overline{B}(z_0, r) \subset \Omega$. Assume that the set

$$\Omega_0 := \big\{ z \in \Omega : \lim_{n \to \infty} f_n(z) \text{ exists} \big\}$$

has a limit point in Ω. Then there exists a holomorphic function $f : \Omega \to X$ such that

$$f^{(k)}(z) = \lim_{n\to\infty} f_n^{(k)}(z)$$

uniformly on all compact subsets of Ω for all $k \in \mathbb{N}_0$.

Proof. Let $l^\infty(X) := \{x = (x_n)_{n\in\mathbb{N}} \subset X : \ \|x\|_\infty := \sup \|x_n\| < \infty\}$. Then $l^\infty(X)$ is a Banach space for the norm $\|\cdot\|_\infty$ and the space $c(X)$ of all convergent sequences is a closed subspace of $l^\infty(X)$. Consider the function $F : \Omega \to l^\infty(X)$ given by $F(z) = (f_n(z))_{n\in\mathbb{N}}$. It follows from Proposition A.3 that F is holomorphic. (One may take N to be the space of all functionals on $l^\infty(X)$ of the form $(x_n)_{n\in\mathbb{N}} \mapsto \langle x_k, x^*\rangle$ where $k \in \mathbb{N}, x^* \in X^*, \|x^*\| \leq 1$). Since $F(z) \in c(X)$ for all $z \in \Omega_0$, it follows from the identity theorem (Proposition A.2) that $F(z) \in c(X)$ for all $z \in \Omega$. Consider the mapping $\phi \in \mathcal{L}(c(X), X)$ given by $\phi((x_n)_{n\in\mathbb{N}}) = \lim_{n\to\infty} x_n$. Then $f = \lim_{n\to\infty} f_n = \phi \circ F : \Omega \to X$ is holomorphic.

Finally, we prove uniform convergence on compact sets. Let $\overline{B}(z_0, r) \subset \Omega$ and $k \in \mathbb{N}_0$. It follows from (A.2) that

$$\frac{1}{k!} f_n^{(k)}(z) = \frac{1}{2\pi i} \int_{|w-z_0|=r} \frac{f_n(w)}{(w-z)^{k+1}}\, dw.$$

Now the dominated convergence theorem implies that $f_n^{(k)}(z)$ converges uniformly on $\overline{B}(z_0, r/2)$ to $f^{(k)}(z)$. Since every compact subset of Ω can be covered by a finite number of discs, the claim follows. □

If in Vitali's theorem (f_n) is a net instead of a sequence, the proof shows that $f(z) = \lim f_n(z)$ exists for all $z \in \Omega$ and defines a holomorphic function $f : \Omega \to X$.

Next we recall a well known theorem from functional analysis.

Theorem A.6 (Krein-Smulyan). *Let X be a Banach space and W be a subspace of the dual space X^*. Denote by B^* the closed unit ball of X^*. Then W is weak* closed if and only if $W \cap B^*$ is weak* closed.*

For a proof, see [Meg98, Theorem 2.7.11].

Now we obtain the following convenient criterion for holomorphy.

Theorem A.7. *Let $\Omega \subset \mathbb{C}$ be open and connected, and let $f : \Omega \to X$ be a locally bounded function. Assume that $W \subset X^*$ is a separating subspace such that $x^* \circ f$ is holomorphic for all $x^* \in W$. Then f is holomorphic.*

Here, W is called *separating* if $\langle x, x^*\rangle = 0$ for all $x^* \in W$ implies $x = 0$ $(x \in X)$.

Proof. Let $Y := \{x^* \in X^* : x^* \circ f \text{ is holomorphic}\}$. Since $W \subset Y$, the subspace Y is weak* dense. It follows from Vitali's theorem (applied to nets if X is not separable) that $Y \cap B^*$ is weak* closed. Now it follows from the Krein-Smulyan theorem that $Y = X^*$. Hence, f is holomorphic by Proposition A.3. □

Notes: Usually, Vitali's theorem is proved with the help of Montel's theorem which is only valid in finite dimensions. A vector-valued version is proved in the book of Hille and Phillips [HP57] by a quite complicated power-series argument going back to Liouville. The very simple proof given here is due to Arendt and Nikolski [AN00] who also proved Theorem A.7 (see also [AN06]).

Appendix B

Closed Operators

Let X be a complex Banach space. An *operator* on X is a linear map $A : D(A) \to X$, where $D(A)$ is a linear subspace of X, known as the *domain* of A. The *range* $\operatorname{Ran} A$, and the *kernel* $\operatorname{Ker} A$, of A are defined by:

$$\begin{aligned} \operatorname{Ran} A &:= \{Ax : x \in D(A)\}, \\ \operatorname{Ker} A &:= \{x \in D(A) : Ax = 0\}. \end{aligned}$$

The operator A is *densely defined* if $D(A)$ is dense in X.

An operator A is *closed* if its graph $G(A)$ is closed in $X \times X$, where

$$G(A) := \{(x, Ax) : x \in D(A)\}.$$

Thus, A is closed if and only if

$$\begin{gathered} \text{Whenever } (x_n) \text{ is a sequence in } D(A),\ x, y \in X, \\ \|x_n - x\| \to 0 \text{ and } \|Ax_n - y\| \to 0, \text{ then } x \in D(A) \text{ and } Ax = y. \end{gathered}$$

It is immediate from this that if A is closed and $\alpha, \beta \in \mathbb{C}$ with $\alpha \neq 0$, then the operator $\alpha A + \beta$ with $D(\alpha A + \beta) = D(A)$ is closed.

An operator A is said to be *closable* if there is an operator $\overline{A}$ (known as the *closure* of A) such that $G(\overline{A})$ is the closure of $G(A)$ in $X \times X$. Thus A is closable if and only if

$$\begin{gathered} \text{Whenever } (x_n) \text{ is a sequence in } D(A),\ y \in X, \\ \|x_n\| \to 0 \text{ and } \|Ax_n - y\| \to 0, \text{ then } y = 0. \end{gathered}$$

When A is closable,

$$\begin{aligned} D(\overline{A}) &= \Big\{ x \in X : \text{there exist } x_n \in D(A) \text{ and } y \in X \\ &\qquad\qquad \text{such that } \|x_n - x\| \to 0 \text{ and } \|Ax_n - y\| \to 0 \Big\}, \\ \overline{A}x &= y. \end{aligned}$$

For an operator A, $D(A)$ becomes a normed space with the *graph norm*

$$\|x\|_{D(A)} := \|x\| + \|Ax\|.$$

The operator $A : D(A) \to X$ is always bounded with respect to the graph norm, and A is closed if and only if $D(A)$ is a Banach space in the graph norm. Note that if A is replaced by $\alpha A + \beta$ where $\alpha \neq 0$, then the space $D(A)$ is unchanged and the graph norm is replaced by an equivalent norm.

Let A be a closed operator on X. A subspace D of $D(A)$ is said to be a *core* of A if D is dense in $D(A)$ with respect to the graph norm. Thus, D is a core of A if and only if A is the closure of $A|_D$, or equivalently for each $x \in D(A)$ there is a sequence (x_n) in D such that $\|x_n - x\| \to 0$ and $\|Ax_n - Ax\| \to 0$.

An operator A on X is said to be *invertible* if there is a bounded operator A^{-1} on X such that $A^{-1}Ax = x$ for all $x \in D(A)$ and $A^{-1}y \in D(A)$ and $AA^{-1}y = y$ for all $y \in X$.

Proposition B.1. *Let A be an operator on X. The following assertions are equivalent:*

(i) *A is invertible.*

(ii) $\operatorname{Ran} A = X$ *and there exists $\delta > 0$ such that $\|Ax\| \geq \delta\|x\|$ for all $x \in D(A)$.*

(iii) *A is closed,* $\operatorname{Ran} A$ *is dense in X, and there exists $\delta > 0$ such that $\|Ax\| \geq \delta\|x\|$ for all $x \in D(A)$.*

(iv) *A is closed,* $\operatorname{Ran} A = X$ *and* $\operatorname{Ker} A = \{0\}$.

Proof. The equivalence of (i) and (ii) is an easy consequence of the definition. Since any bounded operator has closed graph, and since

$$G(A^{-1}) = \{(y, x) : (x, y) \in G(A)\},$$

any invertible operator is closed. Thus, (i) and (ii) imply (iii) and (iv). When (iii) holds, $G(A)$ is complete, and the map $(x, Ax) \mapsto Ax$ is an isomorphism of $G(A)$ onto $\operatorname{Ran} A$, so $\operatorname{Ran} A$ is complete and (ii) follows. When (iv) holds, the inverse mapping theorem can be applied to the map A from $D(A)$ (with the graph norm) to X, showing that A^{-1} exists as a bounded map from X to $D(A)$ and hence to X. □

Let $\lambda \in \mathbb{C}$. Then λ is said to be in the *resolvent set* $\rho(A)$ of A if $\lambda - A$ is invertible, and we write $R(\lambda, A) := (\lambda - A)^{-1}$. The remarks in the previous paragraphs show that if $\rho(A)$ is non-empty, then A is closed. The function $R(\cdot, A) : \rho(A) \to \mathcal{L}(X)$ is the *resolvent* of A. The *spectrum* of A is defined to be:

$$\sigma(A) := \mathbb{C} \setminus \rho(A),$$

and the *spectral bound* is:

$$s(A) := \sup\{\operatorname{Re}\lambda : \lambda \in \sigma(A)\}$$

if the supremum exists ($s(A) := -\infty$ if $\sigma(A)$ is empty). The *point spectrum* $\sigma_p(A)$, and *approximate point spectrum* $\sigma_{ap}(A)$, of A are defined by:

$$\begin{aligned}\sigma_p(A) &:= \{\lambda \in \mathbb{C} : \operatorname{Ker}(\lambda - A) \neq \{0\}\},\\ \sigma_{ap}(A) &:= \Big\{\lambda \in \mathbb{C} : \text{there exist } x_n \in D(A) \text{ such that}\\ &\qquad \|x_n\| = 1 \text{ and } \lim_{n\to\infty} \|(\lambda - A)x_n\| = 0\Big\}.\end{aligned}$$

Thus, $\sigma_p(A)$ and $\sigma_{ap}(A)$ consist of the *eigenvalues* and *approximate eigenvalues* of A, respectively. It is clear that $\sigma_p(A) \subset \sigma_{ap}(A) \subset \sigma(A)$.

Proposition B.2. *Suppose that A has non-empty resolvent set, and let $\mu \in \rho(A)$. Let $\lambda \in \mathbb{C}$, $\lambda \neq \mu$. Then*

a) *$\lambda \in \rho(A)$ if and only if $(\mu - \lambda)^{-1} \in \rho(R(\mu, A))$. In that case,*

$$R(\lambda, A) = (\mu - \lambda)^{-1}\left((\mu - \lambda)^{-1} - R(\mu, A)\right)^{-1} R(\mu, A). \tag{B.1}$$

b) *$\lambda \in \sigma_p(A)$ if and only if $(\mu - \lambda)^{-1} \in \sigma_p(R(\mu, A))$.*

c) *$\lambda \in \sigma_{ap}(A)$ if and only if $(\mu - \lambda)^{-1} \in \sigma_{ap}(R(\mu, A))$.*

d) *The topological boundary of $\sigma(A)$ is contained in $\sigma_{ap}(A)$.*

Proof. Parts a), b) and c) follow immediately from the identity

$$\lambda - A = (\mu - \lambda)\left((\mu - \lambda)^{-1} - R(\mu, A)\right)(\mu - A).$$

Part d) follows from a), c) and the corresponding result for bounded operators. Alternatively, d) may be proved directly in exactly the same way as for bounded operators. □

Corollary B.3. *For any operator A, $\rho(A)$ is open and $\sigma(A)$ is closed in $\mathbb{C}$. Moreover, if $\mu \in \rho(A)$, $\lambda \in \mathbb{C}$ and $|\lambda - \mu| < \|R(\mu, A)\|^{-1}$, then $\lambda \in \rho(A)$, and*

$$R(\lambda, A) = \sum_{n=0}^{\infty} (\mu - \lambda)^n R(\mu, A)^{n+1},$$

where the series is norm-convergent. Hence,

$$\|R(\lambda, A)\| \le \frac{\|R(\mu, A)\|}{1 - |\lambda - \mu|\,\|R(\mu, A)\|}.$$

Moreover, $R(\cdot, A)$ is holomorphic on $\rho(A)$ with values in $\mathcal{L}(X)$ and

$$\frac{R(\mu, A)^{(n)}}{n!} = (-1)^n R(\mu, A)^{n+1} \qquad (n \in \mathbb{N}).$$

Proof. This is immediate from (B.1) and the Neumann expansion, $(I-T)^{-1} = \sum_{n=0}^{\infty} T^n$, when T is a bounded operator with $\|T\| < 1$. □

Proposition B.4. *Let A be an operator on X, and let $\lambda, \mu \in \rho(A)$. Then*

$$R(\lambda, A) - R(\mu, A) = (\mu - \lambda)R(\lambda, A)R(\mu, A). \tag{B.2}$$

Proof. The identity (B.2) follows by rearranging (B.1). □

Proposition B.5. *Let A be an operator on X, and U be a connected open subset of $\mathbb{C}$. Suppose that $U \cap \rho(A)$ is nonempty and that there is a holomorphic function $F : U \to \mathcal{L}(X)$ such that $\{\lambda \in U \cap \rho(A) : F(\lambda) = R(\lambda, A)\}$ has a limit point in U. Then $U \subset \rho(A)$ and $F(\lambda) = R(\lambda, A)$ for all $\lambda \in U$.*

Proof. Let $V = \{\lambda \in U \cap \rho(A) : F(\lambda) = R(\lambda, A)\}$, $\mu \in \rho(A)$, $x \in D(A)$, $y \in X$. For $\lambda \in V$,

$$F(\lambda)(\lambda - A)x = x, \tag{B.3}$$

$$F(\lambda)y = R(\mu, A)y - (\lambda - \mu)R(\mu, A)F(\lambda)y, \tag{B.4}$$

using (B.2). By uniqueness of holomorphic extensions (Proposition A.2), (B.3) and (B.4) are valid for all $\lambda \in U$. Now, (B.4) implies that $F(\lambda)y \in D(A)$ and

$$\begin{aligned} R(\mu, A)(\lambda - A)F(\lambda)y &= F(\lambda)y + (\lambda - \mu)R(\mu, A)F(\lambda)y \\ &= R(\mu, A)y. \end{aligned}$$

Since $R(\mu, A)$ is injective, $(\lambda - A)F(\lambda)y = y$ for all $\lambda \in U$. This and (B.3) imply that $\lambda \in \rho(A)$ and $F(\lambda) = R(\lambda, A)$. □

The equation (B.2) is known as the *resolvent equation* or *resolvent identity*. A function $R : U \to \mathcal{L}(X)$, defined on a subset U of $\mathbb{C}$, is said to be a *pseudo-resolvent* if it satisfies the resolvent equation; i.e., if

$$R(\lambda) - R(\mu) = (\mu - \lambda)R(\lambda)R(\mu) \quad (\lambda, \mu \in U).$$

The following proposition is easy to prove.

Proposition B.6. *Let $R : U \to \mathcal{L}(X)$ be a pseudo-resolvent. Then*

a) $\operatorname{Ker} R(\lambda)$ *and* $\operatorname{Ran} R(\lambda)$ *are independent of $\lambda \in U$.*

b) *There is an operator A on X such that $R(\lambda) = R(\lambda, A)$ for all $\lambda \in U$ if and only if* $\operatorname{Ker} R(\lambda) = \{0\}$.

An operator A is said to have *compact resolvent* if $\rho(A) \neq \emptyset$ and $R(\lambda, A)$ is a compact operator on X. Since the compact operators form an ideal of $\mathcal{L}(X)$, it is immediate from (B.2) that this property is independent of $\lambda \in \rho(A)$. When A has compact resolvent, then $\sigma(A)$ is a discrete subset of $\mathbb{C}$. This follows from (B.1) and the fact that the spectrum of a compact operator has 0 as its only limit point.

The following is easy to prove.

Proposition B.7. *Let A be an operator on X with non-empty resolvent set, and let $T \in \mathcal{L}(X)$. The following are equivalent:*

(i) $R(\lambda, A)T = TR(\lambda, A)$ *for all* $\lambda \in \rho(A)$.

(ii) $R(\lambda, A)T = TR(\lambda, A)$ *for some* $\lambda \in \rho(A)$.

(iii) *For all* $x \in D(A)$, $Tx \in D(A)$ *and* $ATx = TAx$.

For an operator A, the powers A^n $(n \geq 2)$ are defined recursively:

$$\begin{aligned} D(A^n) &:= \left\{x \in D(A^{n-1}) : A^{n-1}x \in D(A)\right\}, \\ A^n x &:= A(A^{n-1}x). \end{aligned}$$

Note that $D((\lambda - A)^n) = D(A^n)$ for all $\lambda \in \mathbb{C}$, $n \in \mathbb{N}$. It is easy to see that A^n is invertible if and only if A is invertible, and then $(A^n)^{-1} = (A^{-1})^n$.

If A is densely defined and $\rho(A) \neq \emptyset$, then $D(A^n)$ is a core for A, for each $n \in \mathbb{N}$. To see this, let $\lambda \in \rho(A)$. Then $R(\lambda, A)$ has dense range $D(A)$. It follows that the range $D(A^{n-1})$ of $R(\lambda, A)^{n-1}$ is dense in X. Let $x \in D(A)$. There is a sequence $(y_m)_{m \in \mathbb{N}}$ in $D(A^{n-1})$ converging to $(\lambda - A)x$. Let $x_m := R(\lambda, A)y_m$. Then $x_m \in D(A^n)$, $\|x_m - x\| \to 0$ and $\|Ax_m - Ax\| \to 0$.

Let A be an operator on X, and let Y be a closed subspace of X. The *part* of A in Y is the operator A_Y on Y defined by

$$\begin{aligned} D(A_Y) &:= \{y \in D(A) \cap Y : Ay \in Y\}, \\ A_Y y &:= Ay. \end{aligned}$$

The following results are easy to prove.

Proposition B.8. *Let A be an operator on X, and let Y be a closed subspace of X.*

a) *If* $D(A) \subset Y$, *then* $\rho(A) \subset \rho(A_Y)$ *and* $R(\lambda, A_Y) = R(\lambda, A)|_Y$ *for all* $\lambda \in \rho(A)$.

b) *Suppose that* $\rho(A) \neq \emptyset$ *and there is a projection* P *of* X *onto* Y *such that* $PR(\lambda, A) = R(\lambda, A)P$ *for some* $\lambda \in \rho(A)$. *Then* A *maps* $D(A) \cap Y$ *into* Y, A_Y *is the restriction of* A *to* $D(A) \cap Y$, $\lambda \in \rho(A_Y)$ *and* $R(\lambda, A_Y) = R(\lambda, A)|_Y$.

One situation where the conditions of Proposition B.8 b) are satisfied is described in the following.

Proposition B.9. *Let A be a closed operator on X with $\rho(A) \neq \emptyset$, and suppose that there are a compact subset E_1 and a closed subset E_2 of $\mathbb{C}$ such that $E_1 \cap E_2 = \emptyset$ and $E_1 \cup E_2 = \sigma(A)$. Then there is a bounded projection P on X such that $R(\lambda, A)P = PR(\lambda, A)$ for all $\lambda \in \rho(A)$, $P(X) \subset D(A)$, $\sigma(A_Y) = E_1$ and $\sigma(A_Z) = E_2$, where $Y := P(X)$, $Z := (I - P)(X)$. Moreover, P is unique, and $A|_Y \in \mathcal{L}(Y)$.*

The projection P is known as the *spectral projection of* A associated with E_1.

Proof. Take $\mu \in \rho(A)$ and consider $R(\mu, A) \in \mathcal{L}(X)$. Then $\sigma(R(\mu, A)) = E_1' \cup E_2'$, where

$$E_1' := \{(\mu - \lambda)^{-1} : \lambda \in E_1\}$$

and

$$E_2' := \begin{cases} \{(\mu - \lambda)^{-1} : \lambda \in E_2\} & \text{if } D(A) = X, \\ \{(\mu - \lambda)^{-1} : \lambda \in E_2\} \cup \{0\} & \text{otherwise.} \end{cases}$$

Then E_1' and E_2' are compact and disjoint. By the functional calculus for bounded operators (see [DS59, p.573]), there is a unique bounded projection P on X such that $R(\lambda, A)P = PR(\lambda, A)$ for all $\lambda \in \rho(A)$, $\sigma(R(\mu, A)|_Y) = E_1'$ and $\sigma(R(\mu, A)|_Z) = E_2'$. Since $0 \notin \sigma(R(\mu, A)|_Y)$, $Y \subset D(A)$ and $A|_Y$ is bounded by the closed graph theorem. The remaining properties follow easily from Proposition B.2 a). □

Suppose that A has compact resolvent, let $\lambda \in \rho(A)$ and $\mu \in \sigma(A)$. Let P be the spectral projection of A associated with $\{\mu\}$. Then there exists $m \in \mathbb{N}$ such that $(R(\lambda, A)P - (\lambda - \mu)^{-1}P)^m = 0$ (see [DS59, Theorem VII.4.5]). Hence, $(A - \mu)^m P = 0$.

Given an operator A on X, let

$$G(A^*) := \{(x^*, y^*) \in X^* \times X^* : \langle Ax, x^* \rangle = \langle x, y^* \rangle \text{ for all } x \in D(A)\},$$

which is a weak* closed subspace of $X^* \times X^*$. If (and only if) A is densely defined, then $G(A^*)$ is the graph of an operator A^* in X^*, known as the *adjoint* of A. For the remainder of this appendix, we shall assume that A is densely defined, and we shall consider properties of A^*.

When A is closed, the operator A can be recovered from A^* in the following way.

Proposition B.10. *Let A be a closed, densely defined operator on X, and let $x, y \in X$. The following assertions are equivalent:*

(i) *$x \in D(A)$ and $Ax = y$.*

(ii) *$\langle x, A^*x^* \rangle = \langle y, x^* \rangle$ for all $x^* \in D(A^*)$.*

Hence, $D(A^*)$ is weak dense in X^*.*

Proof. The implication (i) ⇒ (ii) is immediate from the definition of A^*. For the converse, suppose that $(x, y) \notin G(A)$. By the Hahn-Banach theorem, there exists $(x^*, y^*) \in X^* \times X^*$ such that $\langle x, x^* \rangle + \langle y, y^* \rangle \neq 0$ but $\langle u, x^* \rangle + \langle Au, y^* \rangle = 0$ for all $u \in D(A)$. The latter condition implies that $y^* \in D(A^*)$ and $A^*y^* = -x^*$. Thus, $\langle x, A^*y^* \rangle = -\langle x, x^* \rangle \neq \langle y, y^* \rangle$, so (ii) is violated.

If $D(A^*)$ is not weak* dense in X^*, then by the Hahn-Banach theorem there exists $y \in X$ such that $y \neq 0$ and $\langle y, x^* \rangle = 0$ for all $x^* \in D(A^*)$. By the previous part, $A0 = y$, which is absurd. □

If A is closable (and densely defined), it is easy to see that $(\overline{A})^* = A^*$, so $D(A^*)$ is weak* dense by Proposition B.10. Conversely, it is easy to see that if $D(A^*)$ is weak* dense, then A is closable.

Proposition B.11. *Let A be a closed, densely defined operator on X. Then*

a) *A^* is invertible if and only if A is invertible, and then $(A^*)^{-1} = (A^{-1})^*$.*

b) *$\sigma(A^*) = \sigma(A)$, and $R(\lambda, A^*) = R(\lambda, A)^*$ for all $\lambda \in \rho(A)$.*

c) *$\sigma(A) = \sigma_{ap}(A) \cup \sigma_p(A^*)$.*

Proof. a) If A is invertible, it is easy to verify that $(A^{-1})^* A^* x^* = x^*$ for all $x^* \in D(A^*)$ and $A^*(A^{-1})^* y^* = y^*$ for all $y^* \in X^*$. Thus, A^* is invertible and $(A^*)^{-1} = (A^{-1})^*$.

Now suppose that A^* is invertible, and let $\delta = \left\|(A^*)^{-1}\right\|^{-1}$. Since $\operatorname{Ker} A^* = \{0\}$, $\operatorname{Ran} A$ is dense in X, by a simple application of the Hahn-Banach theorem. For $x \in X$, there exists $x^* \in X^*$ such that $\|x^*\| = 1$ and $\langle x, x^* \rangle = \|x\|$. Let $y^* = (A^*)^{-1} x^* \in D(A^*)$, so that $\|y^*\| \leq \delta^{-1}$ and $A^* y^* = x^*$. Hence,

$$\|Ax\| \geq \delta \, |\langle Ax, y^* \rangle| = \delta \, |\langle x, A^* y^* \rangle| = \delta \|x\|.$$

It follows from Proposition B.1 that A is invertible.

b) This follows from a) by replacing A by $\lambda - A$.

c) This follows from applying Proposition B.1 and the fact that $\operatorname{Ran} A$ is dense in X if and only if $\operatorname{Ker} A^* = \{0\}$ (by the Hahn-Banach theorem), with A replaced by $\lambda - A$. □

Now, let H be a Hilbert space with inner product $(\cdot|\cdot)_H$. Identifying H^* with H by means of the Riesz-Fréchet lemma, we obtain the following. If A is a densely defined operator on H, the *adjoint* A^* of A is defined by

$$\begin{aligned} D(A^*) \quad &:= \quad \Big\{ x \in H : \text{there exists } y \in H \text{ such that} \\ &\qquad\qquad (Au|x)_H = (u|y)_H \text{ for all } u \in D(A) \Big\}, \\ A^* x \quad &= \quad y. \end{aligned}$$

We say that A is *selfadjoint* if $A = A^*$.

Example B.12 (Multiplication operators). Let (Ω, μ) be a measure space, $H := L^2(\Omega, \mu)$, $m : \Omega \to \mathbb{R}$ a measurable function. Define the operator M_m on H by

$$\begin{aligned} D(M_m) \quad &:= \quad \{ f \in H : \, mf \in H \}, \\ M_m f \quad &:= \quad mf. \end{aligned}$$

It is easy to see that M_m is selfadjoint. □

Let H, $\tilde{H}$ be Hilbert spaces. Two operators A on H and $\tilde{A}$ on $\tilde{H}$ are called *unitarily equivalent* if there exists a unitary operator $U : \tilde{H} \to H$ such that

$$\begin{aligned} D(\tilde{A}) &= U^{-1}D(A), \\ \tilde{A}x &= U^{-1}AUx. \end{aligned}$$

It is easy to see that, in that case, $\tilde{A}$ is selfadjoint whenever A is.

Now we can formulate the spectral theorem as follows; we refer to [RS72, Theorem VIII.4] for a proof.

Theorem B.13 (Spectral Theorem). *Each selfadjoint operator is unitarily equivalent to a real multiplication operator.*

Thus, selfadjoint and real multiplication operators are effectively the same thing. In proofs, we frequently regard an arbitrary selfadjoint operator as being a real multiplication operator.

A selfadjoint operator A is always *symmetric*; i.e., $(Ax|y)_H = (x|Ay)_H$ for all $x, y \in D(A)$. In particular, $(Ax|x)_H \in \mathbb{R}$ for all $x \in D(A)$. We say that A is *bounded above* if there exists $\omega \in \mathbb{R}$ such that

$$(Ax|x)_H \le \omega(x|x)_H \qquad (x \in D(A)).$$

In that case, ω is called an *upper bound* of A. If A is a multiplication operator M_m, then this is equivalent to saying that

$$m(y) \le \omega \quad \text{for almost all } \ y \in \Omega.$$

It is easy to see (for example, from the spectral theorem) that for any selfadjoint operator A, we have $\sigma(A) \subset \mathbb{R}$ and ω is an upper bound for A if and only if $\sigma(A) \subset (-\infty, \omega]$; i.e., $\omega \ge s(A)$. Similarly, we say that A is *bounded below* by ω if

$$(Ax|x)_H \ge \omega(x|x)_H \qquad (x \in D(A)).$$

The definition of selfadjointness is not easy to verify in practice. Here is a handy criterion, for a proof of which we refer to [RS72, Theorem X.1].

Theorem B.14. *Let A be an operator on H and let $\omega \in \mathbb{R}$. The following are equivalent:*

(i) *A is selfadjoint with upper bound ω.*

(ii) a) $(Ax|y)_H = (x|Ay)_H \quad (x, y \in D(A))$,
b) $(Ax|x)_H \le \omega(x|x)_H \quad (x \in D(A))$, *and*
c) *there exists $\lambda > \omega$ such that* $\mathrm{Ran}(\lambda - A) = X$.

Finally, we mention one or two topics concerning bounded operators. By the closed graph theorem, an operator T on a Banach space X is bounded if T is closed and $D(T) = X$. Conversely, a densely defined, closed, bounded operator is everywhere defined. By convention, a bounded operator T on a Banach space X will be assumed to be defined on the whole of X. The *spectral radius* of T will be denoted by $r(T)$, so

$$r(T) = \sup\{|\lambda| : \lambda \in \sigma(T)\} = \inf\left\{\|T^n\|^{1/n} : n \in \mathbb{N}\right\}.$$

In order to allow a convenient citation in the book, we state the following standard fact whose proof is straightforward. Note that a family of bounded linear operators is equicontinuous if and only if it is bounded.

Proposition B.15. *Let X, Y be Banach spaces, $T_n \in \mathcal{L}(X,Y)$ $(n \in \mathbb{N})$ such that $\sup_{n\in\mathbb{N}} \|T_n\| < \infty$. The following are equivalent:*

(i) *$(T_n x)_{n\in\mathbb{N}}$ converges for all x in a dense subspace of X.*

(ii) *$(T_n x)_{n\in\mathbb{N}}$ converges for all $x \in X$.*

(iii) *$(T_n x)_{n\in\mathbb{N}}$ converges uniformly in $x \in K$ for all compact subsets K of X.*

Notes: The material of this appendix is standard, and can be found in various books, for example [Kat66, Chapter 3].

Appendix C

Ordered Banach Spaces

Let X be a real Banach space. By a *positive cone* in X we understand a closed subset X_+ of X such that

$$X_+ + X_+ \subset X_+; \tag{C.1}$$
$$\mathbb{R}_+ \cdot X_+ \subset X_+; \tag{C.2}$$
$$X_+ \cap (-X_+) = \{0\}; \quad \text{and} \tag{C.3}$$
$$X_+ - X_+ = X. \tag{C.4}$$

Then an ordering on X is introduced by setting

$$x \leq y \iff y - x \in X_+.$$

The space X together with the positive cone is called a *real ordered Banach space.* The elements of X_+ are called *positive.*

Remark C.1. Property (C.3) is frequently expressed by saying that X_+ is a *proper cone*, and (C.4) says that X_+ is *generating.* We assume these properties throughout without further notice. □

If $x^* \in X^*$, then we say that x^* is *positive* and write $x^* \geq 0$ if

$$\langle x, x^* \rangle \geq 0 \quad \text{for all} \quad x \in X_+.$$

The set $X_+^* := \{x^* \in X^* : x^* \geq 0\}$ is closed and satisfies (C.1), (C.2) and (C.3).

For $x, y \in X$ such that $x \leq y$ we denote by

$$[x, y] := \{z \in X : x \leq z \leq y\}$$

the *order interval* defined by x and y. One says that the cone X_+ is *normal* if all order intervals are bounded.

Proposition C.2. *The cone X_+^* is normal. The cone X_+ is normal if and only if $X_+^* - X_+^* = X^*$.*

Thus, if X_+ is normal then (X^*, X_+^*) is also an ordered Banach space with normal cone. We call X_+^* *the dual cone* of X_+.

If the cone X_+ is normal then there is a constant $c \geq 0$ such that

$$y \leq x \leq z \Longrightarrow \|x\| \leq c \max(\|y\|, \|z\|). \tag{C.5}$$

Indeed, passing to an equivalent norm one can even arrange that $c = 1$.

If X is a real ordered Banach space we tacitly consider the complexification of X. So in this book an ordered Banach space is always the complexification of a real ordered Banach space. Thus, any C^*-algebra is an ordered Banach space with normal cone.

Let X be an ordered Banach space. A linear mapping $T : X \to X$ is called *positive* if

$$Tx \in X_+ \qquad \text{for all } \ x \in X_+.$$

Then we write $T \geq 0$. If $S, T : X \to X$ are linear, we write $S \leq T$ if $T - S \geq 0$.

If X_+ is normal, every positive linear mapping $T : X \to X$ is continuous. Moreover, there is a constant $k \geq 0$ such that

$$\pm S \leq T \Longrightarrow \|S\| \leq k\|T\| \tag{C.6}$$

if $S, T : X \to X$ are linear.

A real ordered Banach space X is a *lattice* if for all $x, y \in X$ there exists a *least upper bound* $x \vee y$ of x and y (i.e., $x \vee y \in X$, $x \vee y \geq x$, $x \vee y \geq y$ and $w \geq x, y$ implies $w \geq x \vee y$). In that case, there also exists a *largest lower bound* $x \wedge y = -((-x) \vee (-y))$. One sets $x^+ = x \vee 0$, $x^- = (-x)^+$, $|x| = x \vee (-x) = x^+ + x^-$. Then X is called a *real Banach lattice* if in addition the following compatibility condition is satisfied:

$$|x| \leq |y| \Longrightarrow \|x\| \leq \|y\| \tag{C.7}$$

for all $x, y \in X$. Thus, the cone of a Banach lattice is always normal.

In this book, a *Banach lattice* is the complexification of a real Banach lattice. Important examples of Banach lattices are the spaces $L^p(\Omega, \mu)$ $(1 \leq p \leq \infty)$, where (Ω, μ) is a measure space, and

$$C(K) := \{f : K \to \mathbb{C} : f \text{ continuous}\},$$

where K is a compact space.

Let X be a real Banach lattice. A subspace Y of X is called a *sublattice* if

$$x \in Y \quad \text{implies} \quad |x| \in Y.$$

The space Y is called an *ideal* if

$$x \in Y, \ y \in X, \ |y| \leq |x| \ \text{ implies } \ y \in Y.$$

Let (Ω, μ) be a σ-finite measure space and $X = L^p(\Omega, \mu)$, where $1 \leq p < \infty$. Then Y is a closed ideal of X if and only if

$$Y = \{f \in X : \; f|_S = 0 \;\text{ a.e.}\}$$

for some measurable subset S of Ω.

If $M \subset X$ is a subset, then

$$M^d := \{x \in X : \; |x| \vee |y| = 0 \;\text{ for all }\; y \in M\}$$

is a closed ideal of X. One says that M is a *band* if $M = M^{dd}$. In that case, $M \oplus M^d = X$.

If X is a complex Banach lattice, then a subspace Y of X is called a sublattice (ideal, band) if

a) $x \in Y \Longrightarrow \operatorname{Re} x \in X$, and

b) $Y \cap X_{\mathbb{R}}$ is a sublattice (ideal, band) of $X_{\mathbb{R}}$,

where $X_{\mathbb{R}}$ denotes the underlying real Banach lattice.

An ordered Banach space has *order continuous norm* if each decreasing positive sequence $(x_n)_{n\in\mathbb{N}}$ converges; i.e.,

$$\text{If }\; x_n \geq x_{n+1} \geq 0 \; (n \in \mathbb{N}), \;\text{ then }\; \lim_{n\to\infty} x_n \;\text{ exists.}$$

The spaces $L^p(\Omega, \mu)$ $(1 \leq p < \infty)$ have order continuous norm, but $L^\infty(\Omega, \mu)$ and $C(K)$ do not if they have infinite dimension. Also, the dual of a C^*-algebra has order continuous norm.

Let X be a Banach lattice. Then the following assertions are equivalent:

(i) If $0 \leq x_n \leq x_{n+1}$ and $\sup_{n\in\mathbb{N}} \|x_n\| < \infty$, then $(x_n)_{n\in\mathbb{N}}$ converges.

(ii) X is a band in X^{**}.

(iii) c_0 is not isomorphic to a closed subspace of X.

In assertion (ii), we identify X with a closed subspace of X^{**} via the canonical evaluation mapping.

A Banach lattice X satisfying the equivalent conditions (i), (ii), and (iii) is called a KB-*space*. Every reflexive Banach lattice and every space of the form $L^1(\Omega, \mu)$ are KB-spaces. Moreover, if X is a KB-space then X has order continuous norm. The space c_0 does have order continuous norm but is not a KB-space. Each closed ideal of a KB-space is a band.

Notes: We refer to the monograph [Sch74] by Schaefer and to the survey article [BR84] for all this and for further information.

Appendix D

Banach Spaces which Contain c_0

We let c_0 be the Banach space of all complex sequences $a = (a_r)_{r\geq 1}$ such that $\lim_{r\to\infty} a_r = 0$, with $\|a\| = \sup_r |a_r|$. For $n \geq 1$, let $e_n := (\delta_{nr})_{r\geq 1}$, so $\|e_n\| = 1$ and

$$\left\| \sum_{n=1}^{m} \alpha_n e_n \right\| = \max_n |\alpha_n|$$

for all $m \in \mathbb{N}$ and $\alpha_1, \dots, \alpha_m \in \mathbb{C}$.

A complex Banach space X is said to *contain* c_0 if there is a closed linear subspace Y of X which is isomorphic (linearly homeomorphic) to c_0. This is equivalent to the existence of a sequence $(x_n)_{n\geq 1}$ in X and strictly positive constants c_1 and c_2 such that

$$c_1 \max_n |\alpha_n| \leq \left\| \sum_{n=1}^{m} \alpha_n x_n \right\| \leq c_2 \max_n |\alpha_n| \tag{D.1}$$

for all $m \in \mathbb{N}$ and $\alpha_1, \dots, \alpha_m \in \mathbb{C}$. Then the map $\sum_{n=1}^{m} \alpha_n x_n \mapsto \sum_{n=1}^{m} \alpha_n e_n$ extends to an isomorphism of the closed linear span of $\{x_n\}$ onto c_0.

Since c_0 is not reflexive, a reflexive Banach space cannot contain c_0. Moreover, for any measure space (Ω, μ), the space $L^1(\Omega, \mu)$ does not contain c_0.

A formal series $\sum_{n=1}^{\infty} x_n$ in X is said to be *unconditionally bounded* if there is a constant M such that

$$\left\| \sum_{j=1}^{m} x_{n_j} \right\| \leq M \tag{D.2}$$

whenever $m \in \mathbb{N}$ and $1 \leq n_1 < n_2 < \cdots < n_m$. The series $\sum_n e_n$ in c_0 is unconditionally bounded (with $M = 1$), but it is divergent. It follows that any Banach space which contains c_0 has a divergent, unconditionally bounded series. In this appendix, we shall give a converse result showing that if X contains a divergent, unconditionally bounded series, then X contains c_0.

Lemma D.1. *Suppose that $\sum_n x_n$ is a divergent, unconditionally bounded series in a complex Banach space X, and let M be as in* (D.2). *Then*

$$\left\| \sum_{n=1}^{m} \alpha_n x_n \right\| \leq 4M \max_{1 \leq n \leq m} |\alpha_n|$$

for all $m \in \mathbb{N}$ and $\alpha_1, \ldots, \alpha_m \in \mathbb{C}$.

Proof. First suppose that $\alpha_n \geq 0$ for $n = 1, 2, \ldots, m$. By rearranging $x_1, x_2, \ldots, x_m$, we may suppose that $0 \leq \alpha_1 \leq \alpha_2 \leq \cdots \leq \alpha_m$. Then

$$\sum_{n=1}^{m} \alpha_n x_n = \alpha_1 \sum_{n=1}^{m} x_n + (\alpha_2 - \alpha_1) \sum_{n=2}^{m} x_n + \cdots + (\alpha_m - \alpha_{m-1}) x_m.$$

Hence,

$$\left\| \sum_{n=1}^{m} \alpha_n x_n \right\| \leq \alpha_1 M + (\alpha_2 - \alpha_1) M + \cdots + (\alpha_m - \alpha_{m-1}) M = \alpha_m M.$$

The general case follows by decomposing each complex number α_n as $\sum_{j=0}^{3} \alpha_{nj} i^j$ where $\alpha_{nj} \geq 0$ and $|\alpha_{nj}| \leq |\alpha_n|$. □

Lemma D.2. *Suppose that X contains a divergent, unconditionally bounded series. Then there is a sequence $(y_j)_{j \geq 1}$ in X such that $\|y_j\| = 1$ for all j and*

$$\left\| \sum_{j=1}^{m} \beta_j y_j \right\| \leq \frac{3}{2} \max_{1 \leq j \leq m} |\beta_j|$$

for all $m \in \mathbb{N}$ and $\beta_1, \ldots, \beta_m \in \mathbb{C}$.

Proof. Let $\sum_n x_n$ be a divergent, unconditionally bounded series, and let

$$\gamma_k := \sup \left\{ \left\| \sum_{n=k+1}^{m} \alpha_n x_n \right\| : m > k, \alpha_n \in \mathbb{C}, |\alpha_n| \leq 1 \right\}.$$

By Lemma D.1, γ_k is finite, and clearly (γ_k) is a decreasing sequence. Let $\gamma := \lim_{k \to \infty} \gamma_k$. Then $\gamma > 0$, since

$$\gamma_k \geq \sup \left\{ \left\| \sum_{n=k+1}^{m} x_n \right\| : m > k \right\}$$

and $\sum_n x_n$ is divergent. Replacing x_n by $(5/4\gamma) x_n$, we may assume that $\gamma = 5/4$.

Choose $k_1 \geq 1$ such that $\gamma_{k_1} < 3/2$. Since $\gamma_{k_1} > 1$, there exist $k_2 > k_1$ and $\alpha_n \in \mathbb{C}$ $(k_1 < n \leq k_2)$ such that $|\alpha_n| \leq 1$ and

$$\nu_1 := \left\| \sum_{n=k_1+1}^{k_2} \alpha_n x_n \right\| > 1.$$

Iterating this, we may choose $k_1 < k_2 < \dots$ and $\alpha_n \in \mathbb{C}$ $(n > k_1)$ such that $|\alpha_n| \leq 1$ and

$$\nu_j := \left\| \sum_{n=k_j+1}^{k_{j+1}} \alpha_n x_n \right\| > 1.$$

Let

$$y_j := \nu_j^{-1} \sum_{n=k_j+1}^{k_{j+1}} \alpha_n x_n.$$

Then $\|y_j\| = 1$. Moreover, if $m \in \mathbb{N}$ and $\beta_j \in \mathbb{C}$ $(j = 1, \dots, m)$ and j_n is chosen so that $k_{j_n} < n \leq k_{j_n+1}$ $(n > k_1)$, then

$$\left\| \sum_{j=1}^{m} \beta_j y_j \right\| = \left\| \sum_{n=k_1+1}^{k_{m+1}} \beta_{j_n} \nu_{j_n}^{-1} \alpha_n x_n \right\| \leq \frac{3}{2} \max_{k_1 < n \leq k_{m+1}} \left| \beta_{j_n} \nu_{j_n}^{-1} \alpha_n \right| \leq \frac{3}{2} \max_{1 \leq j \leq m} |\beta_j|.$$

□

Theorem D.3. *Suppose that X contains a divergent, unconditionally bounded series $\sum_n x_n$. Then X contains c_0.*

Proof. Let (y_j) be as in Lemma D.2. Let $m \in \mathbb{N}$ and $\beta_j \in \mathbb{C}$ $(j = 1, \dots, m)$. Then $\left\| \sum_{j=1}^m \beta_j y_j \right\| \leq \frac{3}{2} \max_j |\beta_j|$. We shall establish that $\left\| \sum_{j=1}^m \beta_j y_j \right\| \geq \frac{1}{2} \max_j |\beta_j|$, so that (y_j) satisfies the condition (D.1), and therefore X contains c_0.

Choose k such that $|\beta_k| = \max_j |\beta_j|$, and choose $x^* \in X^*$ such that $\|x^*\| = 1$ and $\beta_k \langle y_k, x^* \rangle = |\beta_k|$. Let

$$\beta_j' := \begin{cases} \beta_j & (j \neq k), \\ -\beta_k & (j = k). \end{cases}$$

Then

$$\begin{aligned} \left\| \sum_{j=1}^{m} \beta_j y_j \right\| &\geq \operatorname{Re} \langle \sum_{j=1}^{m} \beta_j y_j, x^* \rangle = 2|\beta_k| + \operatorname{Re} \langle \sum_{j=1}^{m} \beta_j' y_j, x^* \rangle \\ &\geq 2|\beta_k| - \left\| \sum_{j=1}^{m} \beta_j' y_j \right\| \geq 2|\beta_k| - \frac{3}{2} \max_j |\beta_j| = \frac{1}{2} \max_j |\beta_j|. \end{aligned}$$

This completes the proof. □

Notes: Theorem D.3 is due to Bessaga and Pelczynski [BP58]. They also showed that X contains c_0 if (and only if) there is a sequence of unit vectors (y_j) in X such that $\sum_j |\langle y_j, x^*\rangle| < \infty$ for all $x^* \in X^*$. Our proof, which is adapted from [LZ82], establishes such a property but in a specific way which eliminates some of the cases considered in [BP58]. Moreover, this proof shows (when constants 5/4 and 3/2 are replaced by constants arbitrarily close to 1) that X contains c_0 "almost isometrically", thereby establishing a positive solution to the "distortion problem" in c_0. This was first proved by James [Jam64].

Another direct proof of Theorem D.3 is given in a paper of Eberhardt and Greiner [EG92]. There are numerous other characterizations of Banach spaces which contain c_0, some of which may be found in the books of Guerre-Delabrière [Gue92], Lindenstrauss and Tzafriri [LT77] and Megginson [Meg98]. Note in particular that a Banach lattice X does not contain c_0 (as a subspace, or equivalently as a sublattice) if and only if X is a KB-*space*, that is, every bounded increasing sequence in X converges [LT77, Theorem II.1.c.4], [Mey91, Theorem 2.4.12].

Appendix E

Distributions and Fourier Multipliers

In this appendix we collect basic facts on distributions and Fourier multipliers. They are needed at various places in the book; those which are essential to understanding Parts I and II are also explained at the appropriate point in the text, while other results from this appendix are needed only for examples in Chapter 3 or for the applications in Part III.

First, we consider distributions on $\mathbb{R}^n$. A *multi-index* is an element $\alpha = (\alpha_1, \dots, \alpha_n) \in \mathbb{N}_0^n$. We write $|\alpha|$ for $\sum_{j=1}^n \alpha_j$, D_j for $\frac{\partial}{\partial x_j}$ and D^α for $D_1^{\alpha_1} \cdots D_n^{\alpha_n}$. We denote by $\mathcal{D}(\mathbb{R}^n)$ (or by $C_c^\infty(\mathbb{R}^n)$ in other contexts) the space of all complex-valued C^∞-functions on $\mathbb{R}^n$ with compact supports (the *test functions*), and by $\mathcal{S}(\mathbb{R}^n)$ the *Schwartz space* of all smooth, rapidly decreasing functions on $\mathbb{R}^n$, i.e.

$$\mathcal{S}(\mathbb{R}^n) := \{\varphi \in C^\infty(\mathbb{R}^n) : \ \|\varphi\|_{m,\alpha} < \infty \quad \text{for all} \ \ m \in \mathbb{N}_0, \ \alpha \in \mathbb{N}_0^n\},$$

where

$$\|\varphi\|_{m,\alpha} := \sup_{x \in \mathbb{R}^n} (1 + |x|)^m |D^\alpha \varphi(x)|.$$

When equipped with the topology defined by the family of all norms $\|\cdot\|_{m,\alpha}$, $\mathcal{S}(\mathbb{R}^n)$ is a Fréchet space, and $\mathcal{D}(\mathbb{R}^n)$ is a dense subspace of $\mathcal{S}(\mathbb{R}^n)$.

We denote by $\mathcal{D}(\mathbb{R}^n)'$ the space of all *distributions*, i.e., linear maps $f : \varphi \mapsto \langle \varphi, f \rangle$ of $\mathcal{D}(\mathbb{R}^n)$ into $\mathbb{C}$ such that for each compact $K \subset \mathbb{R}^n$ there exist $m \in \mathbb{N}$ and $C > 0$ such that

$$|\langle \varphi, f \rangle| \le C \sup_{|\alpha| \le m} \sup_{x \in \mathbb{R}^n} |D^\alpha \varphi(x)|$$

for all $\varphi \in \mathcal{D}(\mathbb{R}^n)$ with $\operatorname{supp} \varphi \subset K$. Let $\mathcal{S}(\mathbb{R}^n)'$ be the space of all *temperate distributions*, i.e., continuous linear maps from $\mathcal{S}(\mathbb{R}^n)$ into $\mathbb{C}$. Then $\mathcal{S}(\mathbb{R}^n)'$ is embedded in $\mathcal{D}(\mathbb{R}^n)'$ in a natural way.

We consider $\mathcal{D}(\mathbb{R}^n)'$ to have the topology arising from the duality with $\mathcal{D}(\mathbb{R}^n)$, so a net (f_α) of distributions converges to 0 in $\mathcal{D}(\mathbb{R}^n)'$ if and only if $\langle \varphi, f_\alpha \rangle \to 0$ for all $\varphi \in \mathcal{D}(\mathbb{R}^n)$.

Any locally integrable $f : \mathbb{R}^n \to \mathbb{C}$ can be identified with a distribution by

$$\langle \varphi, f \rangle := \int_{\mathbb{R}^n} \varphi(x) f(x)\, dx \qquad (\varphi \in \mathcal{S}(\mathbb{R}^n)). \tag{E.1}$$

We shall make such identifications freely.

A function $f : \mathbb{R}^n \to \mathbb{C}$ is said to be *absolutely regular* if there exists $k \in \mathbb{N}_0$ such that $x \mapsto (1 + |x|)^{-k} f(x)$ is Lebesgue integrable on $\mathbb{R}^n$. For an absolutely regular function f, the corresponding distribution is temperate.

Any continuous linear map $T : \mathcal{S}(\mathbb{R}^n) \to \mathcal{S}(\mathbb{R}^n)$ induces an adjoint. We now describe how this enables operators of multiplication, differentiation, Fourier transform and convolution to be extended from functions to distributions.

Let $g : \mathbb{R}^n \to \mathbb{C}$ be a C^∞-function. Then $\varphi \cdot g \in \mathcal{D}(\mathbb{R}^n)$ for all $\varphi \in \mathcal{D}(\mathbb{R}^n)$. Given a distribution $f \in \mathcal{D}(\mathbb{R}^n)'$, we can define $g \cdot f$ by

$$\langle \varphi, g \cdot f \rangle := \langle \varphi \cdot g, f \rangle \qquad (\varphi \in \mathcal{D}(\mathbb{R}^n)). \tag{E.2}$$

If, for each multi-index α, there exists $m_\alpha \in \mathbb{N}$ and $c_\alpha > 0$ such that

$$|(D^\alpha g)(x)| \le c_\alpha (1 + |x|)^{m_\alpha} \qquad (x \in \mathbb{R}^n), \tag{E.3}$$

then the map $\varphi \mapsto \varphi \cdot g$ is continuous from $\mathcal{S}(\mathbb{R}^n)$ into $\mathcal{S}(\mathbb{R}^n)$, and therefore $g \cdot f \in \mathcal{S}(\mathbb{R}^n)'$ whenever $f \in \mathcal{S}(\mathbb{R}^n)'$.

Given a distribution $f \in \mathcal{D}(\mathbb{R}^n)'$, the derivatives $D_j f$ $(j = 1, \dots, n)$ are defined in $\mathcal{D}(\mathbb{R}^n)'$ by

$$\langle \varphi, D_j f \rangle := -\langle D_j \varphi, f \rangle \qquad (\varphi \in \mathcal{D}(\mathbb{R}^n)). \tag{E.4}$$

Then D_j maps $\mathcal{S}(\mathbb{R}^n)'$ into itself. Integration by parts shows that this notation is consistent when differentiable functions are identified with distributions, and the product law extends to derivatives of products of differentiable functions and distributions as discussed above. For higher order derivatives, (E.4) becomes

$$\langle \varphi, D^\alpha f \rangle = (-1)^{|\alpha|} \langle D^\alpha \varphi, f \rangle \qquad (\varphi \in \mathcal{D}(\mathbb{R}^n)). \tag{E.5}$$

Next, we consider convolutions. For functions f and g, the *convolution* $f * g$ is defined by

$$(f * g)(x) := \int_{\mathbb{R}^n} f(x - y) g(y)\, dy$$

whenever the integral exists. For $\varphi, \psi \in \mathcal{S}(\mathbb{R}^n)$, $\psi * \varphi \in \mathcal{S}(\mathbb{R}^n)$ and the map $\psi \mapsto \psi * \varphi$ is continuous. Hence, the *convolution* $\varphi * f$ of $\varphi \in \mathcal{S}(\mathbb{R}^n)$ and $f \in \mathcal{S}(\mathbb{R}^n)'$ can be defined by

$$\langle \psi, \varphi * f \rangle := \langle \psi * \check{\varphi}, f \rangle \qquad (\psi \in \mathcal{S}(\mathbb{R}^n)), \tag{E.6}$$

where $\check{\varphi}(x) := \varphi(-x)$, and then $\varphi * f \in \mathcal{S}(\mathbb{R}^n)'$. An easy calculation shows that this is consistent when functions are identified with distributions.

An alternative way to define $\varphi * f$ is as follows. For $x \in \mathbb{R}^n$ and $\psi \in \mathcal{S}(\mathbb{R}^n)$, let $\tau_x\psi(y) := \psi(y-x)$. For $\varphi \in \mathcal{S}(\mathbb{R}^n)$, $\tau_x\check{\varphi} \in \mathcal{S}(\mathbb{R}^n)$ and the map $x \mapsto \tau_x\check{\varphi}$ is continuous on $\mathcal{S}(\mathbb{R}^n)$. For $f \in \mathcal{S}(\mathbb{R}^n)'$, let

$$(\varphi * f)(x) := \langle \tau_x\check{\varphi}, f\rangle \qquad (x \in \mathbb{R}^n). \tag{E.7}$$

Then $\varphi * f$ is a continuous, bounded function.

These two definitions of $\varphi * f$ are consistent when functions are identified with distributions. Moreover,

$$D_j(\varphi * f) = (D_j\varphi) * f. \tag{E.8}$$

For $f \in L^1(\mathbb{R}^n)$, the *Fourier transform* $\mathcal{F}f$ of f is defined by:

$$(\mathcal{F}f)(\xi) = \int_{\mathbb{R}^n} e^{-ix\cdot\xi} f(x)\,dx \qquad (\xi \in \mathbb{R}^n), \tag{E.9}$$

where $x\cdot\xi := \sum_{j=1}^n x_j\xi_j$. The Fourier inversion theorem [Hör83, Theorem 7.1.5] shows that $\mathcal{F}$ is a linear and topological isomorphism of $\mathcal{S}(\mathbb{R}^n)$, and

$$(\mathcal{F}^{-1}\varphi)(\xi) = (2\pi)^{-n}(\mathcal{F}\varphi)(-\xi) \qquad (\varphi \in \mathcal{S}(\mathbb{R}^n), \xi \in \mathbb{R}^n).$$

The Fourier transform therefore induces an isomorphism of $\mathcal{S}(\mathbb{R}^n)'$, also denoted by $\mathcal{F}$:

$$\langle \varphi, \mathcal{F}f\rangle := \langle \mathcal{F}\varphi, f\rangle \qquad (\varphi \in \mathcal{S}(\mathbb{R}^n),\ f \in \mathcal{S}(\mathbb{R}^n)'). \tag{E.10}$$

A simple application of Fubini's theorem shows that this notation is consistent when $f \in L^1(\mathbb{R}^n)$ and f is identified with a distribution in $\mathcal{S}(\mathbb{R}^n)'$.

The following relations, which are elementary for functions, extend to distributions f:

$$\mathcal{F}^{-1}f = (2\pi)^{-n}(\mathcal{F}f)\check{} = (2\pi)^{-n}\mathcal{F}\check{f}, \quad \text{where } \langle\varphi, \check{f}\rangle := \langle\check{\varphi}, f\rangle, \tag{E.11}$$

$$\mathcal{F}D_jf = i\xi_j\cdot\mathcal{F}f, \tag{E.12}$$

$$\mathcal{F}(\varphi * f) = (\mathcal{F}\varphi)\cdot(\mathcal{F}f) \qquad (\varphi \in \mathcal{S}(\mathbb{R}^n)). \tag{E.13}$$

Plancherel's theorem states that

$$\langle \mathcal{F}\varphi, \mathcal{F}\bar{\psi}\rangle = (2\pi)^n\langle\varphi, \bar{\psi}\rangle \qquad (\varphi, \psi \in \mathcal{S}(\mathbb{R}^n)), \tag{E.14}$$

where $\bar{\psi}$ is the complex conjugate of ψ, and hence $\mathcal{F}$ extends by continuity to a linear operator $\mathcal{F}$ on the Hilbert space $L^2(\mathbb{R}^n)$ such that $(2\pi)^{-n/2}\mathcal{F}$ is unitary. This also says that, for each $f \in L^2(\mathbb{R}^n)$, the distribution $\mathcal{F}f$ belongs to $L^2(\mathbb{R}^n)$.

Many of the concepts above can be extended to the case of distributions on an open subset Ω of $\mathbb{R}^n$. Let $\mathcal{D}(\Omega)$ be the space of *test functions* on Ω, i.e., C^∞-functions of compact support in Ω, and $\mathcal{D}(\Omega)'$ be the space of *distributions* on Ω, i.e., linear functionals f on $\mathcal{D}(\Omega)$ such that for each compact $K \subset \Omega$ there exist $m \in \mathbb{N}$ and $C > 0$ such that

$$|\langle \varphi, f \rangle| \leq C \sup_{|\alpha| \leq m} \sup_{x \in \Omega} |D^\alpha \varphi(x)|$$

for all $\varphi \in \mathcal{D}(\Omega)$ with $\operatorname{supp} \varphi \subset K$. Locally integrable functions on Ω can be identified with distributions, and the derivatives D_j of a distribution f are defined by

$$\langle \varphi, D_j f \rangle := -\langle D_j \varphi, f \rangle \qquad (\varphi \in \mathcal{D}(\Omega)).$$

For $m \in \mathbb{N}$ and $1 \leq p \leq \infty$, the *Sobolev space* $W^{m,p}(\Omega)$ is defined by

$$W^{m,p}(\Omega) := \{f \in L^p(\Omega) : \ D^\alpha f \in L^p(\Omega) \ \text{ for all } \ \alpha \in \mathbb{N}_0^n \ \text{ with } \ |\alpha| \leq m\},$$

where $D^\alpha f$ is understood in the sense of distributions. Thus, $f \in W^{m,p}(\Omega)$ if and only if for each $\alpha \in \mathbb{N}_0^n$ with $|\alpha| \leq m$ there exists $f_\alpha \in L^p(\Omega)$ such that

$$\int_\Omega \varphi f_\alpha \, dx = (-1)^{|\alpha|} \int_\Omega (D^\alpha \varphi) f \, dx \qquad (\varphi \in \mathcal{D}(\Omega)).$$

In the special case when $n = 1$, $f \in W^{m,p}(\Omega)$ if and only if $f \in C^{m-1}(\Omega)$, $f^{(m-1)}$ is absolutely continuous, and $f^{(j)} \in L^p(\Omega)$ for $j = 0, 1, \dots, m$. Equipped with the norm

$$\|f\|_{W^{m,p}(\Omega)} := \sum_{|\alpha| \leq m} \|D^\alpha f\|_p,$$

$W^{m,p}(\Omega)$ becomes a Banach space. The closure of $\mathcal{D}(\Omega)$ in $W^{m,p}(\Omega)$ is denoted by $W_0^{m,p}(\Omega)$. For $p = 2$, we also use the notation

$$H^m(\Omega) := W^{m,2}(\Omega) \quad \text{and} \quad H_0^m(\Omega) := W_0^{m,2}(\Omega).$$

Equipped with the equivalent norm

$$\|f\|_{H^m(\Omega)} := \left(\sum_{|\alpha| \leq m} \|D^\alpha f\|_2^2 \right)^{1/2},$$

$H^m(\Omega)$ is a Hilbert space with the inner product

$$(f|g)_{H^m(\Omega)} = \sum_{|\alpha| \leq m} \int_\Omega D^\alpha f \overline{D^\alpha g} \, dx.$$

Note that Plancherel's theorem and (E.12) show that

$$H^m(\mathbb{R}^n) = \{f \in L^2(\mathbb{R}^n) : \ \xi^\alpha \cdot \mathcal{F} f \in L^2(\mathbb{R}^n) \ \text{ for all } \ \alpha \in \mathbb{N}_0^n \ \text{ with } \ |\alpha| \leq m\},$$

where ξ^α is the function $\xi \mapsto \xi_1^{\alpha_1} \xi_2^{\alpha_2} \cdots \xi_n^{\alpha_n}$. Hence, $f \in H^m(\mathbb{R}^n)$ if and only if $\xi \mapsto (1 + |\xi|^2)^{m/2} (\mathcal{F} f)(\xi)$ belongs to $L^2(\mathbb{R}^n)$.

Now we consider Fourier multipliers. If g is a C^∞-function satisfying the estimates (E.3), then the map $\varphi \mapsto \mathcal{F}^{-1}g\mathcal{F}\varphi := \mathcal{F}^{-1}(g \cdot (\mathcal{F}\varphi))$ is a continuous linear map on $\mathcal{S}(\mathbb{R}^n)$. It is a classical problem to seek conditions on a function such that such a map becomes continuous on a function space $X = L^p(\mathbb{R}^n)$ $(1 \le p \le \infty)$ or $C_0(\mathbb{R}^n)$. Let $m : \mathbb{R}^n \to \mathbb{C}$ be an absolutely regular function. For $\varphi \in \mathcal{S}(\mathbb{R}^n)$, we define $m \cdot (\mathcal{F}\varphi) \in \mathcal{S}(\mathbb{R}^n)'$ by

$$\langle \psi, m \cdot (\mathcal{F}\varphi) \rangle := \int_{\mathbb{R}^n} \psi m \cdot (\mathcal{F}\varphi)\, dx \qquad (\psi \in \mathcal{S}(\mathbb{R}^n)).$$

Then we consider the distribution $\mathcal{F}^{-1}(m \cdot (\mathcal{F}\varphi)) \in \mathcal{S}(\mathbb{R}^n)'$. We call m a *Fourier multiplier* for X if $\mathcal{F}^{-1}(m \cdot (\mathcal{F}\varphi)) \in X$ for all $\varphi \in \mathcal{S}(\mathbb{R}^n)$ and there exists a constant C such that

$$\|\mathcal{F}^{-1}(m \cdot (\mathcal{F}\varphi))\|_X \le C\|\varphi\|_X \qquad (\varphi \in \mathcal{S}(\mathbb{R}^n)).$$

Then the map $\varphi \mapsto \mathcal{F}^{-1}(m \cdot (\mathcal{F}\varphi))$ extends to a bounded linear operator $T_m : f \mapsto \mathcal{F}^{-1}m\mathcal{F}f$ on X (in the case when $X = L^\infty(\mathbb{R}^n)$, the extension is weak* continuous). When m is a C^∞-function, $T_m f$ agrees with the distribution $\mathcal{F}^{-1}(m \cdot \mathcal{F}f)$ defined earlier.

We denote the space of all Fourier multipliers for X by $\mathcal{M}_X(\mathbb{R}^n)$, or by $\mathcal{M}_p(\mathbb{R}^n)$ when $X = L^p(\mathbb{R}^n)$, with the usual identification of functions which coincide a.e. We put

$$\|m\|_{\mathcal{M}_X(\mathbb{R}^n)} := \|T_m\|_{\mathcal{L}(X)}.$$

Fourier multipliers are bounded functions, and $\|m\|_{\mathcal{M}_X(\mathbb{R}^n)} \ge \|m\|_\infty$ (see Proposition E.2). It follows easily that $\mathcal{M}_X(\mathbb{R}^n)$ is a Banach space. Note also that $\mathcal{M}_{C_0}(\mathbb{R}^n) \subset \mathcal{M}_\infty(\mathbb{R}^n)$.

For $a \in \mathbb{R}^n$, define $\tau_a \in \mathcal{L}(X)$ by $\tau_a f(x) := f(x-a)$. If $m \in \mathcal{M}_X(\mathbb{R}^n)$, it is easy to see that

$$T_m \tau_a = \tau_a T_m \qquad (a \in \mathbb{R}^n). \tag{E.15}$$

Conversely, we have the following result.

Proposition E.1. *Let $X = L^p(\mathbb{R}^n)$ $(1 \le p \le \infty)$ or $C_0(\mathbb{R}^n)$, and assume that $T \in \mathcal{L}(X)$ satisfies* (E.15). *Then there exists $m \in \mathcal{M}_X(\mathbb{R}^n)$ such that*

$$Tf = \mathcal{F}^{-1}m\mathcal{F}f \qquad (f \in X).$$

For a proof of Proposition E.1, see [Hör60].

For $N \in \mathbb{N}$, we let $\mathcal{M}_p^N(\mathbb{R}^n)$ be the space of all matrices $m = (m_{ij})_{1\le i,j\le N}$, where $m_{ij} \in \mathcal{M}_p(\mathbb{R}^n)$. Each such matrix m defines a bounded operator $\mathcal{F}^{-1}m\mathcal{F}$ on $L^p(\mathbb{R}^n)^N$, where $\mathcal{F} : L^p(\mathbb{R}^n)^N \to L^p(\mathbb{R}^n)^N$ acts on each coordinate function, and matrix multiplication operates as usual. The norm on $\mathcal{M}_p^N(\mathbb{R}^n)$ is taken to be the norm of the operator $\mathcal{F}^{-1}m\mathcal{F}$ when $L^p(\mathbb{R}^n)^N$ is given the norm of $L^p(\mathbb{R}^n \times \{1,\dots,N\})$. Note that $\mathcal{M}_p^1(\mathbb{R}^n) = \mathcal{M}_p(\mathbb{R}^n)$.

Proposition E.2. *Let* $1 \le p \le \infty$, $N \in \mathbb{N}$. *Then the following hold true:*

a) $\mathcal{M}_p^N(\mathbb{R}^n)$ *is a Banach algebra.*

b) $\mathcal{M}_2^N(\mathbb{R}^n) = L^\infty(\mathbb{R}^n, \mathcal{L}(\mathbb{C}^N)) := \{(m_{ij})_{1\le i,j\le N} : m_{ij} \in L^\infty(\mathbb{R}^n)^N\}$.

c) $\mathcal{M}_p^N(\mathbb{R}^n) = \mathcal{M}_{p'}^N(\mathbb{R}^n)$, *where* $1/p + 1/p' = 1$.

d) $\mathcal{M}_1^N(\mathbb{R}^n) \subset \mathcal{M}_p^N(\mathbb{R}^n) \subset \mathcal{M}_2^N(\mathbb{R}^n)$. *Moreover, for* $m \in \mathcal{M}_1^N(\mathbb{R}^n)$,

$$\|m\|_{\mathcal{M}_p^N(\mathbb{R}^n)} \le \|m\|^{\theta}_{\mathcal{M}_1^N(\mathbb{R}^n)} \|m\|^{1-\theta}_{\mathcal{M}_2^N(\mathbb{R}^n)}, \tag{E.16}$$

where $\theta := 2\left|\frac{1}{p} - \frac{1}{2}\right|$.

e) *Given* $a \in \mathcal{M}_p^N(\mathbb{R}^n)$ *define* a_t *by* $a_t(\xi) := a(t\xi)$ *for* $t > 0$, $\xi \in \mathbb{R}^n$. *Then* $a_t \in \mathcal{M}_p^N(\mathbb{R}^n)$ *for all* $t > 0$ *and*

$$\|a_t\|_{\mathcal{M}_p^N(\mathbb{R}^n)} = \|a\|_{\mathcal{M}_p^N(\mathbb{R}^n)} \qquad (t > 0).$$

f) *Let* $(a_j)_{j\in\mathbb{N}} \subset \mathcal{M}_p^N(\mathbb{R}^n)$. *Assume that there exists a constant* $C > 0$ *such that* $\|a_j\|_{\mathcal{M}_p^N(\mathbb{R}^n)} \le C$ *for* $j \in \mathbb{N}$. *Let* $a \in L^\infty(\mathbb{R}^n)$ *such that* $a_j(x) \to a(x)$ *for almost all* $x \in \mathbb{R}^n$ *as* $j \to \infty$. *Then* $a \in \mathcal{M}_p^N(\mathbb{R}^n)$ *and* $\|a\|_{\mathcal{M}_p^N(\mathbb{R}^n)} \le C$.

Proof. We give only sketches of the proofs; details may be found in [Hör60] or [Ste93].

a) follows from the formal identity $\mathcal{F}^{-1}(m_1m_2)\mathcal{F} = (\mathcal{F}^{-1}m_1\mathcal{F})(\mathcal{F}^{-1}m_2\mathcal{F})$, b) is an easy consequence of Plancherel's theorem and c) is easily proved by duality, showing even that the equalities are isometric.

For d), we can assume by c) that $1 \le p \le 2$. Let $m \in \mathcal{M}_p^N(\mathbb{R}^n)$. By c), $\mathcal{F}^{-1}m\mathcal{F}$ is bounded on $L^p(\mathbb{R}^n)^N$ and on $L^{p'}(\mathbb{R}^n)^N$. Moreover the two versions of the map agree on $L^p(\mathbb{R}^n)^N \cap L^{p'}(\mathbb{R}^n)^N$. By the Riesz-Thorin theorem [Hör83, Theorem 7.1.12], $\mathcal{F}^{-1}m\mathcal{F}$ extends to a bounded linear operator on $L^2(\mathbb{R}^n)^N$. This shows that $\mathcal{M}_p^N(\mathbb{R}^n) \subset \mathcal{M}_2^N(\mathbb{R}^n)$. The inclusion $\mathcal{M}_1^N(\mathbb{R}^n) \subset \mathcal{M}_p^N(\mathbb{R}^n)$ and the inequality (E.16) also follows from the Riesz-Thorin theorem.

e) follows from the fact that $\mathcal{F}^{-1}a_t\mathcal{F} = J_t^{-1}\mathcal{F}^{-1}a\mathcal{F}J_t$, where J_t is the isometry, $(J_tf)(\xi) := t^{-n/p}f(t\xi)$, on $L^p(\mathbb{R}^n)$, and f) is proved by taking limits through the definitions of Fourier multipliers. □

An extremely useful sufficient condition for a function m to belong to $\mathcal{M}_p^1(\mathbb{R}^n)$ for $1 < p < \infty$ is given by the Mikhlin multiplier theorem. Let $j := \min\{k \in \mathbb{N} : k > \frac{n}{2}\}$. Define the Banach space $\mathcal{M}_M$ by

$$\mathcal{M}_M := \left\{m : \mathbb{R}^n \to \mathbb{K} : m \in C^j(\mathbb{R}^n \setminus \{0\}), |m|_M < \infty\right\}, \tag{E.17}$$

where the norm $|\cdot|_M$ is defined by

$$|m|_M := \max_{|\alpha|\le j} \sup_{\xi\in\mathbb{R}^n\setminus\{0\}} |\xi|^{|\alpha|}|D^\alpha m(\xi)|. \tag{E.18}$$

We then have the following result.

Theorem E.3 (Mikhlin). *Let $1 < p < \infty$. Then $\mathcal{M}_M \hookrightarrow \mathcal{M}_p(\mathbb{R}^n)$.*

For a proof of Mikhlin's theorem, we refer to [Ste93, Theorem VI.4.4].
The following results on Fourier multipliers will be useful in Chapter 8.

Theorem E.4. *Let $1 \le p \le \infty$. Then the following hold true:*

a) *Let a be a real homogeneous polynomial on $\mathbb{R}^n$ of degree $m > 1$. Then $e^{ia} \in \mathcal{M}_p(\mathbb{R}^n)$ if and only if $p = 2$.*

b) *Let $a \in C^\infty(\mathbb{R}^n)$ satisfy*

$$a(\xi) := \begin{cases} |\xi|^{-\beta} e^{-i|\xi|^\alpha} & (|\xi| \ge 2), \\ 0 & (|\xi| \le 1), \end{cases}$$

where $\alpha > 0$ and $\beta \ge 0$.

(i) *If $\alpha \ne 1$ and $1 < p < \infty$ (respectively, $p = 1$), then $a \in \mathcal{M}_p(\mathbb{R}^n)$ if and only if $n\left|\frac{1}{2} - \frac{1}{p}\right| \le \frac{\beta}{\alpha}$ (respectively, $\frac{n}{2} < \frac{\beta}{\alpha}$).*

(ii) *If $\alpha = 1$ and $1 < p < \infty$ (respectively, $p = 1$) then $a \in \mathcal{M}_p(\mathbb{R}^n)$ if and only if $(n-1)\left|\frac{1}{2} - \frac{1}{p}\right| \le \beta$ (respectively, $\frac{n-1}{2} < \beta$).*

c) *Define $a_1 : \mathbb{R}^3 \to \mathbb{C}$ and $a_2 : \mathbb{R}^3 \to \mathbb{C}$ by*

$$\begin{aligned} a_1(\xi) &:= (-i)(\xi_1 + \xi_2^2 + \xi_3^2 - i), \\ a_2(\xi) &:= \xi_1 + \xi_2^2 + \xi_3^2 + i. \end{aligned}$$

(i) *If $p \ne 2$, then $a_1^{-1} \notin \mathcal{M}_p(\mathbb{R}^3)$.*

(ii) *Let $a := a_1 a_2$. Define the operator $\mathcal{A}_p$ on $L^p(\mathbb{R}^3)$ by $\mathcal{A}_p f := \mathcal{F}^{-1}(a\mathcal{F}f)$ with $D(\mathcal{A}_p) := \{f \in L^p(\mathbb{R}^3) : \mathcal{F}^{-1}(a\mathcal{F}f) \in L^p(\mathbb{R}^3)\}$. If $\left|\frac{1}{2} - \frac{1}{p}\right| > \frac{3}{8}$, then $\sigma(\mathcal{A}_p) = \mathbb{C}$.*

For a proof of the assertions of Theorem E.4 we refer to [Hör60] (assertion a)), [FS72], [Miy81] and [Per80] (assertion b)), [KT80] (assertion c)i)) and [IS70] (assertion c)ii)).

Finally, we note one instance of Mikhlin's Theorem.

For $x \in \mathbb{R}$, define

$$\operatorname{sign} x := \begin{cases} 1 & (x > 0), \\ 0 & (x = 0), \\ -1 & (x < 0). \end{cases}$$

Then $\operatorname{sign} \in \mathcal{M}_M$. By Mikhlin's theorem, $\operatorname{sign} \in \mathcal{M}_p(\mathbb{R})$ for $1 < p < \infty$.

For $\varphi \in \mathcal{S}(\mathbb{R})$, one finds that $\mathcal{F}^{-1}(-i\operatorname{sign})\mathcal{F}\varphi$ is a function given by

$$(\mathcal{F}^{-1}(-i\operatorname{sign})\mathcal{F}\varphi)(x) = \lim_{\varepsilon \downarrow 0} \frac{1}{\pi} \int_{|x-y| \ge \varepsilon} \frac{\varphi(y)}{x - y}\, dy. \tag{E.19}$$

This is known as the *Hilbert transform* of φ. Thus, we have the following.

Proposition E.5. *Let* $1 < p < \infty$. *Then the Hilbert transform is a bounded linear operator on* $L^p(\mathbb{R})$.

Notes: The material on distributions is very standard and can be found in many books, for example [Hör83]. The basic material on Fourier multipliers can be found in [Ste93].

Bibliography

[Abe26] N. H. Abel. Untersuchungen über die Reihe $1 + \frac{m}{1}x + \frac{m(m-1)}{1 \cdot 2}x^2 + \frac{m(m-1)(m-2)}{1 \cdot 2 \cdot 3}x^3 + \ldots$ u.s.w. *J. Reine Angew. Math.* 1 (1826), 311–339.

[Alb81] E. Albrecht. Spectral decomposition for systems of commuting operators. *Proc. Roy. Irish Acad. Sect. A* 81 (1981), 81–98.

[AB85] C. D. Aliprantis and O. Burkinshaw. *Positive Operators*, Academic Press, London, 1985.

[AOR87] G. R. Allan, A. G. O'Farrell and T. J. Ransford. A Tauberian theorem arising in operator theory. *Bull. London Math. Soc.* 19 (1987), 537–545.

[AR89] G. R. Allan and T. J. Ransford. Power-dominated elements in a Banach algebra. *Studia Math.* 94 (1989), 63–79.

[Ama95] H. Amann. *Linear and Quasilinear Parabolic Problems.* Vol. I. Birkhäuser, Basel, 1995.

[AP71] L. Amerio and G. Prouse. *Almost-periodic Functions and Functional Equations.* Van Nostrand, New York, 1971.

[Are84] W. Arendt. Resolvent positive operators and integrated semigroups. *Semesterbericht Funktionalanalysis*, Univ. Tübingen (1984), 73–101.

[Are87a] W. Arendt. Resolvent positive operators. *Proc. London Math. Soc.* 54 (1987), 321–349.

[Are87b] W. Arendt. Vector–valued Laplace transforms and Cauchy problems. *Israel J. Math.* 59 (1987), 327–352.

[Are91] W. Arendt. Sobolev imbeddings and integrated semigroups. *Semigroup Theory and Evolution Equations*, Proc. Delft 1989, P. Clément et al. eds., Marcel-Dekker, New York (1991), 29–40.

[Are94a] W. Arendt. Vector-valued versions of some representation theorems in real analysis. *Functional Analysis*, Proc. Essen 1991, Marcel-Dekker, New York (1994), 33–50.

[Are94b] W. Arendt. Spectrum and growth of positive semigroups. *Evolution Equations*, Proc. Baton Rouge 1992, G. Ferreyra, G. Goldstein, F. Neubrander eds., Marcel-Dekker, New York (1994), 21–28.

[Are00] W. Arendt. Resolvent positive operators and inhomogeneous boundary value problems. *Ann. Scuola Norm. Sup. Pisa (4)*, 29 (2000), 639–670.

[Are01] W. Arendt. Approximation of degenerate semigroups. *Taiwanese J. Math.* 5 (2001), 279–295.

[Are04] W. Arendt. Semigroups and evolution equations: functional calculus, regularity and kernel estimates. *Evolutionary equations.* Vol. I, Handb. Differ. Equ., North-Holland, Amsterdam (2004), 1–85.

[Are08] W. Arendt. Positive semigroups of kernel operators. *Positivity* 12 (2008), 25–44.

[AB88] W. Arendt and C. J. K. Batty. Tauberian theorems and stability of one-parameter semigroups. *Trans. Amer. Math. Soc.* 306 (1988), 837–852.

[AB92a] W. Arendt and C. J. K. Batty. Domination and ergodicity for positive semigroups. *Proc. Amer. Math. Soc.* 114 (1992), 743–747.

[AB95] W. Arendt and C. J. K. Batty. A complex Tauberian theorem and mean ergodic semigroups. *Semigroup Forum* 50 (1995), 351–366.

[AB97] W. Arendt and C. J. K. Batty. Almost periodic solutions of first- and second-order Cauchy problems. *J. Differential Equations* 137 (1997), 363–383.

[AB99] W. Arendt and C. J. K. Batty. Asymptotically almost periodic solutions of inhomogeneous Cauchy problems on the half-line. *Bull. London Math. Soc.* 31 (1999), 291–304.

[AB00] W. Arendt and C. J. K. Batty. Slowly oscillating solutions of Cauchy problems with countable spectrum. *Proc. Roy. Soc. Edinburgh Sect. A* 130 (2000), 471–484.

[AB06] W. Arendt and C. J. K. Batty. Rank-1 perturbations of cosine functions and semigroups. *J. Funct. Anal.* 238 (2006), 340–352.

[AB07] W. Arendt and C. J. K. Batty. Forms, functional calculus, cosine functions and perturbation. *Perspectives in Operator Theory.* Banach Center Publ. 75, Polish Acad. Sci., Warsaw (2007), 17–38.

[AB92b] W. Arendt and Ph. Bénilan. Inégalités de Kato et semi-groupes sous-markoviens. *Rev. Mat. Univ. Complut. Madrid* 5 (1992), 279–308.

[AB98] W. Arendt and Ph. Bénilan. Wiener regularity and heat semigroups on spaces of continuous functions. *Topics in Nonlinear Analysis*, J. Escher, G. Simonett eds., Birkhäuser, Basel (1998), 29–49.

[AB04] W. Arendt and S. Bu. The operator-valued Marcinkiewicz multiplier theorem and maximal regularity. *Math. Z.* 240 (2002), 311-343.

[ACK82] W. Arendt, P. Chernoff and T. Kato. A generalization of dissipativity and positive semigroups. *J. Operator Theory* 8 (1982), 167–180.

[AD08] W. Arendt and D. Daners. Varying domains: stability of the Dirichlet and the Poisson problem. *Discrete Contin. Dyn. Syst.* 21 (2008), 21–39.

[AEH97] W. Arendt, O. El-Mennaoui and M. Hieber. Boundary values of holomorphic semigroups. *Proc. Amer. Math. Soc.* 125 (1997), 635–647.

[AEK94] W. Arendt, O. El-Mennaoui and V. Keyantuo. Local integrated semigroups: evolution with jumps of regularity. *J. Math. Anal. Appl.* 186 (1994), 572–595.

[AF93] W. Arendt and A. Favini. Integrated solutions to implicit differential equations. *Rend. Sem. Mat. Univ. Politec. Torino* 51 (1993), 315–329.

[AK89] W. Arendt and H. Kellermann. Integrated solutions of Volterra integrodifferential equations and applications. *Volterra Integrodifferential Equations in Banach Spaces and Applications*, Proc. Trento 1987, G. Da Prato, M. Iannelli eds., Pitman Res. Notes Math. 190, Longman, Harlow (1989), 21–51.

[ANS92] W. Arendt, F. Neubrander and U. Schlotterbeck. Interpolation of semigroups and integrated semigroups. *Semigroup Forum* 45 (1992), 26–37.

[AN00] W. Arendt and N. Nikolski. Vector-valued holomorphic functions revisited. *Math. Z.* 234 (2000), 777–805.

[AN06] W. Arendt and N. Nikolski. Addendum: "Vector-valued holomorphic functions revisited". *Math. Z.* 252 (2006), 687–689.

[AP92] W. Arendt and J. Prüss. Vector-valued Tauberian theorems and asymptotic behavior of linear Volterra equations. *SIAM J. Math. Anal.* 23 (1992), 412–448.

[AR91] W. Arendt and A. Rhandi. Perturbation of positive semigroups. *Arch. Math. (Basel)* 56 (1991), 107–119.

[AS99] W. Arendt and S. Schweiker. Discrete spectrum and almost periodicity. *Taiwanese J. Math.* 3 (1999), 475–490.

[Arv82] W. Arveson. The harmonic analysis of automorphism groups. *Operator Algebras and Applications.* Vol. I. Proc. Symp. Pure Math. 38, Amer. Math. Soc., Providence (1982), 199–269.

[Atz84] A. Atzmon. On the existence of hyperinvariant subspaces. *J. Operator Theory* 11 (1984), 3–40.

[BV62] I. Babuŝka and R. Výborný. Reguläre und stabile Randpunkte für das Problem des Wärmeleitungsgleichung. *Ann. Polon. Math.* 12 (1962), 91–104.

[BB05] B. Baillard and H. Bourget (eds.). *Correspondance d'Hermite et de Stieltjes.* Gauthier Villars, Paris, 1905.

[BE79] M. Balabane and H. A. Emami-Rad. Smooth distribution group and Schrödinger equation in $L^p(\mathbb{R}^N)$. *J. Math. Anal. Appl.* 70 (1979), 61–71.

[BE85] M. Balabane and H. A. Emami-Rad. L^p estimates for Schrödinger evolution equations. *Trans. Amer. Math. Soc.* 292 (1985), 357–373.

[BEJ93] M. Balabane, H. A. Emami-Rad and M. Jazar. Spectral distributions and generalization of Stone's theorem. *Acta. Appl. Math.* 31 (1993), 275–295.

[Bal60] A. V. Balakrishnan. Fractional powers of closed operators and the semigroups generated by them. *Pacific J. Math.* 10 (1960), 419–437.

[BLR89] C. Bardos, G. Lebeau and J. Rauch. Un exemple d'utilisation des notions de propagation pour le contrôle et la stabilisation de problèmes hyperboliques. *Rend. Sem. Mat. Univ. Politec. Torino 1988*, Special Issue (1989), 11–31.

[Bas95] B. Basit. Some problems concerning different types of vector valued almost periodic functions. *Dissertationes Math.* 338, 1995.

[Bas97] B. Basit. Harmonic analysis and asymptotic behavior of solutions to the abstract Cauchy problem. *Semigroup Forum* 54 (1997), 58–74.

[Bas78] A. G. Baskakov. Spectral criteria for almost periodicity of solutions of functional equations. *Math. Notes* 24 (1978), 606–612.

[Bas85] A. G. Baskakov. Harmonic analysis of cosine and exponential operator-valued functions. *Math. USSR-Sb.* 52 (1985), 63–90.

[BEPS06] A. Bátkai, K.-J. Engel, J. Prüss and R. Schnaubelt. Polynomial stability of operator semigroups. *Math. Nachr.* 279 (2006), 1425–1440.

[Bat78] C. J. K. Batty. Dissipative mappings and well-behaved derivations. *J. London Math. Soc. (2)* 18 (1978), 527–533.

[Bat90] C. J. K. Batty. Tauberian theorems for the Laplace-Stieltjes transform. *Trans. Amer. Math. Soc.* 322 (1990), 783–804.

[Bat94] C. J. K. Batty. Asymptotic behaviour of semigroups of operators. *Functional Analysis and Operator Theory.* Banach Center Publ. 30, Polish Acad. Sci., Warsaw (1994), 35–52.

[Bat96] C. J. K. Batty. Spectral conditions for stability of one-parameter semigroups. *J. Differential Equations* 127 (1996), 87–96.

[Bat03] C. J. K. Batty. Bounded Laplace transforms, primitives and semigroup orbits. *Arch. Math. (Basel)* 81 (2003), 72–81.

[BB00] C. J. .K. Batty and M. D. Blake. Convergence of Laplace integrals. *C. R. Acad. Sci. Paris Sér. I Math.* 330 (2000), 71–75.

[BBG96] C. J. K. Batty, Z. Brzeźniak and D. A. Greenfield. A quantitative asymptotic theorem for contraction semigroups with countable unitary spectrum. *Studia Math.* 121 (1996), 167–183.

[BC99] C. J. K. Batty and R. Chill. Bounded convolutions and solutions of inhomogeneous Cauchy problems. *Forum Math.* 11 (1999), 253–277.

[BCN98] C. J. K. Batty, R. Chill and J. M. A. M. van Neerven. Asymptotic behaviour of C_0-semigroups with bounded local resolvents. *Math. Nachr.* 219 (2000), 65–83.

[BCT02] C. J. K. Batty, R. Chill and Y. Tomilov. Strong stability of bounded evolution families and semigroups. *J. Funct. Anal.* 193 (2002), 116–139.

[BD82] C. J. K. Batty and E. B. Davies. Positive semigroups and resolvents. *J. Operator Theory* 10 (1982), 357–363.

[BD08] C. J. K. Batty and T. Duyckaerts. Non-uniform stability for bounded semi-groups in Banach spaces. *J. Evolution Equations* 8 (2008), 765–780.

[BHR99] C. J. K. Batty, W. Hutter and F. Räbiger. Almost periodicity of mild solutions of inhomogeneous periodic Cauchy problems. *J. Differential Equations* 156 (1999), 309–327.

[BNR98a] C. J. K. Batty, J. M. A. M. van Neerven, and F. Räbiger. Local spectra and individual stability of uniformly bounded C_0-semigroups. *Trans. Amer. Math. Soc.* 350 (1998), 2071–2085.

[BNR98b] C. J. K. Batty, J. M. A. M. van Neerven, and F. Räbiger. Tauberian theorems and stability of solutions of the Cauchy problem. *Trans. Amer. Math. Soc.* 350 (1998), 2087–2103.

[BR84] C. J. K. Batty and D. W. Robinson. Positive one-parameter semigroups on ordered Banach spaces. *Acta Appl. Math.* 2 (1984), 221–296.

[BV90] C. J. K. Batty and Q. P. Vũ. Stability of individual elements under one-parameter semigroups. *Trans. Amer. Math. Soc.* 322 (1990), 805–818.

[BV92] C. J. K. Batty and Q. P. Vũ. Stability of strongly continuous representations of abelian semigroups. *Math. Z.* 209 (1992), 75–88.

[BY00] C. J. K. Batty and S. B. Yeates. Weighted and local stability of semigroups of operators. *Math. Proc. Cambridge Phil. Soc.* 129 (2000), 85–98.

[Bäu97] B. Bäumer. *Vector-Valued Operational Calculus and Abstract Cauchy Problems.* Dissertation, Louisiana State Univ., Baton Rouge, 1997.

[Bäu01] B. Bäumer. Approximate solutions to the abstract Cauchy problem. *Evolution equations and their applications in physical and life sciences.* Proc. Bad Herrenalb 1998, Lecture Notes in Pure and Appl. Math., 215, Marcel-Dekker, New York (2001), 33–41.

[Bäu03] B. Bäumer. On the inversion of the convolution and Laplace transform. *Trans. Amer. Math. Soc.* 355 (2003), 1201-1212.

[BLN99] B. Bäumer, G. Lumer and F. Neubrander. Convolution kernels and generalized functions. *Generalized Functions, Operator Theory, and Dynamical Systems.* Proc. Brussels 1997, Chapman & Hall, Boca Raton (1999), 68–78.

[BN94] B. Bäumer and F. Neubrander. Laplace transform methods for evolution equations. *Confer. Sem. Mat. Univ. Bari* 259 (1994), 27–60.

[BN96] B. Bäumer and F. Neubrander. Existence and uniqueness of solutions of ordinary linear differential equations in Banach spaces. Unpublished manuscript, Baton Rouge, 1996.

[Bea72] R. Beals. On the abstract Cauchy problem, *J. Funct. Anal.* 10 (1972), 281–299.

[BCP88] Ph. Bénilan, M. G. Crandall and A. Pazy. "Bonnes solutions" d'un problème d'évolution semi-linéaire. *C. R. Acad. Sci. Paris Sér. I Math.* 306 (1988), 527–530.

[BCP90] Ph. Bénilan, M. G. Crandall and A. Pazy. Evolution problems governed by accretive operators. Unpublished manuscript, Besançon, 1990.

[BB65] H. Berens and P. L. Butzer. Über die Darstellung vektorwertiger holomorpher Funktionen durch Laplace-Integrale. *Math. Ann.* 158 (1965), 269–283.

[Ber28] S. Bernstein. Sur les fonctions absolument monotones. *Acta Math.* 52 (1928), 1–66.

[BP58] C. Bessaga and A. Pelczynski. On bases and unconditional convergence of series in Banach spaces. *Studia Math.* 17 (1958), 329–396.

[Beu47] A. Beurling. Sur une classe de fonctions presque-périodiques. *C. R. Acad. Sci. Paris* 225 (1947), 326–328.

[Bid33] Sylvia Martis in Biddau. Studio della transformazione di Laplace e della sua inversa dal punto di vista dei funzionali analitici. *Rend. Circ. Mat. Palermo* 57 (1933), 1–70.

[Bla99] M. D. Blake. *Asymptotically norm-continuous semigroups of operators.* DPhil Thesis, Oxford, 1999.

[Bla01] M. D. Blake. A spectral bound for asymptotically norm-continuous semigroups. *J. Operator Theory* 45 (2001), 111–130.

[BM96] O. Blasco and J. Martinez. Norm continuity and related notions for semigroups on Banach spaces. *Arch. Math. (Basel)* 66 (1996), 470–478.

[Blo49] P. H. Bloch. Über den Zusammenhang zwischen den Konvergenzabszissen, der Holomorphie- und der Beschränktheitsabszisse bei der Laplace-Transformation. *Comment. Math. Helv.* 22 (1949), 34–47.

[Bob97a] A. Bobrowski. On the Yosida approximation and the Widder-Arendt representation theorem. *Studia Math.* 124 (1997), 281–290.

[Bob97b] A. Bobrowski. The Widder-Arendt theorem on inverting of the Laplace transform, and its relationships with the theory of semigroups of operators. *Methods Funct. Anal. Topology* 3 (1997), 1–39.

[Bob01] A. Bobrowski. Inversion of the Laplace transform and generation of Abel summable semigroups. *J. Funct. Anal.* 186 (2001), 1–24.

[Boc42] S. Bochner. Completely monotone functions in partially ordered spaces. *Duke Math. J.* 9 (1942), 519–526.

[Boh25] H. Bohr. Zur Theorie der fastperiodischen Funktion, I and II. *Acta Math.* 45 (1925), 29–127 and 46 (1925), 101–214.

[Boh47] H. Bohr. *Almost Periodic Functions.* Chelsea, New York, 1947.

[BCT07] A. Borichev, R. Chill and Y. Tomilov. Uniqueness theorems for (sub-) harmonic functions with applications to operator theory. *Proc. London Math. Soc.* 95 (2007), 687–708.

[BT10] A. Borichev and Y. Tomilov. Optimal polynomial decay of functions and operator semigroups. *Math. Ann.* 347 (2010), 455-478.

[BY84] J. M. Borwein and D. T. Yost. Absolute norms on vector lattices. *Proc. Edinburgh Math. Soc.* 27 (1984), 215–222.

[Bou82] J. Bourgain. A Hausdorff-Young inequality for B-convex Banach spaces. *Pacific J. Math.* 101 (1982), 255–262.

[Bou83] J. Bourgain. Some remarks on Banach spaces in which martingale difference sequences are unconditional. *Ark. Math.* 21 (1983), 163–168.

[Bou88] J. Bourgain. Vector-valued Hausdorff-Young inequalities and applications. *Geometric Aspects of Functional Analysis.* Lect. Notes in Math. 1317, Springer-Verlag, Berlin (1988), 239–249.

[BD92] K. Boyadzhiev and R. deLaubenfels. Semigroups and resolvents of bounded variation, imaginary powers and H^∞ functional calculus. *Semigroup Forum* 45 (1992), 372–384.

[Bre66] Ph. Brenner. The Cauchy problem for symmetric hyperbolic systems in L_p. *Math. Scand.* 19 (1966), 27–37.

[Bre73] Ph. Brenner. The Cauchy problem for systems in L_p and $L_{p,\alpha}$. *Ark. Mat.* 11 (1973), 75–101.

[BT79] Ph. Brenner and V. Thomée. On rational approximations of semigroups. *SIAM J. Numer. Anal.* 16 (1979), 683-694.

[Bre83] H. Brézis. *Analyse Fonctionelle.* Masson, Paris, 1983.

[Buk81] A. V. Bukhvalov. Hardy spaces of vector-valued functions. *J. Soviet Math.* 16 (1981), 1051–1059.

[BD82] A. V. Bukhvalov and A. A. Danilevich. Boundary properties of analytic and harmonic functions with values in Banach space. *Math. Notes* 31 (1982), 104–110.

[Bur81] D. L. Burkholder. A geometrical characterization of Banach spaces in which martingale difference sequences are unconditional. *Ann. Probab.* 9 (1981), 997–1011.

[Bur98] N. Burq. Décroissance de l'énergie locale de l'équation des ondes pour le problème extérieur et absence de résonance au voisinage du réel. *Acta Math.* 180 (1998), 1–29.

[Cas85] J. A. van Casteren. *Generators of Strongly Continuous Semigroups.* Pitman Res. Notes Math. 115, Longman, Harlow, 1985.

[Cha84] S. D. Chatterji. Tauber's theorem—a few historical remarks. *Jahrb. Überbl. Math.*, 1984, Bibliographisches Institut, Mannheim (1984), 167–175.

[Cha71] J. Chazarain. Problèmes de Cauchy abstraits et applications à quelques problèmes mixtes, *J. Funct. Anal.* 7 (1971), 346–446.

[Che68] P. R. Chernoff. Note on product formulas for operator semigroups. *J. Funct. Anal.* 2 (1968), 238–242.

[Che74] P. R. Chernoff. *Product Formulas, Nonlinear Semigroups and Addition of Unbounded Operators.* Mem. Amer. Math. Soc. 140, 1974.

[CL99] C. Chicone and Y. Latushkin. *Evolution Semigroups in Dynamical Systems and Differential Equations.* Amer. Math. Soc., Providence, 1999.

[Chi98a] R. Chill. *Fourier Transforms and Asymptotics of Evolution Equations.* PhD Thesis, Ulm, 1998.

[Chi98b] R. Chill. Tauberian theorems for vector-valued Fourier and Laplace transforms. *Studia Math.* 128 (1998), 55–69.

[CP01] R. Chill and J. Prüss. Asymptotic behaviour of linear evolutionary integral equations. *Integral Equations Operator Theory* 39 (2001), 193–213.

[CT03] R. Chill and Y. Tomilov. Stability of C_0-semigroups and geometry of Banach spaces. *Math. Proc. Cambridge Philos. Soc.* 135 (2003), 493–511.

[CT04] R. Chill and Y. Tomilov. Analytic continuation and stability of operator semigroups. *J. Anal. Math.* 93 (2004), 331–357.

[CT07] R. Chill and Y. Tomilov. Stability of operator semigroups: ideas and results. *Perspectives in Operator Theory.* Banach Center Publ. 75, Polish Acad. Sci., Warsaw (2007), 71–109.

[CT09] R. Chill and Y. Tomilov. Operators $L^1(\mathbb{R}_+) \to X$ and the norm continuity problem for semigroups. *J. Funct. Anal.* 256 (2009), 352–384.

[Cho02] W. Chojnacki. A generalization of the Widder-Arendt theorem. *Proc. Edinb. Math. Soc. (2)* 45 (2002), 161–179.

[CL03] I. Cioranescu and C. Lizama. On the inversion of the Laplace transform for resolvent families in UMD-spaces. *Arch. Math. (Basel)* 81 (2003), 182-192.

[CL94] I. Cioranescu and G. Lumer. Problèmes d'évolution régularisés par un noyau général $K(t)$. Formule de Duhamel, prolongements, théorèmes de génération. *C.R. Acad. Sci. Paris Sér. I Math.* 319 (1994), 1273–1278.

[CHA87] Ph. Clément, H. J. A. M. Heijmans, S. Angenent, C. J. van Duijn and B. de Pagter. *One-parameter Semigroups*. North-Holland, Amsterdam, 1987.

[Con73] J. B. Conway. *Functions of One Complex Variable*. Springer-Verlag, Berlin, 1973.

[CW77] R. R. Coifman and G. Weiss. *Transference Methods in Analysis*. Conf. Board Math. Sci. Reg. Conf. Series Math. 31, Amer. Math. Soc., Providence, 1977.

[Cou84] T. Coulhon. Suites d'opérateurs sur un espace $C(K)$ de Grothendieck. *C. R. Acad. Sci. Paris Sér. I Math.* 298 (1984), 13–15.

[Cro04] M. Crouzeix. Operators with numerical range in a parabola. *Arch. Math. (Basel)* 82 (2004), 517–527.

[Cro07] M. Crouzeix. Numerical range and functional calculus in Hilbert space. *J. Funct. Anal.* 244 (2007), 668–690.

[Cro08] M. Crouzeix. A functional calculus based on the numerical range: applications. *Linear Multilinear Algebra* 56 (2008), 81–103.

[DaP66] G. Da Prato. Semigruppi regolarizzabili. *Ricerche Mat.* 15 (1966), 223–248.

[DG67] G. Da Prato and F. Giusti. Una caratterizzazione dei generatori di funzioni coseno astratte. *Bull. Un. Mat. Ital.* 22 (1967), 357–362.

[DS87] G. Da Prato and E. Sinestrari. Differential operators with nondense domain. *Ann. Scuola Norm. Sup. Pisa Cl. Sci. (4)* 14 (1987), 285–344.

[DK74] Y. L. Daletskii and M. G. Krein. *Stability of Solutions of Differential Equations in Banach Space*. Amer. Math. Soc., Providence, 1974.

[Dat70] R. Datko. Extending a theorem of A. M. Liapunov to Hilbert space. *J. Math. Anal. Appl.* 32 (1970), 610–616.

[Dat72] R. Datko. Uniform asymptotic stability of evolutionary processes in a Banach space. *SIAM J. Math. Anal.* 3 (1972), 428–445.

[DM95] C. Datry and G. Muraz. Analyse harmonique dans les modules de Banach, I: propriétés générales. *Bull. Sci. Math.* 119 (1995), 299–337.

[DM96] C. Datry and G. Muraz. Analyse harmonique dans les modules de Banach, II: presque-périodicité et ergodicité. *Bull. Sci. Math.* 120 (1996), 493–536.

[DL90] R. Dautray and J. L. Lions. *Mathematical Analysis and Numerical Methods for Science and Technology.* Vol. 1–3. Springer-Verlag, Berlin, 1990.

[Dav80] E. B. Davies. *One-parameter Semigroups.* Academic Press, London, 1980.

[Dav90] E. B. Davies. *Heat Kernels and Spectral Theory.* Cambridge Univ. Press, Cambridge, 1990.

[Dav95] E. B. Davies. *Spectral Theory and Differential Operators.* Cambridge Univ. Press, Cambridge, 1995.

[Dav05] E. B. Davies. Triviality of the peripheral point spectrum. *J. Evol. Equ.* 5 (2005), 407–415.

[Dav07] E. B. Davies. *Linear Operators and their Spectra.* Cambridge Univ. Press, Cambridge, 2007.

[DP87] E. B. Davies and M. M. Pang. The Cauchy problem and a generalization of the Hille-Yosida theorem. *Proc. London Math. Soc.* 55 (1987), 181–208.

[Dea81] M. A. B. Deakin. The development of the Laplace transform, 1737-1937. I. Euler to Spitzer, 1737-1880. *Arch. Hist. Exact Sci.* 25 (1981), 343–390.

[Dea82] M. A. B. Deakin. The development of the Laplace transform, 1737-1937. II. Poincaré to Doetsch, 1880-1937. *Arch. Hist. Exact Sci.* 26 (1982), 351–381.

[deL90] R. deLaubenfels. Integrated semigroups and integrodifferential equations. *Math. Z.* 204 (1990), 501–514.

[deL94] R. deLaubenfels. *Existence Families, Functional Calculi and Evolution Equations.* Lecture Notes in Math. 1570, Springer-Verlag, Berlin, 1994.

[DHW97] R. deLaubenfels, Z. Huang, S. Wang, and Y. Wang. Laplace transforms of polynomially bounded vector-valued functions and semigroups of operators. *Israel J. Math.* 98 (1997), 189–207.

[DVW02] R. deLaubenfels, Q. P. Vũ, and S. Wang. Laplace transforms of vector-valued functions with growth ω and semigroups of operators. *Semigroup Forum* 64 (2002), 355–375.

[Der80] R. Derndinger. Über das Spektrum positiver Generatoren. *Math. Z.* 172 (1980), 281–293.

[DP93] W. Desch and J. Prüss. Counterexamples for abstract linear Volterra equations. *J. Integral Equations Appl.* 5 (1993), 29–45.

[DS88] W. Desch and W. Schappacher. Some perturbation results for analytic semigroups. *Math. Ann.* 281 (1988), 157–162.

[DSS09] W. Desch, G. Schappacher and W. Schappacher. Relatively bounded rank one perturbations of non-analytic semigroups can generate large point spectrum. *Semigroup Forum* 75 (2007), 470–476.

[DU77] J. Diestel and J. J. Uhl. *Vector Measures.* Amer. Math. Soc., Providence, 1977.

[Doe37] G. Doetsch. *Theorie und Anwendung der Laplace-Transformation.* Verlag Julius Springer, Berlin, 1937.

[Doe50] G. Doetsch. *Handbuch der Laplace-Transformation.* Vol. I, II, III. Birkhäuser, Basel, 1950, 1955, 1956.

[DE99] A. Driouich and O. El-Mennaoui. On the inverse formula for Laplace transforms. *Arch. Math. (Basel)* 72 (1999), 56–63.

[DS59] N. Dunford and J. Schwartz. *Linear Operators.* Vol. I. Interscience, New York, 1958.

[Dur70] P. L. Duren. *Theory of H^p Spaces.* Academic Press, New York, 1970.

[EG92] B. Eberhardt and G. Greiner. Baillon's theorem on maximal regularity. *Acta Appl. Math.* 27 (1992), 47–54.

[Eis10] T. Eisner. *Stability of Operators and Operator Semigroups.* Birkhäuser, Basel, 2010.

[Elm92] O. El-Mennaoui. *Traces des semi-groupes holomorphes singuliers à l'origine et comportement asymptotique.* PhD Thesis, Besançon, 1992.

[Elm94] O. El-Mennaoui. Asymptotic behaviour of integrated semigroups. *J. Comput. Appl. Math.* 54 (1994), 351–369.

[EE94] O. El-Mennaoui and K. J. Engel. On the characterization of eventually norm continuous semigroups in Hilbert spaces. *Arch. Math. (Basel)* 63 (1994), 437–440.

[EK96a] O. El-Mennaoui and V. Keyantuo. On the Schrödinger equation in L^p-spaces. *Math. Ann.* 304 (1996), 293–302.

[EK96b] O. El-Mennaoui and V. Keyantuo. Trace theorems for holomorphic semigroups and the second order Cauchy problem. *Proc. Amer. Math. Soc.* 124 (1996), 1445–1458.

[EN00] K. J. Engel and R. Nagel. *One-parameter Semigroups for Linear Evolution Equations.* Springer-Verlag, Berlin, 2000.

[EN06] K. J. Engel and R. Nagel. *A Short Course on Operator Semigroups.* Springer-Verlag, Berlin, 2006.

[EW85] I. Erdelyi and S. W. Wang. *A Local Spectral Theory for Closed Operators.* London Math. Soc. Lecture Note 105, Cambridge Univ. Press, Cambridge, 1985.

[ESZ92] J. Esterle, E. Strouse, and F. Zouakia. Stabilité asymptotique de certains semi-groupes d'opérateurs et idéaux primaires de $L^1(\mathbb{R}^+)$. *J. Operator Theory* 28 (1992), 203–227.

[Eva98] L. C. Evans. *Partial Differential Equations.* Amer. Math. Soc., Providence, 1998.

[FP01] E. Fasangova and J. Prüss. Asymptotic behaviour of a semilinear viscoelastic beam model. *Arch. Math. (Basel)* 77 (2001), 488–497

[Fat69] H. O. Fattorini. Ordinary differential equations in linear topological spaces, II. *J. Differential Equations* 6 (1969), 50–70.

[Fat83] H. O. Fattorini. *The Cauchy Problem.* Addison-Wesley, Reading, 1983.

[Fat85] H. O. Fattorini. *Second Order Linear Differential Equations in Banach Spaces.* North-Holland, Amsterdam, 1985.

[FY99] A. Favini and A. Yagi. *Degenerate Differential Equations in Banach Spaces.* Marcel-Dekker New York, 1999.

[FS72] C. Fefferman, E. M. Stein. H^p spaces of several variables. *Acta Math.* 129 (1972), 137–193.

[Fin74] A. M. Fink. *Almost Periodic Differential Equations.* Lecture Notes in Math. 377, Springer-Verlag, Berlin, 1974.

[Fra86] J. L. Rubio de Francia. Martingale and integral transforms of Banach space valued functions. *Probability and Banach Spaces*, Proc. Zaragoza 1985, Lecture Notes in Math. 1221, Springer-Verlag, Berlin (1986), 195–222.

[Ful56] W. Fulks. A note on the steady state solutions of the heat equation. *Proc. Amer. Math. Soc.* 7 (1956), 766–770.

[Ful57] W. Fulks. Regular regions for the heat equation. *Pacific J. Math.* 7 (1957), 867–877.

[Gea78] L. Gearhart. Spectral theory for contraction semigroups on Hilbert space. *Trans. Amer. Math. Soc.* 236 (1978), 385–394.

[GT83] D. Gilbarg and N. S. Trudinger. *Elliptic Partial Differential Equations of Second Order.* Springer-Verlag, Berlin, 1983.

[GL61] I. Glicksberg and K. de Leeuw. Applications of almost periodic compactifications. *Acta Math.* 105 (1961), 63–97.

[GW99] V. Goersmeyer and L. Weis. Norm continuity of C_0-semigroups. *Studia Math.* 134 (1999), 169–178.

[Gol85] J. A. Goldstein. *Semigroups of Linear Operators and Applications.* Oxford Univ. Press, Oxford, 1985.

[Gom99] A. M. Gomilko. On conditions for the generating operator of a uniformly bounded C_0-semigroup of operators. *Funct. Anal. Appl.* 33 (1999), 294–296.

[GN89] A. Grabosch and R. Nagel. Order structure of the semigroup dual: a counterexample. *Nederl. Akad. Wetensch. Indag. Math.* 51 (1989), 199–201.

[Gre94] D. A. Greenfield. *Semigroup Representations: an Abstract Approach.* DPhil Thesis, Oxford, 1994.

[Gre82] G. Greiner. Spektrum und Asymptotik stark stetiger Halbgruppen positiver Operatoren. *Sitzungsber. Heidelb. Akad. Wiss. Math.-Natur. Kl.* 1982, 55–80.

[Gre84] G. Greiner. Some applications of Fejér's theorem to one-parameter semigroups. *Semesterbericht Funktionalanalysis*, Tübingen (1984/1985), 33–50.

[Gre87] G. Greiner. Perturbing the boundary conditions of a generator. *Houston J. Math.* 13 (1987), 213–229.

[GM93] G. Greiner and M. Müller. The spectral mapping theorem for integrated semigroups. *Semigroup Forum* 47 (1993), 115–122.

[GN83] G. Greiner and R. Nagel. On the stability of strongly continuous semigroups of positive operators on $L^2(\mu)$. *Ann. Scuola Norm. Sup. Pisa Cl. Sci. (4)* 10 (1983), 257–262.

[GVW81] G. Greiner, J. Voigt and M. Wolff. On the spectral bound of the generator of semigroups of positive operators. *J. Operator Theory* 5 (1981), 245–256.

[GN81] U. Groh and F. Neubrander. Stabilität starkstetiger, positiver Operatorhalbgruppen auf C^*-Algebren. *Math. Ann.* 256 (1981), 509–516.

[Gue92] S. Guerre-Delabrière. *Classical Sequences in Banach Spaces.* Marcel-Dekker, New York, 1992.

[Haa06] M. Haase. *The Functional calculus for Sectorial Operators.* Birkhäuser, Basel, 2006.

[Haa07a] M. Haase. Functional calculus for groups and applications to evolution equations. *J. Evol. Equ.* 7 (2007), 529–554.

[Haa07b] M. Haase. Convexity inequalities for positive operators. *Positivity* 11 (2007), 57–68.

[Haa08] M. Haase. The complex inversion formula revisited. *J. Aust. Math. Soc.* 84 (2008), 7383.

[Haa09] M. Haase. The group reduction for bounded cosine functions on UMD spaces. *Math. Z.* 262 (2009), 281–299.

[Har91] A. Haraux. *Systèmes Dynamiques Dissipatifs et Applications.* Masson, Paris, 1991.

[Hea93] O. Heaviside. *Electromagnetic Theory.* Vol. I–III. Benn, London, 1893–1899.

[Hel69] L. L. Helms. *Introduction to Potential Theory.* Wiley, New York, 1969.

[HN93] B. Hennig and F. Neubrander. On representations, inversions, and approximations of Laplace transforms in Banach spaces. *Appl. Anal.* 49 (1993), 151–170.

[Her83] W. Herbst. The spectrum of Hilbert space semigroups. *J. Operator Theory* 10 (1983), 87–94.

[HK79] R. Hersh and T. Kato. High-accuracy stable difference schemes for well-posed initial value problems. *SIAM J. Numer. Anal.* 16 (1979), 670-682.

[Hes70] P. Hess. Zur Störungstheorie linearer Operatoren in Banachräumen. *Comment. Math. Helv.* 45 (1970), 229–235.

[Hie91a] M. Hieber. Integrated semigroups and differential operators on L^p spaces. *Math. Ann.* 291 (1991), 1–16.

[Hie91b] M. Hieber. Laplace transforms and α-times integrated semigroups. *Forum Math.* 3 (1991), 595–612.

[Hie91c] M. Hieber. Integrated semigroups and the Cauchy problem for systems in L^p-spaces. *J. Math. Ann. Appl.* 162 (1991), 300–308.

[Hie91d] M. Hieber. An operator-theoretical approach to Dirac's equation on L^p-spaces. *Semigroup Theory and Evolution Equations*, Proc. Delft 1989, P. Clément et al. eds., Marcel-Dekker, New York (1991), 259-265.

[Hie95] M. Hieber. L_p spectra of pseudodifferential operators generating integrated semigroups. *Trans. Amer. Math. Soc.* 347 (1995), 4023–4035.

[HHN92] M. Hieber, A. Holderieth and F. Neubrander. Regularized semigroups and systems of partial differential equations. *Ann. Scuola Norm. Sup. Pisa Cl. Sci. (4)* 19 (1992), 263–379.

[Hil48] E. Hille. *Functional Analysis and Semi-groups.* Amer. Math. Soc., Providence, 1948.

[HP57] E. Hille and R. S. Phillips. *Functional Analysis and Semi-groups.* Amer. Math. Soc., Providence, 1957.

[Hör60] L. Hörmander. Estimates for translation invariant operators in L^p spaces. *Acta Math.* 104 (1960), 93–139.

[Hör83] L. Hörmander. *The Analysis of Linear Partial Differential Operators.* Vol. I,II. Springer-Verlag, Berlin, 1983.

[How84] J. S. Howland. On a theorem of Gearhart. *Integral Equations Operator Theory* 7 (1984), 138–142.

[Hua83] F. L. Huang. Asymptotic stability theory for linear dynamical systems in Banach spaces. *Kexue Tongbao* 28 (1983), 584–586.

[Hua85] F. L. Huang. Characteristic conditions for exponential stability of linear dynamical systems in Hilbert spaces. *Ann. Differential Equations* 1 (1985), 43–56.

[Hua93a] F. L. Huang. Spectral properties and stability of one-parameter semigroups. *J. Differential Equations* 104 (1993), 182–195.

[Hua93b] F. L. Huang. Strong asymptotic stability of linear dynamical systems in Banach spaces. *J. Differential Equations* 104 (1993), 307–324.

[Hua99] S. Z. Huang. A local version of Gearhart's theorem. *Semigroup Forum* 58 (1999), 323–335.

[HN99] S. Z. Huang and J. M. A. M. van Neerven. B-convexity, the analytic Radon-Nikodym property, and individual stability of C_0-semigroups. *J. Math. Anal. Appl.* 231 (1999), 1–20.

[HR94] S. Z. Huang and F. Räbiger. Superstable C_0-semigroups on Banach spaces. *Evolution equations, control theory and biomathematics*, Proc. Han sur Lesse 1991, Marcel-Dekker, New York (1994), 291–300.

[IS70] F. T. Iha and C. F. Schubert. The spectrum of partial differential operators on $L^p(\mathbb{R}^n)$. *Trans. Amer. Math. Soc.* 152 (1970), 215–226.

[Ile07] P. Iley. Perturbations of differentiable semigroups. *J. Evol. Equ.* 7 (2007), 765–781.

[Ing35] A. E. Ingham. On Wiener's method in Tauberian theorems. *Proc. London Math. Soc.* 38 (1935), 458–480.

[Jac76] S. V. D'Jacenko. Semigroups of almost negative type and their applications. *Soviet Math. Dokl.* 17 (1976), 1189–1193.

[Jac56] K. Jacobs. Ergodentheorie und fastperiodische Funktionen auf Halbgruppen. *Math. Z.* 64 (1956), 298–338.

[Jam64] R. C. James. Uniformly non-square Banach spaces. *Ann. of Math.* 80 (1964), 542–550.

[Jar08] P. Jara. Rational approximation schemes for bi-continuous semigroups. *J. Math. Anal. Appl.* 344 (2008), 956–968.

[JNO08] P. Jara, F. Neubrander and K. Ozer. Rational inversion of the Laplace transform. Preprint, 2008.

[Jaz95] J. M. Jazar. Fractional powers of momentum of a spectral distribution. *Proc. Amer. Math. Soc.* 123 (1995), 1805–1813.

[Kad69] M. I. Kadets. On the integration of almost periodic functions with values in a Banach space. *Funct. Anal. Appl.* 3 (1969), 228–230.

[KW03] C. Kaiser and L. Weis. Perturbation theorems for α-times integrated semigroups. *Arch. Math. (Basel)* 81 (2003), 215–228.

[Kat59] T. Kato. Remarks on pseudo-resolvents and infinitesimal generators of semi-groups. *Proc. Japan Acad.* 35 (1959), 467–468.

[Kat66] T. Kato. *Perturbation Theory for Linear Operators.* Springer-Verlag, Berlin, 1966.

[Kat82] T. Kato. *A Short Introduction to Perturbation Theory for Linear Operators.* Springer-Verlag, Berlin, 1982.

[Kat68] Y. Katznelson. *An Introduction to Harmonic Analysis.* Wiley, New York, 1968.

[KT86] Y. Katznelson and L. Tzafriri. On power bounded operators. *J. Funct. Anal.* 68 (1986), 313–328.

[Kei06] V. Keicher. On the peripheral spectrum of bounded positive semigroups on atomic Banach lattices. *Arch. Math. (Basel)* 87 (2006), 359–367.

[Kel86] H. Kellermann. *Integrated Semigroups*. PhD Thesis, Tübingen, 1986.

[KH89] H. Kellermann and M. Hieber. Integrated semigroups. *J. Funct. Anal.* 84 (1989), 160–180.

[Kel67] O. D. Kellogg. *Foundations of Potential Theory*. Springer-Verlag, Berlin, 1967.

[KT80] C. E. Kenig and P. A. Tomas. L^p behavior of certain second order partial differential operators. *Trans. Amer. Math. Soc.* 262 (1980), 521–531.

[Kér97] L. Kérchy. Operators with regular norm-sequences. *Acta Sci. Math. (Szeged)* 63 (1997), 571–605.

[Kér99] L. Kérchy. Representations with regular norm-behaviour of discrete abelian semigroups. *Acta Sci. Math. (Szeged)* 65 (1999), 701–726.

[Key95a] V. Keyantuo. A note on interpolation of semigroups. *Proc. Amer. Math. Soc.* 123 (1995), 2123–2132.

[Key95b] V. Keyantuo. The Laplace transform and the ascent method for abstract wave equations. *J. Differential Equations* 122 (1995), 27–47.

[KMV03] V. Keyantuo, C. Müller and P. Vieten. Finite and local Laplace transforms in Banach spaces. *Proc. Edinb. Math. Soc. (2)* 46 (2003), 357–372.

[KR81] A. Kishimoto and D. W. Robinson. Subordinate semigroups and order properties. *J. Austral. Math. Soc. Ser. A* 31 (1981), 59–76.

[Kis72] J. Kisyński. On cosine operator functions and one-parameter groups of operators. *Studia Math.* 44 (1972), 93–105.

[Kis76] J. Kisyński. Semi-groups of operators and some of their applications to partial differential equations. *Control Theory and Topics in Functional Analysis*, Proc. Trieste 1974, Vol. III, International Atomic Energy Agency, Vienna (1976), 305–405.

[Kis00] J. Kisyński. Around Widder's characterization of the Laplace transform of an element of $L^\infty(\mathbb{R}^+)$. *Ann. Polon. Math.* 74 (2000), 161–200.

[KN94] C. Knuckles and F. Neubrander. Remarks on the Cauchy problem for multi-valued linear operators. *Partial Differential Equations*, Proc. Han-sur-Lesse 1993, Akademie Verlag, Berlin (1994), 174–187.

[Kom66] H. Komatsu. Fractional powers of operators. *Pacific J. Math.* 19 (1966), 285–346.

[Kön60] H. König. Neuer Beweis eines klassischen Tauber-Satzes. *Arch. Math.*, 11 (1960), 278–279.

[Koo80] P. Koosis. *Introduction to H_p Spaces.* London Math. Soc. Lecture Note 40. Cambridge Univ. Press, 1980.

[Kor82] J. Korevaar. On Newman's quick way to the prime number theorem. *Math. Intelligencer* 4 (1982), 108–115.

[Kor04] J. Korevaar. *Tauberian theory. A century of developments.* Springer-Verlag, Berlin, 2004.

[Kov07] M. Kovács. On the convergence of rational approximations of semigroups on intermediate spaces. *Math. Comp.* 76 (2007), 273-286.

[KS90] W. Kratz and U. Stadtmüller. Tauberian theorems for Borel-type methods of summability. *Arch. Math. (Basel)* 55 (1990), 465–474.

[Kre71] S. G. Krein. *Linear Differential Equations in Banach Spaces.* Amer. Math. Soc.. Providence, 1971.

[Kre59] H. O. Kreiss. Über Matrizen die beschränkte Halbgruppen erzeugen. *Math. Scand.* 7 (1959), 71–80.

[Kre85] U. Krengel. *Ergodic Theorems.* De Gruyter, Berlin, 1985.

[KW04] P.C. Kunstmann and L. Weis. Maximal L_p-regularity for parabolic equations, Fourier multiplier theorems and H^∞-functional calculus. *Functional analytic methods for evolution equations*, Lecture Notes in Math. 1855, Springer-Verlag, Berlin (2004), 65–311.

[Kur69] T. G. Kurtz. Extensions of Trotter's operator semigroup approximation theorems. *J. Funct. Anal.* 3 (1969), 354–375.

[Kwa72] S. Kwapień. Isomorphic characterizations of inner product spaces by orthogonal series with vector valued coefficients. *Studia Math.* 44 (1972), 583–595.

[Lan68] E. Lanconelli. Valutazioni in $L^p(\mathbb{R}^n)$ della soluzione del problema di Cauchy per l'equazione di Schrödinger. *Boll. Un. Mat. Ital. (4)* 1 (1968), 591–607.

[Lan72] N. S. Landkof. *Foundation of Modern Potential Theory.* Springer-Verlag, Berlin, 1972.

[LM95] Y. Latushkin and S. Montgomery-Smith. Evolutionary semigroups and Lyapunov theorems in Banach spaces. *J. Funct. Anal.* 127 (1995), 173–197.

[LP67] P. D. Lax and R. S. Phillips. *Scattering Theory.* Academic Press, New York, 1967.

[Leb96] G. Lebeau. Équation des ondes amorties. *Algebraic and geometric methods in mathematical physics*, Proc. Kaciveli 1993, Math. Phys. Stud. 19, Kluwer Acad. Publ., Dordrecht (1996), 73–109.

[Leb72] N. N. Lebedev. *Special Functions and their Applications.* Dover, New York, 1972.

[LS97] H. G. Leopold and E. Schrohe. Invariance of the L_p spectrum for hypoelliptic operators. *Proc. Amer. Math. Soc.* 125 (1997), 3679-3687.

[Lev69] D. Leviatan. On the representation of functions as Laplace integrals. *J. London Math. Soc.* 44 (1969), 88–92.

[Lev66] B. M. Levitan. Integration of almost periodic functions with values in a Banach space. *Izv. Akad. Nauk SSSR Ser. Mat.* 30 (1966), 1101–1110.

[LZ82] B. M. Levitan and V. V. Zhikov. *Almost Periodic Functions and Differential Equations.* Cambridge Univ. Press, Cambridge, 1982.

[LT77] J. Lindenstrauss and L. Tzafriri. *Classical Banach Spaces.* Vol. I,II. Springer-Verlag, Berlin, 1977.

[Lio60] J.-L. Lions. Les semi groupes distributions. *Portugal. Math.* 19 (1960) 141–164.

[Lit63] W. Littman. The wave operator and L_p norms. *J. Math. Mech.* 12 (1963), 55–68.

[LR05] Z. Liu and B. Rao. Characterization of polynomial decay rate for the solution of linear evolution equation. *Z. Angew. Math. Phys.* 56 (2005), 630–644.

[Liz94] C. Lizama. On the convergence and approximation of integrated semigroups. *J. Math. Anal. Appl.* 181 (1994), 89–103.

[Liz00] C. Lizama. Regularized solutions for abstract Volterra equations. *J. Math. Anal. Appl.* 243 (2000), 278–292.

[Loo60] L. H. Loomis. On the spectral characterization of a class of almost periodic functions. *Ann. of Math.* 72 (1960), 362–368.

[LB07] L. Lorenzi and M. Bertoldi. *Analytical Methods for Markov Semigroups.* Chapman & Hall/CRC, Boca Raton, 2007.

[Lot85] H. P. Lotz. Uniform convergence of operators on L^∞ and similar spaces. *Math. Z.* 190 (1985), 207–220.

[Lum75] G. Lumer. Problème de Cauchy avec valeurs au bord continues. *C. R. Acad. Sci. Paris Sér. A* 281 (1975), 805–807.

[Lum90] G. Lumer. Solutions généralisées et semi-groupes intégrés. *C. R. Acad. Sci. Paris Sér. I Math.* 310 (1990), 577–582.

[Lum92] G. Lumer. Semi-groupes irréguliers et semi-groupes intégrés: application à l'identification de semi-groupes irréguliers analytiques et résultats de génération. *C. R. Acad. Sci. Paris Sér. I Math.* 314 (1992), 1033–1038.

[Lum94] G. Lumer. Evolution equations. Solutions for irregular evolution problems via generalized solutions and generalized initial values. Applications to periodic shocks modes. *Ann. Univ. Sarav. Ser. Math.* 5 (1994), 1–102.

[Lum97] G. Lumer. Singular evolution problems, regularization, and applications to physics, engineering, and biology. *Linear Operators*, Banach Center Publ. 34, Polish Acad. Sci., Warsaw (1997), 205–216.

[LN97] G. Lumer and F. Neubrander. Signaux non-détectables en dimension N dans des systèmes gouvernés par des équations de type paraboliques. *C. R. Acad. Sci. Paris Sér. I Math.* 324 (1997), 731–736.

[LN99] G. Lumer and F. Neubrander. Asymptotic Laplace transforms and evolution equations. *Evolution Equations, Feshbach Resonances, Singular Hodge Theory*, Wiley-VCH, Berlin (1999), 37–57.

[LN01] G. Lumer and F. Neubrander. The asymptotic Laplace transform: new results and relation to Komatsu's Laplace transform of hyperfunctions. *Partial Differential Equations on Multistructures*, Marcel-Dekker, New York (2001), 147–162.

[LP79] G. Lumer and L. Paquet. Semi-groupes holomorphes, produit tensoriel de semi-groupes et équations d'évolution. *Sém. Theorie du Potentiel 4 (Paris, 1977/78)*, Lecture Notes in Math. 713, Springer-Verlag, Berlin (1979), 156–177.

[LS99] G. Lumer and R. Schnaubelt. Local operator methods and time dependent parabolic equations on non-cylindrical domains. *Evolution Equations, Feshbach Resonances, Singular Hodge Theory*, Wiley-VCH, Berlin (1999), 58–130.

[Lun95] A. Lunardi. *Analytic Semigroups and Optimal Regularity in Parabolic Equations.* Birkhäuser, Basel, 1995.

[Lun86] J. van de Lune. *An Introduction to Tauberian Theory: from Tauber to Wiener.* Stichting Math. Centrum, Centrum Wisk. Inform., Amsterdam, 1986.

[Lyu66] Yu. I. Lyubich. The classical and local Laplace transformation in an abstract Cauchy problem. *Russian Math. Surveys* 21 (1966), 1–52.

[LV88] Y. I. Lyubich and Q. P. Vũ. Asymptotic stability of linear differential equations in Banach spaces. *Studia Math.* 88 (1988), 37–42.

[LV90a] Y. I. Lyubich and Q. P. Vũ. A spectral criterion for the almost periodicity of one-parameter semigroups. *J. Soviet. Math.* 48 (1990), 644–647.

[LV90b] Y. I. Lyubich and Q. P. Vũ. A spectral criterion for asymptotic almost periodicity of uniformly continuous representations of abelian semigroups. *J. Soviet. Math.* 49 (1990), 1263–1266.

[MS01] C. Martínez Carracedo and M. Sanz Alix. *The Theory of Fractional Powers of Operators.* North-Holland, Amsterdam, 2001.

[MM96] J. Martinez and J. M. Mazon. C_0-semigroups norm continuous at infinity. *Semigroup Forum* 52 (1996), 213–224.

[Mar11] M. Martínez. Decay estimates of functions through singular extensions of vector-valued Laplace transforms. *J. Math. Anal. Appl.* 375 (2011), 196–206.

[Mat08] T. Mátrai. Resolvent norm decay does not characterize norm continuity. *Israel J. Math.* 168 (2008), 1–28.

[Meg98] R. E. Megginson. *An Introduction to Banach Space Theory.* Springer-Verlag, Berlin, 1998.

[MF01] I.V. Melnikova and A. Filinkov. *Abstract Cauchy Problems: Three Approaches.* Chapman & Hall, Boca Raton, 2001.

[MS78] R. B. Melrose and J. Sjöstrand. Singularities of boundary value problems. I. *Comm. Pure Appl. Math.* 31 (1978),593–617.

[Mey91] P. Meyer-Nieberg. *Banach Lattices.* Springer-Verlag, Berlin, 1991.

[Mih09] C. Mihai. Zeros of the Laplace transform. *Int. J. Pure Appl. Math.* 56 (2009), 49–55.

[MB87] J. Mikusiński and T. Boehme. *Operational Calculus*, Vol. II. 2nd ed., Pergamon Press, Oxford, 1987.

[Mil80] P. Milnes. On vector-valued weakly almost periodic functions. *J. London Math. Soc. (2)* 22 (1980), 467–472.

[Miy81] A. Miyachi. On some singular Fourier multipliers. *J. Fac. Sci. Univ. Tokyo Sect. 1A Math.* 28 (1981), 267–315.

[Miy56] I. Miyadera. On the representation theorem by the Laplace transformation of vector-valued functions. *Tôhoku Math. J.* 8 (1956), 170–180.

[Mon96] S. Montgomery-Smith. Stability and dichotomy of positive semigroups on L_p. *Proc. Amer. Math. Soc.* 124 (1996), 2433-2437.

[MRS77] C. S. Morawetz, J. V. Ralston and W. A. Strauss. Decay of solutions of the wave equation outside nontrapping obstacles. *Comm. Pure Appl. Math.* 30 (1977), 447–508.

[Nag86] R. Nagel (ed.). *One-parameter Semigroups of Positive Operators.* Lecture Notes in Math. 1184. Springer-Verlag, Berlin, 1986.

[NP00] R. Nagel and J. Poland. The critical spectrum of a strongly continuous semigroup. *Adv. Math.* 152 (2000), 120–133.

[NR93] R. Nagel and F. Räbiger. Superstable operators on Banach spaces. *Israel J. Math.* 81 (1993), 213–226.

[NS94] R. Nagel and E. Sinestrari. Inhomogeneous Volterra integrodifferential equations for Hille-Yosida operators. *Functional Analysis*, Proc. Essen 1991, Marcel-Dekker, New York (1994), 51–70.

[Nee92] J. M. A. M. van Neerven. *The Adjoint of a Semigroup of Linear Operators.* Lecture Notes in Math. 1529, Springer-Verlag, Berlin, 1992.

[Nee96a] J. M. A. M. van Neerven. Characterization of exponential stability of a semigroup of operators in terms of its action by convolution on vector-valued function spaces over $\mathbb{R}_+$. *J. Differential Equations* 124 (1996), 324–342.

[Nee96b] J. M. A. M. van Neerven. Individual stability of C_0-semigroups with uniformly bounded local resolvent. *Semigroup Forum* 53 (1996), 155–161.

[Nee96c] J. M. A. M. van Neerven. *The Asymptotic Behaviour of Semigroups of Linear Operators.* Birkhäuser, Basel, 1996.

[Nee96d] J. M. A. M. van Neerven. Inequality of spectral bound and growth bound for positive semigroups in rearrangement invariant Banach function spaces. *Arch. Math. (Basel)* 66 (1996), 406–416.

[Nee00] J. M. A. M. van Neerven. The vector-valued Loomis theorem for the half-line and individual stability of C_0-semigroups: a counterexample. *Semigroup Forum* 60 (2000), 271–283.

[Nee09] J. M. A. M. van Neerven. Asymptotic behaviour of C_0-semigroups and γ-boundedness of the resolvent. *J. Math. Anal. Appl.* 358 (2009), 380–388.

[NSW95] J. M. A. M. van Neerven, B. Straub and L. Weis. On the asymptotic behaviour of a semigroup of linear operators. *Indag. Math. (N.S.)* 6 (1995), 453–476.

[Neu86] F. Neubrander. Laplace transform and asymptotic behavior of strongly continuous semigroups. *Houston J. Math.* 12 (1986), 549–561.

[Neu88] F. Neubrander. Integrated semigroups and their applications to the abstract Cauchy problem. *Pacific J. Math.* 135 (1998), 111–155.

[Neu89a] F. Neubrander. Integrated semigroups and their applications to complete second order Cauchy problems. *Semigroup Forum* 38 (1989), 233–251.

[Neu89b] F. Neubrander. Abstract elliptic operators, analytic interpolation semigroups, and Laplace transforms of analytic functions. *Semesterbericht Funktionalanalysis*, Univ. Tübingen (1988/89), 163–185.

[Neu94] F. Neubrander. The Laplace-Stieltjes transform in Banach spaces and abstract Cauchy problems. *Evolution Equations, Control Theory, and Biomathematics*, Proc. Han sur Lesse 1991, Marcel-Dekker, New York (1994), 417–431.

[New80] D. J. Newman. Simple analytic proof of the prime number theorem. *Amer. Math. Monthly* 87 (1980), 693–696.

[New98] D. J. Newman. *Analytic Number Theory.* Springer-Verlag, Berlin, 1998.

[Nic93] S. Nicaise. The Hille-Yosida and Trotter-Kato theorems for integrated semigroups. *J. Math. Anal. Appl.* 180 (1993), 303–316.

[Oha71] S. Oharu. Semigroups of linear operators in a Banach space. *Publ. Res. Inst. Math. Sci.* 7 (1971), 205–260.

[Ouh05] E.-M. Ouhabaz. *Analysis of Heat Equations on Domains.* Princeton Univ. Press, Princeton, 2005.

[Pag89] B. de Pagter. A characterization of sun-reflexivity. *Math. Ann.* 283 (1989), 511–518.

[Pas19] J. Rey Pastor. La investigación matemática. Boletin de critica, pedagogia, historia y bibliografia 1 (1919), 97 - 108.

[Paz72] A. Pazy. On the applicability of Lyapunov's theorem in Hilbert space. *SIAM J. Math. Anal.* 3 (1972), 291–294.

[Paz83] A. Pazy. *Semigroups of Linear Operators and Applications to Partial Differential Equations.* Springer-Verlag, Berlin, 1983.

[Ped89] G. K. Pedersen. *Analysis Now.* Springer-Verlag, Berlin, 1989.

[PC98] Jigen Peng and Si-Kit Chung. Laplace transforms and generators of semigroups of operators. *Proc. Amer. Math. Soc.* 126 (1998), 2407–2416.

[Per80] J. C. Peral. L^p estimates for the wave equation. *J. Funct. Anal.* 36 (1980), 114–145.

[Phr04] E. Phragmén. Sur une extension d'un théorème classique de la théorie des fonctions. *Acta Math.* 28 (1904), 351–368.

[Pis86] G. Pisier. *Factorization of Linear Operators and Geometry of Banach Spaces.* Conf. Board Math. Sci. Reg. Conf. Series Math. 60, Amer. Math. Soc., Providence, 1986.

[Pit58] H. R. Pitt. *Tauberian Theorems.* Oxford Univ. Press, Oxford, 1958.

[Prü84] J. Prüss. On the spectrum of C_0-semigroups. *Trans. Amer. Math. Soc.* 284 (1984), 847–857.

[Prü93] J. Prüss. *Evolutionary Integral Equations and Applications.* Birkhäuser, Basel, 1993.

[RW95] F. Räbiger and M. Wolff. Superstable semigroups of operators. *Indag. Math. (N.S.)* 6 (1995), 481–494.

[RS72] M. Reed and B. Simon. *Methods of Modern Mathematical Physics.* Vol. I,II. Academic Press, New York, 1972, 1975.

[Rei52] H. J. Reiter. Investigations in harmonic analysis. *Trans. Amer. Math. Soc.* 73 (1952), 401–427.

[Rob91] D.W. Robinson. *Elliptic Operators and Lie Groups.* Oxford Univ. Press, Oxford, 1991.

[RRS91] J. Rosenblatt, W. M. Ruess, F. D. Sentilles. On the critical part of a weakly almost periodic function. *Houston J. Math.* 17 (1991), 237–249.

[Rud62] W. Rudin. *Fourier Analysis on Groups.* Wiley, New York, 1962.

[Rud76] W. Rudin. *Principles of Mathematical Analysis.* 3rd ed., McGraw-Hill, New York, 1976.

[Rud87] W. Rudin. *Real and Complex Analysis.* 3rd ed., McGraw-Hill, New York, 1987.

[Rud91] W. Rudin. *Functional Analysis.* 2nd ed., McGraw-Hill, New York, 1991.

[Rue91] W. M. Ruess. Almost periodicity properties of solutions to the nonlinear Cauchy problem in Banach spaces. *Semigroup Theory and Evolution Equations*, Proc. Delft 1989, P. Clément et al. eds., Marcel-Dekker, New York (1991), 421–440.

[Rue95] W. M. Ruess. Purely imaginary eigenvalues of operator semigroups. *Semigroup Forum* 51 (1995), 335–341.

[RS86] W. M. Ruess and W. H. Summers. Asymptotic almost periodicity and motions of semigroups of operators. *Linear Algebra Appl.* 84 (1986), 335–351.

[RS87] W. M. Ruess and W. H. Summers. Presque-périodicité faible et théorème ergodique pour les semi-groupes de contractions non linéaires. *C. R. Acad. Sci. Paris Sér. I Math.* 305 (1987), 741–744.

[RS88a] W. M. Ruess and W. H. Summers. Weak almost periodicity and the strong ergodic limit theorem for contraction semigroups *Israel J. Math.* 64 (1988), 139–157.

[RS88b] W. M. Ruess and W. H. Summers. Compactness in spaces of vector valued continuous functions and asymptotic almost periodicity. *Math. Nachr.* 135 (1988), 7–33.

[RS89] W. M. Ruess and W. H. Summers. Integration of asymptotically almost periodic functions and weak asymptotic almost periodicity. *Dissertationes Math.* 279, 1989.

[RS90a] W. M. Ruess and W. H. Summers. Weak almost periodicity and the strong ergodic limit theorem for periodic evolution systems. *J. Funct. Anal.* 94 (1990), 177–195.

[RS90b] W. M. Ruess and W. H. Summers. Weakly almost periodic semigroups of operators. *Pacific J. Math.* 143 (1990), 175–193.

[RS92a] W. M. Ruess and W. H. Summers. Ergodic theorems for semigroups of operators. *Proc. Amer. Math. Soc.* 114 (1992), 423–432.

[RS92b] W. M. Ruess and W. H. Summers. Weak asymptotic almost periodicity for semigroups of operators. *J. Math. Anal. Appl.* 164 (1992), 242–262.

[RV95] W. M. Ruess and Q. P. Vũ. Asymptotically almost periodic solutions of evolution equations in Banach spaces. *J. Differential Equations* 122 (1995), 282–301.

[San75] N. Sanekata. Some remarks on the abstract Cauchy problem. *Publ. Res. Inst. Math. Sci.* 11 (1975), 51–65.

[Sch74] H. H. Schaefer. *Banach Lattices and Positive Operators.* Springer-Verlag, Berlin, 1974.

[Sch71] M. Schechter. *Spectra of Partial Differential Operators.* North Holland, Amsterdam, 1971.

[SV98] E. Schüler and Q. P. Vũ. The operator equation $AX - XB = C$, admissibility, and asymptotic behavior of differential equations. *J. Differential Equations* 145 (1998), 394–419.

[See68] G. L. Seever. Measures on F-spaces. *Trans. Amer. Math. Soc.* 133 (1968), 267–280.

[She47] Yu-Cheng Shen. The identical vanishing of the Laplace integral. *Duke Math. J.* 14 (1947), 967-973.

[SF00] D.-H. Shi and D.-X. Feng. Characteristic conditions of the generation of C_0 semigroups in a Hilbert space. *J. Math. Anal. Appl.* 247 (2000), 356–376.

[Sim79] B. Simon. *Functional Integration and Quantum Physics.* Academic Press, London, 1979.

[Sin85] E. Sinestrari. On the abstract Cauchy problem of parabolic type in spaces of continuous functions. *J. Math. Anal. Appl.* 107 (1985), 16–66.

[Sjo70] S. Sjöstrand. On the Riesz means of the solution of the Schrödinger equation. *Ann. Scuola Norm. Sup. Pisa (3)* 24 (1970), 331–348.

[SS82] G. M. Skylar and V. Ya. Shirman. On the asymptotic stability of a linear differential equation in a Banach space. *Teor. Funktsiĭ Funktsional. Anal. i Prilozhen.* 37 (1982), 127–132.

[Sle76] M. Slemrod. Asymptotic behavior of C_0-semigroups as determined by the spectrum of the generator. *Indiana Univ. Math. J.* 25 (1976), 783–792.

[Sov66] M. Sova. Cosine operator functions. *Rozprawy Mat.* 49, 1966.

[Sov68] M. Sova. Problème de Cauchy pour équations hyperboliques opérationnelles à coefficients constants non-bornés. *Ann. Scuola Norm. Sup. Pisa (3)* 22 (1968), 67–100.

[Sov77] M. Sova. On inversion of Laplace transform, I. *Časopis Pěst. Mat.* 102 (1977), 166–172.

[Sov79a] M. Sova. The Laplace transform of analytic vector-valued functions (real conditions). *Časopis Pěst. Mat.* 104 (1979), 188–199.

[Sov79b] M. Sova. The Laplace transform of analytic vector-valued functions (complex conditions). *Časopis Pěst. Mat.* 104 (1979), 267–280.

[Sov79c] M. Sova. Laplace transform of exponentially lipschitzian vector-valued functions. *Časopis Pěst. Mat.* 104 (1979), 370–381.

[Sov80a] M. Sova. The Laplace transform of exponentially bounded vector-valued functions (real conditions). *Časopis Pěst. Mat.* 105 (1980), 1–13.

[Sov80b] M. Sova. Relation between real and complex properties of the Laplace transform. *Časopis Pěst. Mat.* 105 (1980), 111–119.

[Sov81a] M. Sova. On the equivalence of Widder-Miyadera's and Leviatan's representability conditions for the Laplace transform of integrable vector-valued functions. *Časopis Pěst. Mat.* 106 (1981), 117–126.

[Sov81b] M. Sova. On a fundamental theorem of the Laplace transform theory. *Časopis Pěst. Mat.* 106 (1981), 231–242.

[Sov82] M. Sova. General representability problem for the Laplace transform of exponentially bounded vector-valued functions. *Časopis Pěst. Mat.* 107 (1982), 69–89.

[ST81] U. Stadtmüller and R. Trautner. Tauberian theorems for Laplace transforms in dimension $d > 1$. *J. Reine Angew. Math.* 323 (1981), 127–138.

[Sta81] O. J. Staffans. On asymptotically almost periodic solutions of a convolution equation. *Trans. Amer. Math. Soc.* 266 (1981), 603–616.

[Ste93] E. M. Stein. *Harmonic Analysis: Real-variable Methods, Orthogonality, and Oscillatory Integrals.* Princeton Univ. Press, Princeton, 1993.

[Tai95] K. Taira. *Analytic Semigroups and Semilinear Initial Boundary Value Problems.* London Math. Soc. Lecture Note 222, Cambridge Univ. Press, Cambridge, 1995.

[TakO90] T. Takenaka and N. Okazawa. Wellposedness of abstract Cauchy problems for second order differential equations. *Israel J. Math.* 69 (1990), 257–288.

[TM92] N. Tanaka and I. Miyadera. C-semigroups and the abstract Cauchy problem. *J. Math. Anal. Appl.* 170 (1992), 196–206.

[TanO90] N. Tanaka and N. Okazawa. Local C-semigroups and local integrated semigroups. *Proc. London Math. Soc.* 61 (1990), 63–90.

[Tau97] A. Tauber. Ein Satz aus der Theorie der unendlichen Reihen. *Monatsh. Math. Phys.* 8 (1897), 273–277.

[Tay73] S. J. Taylor. *Introduction to Measure and Integration.* Cambridge Univ. Press, Cambridge, 1973.

[Thi98a] H. R. Thieme. Remarks on resolvent positive operators and their perturbation. *Discrete Contin. Dynam. Systems* 4 (1998), 73–90.

[Thi98b] H. R. Thieme. Positive perturbation of operators semigroups: growth bounds, essential compactness, and asynchronous exponential growth. *Disrete Contin. Dynam. Systems* 4 (1998), 735–764.

[Tom01] Y. Tomilov. Resolvent approach to stability of operator semigroups. *J. Operator Theory* 46 (2001), 63–98.

[TE05] L. N. Trefethen and M. Embree. *Spectra and pseudospectra. The behavior of nonnormal matrices and operators.* Princeton University Press, Princeton, 2005.

[Tyc38] A. Tychonoff. Sur l'équation de la chaleur à plusieurs variables. *Bull. Univ. Moscow Ser. Int. Sect.* A1 (1938), 1–44.

[Ulm99] M. Ulm. The interval of resolvent-positivity for the biharmonic operator. *Proc. Amer. Math. Soc.* 127 (1999), 481–489.

[Vie95] P. Vieten. *Holomorphie und Laplace Transformation banachraumwertiger Funktionen.* Dissertation, Universität Kaiserslautern, 1995.

[Vig39] J. C. Vignaux. Sugli integrali di Laplace asintotici. *Atti Accad. Naz. Lincei Rend. Cl. Sci. Fis. Mat. (6)* 29 (1939), 396–402.

[VC44] J. C. Vignaux and M. Cotlar. Asymptotic Laplace-Stieltjes integrals. *Univ. Nac. La Plata. Publ. Fac. Ci. Fisicomat. (2)* 3(14) (1944), 345–400.

[Voi89] J. Voigt. On resolvent positive operators and positive C_0-semigroups on AL-spaces. *Semigroup Forum* 38 (1989), 263–266.

[Vu91] Q. P. Vũ. The operator equation $AX - XB = C$ with unbounded operators A and B and related abstract Cauchy problems. *Math. Z.* 208 (1991), 567–588.

[Vu92] Q. P. Vũ. Theorems of Katznelson-Tzafriri type for semigroups of operators. *J. Funct. Anal.* 103 (1992), 74–84.

[Vu93] Q. P. Vũ. Semigroups with nonquasianalytic growth. *Studia Math.* 104 (1993), 229–241.

[Vu97] Q. P. Vũ. Almost periodic and strongly stable semigroups of operators. *Linear Operators*, Banach Center Publ. 38, Polish Acad. Sci., Warsaw (1997), 401–426.

[Wei93] L. Weis. Inversion of the vector-valued Laplace transform in $L_p(X)$-spaces. *Differential Equations in Banach Spaces*, Proc. Bologna 1991, North- Holland, Amsterdam, 1993.

[Wei95] L. Weis. The stability of positive semigroups on L_p spaces. *Proc. Amer. Math. Soc.* 123 (1995), 3089–3094.

[Wei96] L. Weis. Stability theorems for semi-groups via multiplier theorems. *Tübinger Berichte Funktionalanalysis* 6 (1996/97), 253–268.

[Wei98] L. Weis. A short proof for the stability theorem for positive semigroups on $L_p(\mu)$. *Proc. Amer. Math. Soc.* 126 (1998), 3253–3256.

[WW96] L. Weis and V. Wrobel. Asymptotic behavior of C_0-semigroups in Banach spaces. *Proc. Amer. Math. Soc.* 124 (1996), 3663–3671.

[Wei88] G. Weiss. Weak L^p-stability of a linear semigroup on a Hilbert space implies exponential stability. *J. Differential Equations* 76 (1988), 269–285.

[Wei90] G. Weiss. The resolvent growth assumption for semigroups on Hilbert spaces. *J. Math. Anal. Appl.* 145 (1990), 154–171.

[Wid36] D. V. Widder. A classification of generating functions. *Trans. Amer. Math. Soc.* 39 (1936), 244–298.

[Wid41] D. V. Widder. *The Laplace Transform.* Princeton Univ. Press, Princeton, 1941.

[Wid71] D. V. Widder. *An Introduction to Transform Theory.* Academic Press, New York, 1971.

[Wil70] S. Willard. *General Topology.* Addison-Wesley, Reading, 1970.

[Woj91] P. Wojtaszczyk. *Banach Spaces for Analysts.* Cambridge Univ. Press, Cambridge, 1991.

[Wol81] M. Wolff. A remark on the spectral bound of the generator of semigroups of positive operators with applications to stability theory. *Functional Analysis and Approximation*, Proc. Oberwolfach 1980, Birkhäuser (1981), 39–50.

[Wol07] M. Wolff. Triviality of the peripheral point spectrum of positive semigroups on atomic Banach lattices. *Positivity* 12 (2008), 185–192.

[Woo74] G. S. Woodward. The generalized almost periodic part of an ergodic function. *Studia Math.* 50 (1974), 103–116.

[Wro89] V. Wrobel. Asymptotic behavior of C_0-semigroups in B-convex spaces. *Indiana Univ. Math. J.* 38 (1989), 101–114.

[XL98] T. J. Xiao and J. Liang. *The Cauchy Problem for Higher-Order Abstract Differential Equations.* Lecture Notes in Math. 1701, Springer-Verlag, Berlin, 1998.

[XL00] T. J. Xiao and J. Liang. Approximations of Laplace transforms and integrated semigroups. *J. Funct. Anal.* 172 (2000), 202–220.

[Yos80] K. Yosida. *Functional Analysis.* 6th ed., Springer-Verlag, Berlin, 1980.

[You92] P. H. You. Characteristic conditions for a C_0-semigroup with continuity in the uniform operator topology for $t > 0$. *Proc. Amer. Math. Soc.* 116 (1992), 991–997.

[Zab75] A. Zabczyk. A note on C_0-semigroups. *Bull. Acad. Polon. Sci.* 23 (1975), 895–898.

[Zab79] A. Zabczyk. Stabilization of boundary control systems. *Systems Optimization and Analysis*, Lecture Notes in Control Information Sci. 14, Springer-Verlag, Berlin (1979), 321–333.

[Zai60] S. Zaidman. Sur un théorème de I. Miyadera concernant la représentation des functions vectorielles par des intégrales de Laplace. *Tôhoku Math. J. (2)* 12 (1960), 47–51.

[Zar93] M. Zarrabi. Contractions à spectre dénombrable et propriétés d'unicité des fermés dénombrables du cercle. *Ann. Inst. Fourier (Grenoble)* 43 (1993), 251–263.

[Zem94] J. Zemánek. On the Gelfand-Hille theorem. *Functional Analysis and Operator Theory.* Banach Center Publ. 30, Polish Acad. Sci., Warsaw (1994), 369–385.

Notation

Function and Distribution Spaces

Dual Spaces and Subspaces

Norms and Dualities

Functions, Integrals and Abscissas

Operators

Spectrum and Resolvent Set

Subsets of $\mathbb{R}^n$ or $\mathbb{C}$

Transformations

Cauchy Problems

Other Notation

Index

M

N

O

P

T